Mechanisms in
HOMOGENEOUS AND HETEROGENEOUS EPOXIDATION CATALYSIS

Mechanisms in
HOMOGENEOUS AND HETEROGENEOUS EPOXIDATION CATALYSIS

Edited by

S. TED OYAMA
Department of Chemical Engineering
Virginia Tech, Blacksburg, USA

ELSEVIER

Amsterdam • Boston • Heidelberg • London • New York • Oxford • Paris
San Diego • San Francisco • Singapore • Sydney • Tokyo

Elsevier
Radarweg 29, PO Box 211, 1000 AE Amsterdam, The Netherlands
Linacre House, Jordan Hill, Oxford OX2 8DP, UK

First edition 2008

Copyright © 2008 Elsevier B.V. All rights reserved

No part of this publication may be reproduced, stored in a retrieval system or transmitted in any form or by any means electronic, mechanical, photocopying, recording or otherwise without the prior written permission of the publisher

Permissions may be sought directly from Elsevier's Science & Technology Rights Department in Oxford, UK: phone (+44) (0) 1865 843830; fax (+44) (0) 1865 853333; email: permissions@elsevier.com. Alternatively you can submit your request online by visiting the Elsevier web site at http://elsevier.com/locate/permissions, and selecting *Obtaining permission to use Elsevier material*

Notice
No responsibility is assumed by the publisher for any injury and/or damage to persons or property as a matter of products liability, negligence or otherwise, or from any use or operation of any methods, products, instructions or ideas contained in the material herein. Because of rapid advances in the medical sciences, in particular, independent verification of diagnoses and drug dosages should be made.

Library of Congress Cataloging-in-Publication Data
A catalog record for this book is available from the Library of Congress

British Library Cataloguing in Publication Data
A catalogue record for this book is available from the British Library

ISBN: 978-0-444-53188-9

For information on all Elsevier publications
visit our website at books.elsevier.com

Printed and Bound in Hungary
08 09 10 11 12   10 9 8 7 6 5 4 3 2 1

**Working together to grow
libraries in developing countries**

www.elsevier.com | www.bookaid.org | www.sabre.org

ELSEVIER   BOOK AID International   Sabre Foundation

DEDICATION

Dedicated with much thanks and affection to my parents with deep gratitude for all their loving support, and my immediate family, my wife, Hideko, our children, Monika and Leo, who, by their love and faith in me, have always been a source of great encouragement to me.

S. Ted Oyama

CONTENTS

Contributors   *xiii*
Preface   *xix*

SECTION 1: Introduction   1

1. Rates, Kinetics, and Mechanisms of Epoxidation: Homogeneous, Heterogeneous, and Biological Routes   **3**

    S. Ted Oyama

    1. Introduction   4
    2. Epoxide Uses and Markets   5
    3. Catalysts and Rates in Commodity and Heterogeneous Epoxidation Processes   12
    4. Catalysts and Rates in Homogeneous Epoxidation Reactions   16
    5. Catalysts and Rates in Biomimetic Epoxidation Reactions   25
    6. Catalysts and Rates in Biological Epoxidation Reactions   28
    7. Summary and Perspective on the Reactivity Results   30
    8. Oxidants for Epoxidation   35
    9. Mechanisms   37
    10. Homogeneous Epoxidation by Early Transition Metals (Lewis Acid Mechanism)   47
    11. Main Group Elements   57
    12. Homogeneous Epoxidation by Late Transition Metals (Redox Mechanism)   59
    13. Biological Systems   70
    14. Perspective and Conclusions   71
    Acknowledgments   72
    References   72

SECTION 2: Homogeneous Catalysis   101

2. Unprecedented Selectivity in the H_2O_2 Epoxidation of Simple Alkenes Imparted by Soft Pt(II) Lewis Acid Catalysts   **103**

    Giorgio Strukul and Alessandro Scarso

    1. Introduction   104
    2. Catalyst Synthesis and Lewis Acid Properties   105
    3. General Epoxidation Activity   106
    4. Regioselectivity   107
    5. Diastereoselectivity   108
    6. Enantioselectivity   109
    7. Reactions in Micellar Media   110

        8. Reaction Mechanism   113
        9. Conclusions   113
    References   115

3. Lewis Acid Catalyzed Epoxidation of Olefins Using Hydrogen Peroxide: Growing Prominence and Expanding Range   119

Daryle H. Busch, Guochuan Yin, and Hyun-Jin Lee

    1. Introduction   120
    2. Industrial Processes for Propylene Oxide Manufacture: Present and Future   123
    3. A New Pressure Intensified Epoxidation Process for Light Olefins   126
    4. Expanding Range of Lewis Acid Catalysis Chemistry   130
    5. Mid- to Late Transition Metal Catalysts also Perform Epoxidation Reactions by the Lewis Acid Mechanism   131
   Acknowledgments   148
   References   149

4. Activation of Hydrogen Peroxide by Polyoxometalates   155

Noritaka Mizuno, Kazuya Yamaguchi, Keigo Kamata, and Yoshinao Nakagawa

    1. Introduction   156
    2. Activation of Hydrogen Peroxide by Polyoxometalates   157
    3. Catalytic Oxidation by Polyoxometalates   166
    4. Conclusions   170
    5. Future View   170
   Acknowledgments   171
   References   171

5. Oxaziridinium Salt-Mediated Catalytic Asymmetric Epoxidation   177

Philip C. Bulman Page and Benjamin R. Buckley

    1. Introduction   178
    2. Page Group Findings   184
   Acknowledgements   214
   References   214

6. Selective Aerobic Radical Epoxidation of α-Olefins Catalyzed by N-Hydroxyphthalimide   217

Carlo Punta, Davide Moscatelli, Ombretta Porta, Francesco Minisci, Cristian Gambarotti, and Marco Lucarini

    1. Introduction   218
    2. Experimental   219
    3. Results and Discussion   220
    4. Conclusions   227
   Acknowledgments   227
   References   227

SECTION 3: Heterogeneous Catalysis   231

7. **Investigation of the Origins of Selectivity in Ethylene Epoxidation on Promoted and Unpromoted Ag/α-Al$_2$O$_3$ Catalysts: A Detailed Kinetic, Mechanistic and Adsorptive Study**   233

    Kenneth C. Waugh and M. Hague

    1. Introduction   234
    2. The Detailed Kinetics of the Adsorption and Desorption of Oxygen on Silver and the Structure of the Oxide Overlayer and Reaction Conditions   235
    3. The Adsorption of Ethylene and Ethylene Oxide on Clean and Oxidised Ag(111) and Ag(110)   240
    4. The Reaction Mechanism   242
    5. Subsurface Oxygen   247
    6. Promotion   249
    7. The Detailed Kinetics of the Adsorption and Desorption of Ethylene on an Unoxidised and Oxidised Ag/α-Al$_2$O$_3$ Catalyst   254
    8. The Detailed Kinetics of the Adsorption and Desorption of Ethylene on an Oxidised Ag/α-Al$_2$O$_3$ Catalyst Subtending the α_1-O State Only   255
    9. The Detailed Kinetics of the Adsorption and Desorption of Ethylene on the α_2-O State Principally   257
    10. Ethylene Desorption from an Unoxidised and Oxidised Cs-Promoted Ag/α-Al$_2$O$_3$ Catalyst   257
    11. The Desorption of Ethylene from an Oxidised Cl-Promoted Cs/Ag/α-Al$_2$O$_3$ Catalyst   258
    12. Overall Conclusions   259
    Acknowledgement   261
    References   261

8. **Computational Strategies for Identification of Bimetallic Ethylene Epoxidation Catalysts**   265

    A. B. Mhadeshwar and M. A. Barteau

    1. Introduction   266
    2. Computational Details   267
    3. Activation Energies on Ag   270
    4. Extension to Ag-Based Bimetallic Catalysts   272
    5. Predictions Using Microkinetic Modeling   276
    6. Conclusions   279
    Acknowledgments   279
    References   279

9. **Effect of Support on Ethylene Epoxidation on Ag, Au, and Au–Ag Catalysts**   283

    Sumaeth Chavadej, Siriphong Rojluechai, Johannes W. Schwank, and Vissanu Meeyoo

    1. Introduction   284
    2. Experimental   285

	3. Results and Discussion	287
	4. Conclusions	295
	Acknowledgments	295
	References	295

10. Epoxidation of Propylene with Oxygen–Hydrogen Mixtures   297

Masatake Haruta and Jun Kawahara

	1. Introduction	298
	2. Liquid-Phase Epoxidation of Propylene	300
	3. Gas-Phase Epoxidation of Propylene	301
	4. Conclusions	310
	References	311

11. Propylene Epoxidation by $O_2 + H_2$ over Au Nanoparticles on Ti-Nanoporous Supports   315

Ajay M. Joshi, Bradley Taylor, Lasitha Cumaranatunge, Kendall T. Thomson, and W. Nicholas Delgass

	1. Introduction	316
	2. Propylene Epoxidation over Au/TS-1 Catalysts	317
	3. Au on Ti-Containing Mesoporous Supports	322
	4. Promoters and Postsynthesis Support Treatments	324
	5. Reaction Kinetics	327
	6. Conclusions and Future Outlook	331
	Acknowledgments	333
	References	333

12. The Epoxidation of Propene over Gold Nanoparticle Catalysts   339

T. Alexander Nijhuis, Elena Sacaliuc, and Bert M. Weckhuysen

	1. Introduction	340
	2. Experimental	341
	3. Results and Discussion	343
	4. Conclusions	352
	Acknowledgments	353
	References	353

13. Propylene Epoxidation via Shell's SMPO Process: 30 Years of Research and Operation   355

J. K. F. Buijink, Jean-Paul Lange, A. N. R. Bos, A. D. Horton, and F. G. M. Niele

	1. Introduction	356
	2. Catalytic Epoxidation	358
	3. SMPO Process Improvement	363
	4. Conclusions	369

Acknowledgment	369
References	369

14. Propylene Epoxidation with Ethylbenzene Hydroperoxide over Ti-Containing Catalysts Prepared by Chemical Vapor Deposition — 373

Kuo-Tseng Li, Chia-Chieh Lin, and Ping-Hung Lin

1. Introduction	374
2. Experimental	375
3. Results and Discussion	376
4. Conclusions	385
Acknowledgment	385
References	385

15. Metal Species Supported on Organic Polymers as Catalysts for the Epoxidation of Alkenes — 387

Ulrich Arnold

1. Introduction	388
2. Supported Manganese Catalysts	389
3. Supported Molybdenum Catalysts	396
4. Supported Ruthenium and Iron Catalysts	398
5. Supported Titanium Catalysts	399
6. Supported Tungsten Catalysts	400
7. Supported Rhenium Catalysts	401
8. Supported Cobalt, Nickel, and Platinum Catalysts	402
9. Supported BINOL-Complexes of Lanthanoids and Calcium	402
10. Conclusion	403
Acknowledgment	407
References	407

SECTION 4: Phase-Transfer Catalysis — 413

16. Fine-Tuning and Recycling of Homogeneous Tungstate and Polytungstate Epoxidation Catalysts — 415

Paul L. Alsters, Peter T. Witte, Ronny Neumann, Dorit Sloboda Rozner, Waldemar Adam, Rui Zhang, Jan Reedijk, Patrick Gamez, Johan E. ten Elshof, and Sankhanilay Roy Chowdhury

1. Introduction	416
2. Characteristics and Preparation of Sandwich POMs	417
3. Benchmarking Sandwich POM-Catalyzed Epoxidations	418
4. Effect of Carboxylic Acids as Cocatalysts in Tungstate-Catalyzed Epoxidations	420
5. Epoxidations that Afford Acid-Sensitive Products	421
6. Sandwich POM Catalyst Recycling	425
7. Conclusions	426

	Acknowledgments	427
	References	427

17. Reaction-Controlled Phase-Transfer Catalysis for Epoxidation of Olefins   429

Shuang Gao and Zuwei Xi

1.	Introduction	430
2.	Reaction-Controlled Phase-Transfer Catalyst Based on Quaternary Ammonium Phosphotungstates	431
3.	Influence of the Composition of the Heteropolyphosphotungstate Anion [40]	432
4.	Influence of Different Quaternary Ammonium Cations [43]	433
5.	Epoxidation of Propylene with *In Situ* Generated H_2O_2 as the Oxidant	435
6.	Epoxidation of Propylene with Aqueous H_2O_2 [45] as the Oxidant	438
7.	Epoxidation of Cyclohexene and Others Olefins	439
8.	Epoxidation of Allyl Chloride	440
9.	Conclusion	444
	Acknowledgments	444
	References	444

SECTION 5: Biomimetic Catalysis   449

18. Bio-Inspired Iron-Catalyzed Olefin Oxidations: Epoxidation Versus *cis*-Dihydroxylation   451

Paul D. Oldenburg, Rubén Mas-Ballesté, and Lawrence Que, Jr

1.	Introduction	452
2.	Structure–Reactivity Correlation of Catalysts	453
3.	Toward Synthetically Useful Applications	457
4.	Mechanistic Landscape	459
	Acknowledgment	466
	References	466

19. Quantum Chemical Analysis of the Reaction Pathway for Styrene Epoxidation Catalyzed by Mn-Porphyrins   471

María C. Curet-Arana, Randall Q. Snurr, and Linda J. Broadbelt

1.	Introduction	472
2.	Methodology	473
3.	Results	475
4.	Conclusions	483
	Acknowledgments	483
	References	484

Index   487

CONTRIBUTORS

Waldemar Adam
Institute of Organic Chemistry, University of Wurzburg, Am Hubland, 97074 Wurzburg, Germany.
Department of Chemistry, Facundo Bueso FB-110, University of Puerto Rico, Rio Piedras, Puerto Rico 00931.

Paul L. Alsters
DSM Pharma Products, Advanced Synthesis, Catalysis, and Development, P.O. Box 18, 6160 MD Geleen, The Netherlands.

Ulrich Arnold
Department of Chemical Engineering (ITC-CPV), Forschungszentrum Karlsruhe GmbH, Hermann-von-Helmholtz-Platz 1, D-76344 Eggenstein-Leopoldshafen, Germany.

M. A. Barteau
Center for Catalytic Science and Technology, Department of Chemical Engineering, University of Delaware, Newark, Delaware 19716.

A. N. R. Bos
Shell Global Solutions International B.V., P.O. Box 38000, 1030 BN Amsterdam, The Netherlands.

Linda J. Broadbelt
Department of Chemical and Biological Engineering and Institute for Catalysis in Energy Processes, Northwestern University, Evanston, Illinois 60208.

Benjamin R. Buckley
Department of Chemistry, Loughborough University, Loughborough LE11 3TU, United Kingdom.

J. K. F. Buijink
Shell Global Solutions International B.V., P.O. Box 38000, 1030 BN Amsterdam, The Netherlands.

Philip C. Bulman Page
Department of Chemistry, Loughborough University, Loughborough LE11 3TU, United Kingdom.

Daryle H. Busch
Department of Chemistry and Center for Environmentally Beneficial Catalysis, University of Kansas, Lawrence, Kansas 66047.

Sumaeth Chavadej
The Petroleum and Petrochemical College, Chulalongkorn University, Bangkok 10330, Thailand.

Sankhanilay Roy Chowdhury
University of Twente, MESA + Institute for Nanotechnology, P.O. Box 217, 7500 AE Enschede, The Netherlands.

Lasitha Cumaranatunge
School of Chemical Engineering, Purdue University, West Lafayette, Indiana 47907.

María C. Curet-Arana
Department of Chemical and Biological Engineering and Institute for Catalysis in Energy Processes, Northwestern University, Evanston, Illinois 60208.

W. Nicholas Delgass
School of Chemical Engineering, Purdue University, West Lafayette, Indiana 47907.

Cristian Gambarotti
Dipartimento di Chimica, Materiali e Ingegneria Chimica "Giulio Natta," Politecnico di Milano, Via Mancinelli 7–I-20131 Milano, Italy.

Patrick Gamez
Leiden Institute of Chemistry, Gorlaeus Laboratories, Leiden University, P.O. Box 9502, 2300 RA, Leiden, The Netherlands.

Shuang Gao
Dalian Institute of Chemical Physics, Chinese Academy of Sciences, Dalian 116023, China.

M. Hague
BP Oil International, 20 Canada Square, Canary Wharf, London E14 5NJ, United Kingdom.

Masatake Haruta
Department of Applied Chemistry, Graduate School of Urban Environmental Sciences, Tokyo Metropolitan University, Hachioji 192–0397, Tokyo, Japan, and Japan Science and Technology Agency, CREST, Kawaguchi 332–0012, Saitama, Japan.

A. D. Horton
Shell Global Solutions International B.V., P.O. Box 38000, 1030 BN Amsterdam, The Netherlands.

Ajay M. Joshi
School of Chemical Engineering, Purdue University, West Lafayette, Indiana 47907.

Keigo Kamata
Department of Applied Chemistry, School of Engineering, The University of Tokyo, 7-3-1 Hongo, Bunkyo-ku, Tokyo 113-8656, Japan.

Jun Kawahara
Department of Applied Chemistry, Graduate School of Urban Environmental Sciences, Tokyo Metropolitan University, Hachioji 192-0397, Tokyo, Japan, and Japan Science and Technology Agency, CREST, Kawaguchi 332–0012, Saitama, Japan.

Jean-Paul Lange
Shell Global Solutions International B.V., P.O. Box 38000, 1030 BN Amsterdam, The Netherlands.

Hyun-Jin Lee
Department of Chemistry and Center for Environmentally Beneficial Catalysis, University of Kansas, Lawrence, Kansas 66047.

Kuo-Tseng Li
Department of Chemical Engineering, Tunghai University, Taichung, Taiwan, ROC.

Chia-Chieh Lin
Department of Chemical Engineering, Tunghai University, Taichung, Taiwan, ROC.

Ping-Hung Lin
Department of Chemical Engineering, Tunghai University, Taichung, Taiwan, ROC.

Marco Lucarini
Dipartimento di Chimica Organica "A. Mangini," Università di Bologna, Via San Giacomo 11–40126 Bologna, Italy.

Rubén Mas-Ballesté
Department of Chemistry and Center for Metals in Biocatalysis, University of Minnesota, Minneapolis, Minnesota 55455.

Vissanu Meeyoo
Department of Chemical Engineering, Mahanakorn University, Bangkok 10530, Thailand.

A. B. Mhadeshwar
Center for Catalytic Science and Technology, Department of Chemical Engineering, University of Delaware, Newark, Delaware 19716.

Francesco Minisci
Dipartimento di Chimica, Materiali e Ingegneria Chimica "Giulio Natta," Politecnico di Milano, Via Mancinelli 7–I-20131 Milano, Italy.

Noritaka Mizuno
Department of Applied Chemistry, School of Engineering, The University of Tokyo, 7-3-1 Hongo, Bunkyo-ku, Tokyo 113-8656, Japan.

Davide Moscatelli
Dipartimento di Chimica, Materiali e Ingegneria Chimica "Giulio Natta," Politecnico di Milano, Via Mancinelli 7–I-20131 Milano, Italy.

Yoshinao Nakagawa
Department of Applied Chemistry, School of Engineering, The University of Tokyo, 7-3-1 Hongo, Bunkyo-ku, Tokyo 113-8656, Japan.

Ronny Neumann
Department of Organic Chemistry, Weizmann Institute of Science, Rehovot 76100, Israel.

F. G. M. Niele
Shell Global Solutions International B.V., P.O. Box 38000, 1030 BN Amsterdam, The Netherlands.

T. Alexander Nijhuis
Laboratory for Chemical Reactor Engineering, Department of Chemical Engineering and Chemistry, Eindhoven University of Technology, P.O. Box 513, 5600 MB Eindhoven, The Netherlands.

Paul D. Oldenburg
Department of Chemistry and Center for Metals in Biocatalysis, University of Minnesota, Minneapolis, Minnesota 55455.

S. Ted Oyama
Department of Chemical Engineering, Virginia Polytechnic Institute & State University, Virginia 24061, USA.

Ombretta Porta
Dipartimento di Chimica, Materiali e Ingegneria Chimica "Giulio Natta," Politecnico di Milano, Via Mancinelli 7–I-20131 Milano, Italy.

Carlo Punta
Dipartimento di Chimica, Materiali e Ingegneria Chimica "Giulio Natta," Politecnico di Milano, Via Mancinelli 7–I-20131 Milano, Italy.

Lawrence Que, Jr
Department of Chemistry and Center for Metals in Biocatalysis, University of Minnesota, Minneapolis, Minnesota 55455.

Jan Reedijk
Leiden Institute of Chemistry, Gorlaeus Laboratories, Leiden University, P.O. Box 9502, 2300 RA, Leiden, The Netherlands.

Siriphong Rojluechai
The Petroleum and Petrochemical College, Chulalongkorn University, Bangkok 10330, Thailand.

Dorit Sloboda Rozner
Department of Organic Chemistry, Weizmann Institute of Science, Rehovot 76100, Israel.

Elena Sacaliuc
Laboratory for Inorganic Chemistry and Catalysis, Department of Chemistry, Utrecht University, Sorbonnelaan 16, 3584 CA Utrecht, The Netherlands.

Alessandro Scarso
Dipartimento di Chimica, Università Ca' Foscari di Venezia, 30123 Venice, Italy.

Johannes W. Schwank
Department of Chemical Engineering, The University of Michigan, Ann Arbor, Michigan 48109.

Randall Q. Snurr
Department of Chemical and Biological Engineering and Institute for Catalysis in Energy Processes, Northwestern University, Evanston, Illinois 60208.

Giorgio Strukul
Dipartimento di Chimica, Università Ca' Foscari di Venezia, 30123 Venice, Italy.

Bradley Taylor
School of Chemical Engineering, Purdue University, West Lafayette, Indiana 47907.

Johan E. ten Elshof
University of Twente, MESA + Institute for Nanotechnology, P.O. Box 217, 7500 AE Enschede, The Netherlands.

Kendall T. Thomson
School of Chemical Engineering, Purdue University, West Lafayette, Indiana 47907.

Kenneth C. Waugh
School of Chemistry, University of Manchester, Manchester M13 9PL, United Kingdom.

Bert M. Weckhuysen
Laboratory for Inorganic Chemistry and Catalysis, Department of Chemistry, Utrecht University, Sorbonnelaan 16, 3584 CA Utrecht, The Netherlands.

Peter T. Witte
DSM Pharma Products, Advanced Synthesis, Catalysis, and Development, P.O. Box 18, 6160 MD Geleen, The Netherlands.

Zuwei Xi
Dalian Institute of Chemical Physics, Chinese Academy of Sciences, Dalian 116023, China.

Kazuya Yamaguchi
Department of Applied Chemistry, School of Engineering, The University of Tokyo, 7-3-1 Hongo, Bunkyo-ku, Tokyo 113-8656, Japan.

Guochuan Yin
Department of Chemistry and Center for Environmentally Beneficial Catalysis, University of Kansas, Lawrence, Kansas 66047.

Rui Zhang
Institute of Organic Chemistry, University of Wurzburg, Am Hubland, 97074 Wurzburg, Germany.

PREFACE

The literature on epoxidation is vast, covering an immense subject that spans industrial heterogeneous catalysts, homogenous complexes, and biomimetic and biological systems. This book was motivated by a desire to organize the information on the subject, which was scattered in a few disparate reviews, many journal articles, and some patents. My interest in kinetics and mechanism made these disciplines the natural center around which to organize the book. In fact, this turned out to be a good choice, as my research led to the finding of many common threads among the topics, as well as unexpected divergence.

This book is different from conventional edited tomes, which collect assorted topics. I composed the first chapter as a comprehensive review of the field to be useful to experts and beginners alike. The chapter presents broad coverage of different catalysts used in epoxidation. In the first part, it presents a review of the rates of reaction expressed as turnover frequencies in order to compare all catalysts on an equal footing. It is found that the turnover frequencies vary over 7 orders of magnitude in the following order:

Biological and biomimetic systems > heterogeneous catalysts > homogeneous complexes

Peculiarities about each catalyst are also presented in an easy-to-understand format with many tables, figures, and summaries. The chapter then introduces in detail all the major systems in epoxidation in which there are solid foundations of knowledge about the mechanism. Effort is made to present the original references as well as the most important and latest articles so that the reader gets a historical perspective as well as up-to-date scientific findings. Topics that are covered are mentioned below.

Lewis acid mechanism

The Lewis acid mechanism occurs with early transition metals (e.g., Mo, V, Re) which generally utilize organic peroxides as oxidizing agents. The mechanism involves the transfer of the oxygen atom next to metal center, the proximal oxygen, in a concerted step that leaves the oxidation state of the metal unchanged.

Main group oxidations

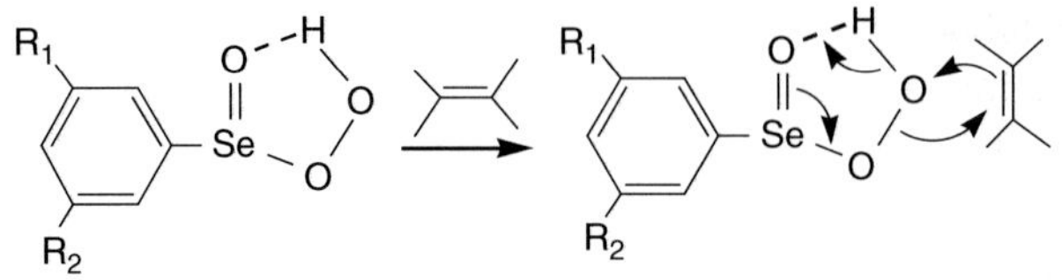

Main group elements such as Se, B, As, Al, and C carry out epoxidations with hydrogen peroxide and other oxidants such as persulfates. In this case, the oxygen atom transferred is the distal oxygen atom.

Hydroperoxide mechanism

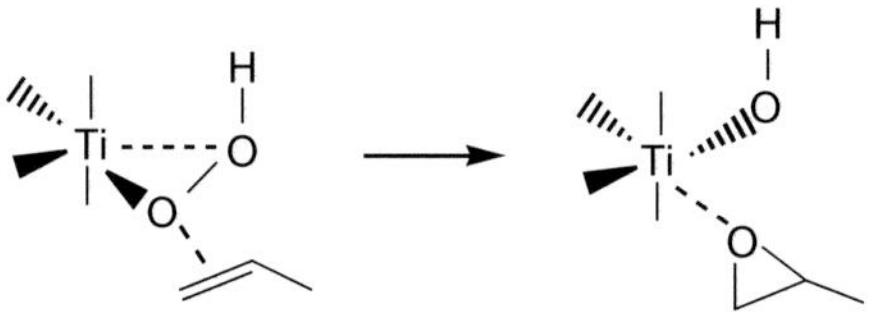

Ti carries out epoxidation with hydrogen peroxide transferring the proximal oxygen atom, like the early transition metals. However, in the case of Ti, the transfer is promoted by protic substances such as alcohols or water, making the use of dilute hydrogen peroxide possible.

Redox mechanism

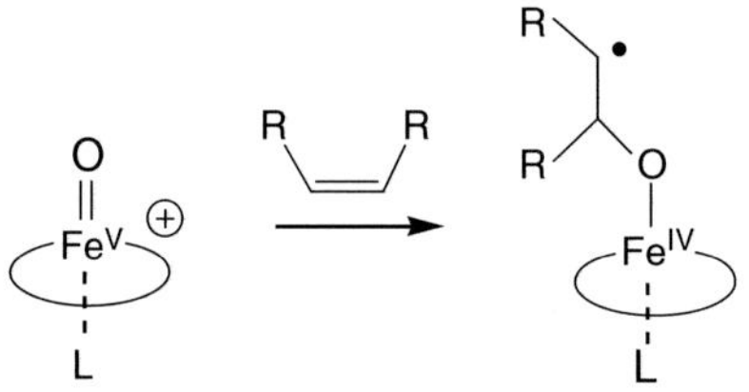

The redox mechanism operates with late transition metals, as exemplified by Mn, Fe, and Ru, which have easily accessible multiple oxidation states. Theoretical studies and trapping experiments suggest that in porphyrins and salen compounds, the mechanism involves the formation of a radical intermediate.

Oxametallacycle mechanism

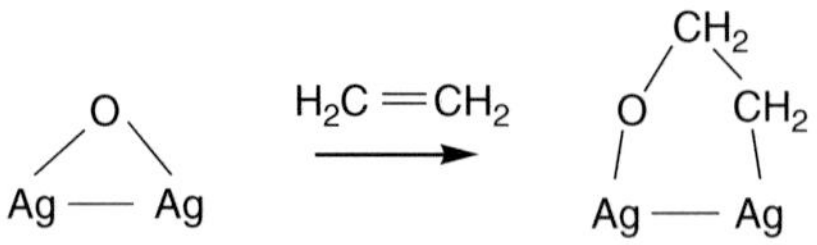

Although suggested to occur with other metals, strong kinetic and spectroscopic evidence for the oxametallacycle mechanism is found only with heterogeneous silver catalysts.

The rest of the book provides detailed and in-depth coverage of a broad selection of specialized topics. These were selectively taken from a major symposium I organized at the 234th National Meeting of the American Chemical Society in Boston, August 19–23, 2007, in the Division of Petroleum Chemistry, Inc. The preprints of the entire symposium are available online at http://membership.acs.org/P/PETR/.

In the book, the section on homogeneous catalysis covers soft Pt(II) Lewis acid catalysts, methyltrioxorhenium, polyoxometallates, oxaziridinium salts, and N-hydroxyphthalimide. The section on heterogeneous catalysis describes supported silver and gold catalysts, as well as heterogenized Ti catalysts, and polymer-supported metal complexes. The section on phase-transfer catalysis describes several new approaches to the utilization of polyoxometallates. The section on biomimetic catalysis covers nonheme Fe catalysts and a theoretical description of the mechanism on porphyrins.

I would like to extend my thanks to my Ph.D. advisor, Prof. Michel Boudart, and to my father-in-law, Prof. Kenzi Tamaru, who instilled in me the appreciation for the subjects of kinetics and mechanism and the importance of turnover frequency, which are a central part of this book. I am also grateful to Prof. Masatake Haruta, who initiated me in the fascinating subject of epoxidation. The very elegant cover illustrations were kindly provided by Prof. Mark Barteau, and show the steps by which ethylene is epoxidized through an oxometallacycle intermediate on a silver surface.

SECTION 1
Introduction

CHAPTER 1

Rates, Kinetics, and Mechanisms of Epoxidation: Homogeneous, Heterogeneous, and Biological Routes

S. Ted Oyama

Contents

1.	Introduction	4
2.	Epoxide Uses and Markets	5
3.	Catalysts and Rates in Commodity and Heterogeneous Epoxidation Processes	12
4.	Catalysts and Rates in Homogeneous Epoxidation Reactions	16
	4.1. Transition metal complexes	16
	4.2. Main group oxidations	21
5.	Catalysts and Rates in Biomimetic Epoxidation Reactions	25
6.	Catalysts and Rates in Biological Epoxidation Reactions	28
7.	Summary and Perspective on the Reactivity Results	30
8.	Oxidants for Epoxidation	35
9.	Mechanisms	37
	9.1. Heterogeneous epoxidation on silver catalysts	37
	9.2. Heterogeneous epoxidation on Ti catalysts	42
	9.3. Heterogeneous epoxidation on Au/titanosilicate catalysts	45
10.	Homogeneous Epoxidation by Early Transition Metals (Lewis Acid Mechanism)	47
	10.1. Molybdenum complexes	48
	10.2. Vanadium complexes	52
	10.3. Titanium complexes (Sharpless Ti tartrate asymmetric epoxidation catalyst)	53
	10.4. Polyoxometallates	56
	10.5. Methyltrioxorhenium	56
11.	Main Group Elements	57

Department of Chemical Engineering, Virginia Polytechnic Institute & State University, Virginia 24061, USA

12.	Homogeneous Epoxidation by Late Transition Metals (Redox Mechanism)	59
	12.1. Porphyrin complexes	60
	12.2. Salen complexes	66
	12.3. Nonheme Fe complexes	69
13.	Biological Systems	70
14.	Perspective and Conclusions	71
Acknowledgments		72
References		72

Abstract This chapter will cover the following topics. First, a survey of the commercial importance of epoxides will be given, including the latest available market figures for commodity products. Second, a description of the different catalysts employed for large-scale and small-scale chemical production will be provided, with an emphasis on the reporting of rates. Third, a compendium of oxidants will be presented. Finally, a detailed description of known aspects of mechanism will be covered for the main catalyst types, including heterogeneous, homogeneous, main-group, biomimetic, and biological catalysts.

Key Words: Ethylene oxide, Propylene oxide, Epoxybutene, Market, Isoamylene oxide, Cyclohexene oxide, Styrene oxide, Norbornene oxide, Epichlorohydrin, Epoxy resins, Carbamazepine, Terpenes, Limonene, α-Pinene, Fatty acid epoxides, Allyl epoxides, Sharpless epoxidation, Turnover frequency, Space time yield, Hydrogen peroxide, Polyoxometallates, Phase-transfer reagents, Methyltrioxorhenium (MTO), Fluorinated acetone, Alkylmetaborate esters, Alumina, Iminium salts, Porphyrins, Jacobsen–Katsuki oxidation, Salen, Peroxoacetic acid, P450 BM-3, *Escherichia coli*, Iodosylbenzene, Oxometallacycle, DFT, Lewis acid mechanism, Metalladioxolane, Mimoun complex, Sheldon complex, Michaelis–Menten, Schiff bases, Redox mechanism, Oxygen-rebound mechanism, Spiro structure.   © 2008 Elsevier B.V.

1. INTRODUCTION

The epoxidation of olefins plays an important role in the industrial production of several commodity compounds, as well as in the synthesis of many intermediates, fine chemicals, and pharmaceuticals. The scale of production ranges from millions of tons per year to a few grams per year. The diversity of catalysts is large and encompasses all the known categories of catalyst type: homogeneous, heterogeneous, and biological.

Great advances have been made in the field of catalytic epoxidation in the last 10 years, and continue to be made as new catalysts and reactions are discovered and new processes are developed. Of particular interest has been the subject of mechanism, which has played an important role not only in improving technology, but also in advancing fundamental knowledge in the area of chemistry. Notable are the high levels of understanding that have been achieved in the areas of heterogeneous epoxidation on Ag catalysts, homogeneous catalysts based on Ti, V, Mo, W, and Re, and biomimetic catalysts based on Fe and Mn. Much insight is provided by the advent of theoretical methods, notably density functional theory (DFT), which provide detailed pictures of the relationship between structure and reactivity, and which can be used to distinguish between competing views on reaction pathways.

The goal of this chapter is to present a detailed overview of the epoxidation area and to organize the advances made in the area of mechanism, so as to give a perspective on similarities and differences between different catalysts. It is often said that mechanisms come and go, but rates remain firm. With this in mind, considerable attention is also placed on the reporting of rates. Because of the large difference in catalyst types, the conditions are not generally the same, but an attempt is made to utilize turnover frequencies (TOFs) so as to permit comparisons between different systems.

An important aspect of epoxidation is the source of the oxygen. The most desirable terminal source is molecular oxygen, which is a plentiful and economical reagent. However, aside from silver, other catalysts almost universally employ activated forms of oxygen, which involve the use of a sacrificial reductant for their production. Hydrogen peroxide has emerged as the most attractive oxidant at laboratory and commercial scales, and the reasons and implications of this will be discussed.

2. EPOXIDE USES AND MARKETS

Major commodity chemicals produced by epoxidation are ethylene oxide and propylene oxide. Epoxybutene is an important intermediate, but is no longer produced on a large scale. The general subject of epoxides has been reviewed, including properties and preparation by stoichiometric methods [1]. Reliable production and price data are not available for most epoxides because they are used captively, and only available market values for some commodity chemicals will be presented.

Ethylene oxide (EO) is an important commodity chemical [2]. It is mainly used to produce ethylene glycol (EG) and surface-active agents such as nonionic alkylphenol ethoxylates and detergent alcohol ethoxylates. Several dozen important fine petroleum and chemical intermediates are derivatives of EO, and it is therefore extensively used in applications such as washing/dyeing, electronics, pharmaceuticals, pesticides, textiles, papermaking, automobiles, oil recovery, and oil refining. The worldwide consumption of EO in 2002 was 1.47×10^7 metric tons year^{-1}, making it the most utilized epoxide species. Usage was divided among North America, 10%; Western Europe, 27%; Japan, 20%; other Asia, 6%; and other regions, 3% [3]. The merchant price of ethylene oxide for 2006 is listed as US$ 1.66–1.88 kg^{-1} [4].

Propylene oxide (PO) is used in the production of polyurethane polyols (60–65%), propylene glycols (20–25%), P-series glycol ethers (3–5%), di- and tri-propylene glycols (3–5%), and other miscellaneous chemicals (10%), which include polyalkylene glycols, allyl alcohol, and isopropylamines [5,6]. Polyurethane polyols are used in the manufacture of polyurethane foams, while propylene glycols are used to make unsaturated polyester resins for the textile and construction industries. Propylene glycols also are employed in drugs, cosmetics, solvents, and emollients in food, plasticizers, heat transfer and hydraulic fluids, and antifreezes. PO also finds use in the sterilization of packaged foods and as a pesticide [7]. The worldwide production of PO is reported to have been 5.78×10^6 metric tons in 1999 [5] and is estimated to have been 6.74×10^6 metric tons in 2003 [8], making PO the second most important epoxide product by amount produced. Notable is the growth in China where usage is predicted to rise to 1.1×10^6 metric tons in 2010. The merchant price of propylene oxide for 2006 is listed as US$ 1.88–2.03 kg^{-1} [4].

Butadiene epoxidation to epoxybutene (EpB^{TM}) was practiced at a semiworks scale of 1.4×10^3 metric tons year^{-1} by Tennessee Eastman [9] between 1997 and 2004 [10]. Epoxybutene is a versatile intermediate [11] that can be used to produce a large variety of different products such as epoxybutane, 1,4-butane diols and alcohols, 1,2-butane diols and alcohols, 2,3-dihydrofuran, 2,5-dihydrofuran, tetrahydrofuran, N-methylpyrrolidone, cyclopropyl carboxyaldehyde (CPCA) derivatives, vinyl ethylene carbonate (addition of CO_2 to EpB), and 3,4-dihydroxy-1-butene (addition of H_2O to EpB).

There are many other epoxide compounds produced in smaller scale for specialized, but important applications. These include aliphatic, alicyclic, aromatic, and heterosubstituted compounds.

Terminal and internal olefin epoxides such as 1-hexene epoxide, 1-octene epoxide, 1-decene epoxide, etc. are used as stabilizers for halogen hydrocarbons, as reactive diluents (e.g., for epoxy resins), as resin modifiers, or as coating materials. The terminal epoxides, as well as their *iso*-derivatives, are also useful for conversion to hydroxyethers by ring-opening of the epoxide with alcohols. These hydroxyethers are important ingredients in creams, lotions, and ointments for use in cosmetic and pharmaceutical applications, serving as oil-soluble bases in which many lipid-soluble solids can be dissolved [12]. The products are valuable because they are nonsticky and have good compatibility with skin, providing smoothness, and suppleness.

Isoamylene oxide (2-methyl-2-butene oxide) is used commercially as a stabilizer for chlorinated hydrocarbons like 1,1,1-trichloroethane and trichloroethylene [13]. These chlorohydrocarbons find application in metal cleaning (including light metal alloys) and must be protected against decomposition and formation of acidic substances during storage and application. The basic stabilizer system also contains either an organic amine or a nitro compound in addition to the low-molecular weight epoxide.

Cyclohexene oxide is useful as a monomer in polymerization and the coating industry. It is used in the synthesis of alicyclic molecules used in pesticides, pharmaceuticals, perfumery, and dyestuffs, and as a monomer in polymerization with CO_2

to yield aliphatic polycarbonates [14]. It can also lead to enantiopure β-aminoalcohols and β-diamines, which are useful as chirality-inducing ligands and as precursors for chiral oxazolidinones, oxazinones, phosphonamides, etc. [15].

Styrene oxide is used industrially as a reactive diluent for epoxy resins, but is also an extremely useful building block for the synthesis of chiral and nonchiral organic compounds [16]. For example, it can be used for the preparation of chiral aziridines with retention of configuration [17], and chiral ring-opened products with alcohols [18] and amines [19]. A general reaction of epoxides is the Meinwald rearrangement [20] which has been used to produce a wide range of aldehydes and ketones [21]. Reaction with group 5 or 6 metal halides produce an eight-membered ring compound, 2,3,6,7-dibenzo-9-oxabicycle[3,3,1]nona-2,6-diene [22]. Reaction with aliphatic ketones catalyzed by Y, USY, and ZSM-5 zeolites yield 4-phenyl-1–1,3-dioxolanes [23].

Norbornene oxide can react with a *bis*-(cyclopentadienyl)(*tert*-butylimido)-zirconium complex ($Cp_2Zr=N-t$-Bu) tetrahydrofuran (THF) to produce the 1,2-amino alcohol [24], and can be transformed enantioselectively by α-deprotonation-rearrangement to (-)-nortricyclanol in up to 52% ee using a nonracemic lithium amide, or an organolithium compound in the presence of (-)-sparteine [25].

Epichlorohydrin or chloromethyloxirane is manufactured from allyl chloride, and, in 2006, had a merchant price of US$ 1.66 kg^{-1} [4]. It is used as a building block in the manufacture of plastics, epoxy resins, phenoxy resins, and other polymers, and as a solvent for cellulose, resins, and paints, and has also found use as an insect fumigant. Epoxy resins (aryl glycidyl ethers) are manufactured successfully in large scale (1.2×10^6 metric tons in 2000) [26] and are widely used in a variety of industrial and commercial applications [27]. These are made by addition reactions of epichlorohydrins or by epoxidation of allyl ethers or esters (Table 1.1). Epichlorohydrin can be reacted with an alkali nitrate to produce glycidyl nitrate, an energetic binder used in explosive and propellant compositions.

Various epoxides have pharmacological applications and their synthesis have been described in a review of oxidation chemistry [28]. The epoxide of *trans*-stilbene has estrogen-mimic properties and induces drug-metabolizing enzymes in rat and mouse livers [29,30]. For example, it is a potent inducer of epoxide hydratase in rat liver which does not significantly alter cytochrome P-450 content [31]. Carbamazepine (CBZ), 5H-dibenzepine-5-carboxamide, is an antiepileptic drug used in clinical practice as a first-line treatment for generalized tonic–clonic and partial seizures [32–34]. The epoxide of this anticonvulsant, carbamazepine 10,11-epoxide (CBZ-EP), is as pharmacologically active as the parent compound in experimental animals [35,36]. Its synthesis by an immobilized Jacobsen–Katsuki Mn(salen) catalyst [37] and a Mn(porphyrin) [32] has been described. A precursor to the antihypertensive agent levcromakalim has been obtained by the highly asymmetric epoxidation with iminium salts [38]. The intermediate, 3,4-epoxytetrahydrofuran, is a useful building block in combinatorial chemistry for pharmaceutical applications [39], for example, as a key synthesis component for the HIV-protease inhibitor nefinavir [40]. The epoxide has been obtained in high yield by the heterogeneous epoxidation of 2,5-dihydrofuran with H_2O_2 over a novel titanosilicate catalyst, Ti-MWW [41].

TABLE 1.1 Epoxy resins: Epoxidation reactions and catalysts

Reactant	Catalyst and conditions	References
[Aryl glycidyl ether structure] R_1 = isopropyl, R_2 = H, R_3 = methyl Aryl glycidyl ether	$Mo(CO)_6$/TBHP; 65 °C, 1 atm, 8 h, $Cl_2CH–CHCl_2$ solvent; conv. 92% TBHP; select. 98%	[27]
[Diaryl glycidyl ether structure] R_1 = R_3 = methyl, Z = isopropylidine Diaryl glycidyl ether	$Mo(CO)_6$/TBHP; 65 °C, 1 atm, 16 h, $Cl_2CH–CHCl_2$ solvent; conv. 81% TBHP; select. 58% diepoxide; select. 36% monoepoxide	[27]

Some epoxides carry functional groups. The compound (1R,2S)-(–)-(1,2)-epoxypropyl phosphonic acid (fosfomycin) is a clinically important drug with wide-spectrum antibiotic activity. It was isolated originally from a fermentation broth of *Streptomyces fradiae* and prepared mainly by epoxidation of cis-1-propenylphosphonic acid (CPPA) [42] followed by optical resolution of the racemic epoxide with chiral amines. Recently, chiral W^{VI}(salen) and Mo^{VI}(salen) complexes have been used in the asymmetric epoxidation of CPPA [43].

A large variety of epoxides with pharmacological and medicinal properties are enumerated by Jacobsen and coworkers [44]. These include fumagillin (antibiotic originally used against fungal Nosema apis infections) [45], ovalicin (potent immunosuppressive agent) [46], coriolin (antitumor, low-toxicity agent) [47], disparlure (sex pheromone of the gypsy moth) [48], triptolide (potent immunosuppressive agent) [49], periplanone (pesticide, it is the specific sex-attractant of cockroaches) [50], neocarzinostatin chromophore (antitumor protein) [51], trapoxins (specific inhibitors of histone deacetylases) [52], epothilones (anticancer drugs which arrest division of cells) [53], and FR901464 (antitumor agent) [54].

The epoxidation of terpenic substrates is of interest in the flavor and fragrance industry [55,56]. Terpenes are derivatives of isoprene, which has formula C_5H_8 (2-methyl-*trans*-butadiene). There are tens of examples of terpenes, including limonene, α-pinene, geraniol, citronellol, myrcene, ocimene, camphene, α-terpineol, menthol, and isopugelol. Limonene is an abundant monoterpene extracted from citrus oil, which can be epoxidized to obtain fragrances, perfumes, and food additives. The terpene α-pinene, when converted to the epoxide, can be used in the one-step synthesis of α-campholenic alcohol, naturanol [57], a valuable fine chemical largely used in the food and perfumery industry due to its particularly well-defined sweet, natural, and berry-like fragrance [58]. Another industrially important reaction is the rearrangement of optically active α-pinene oxide to campholene aldehyde (e.g., in the presence of zinc bromide), which can undergo aldol condensation with lower aliphatic aldehydes or ketones (e.g., propionaldehyde, butyraldehyde, or acetone) to form unsaturated aldehydes or ketones that can be reduced to the corresponding saturated alcohols [59]. These are used in the fragrance industry as a sandalwood scent [60]. The hydrolysis and rearrangement of α-pinene oxide gives therapeutically active substances [61]. Many catalyst systems are utilized for the epoxidation of terpenic substances, including immobilized phosphotungstates [62,63] and mesoporous materials [55] (Table 1.2). The challenge is that very often the terpenes carry a second oxidizable group, such as an alcohol.

Fatty acid epoxides have numerous uses. In particular, oils and fats of vegetable and animal origin represent the greatest proportion of current consumption of renewable raw materials in the chemical industry, providing applications that cannot be met by petrochemicals [64]. Polyether polyols produced from methyl oleate by the Prileshajev epoxidation (using peracetic acid) are an example. Epoxidized soybean oil (ESBO) is a mixture of the glycerol esters of epoxidized linoleic, linolenic, and oleic acids. It is used as a plasticizer and stabilizer for poly(vinyl chloride) (PVC) [1] and as a stabilizer for PVC resins to improve flexibility, elasticity, and toughness [65]. The ESBO market is second to that of epoxy resins and its worldwide production was 2×10^5 tons year^{-1} in 1999 [66]. High-temperature greases are prepared from special hydroxylated fatty acids obtained from the splitting of castor oil. Fatty acid epoxides may be used to produce polyols with high viscosity for grease preparations [67]. Fatty acid epoxides from the C18 family are used as defense compounds in infected plants [68,69] (e.g., to combat fungal aggression). Fatty acid epoxides are also used to form coating compositions by reactions with urethanes and alcohols [70], plasticizers by reaction with triacetin to form epoxidized glyceride acetate [71], and antifoaming compounds by reactions with alcohols or alkylamides [72]. The best-known epoxy acid from natural sources is vernolic acid (12,13-epoxy-9-octadecenoic acid) and is an epoxy oleic acid [73]. Epoxyeicosatrienoic acid is an important signalling compound in the endothelium which induces dilation of coronary arteries and may be useful for treatment of coronary artery disease [74].

Allyl epoxides are produced by the acclaimed Sharpless asymmetric epoxidation reaction [75], and are important intermediates and products. For example, an allyl epoxide is a vital part of the structure of amphidinolides, a series of unique macrolides isolated from dinoflagellates (*Amphidinium* sp.). Amphidinolide H (AmpH) is a potent cytotoxic 26-membered macrolide with potent cytotoxicity for several carcinoma cell lines [76]. An allyl epoxide is involved in the total synthesis of prostaglandin A2 with a cuprate reagent [77]. Allyl epoxides derived from Sharpless chemistry are a practical method for construction of polypropionate structures by Lewis acid-induced rearrangement [78,79]. Other allyl epoxides such as 1,2-epoxy-3-methyl-3-butanol are useful organic intermediates for the production of α-hydroxyketones, which are used for the synthesis of various natural

TABLE 1.2 Terpenic compounds: Epoxidation reactions and catalysts

Reactant	Catalyst and conditions	References
Limonene	$[PW_4O_{24}]^{3-}$-Amberlite IRA-900/H_2O_2; 33 °C, 1 atm, CH_3CN solvent, 24 h; conv. 77%, select. 93% (endo epoxide); 59% H_2O_2 select., TOF = 5.8×10^{-4} s^{-1}	[62,63]
Limonene, Isopulegol	Ti-MCM-41/H_2O_2; 90 °C, 1 atm, CH_3CN solvent, 24 h. Limonene: conv. 59%; select. 90% (endo); TOF = 3.1×10^{-4} s^{-1}. Isopulegol: conv. 71%; select. 80%; TOF = 3.3×10^{-4} s^{-1}	[55]
α-Terpineol, Terpinen-4-ol	Ti-MCM-41/H_2O_2; 90 °C, 1 atm, CH_3CN solvent, 24 h. α-Terpineol: conv. 86%; select. 51%; TOF = 2.6×10^{-4} s^{-1}. Terpinen-4-ol: conv. 74%; select. 61%; TOF = 2.7×10^{-4} s^{-1}	[55]
Carveol, Carvotanacetol	Ti-MCM-41/H_2O_2; 90 °C, 1 atm, CH_3CN solvent, 24 h. Carveol: conv. 78%; select. 73%; TOF = 3.4×10^{-4} s^{-1}. Carvotanacetol: conv. 84%; select. 64%; TOF = 3.2×10^{-4} s^{-1}	[55]

products including biologically active compounds sugars, and β-hydroxy-α-amino acids [80,81]. The ketones are difficult to obtain by conventional means.

In summary, epoxides are produced not only as endproducts, but also as intermediates because they are valuable building blocks in synthetic organic chemistry [82–84] (Table 1.3). Until recently, epoxide intermediates were produced by direct oxygen transfer to olefins by a variety of stoichiometric methods. Recently, considerable efforts have been made to conduct the transformations selectively under catalytic conditions. Because epoxides are reactive substances, they can undergo diverse transformations by reactions with acids and bases, and their reactivity has been exploited to form a diverse range of products by so-called click chemistry [85,86], which combines the breadth of combinatorial methods with the precise synthesis of organic chemistry.

TABLE 1.3 Some epoxides and their endproducts

Olefin starting materials	Epoxides	Product	References
1-Hexene, 1-Octene		Hydroxyethers (R^1, OR^3, R^2, OH)	[12]
2-Methyl-2-butene		Stabilizer for chlorinated solvents	[13]
Cyclohexene		β-Aminoalcohols, β-Diamines	[15]
(styrene)		2,3,6,7-Dibenzo-9-oxabicyclo[3,3,1]nona-2,6-diene	[22]
Norbornene		Nortricyclanol	[25]
Allyl chloride		Epoxy resin (with bisphenol A)	[27]
Allyl alcohols		Proprionates	[78,79]
2-Methyl-3-buten-2-ol		α-Hydroxyketones	[80,81]
Limonene		Fragrances, perfumes, and food additives	[56]
α-Pinene		Fragrances, pharmaceuticals	[59,60]

(*continued*)

TABLE 1.3 (continued)

Olefin starting materials	Epoxides	Product	References
trans-Stilbene		Enzyme inducer, artificial hormone	[29,30]
Carbamazepine		Anticonvulsant	[37]
6-Cyano-2,2-dimethylbenzopyran		Levcromakalim precursor (antihypertensive agent)	[38]
2,5-dihydrofuran		Nefinavir precursor (HIV-protease inhibitor)	[41]
cis-1-Propenyl-phosphonic acid		Fosfomycin (antibiotic)	[43]
cis-9-Octadecenoic acid (oleic acid)		High-temperature greases, defoaming compounds, antifungal compounds	[67,68,72]

3. CATALYSTS AND RATES IN COMMODITY AND HETEROGENEOUS EPOXIDATION PROCESSES

Various catalysts are used in different commodity epoxidation processes at diverse conditions. Important characteristics are the space time yield based on the weight of epoxide formed per kilogram of catalyst, and the TOF as a measure of the intrinsic activity per active surface metal atom (Table 1.4). In Table 1.4 and other tables in this chapter, conversion is always based on the reactant olefin and not on the oxidant, unless specified. Selectivity refers to the epoxide product, and yield to the product of conversion and selectivity, unless defined otherwise.

EO is mainly produced by the direct oxidation of ethylene with air or oxygen in a packed-bed, multitubular reactor with recycle [2]. Catalysts for EO production

TABLE 1.4 Commodity epoxidation reactions and catalysts

Reactant	Catalysts and conditions	References
Ethylene	Shell process; Ag/α-Al$_2$O$_3$ + Re + Cs promoters/O$_2$; 230–280 °C, 10–35 atm, 3 ppmv ethylchloride; conv. ~10%, select. ~85%. Yield^a = 44–440 g kg$_{cat}^{-1}$ h^{-1}, TOF = 0.014–0.15 s^{-1}	[99]
Propylene	Halcon styrene monomer process; Mo naphthenate + K naphthenate promoter/EBHP; 90 °C, 10 bar; conv. hydroperoxide 92%, select. 90%. Yield^a = 72 g kg Mo^{-1} h^{-1}, TOF = 0.11 s^{-1}	[101]
Propylene	Eniricerche process; TS-1 (Si/Ti = 46)/H$_2$O$_2$; 30 °C, 4 atm, MeOH solvent, *tert*-butylmethyl ether, 10 min; conv. (H$_2$O$_2$) 90%, (C$_3$H$_6$) 5%, select. 97%. Yield^a = 7,000 g kg$_{cat}^{-1}$ h^{-1}, TOF = 9.2 × 10^{-2} s^{-1}	[105,106]
Propylene	Degussa–Huls–Headwaters hydrogen peroxide process; TS-1 + liquid base promoter (ammonia)/H$_2$O$_2$; 50 °C, 15 bar, pH = 8.5, solvent H$_2$O, MeOH, MTBE; conv. (H$_2$O$_2$) 94%, (C$_3$H$_6$) 19%, select. 95%. Yield^a = 770 g kg$_{cat}^{-1}$ h^{-1}, TOF = 2.2 × 10^{-2} s^{-1}	[108]
Butadiene	Eastman Chemical process; 12 wt% Ag–Cs/α-Al$_2$O$_3$/O$_2$; 210 °C, 1 atm, C$_4$H$_6$/O$_2$/He = 1/1/4, SV = 5,400 h^{-1}. Yield^a = 250 g kg$_{cat}^{-1}$ h^{-1}, TOF = 0.12 s^{-1}	[9]

^{a} Space time yield (g of epoxide per kg catalyst per hour).

are generally composed of silver (10–20 wt%) supported on a low–surface area (1–2 m^2 g^{-1}) α-Al$_2$O$_3$ support with alkali metal (500–1,200 wppm) promoters [87–89], especially Cs [90–92], and may contain fluoride [93], Re [94], and other promoters (see Chapter 8). Chlorine is also added as a promoter at the ppm level [91,95,96,97] in the form of a gas-phase additive such as dichloroethane (C$_2$H$_4$Cl$_2$) or vinyl chloride (C$_2$H$_3$Cl). High–surface area aluminas are poor support materials for EO [98] (see Chapter 9). Typical conditions [89,99] are 10–35 atm, 230–280 °C with gas hourly space velocity (GHSV) 1,500–10,000 h^{-1}, and contact time 0.1–5 s. The space time yield of industrial EO production is given as 0.032–0.32 g$_{EO}$ h^{-1} (cm^3 cat)$^{-1}$ [99], 330 kg$_{EO}$ m^{-3} h^{-1} [93], and 2–25 lbs$_{EO}$ ft^{-3} h^{-1} [88]. Given that the density of typical catalysts is about 45 lb ft^{-3} (0.72 g cm^{-3}) [88,89], these values translate to a space time yield of about 44–440 g$_{EO}$ kg$_{cat}^{-1}$ h^{-1}. Assuming full coverage of the support with silver with a site density of 10^{19} m^{-2}, a TOF of 0.014–0.15 s^{-1} can be calculated (Table 1.4).

PO is produced by several indirect, multistep processes (Fig. 1.1). A review describing the most important processes and recent advances is available [100]. Historically, the first process developed was the chlorohydrin process in which

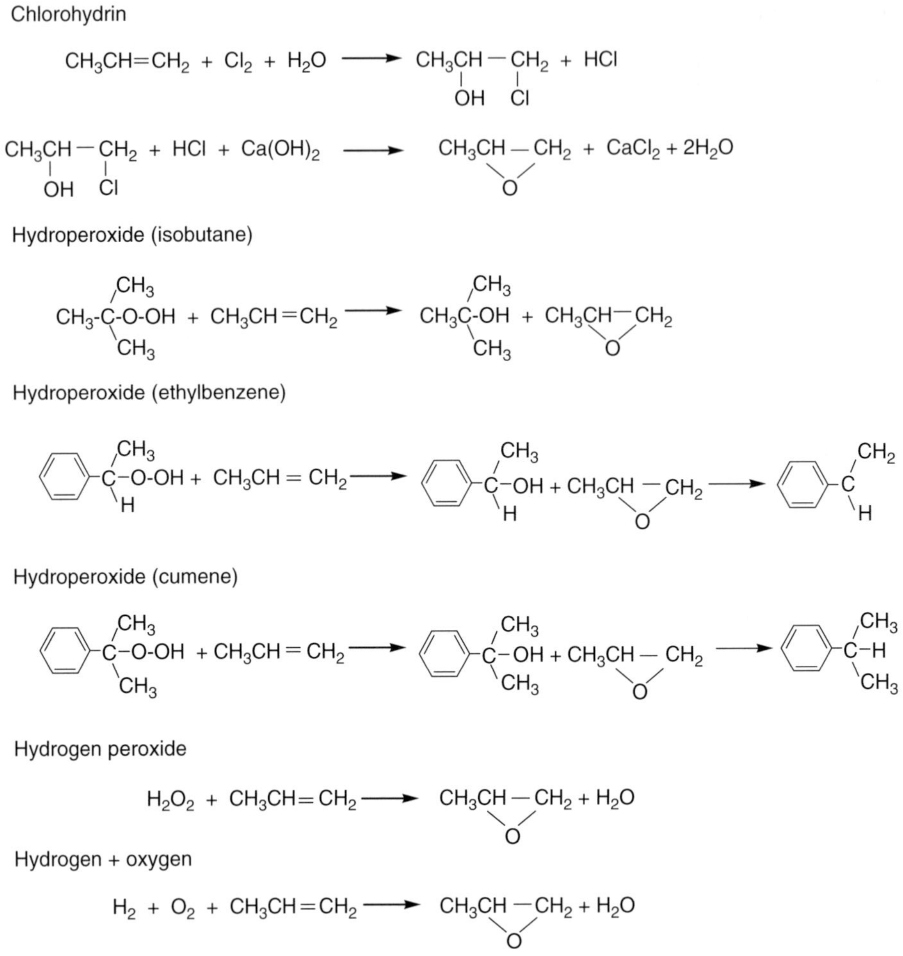

Chlorohydrin

$$CH_3CH=CH_2 + Cl_2 + H_2O \longrightarrow \underset{\underset{OH}{|}\underset{Cl}{|}}{CH_3CH-CH_2} + HCl$$

$$\underset{\underset{OH}{|}\underset{Cl}{|}}{CH_3CH-CH_2} + HCl + Ca(OH)_2 \longrightarrow CH_3CH\underset{O}{\overset{}{\diagdown\diagup}}CH_2 + CaCl_2 + 2H_2O$$

Hydroperoxide (isobutane)

$$\underset{\underset{CH_3}{\diagup}\overset{CH_3}{\diagdown}}{CH_3-C-O-OH} + CH_3CH=CH_2 \longrightarrow \underset{\underset{CH_3}{\diagup}\overset{CH_3}{\diagdown}}{CH_3C-OH} + CH_3CH\underset{O}{\overset{}{\diagdown\diagup}}CH_2$$

Hydroperoxide (ethylbenzene)

$$Ph\underset{\underset{H}{\diagup}\overset{CH_3}{\diagdown}}{C-O-OH} + CH_3CH=CH_2 \longrightarrow Ph\underset{\underset{H}{\diagup}\overset{CH_3}{\diagdown}}{C-OH} + CH_3CH\underset{O}{\overset{}{\diagdown\diagup}}CH_2 \longrightarrow Ph\underset{\underset{H}{\diagup}\overset{CH_2}{\diagdown\diagdown}}{C}$$

Hydroperoxide (cumene)

$$Ph\underset{\underset{CH_3}{\diagup}\overset{CH_3}{\diagdown}}{C-O-OH} + CH_3CH=CH_2 \longrightarrow Ph\underset{\underset{CH_3}{\diagup}\overset{CH_3}{\diagdown}}{C-OH} + CH_3CH\underset{O}{\overset{}{\diagdown\diagup}}CH_2 \longrightarrow Ph\underset{\underset{CH_3}{\diagup}\overset{CH_3}{\diagdown}}{C-H}$$

Hydrogen peroxide

$$H_2O_2 + CH_3CH=CH_2 \longrightarrow CH_3CH\underset{O}{\overset{}{\diagdown\diagup}}CH_2 + H_2O$$

Hydrogen + oxygen

$$H_2 + O_2 + CH_3CH=CH_2 \longrightarrow CH_3CH\underset{O}{\overset{}{\diagdown\diagup}}CH_2 + H_2O$$

FIGURE 1.1 Technologies for production of propylene oxide.

propylene is reacted with chlorine and water to form the chlorohydrin, which in turn is reacted with a base such as calcium hydroxide to form PO and calcium chloride. The calcium chloride requires disposal.

Although the chlorohydrin process still accounts for almost half of worldwide PO production, new plants in the last 20 years have exclusively used epoxidation with organic hydroperoxides produced by homogeneous oxidation of isobutane, ethylbenzene, and cumene (Fig. 1.1). These are the bases of the so-called Oxirane, Halcon, ARCO (isobutane and ethylbenzene), Shell (ethylbenzene), and Sumitomo (cumene) processes. The early work of Kollar [101] showed that Mo was the best soluble catalyst, that various hydroperoxides, including *tert*-butyl hydroperoxide, ethylbenzene hydroperoxide, and cumene hydroperoxide, were effective epoxidizing agents, and that the addition of an alkali metal improved selectivity. Mo is an effective catalyst for epoxidation with alkylperoxides because it withdraws electrons from a coordinated alkylperoxo moiety, thereby increasing the electrophilic character of the peroxidic oxygens [102]. In this manner, Mo acts as a Lewis acid. Shell has developed a heterogeneous route to PO using Ti/SiO$_2$ catalysts and ethylbenzene hydroperoxide [103] (see Chapters 13 and 14). Homogeneous Ti catalysts are not very active for oxidations with RO$_2$H hydroperoxides because of facile oligomerization of oxotitanium (IV) species to unreactive μ-oxotitanium oligomers. The Shell Ti(IV)/SiO$_2$ catalyst is effective because of the formation of site-isolated Ti species on the surface of the support [104], and because of the increased Lewis acidity of the Ti(IV) due to electron withdrawal by the silanoxy ligands [102].

All these processes suffer the drawback of producing a coproduct that needs to be sold separately (*tert*-butyl alcohol, styrene) or recycled (cumyl alcohol). They are also multistep, and require complex facilities.

An emerging, simpler technology for PO production is the epoxidation of propylene with hydrogen peroxide, which does not produce organic coproducts. The reaction is run with a titanosilicate catalyst like TS-1 in a methanol solvent in the liquid phase with propylene bubbling through. The reaction was discovered by researchers at Eniricerche [105–107]. In a patent by Thiele of Degussa, the reaction is run in a mixed aqueous-organic solvent [methanol and an ether like methyl *tert*-butyl ether (MTBE)] with an added base in solution [108]. The base, which can be an alkali compound or ammonia, generally increases the pH from a level of 2–5 to 8–9 and results in a slight decrease in H$_2$O$_2$ conversion, but increases PO selectivity. For example, at 50 °C and 15 bar using 68 g of TS-1 with a feed of 600 g h^{-1} of 6.3% H$_2$O$_2$, 15.9% H$_2$O, 77.4% MeOH, 0.3% MTBE, mixed with 200 g h^{-1} of propylene, increasing the pH from 4.8 to 8.5 causes the H$_2$O$_2$ conversion to decrease from 97 to 94% but the PO selectivity to increase from 82 to 95%. At the latter conditions, assuming a utilization of H$_2$O$_2$ of 80% [106], which would result in a propylene conversion of 19%, the space time yield is 770 g$_{PO}$ kg$_{cat}^{-1}$ h^{-1}, a very high value. Further assuming that the Ti sites in TS-1 are responsible for the epoxidation and that the Ti/Si ratio is 1/100, a TOF of 0.022 s^{-1} can be calculated (Table 1.4).

On a stoichiometric basis, 1 kg of H$_2$O$_2$ is required to produce 1.7 kg of propylene oxide, and thus requires a substantial production capacity of H$_2$O$_2$ [109]. For this reason, it would be advantageous to have a direct means of producing PO from H$_2$ and O$_2$. The first report on epoxidation of propylene using *in situ* generated H$_2$O$_2$ from H$_2$/O$_2$ mixtures was in a 1992 patent to Sato and Miyake [110] using a catalyst comprising a group 10 metal like Pd supported on a crystalline titanosilicate and an optional solvent. The use of H$_2$/O$_2$ mixtures in the gas phase first appeared in the open literature in 1995 with reports by the group of Haruta [111,112] with Au/TiO$_2$, and subsequently by the group of Hölderich [113–116] with Pd–Pt/TS-1 catalysts. A patent in 1996 by Haruta *et al.* [117] describes a gold catalyst supported on titania. Since then, this area has seen considerable activity (see Chapters 10–12). Many of the catalysts reported suffered from deactivation. Substantial improvements were obtained with catalysts in which the Ti centers were isolated. The highest PO yield in the open literature is reported by Delgass and collaborators and was obtained on a stable Au/TS-1 catalyst at 200 °C and 1 atm [118]. They obtained a propylene conversion of 8.8% with PO selectivity of 81%, and a TOF based on Au of 0.33 s^{-1}, with the reasonable assumption of 100%

dispersion of Au given the low loading of 0.01 wt% Au (Table 1.4). Another stable catalyst is Au–Ba/Ti-TUD, although it tends to produce lower TOFs because it operates at lower temperatures [119].

Epoxybutene can be obtained by the epoxidation of 1,3-butadiene using silver catalysts supported on α-Al$_2$O$_3$. The catalyst was developed by Monnier and coworkers, and typically contains 10–20 wt% silver and is promoted by CsCl [120,121], RbCl [120,121], or TlCl [122] at optimal levels of 6–8 μmol promoter cation gcat^{-1} [9], and can be used in the presence of a gas-phase fluorinated hydrocarbon such as C$_2$F$_6$ [123] and C5–C10 hydrocarbon diluents [124]. For example [9], for a 12 wt% Ag–Cs/α-Al$_2$O$_3$ catalyst at 210 °C and 1 bar using a feed composition of He/C$_4$H$_6$/O$_2$ = 4/1/1 at a space velocity of 5,400 h^{-1} the butadiene conversion is 8.5% and the epoxybutene selectivity is 94%. Assuming a packing density for the support of 0.9 g cm^{-3}, this gives a space time yield of 250 g$_{EB}$ kg$_{cat}^{-1}$ h^{-1}, which is reasonable. A typical surface area used in the preparations is 0.5 m^2 g^{-1}, which, if assumed is totally covered by silver with a site density of 10^{19} m^{-2}, can be used to calculate a TOF of 0.12 s^{-1} (Table 1.4).

The commodity processes all target small molecules, as these are versatile intermediates in the chemical industry. With the exception of the Mo-based hydroperoxide process for PO production, most of the catalysts used in commodity chemical manufacture are solid catalysts. Other heterogeneous catalysts are used in the production of large molecules (Table 1.5). Thomas *et al.* compare the TOFs in the epoxidation of cyclohexene with *tert*-butyl hydroperoxide (TBHP) of various titanosilicates including Ti-MCM-41, and report a highly active material containing Ge, and find that the rates are comparable to those of analogous homogeneous materials [125]. The group of Baiker has studied titania–silica mixed aerogel oxides for the epoxidation of larger sized olefins [126] and functionalized olefins [127]. Caps and coworkers have investigated Au/TiO$_2$ catalysts for the epoxidation of *trans*-stilbene and found that the catalyst induces the generation of radicals and one-electron chain reactions [128].

4. CATALYSTS AND RATES IN HOMOGENEOUS EPOXIDATION REACTIONS

4.1. Transition metal complexes

The vast majority of homogeneous catalysts are transition metal complexes and many systems have been reported, for example, Ru(III) [129], W(VI) [130], polyoxometallates [131], Re(V) [132], Fe(III) [133], and Pt(II) [134] with hydrogen peroxide, Mn(II) [135–137] with peracetic acid, and Ti-tartrate with alkyl hydroperoxides [75]. The subject of epoxidation by H$_2$O$_2$ has been reviewed [138–140].

A summary of typical homogeneous catalysts, oxidants used, conditions of use, conversions and yields based on the olefin reactant (unless specified), and TOF is provided at the end of this section (Table 1.6). It is lamented that in general, in the homogeneous epoxidation catalysis field, TOFs are not often reported, as more emphasis is placed on selectivity than on rate. Many times, the reactions are run with hydrogen peroxide as oxidant, in excess because of its tendency to decompose, and the addition is not controlled carefully. The reported TOFs are

TABLE 1.5 Heterogeneous epoxidation reactions and catalysts

Reactant	Catalysts and conditions	References
Cyclohexene	TiO_2–SiO_2/TBHP; 30 °C, 1 atm, CH_3CN solvent, 1 h, select. > 95%; TOF = 7.2 × 10^{-3} s^{-1}	[125]
Cyclohexene	Ti-MCM-41/TBHP; 30 °C, 1 atm, CH_3CN solvent, 1 h, select. > 95%; TOF = 9.4 × 10^{-3} s^{-1}	[125]
Cyclohexene	Ti-Ge-MCM-41/TBHP; 30 °C, 1 atm, CH_3CN solvent, 1 h, select. > 95%; TOF = 1.1 × 10^{-2} s^{-1}	[125]
Cyclohexene	TiO_2–SiO_2 aerogel/cumene hydroperoxide; 90 °C, 1 atm, cumene solvent, 0.25 h, select. 100%; TOF = 3.4 × 10^{-2} s^{-1}	[126]
Norbornene	TiO_2–SiO_2 aerogel/cumene hydroperoxide; 90 °C, 1 atm, cumene solvent, 1 h, select. 97%; TOF = 8.3 × 10^{-3} s^{-1}	[126]
Limonene	TiO_2–SiO_2 aerogel/cumene hydroperoxide; 90 °C, 1 atm, cumene solvent, 0.5 h, select. 87%; TOF = 1.5 × 10^{-2} s^{-1}	[126]
Propylene	Au/TS-1/H_2 + O_2; 200 °C, 1 atm, $C_3^=$/H_2/O_2/He = 10/10/10/70; SV 7,000 cm^3 g_{cat}^{-1} h^{-1}, conv. 8.8%, select. 81%. Yield^a = 116 g kg_{cat}^{-1} h^{-1}, TOF = 0.33 s^{-1}	[118]
Propylene	Au–Ba/Ti-TUD/H_2 + O_2; 150 °C, 1 atm, $C_3^=$/H_2/O_2/Ar = 10/10/10/70; SV 7,000 cm^3 g_{cat}^{-1} h^{-1}, conv. 1.4%, select. 99.6%. Yield^a = 25 g kg_{cat}^{-1} h^{-1}, TOF = 2.2 × 10^{-2} s^{-1}	[119]

^{a} Space time yield (g of epoxide per kg catalyst per hour).

mostly calculated by this author using the reported yield, and the time of reaction, so they must be considered as an average TOF.

Hydrogen peroxide is normally available in aqueous solution at concentrations between 30 and 60 wt%, and even higher. The aqueous solution constitutes a problem as the majority of metal catalysts are inhibited by water. A significant breakthrough was achieved by Venturello and coworkers who used phase-transfer reagents (PTRs) in addition to tungstate catalysts based on mixtures of Na_2WO_4/H_3PO_4 to overcome the problem of H_2O inhibition in H_2O_2 epoxidations of olefins [141–143] (Table 1.6). Phase-transfer reagents are chemical compounds, often quarternary ammonium salts, which facilitate the migration of an anionic chemical

TABLE 1.6 Homogeneous epoxidation reactions and catalysts

Reactant	Catalysts and conditions	References
1-Octene	Venturello system: $[PWO_4O_{24}]^{3-}$, $[(C_8H_{17})_3NCH_3]^+/H_2O_2$; 70 °C, CH_2Cl_2, 0.75 h. Yield 88% (H_2O_2); yield 53%; TOF = 6.3×10^{-2} s^{-1}	[141]
Cyclohexene	Venturello system: $[PWO_4O_{24}]^{3-}$, $[(C_{18}H_{37})_{0.76}(C_{16}H_{33})_{0.24}N(CH_3)_2]^+/H_2O_2$; 70 °C, CH_2Cl_2, 1 h. Yield 88% (H_2O_2); yield 53%; TOF = 7.3×10^{-2} s^{-1}	[141]
Propylene	Mizuno system: $[\gamma\text{-SiW}_{10}O_{34}(H_2O)_2]^{4-}/H_2O_2$; 32 °C, CH_3CN, 8 h. Yield 90%, H_2O_2 utilization 99%; TOF = 2.0×10^{-2} s^{-1}	[131]
1-Octene	Mizuno system: $[\gamma\text{-SiW}_{10}O_{34}(H_2O)_2]^{4-}/H_2O_2$; 32 °C, CH_3CN, 10 h. Yield 90%, H_2O_2 utilization 99%; TOF = 1.6×10^{-2} s^{-1}	[131]
Propylene	Busch system: CH_3ReO_3, pyridine-N-oxide/H_2O_2; 30 °C, 20 bar N_2, solvent CH_3OH; TOF = 5.7×10^{-3} s^{-1}	[166]
1-Hexene	Herrmann system: CH_3ReO_3, pyrazole/H_2O_2; 25 °C, 1 atm, solvent CH_2Cl_2. Yield 99%; TOF = 6.6×10^{-2} s^{-1}	[161]
Cyclohexene	Herrmann system: CH_3ReO_3, pyrazole/H_2O_2; 25 °C, 1 atm, solvent CH_2Cl_2. Yield 99%; TOF = 5.5×10^{-2} s^{-1}	[161]
cis-Cyclooctene	Kühn and Romào system: $\eta^5\text{-}(C_5Bz_5)MoO_2Cl$/TBHP; 55 °C, 1 atm, CH_2Cl_2 solvent, 4 h. Yield 100%; TOF = 6.9×10^{-3} s^{-1}	[174]
1-Octene	Noyori system: $WO_4^{2-}/H_2NCH_2PO_3H_2/(C_8H_{17})_3MeN^+HSO_4^-/H_2O_2$; 90 °C, 1 atm, solvent toluene. Yield 81%; TOF = 3.3×10^{-3} s^{-1}	[145]
1-Hexene	Jacobs system: Mn(tmtacn), oxalate buffer/H_2O_2; 5 °C, 1 atm, solvent CH_3CN, 0.6 h, select. 99%. Yield 95%; TOF = 0.79 s^{-1}	[167]

(continued)

TABLE 1.6 (continued)

Reactant	Catalysts and conditions	References
Cyclohexene	Jacobs system: Mn(tmtacn), oxalate buffer/ H_2O_2; 0 °C, 1 atm, solvent acetone, select. 89%; TOF = 8.2×10^{-2} s^{-1}	[168]
Cyclohexene	Noyori system: WO_4^{2-}/$H_2NCH_2PO_3H_2$/ $(C_8H_{17})_3MeN^+HSO_4^-$/$H_2O_2$; 90 °C, 1 atm, solvent toluene. Yield 0%; gives adipic acid	[146]
Cyclohexene	Reyes system: Mo(acac)$_2$/TBHP; 70 °C, 1 atm, solvent 1,2-dichloroethane, 3.5 h; conv. 65%; select. 100%; TOF = 9.0×10^{-2} s^{-1}	[172]
1-Octene	Strukul system: [(P–P)Pt(C$_6$F$_5$)(H$_2$O)]/H$_2$O$_2$; 20 °C, 1 atm, 1,2 dichloroethane solvent, 4 h. Yield 80%, TOF = 2.8×10^{-3} s^{-1}	[134]

species from one phase to another. They consist of a positively charged onium core, Q^+, long-chain alkyl groups, R, and a counterion, and work by encapsulating the anionic species. In this case, an active peroxotungstate species is transferred from the aqueous phase containing the H_2O_2 to the organic phase containing the olefin. After it is spent, the tungstate returns to the aqueous phase to be reoxidized. The use of PTRs has greatly influenced epoxidation research. The reactivity of the system is affected not only by the PTR cation, but also by the counterion for example, phosphates give better results than do carbonates [144].

The group of Noyori developed an improved version, consisting of a W complex derived from Na_2WO_4 and aminomethylphosphonic acid with methyl-tri-n-octyl ammonium hydrogen sulfate as PTR (WO_4^{2-}/$H_2NCH_2PO_3H_2$/$(C_8H_{17})_3$ $MeN^+HSO_4^-$) [145] (Table 1.6). This catalyst system was attractive because it could be used to epoxidize unreactive terminal alkenes with H_2O_2, and because it could operate without chlorinated solvents, which can be toxic and carcinogenic. A drawback was that the system was not effective in the epoxidation of styrene, because of hydrolytic decomposition of the epoxide, and of cyclohexene, because of the production of adipic acid [146].

The group of Ishi described a similar system to that of Venturello consisting of $H_3[PW_{12}O_{40}]$ in combination with H_2O_2 and a PTR [147,148]. However, it has been demonstrated that the two catalyst systems involve the oxoperoxo anion $[PO_4\{WO(O_2)_2\}_4]^{3-}$ [149,150] derived from polyoxometallates.

Polyoxometallates (POMs) have attracted interest because they are fully inorganic mimics of biological-type catalysts such as porphyrins or salen compounds. POMs are oxygen anion multimetallic clusters containing early transition metals such as V, Nb, Ta, W, and Mo [151–153]. Recent reviews are provided by Brégeault *et al.* [154] and Mizuno *et al.* [131] (see Chapters 4 and 16). A very active species is the lacunary POM [γ-SiW$_{10}$O$_{34}$(H$_2$O)$_2$]$^{4-}$, whose tetra-*n*-butylammonium salt derivative is able to epoxidize various olefins including nonreactive terminal olefins such as propylene with H$_2$O$_2$ at mild conditions. Also noteworthy is a *bis*-(μ-hydroxo)-bridged di-vanadium species implemented in a peroxotungstate framework which shows high selectivity to epoxidation with H$_2$O$_2$, in particular toward terminal alkenes [151]. Another interesting POM system is described by Xi *et al.* [155] and consists of the initially insoluble [π-C$_5$H$_5$NC$_{16}$H$_{33}$]$_3$[PO$_4$(WO$_3$)$_4$] which forms the soluble [π-C$_5$H$_5$NC$_{16}$H$_{33}$]$_3$[PO$_4${WO$_2$(O$_2$)}$_4$] by reaction with *in situ*–generated hydrogen peroxide by a coupled anthraquinone system (see Chapter 17). When the H$_2$O$_2$ is used up, the catalyst became insoluble again, and permits catalyst recovery.

The kinetics of epoxidation of limonene were studied on a phosphotungstate peroxo complex [PW$_4$O$_{24}$]$^{-3}$ heterogenized on the ion-exchange resin amberlite and denoted as PW-Amberlite [62,63] (Table 1.2). The reaction was carried out at triphasic conditions (solid, aqueous, organic phases) with an acetonitrile solvent, and aqueous hydrogen peroxide. The authors pay good attention to the elimination of mass transfer limitations. The catalyst underwent loss of activity due to limonene oxide (LO) adsorption, but could be regenerated easily by washing with acetone. The space time yield of LO after 24 h was 425 g$_{LO}$ kg$_{cat}^{-1}$ h^{-1} (2.8 mol kg$_{cat}^{-1}$ h^{-1}).

The rhenium compound methyltrioxorhenium (MTO) was first reported as an active and stable epoxidation catalyst by Herrmann *et al.* [156], and has been the subject of extensive investigations [132,157–160]. A problem with the catalyst is that it tends to form diols, but the addition of Lewis bases such as pyridine and particularly pyrazole hampers the formation of diols without affecting the rate. The reaction of CH$_3$ReO$_3$/pyrazole with a CH$_2$Cl$_2$ solvent and a pyrazole PTR in the epoxidation with H$_2$O$_2$ of various olefins including 1-alkenes and cyclohexene has been reported [161] (Table 1.6). The reaction is greatly accelerated by the use of CF$_3$CH$_2$OH as a solvent [162], likely because it polarizes the hydrogen peroxide molecule, which renders it more reactive. The subject of MTO has been reviewed [163–165] (see Chapter 3). The epoxidation of propylene in a phase-transfer system with MTO has been reported in a patent application [166].

The group of Jacobs reports a Mn complex with a tridentate nitrogen (N3) ligand (trimethyltriazacyclonane) catalyst that can be used for the epoxidation with H$_2$O$_2$ of various terminal olefins [167] and cyclohexene [168] (Table 1.6). It was found that cocatalysts such as oxalic acid suppressed solvent oxidation and the ensuing generation of radicals.

Complexes of Mo and V as naphthenates [126], carbonyls [169], and acetylacetonates [170] have been studied extensively [101,171–173] (Table 1.6). Mo and V are the most active metals for epoxidation with hydroperoxides, and are the basis for one of the industrial production methods of PO. An organomolybdenum compound, η^5-(C$_5$Bz$_5$)MoO$_2$Cl has been reported with high activity for epoxidation of *cis*-cyclooctene with TBHP [174]. The authors report a TOF of 0.33 s^{-1} at 55 °C, but this may have been an initial rate. Using quantities from the paper, an average TOF of 6.9 × 10^{-3} s^{-1} in 4 h is calculated. The catalyst has similar activity for styrene epoxidation, but lower for 1-octene.

The group of Strukul [134,175] has developed a class of electron-poor Pt(II) complexes which are efficient catalysts for the epoxidation of terminal alkenes with H_2O_2 (Table 1.6). The complexes have general structure [(P–P)Pt(C_6F_5)(H_2O)][X] where (P–P) is a diphosphine and X is BF_4^- or OTf. Kinetic studies showed that the complexes owed their reactivity to their ability to increase the nucleophilicity of the olefin by coordination, thereby changing the traditional electrophile/nucleophile roles of the system [176] (see Chapter 2).

Sharpless and coworkers developed the first effective asymmetric epoxidation catalyst consisting of Ti(*i*-O-Pr)$_4$ in combination with a chiral tartrate diester and using an alkyl hydroperoxide such as TBHP [75]. The method is used with allyl alcohol derivatives. A modified procedure employing 3A and 4A molecular sieves [177] reduces the amount of catalyst required to 5–10% of the originally reported [75] (Table 1.7). The purpose of the molecular sieves is to remove water from the system.

As an alternative to Ti-based systems for allylic oxidation, the groups of Brégeault [81] and Yamamoto [178] have reported V-based catalysts. Brégeault uses a O=V(OC$_3$H$_7$)$_3$ catalyst which achieves high TOF. Yamamoto employs V catalysts with chiral *bis*-hydroxamic acid (*bis*-hydroxam) ligands of C$_2$ symmetry which allow use of lower catalyst concentrations and the use of aqueous TBHP rather than the use of anhydrous reagent (Table 1.7).

4.2. Main group oxidations

Main group elements and organic compounds have catalytic activity in epoxidation with activated oxidants. They could be alternatives to catalysts containing expensive or toxic transition metals. The earliest work was carried out with fluorinated acetone, alkyl metaborate esters, and organic derivatives of arsenic and selenium (Fig. 1.2a–d) (Table 1.8).

Fluorinated acetone [179,180] forms adducts with hydrogen peroxide that are active in epoxidation. The adducts function at relatively high temperatures and use halogenated solvents. The use of fluorinated alcohols such as trifluoroethanol improves the performance [181].

Alkyl metaborate esters in combination with organic hydroperoxides are reported to be active for epoxidation [182]. The borate esters are cyclic and have general formula (ROBO)$_3$, where R is an alkyl group like cyclohexyl (C$_6$H$_{11}$). They are six-membered ring compounds formed from O–B units with a pendant RO, and are 1:1 dehydro adducts of alcohol and boric acid.

Organic derivatives of arsenic and selenium have been described in a review by Arends and Sheldon [138]. Hydrogen peroxide adducts of arsine oxide have been found to carry out epoxidation [183,184]. The organic derivative, dibutylphenylarsine, PhAsBu$_2$, gave better results, producing even sensitive epoxides like cyclohexene oxide [185]. The fluorinated alcohol hexafluoroisopropanol (HFIP, 1,1,1,3,3,3-hexafluoro-2-propanol) mixed with dioxane gave extremely high (10^5) enhancements of epoxidation rate of *cis*-cyclooctene and 1-octene by a phenyl arsenic acid catalyst [186]. The compound Sm(BINOL)/Ph$_3$AsO is effective in the asymmetric epoxidation of α-β-unsaturated amides [187] (BINOL = 1,1'-*bis*-2-napthol).

TABLE 1.7 Allylic alcohols: Epoxidation reactions and catalysts

Reactant	Catalysts and conditions	References
C$_7$H$_{15}$⁀⁀OH (E)-2-Decen-1-ol	Sharpless system: Ti(O-i-Pr)$_4$, L-(+)-diethyltartrate, 4A sieves/TBHP; −23 °C, 1 atm, CH$_2$Cl$_2$ solvent, 2.5 h. Yield 85%, ee 96% (2S-$trans$ product). Yield^a = 3,700 g kg$_{cat}^{-1}$ h^{-1}, TOF = 1.9 × 10^{-3} s^{-1}	[177]
Geraniol	Sharpless system: Ti(O-i-Pr)$_4$, L-(+)-diethyltartrate, 4A sieves/TBHP; −23 °C, 1 atm, CH$_2$Cl$_2$ solvent, 0.75 h. Yield 95%, ee 91% (2S-$trans$ product). Yield^a = 14,000 g kg$_{cat}^{-1}$ h^{-1}, TOF = 1.4 × 10^{-3} s^{-1}	[177]
1-Cyclohexenylmethanol	Sharpless system: Ti(O-i-Pr)$_4$, L-(+)-diethyltartrate, 4A sieves/TBHP; −40 °C, 1 atm, CH$_2$Cl$_2$ solvent, 3 h. Yield 77%, ee 77% (2S-$trans$ product). Yield^a = 200 g kg$_{cat}^{-1}$ h^{-1}, TOF = 7.1 × 10^{-3} s^{-1}	[177]
Allyl alcohol	Sharpless system: Ti(O-i-Pr)$_4$, L-(+)-diethyltartrate, 3A sieves; cumene hydroperoxide; 0 °C, 1 atm, CH$_2$Cl$_2$ solvent, 5 h. Yield 65%, ee 90% (2S-$trans$ product). Yield^a = 540 g kg$_{cat}^{-1}$ h^{-1}, TOF = 4.2 × 10^{-5} s^{-1}	[177]
2-Methyl-3-buten-2-ol	Brégeault system: O=W(OC$_2$H$_5$)$_4$/TBHP; 25 °C, 1 atm, benzene or toluene solvent, 24 h; conv. 5%, select. 100%; TOF = 1.2 × 10^{-5} s^{-1}	[81]
2-Methyl-3-buten-2-ol	Brégeault system: Ti(OC$_3$H$_7$)$_4$/TBHP; 40 °C, 1 atm, benzene or toluene solvent, 24 h; conv. 10%, select. 99%; TOF = 8.0 × 10^{-5} s^{-1}	[81]
2-Methyl-3-buten-2-ol	Brégeault system: O=V(OC$_3$H$_7$)$_3$/TBHP; 25 °C, 1 atm, benzene or toluene solvent, 4 h; conv. 99%, select. 98%; TOF = 1.1 × 10^{-2} s^{-1}	[81]

($continued$)

TABLE 1.7 (continued)

Reactant	Catalysts and conditions	References
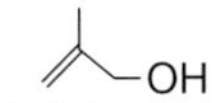 2-Methyl-2-propen-1-ol	Yamamoto system: V(O^{i}Pr)(bis-hydroxam); 0 °C, 1 atm, CH$_2$Cl$_2$ solvent, 12 h. Yield 78%, ee 99%, TOF = 1.8 × 10^{-3} s^{-1}	[178]
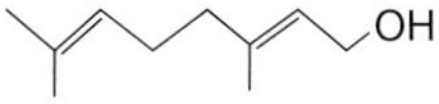 Geraniol	Yamamoto system: V(O^{i}Pr)(bis-hydroxam); −20 °C, 1 atm, CH$_2$Cl$_2$ solvent, 48 h. Yield 68%, ee 95%, TOF = 4.0 × 10^{-4} s^{-1}	[178]

^{a} Space time yield (g of epoxide per kg catalyst per hour).

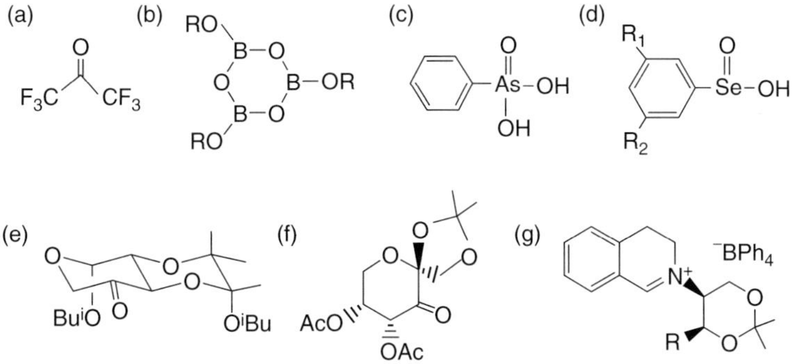

FIGURE 1.2 Main group epoxidation catalysts. (a) Fluoroacetone; (b) borate esters; (c) phenylarsenic acid; (d) organoselenic acid; (e) arabinose-derived ketone; (f) Shi ketone catalyst; and (g) Page iminium salt.

Aryl selenium compounds also form adducts with hydrogen peroxide, ArSe(O)OOH, that are able to carry out epoxidation [188,189], albeit at low rates [190,191]. Improvements can be made by substitution of the aromatic ring with electron-withdrawing trifluoromethyl groups [192], for example, to form the bis-[3,5-bis-(trifluoromethyl)phenyl]diselenide.

It was reported early that alumina with H$_2$O$_2$ and silyl hydroperoxides, like Ph$_3$SiOOH, were active for epoxidation of substituted olefins [193]. Alumina has been studied further, and it is believed that it forms a surface hydroperoxide species by the interaction of H$_2$O$_2$ with surface OH groups of low acidity [194,195]. A sol–gel-derived material has good performance [196], but the catalyst deactivates slowly by an apparent structural change of the surface induced by the condensation of adjacent OH groups [197].

Recent work with main group catalysts has concentrated on the use of Oxone^{TM} (potassium peroxymonosulfate) as co-oxidant with organic ketone derivatives. Shing et al. have described an arabinose ketone catalyst containing a tuneable butanediacetal functionality (Fig. 1.2e) which can be used for asymmetric epoxidation with up to 90% ee [198]. The group of Shi reports on a range of ketones bearing

TABLE 1.8 Main group epoxidation reactions and catalysts

Reactant	Catalysts and conditions	References
2-Penten-4-ol	CF_3COCF_3/H_2O_2; 61 °C, 1 atm, $CDCl_3$ solvent, 5.5 h; conv. 54%, select. 95%, threo:erythro = 62:38; TOF = 0.22 s^{-1}	[179]
Cyclohexene	$(RO–BO)_3$ R = C_6H_{11}/tetralin hydroperoxide; 80 °C, 1 atm, cyclohexane solvent, 3,300 s. Yield 90%; TOF = 5.2 × 10^{-4} s^{-1}	[182]
Cyclohexene	$PhAsBu_2/H_2O_2$; 78 °C, 1 atm, CF_3CH_2OH solvent, 1 h. Yield 94%; TOF = 1.3 × 10^{-2} s^{-1}	[185]
Cyclohexene	$((CF_3)_2PhSe)_2/H_2O_2$; 30 °C, 1 atm, CF_3CH_2OH solvent, 1 h, select. 90%. Yield 88%; TOF = 6.9 × 10^{-2} s^{-1}	[192]
Limonene	Al_2O_3/H_2O_2; 77 °C, 1 atm, ethyl acetate solvent, 2 h, conv. 74%, select. 91%; TOF = 3.9 × 10^{-4} s^{-1}	[196]
Phenylcyclohexene	$[R_2N^+=C]\ BPh_4^-/TPPP$ (peroxysulfate); −40 °C, 1 atm, CH_3CN, 0.05 h, conv. 100%, ee 67%; TOF = 5.6 × 10^{-2} s^{-1}	[204]
cis-β-Methylstyrene	$[R_2N^+=C]\ BPh_4^-/TPPP$ (peroxysulfate); −40 °C, 1 atm, $CHCl_3$, 24 h. Yield 85%, ee 70%; TOF = 9.8 × 10^{-5} s^{-1}	[38]

a spirocyclic oxazolidinone in the α-position (Fig. 1.2f) to probe the effect of structure on enantioselectivity [199], and proposed a structure for the transition state [200]. Theoretical studies on epoxidation by dioxiranes derived from chiral ketones support the experimental results and provides better understanding of the process [201]. In particular, the study shows that asynchronicity in the formation of the two epoxide C–O bonds needs to be considered in the transition state.

A related epoxidation system consists of iminium salts also with Oxone™ as the oxygen source, an area first started by Lusinchi [202], and recently reviewed by Lacour and coworkers [203]. The group of Page (see Chapter 5) has developed a nonaqueous oxidation system employing tetraphenylphosphonium monoperoxysulfate or bisulfate (TPPP) as co-oxidant, and applied it with a dihydroisoquinolinium salt as catalyst (Fig. 1.2g) to obtain good yields of phenyl-substituted olefins, including cis-isomers [38,204] (Table 1.8). A catalyst system not using Oxone™ but employing N-hydroxyphthalimide (NHPI) to produce a radical in a co-oxidation step has been described by the group of Ishii [205]. It has been applied in epoxidation and other reactions (see Chapter 6) [206].

Another amine-mediated epoxidation system, discovered by the group of Jorgensen, consists of proline-derived catalysts and is mechanistically distinct from the iminium salt system [207]. These catalysts are able to epoxidize α-β-unsaturated aldehydes with excellent enantioselectivity and diastereoselectivity for *trans*-aliphatic and aromatic enals.

5. CATALYSTS AND RATES IN BIOMIMETIC EPOXIDATION REACTIONS

Biomimetic oxidations refer to oxidations carried out by relatively low molecular weight complexes which mimic biological systems in their catalytic efficiency. The subject is covered in recent tomes edited by Meunier [208] and van Eldik and Reedijk [209].

Porphyrins are tetrapyrolic macrocycles related to iron-chelating heme systems, and have been used as models of cytochrome P450. Mansuy provides a recent historic account [210]. The groups of Collman and Brauman report the oxidation of a variety of olefins by an Fe porphyrin [211]. Specifically, they used an Fe(III) tetrakis(2,6-dichlorophenyl)porphyrin chloride with pentafluoroiodosylbenzene (F_5PhIO). The latter oxidant is more powerful than iodosylbenzene, and helps ensure that the rate-determining step is oxygen transfer to the olefin. Moreover, conditions were chosen to avoid formation of a porphyrin dimer. The paper reports very high TOFs (10 s^{-1} for *cis*-cyclooctene and 31 s^{-1} for both *trans*-β-methylstyrene and 2-methyl-2-pentene). More recent results show lower rates [212–216] (Table 1.9).

Corroles are also tetrapyrrolic macrocycles related to the cobalt-chelating corrin in vitamin B_{12} and have received considerable interest recently [217]. They are related to porphyrins by ring contraction. Corrolazines are derivative compounds obtained by aza substitution of corroles [218]. Related compounds are phthalocyanines (tetrabenzotetraazaporphyrin) [219]. Although these are promising systems, so far they seem to have lower activity and selectivity than that of the corresponding porphyrins.

The well-known Jacobsen–Katsuki enantioselective oxidation [220,221] uses Mn^{III}(salen) compounds with usually PhIO or NaOCl as the terminal oxidants (TOs) [222,223], and also peracids like *m*-chloroperbenzoic acid (*m*-CPBA) [224,225] (Table 1.9). There are also Mo(salen) and W(salen) complexes reported [43]. The salen group is partly an oxygen atom donor, and so the epoxidation generally proceeds with relatively low turnover numbers, and most are unstable to prolonged oxidative conditions and thus are not recyclable [226]. Much effort has been expended in improving the catalyst by modifying the salen ligand [227], for example, by cyclization to improve stability [228], or by use of related macrocycles such as salan ligands with Ti by the group of Katsuki to improve yields [229].

Epoxidation by Fe-containing complexes with H_2O_2 is difficult because H_2O_2 is readily decomposed by Fe [230–232]. However, there are several examples of workable nonheme systems from the groups of Beller [233], Que [234,235], and Jacobsen [236] (Table 1.9).

Beller and coworkers have reported a complex of $FeCl_3 \cdot 6H_2O$ with pyridine-2,6-dicarboxylic acid (H_2pydic) (Fig. 1.3a) together with an inorganic base such as

TABLE 1.9 Biomimetic epoxidation reactions and catalysts

Reactant	Catalysts and conditions	References
cis-Cyclooctene	Mn^{III} porphyrin (MnTPPCl)/NaOCl: 25 °C, 1 atm, CH_2Cl_2 solvent, select. 100%; TOF = 0.54 s^{-1}	[212]
trans-β-Methylstyrene	Mn^{III} porphyrin (MnTPPCl)/NaOCl: 25 °C, 1 atm, CH_2Cl_2 solvent, select. 100% (trans); TOF = 0.62 s^{-1}	[212]
cis-Cyclooctene	Mn^{III} porphyrin (MnTDCPPCl)/HOCl: 25 °C, 1 atm, CH_2Cl_2 solvent, select. 100% (trans); TOF = 0.17 s^{-1}	[213]
cis-Cyclooctene	Mn^{III} porphyrin (MnTPFPPCl)/(t-Bu-SO_2)PhIO: 0 °C, 1 atm, CH_2Cl_2 solvent, select. 100% (cis); TOF = 0.25 s^{-1}	[214]
Cyclohexene	Fe^{III} porphyrin (FeTMpyCl)/PhIO: 25 °C, 1 atm, MeOH solvent. Yield 30%; TOF = 0.06 s^{-1}	[215]
Norbornene	Fe^{III} porphyrin (FeTDCPPCl)/PFPhIO: 25 °C, 1 atm, CH_2Cl_2-MeOH solvent, select. 100%; TOF = 1.4 s^{-1}	[216]
cis-β-Methylstyrene	Jacobsen–Katsuki system: Mn^{III}(salen)/PhIO: 0 °C, 1 atm, 3 h, CH_3CN solvent, pyridine-N-oxide. Yield 25%, ee 92%, TOF = 9.3 × 10^{-4} s^{-1}	[222]
cis-β-Methylstyrene	Jacobsen–Katsuki system: Mn^{III}(salen)/NaOCl: 0 °C, 1 atm, 3 h, CH_2Cl_2 solvent, Cl^-, pH 10. Yield 74%, ee 99%, TOF = 1.7 × 10^{-3} s^{-1}	[223]
1,2-Dihydronaphthalene	Katsuki system: Ti^{IV}(salan)/H_2O_2: 25 °C, 1 atm, 6 h, CH_2Cl_2 solvent. Yield 87%, ee 96%, TOF = 8.0 × 10^{-4} s^{-1}	[229]
trans-β-Methylstyrene	Beller system: Fe^{III}(H_2pydic)(pyrrolidine)/H_2O_2: 25 °C, 1 atm, pentanol solvent, 1 h; select. 95%; TOF = 5.3 × 10^{-3} s^{-1}	[233]

(continued)

TABLE 1.9 (*continued*)

Reactant	Catalysts and conditions	References
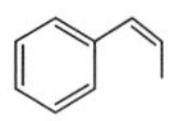 *cis*-β-Methylstyrene	Beller system: $Fe^{III}(H_2pydic)(pyrrolidine)/$ H_2O_2: 25 °C, 1 atm, pentanol solvent, 1 h, select. 75%; TOF = 3.1×10^{-3} s^{-1}	[233]
 *cis*-Cyclooctene	Que system: Fe^{III} bispidine/H_2O_2: 25 °C, 1 atm, CH_3CN solvent, 0.5 h, select. 100%; TOF = 2.8×10^{-2} s^{-1}	[242]
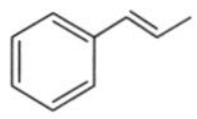 *trans*-β-Methylstyrene	Beller system: $Ru^{II}(Ph_2\text{-pybox})(pydic)/$ TBHP: 25 °C, 1 atm, *t*-BuOH solvent, 12 h. Yield 90%, ee 51%, TOF = 4.2×10^{-4} s^{-1}	[243]
 *cis*-Cyclooctene	Jacobsen system: $Fe^{III}(mep)/CH_3CO_3H$ ($H_2O_2 + CH_3COOH$): 4 °C, 1 atm, CH_3CN solvent, 5 min, conv. 100%. Yield 86%, TOF = 9.8×10^{-2} s^{-1}	[236]
 *cis*-Cyclooctene	Stack system: $((phen)_2(H_2O)Fe^{III})_2(\mu\text{-O})/$ CH_3CO_3H: 0 °C, 5 min, CH_3CN solvent. Yield 100%; TOF = 1.3 s^{-1}	[245]
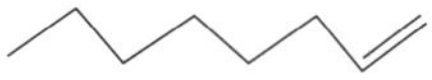 1-Octene	Stack system: $((phen)_2(H_2O)$ $Mn^{II})_2(CF_3SO_3)_2/CH_3CO_3H$: 25 °C, 5 min, CH_3CN solvent, conv. 99%, select. 93%; TOF = 0.31 s^{-1}	[137]

TPPCl, tetraphenylporphyrin chloride; TDCPPL, tetrakis-(2,6-dichlorophenyl)porphyrin chloride; TPFPPCl, tetrakis-(pentafluorophenyl)porphyrin chloride; TmpyCl, tetra(4-*N*-methylpyridyl)porphyrin chloride.

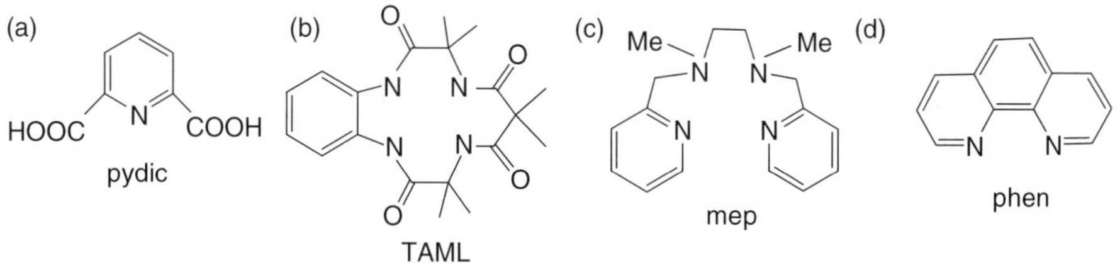

FIGURE 1.3 Ligands used in various biomimetic catalytic systems. (a) Pydic = 2,6-pyridinedicarboxylic acid (Beller); (b) TAML = tetraamido macrocyclic ligand = 3,3,6,6,9,9-hexamethyl-3,4,8,9-tetrahydro-1H-1,4,8,11-benzotetraazacyclotridecine-2,5,7,10(6H,11H)-tetraone (Que) (c) mep = *N,N*′-dimethyl-*N,N*′-bis(2-pyridylmethyl)-ethane (Jacobsen); and (d) phen = phenanthroline (Stack).

benzylamine that is active in the epoxidation of various olefins with H_2O_2 [233]. The complex that is formed is unique in operating at neutral pH.

The group of Que has studied a number of nonheme Fe complexes that are analogues of Rieske dioxygenase enzymes [237,238] that are known for the cis-dihydroxylation of arene double bonds. The most extensively studied complexes thus far have tetradentate nitrogen (N4) ligands (TAML = tetraamido macrocyclic ligand) (Fig. 1.3b) which make available cis-oriented coordination sites, analogous to those in the Rieske enzymes [234,235,239,240]. These complexes catalyze both the epoxidation and cis-dihydroxylation of olefins with H_2O_2, with the selectivity controlled by diverse factors, including the α-methylation of the pyridyl ring ligands [241] (see Chapter 18). A bispidine Fe complex is found to be highly selective to epoxide [242].

Beller and coworkers have also studied Ru complexes [243] such as Ru^{II}(Ph_2-pybox)(pydic) ((Ph_2-pybox)=S,S-diphenyl-pyridinebisoxazoline, (pydic) = pyridinecarboxylate) with the ligands forming a hexadentate coordination sphere of four nitrogen and two oxygen atoms. These have lower activity than that of Fe complexes.

Other Fe complexes are known to epoxidize olefins with peracetic acid [244]. For example, Jacobson and coworkers report a Fe(mep) complex (mep=N,N'-dimethyl-N,N'-bis(2-pyridylmethyl)-ethane) (Fig. 1.3c) which self-assembles under reaction conditions to form a μ-oxo, carboxylate-bridged diiron complex similar to that found in the core of the hydroxylase active site of oxidized methane monooxygenase (MMO) [236].

The group of Stack reports epoxidations with peroxoacetic acid using an Fe (phen) complex (phen = phenanthroline) (Fig. 1.3d) where two phen ligands ligate iron with cis-coordination in a tetradentate fashion (N4) to form a ferric μ-oxo dimer of formula [((phen)$_2$(H_2O)Fe^{III})$_2$(μ-O)](ClO_4)$_4$ [245] (Table 1.9). The complex is readily assembled from common chemicals and has very high activity for epoxidation of a wide variety of olefins, including terminal, aromatic, and cyclic olefins. A similar Mn^{II}(phen) complex is active in the epoxidation of terminal olefins [137].

6. CATALYSTS AND RATES IN BIOLOGICAL EPOXIDATION REACTIONS

Biological enzymes are well known to carry out epoxidation. For example, MMO is an efficient and selective catalyst for epoxidation of small terminal olefins such as ethylene, propylene, and 1-butene [246,247]. Lipases have been used to generate peroxoacids which in turn are used for epoxidation reactions [248,249]. The subject has been reviewed [250]. Biocatalytic systems are of interest not only because they can carry out enantioselective epoxidation of substrates, but also because they offer the exciting possibility of being engineered for specific transformations of nonnatural reagents.

There are two approaches for the use of enzymes. They may be extracted from living organisms and used in (partially) purified form, or they may be used within the organism as whole cell biocatalysts. Growing cells are favored when regeneration of reduced cofactors is required, for example, in reactions carried out by monooxygenases [251].

An example of the use of whole cells is provided for the production of (S)-styrene oxide by recombinant *Escherichia coli* synthesizing styrene monooxygenase [252] (Table 1.10). The work was based on the construction of the plasmid pSPZ10 containing the genes *sty*AB for the styrene monooxygenase of *Pseudonomas* sp. strain VLB120 under the control of the *alk* regulatory system of *P. oleovorans* GPol and the transformation of *E. coli* JM101 with sSPZ10 [253]. After cultivation of the cells, they were charged into a stirred tank reactor (13.5 L) with glucose solution as a nutrient, with the pH maintained at 7.0 by the addition of 25% ammonia or 30% phosphoric acid solutions at a temperature of 30 °C. The reaction was started by adding an organic phase (14.8 L) containing bis(2-ethylhexyl)phthalate as the apolar carrier solvent, 2 vol% styrene and 1 vol% *n*-octane (inducer of the *alk* regulatory system), and using an aeration rate increasing from 15 to 50 L min^{-1} and a stirring speed increasing from 400 to 950 rpm. In total, the biocatalyst produced 388 g of styrene oxide in 12 L of organic phase in 16 h, with about 36 g of 2-phenylethanol side product. The average volumetric productivity was 1 g L^{-1} h^{-1} (8.3 mmol L^{-1} h^{-1}) of styrene oxide and the cell dry weight concentration increased from 8.6 to around 12 g L^{-1}. This corresponds to an average space time yield of 100 g$_{SO}$ kg$_{cat}^{-1}$ h^{-1}. Assuming a cell concentration of 1 × 10^7 cm^{-3} and 10^6 enzymes per cell, this gives a TOF of 1.4 × 10^2 s^{-1}.

An example of the use of isolated enzymes is the work on directed evolution by random mutagenesis of the fatty acid hydroxylase cytochrome P450 BM-3 by the group of Arnold [254] (Table 1.10). The target reaction was the enantioselective

TABLE 1.10 Biological epoxidation reactions and catalysts

Reactant	Catalysts and conditions	References
Styrene	*Escherichia coli* JM101 with sPZ10 plasmid/O$_2$; 30 °C, 1 atm, pH 7, 16 h, glucose; select. 91%, ee% >99(S); average styrene oxide (SO) rate = 8.3 mmol L^{-1} h^{-1}; average cell dry weight 10 g L^{-1}; average yield^a = 100 g kg$_{cat}^{-1}$ h^{-1}; TOF = 1.4 × 10^2 s^{-1} (assuming 10^7 cells cm^{-3}, 10^6 enzymes/cell)	[252]
1-Pentene	P450 BM-3 variant SH-44/O$_2$; 20 °C, 1 atm, 3 h, NADPH; TON 1370, select. 93%, ee% 73(S); TOF = 0.13 s^{-1}	[254]
Styrene	P450 BM-3 variant 139–3/O$_2$; 25 °C, 1 atm, pH 8, NADPH, initial rate, select. 100%; TOF = 6.4 s^{-1}	[264]
Propylene	P450 BM-3 variant 139–3/O$_2$; 25 °C, 1 atm, pH 8, NADPH, initial rate, select. 100% TOF = 12 s^{-1}	[264]

^{a} Space time yield (g of epoxide per kg catalyst per hour).
TON, turnover number.

epoxidation of terminal alkenes, a transformation that poses synthetic challenges [255] which have not been fully addressed by catalytic systems such as Ti (tartrate) [75], Mn(salen) [220,221], and Fe(porphyrin) complexes [256]. Arnold and coworkers point out that several enzymes including cytochrome 450 [257], MMO [258], toluene monooxygenase [259], styrene monooxygenase [260], chloroperoxidase [261], as well as microbial whole-cell catalysts such as *Rhodococcus rhodochrous* [262], do catalyze the enantioselective epoxidation of terminal olefins, but although high optical purity is achieved in some cases, these biocatalytic systems generally only produce a single enantiomer, and can only accept a limited range of alkene substrates [263]. In their work, Arnold and coworkers [254] isolated cytochrome P450 BM-3 from *Bacillus megaterium* that has been made to undergo saturation mutagenesis of key residues close to the active site, to generate 11 libraries, each containing all possible amino acid substitutions. They obtain two P450 variants that are able to convert a range of terminal alkenes to either (*R*)- or (*S*)-epoxide (up to 83% ee) with high catalytic turnovers (up to 1,370) and high epoxidation selectivities (up to 95%). Using nicotinamide dinucleotide phosphate (NADPH) as a sacrificial reductant and accounting for part of the NADPH forming H_2O (65% loss), they obtain a TOF for styrene epoxidation of 6.4 s^{-1} and a TOF for propylene epoxidation of 12 s^{-1} (assuming 0% loss) [264].

Although the field of enzymatic catalysis for oxidations is relatively young, there is considerable potential in the area. The impact of modern molecular biology on catalyst development is expected to result in new opportunities for recombinant whole cell transformation. Some of the problems that need to be solved are enzyme immobilization and recovery (similar to homogeneous catalysis) and in the case of living cell catalysis, toxicity by the reactants, products, or extractant media, energy transfer challenges in mixing and heat exchange, and mass transfer issues related to the provision of nutrients and removal of wastes.

7. SUMMARY AND PERSPECTIVE ON THE REACTIVITY RESULTS

The synthetic value of epoxidation systems greatly depends upon a number of factors, including the existence of nonselective pathways, the reactivity of the alkene, and the stability at reaction conditions of the epoxide formed, and these must be taken into account when comparing different catalysts. Considering pathways, the occurrence of competing reactions such as allylic oxidation, common with heterogeneous catalysts, must be considered in assessing epoxidation systems. As for reactivity, a wide variety of factors must be considered including properties of the catalyst, substrate, and solvent. Finally, regarding the stability of epoxides, the influence of both steric and electronic effects must be noted. For example, cyclooctene epoxide is one of the most stable epoxides known because side reactions of the epoxide group are retarded by the steric hindrance of the ring. In contrast, cyclohexene epoxide is highly reactive because in this case, steric factors of the cyclohexane ring favor ring opening of the epoxide [138]. Terminal epoxides tend to be stable towards ring opening, but styrene oxide is quite reactive owing to the electronic stabilization of intermediate products by the aromatic ring. Epoxides of bicyclic alkenes (e.g., terpenes) are also very reactive [138].

Previous tables concentrated on reactivity results with typical molecules for comparison purposes. A summary of the reactivity patterns and general conditions is provided in the following (Table 1.11).

Some generalized comments can be made concerning the various systems. Deviations from these generalizations, which can be seen in some of the results in the table, indicate exceptional behavior and are noteworthy.

In general, since epoxidation involves electrophilic reagents, double bonds with electron-donating substituents are more reactive. Thus, internal or cyclic alkenes give higher conversions than do terminal olefins. Conversely, in general, electron-withdrawing substituents increase the activity of electrophilic catalysts.

In general, planar-type catalysts like Mn^{III}(salen) or porphyrins favor epoxidation of *cis*-substituted olefins, while nonplanar complexes can carry out epoxidation of *trans*-substituted olefins, even when four-coordinate.

In general, halogenated solvents are undesirable because they are toxic, and sometimes carcinogenic, and pose safety problems in handling and disposal. However, halogenated solvents are effective with H_2O_2 and hydroperoxides because they polarize the molecules, weakening the O–O bond and making them more reactive.

The most effective solvents for H_2O_2 are polar, noncoordinating, nonbasic, and inert under oxidizing conditions [192]. Water-immiscible solvents have the advantage of minimizing hydrolysis of epoxides, but can result in mass transfer limitations with aqueous hydrogen peroxide.

Highly acidic or basic aqueous solutions result in hydrolytic cleavage of oxirane rings. The use of biphasic systems in combination with phase-transfer reagents give high selectivity to epoxides.

PTRs have received considerable attention as effective agents to facilitate the use of hydrogen peroxide, but require many parameters to be adjusted to obtain good selectivity. Among them are the nature and concentration of the PTRs, where in general, the length of the alkyl groups determines the extraction efficiency, the solvent, the pH of the aqueous phase, and the presence of salts like NaCl, which modifies the distribution of the species [154].

Multidentate nitrogen species are effective ligands for transition metal ions and also play the role of Brønsted bases acting like "proton sponges" to prevent side reactions and hence improve epoxide selectivity (Fig. 1.3). Particularly effective in the latter role is bipyridine or preferably 2,2'-bipyridyl-*N,N'*-dioxide [81,149].

A final caveat [265] with all these catalysts operating in the liquid phase with hydrogen peroxide is the possibility for the formation of free-radical species that can lead to parallel homolytic pathways for epoxidation. This was noted for epoxidation with hydroperoxides with Groups 4–6 metal oxides by the group of Sheldon [266] and others [267], and remains a concern with materials containing V, Fe, and other redox metals [102].

In general, liquid-phase systems, encompassing homogeneous and biomimetic complexes, suffer from catalyst degradation in the oxidative environment, catalyst recovery problems, and contamination of the product. For these reasons, considerable efforts have been expended on immobilization of these systems on

TABLE 1.11 Reactivity patterns of various catalyst systems

Catalyst/system	Reactive olefins	Unreactive olefins	Conditions, remarks	References
TS-1 Clerici, Ingallina, Notari	Terminal C2–C4 olefins, allyl alcohol	Branched, cyclic olefins	H_2O_2, basic, alcohol solvent, shape selectivity	[105]
TiO_2–SiO_2 aerogel Baiker	Alicyclic, tertiary olefins	Terminal olefins	Hydroperoxides, H_2O_2	[126]
Ti^{IV}(tartrate) Sharpless	Trisubstituted, *trans*-1,2-disubstituted allyl alcohols	*cis*-1,2-Disubstituted allyl alcohols	Anhydrous TBHP, CH_2Cl_2, moderate activity for *cis*-1,2-disubstituted allyl alcohols	[177]
V^V(O^iPr) (bis-hydroxam) Yamamoto	Trisubstituted, *trans*-1,2-disubstituted allyl alcohols	*cis*-1,2-Disubstituted allyl alcohols	Aqueous TBHP, CH_2Cl_2	[178]
Mn(tmtacn) Jacobs	Terminal olefins	Substituted olefins	H_2O_2, CH_3CN	[167]
[γ-$SiW_{10}O_{34}$($H_2O)_2$]$^{4-}$ Mizuno	Terminal, alicyclic, diene olefins, styrene	*Trans*-olefins	H_2O_2, CH_3CN, high H_2O_2 effic., stable	[131]
Fe^{III}(H_2pydic) (pyrrolidine) Beller	Aromatic, dienes, *cis*-, *trans*-olefins	Unreported	H_2O_2, neutral, alcohol solvent	[233]
Ru^{II}(Ph_2-pybox)(pydic) Beller	1,2-Disubstituted and trisubstituted aromatic olefins,	Styrenes and halogen-substituted styrenes, aliphatic olefins	TBHP	[243]
Fe^{III}(phen) Stack	Terminal, cyclic, aromatic, substituted olefins	Electron poor olefins: butene oxide, allyl acetate	CH_3CO_3H, acid	[245]
Fe^{II}(mep) Jacobsen	Terminal, *trans*-substituted olefins		CH_3CO_3H, acidic	[236]

Catalyst	Substrates	Conditions	Ref
Mn^{III}(salen)$^+$ Jacobsen	Aromatic, cis-substituted olefins	Terminal, trans-substituted olefins	m-CPBA, acidic, high enantioselectivity, deactivates	[220, 221]
CH_3ReO_3 Herrmann	Terminal, cyclic olefins		H_2O_2, halogenated solvent, tendency to form diols	[161]
[(P-P)Pt(C_6F_5)(H_2O)] Strukul	Terminal olefins		H_2O_2, halogenated solvent	[134]
Mn,Fe(porphyrin) Collman, Brauman	Cyclic olefins, styrene	Terminal, electron poor olefins	PhIO, NaOCl, CH_2Cl_2 solvent, oxidative degradation, H_2O_2 decomposes	[212, 215]
Jacobsen–Katsuki system Mn^{III}(salen)	Aromatic, cis-substituted olefins	Terminal, trans-substituted olefins	PhIO, NaOCl, basic, halogenated solvent, high enantioselectivity	[222, 223]
CF_3COCF_3	cis-, trans-substituted olefins, cyclic, substituted allyl alcohols	Terminal olefins, allyl alcohol	H_2O_2, neutral, halogenated solvent	[179, 180]
$PhAsBu_2$	Alicyclic olefins, substituted olefins	Terminal olefins	H_2O_2, fluorinated solvent, toxicity	[185]
((CF_3)$_2$PhSe)$_2$/H_2O_2	di- and tri-substituted olefins	Terminal olefins, styrene	H_2O_2, fluorinated solvent, toxicity	[192]

solid supports. Many types of supports have been employed including silica [268–271], magnesium oxide [272], montmorillonite [273,274], layered double hydroxides [275,276,], zeolites, mesoporous materials [277], ion-exchange resins [278], and polymers [279]. The latter has been recently reviewed [280] and is covered in detail (see Chapter 15). With a few exceptions [277,281–284], a general problem has been significant loss of activity. Another recurring problem is leaching [285,286]. However, there are cases where immobilization imparts unique properties to a catalyst system. For example, normally inactive Cu(salen) complexes are reported to be efficient catalysts when supported on MCM-41 [287]. Attachment of the Sharpless Ti(tartrate) catalyst on poly(ethyleneglycol) monomethyl ether (MPEG) results in the reversal of enantioselectivity depending on the molecular weight of the polymer due to the formation of a 2:1 Ti:ligand stoichiometry rather than a 2:2 ratio [288]. Confinement of a chiral Mn(salen) complex in a metal–organic framework imparts substrate size selectivity while enhancing enantiomeric excess and stability [289]. The subject of immobilization is very broad, and will only be mentioned occasionally with specific systems.

The data on TOFs compiled in the previous tables is given in graphical form (Fig. 1.4) and is seen to fall into three regions. The top region comprises biomimetic or biological systems operating in the liquid phase. They have the highest rates, with over several orders of magnitude advantage over other systems, and are very promising. Their drawbacks are the difficult handling (microorganisms) and the requirements for cofactors (enzymes) to produce the

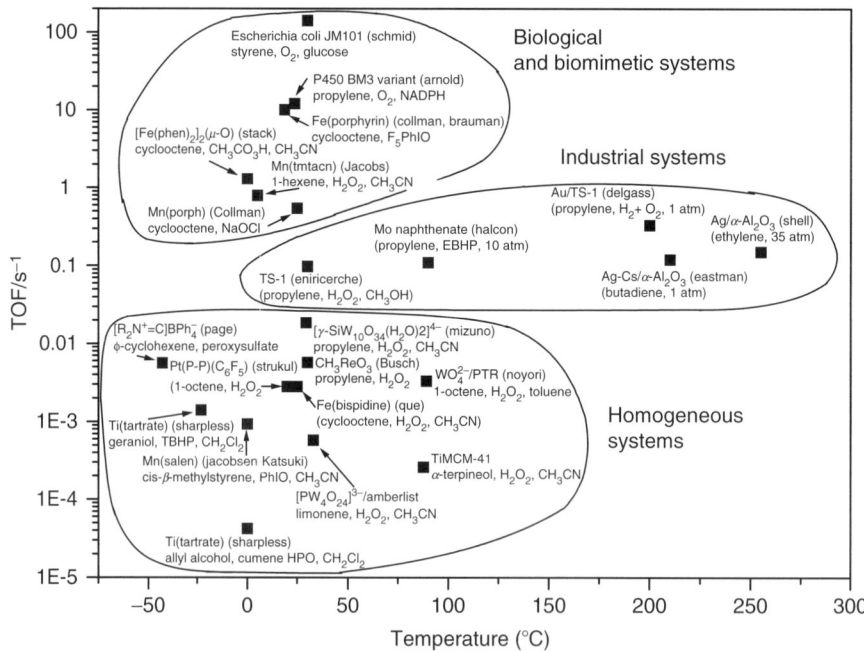

FIGURE 1.4 Comparison of rates for different catalyst systems.

reductant NADPH (see next section). The middle region encompasses catalyst systems that are commercial or close to commercial and includes heterogeneous silver catalysts for ethylene and butadiene epoxidation and liquid-phase catalysts for propylene epoxidation. The bottom region includes predominantly homogeneous catalyst systems that are employed for fine chemicals synthesis. These operate at low temperatures with low TOFs, but give rise to high-value-added products.

In terms of space time yields, gas-phase processes with heterogeneous catalysts operate commercially in the range of 100–200 $g_{epoxide}$ kg_{cat}^{-1} h^{-1}. For liquid-phase processes with homogeneous catalysts, the space time yields increase to 500–14,000 $g_{epoxide}$ kg_{cat}^{-1} h^{-1}, because of the higher dispersion of the catalysts.

8. OXIDANTS FOR EPOXIDATION

Clearly, molecular oxygen is the most desirable oxidant for epoxidation because of its ready availability and low cost. However, O_2 cannot readily be used in practice because of the occurrence of autooxidation radical pathways, which are unselective. Good selectivity can be obtained by co-autooxidation with aldehydes [290–292], but the process produces acid coproducts. Adolfsson gives a list of common oxidants, their active oxygen contents and waste products [293].

It would seem that the use of both oxygen atoms in diatomic oxygen should be possible by the design of suitable catalytic complexes to activate the molecule. This has been found to be more difficult than appears superficially. Although the molecule is symmetric, use of one of the oxygen atoms leaves behind the other and imparts an inherent asymmetry to the activation process. The only catalytic system that can use molecular oxygen directly with any efficiency is bulk silver in heterogeneous catalysts, and it achieves this by dissociative adsorption into atomic oxygen species, which are essentially equivalent on its surface. In other catalytic systems, the oxygen molecule needs to be activated in some form, and this can be carried out directly or indirectly. Examples of products of direct activation are hydrogen peroxide and organic peroxides [294], which retain the O–O bond. Hydrogen peroxide is a particularly attractive oxidant because its only by-product is water, and thus constitutes an environmentally friendly oxidant. Moreover, it is easy to use and has relatively high active oxygen content, so is an atom-efficient oxidant. Its use in epoxidation has been covered in several reviews [138–140]. Examples of indirect activation are molecules such as N_2O [295], iodosyl aromatics, and hypochlorite, which contain a single oxygen atom, but in a highly activated state. Molecules such as persulfate, NO_2 or O_3 can be thought of as belonging to the latter category, because after delivery of an oxygen equivalent, the remnant, sulfate, NO or O_2, is kinetically incompetent for further reaction.

In all these systems where molecular oxygen is activated directly or indirectly, the overall process of oxygen utilization may be considered as involving a sacrificial reductant (red) which reacts with O_2 to form an active oxygen and an oxidized form of reductant (oxid). The process may be written schematically as shown below.

$$O_2 + \text{red} \rightarrow \text{oxid} + [O] \tag{1.1}$$

$$C=C + [O] \rightarrow \text{epoxide} \tag{1.2}$$

For example, in the case of hydrogen peroxide, the reductant is H_2 and the oxidized species is H_2O; in the case of alkyl hydroperoxides, the reductant is the hydrocarbon and the oxidized species is the resultant alcohol.

Similarly, for an oxidant that contains an indirectly activated oxygen, the preparation method can always be used to determine the oxidant. For example, iodosylbenzene (PhIO), a common oxidant, is made by treatment of the hypervalent iodine reagent $PhI(OAc)_2$ with sodium hydroxide. The diacetate is made from the action of peracetic acid and acetic acid on iodobenzene. Thus, the iodosylbenzene oxidant can be traced to hydrogen peroxide, which is used to make the peracetic acid.

Iodosylbenzene (PhIO) has been widely employed as the sacrificial TO in oxygenation reactions catalyzed by metalloporphyrins [296] as well as salen complexes [220]. PhIO (and other iodosylarenes) is soluble in reactive solvents such as methanol, which forms hydromethoxy and dimethoxy derivatives [297], but is insoluble in most organic solvents because of its polymeric structure resulting from a strong intermolecular secondary I⋀O bond [298]. Thus, its exact concentration in the reaction phase is variable, and depends on factors such as its state of dispersion, the temperature, and the stirring speed. For kinetic studies, it is desirable to have soluble species [212]. Derivatives such as iododimethoxybenzene [215] are soluble, but are expected to have different behavior from PhIO. However, a sulfone compound is soluble in CH_2Cl_2 and retains the iodosyl functionality and is useful for mechanistic studies [299]. The fluoroderivative pentafluoroiodosylbenzene (PFIB, PFPhIO) is even more reactive than PhIO [216,300], and reacts readily with porphyrins even though its not soluble in CH_2Cl_2.

Montanari and coworkers demonstrated that by lowering the pH of the aqueous NaOCl solutions from 12.7 to 10.0 ± 0.5, HOCl can be extracted into the organic phase as a soluble TO [301].

In certain cases, the sacrificial reductant is delivered directly, such as hydrogen in H_2/O_2 processes, $NaBH_4/O_2$ or NADPH in biological processes. The use of sodium borohydride together with O_2 was reported by Tabushi and Koga [302] and of tetraalkylammonium borohydride by Perrée-Fauvet and Gaudemer [303]. The borohydride tends to be too strong a reductant and the product epoxides are transformed to alcohols. The NADPH itself can be produced by oxidation of another hydrogen donor using a cofactor enzyme such as alcohol dehydrogenase, isocitric dehydrogenase, or formate dehydrogenase [304]. The group of Arnold demonstrate this with an alcohol dehydrogenase from *Thermoanaerobium brockii* [254].

Often different oxidants can be used with the same system. For example, methane monooxygenase mimics have been used with H_2O_2 [305], TBHP [306],

PhIO [307], and O_2 [308,309]. Porphyrins have been used with iodosyl aromatics, and alkyl aromatics, but not in general with H_2O_2 because of competing decomposition.

As stated initially, the use of molecular oxygen for epoxidation with metal complexes is difficult. Nevertheless, there have been some reports of its use in dihydroxylation [310–312], including a general review on general oxidation with O_2 [313]. Most of the work with molecular oxygen involves co-oxidation. A report of propylene oxide production on Ti silicates [314] is probably due to gas-phase chain reactions as considerable amounts of other oxygenates are reported. This was substantiated in a later study which showed the considerable influence of the post-catalyst bed volume [315]. The group of Herrmann reports oxovanadium(IV) compounds with bidentate N,O-ligands consisting of pyridyl alcohols that can be used for the epoxidation of 1-octene with O_2 [316]. However, only O_2 consumption is reported, and no product analysis, so the occurrence of autooxidation cannot be discounted.

There are other oxidants reported. The combination of *m*-CPBA and *N*-methylmorpholine-*N*-oxide is an effective anhydrous oxidant system for enantioselective oxidation with Mn(salen) compounds [224,225]. Dimethyldioxirane is prepared by the reaction of Oxone^{TM} (potassium monoperoxysulfate) with acetone [317] and is a member of the smallest cyclic peroxide system. It is an active oxidant for a variety of olefins [318,319].

9. MECHANISMS

9.1. Heterogeneous epoxidation on silver catalysts

9.1.1. Ethylene

The silver-catalyzed oxidation of ethylene was first reported in patents by Lefort in 1931 and 1935 [320,321] which, interestingly, also described the effectiveness of gold, iron, and copper promoters. These are elements that have been rediscovered in recent studies [322,323] (see Chapters 8 and 9). Due to the importance of the commercial process, which has seen phenomenal growth since 1940, the mechanism of the reaction has been investigated extensively.

A central question that has been broadly addressed is the nature of the oxygen responsible for the epoxide formation [324] (see Chapter 7). Surface characterization studies have shown that there are at least three types of oxygen species in the surface region: monoatomic chemisorbed oxygen, diatomic (molecular) oxygen, and subsurface oxygen. Atomically adsorbed oxygen results from dissociative oxygen adsorption.

$$O_2 + 4\,Ag \rightarrow 2O^{2-}_{ads} + 4\,Ag^+ \quad (1.3)$$

Molecularly adsorbed oxygen results from nondissociative adsorption and is more weakly held.

$$O_2 + Ag \rightarrow O^{-}_{2ads} + Ag^+ \quad (1.4)$$

Adsorbed oxygen has been suggested to give selective epoxidation of ethylene by Kazanski et al. [325] and in more recent works detailing the existence of nucleophilic (nonselective) and electrophilic (selective) adsorbed atomic oxygen [326,327]. Subsurface oxygen is formed at temperatures higher than 420 K by migration of atomically adsorbed oxygen to an underlying layer. Subsurface oxygen was first suggested by van Santen and Kuipers [328] and has been utilized to explain a variety of results in the EO literature such as isotopic exchange, effect of promoters, transient use of oxygen inventories, and results of microkinetic analysis [329–331]. Although it has been claimed that subsurface oxygen has no effect on the desorption kinetics of oxygen held on the Ag(111) surface nor on the reaction to form EO [91], other studies report that it leads to a decrease in the activation energy for oxygen adsorption [332] and enhanced reactivity of ethylene with adsorbed oxygen [333].

Considerable evidence has accumulated showing that the oxygen responsible for epoxidation is atomic oxygen. One argument against the involvement of molecular oxygen is based on the primary selectivity to ethylene oxide. If the selective oxygen species leading to EO is molecular, every epoxidation event will leave behind an oxygen atom. In the absence of recombination, these oxygen atoms must be utilized to react with ethylene to form CO_2.

$$6O^{2-}_{ads} + C_2H_4 \rightarrow 2CO_2 + 2H_2O \tag{1.5}$$

Thus for every six EO molecules formed, one C_2H_4 molecule will be combusted and the maximum selectivity to EO should be 6/7 or 85.7% [324]. This so-called 6/7 rule has been violated in a number of cases [334] leading to the conclusion that molecular oxygen is not the active species. Moreover, there have been considerable studies indicating that atomic oxygen can form EO [335].

It was traditionally believed that the ethylene epoxidation network was triangular with both ethylene and ethylene oxide contributing to CO_2 (Fig. 1.5, Scheme 1). Cant and Hall studied the epoxidation of deuterium-labeled and unlabeled ethylene on silver catalysts and found an inverse isotope effect in

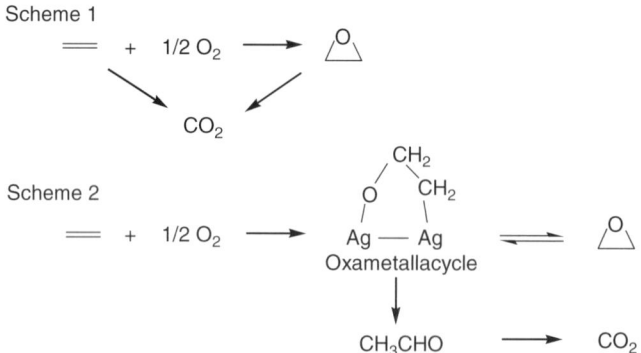

FIGURE 1.5 Reaction networks in ethylene oxidation.

which the heavier ethylene produced more EO [336]. This unexpected result was explained as occurring from the formation of a common surface intermediate for EO and CO_2. The reaction pathway leading to CO_2 involved C–H bond breaking and was slowed by the isotopic substitution giving rise to enhanced EO production. Furthermore, they found *cis–trans* isomerization in the products, strongly implicating an intermediate that could rotate freely around the C–C axis.

The group of Barteau has carried out extensive studies on the kinetics and mechanism of the reaction [337]. On the basis of spectroscopic and quantum mechanical calculations [338], they have presented strong evidence for a surface oxametallacycle as the key intermediate in the reaction. This intermediate is believed to react by two pathways, a selective pathway which results in the formation and release of EO, and an unselective pathway which results in the formation of acetaldehyde, and hence CO_2 (Fig. 1.5, Scheme 2) (see Chapters 7 and 8). Acetaldehyde is known to react to form carbon dioxide on silver [328,339]. The essential features of the mechanism can be represented by four steps [340].

$$O_2 + 2* \xrightarrow{} 2\,O^* \tag{1.6}$$

$$C_2H_4 + * \rightleftharpoons C_2H_4^* \tag{1.7}$$

$$C_2H_4^* + O^* \xrightarrow{} CH_2CH_2O^* + * \tag{1.8}$$

$$CH_2CH_2O^* \rightleftharpoons C_2H_4O + * \tag{1.9}$$

In the steps above, the symbol * indicates empty sites or adsorbed intermediates, the symbol $\xrightarrow{}$ indicates a rate-determining step, and the symbol $\rightleftharpoons$ reversible steps. The mechanism involves dissociative adsorption of oxygen, adsorption of ethylene, a surface reaction between adsorbed ethylene and adsorbed oxygen to form the species $CH_2CH_2O^*$, which denotes the oxametallacycle, and finally a reaction/desorption step where the oxametallacycle forms ethylene oxide which is released to the gas phase.

Linic and Barteau [340] and Stegelmann *et al.* [331,341] have developed microkinetic models to describe the process. The mechanism has two kinetically significant steps, the adsorption of oxygen and the surface reaction to form the oxametallacycle, and the reaction conditions dictate which dominates. For ethylene-rich feeds, the rate-determining step is oxygen adsorption, and the rate increases with oxygen partial pressure. In this regime, as ethylene partial pressure is increased, the rate remains flat or decreases slightly and the selectivity to EO increases. Ethylene here acts to block sites, including those that lead to combustion to CO_2. For oxygen-rich feeds, the rate-determining step shifts to the surface reaction, and an increase in rate is observed with increasing ethylene partial pressure. In this regime, oxygen partial pressure has a strong positive effect on both rate and selectivity. This is attributed to the formation of subsurface oxygen, which decreases the activation energy for oxygen adsorption and also reduces the decomposition pathway of the oxametallacycle that leads to acetaldehyde, and then to CO_2. Calculations by the group of Mavrikakis [332] using DFT have shown that subsurface oxygen decreases the barrier for oxygen adsorption on Ag(111). Earlier work had also shown that the presence of subsurface oxygen could make adsorbed oxygen more susceptible to electrophilic attack by olefins [333].

The studies on the oxametallacycle intermediate have led to theoretical work with DFT that predicts that Cu–Ag alloys will have enhanced activity [342] (see Chapter 8), and this has been experimentally verified [323,343], validating the initial reports by Lefort [320,321].

9.1.2. Propylene

The direct epoxidation of propylene is highly desirable and catalysts for the transformation have been sought extensively in numerous screening [9,344–347] and combinatorial [348] studies. The latter work yielded Rh as a promising candidate, but it has been suggested that this was a result of the method of analysis (mass spectrometry) which could not properly distinguish between products of the same parent mass as PO (propionaldehyde and acetone) [349]. The best catalysts that have emerged for the direct epoxidation of propylene are high-loading $Ag/CaCO_3$ catalysts developed by the group of Gaffney at ARCO [344,350–353] and a Cu/SiO_2 catalyst reported by the group of Lambert [354]. Interestingly, a supported Cu catalyst for PO was patented by ICI in 1976 [355]. Although for both Cu/SiO_2 and $Ag/CaCO_3$ a selectivity to PO as high as 60% could be obtained [355,356], this was for low conversions of about 1%, and the production of PO decreased rapidly with conversion.

The main problem with propylene, which is common to other similar molecules, is the presence of reactive allylic hydrogens which are prone to attack by nucleophilic oxygen species [357–359]. This has been supported in studies with molecules not possessing reactive allylic hydrogens such as butadiene [360,361], norbornene [362], 3,3-dimethylbutene [363], and styrene [364–366], which can undergo selective oxidation. Interestingly though, there is still a considerable structure-sensitivity. Although styrene oxide is observed on Ag(111) and Ag(100), its formation is inhibited on Ag(110) [367].

A theoretical study of the mechanism of propylene epoxidation on Ag(111) and Cu(111) surfaces has been carried out using DFT by the groups of Lopez and Lambert [368]. The reaction between propylene and oxygen on the metal surface can follow either of two pathways, one pathway forms an allylic species and a hydroxyl group by dehydrogenation of propylene and leads to total oxidation products, the other pathway forms an intermediary propylene oxametallacycle, denoted as (OMMP), involving two adjacent surface metal atoms (MM). The latter follows the mechanism of Linic and Barteau [337,338,340] for ethylene epoxidation, with the OMMP rearranging to form PO or propionaldehyde, which further reacts to form CO_2. Oxygen adatoms (O_a) are preferentially located at threefold sites [369], and can react with neighboring adsorbed propylene molecules by abstraction of an allyl hydrogen or by insertion between the secondary carbon and the metal to form the OMMP intermediate. The fundamental difference between silver and copper is that silver energetically prefers the abstraction pathway while copper the insertion pathway. The difference is related to the greater basicity of O_a in silver which favors interaction with the acidic allylic hydrogens [368].

9.1.3. Butadiene

Medlin *et al.* studied the epoxidation of deuterium-labeled 1,3-butadiene to epoxybutene (EpB) on unpromoted and cesium-promoted silver catalysts and found parallel results to those obtained by Cant and Hall for ethylene epoxidation [370]. Again, it was observed that deuterium-labeling enhanced epoxide formation and this suggested the involvement of a common surface intermediate leading to both EpB and CO_2. In particular, it was found that the kinetic isotope effect was significant only when D was incorporated in the 1- and 4-positions, suggesting that combustion was initiated by the cleavage of a terminal C–H bond.

A further study by Monnier *et al.* using oxygen-18 led to the conclusion that the rate-limiting step in the epoxidation of butadiene was dissociation of a molecular oxygen species [371]. It was also noted that if molecular oxygen were involved, the maximum selectivity should follow an 11/12 rule corresponding to a selectivity of 91.7%, in analogy to the ethylene oxide 6/7 rule. In fact, selectivities well in excess of the maximum predicted are routinely observed. On the basis of these facts, the authors suggested the following sequence of steps (Fig. 1.6).

$$O_2 + 2* \xrightarrow{k_1} 2O* \qquad (1.10)$$

$$B + O* \xrightarrow{k_2} EpB* \qquad (1.11)$$

$$EpB* \xrightleftharpoons{k_3} EpB + * \qquad (1.12)$$

The first step, the dissociative chemisorption of oxygen, was taken to be the rate-determining step. This was followed by an irreversible reaction to form

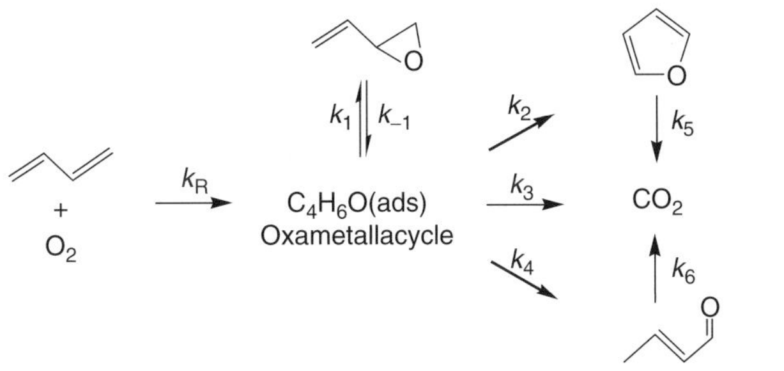

FIGURE 1.6 Proposed reaction network in butadiene epoxidation.

adsorbed EpB*, which desorbed in a final equilibrated step. This intermediate is believed to have an oxametallacycle structure. Assuming adsorption saturation by EpB led to the following rate expression

$$r = \frac{2(L)^2 k_1 [O_2]}{(K_3)^2 [\text{EpB}]^2} \qquad (1.13)$$

The rate was predicted to be first order in oxygen and strongly inhibited by the epoxybutene product, in accordance with experimental results.

9.2. Heterogeneous epoxidation on Ti catalysts

The heterogeneous Ti(IV)/SiO$_2$ catalysts developed by Shell for the epoxidation of propylene with ethylbenzene hydroperoxide constituted the first major application of a solid catalyst in a homogeneous medium. Solid titanosilicates for epoxidation with hydrogen peroxide have been reviewed by Langhendries et al. [265], Sheldon et al. [102,372], and Clerici [373]. Until the late seventies, efforts to develop redox molecular sieves were limited to introducing redox ions through ion exchange. The resulting materials suffered from loss of the ions by leaching. A major development occurred in the eighties when scientists from SnamProgetti (Eniricerche) discovered that a framework-substituted titanium silicalite (TS-1) catalyst was effective in selective oxidations with dilute (30%) hydrogen peroxide, and was not inhibited by water [107,374–376]. Conventional Ti(IV)/SiO$_2$ catalysts were ineffective for epoxidation with aqueous hydrogen peroxide because of strong inhibition of the reaction by water [102]. These systems required > 95% hydrogen peroxide to ensure high selectivity to the epoxide and to minimize side products [377]. The active sites are different in both catalysts, likely consisting of titanyl species (Ti=O) in Ti(IV)/SiO$_2$ and tetrahedral Ti in TS-1.

The finding of an active solid redox system resulted in a flourish of activity in the development and application of diverse redox molecular sieves containing titanium (IV) and other metal ions [378–380]. Like the earlier ion-exchanged zeolites, many of the resulting catalysts, however, also suffered from loss by leaching, even when the redox element was substituted in the framework [102]. Ti-substituted zeolites remain special because of then stability.

Materials with larger porosity such as Ti-Beta [381], Ti-MCM-41 [382,383], and titania–silica aerogels [384] were less effective for PO production, but have allowed the catalytic epoxidation of more bulky olefins. The basic order of reactivity for PO production with H$_2$O$_2$ is TS-1 > Ti-Beta > [Ti,Al]-Beta > Ti-MCM-41. There is no proof that the decrease in rate is linked to a decrease in activity of the PO sites [373], but it does match the decrease in hydrophobicity in the same order. This may be due to the adsorption of the water at the expense of the olefin [373]. It is reported that for cyclohexene epoxidation with cumene hydroperoxide, the order of reactivity is Ti–Si aerogel > TiO$_2$/SiO$_2$ > amorphous TiO$_2$ > TS-1 > crystalline TiO$_2$ > silica, and that the aerogel is more active than Ti-Beta and

Ti-MCM-41 [126]. The high activity of the aerogel was attributed to isolated Ti centers.

Much work has been done on the mechanism of epoxidation with hydrogen peroxide in TS-1 [385,386]. The original patents [387–389] and early studies by Clerici and coworkers [105,390] indicated that the reaction was fast, and could be carried out with dilute H_2O_2 (~1–10%) in solvents (alcohols, acetone, water) at close to room temperature and moderate pressure. The reaction of propylene with Ti-OOH preformed in TS-1 occurs at room temperature [391]. The best results were obtained with alcohols with the rate of epoxidation found to be highest in the most polar alcohol following the order: methanol > ethanol > isopropanol > tert-butanol, with an order of magnitude difference in rates between the members [105,390,392,393]. The presence of water reduced the rate, possibly because of faster catalyst deactivation or the lower solubility of propylene in the alcohol when water is present [373]. These results were largely confirmed in subsequent studies, which extended the range of catalysts and solvents, for propylene [393] and other olefins [394–397].

It was originally proposed that the lower activity with the higher alcohols was due to coadsorption on the Ti centers, and steric hindrance of the epoxidation step [390]. In addition, increased electron donation to the Ti centers was suggested to decrease their activity. However, later work from the group of Jacobs showed that the alcohol medium had a role in the partition of the reagents between the catalyst pores and the solvent [398]. In particular, methanol gave a higher intraporous concentration of the olefin than did the higher alcohols or acetone or acetonitrile, explaining the higher rate of epoxidation.

Solvent effects differ with large size zeolites which have lesser hydrophobicity than TS-1, and these are discussed by Clerici [373]. Larger sized zeolites, such as Ti-beta, have been found to be more effective for bulkier olefins such as norbornene, limonene, and α-terpineol [381,399]. The subject has been reviewed by Baiker et al. [127].

Studies of the behavior of TS-1 in the epoxidation of propylene with H_2O_2 have been reported by Thiele and Roland of Degussa [400]. The active site was suggested to be tetrahedrally coordinated Ti with Lewis acid character, capable of coordinating two nucleophilic molecules. The rate was reported to be pseudo first-order in hydrogen peroxide ($dC_{H_2O_2}/dt = kC_{H_2O_2}$), with the rate constant following the order methanol > methyl acetate > acetone > acetonitrile > t-butanol > 2-butanone > tetrahydrofuran. The major by-products in order of importance were methyl ethers, propylene glycol, dimers, and hydroperoxides [400]. These products were formed from PO by ring opening and reaction with a nucleophile. From the ratio of 1-methoxy-2-propanol and 2-methoxy-1-propanol, it was concluded that ring opening was acid catalyzed. It was found that treatment of the catalyst with a weak base like sodium acetate and calcination produced an active catalyst with reduced production of by-products. However, strong bases reduced activity significantly, and it was suggested that they deactivated the Ti by causing the dissociation of coordinated water to form OH^- groups with a coordinated counterion ($[SiO]_4Ti-OH^-M^+$). The TS-1 catalyst deactivated in the span of about a day, probably by the deposition of PO oligomers. The catalyst could be regenerated by calcination at 550 °C, but it was also found that treatment with dilute hydrogen peroxide restored activity. The presence of mostly PO dimers and some oligomers was demonstrated by FTIR spectroscopy [393].

The structure of the oxidant formed by the interaction of hydrogen peroxide with the Ti site in TS-1 has been debated extensively [373]. Possible structures include $\eta^2(O_2)$, $\eta^1(OOH)$, $\eta^2(OOH)$, $\eta^1(OOH)(ROH)$, $\eta^2(OOH)(ROH)$ [171], where ROH is a coordinated water or alcohol ligand (Fig. 1.7). An overview of the literature up to 1995 has been given by Notari [401] and a recent comprehensive review is provided by Ratnasamy et al. [402]. An early proposal based on work on the oxidation of alkanes was that the oxidant was a side-on $Ti(\eta^2-O_2)$ peroxide species [403]. Subsequent work suggested a hydroperoxide (–OOH) species, based on analogy to the structure of inorganic and organic peroxo compounds [404]. Although the hydroperoxide species was accepted, its exact structure was controversial. Despite infrared and NMR characterization, it was unsure whether it was a $Ti(\eta^2-OOH)$ species or an end-on $Ti(\eta^1-OOH)$ species [405,406] stabilized by the coadsorption of a protic molecule such as water or an alcohol. Theoretical calculations using DFT suggested an η^2 coordination without the adsorption of a protic species [407,408]. Sinclair and Catlow presented a compromise structure $Ti(\eta^2-OOH)(ROH)$ with a coordinating alcohol molecule [409]. However, another DFT study indicated that a cyclic $Ti(\eta^1-OOH)$ structure is lower in energy than $Ti(\eta^2-OOH)$ [410].

The epoxidation mechanism for the $Ti(\eta^1-OOH)$ species is thought to begin with reversible splitting of a Ti–O–Si bond by H_2O_2, to form the Ti-OOH species (and a Si-OH), followed by the coadsorption of an alcohol or water molecule to stabilize the hydroperoxide through a five-membered ring [390, 411] (Fig. 1.8a). It was also suggested that water could react with the tetrahedral Ti site at room temperature to produce Ti-OH and Si-OH groups. The epoxidation is carried out with the peroxy oxygen *vicinal* to Ti with the concomitant formation of a water molecule and a Ti-alkoxide. Theoretical calculations for ethylene epoxidation indicated that the double bond attacks the vicinal oxygen next to the Ti in the hydroperoxide complex because of substantially reduced electrostatic repulsion [411]. The catalytic cycle is completed by desorption of the epoxide and the reaction of the Ti-OR species with H_2O_2 (not shown) to form the active species [373].

Clerici cites considerable support for the mechanism involving the Ti (η^1-OOH) species [373]. In TS-1, numerous spectroscopic studies [412] show the adsorption of protic molecules on Ti sites, for example, as observed by XANES by the decrease in intensity of the tetrahedral Ti pre-edge peak on adsorption of water and ammonia [413]. Rate laws suggest that one molecule of alcohol solvent is adsorbed on the active species in the oxidation of alcohols [414] and the

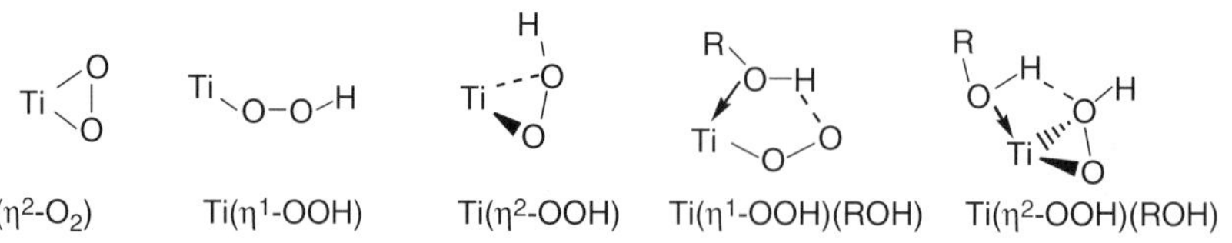

FIGURE 1.7 Structure of Ti peroxides.

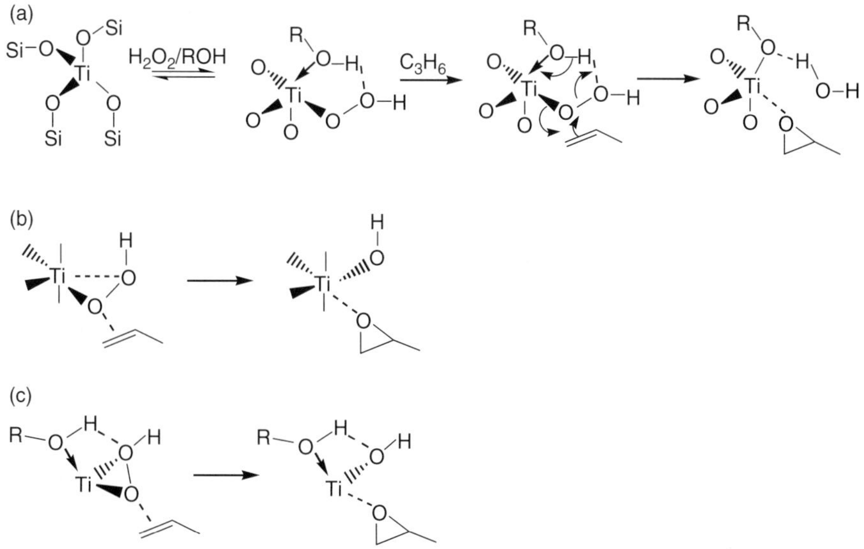

FIGURE 1.8 Epoxidation mechanisms of (a) Ti(η^1-OOH); (b) Ti(η^2-OOH); and (c) Ti(η^2-OOH)(ROH).

epoxidation of allyl chloride [415]. In Ti,Al-β, the presence of water in acetonitrile solvent accelerates the epoxidation of olefins [394]. In TS-1 and Ti-Beta catalysts, large olefins experience steric effects consistent with repulsions in the approach of the double bond to the peroxide group [416–418]. In TS-1 and Ti-MOR, the adsorption capacity and diffusivity of aromatic substrates were decreased by the simultaneous presence of H_2O_2 and H_2O [419].

An alternative to the involvement of Ti(η^1-OOH) is the reaction of Ti(η^2-OOH) or Ti(η^1-OOH)(ROH) species (Fig. 1.8b,c). Much of the same evidence in support for the η^1 route can be applied to the η^2 pathway [409].

Although these mechanisms appear similar to that suggested for Mo catalysts (See next section), it must be remembered [373] that soluble Mo catalysts are strongly inhibited by protic compounds that compete with the oxidant for the active sites. On the other hand, these compounds promote the activity of TS-1.

9.3. Heterogeneous epoxidation on Au/titanosilicate catalysts

Due to the high selectivity to PO (> 90%) and the lower cost of the feedstocks, the hydrogen–oxygen route over Au/Ti-SiO$_2$ catalysts has attracted great attention. The subject has been recently reviewed briefly with a concentration on the chemistry of gold [420], and the role of the titanosilicate support (see Chapters 10–12). In particular, two major catalysts have been developed for PO synthesis with H_2 and O_2 consisting of Au supported on microporous (TS-1, Ti/Si = 1/100) [118,421,422] and Au on mesoporous (amorphous Ti-SiO$_2$, Ti/Si = 3/100) titanosilicates [423–425]. Catalytic activity tests have resulted in space time yields for the

Au/TS1 and the Au/(mesoporous)Ti-SiO$_2$ of 116 and 92 g$_{PO}$ kg$_{cat}^{-1}$ h^{-1}, respectively, at propylene conversions close to 10%, PO selectivities over 80%, and H$_2$ efficiencies over 20%. These results are quite remarkable since it has been estimated that a commercially viable process would require values of C$_3$H$_6$ conversion > 10%, PO selectivity > 90%, and H$_2$ efficiency > 50% [423].

In spite of the major advances in catalyst development for PO synthesis through the hydrogen–oxygen route, not as much work has been carried out to understand the reaction pathways. Some theoretical work has been carried out in this area [426–428], and the kinetics of PO synthesis over Au/TS-1 and Au–Ba/Ti-TUD (mesoporous Ti-SiO$_2$) have been recently investigated. The similarity in the power-rate law expressions for PO synthesis on the Au/(microporous) TS1 [429]:

$$r_{PO} = k(H_2)^{0.60}(O_2)^{0.31}(C_3H_6)^{0.18} \quad (1.14)$$

and the Au–Ba/(mesoporous) Ti-TUD [430]:

$$r_{PO} = k(H_2)^{0.54}(O_2)^{0.24}(C_3H_6)^{0.36} \quad (1.15)$$

suggests that the sequence of steps occurring on both catalysts is similar. From these reports, there seems to be agreement that the important steps during PO synthesis consist of the following sequence: (a) synthesis of hydrogen peroxide from hydrogen and oxygen on gold nanoparticles; (b) formation of Ti-hydroperoxo or peroxo species from hydrogen peroxide on tetrahedral Ti centers; (c) reaction of propylene with the Ti-hydroperoxide species to form PO; and (d) decomposition of hydrogen peroxide to water.

The reaction rate appears to be determined by two irreversible steps: the production of hydrogen peroxide on a gold site, and the epoxidation of propylene by a hydroperoxide species on a Ti site. Despite these studies, direct experimental evidence supporting this sequence of steps has been lacking. For example, Delgass and coworkers suggested the involvement of hydroperoxide species based on a D$_2$ kinetic isotope effect found for the PO reaction, but these species were not directly observed [431]. Using inelastic neutron scattering (INS), Goodman and coworkers found the presence of hydroperoxide species on an Au/TiO$_2$ catalyst; however, the measurement conditions (20 K) were far from reaction conditions [432]. More recently, Chowdhury et al. reported the presence of Ti-hydroperoxo species on an Au/(mesoporous) Ti-SiO$_2$ catalyst during in situ UV–vis measurements at PO synthesis conditions, but did not confirm that it was a reactive species [433]. Although these results support the formation of Ti-hydroperoxide species in the previously mentioned sequence, the sole detection of these species is not sufficient proof that they are true intermediates rather than spectators during the actual reaction [434–437]. To demonstrate that these spectroscopically detected species are true intermediates, it is necessary to show that they are reacting at a rate similar to that of the overall rate of reaction [438]. This has been demonstrated for the hydroperoxide intermediate on Au–Ba/Ti-TUD catalyst [439].

The reaction pathways can be envisaged to involve isolated tetrahedral Ti sites embedded in an amorphous silica network (Fig. 1.9). In the bulk, the Ti atoms are coordinated by four Si–O ligands (tetrapodal Ti site), but on the surface, the active

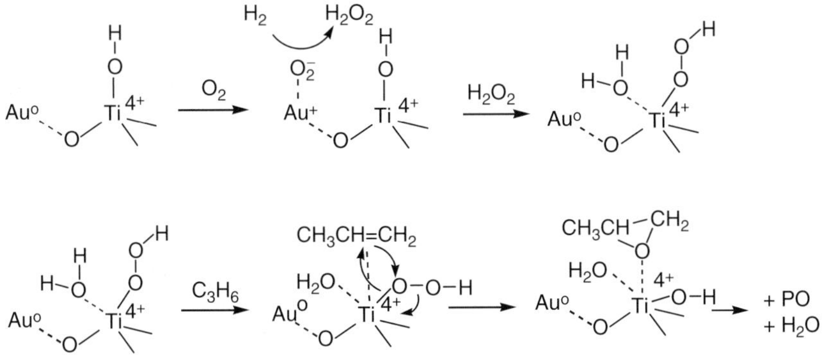

FIGURE 1.9 Epoxidation mechanism with $H_2 + O_2$ on Au–Ba/Ti-TUD.

Ti centers are likely to be tripodally held to Si–O as Ti-OH species in order to be sterically accessible [440]. This tripodal Ti site can be readily obtained by hydrolysis of a Ti–O–Si bond in a tetrapodal Ti site, and evidence for this occurrence has been presented by Sinclair et al. [441]. Gold particles are found in the vicinity of these Ti centers due to its affinity for oxidized Ti [442], although not necessarily bonded directly to the Ti [118].

The first step in the proposed sequence is the formation of hydrogen peroxide from adsorbed H_2 and O_2 on gold sites. Adsorption of H_2 and O_2 has been theoretically considered as an intermediate step in the production of H_2O_2 [443,444]. From electron paramagnetic resonance (EPR) measurements, it has been reported that oxygen adsorbed on Au/TiO_2 and $Au/Ti-SiO_2$ may form (O_2^-) adsorbed species on Au, Ti^{4+}, or more likely at the Au–Ti^{4+} interface [433,445]. A subsequent step in the sequence is the formation of Ti-hydroperoxo species from the reaction of H_2O_2 and tetrahedral Ti sites, more likely Ti tripodal sites. This Ti-hydroperoxo species has been proposed as an intermediate in the gas-phase epoxidation of propylene by analogy with the well-known chemistry for oxidations in the liquid phase with H_2O_2 and TS-1 [105,106,401,446]. In gas-phase reactions, hydroperoxide species have been inferred by D_2 isotopic experiments [431] and detected by *ex situ* INS [432] and *in situ* UV–vis measurements [433]. Other species in this simplified sequence include adsorbed propylene on a Ti-hydroperoxo site and adsorbed PO on a Ti tripodal site. Desorption of PO and water results in the original Ti species, which closes the catalytic cycle.

10. HOMOGENEOUS EPOXIDATION BY EARLY TRANSITION METALS (LEWIS ACID MECHANISM)

Early transition metal ions in their highest oxidation states, such as Ti(IV), V(V), W(VI), and Mo(VI), tend to be stable toward changes in their oxidation states. Consequently, in epoxidation reactions with hydrogen peroxide or alkyl hydroperoxides they form adducts (M-OOH and M-OOR) that are the key intermediates in the epoxidation, and the role of the metal ion is that of a Lewis acid [75,447–450]. The metal center acts as a Lewis acid by removing charge from the O–O bond, facilitating its dissociation, and activating the nearest oxygen atom (the *proximal* oxygen) for insertion into the olefin double bond. Thus, the oxidation is commonly refered to as an electrophilic oxidation by the positively charged proximal oxygen. The more distant oxygen (the *distal* oxygen) constitutes a good leaving group (LG) in the form of OH or OR. The metal center does not undergo a change in oxidation state. The most effective metals are strong Lewis acids but are relatively weak oxidants (to avoid one-electron oxidation of the peroxide) in their highest oxidation state [138].

In many ways, the behavior of Re(VII) is similar to that of the other metals, but in reactions with hydrogen peroxide, it prefers to form peroxo complexes, as does in certain cases, Mo(VI). Ti(IV) forms exclusively hydroperoxo intermediates.

Although, as stated above, olefin epoxidation is commonly referred to as an electrophilic oxidation, recent theoretical calculations suggest that the electronic character of the oxygen transfer step needs to be considered to fully understand the mechanism [451]. The electronic character, that is, whether the oxidant acts as an electrophile or a nucleophile is studied by charge decomposition analysis (CDA) [452,453]. This analysis is a quantitative interpretation of the Dewar–Chatt–Duncanson model and evaluates the relative importance of the orbital interactions between the olefin (donor) and the oxidant (acceptor) and vice versa [451]. For example, dimethyldioxirane (DMD) is described as a chameleon oxidant because in the oxidations of acrolein and acrylonitrile, it acts as a nucleophile [454]. In most cases though, epoxidation with peroxides occurs predominantly by electron donation from the π orbital of the olefin into the σ* orbital of the O–O bond in the transition state [455,456] (Fig. 1.10), so the oxidation is justifiably called an electrophilic process.

10.1. Molybdenum complexes

Epoxidation with hydroperoxides is the basis for the large-scale indirect production of propylene oxide by a process that has been called the Oxirane or Halcon processes. Early work was reported by Smith in a patent issued in 1956 [457], which described soluble heteropoly acids containing transition metals such as chromium, molybdenum, and tungsten that could be employed as homogeneous catalysts for the reaction of olefins with organic hydroperoxides and hydrogen peroxide.

The work of Kollar [101] showed that Mo was the best soluble catalyst, that various hydroperoxides, including TBHP, ethylbenzene hydroperoxide, and

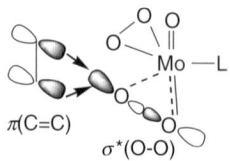

FIGURE 1.10 Predominant orbital interactions in the transition state of epoxidation with peroxo complexes.

cumene hydroperoxide, were effective epoxidizing agents, and that the addition of an alkali metal improved selectivity. Interestingly, for developments in later decades, it was found that hydrogen peroxide was a much less competent epoxidizing agent with the soluble catalysts employed, and that Ti was less active than Mo. The process was commercialized together with ARCO using Mo catalysts [458–463]. Two variants exist which utilize isobutane and ethylbenzene as sources for the hydroperoxides.

The reaction of Mo species with H_2O_2 forms peroxide complexes known as Mimoun complexes (Fig. 1.11a). These are not as active as organic hydroperoxides Mo-OOR in epoxidation. Peroxo complexes have been reviewed [464].

The Lewis acid catalyzed epoxidation reaction with organic hydroperoxides belongs to a class of reactions known as heterolytic reactions which involve two-electron transfer processes. An additional feature of the epoxidation reaction is that the catalytic center does not undergo a change in oxidation state. This occurs because the electron-transfer steps involving the metal are concerted and there is no net change in valence in the metal. The role of the metal center is to activate an organic hydroperoxide (ROOH) so that an oxygen atom from it can be transferred to an olefin. The reaction was initially described by Brill [173] and Kollar [101]. The epoxidation does not proceed to any appreciable rate without a catalyst because the hydroperoxide alone is not electrophilic enough to attack the double bond [465]. It should be noted that peroxyacids (R(C=O)OOH) are more electrophilic and do carry out the epoxidation in the absence of a catalyst [466].

The most active elements for the oxygen transfer are the transition metals to the left of the Periodic Table. The order of activity of these is Mo > W > Ti > V > U > Th > Zr, Nb [465]. In addition, several nontransition metal compounds are effective in the reaction, most notably SeO_2 and borate esters (See Section 11). The catalytic elements are typically in their highest attainable oxidation state, and have the essential feature of not having a readily accessible lower oxidation state. This is necessary in order not to promote the metal-catalyzed decomposition of the peroxides, which could initiate radical chain reactions. Elements such as Mn, Fe, Co, Rh, Ni, Pt, and Cu are ineffective for this reason.

Among the most active transition metals are Mo and V, and both are effective in their highest oxidation state during reaction. Linden and Farona have found that Mo(V) is inactive for the epoxidation reaction and that V(IV) is converted to V(V) when contacted with a hydroperoxide [467]. The metals are added as compounds soluble in the reaction mixture, for example, $Mo(CO)_6$, $MoO_2(acac)$, and $VO(acac)_2$ (acac = monoanion of acetylacetone). Sheldon and Van Doorn [266] have found that irrespective of the starting material, all molybdenum catalysts give rise to a common compound, a 1,2-diol complex (Fig. 1.11b). This is formed

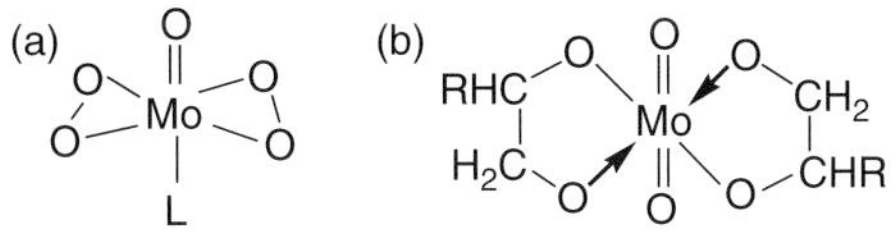

FIGURE 1.11 Mo complexes involved in epoxidation. (a) Mimoun peroxo complex and (b) Sheldon complex.

from ring opening of the product epoxide. Heterogeneous Mo catalysts are also effective in the reaction but have a tendency to leach into the solution and are not used [468–471]. The most promising application of immobilized catalysts are as polymer-supported epoxidation catalysts for fine chemical and pharmaceutical applications [472].

For Mo catalysts Chong and Sharpless [473] have proposed a reaction sequence (Fig. 1.12) consistent with the observed epoxidation kinetics:

$$r = \frac{k_2 K_1 [\text{Mo}]_o [\text{olefin}][\text{RO}_2\text{H}]}{1 + K_1 [\text{RO}_2\text{H}]} \tag{1.16}$$

The hydroperoxide is suggested to coordinate to the metal center through the *distal* or terminal oxygen, rather than the *proximal* oxygen (Fig. 1.12). The olefin then undergoes complexation and oxygen transfer. The coordination of the distal oxygen atom is reasonable from steric considerations and the fact that its complexation to the metal center will activate it. Formation of peroxo complexes $M\overset{O}{\underset{O}{\lessgtr}}$ was ruled out from ^{18}O-labeling experiments which showed that the hydroperoxide remained intact during the reaction [473]. The scheme also suggests that the stereochemistry of the olefin would be retained in the formation of the epoxide. Indeed, the epoxidation reaction occurs without *cis–trans* isomerization [474].

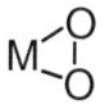

Mimoun has given an alternative possibility, suggesting the formation of peroxo metallacyclic adducts following complexation of the olefin to the metal [171,475]. The reaction sequence (Fig. 1.13) involves a reactive species denoted as Ln–Mo, a Mo(VI) ion with a set of alkoxy ligands. The reaction starts by formation of an alkylperoxo Mo complex by ligand exchange of an alkylperoxide with an alcohol. This is followed by complexation of the olefin by a coordination bond. A subsequent peroxy metallation of the olefin produces a five-membered

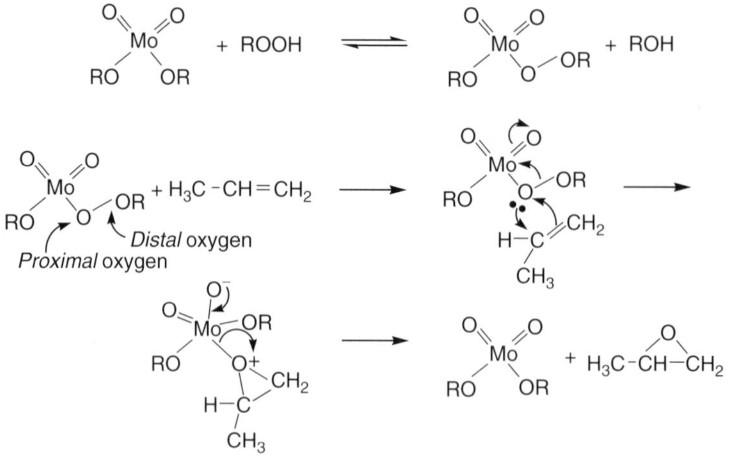

FIGURE 1.12 Accepted alkylperoxo mechanism of molybdenum-catalyzed epoxidation with hydroperoxides.

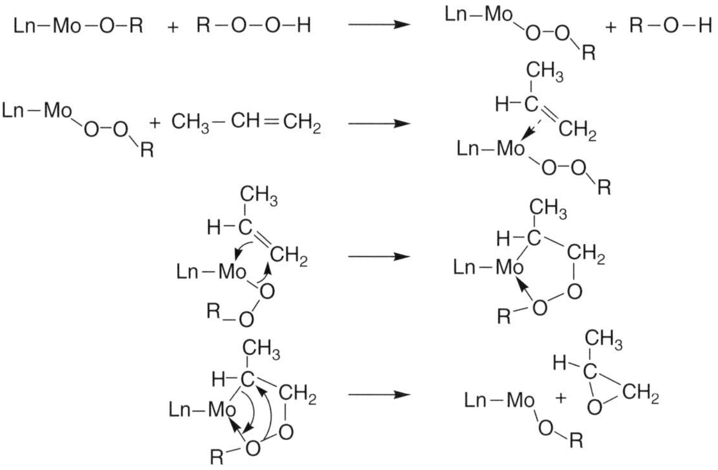

FIGURE 1.13 Disfavored metalladioxolane mechanism of molybdenum-catalyzed epoxidation with hydroperoxides.

metalladioxolane (peroxometallacycle) intermediate, which then subsequently decomposes to produce PO and a molybdenum alkoxide.

Evidence in favor of this mechanism is summarized by Mimoun [171]. (a) It is found that the epoxidation proceeds with any olefin/hydroperoxide ratio [266,476], indicating that the olefin and the hydroperoxide do not compete for sites. This is because the alkylhydroperoxide needs an anionic position on the metal and the olefin needs a vacant coordination site. (b) The rate of epoxidation increases with increasing substitution of the olefin with electron-donating groups, which is expected to strengthen the binding of the olefin to the Mo center [477]. On the other hand, the rate is strongly inhibited by coordinating α-donor solvents or ligands which compete with olefins for vacant sites on the metal [474]. This indicates that the olefin is coordinated to the metal before the epoxidation step. (c) The mechanism explains the retardation of the rate by the alcohol product, which occurs by competition with the hydroperoxide for the anionic site. (d) The Mo(VI) center involved in the reaction has no d electrons for back-bonding, so the coordination of the olefin to the metal is by a Lewis acid–Lewis base interaction. By analogy, it is expected that Lewis acid centers having alkylperoxo groups should be active for epoxidation. This is found for boron and boron alkylperoxides [266,478] (See Section 11). In fact, this evidence is also compatible with the mechanism suggested by Chong and Sharpless. The long-standing question [479] of the correct mechanism is finally being resolved by theoretical calculations (see also paragraph below) which suggest that direct oxygen transfer from a hydroperoxide species by the mechanism of Chong and Sharpless is more energetically favorable than the formation of the metalladioxolane [480].

As stated earlier, H_2O_2 is not as effective as organic hydroperoxides in epoxidation with Mo complexes. Nevertheless, the subject has been studied, and again

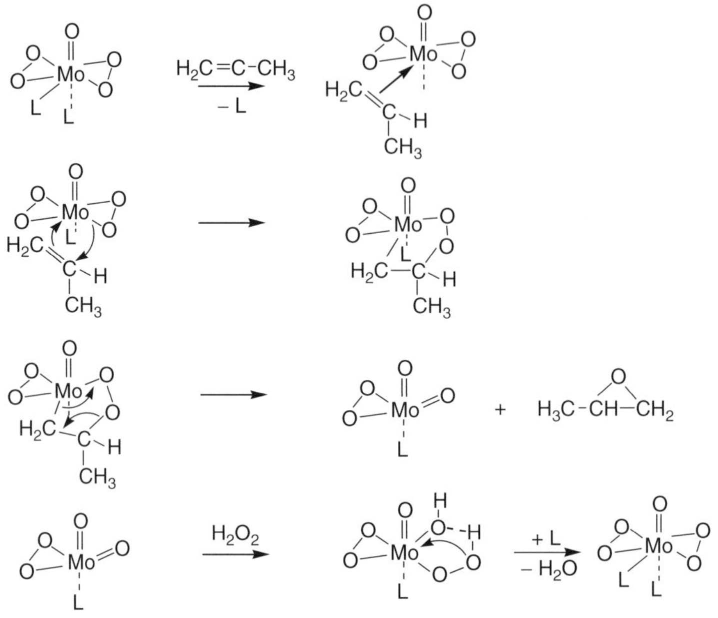

FIGURE 1.14 Disfavored metalladioxolane mechanism of Mo-catalyzed epoxidation with H_2O_2.

there are two viewpoints concerning the mechanism. Mimoun suggests a similar mechanism as occurring with hydroperoxides involving the formation of a metalladioxolane [171,475] (Fig. 1.14), while Sharpless suggests a direct oxygen transfer [481,482] (Fig. 1.15).

Recent experimental work by the group of Shi [483] and extensive theoretical work with DFT [480,484,485] that includes the formation of hydroperoxides [486] give overwhelming support to the direct oxygen transfer mechanism suggested by Sharpless. Calculations of the geometry involved for Cr, Mo, and W complexes indicate that a spiro (nonplanar) geometry is favored [487] (Fig. 1.16). The insertion mechanism (Fig. 1.16a) is unfavorable energetically. Although the details depend on the exact geometry of the peroxo system, in general, attack on the front oxygen (Fig. 1.16b) yields lower activation barriers than attack on the back oxygen (Fig. 1.16c). A clear recent account of the contributions of theory to the understanding of these systems is given by Rösch and coworkers [451].

10.2. Vanadium complexes

The mechanism of epoxidation of propylene by *tert*-butylhydroperoxide on V(V) complexes has been thoroughly investigated by Mimoun [488] and bears many similarities to epoxidation by Mo(VI) complexes. Notable conclusions are the following: (a) The reaction is highly stereoselective, *cis* olefins give *cis* epoxides and *trans* olefins give *trans* epoxides. (b) The reactivity of olefins increases with

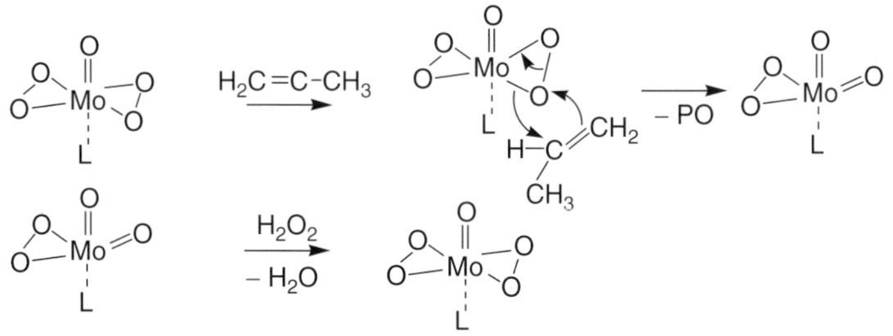

FIGURE 1.15 Accepted direct oxygen transfer mechanism of Mo-catalyzed epoxidation with H_2O_2.

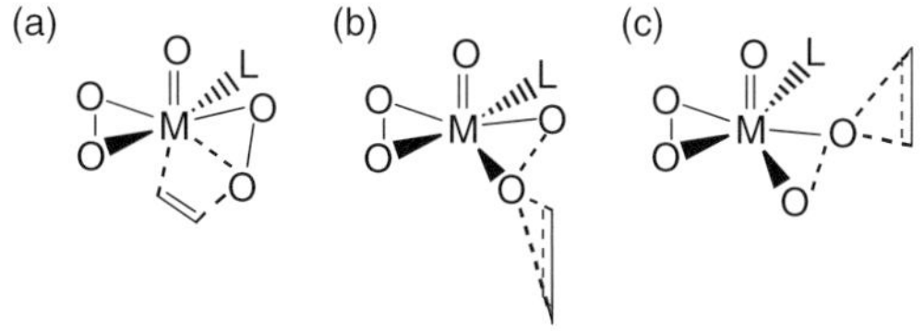

FIGURE 1.16 Spiro structures of Mo complexes. (a) Insertion; (b) front spiro (favored); and (c) back spiro.

their nucleophilicity and is affected by steric effects. Since V(V) is a d^0 metal it does not have electrons to backbond to the olefin, and donor groups on the olefin will increase its ability to complex with the metal. (c) The reaction is strongly inhibited by water, alcohol, and basic solvents and is accelerated by nonpolar solvents. (d) Competitive epoxidation of several olefins dramatically shows the importance of precomplexation of the olefins. (e) The rate-determining step is believed to be the insertion of the olefin to the V–O bond in the bound alkylperoxide complex (V–O–OR) to form a metalladioxolane (dioxometallacyclopentane) ring. This is favored by the polarization of the bond to $V^{\delta+}-O^{\delta-}$.

As in the case with Mo(VI) complexes, the current view is that the epoxidation occurs via direct oxygen insertion as suggested by Chong and Sharpless [473].

10.3. Titanium complexes (Sharpless Ti tartrate asymmetric epoxidation catalyst)

Tartaric acid is the least expensive chiral starting material with twofold symmetry available from natural sources. The Sharpless Ti tartrate asymmetric epoxidation catalyst consists of titanium(IV) tetraisopropoxide (Ti(OiPr)$_4$) in combination with a chiral tartrate diester to induce asymmetry in the reaction of allylic alcohols. It is used with an alkyl hydroperoxide such as TBHP in the presence of 3A or 4A molecular sieves (to remove water) for the epoxidation of allylic alcohols [75]. It was the first effective asymmetric epoxidation catalyst reported.

The mechanism of epoxidation has been studied in depth [489,490]. The mixing of 1 equiv. of dialkyl tartrate with 1 equiv. of titanium tetraisopropoxide produces 2 equiv. of alcohol in accordance with the following reaction:

$$n[\text{Ti}(\text{OR})_4] + n[\text{tartrate}] \rightarrow n/2[\text{Ti}(\text{tartrate})(\text{OR})_2]_2 + 2n[\text{ROH}] \tag{1.17}$$

Molecular weight determinations by a technique related to vapor-phase osmometry indicated that the Ti(tartrate) is dimeric [490], so $n = 2$. The kinetic rate expression is as follows [489]:

$$r = k \frac{[\text{allyl alcohol}][\text{Ti}(\text{tartrate})][\text{ROOH}]}{[\text{ROH}]^2} \tag{1.18}$$

The rate is first order with respect to allyl alcohol, Ti(tartrate), and the hydroperoxide oxidizing agent, and is inhibited by alcohol. The rate expression is consistent with the reaction sequence (Fig. 1.17). The Ti(tartrate) complex is formed by removal of two alkoxide ligands, and then the remaining two alkoxide ligands are displaced by TBHP and the allyl alcohol. The order of displacement is immaterial so the "loaded" complex can be reached by either pathway shown.

The rate-determining step is oxygen transfer from the hydroperoxide to the olefin and yields the epoxy alkoxide and *tert*-butoxide, with all the reactants and products coordinated. The inverse-squared dependence on alcohol is due to the need to replace the two alkoxide ligands with the hydroperoxide and the allylic alcohol. It is found that allyl alcohols with electron-donating groups increase the rate, while alcohols with electron-withdrawing groups decrease the rate, indicating that the olefinic moiety acts like a nucleophile and providing support for its prior coordination to the titanium center. Variation of the Ti-tartrate stoichiometry indicates that more than one Ti-tartrate species is active, but that one species is

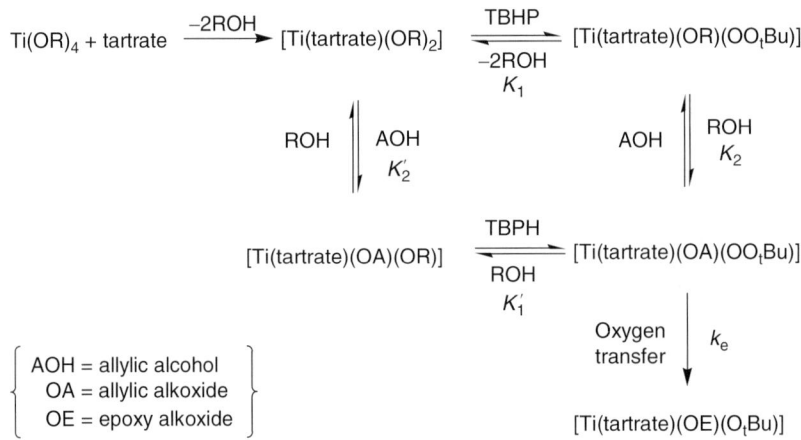

FIGURE 1.17 Ligand exchange pathway in the Jacobsen Ti-tartrate epoxidation mechanism.

dominant by virtue of its enhanced reactivity or its presence in excess [489]. An alternative ion pairing model does not agree with the observed kinetics [491].

The dominant Ti(tartrate) species is a dimer (Fig. 1.18a), as deduced on the basis of FTIR and NMR measurements [490]. The hydroperoxide is coordinated to Ti in η^2 bidentate fashion (Fig. 1.18b) [492] as indicated by the observation that K_{eq} for TBHP is less than 1.0 with both Ti(DIPT)(O-i-Pr)$_2$ and Ti(DIPT)(O-t-Bu)$_2$, and supported by DFT calculations [493,494] (DIPT = diisopropyltartrate). This implies that TBHP is sterically more demanding than *iso*-propoxide or *tert*-butoxide ligands, unlikely unless bidentate coordination of the alkyl peroxide were important [489].

The overall mechanism of the reaction (Fig. 1.19) is supported by theoretical DFT calculations which suggest that the approach of the Ti–O–O to the allyl C=C bond is in a spiro fashion. The outer C–O forming bond is about 0.01–0.02 nm shorter than the inner C–O forming bond, in agreement with the secondary isotope

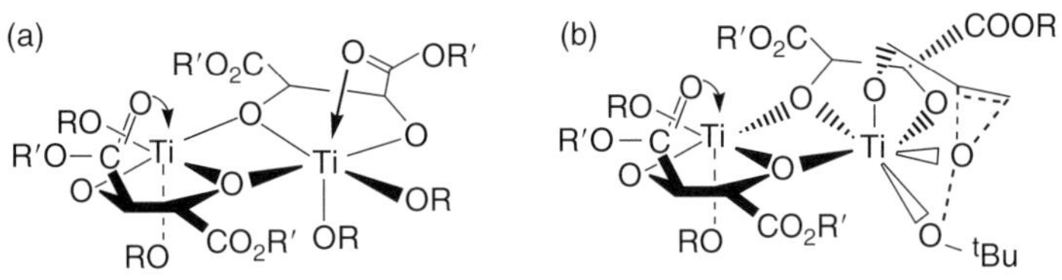

FIGURE 1.18 Structure of the Ti-tartrate catalyst and intermediate. (a) [Ti(tartrate)(OR)$_2$]$_2$ and (b) transition state.

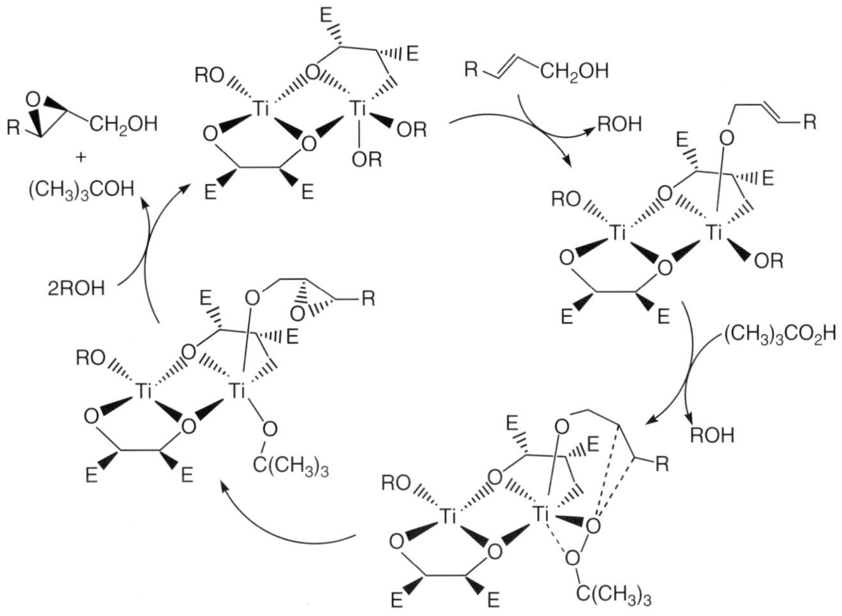

FIGURE 1.19 Overall mechanism of the Jacobsen Ti-tartrate epoxidation.

effect observed by Sharpless. The ester groups favor the equatorial positions rather than the axial positions in the transition state and do not interact with the Ti center.

10.4. Polyoxometallates

The mechanism of the $[\gamma\text{-}H_2SiV_2W_{10}O_{40}]^{4-}$ (Species I) epoxidation of alkenes with H_2O_2 has been studied in detail by the group of Mizuno [152]. Kinetic measurements backed by ^{51}V NMR, ^{183}W NMR, mass spectrometry [151], and quantum chemical calculations suggest that the vanadotungstate ion reversibly forms a hydroperoxide intermediate $[\gamma\text{-}HSiV_2W_{10}O_{39}OOH]^{4-}$ (Species II), which upon dehydration forms a species with a $\mu\text{-}\eta^2\text{:}\eta^2$-peroxo group (Species III) which is the active species for epoxidation. This dimeric V group $V\overset{O}{\underset{O}{\lessgtr}}V$ is notable because V(V) compounds are generally not effective compounds for the epoxidation of nonfunctionalized olefins with H_2O_2 due to the formation of radicals [495]. A sequence of steps has been suggested.

$$I + H_2O_2 \underset{k_2}{\overset{k_1}{\rightleftarrows}} II + H_2O \tag{1.19}$$

$$II \underset{k_4}{\overset{k_3}{\rightleftarrows}} III + H_2O \tag{1.20}$$

$$III + \text{alkene} \overset{k_5}{\rightarrow} I + \text{epoxide} \tag{1.21}$$

The rate expression corresponding to this sequence is

$$r = \frac{k_1 k_3 k_5 [\text{catalyst}][H_2O_2][\text{alkene}]}{k_2 k_4 [H_2O]^2 + (k_2[H_2O] + k_3)k_5[\text{alkene}]} \tag{1.22}$$

10.5. Methyltrioxorhenium

Epoxidation with MTO has been studied thoroughly by the group of Herrmann [156–161] and its mechanism of epoxidation with H_2O_2 has been discussed [163]. Two important intermediates have been isolated, a bisperoxo complex of stoichiometry $(CH_3)Re(O_2)_2O\cdot H_2O$ [496], present with excess H_2O_2, and a monoperoxo complex of composition $(CH_3)Re(O_2)O_2$ [159,497], obtained by reaction of MTO with 1 equiv. of H_2O_2. Experiments with the isolated complexes indicate that they are active in epoxidation [496,498], and have rate constants of reaction that are of a similar order of magnitude [159,497,499]. These findings have been supported by density functional calculations [480,500,501]. The activation parameters for the coordination of H_2O_2 to MTO indicate a mechanism involving nucleophilic attack by H_2O_2. The protons lost in converting H_2O_2 to a coordinated peroxo ligand, O_2^{2-}, are transferred to one of the terminal oxygen atoms, which remains on the Re as an aqua ligand. The rate of this reaction is not pH dependent [502].

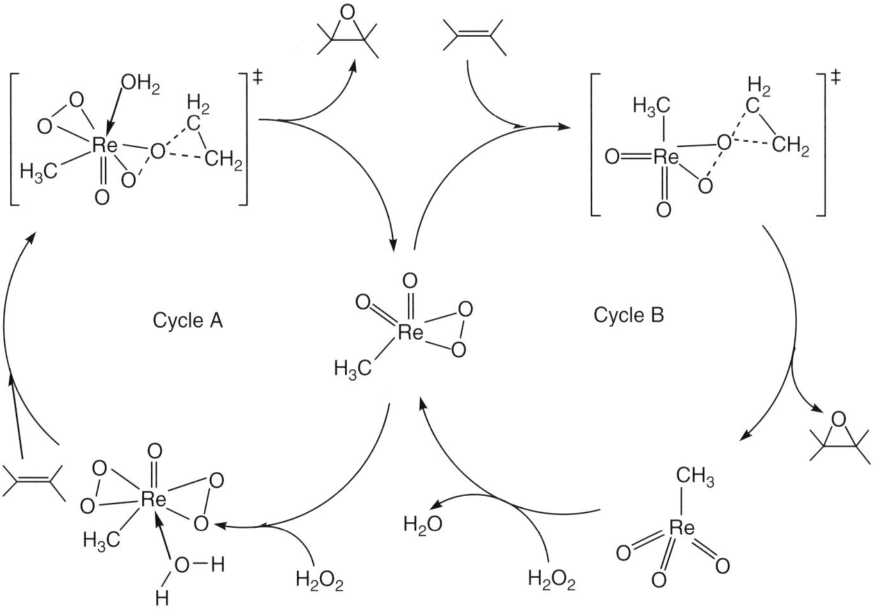

FIGURE 1.20 Suggested mechanism of methyltrioxorhenium (MTO) epoxidation.

The epoxidation reaction can be described as proceeding through two pathways, depending on the concentration of the hydrogen peroxide used (Fig. 1.20) [163]. At high H_2O_2 concentration (85 wt%), the bisperoxo complex is responsible for the epoxidation activity (Cycle A), while at low concentration (<30 wt%), the monoperoxo complex also takes part in the epoxidation (Cycle B). For both cycles, a concerted mechanism is suggested in which the electron-rich double bond of the alkene electrophilically attacks a peroxidic oxygen of the complex possibly through a spiro arrangement [500–503].

Because of the success in producing chiral epoxidation catalysts such as the Sharpless Ti(tartrate) [75] or the Jacobsen–Katsuki Mn(salen) catalysts [220,221] by the addition of chiral organic ligands to a metal, several attempts were made to transform MTO (and also the $(MoO_2)^{2+}$ moiety) into enantioselective epoxidation catalysts by the addition of chiral Lewis base ligands [504]. Unfortunately, these efforts were not successful because of the weak coordination of the base ligands to the metal center, which led either to high ees only at the very beginning of the catalytic reaction (low conversion) or to generally low enantiomeric excesses.

11. MAIN GROUP ELEMENTS

The structure of intermediates in epoxidation by main group elements is similar to that formed with organic peroxoacids (Fig. 1.21) [505]. The oxygen atom that is transferred is the *distal* atom. The intermediate likely has a spiro structure as in the

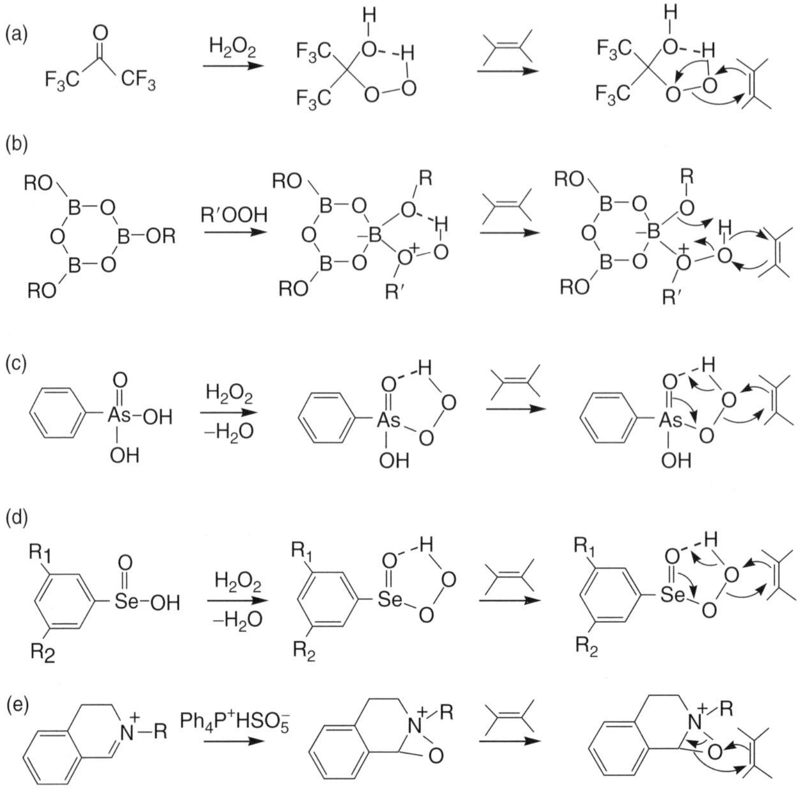

FIGURE 1.21 Intermediates in epoxidation by main group elements. (a) Fluoroacetone; (b) boron esters; (c) arylarsenic acid; (d) arylselenic acid; and (e) iminium salt.

case of the peroxoacids [506]. Because there are no metals involved, there is no change in oxidation state.

The transition state in epoxidation with fluorinated acetone has been probed using substituted allyl alcohol substrates [180]. This organic compound catalyst is capable of regiospecific and diastereoselective epoxidation because of the interaction of the allylic hydroxy group with the coordinated hydrogen peroxide. Its structure is related to that of peroxycarboxylic acids and DMD (Fig. 1.22 Structures **1–3**) The syn:anti ratio has been related to the dihedral angle, α, between the allylic hydroxyl group and the plane of the π-bond. The transition state is similar to that obtained with stoichiometric oxidants such as percarboxylic acids but has a planar character due to the small α, unlike the spiro geometry [318,319] of DMD [507,508,] (Fig. 1.22 Structures **4–6**). A possible overall mechanism of reaction has been given by Arends and Sheldon [138].

Berkessel and Adrio report that the epoxidation rate by phenylarsonic acid using H_2O_2 is greatly accelerated by 10^5 by the use of hexafluoroisopropanol (HFIP) as a solvent for substrates like cyclooctene or 1-octene [186]. The

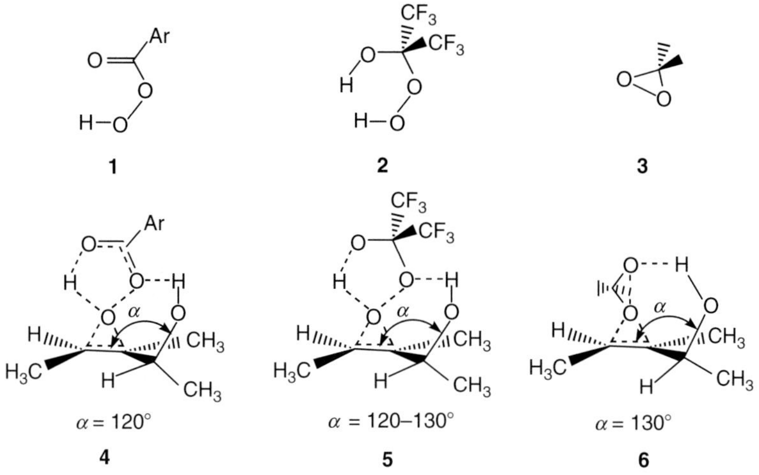

FIGURE 1.22 Structures of peroxycarboxylic acid (**1**), fluorinated acetone (**2**), dimethyldioxirane (DMD) (**3**), and their respective transition states (**4–6**).

enhancement is believed to occur through the formation of an ester complex, which activates the H_2O_2 molecule. Interestingly, the rate shows a 12th-order dependence on HFIP concentration, which is rationalized by the formation of a coordination sphere of 12 HFIP molecules. This interpretation is given credence by the finding by molecular dynamics simulations that HFIP forms dimers and a second coordination sphere containing 13 HFIP molecules [509]. It is suggested that the pocket surrounding the arsenic acid catalytic entity bears resemblance to that formed during enzymatic reactions in a protein matrix.

In the case of iminium salts [38,204], the group of Page has shown that the intermediate is an oxaziridinium salt generated by a single oxygen atom transfer from a TPPP as determined by ^{1}H NMR [510] (see Chapter 5). Calculations of the transition state indicate that the transition state is synchronous [511].

12. HOMOGENEOUS EPOXIDATION BY LATE TRANSITION METALS (REDOX MECHANISM)

Oxygen transfer processes in cytochrome P-450, salen complexes, and related biomimetic systems, such as certain coordination compounds of Fe, Mn, and Cr, take place by a redox mechanism in which the metal undergoes successive oxidation state changes as it incorporates oxygen and then transfers it to the olefin. For porphyrins, this mechanism was denoted by Groves [512] as the oxygen-rebound mechanism. In this mechanism, a critical oxygen atom is transferred from a TO to a metal ion to form a high–oxidation state metal-oxo species (M=O), which in turn delivers the same oxygen atom to the substrate. In this manner, highly efficient use of oxidants and high yields in the epoxidation of olefins are obtained [136,223,231,513–520].

12.1. Porphyrin complexes

Porphyrins are found in the nucleus of various oxidation enzymes such as cytochrome P450, catalases, and peroxidases, and have been used as model catalysts. Porphyrins are heterocyclic macrocycles derived from four pyrrole-like subunits linked by their α carbon atoms through methine bridges (=CH–) to form a planar structure (Fig. 1.23). Excellent descriptions are available in reviews by Mansuy [210] and Meunier [521], as well as in books [522,523], and an overview of enantioselective epoxidation using chiral metalloporphyrins [524]. These macrocycles are highly conjugated, and consequently deeply colored, hence the name porphyrin, from the Greek word for purple. The macrocycles possess the ability to coordinate catalytically active metal ions, such as Fe, Mn, or Cr, in a planar central cavity formed by the pyrrole units. The metal ions are held by the four N atoms of the pyrrole units, but possess the ability to coordinate other species above and below the plane of the macrocycle. This gives rise to their unique reactivity. Usually, one side is occupied by a coordinating ligand, which is able to affect reactivity of the metal ion by electron donation [525,526].

The porphyrins have excellent activity, but practical use has been hampered by their multistep synthesis and tendency to deactivate, usually by oxidative degradation from free-radical side reactions, or by alkylation of the pyrrole unit. The porphyrins also have a tendency to undergo comproportionation where the active epoxidizing intermediate forms a less active dimer through axial bonding as shown by Bruice [527]. Stability has been improved by the addition of bulky groups on the sides of the porphyrin [528,529], the synthesis of elaborate

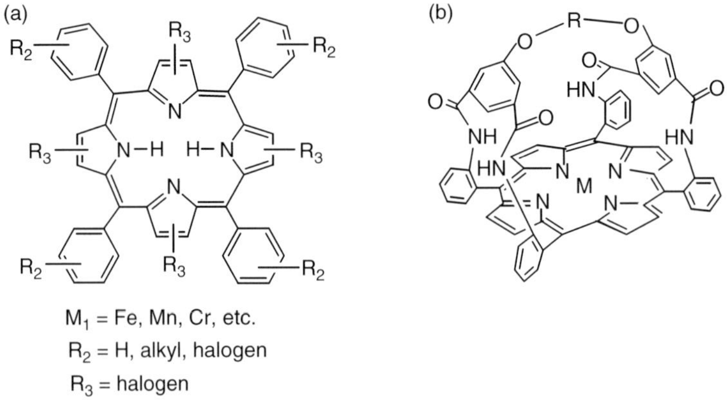

M_1 = Fe, Mn, Cr, etc.
R_2 = H, alkyl, halogen
R_3 = halogen

FIGURE 1.23 Four generations of derivatives of metal porphyrins. (a) First generation (M_1): different metal derivatives (e.g., Fe, Mn, Cr). Second generation (R_2): alkyl or halogen substitution in the phenyl groups of the macrocycle meso positions. Third generation (R_3): substitution of halogens in the β-positions of pyrroles. (b) Fourth generation (picket-fence, picnic basket).

structures such as the picket-fence or picnic-basket porphyrins created by Collman, Brauman, and coworkers [530] (Fig. 1.23), and by immobilization on solid supports [212,531], or in self-assembled systems [532].

The first use of a porphyrin as a functional model for cytochrome P-450 was the seminal work of the group of Groves [296] in which an Fe(III) porphyrin was used with iodosylbenzene. The success of this early work led to the development of what has been described as three generations of catalysts [521], and here expanded to four generations, based on metallotetraarylporphyrins which differ on the degree of substitution in the rings (Fig. 1.23).

Most kinetic work is carried out on biomimetic hemins having a chelating imidazole ligand, or a synthetic, tetraaryl porphyrin. A general scheme of reaction is illustrated for an iron porphyrin below [513] where the terminal ligand X is Cl^- or ClO_4^-.

$$(Porph)Fe^{III}(X) + TO \rightarrow (^+Porph)Fe^{IV}(O)(X) + (T) \qquad (1.23)$$

$$(^+Porph)Fe^{IV}(O)(X) + C{=}C \rightarrow (Porph)Fe^{III}(X) + epoxide \qquad (1.24)$$

The reaction starts by the reaction of a reduced form of the porphyrin with a TO to form a high valent oxo center, which for iron is sometimes denoted as an $Fe^V{=}O$ complex and is called an oxene or a hypervalent iron complex, but in actuality is the Fe^{IV}oxo π radical cation [526,533–535] with the porphyrin delocalizing the charge ($^+$Porph). The TO may be any single-oxygen atom transfer agent such as iodosyl aromatics [296,536], N-oxides [537,538], or hypochlorite [539], or may be organic or inorganic peroxides such as peroxycarboxylic acids [540,541], hydroperoxides [231,525,542], hydrogen peroxide [543], or potassium persulfate ($KHSO_5$) [544]. The hypervalent metal center then transfers the oxygen to an olefin in a subsequent step. This will be discussed presently.

The function of single-oxygen atom transfer agents is straightforward; in fact, some like the oxidant pentafluoroiodosylbenzene (F_5PhIO) are kinetically extremely fast. In contrast, the action of some of the peroxide species is more controversial because they have been associated with radical generation. Although it is well established that peracids are cleaved heterolytically [541,545–548], the reactions of hydroperoxides have been less clear. The groups of Groves and Traylor have suggested that the oxene forms by a heterolytic process [231, 534,542,549] where X = RO.

$$Fe^+ + XOH \rightleftharpoons H^+ + FeOX \xrightarrow{BH^+} Fe^+ = O + HX + B \qquad (1.25)$$

The process involves deprotonation of the peroxy reagent and proton transfer to assist loss of a poor LG. Consistent with this heterolytic process, the reactivity of Fe porphyrins is found to be retarded by electron-withdrawing substitution when the oxidant is pentafluoroiodosylbenzene, but is accelerated with hydroperoxides [534].

A problem with hydroperoxides is that they can also form peroxidic radicals (RO) by homolytic cleavage of the peroxidic O–O bond or other pathways as advocated by the groups of Mansuy and Bruice [550,551]. This can lead to unwanted side reactions, including oxidative degradation of the porphyrin.

$$Fe^+ + XOH \leftrightharpoons H^+ + FeOX \to Fe = O + X \quad (1.26)$$

The formation of radicals is even more facile with H_2O_2, so it is not often used as an oxidant, even though it is extremely reactive [552–554]. In the case of X = HO, the decomposition above is reminiscent of Fenton chemistry.

The processes involved with hydroperoxides have been recently resolved by Nam and coworkers [555]. They have shown that hydroperoxide O–O bonds can be cleaved both heterolytically and homolytically, with their partitioning governed by the electronic nature of the iron porphyrin complexes (i.e., electronic properties of porphyrin and axial ligands). Electron-deficient iron porphyrin complexes tend to cleave the hydroperoxide O–O bonds heterolytically, whereas electron-rich iron porphyrin complexes cleave the hydroperoxide O–O bonds homolytically. The O–O bond cleavage was also found to be significantly affected by the substituent, R, of the hydroperoxides, ROOH, in which the tendency of O–O bond heterolysis was in the order of *m*-CPBA > H_2O_2 > *t*-BuOOH > 2-methyl-1-phenyl-2-propyl hydroperoxide (MPPH). These results indicate that the O–O bond of hydroperoxides containing electron-donating *tert*-alkyl R groups such as *t*-BuOOH and MPPH tends to be cleaved homolytically, whereas those containing electron-withdrawing substituents such as the acyl group in *m*-CPBA tend to undergo O–O bond heterolysis. Since it is observed that homolytic O–O bond cleavage prevails with electron-rich iron porphyrin complexes and with hydroperoxides containing electron-donating substituents such as the *tert*-alkyl group, Nam et al. [555] suggest that the homolytic O–O bond cleavage is facilitated when more electron density resides on the O–O bond of (Porp)Fe^{III}-OOR intermediates.

The kinetics for a number of Mn and Fe porphyrin systems [211,213,556] follow a simple Michaelis–Menten-type form for enzymes [557]. This has been recently confirmed by studies with a soluble iodosylbenzene derivative [212].

$$E + S \underset{}{\overset{k_1}{\rightleftharpoons}} ES \overset{k_2}{\to} E + P \quad (1.27)$$

where E is the enzyme, S is the substrate, ES is the enzyme–substrate complex, and P is the product. In this case, E is the porphyrin, S the olefin, and P the epoxide. This gives rise to the following expression for the dependence of epoxide formation on substrate.

$$\frac{d[P]}{dt} = \frac{V_{max}[S]}{K_m + [S]} \quad (1.28)$$

where V_{max} is equal to k_2 times the total porphyrin concentration and gives a direct measure of the rate-determining step. K_m is defined as $(k_{-1} + k_2)/k_1$ and in the limit that $k_{-1} \gg k_2$ gives a measure of the binding affinity of the substrate.

(A large binding affinity gives rise to a low K_m.) In the case where there are competing substrates, a and b, the expression becomes.

$$\frac{d[P]}{dt} = \frac{V_{max,a}[S]_a}{K_{m,a}(1 + [S]_b/K_{m,b}) + [S]_a} \tag{1.29}$$

The dependence on porphyrin [P] and oxidant [O] is generally first-order [534], so that the overall rate is given as:

$$r_E = \frac{\alpha[P][S][O]}{1 + \beta[S]} \tag{1.30}$$

The mechanism for epoxidation has a "short route" which utilizes a single oxygen atom donor, and a "long route," which employs molecular oxygen, and two electrons and two protons [521] (Fig. 1.24). The scheme describes competitive adsorption well when different olefins are reacted together [211,556], but although the results provide evidence for the formation of a catalyst–enzyme complex, ES, they could not be used by themselves to specify the nature of the intermediate ES. In fact, the original work suggested that an oxametallacycle was the intermediate [211,556], but subsequent work (vide infra) provided evidence that a radical intermediate is more likely.

In cases where unsubstituted porphyrins are used, the reactions are complicated by dimerization and disproportionation processes of the porphyrins (P), as illustrated for a Cr complex in the epoxidation of norbornene (Nb) [513].

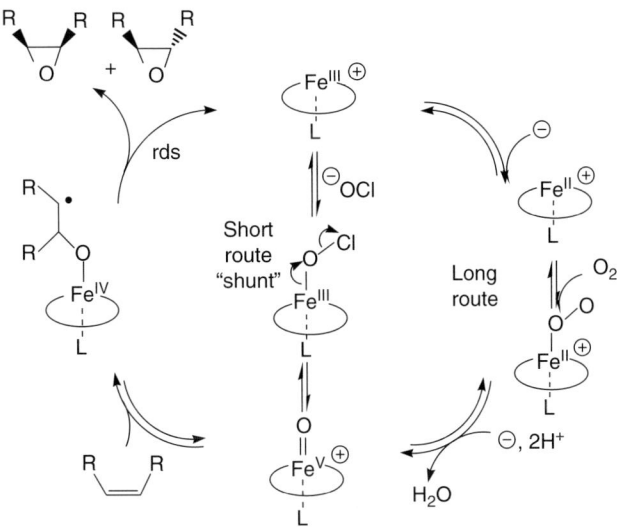

FIGURE 1.24 Scheme for the epoxidation of olefins by porphyrins.

$$(P)Cr^V(O)(X) + Nb \xrightarrow{k_1} (P)Cr^{III}(Cl) + NbO \qquad (1.31)$$

$$(P)Cr^V(O)(X) + (P)Cr^{III}(Cl) \xrightarrow{k_2} (P)Cr^{IV} - O - Cr^{IV}(P)(Cl)(X) \qquad (1.32)$$

$$(P)Cr^{IV} - O - Cr^{IV}(P)(Cl)(X) \xrightarrow{k_3} (P)Cr^V(O)(X) + (P)Cr^{III}(Cl) \qquad (1.33)$$

$$(P)Cr^V(O)(X) \xrightarrow{k_4} (P)Cr^{IV}(O) \qquad (1.34)$$

The reaction of [5,10,15,20-tetrakis(2,6-dichlorophenyl)-porphyrinato]Cr(V) oxide ((Cl_8TPP)$Cr^V(O)(X)$, 2.5×10^{-5} M) with norbornene (9×10^{-3} M) and cis-cyclooctene (2.3×10^{-2} M) was followed by ultraviolet–visible (UV–vis) spectroscopy [513]. Fits to the time course of absorbance gave $k_1 = 0.95$ s^{-1} M^{-1}, $k_2 = 8.0 \times 10^2$ s^{-1} M^{-1}, $k_3 = 8.0 \times 10^1$ s^{-1} M^{-1}, and $k_4 = 3.5 \times 10^{-4}$ s^{-1}. The yield of norbonene oxide was 77% based on (Cl_8TPP)$Cr^V(O)(X)$.

A number of intermediates have been suggested in the epoxidation by hypervalent Fe oxo [296,526,533,538,558,559] and Mn oxo [536,537,539,558,559] porphyrins (Fig. 1.25): **1** cation radical, **2** metal oxo carbon radical, **3** metal oxo carbocation, and **4** metallaoxetane. Another possibility is the direct oxygen transfer by concerted bond shifts that does not involve an intermediate.

The various possibilities have been discussed by Garrison and Bruice [513]. The basis for intermediate **1** is the observation that hypervalent Fe oxo porphyrins can undergo 1e oxidation [560] and the O_2-dependent formation of benzaldehyde in the epoxidation of cis-stilbene [561]. However, calculations indicate that the rate constant for its reaction must be inordinately high [513]. Intermediate **2** is considered unlikely by trapping experiments [562], although theoretical calculations by the groups of Snurr and Broadbelt suggest that it is a possible intermediate [563] (see Chapter 19). The basis for **3** is the observation of rearranged products including cis–trans isomers of cis-stilbene [561] and consistency with kinetics [513]. The suggestion of **4** for both hypervalent Fe and Mn porphyrins [211,556,564] was partly on the basis of the earlier proposal for such an intermediate in the epoxidation by chromyl chloride [565]. However, the lack of UV–vis evidence for its formation [513,533], molecular modeling studies which show high steric hindrance [566], and the low exo/endo ratios obtained in the epoxidation of norbornene [567] suggests that such a metallaoxetane is not formed. Overall, the most likely intermediate is the radical species **2**.

The activation of O_2 by Co, Mn, and Mo porphyrins has been studied theoretically by Witko and coworkers [568]. In the case of the Co porphyrin, the active

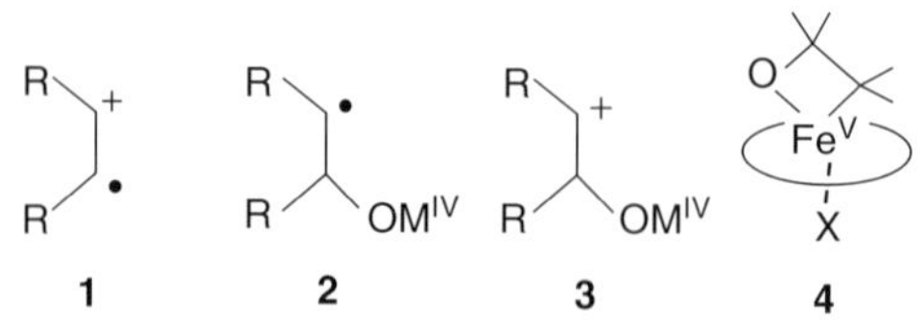

FIGURE 1.25 Proposed intermediates in epoxidation by hypervalent Fe and Mn porphyrin complexes.

forms of the catalyst are found to include an end-on complex with dioxygen and the hydroperoxo form, but not an oxo O=Co(porph) complex because water will not easily dissociate. In the case of the Mn porphyrin, the side-on, hydroperoxo, and oxo types of ligands are possible. In the case of the Mo porphyrin, all forms, the side-on, end-on, hydroperoxo, and oxo forms, are possible.

Laser flash photolysis has been used to generate a hypervalent $Mn^V=O$ species, which has an extremely high reactivity [569], related to that observed earlier in stopped-flow spectrophotometry experiments [32]. In contradiction to the Michaelis–Menten kinetics, it was found to react by second-order kinetics and to yield a TOF of 5.2×10^4 s^{-1}, which is orders-of-magnitude higher than that obtained by standard methods with a Mn porphyrin [214]. It is concluded that the species obtained by flash photolysis is different from the intermediate involved in standard epoxidation with PhIO [214].

Although hypervalent Fe porphyrins are largely believed to operate by the oxygen-rebound mechanism, evidence exists that they can also react by the Lewis acid mechanism. A study from the group of Busch [570] points out that Fe(III) also can form certain adducts of hydrogen peroxide or alkyl hydroperoxides that serve as the reactive species in oxygen-transfer processes [230,571–575], and isotopic exchange experiments confirm the existence of multiple reaction channels in oxidation with these species [576]. Nam and coworkers have studied the mechanism of epoxidation by Fe(porph) complexes [555,577,578,583]. When peroxidic oxygen donors such as hydrogen peroxide, TBHP, and m-CPBA initially an FeOOR(porph) adduct was formed which could react in three ways (Fig. 1.26). The adduct could epoxidize olefins directly by a Lewis acid mechanism (path A), alternatively the O–O bond of the oxygen donor ROOH could be cleaved either heterolytically (path B) to form an Fe(V)oxo species (actually the Fe(IV)oxo radical cation) or homolytically (path C) to yield the Fe(IV)oxo complex. The Fe(V)oxo species gives high yields of epoxides, but the Fe(IV)oxo complex carries out allylic oxidation. Different oxygen donors and axial ligands L give different amounts of homolytic and heterolytic cleavage. To summarize, studies have shown that Fe porphyrin systems can carry out epoxidation by both the Lewis acid mechanism and the rebound oxygen mechanism. The precise nature of the intermediate is the subject of continuing studies.

As stated earlier, most work with porphyrins have employed iodosylbenzenes or hypochlorite because the use of H_2O_2 results in the generation of OH radicals

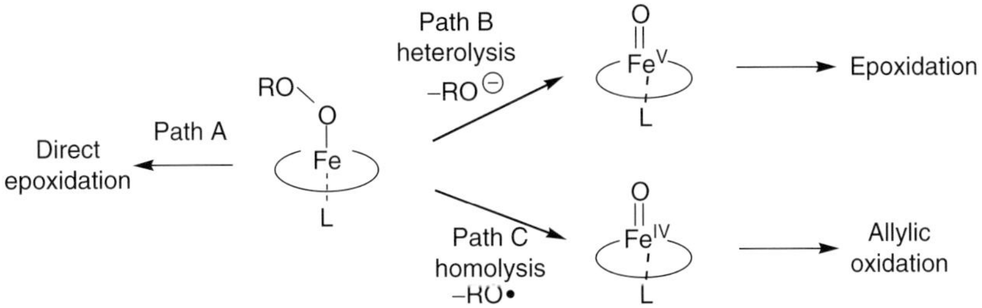

FIGURE 1.26 Three pathways for the reaction of the FeOOR(porph) adduct in epoxidation.

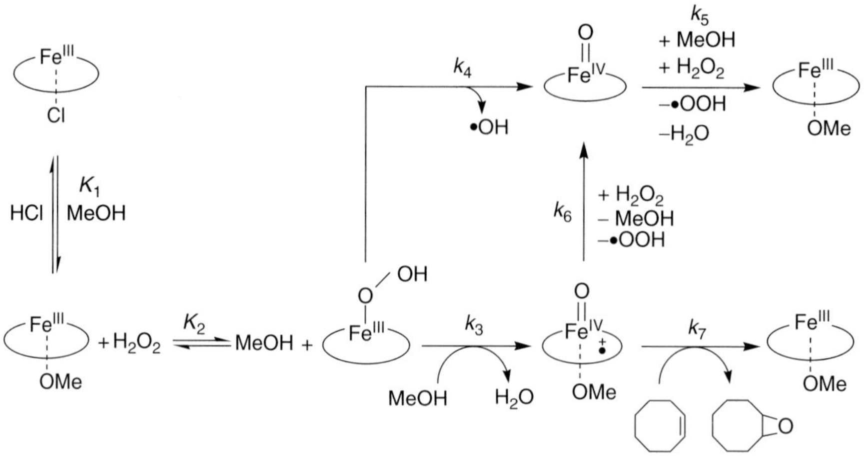

FIGURE 1.27 Kinetic pathway for epoxidation on F_{20}TPPFe(III) with H_2O_2.

which leads to deactivation of the porphyrin. A kinetic analysis for epoxidation with H_2O_2 has been carried out by Stephenson and Bell [579–582] (Fig. 1.27). Porphyrins with strong ligands (e.g., OH^-, OAc^-, Cl^-) do not catalyze the epoxidation of olefins with H_2O_2, but those with weak ligands (e.g., $CF_3SO_3^-$, ClO_4^-, NO_3^-) are able to carry out epoxidation [583]. Stephenson and Bell show that an Fe(III)porphyrin chloride becomes active in methanol containing solvents by replacement of the chloride by a methoxide.

12.2. Salen complexes

The first reports of a reaction of an amine with an aldehyde by Schiff [584] led to the establishment of a large class of ligands called Schiff bases. Among the most important of the Schiff bases are the tetradentate salen ligands (N,N'-bis(salicylaldehydo)ethylenediamine), which were studied extensively by Kochi and co-workers, who observed their high potential in chemoselective catalytic epoxidation reactions [585]. The best known method to epoxidize unfunctionalized olefins enantioselectively is the Jacobsen–Katsuki epoxidation reported independently by these researchers in 1990 [220,221]. In this method [515,586–589], optically active Mn^{III}(salen) compounds are used as catalysts, with usually PhIO or NaOCl as the terminal oxygen sources, and with a O=Mn^V(salen) species as the active [590,591] oxidant [586–594]. Despite the undisputed synthetic value of this method, the mechanism by which the reaction occurs is still the subject of considerable research [514,586,591]. The subject has been covered in a recent extensive review [595], which also discusses the less-studied Cr^{III}(salen) complexes, which can display different, and thus useful selectivity [596]. Computational and ^{1}H NMR studies have related observed epoxide enantioselectivities to the conformation of the salen groups [597]. The salen ligand has been described as "privileged" because of its wide use, manipulability, and broad scope in a variety of reaction systems [598]. Other ligands (e.g., BINOL = 1,1′-bi-2-napthol) are not as effective [599].

It is generally accepted that the Jacobsen–Katsuki epoxidation is initiated by the formation from a terminal oxygen donor (OxD) of a O=Mn^V oxo species (Fig. 1.28). The finding by Collman *et al.* that in Mn^{III}(salen) oxidation with various iodosylbenzene derivatives (PhIO, C$_6$F$_5$IO, and MesIO), the *cis/trans* ratios of stilbene oxides were strongly dependent on the oxidant and the reaction conditions suggest that the oxidant is complexed with the Mn [600]. Otherwise, a single active intermediate, O=Mn^V(salen), would produce similar *cis/trans* ratios of stilbene oxides with the different iodosylbenzenes. Adam *et al.* found that the *cis/trans* ratio is affected by the nature of the ligand [514]. Mn(salen)X bearing ligating counterions (e.g., X$^-$ = Cl$^-$) gave *trans*-stilbene oxide as the major product in the epoxidation of *cis*-stilbene, whereas Mn(salen)X bearing nonligating counterions (X$^-$ = PF$_6^-$) yielded *cis*-stilbene oxide as the major product. Similar results have been found for porphyrins [601]. These oxidant and counterion effects have been rationalized by the participation of different spin states of oxomanganese(V) salen intermediates and/or multiple active oxidants in oxygen atom transfer reactions [514,600,602–605].

Enantioselectivity in Mn(salen) asymmetric epoxidation correlates directly with the electronic properties of the ligand substituents, with complexes bearing electron-donating substituents giving the highest ees. In the epoxidation of *cis*-deuteriostyrene, electron-rich catalysts display a more pronounced secondary inverse isotope effect than electron-deficient catalysts. It is concluded that enantioselectivity is tied to the position of the transition state along the reaction coordinate. Electron-withdrawing groups produce a more reactive Mn oxo species, which adds an olefin in an early transition state and affords lower enantioselectivity, while electron-donating groups attenuate the reactivity of the oxo species, leading to a late transition state and higher enantioselectivity [586]. This is an important deduction for reactions without substrate precoordination, and is to be contrasted with enzymatic processes in which substrate precoordination

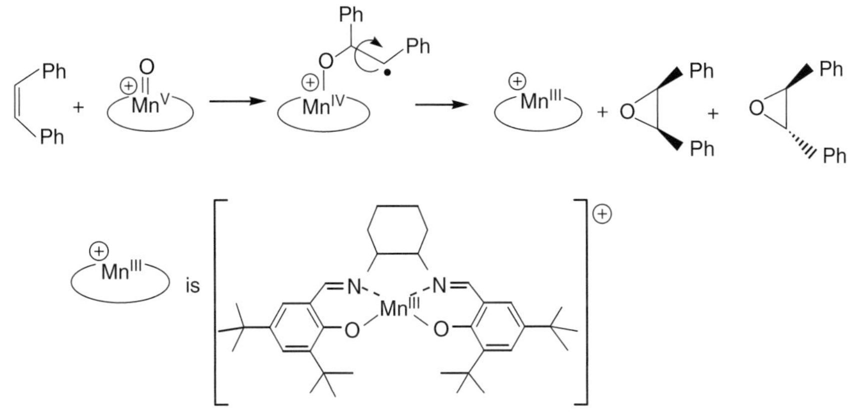

FIGURE 1.28 Likely mechanism of the isomerization in the Jacobsen–Katsuki epoxidation of *cis*-stilbene.

results in substantial selectivity-determining interactions between the catalyst's asymmetric environment and the substrate.

The formation of the $O=Mn^V$ oxo species is followed by addition of an olefin, and here the nature of the intermediate, as in the case of the porphyrin chemistry (Fig. 1.28), is in question [515,606,607]. The observation of variable *cis/trans* ratios with olefins such as *cis*-stilbene suggests that a radical intermediate is involved that can undergo rotation around the C–C axis (Fig. 1.28), but a cation intermediate is also possible, as has been suggested for the Cr(V) porphyrin case [513]. Another possibility is an oxametallacycle (metallaoxetane) intermediate, and evidence for this was presented by experiments with a vinylcyclopropane species which did not result in ring-opened products as expected for a radical intermediate [606]. These experiments were superceded by experiments by the group of Roschmann [514] which used a substituted vinylcyclopropane as a probe for distinguishing between radical and cation intermediates (Fig. 1.29). This probe can react to form different derivatives, and in this manner, it was shown for $Mn^V(oxo)$ that the intermediate was a radical. Furthermore, it was noted that if the $Mn^V(oxo)$ complex were the only oxidant in the epoxidation, irrespective of which oxygen donor was used to generate the reagent, the *cis/trans* ratio should be the same.

The study by the group of Roschmann further suggested the involvement of parallel Lewis acid and redox-type paths for the stereoselective epoxidation to account for the observation of different *cis/trans* ratios with different oxidants (Fig. 1.30). A $Mn^V(oxo)$ species can react with olefins via the radical intermediate mechanism discussed above to produce a mixture of *cis*- and *trans*-epoxides and release a spent Mn(III) species (path 1). Ligation of the oxygen donor to the spent Mn(III) catalyst would result in a $Mn^{III}(OLG)$ adduct. This adduct could release the LG to regenerate the active $Mn^V(oxo)$ oxidant in the Groves-type rebound oxygen mechanism. Alternatively, the adduct could also react by concerted oxygen transfer without an oxidation state change to give a *cis*-epoxide by the Lewis acid mechanism (path 2). Thus, the combination of the stepwise epoxidation by the established $Mn^V(oxo)$ species and the concerted epoxidation (path 2) accounts for the various stereoisomers produced by different oxidants and ligands.

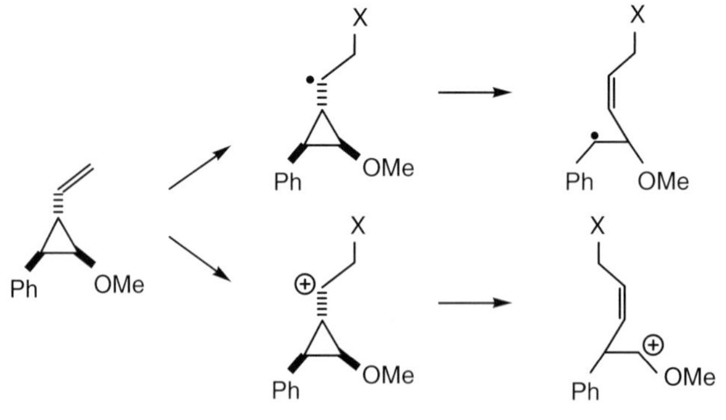

FIGURE 1.29 Vinylcyclopropane probe to distinguish between radical and cation intermediates.

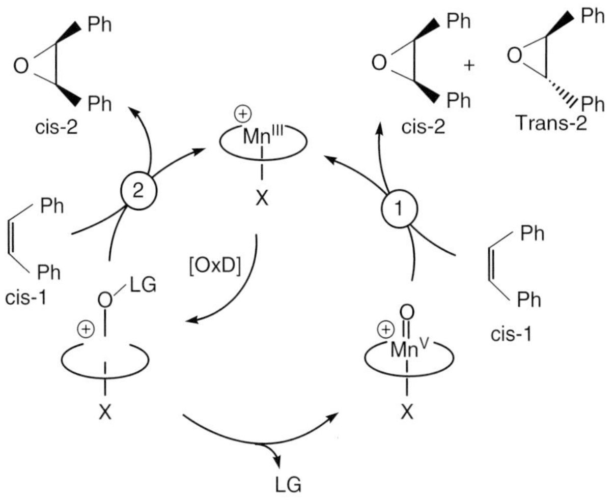

FIGURE 1.30 Lewis acid and redox pathways in epoxidation in the Mn(salen) system.

Research with salen compounds continues actively with the invention of new ligand derivatives and the employment of metals other than Mn such as Ni or Cu, and even the creation of bimetallic systems, for example, containing Ni and Zr [608].

The group of Busch also presented evidence that the Mn(II) complex of a cross-bridged cyclam ligand, 4,11-dimethyl-1,4,8,11-tetraazabicyclo[6.6.2]hexadecane, denoted as $Mn^{II}(Me_2EBC)Cl_2$, forms a Mn(IV) adduct with iodosylbenzene which is a new active intermediate in epoxidation reactions [600,609,610]. These examples with Mn, and in the previous section with Fe, are instances of late transition metals catalyzing epoxidation reactions by both the redox and the Lewis acid mechanisms (see Chapter 3).

12.3. Nonheme Fe complexes

The reaction of nonheme Fe complexes that are analogues of Rieske dioxygenase enzymes have been studied by the group of Que [234,235] (see Chapter 18). A possible mechanism has been suggested [241] (Fig. 1.31) for a bispidine ligand. An [Fe^{III}-OOH] intermediate is formed initially, which undergoes O–O bond homolysis to form a ferryl [Fe^{IV}=O] oxidant and HO•. The oxidant was formed independently by reaction of the complex with iodosylbenzene as indicated by the appearance of an expected electronic transition in the near infrared of $\varepsilon = 400$ M^{-1} cm^{-1}. This was able to epoxidize cis-cyclooctene in an Ar atmosphere.

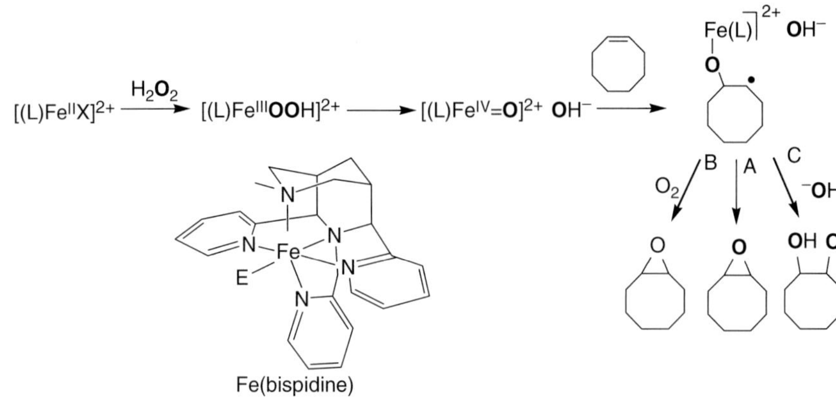

FIGURE 1.31 Proposed mechanism for formation of epoxides and diols with Fe(bispidine).

13. BIOLOGICAL SYSTEMS

The mechanism of enzymatic action differs from that of their homogeneous homologues in two significant ways, both of which are related to the existence of the protein scaffold surrounding the active site that gives rise to the secondary, tertiary, and quaternary structure. The first is the well-known ability of enzymes to form docking cavities or pockets with precise hydrophobic or hydrophilic interaction properties, which gives rise to exquisite selectivity properties. The second involves the enzyme's ability, by dynamic motion of its structure, to bring into play various groups involved in polarization effects and proton or electron transfer to carry out fast reactions. These aspects will be illustrated through the example of cytochrome P-450, a naturally occurring enzyme of size from 40–50 kDa that, like peroxidases and catalases, has at its nucleus a single heme group.

Cytochrome P-450 is unique in being able to heterolytically cleave the O–O bond of putative Fe(III) hydroperoxide intermediates [578], and to activate a variety of electron-rich double bonds. Its selectivity for internal olefins, but not terminal olefins, has been related to its substrate-binding pocket [264], which resembles a long funnel, at the end of which a small hydrophobic pocket sequesters the substrate terminus, making it unavailable for oxidation [611–613].

Enzymes are also able to position terminal ligands that play a key role in the reactivity of the heme. These ligands are thiolate in cytochrome P-450, imidazolate in peroxidase, and phenolate in catalase. The ligands facilitate the O–O bond cleavage by serving as internal electron donors [614–616], in a similar manner as the axial ligands in porphyrin models [525,526], but importantly, function in a more dynamic mode as they are not rigidly bound. As shown by the group of Poulos, keys to enzyme function are also proton transfer [617] and electron transfer, the latter involving complex formation between an electron donor protein and the P-450 [618]. In microsomal P-450s, the electron donor is P-450 reductase which contains flavin mononucleotide (FMN) and flavin adenine dinucleotide (FAD). Electrons are transfered from NADPH to the flavins and finally to the P-450 heme iron. P450BM-3 is an unusual bacterial P-450 since the P-450 and P-450 reductase are linked together to form a large single polypeptide chain. The reductase domain contains FMN and FAD, and the electron flow is from NADPH-to-FAD-to-FMN-to-heme.

14. PERSPECTIVE AND CONCLUSIONS

The epoxidation of olefins is used in the production of a wide variety of chemical products, intermediates, fine chemicals, and pharmaceuticals with myriads of uses. The scale of production ranges from millions of tons for chemical commodities such as ethylene oxide and propylene oxide to a few grams per year for synthetic intermediates. The catalysts employed in epoxidation are extremely diverse and encompass the gamut of homogeneous, heterogeneous, and biological systems, including hybrid materials combining several functions such as immobilized homogeneous complexes or phase-transfer–mediated materials.

The advances made in the field of catalytic epoxidation in the last 10 years have been enumerated, and broad comparisons are made on the basis of TOF. It is found that biological systems have a considerable advantage over conventional systems, although their implementation in practice still faces tremendous challenges. For example, with microorganisms, substantial efforts are required to provide life-supporting systems and to minimize product toxicity problems, while with enzymes, the supply of cofactors and stability issues remain largely unresolved. Heterogeneous systems are the basis for the most important industrial commodity compound production technologies because of their robustness, and ease in product separation. They set as a benchmark a required TOF of about $1 \times 10^{-1}\, s^{-1}$ for large-scale production. Homogeneous systems find considerable use in smaller scale, higher value-added products. In the latter area, much emphasis has been placed on selectivity rather than rate, yet both are important in practice. It is hard to imagine that a reaction with a TOF of $1 \times 10^{-2}\, s^{-1}$ or less will be of substantial commercial significance, no matter if it has 100% selectivity and 100% ee.

A subject of great interest has been the study of mechanism, a subject which has played an important role not only in improving technology, but also in advancing fundamental knowledge in the area of chemistry. Notable are the high levels of understanding that have been achieved in the areas of heterogeneous epoxidation on Ag and Ti catalysts, homogeneous catalysts based on Mo and V, and biomimetic catalysts based on Cr, Mn, and Fe. These are discussed in detail, including results from theoretical studies on both heterogeneous and homogeneous catalysts, which have been used to resolve long-standing issues in these fields.

Advances made in the area of mechanism have been presented so as to give a perspective on similarities and differences between different catalysts. Both classical and novel systems have been covered with considerable attention placed on comparisons. Less detail is given for certain homogeneous and heterogeneous catalyst systems reported elsewhere in this book, such as Au/titanosilicates,

TiO$_2$/SiO$_2$, Ag–Cu alloys, soft Pt(II) Lewis acids, MTO, oxaziridinium salts, NHPI, polyoxometallates, phase-transfer systems, polymer-supported complexes, calixarene complexes, and biologically based systems.

Early transition metal ions in their highest oxidation states, such as Ti(IV), V(V), W(VI), Mo(VI), and Re(VIII), operate by a Lewis acid mechanism of epoxidation. They act as Lewis acids coordinating hydrogen peroxide or alkyl hydroperoxides as adducts (M-OOH or M-OOR), which transfer the *proximal* oxygen atom to an olefin double bond. The *distal* oxygen atom and its cohort shifts to the metal (M-OH or M-OR) leaving it unchanged in oxidation state. The most effective metals are strong Lewis acids with open coordination sites that are not strong oxidants (to avoid one-electron oxidations).

Late transition metal ions that can accommodate a two-electron rise in their oxidation state, like Cr(III), Mn(III), and Fe(III), and likely Ru(I), operate by a redox mechanism of epoxidation. They receive an oxygen atom from a TO to form an oxene species (M=O) which then transfers the oxygen to an olefin by the intermediacy of a metallacycle, or a radical or cation species. Interestingly, these systems are not inhibited by water or alcohol as are the Lewis acid metals.

There are several movements that can be detected in research in epoxidation as well as in its practice. One aspect is the shift from alkyl hydroperoxides to the more atom-efficient and environmentally benign oxidant H$_2$O$_2$. Another aspect is the substitution of chlorinated solvents, which are carcinogenic and polluting, by aqueous solvents. These shifts have required compensating developments, as H$_2$O$_2$ is available only in aqueous solution, and water is a strong inhibitor of metal catalysts. One such development is the use of PTRs. A further movement is the erosion of the dichotomy of Lewis acid and redox mechanisms by the discovery of systems that utilize both pathways. As our understanding of mechanisms improves, it is expected that further development of new catalyst systems will take place.

ACKNOWLEDGMENTS

The author is grateful for support from the National Institute for Advanced Industrial Science and Technology (AIST), the National Science Foundation, and the Japan Society for the Promotion of Science (JSPS) through the Invited Fellow Program.

REFERENCES

[1] G. Sienel, R. Rieth, K. T. Rowbottom, Epoxides, in: *Ulmann's Encyclopedia of Industrial Chemistry*, 6th ed., Verlag Chemie, Weinheim, 2003, p. 269.
[2] J. P. Dever, K. F. George, W. C. Hoffman, H. Soo, Ethylene oxide, in: *Kirk-Othmer Encyclopedia of Chemical Technology*, John Wiley & Sons, New York, 2001, (on-line edition updated March 227), pp. 632–673.
[3] J. Lacson, Ethylene oxide, in: *Chemical Economics Handbook*, SRI International, Menlo Park, CA, Oct. 2003.
[4] *Chemical Market Reporter*, Aug. 8, 2006 (currently ICIS Chemical Business Americas, www.icis.com).

[5] D. L. Trent, Propylene oxide, in: *Kirk Othmer Encyclopedia of Chemical Technology*, on-line edition, John Wiley & Sons, New York, 2001.
[6] Chemexpo.com, http://www.chemexpo.com, Chemical Profile, 9/10/2001.
[7] Hazardous Substance Data Bank, Online database produced by the National Library of Medicine, 1,2-Propylene Oxide Profile last updated, (October 10, 2001).
[8] M. Ishino, J. Yamamoto, Propylene oxide manufacturing processes, *Shokubai* 48 (2006) 511–515.
[9] J. R. Monnier, The direct epoxidation of higher olefins using molecular oxygen, *Appl. Catal. A: Gen.* 221 (2001) 73.
[10] J. R. Monnier, Private communication.
[11] J. R. Monnier, The selective epoxidation of non-allylic olefins over supported silver catalysts, in: R. K. Grasselli, S. T. Oyama, A. M. Gaffney, J. E. Lyons (Eds.), 3rd World Congress on Oxidation Catalysis, Elsevier, Amsterdam, 1997, *Stud. Surf. Sci. Catal.* 110, (1997)135.
[12] A. Ansmann, R. Kawa, M. Neuss, Cosmetic composition containing hydroxyethers, *US Patent* 7,083,780 B2, Aug. 1, 2006, To Cognis Deutschland, Gmbh & Co. KG.
[13] M. Servais, R. Crochet, Stabilised composition of 1,1,1,-trichloroethane, *European Patent EP* 62,952, Oct. 20, 1982, To Solvay (BE).
[14] I. Kim, S. M. Kim, C.-S. Ha, D.-W. Park, Synthesis and cyclohexene oxide/carbon dioxide copolymerizations of zinc acetate complexes bearing bidentate pyridine-alkoxide ligands, *Macromolec. Rapid Commun.* 25 (2004) 888.
[15] C. Anaya de Parrodi, E. Juaristi, Chiral 1,2-amino alcohols and 1,2-diamines derived from cyclohexene oxide: Recent applications in asymmetric synthesis, *Synlett* (2006) 2699.
[16] H. H. Szmant, *Organic Building Blocks of the Chemical Industry*, Wiley, New York, 1989, p. 4.
[17] A. Toshimitsu, H. Abe, C. Hirosawa, S. Tanimoto, Preparation of chiral aziridines from chiral oxiranes with retention of configuration, *J. Chem. Soc. Chem. Commun.* (1992) 284.
[18] Y. Niibo, T. Nakata, J. Otera, H. Nozaki, Stereospecific ring opening at the benzylic carbon of phenyloxirane derivatives by alcohols, *Synlett* (1991) 97.
[19] M. Chini, P. Crotti, F. Macchia, Regioalternating selectivity in the metal salt catalyzed aminolysis of styrene oxide, *J. Org. Chem.* 56 (1991) 5939.
[20] J. Meinwald, S. S. Labana, M. S. Chadha, Peracid reactions. III. The oxidation of bicyclo [2.2.1] heptadiene, *J. Am. Chem. Soc.* 85 (1963) 582.
[21] K. Miyamotoa, K. Okuroa, H. Ohta, Substrate specificity and reaction mechanism of recombinant styrene oxide isomerase from Pseudomonas putida, *Tetrahedron Lett.* 48 (2007) 3255.
[22] Q. Guo, K. Nakajima, T. Takahashi, Formation of 8-membered ring compounds by the reaction of styrene oxide with $MoCl_5$, *Chem. Lett.* 32 (2003) 1044.
[23] L. W. Zatorski1, P. T. Wierzchowski, Zeolite-catalyzed synthesis of 4-phenyl-1,3-dioxolanes from styrene oxide, *Catal. Lett.* 10 (1991) 211.
[24] S. A. Blum, V. A. Rivera, R. T. Ruck, F. E. Michael, R. G. Bergman, Synthetic and mechanistic studies of strained heterocycle opening reactions mediated by zirconium(IV) imido complexes, *Organometallics* 24 (2005) 1647.
[25] D. M. Hodgson, R. Wisedale, Enantioselective rearrangement of *exo*-norbornene oxide to nortricyclanol, *Tetrahedron Asymm.* 7 (1996) 1275.
[26] H. Q. Pham, M. J. Marks, Epoxy Resins, in: *Ulmann's Encyclopedia of Industrial Chemistry*, 6th ed., Verlag Chemie, Weinheim, 2003, On-line edition.
[27] Z. K. Liao, C. J. Boriack, Epoxidation process for aryl allyl ethers, *US Patent* 6,087,513, July 11, 2000, To the Dow Chemical Company.
[28] J. Bernadou, B. Meunier, Biomimetic chemical catalysts in the oxidative activation of drugs, *Adv. Synth. Catal.* 346 (2004) 171.
[29] A. L. Slitt, N. J. Cherrington, M. Z. Dieter, L. M. Aleksunes, G. L. Scheffer, W. Huang, D. D. Moore, C. D. Klaassen, *Trans*-stilbene oxide induces expression of genes involved in metabolism and transport in mouse liver via CAR and Nrf2 transcription factors, *Mol. Pharmacol.* 69 (2005) 1554.
[30] A. L. Slitt, N. J. Cherrington, C. D. Fisher, M. Negishi, C. D. Klaassen, Induction of genes for metabolism and transport by *trans*-stilbene oxide in livers of Sprague-Dawley and Wistar-Kyoto rats, *Drug Metab. Dispos.* 34 (2006) 1190.

[31] M. Bücker, M. Golan, H. U. Schmassmann, H. R. Glatt, P. Stasiecki, F. Oesch, The epoxide hydratase inducer *trans*-stilbene oxide shifts the metabolic epoxidation of benzo(a)pyrene from the bay- to the k-region and reduces its mutagenicity, *Mol. Pharmacol.* 16 (1979) 656.

[32] J. T. Groves, J. Lee, S. S. Marla, Detection and characterization of an oxomanganese(V) porphyrin complex by rapid-mixing stopped-flow spectrophotometry, *J. Am. Chem. Soc.* 119 (1997) 6269.

[33] T. J. Hubin, J. M. McCormick, S. R. Collinson, M. Buchalova, C. M. Perkins, N. W. Alcock, P. K. Kahol, A. Raghunathan, D. H. Busch, New iron(II) and manganese(II) complexes of two ultra-rigid, cross-bridged tetraazamacrocycles for catalysis and biomimicry, *J. Am. Chem. Soc.* 122 (2000) 2512.

[34] H. Breton, M. Cociglio, F. Bressolle, H. Peyriere, J. P. Blayac, D. H. Buys, Liquid chromatography–electrospray mass spectrometry determination of carbamazepine, oxcarbazepine and eight of their metabolites in human plasma, *J. Chromatogr. B* 828 (2005) 80.

[35] K. Lertratanangkoon, M. G. Horing, Metabolism of carbamazepine, *Drug. Metab. Dispos.* 10 (1982) 1.

[36] Y. Zhu, H. Chiang, M. W. Radcliffe, R. Hilt, P. Wong, C. B. Kissinger, P. T. Kissinger, Liquid chromatography/tandem mass spectrometry for the determination of carbamazepine and its main metabolite in rat plasma utilizing an automated blood sampling system, *J. Pharm. Biomed. Anal.* 38 (2005) 119.

[37] T. C. O. Mac Leod, V. P. Barros, A. L. Faria, M. A. Schiavon, I. V. P. Yoshida, M. E. C. Queiroz, M. D. Assis, Jacobsen catalyst as a P450 biomimetic model for the oxidation of an antiepileptic drug, *J. Mol. Catal. A: Chem.* 273 (2007) 259.

[38] P. C. B. Page, B. R. Buckley, H. Heaney, A. J. Blacker, Asymmetric epoxidation of *cis*-alkenes mediated by iminium salts: Highly enantioselective synthesis of levcromakalim, *Org. Lett.* 7 (2005) 375.

[39] G. F. Lai, A convenient preparation of tetrahydrofuran-based diamines, *Synth. Commun.* 34 (2004) 1981.

[40] S. E. Zook, J. K. Busse, B. C. Borer, A concise synthesis of the HIV-protease inhibitor nelfinavir via an unusual tetrahydrofuran rearrangement, *Tetrahedron Lett.* 41 (2000) 7017.

[41] H. Wu, L. Wang, H. Zhang, Y. Liu, P. Wu, M. He, Highly efficient and clean synthesis of 3,4-epoxytetrahydrofuran over a novel titanosilicate catalyst, Ti-MWW, *Green Chem.* 8 (2006) 78.

[42] D. Hendlin, E. O. Stapley, M. Jackson, H. Wallick, A. K. Miller, F. J. Wolf, T. W. Miller, L. Chaiet, F. M. Kahan, E. L. Foltz, H. B. Woodruff, J. M. Mata, S. Hernandez, S. Mochales, Phosphonomycin, a new antibiotic produced by strains of streptomyces, *Science* 166 (1969) 122.

[43] X. Y. Wang, H. C. Shi, C. Sun, *et al.*, Asymmetric epoxidation of *cis*-1-propenylphosphonic acid (CPPA) catalyzed by chiral tungsten(VI) and molybdenum(VI) complexes, *Tetrahedron* 60 (2004) 10993.

[44] S. E. Schaus, B. D. Brandes, J. F. Larrow, M. Tokunaga, K. B. Hansen, A. E. Gould, M. E. Furrow, E. N. Jacobsen, Highly selective hydrolytic kinetic resolution of terminal epoxides catalyzed by chiral (salen)Co^{III} complexes. Practical synthesis of enantioenriched terminal epoxides and 1,2-diols, *J. Am. Chem. Soc.* 124 (2002) 1307.

[45] D. S. Tarbell, R. M. Carman, D. D. Chapman, S. E. Cremer, A. D. Cross, K. R. Huffman, M. Kuntsmann, N. J. McCorkindale, J. G. McNally, A. Rosowsky, F. H. L. Varino, R. L. West, *J. Am. Chem. Soc.* 83 (1961) 3096.

[46] H. P. Sigg, H. P. Weber, Isolierung und Strukturaufklärung von Ovalicin, *Helv. Chim. Acta* 51 (1968) 1395.

[47] T. Takeuchi, H. Iinuma, J. Iwanaga, S. Takahashi, T. Takita, H. Umezawa, Coriolin, a new Basidiomycetes antibiotic, *J. Antibiot.* 22 (1969) 215.

[48] B. A. Bierl, M. Beroza, C. W. Collier, Potent sex attractant of the gypsy moth: Its isolation, identification, and synthesis, *Science* 170 (1970) 87.

[49] S. M. Kupchan, W. A. Court, R. G. Dailey, C. J. Gilmore, R. F. Bryan, Tumor inhibitors. LXXIV. Triptolide and tripdiolide, novel antileukemic diterpenoid triepoxides from Tripterygium wilfordii, *J. Am. Chem. Soc.* 94 (1972) 7194.

[50] B. C. J. Persoons, P. E. J. Verwiel, F. J. Ritter, E. Talman, P. J. Nooijen, W. J. Nooijen, Sex pheromones of the american cockroach, *Periplaneta americana*: A tentative structure of periplanone-B, *Tetrahedron Lett.* 17 (1976) 2055.

[51] K. Edo, M. Mizugaki, Y. Koide, H. Seto, K. Furihata, N. Otake, N. Ishida, The structure of neocarzinostatin chromophore possessing a novel bicyclo-[7,3,0]dodecadiyne system, *Tetrahedron Lett.* 26 (1985) 331.
[52] H. Itazaki, K. Nagashima, K. Sugita, H. Yoshida, Y. Kawamura, Y. Yasuda, K. Matsumoto, K. Ishii, N. Uotani, H. Nakai, A. Terui, S. Yoshimatsu, Isolation and structural elucidation of new cyclotetrapeptides, trapoxins A and B, having detransformation activities as antitumor agents, *J. Antibiot.* 43 (1990) 1524.
[53] D. M. Bollag, P. A. McQueney, J. Zhu, O. Hensens, L. Koupal, J. Liesch, M. Goetz, E. Lazarides, C. M. Woods, Epothilones, a new class of microtubule-stabilizing agents with a taxol-like mechanism of action, *Cancer Res.* 55 (1995) 2325.
[54] H. Nakajima, B. Sato, T. Fujita, S. Takase, H. Terano, M. Okuhara, New antitumor substances, FR901463, FR901464 and FR901465. I. Taxonomy, fermentation, isolation, physico-chemical properties and biological activities, *J. Antibiot.* 49 (1996) 1196.
[55] M. Guidotti, N. Ravasio, R. Psaro, G. Ferraris, G. Moretti, Epoxidation on titanium-containing silicates: Do structural features really affect the catalytic performance? *J. Catal.* 214 (2003) 242.
[56] C. S. Sell, Terpenoids, in: *Kirk-Othmer Encyclopedia of Chemical Technology*, John Wiley & Sons, New York, 2001(on-line edition updated Sept. 2006).
[57] G. Neri, G. Rizzo, A. S. Arico, C. Crisafulli, L. De Luca, A. Donato, M. G. Musolino, and R. Pietropaolo, One-pot synthesis of naturanol from α-pinene oxide on bifunctional Pt-Sn/SiO$_2$ heterogeneous catalysts: Part I. The catalytic system, *Appl. Catal. A: Gen.* 325 (2007) 15.
[58] M. Rohr, R. H. Potter, R. E. Naipawer, Flavoring with α-campholenic alcohol, US Patent 4,766,002, Aug. 23, 1988, To Givaudan Corporation.
[59] E. J. Brunke, E. Klein, Polysubstituted cyclopentene derivatives, German Patent DE 2,827,957, Jan. 10, 1980, To Dragoco Gerberding, Co. GBMH.
[60] M. Mühlstädt, W. Dollase, M. Herrmann, G. Feustel, Riechstoffe und riechstoffkompositionen, German Patent DE 1,922,391, Aug. 27, 1970, To Chem Fab Miltitz Veb.
[61] C. Corvi-Mora, Method of preparing sobrerol and the pharmaceutical application of the sobrerol thus obtained, US Patent 4,639,469, Sep. 30, 1982, To C. Corvi-Mora.
[62] R. Barrera Zapata, A. L. Villa, C. Montes de Correa, Limonene epoxidation: Diffusion and reaction over PW-amberlite in a triphasic system, *Ind. Eng. Chem. Res.* 45 (2006) 4589.
[63] A. L. Villa, Personal communication.
[64] G. Lligadas, J. C. Ronda, M. Galià, U. Biermann, J. O. Metzer, Synthesis and characterization of polyurethanes from epoxidized methyl oleate based polyether polyols as renewable resources, *J. Pol. Sci. Part A: Pol. Chem.* 44 (2005) 634.
[65] V. V. Goud, A. V. Patwardhan, N. C. Pradhan, Studies on the epoxidation of manua oil (Madhumica indica) by hydrogen peroxide, *Biores. Technol.* 97 (2006) 1365.
[66] M. R. Klaas, S. Warwel, Complete and partial epoxidation of plant oils by lipase-catalysed perhydrolysis, *Ind. Crops. Prod.* 9 (1999) 125.
[67] G. J. Piazza, T. A. Foglia, One-pot synthesis of fatty acid epoxides from triacylglycerols using enzymes present in oat seeds, *J. Am. Oil Chem. Soc.* 83 (2006) 1021.
[68] T. Kato, Y. Yamaguchi, T. Hirano, T. Yokoyama, T. Uyehara, T. Namai, S. Yamanaka, N. Harada, Unsaturated hydroxyl fatty acids, the self defensive substances in rice plant against rice blast disease, *Chem. Lett.* 26 (1984) 409.
[69] P. Schweizer, A. Jeanguenat, D. Whitacre, J. P. Métraux, E. Mösinger, Induction of resistance in barley against Erysiphe graminis f.sp. hordei by free cutin monomers, *Physiol. Mol. Plant Pathol.* 49 (1996) 103.
[70] W. Gress, R. Hoefer, R. Gruetzmacher, U. Nagorny, A. Heidbreder, B. Hirschberger, Fatty chemical polyalcohols as reagent thinners, US Patent 6,433,125, Aug. 13, 2002, To Henkel Kommanditgesellschaft auf Aktien.
[71] P. Daute, R. Picard, J.-D. Klamann, P. Wedl, A. Peters, Method for producing glyceride acetates, US Patent 7,071,343 B2, July 4, 2006, To Cognis Deutschland GmbH & Co. KG.
[72] A. B. Cook, J. J. Palmer, J. M. Rodriguez, Defoamer composition and method of using the same, US Patent 5,645,762, July 8, 1997, To Henkel Corporation.
[73] F. D. Gunstone, Fatty acids. Part II. The nature of the oxygenated acid present in *Vernonia anthelmintica* (Willd.) seed oil, *J. Chem. Soc.* (1954) 1611.

[74] B. T. Larsen, D. D. Gutterman, O. A. Hatoum, Emerging role of epoxyeicosatrienoic acids in coronary vascular function, *Eur. J. Clin. Inv.* 36 (2006) 293.
[75] T. Katsuki, K. B. Sharpless, The first practical method for asymmetric epoxidation, *J. Am. Chem. Soc.* 102 (1980) 5974.
[76] T. Usui, S. Kazami, N. Dohmae, Y. Mashimo, H. Kondo, M. Tsuda, A. Terasaki, K. Ohashi, J. Kobayashi, H. Osada, Amphidinolide H, a potent cytotoxic macrolide, covalently binds on actin subdomain 4 and stabilizes actin filament, *Chem. Biol.* 11 (2004) 1269.
[77] C. B. Chapleo, M. A. W. Finch, T. V. Lee, S. M. Roberts, R. F. Newton, Total synthesis of prostaglandin A2 involving the reaction of a heterocuprate reagent with an allyl epoxide, *J. Chem. Soc., Perkin Trans.* 1 (1980) 2084.
[78] M. E. Jung, D. Sun, Stereoselective production of β-amino alcohols and β-thioacyl alcohols via an application of the non-aldol aldol process, *Tetrahedron Lett.* 40 (1999) 8343.
[79] M. E. Jung, A. van den Heuvel, Diastereoselectivity in non-aldol aldol reactions: Silyl triflate-promoted Payne rearrangements, *Tetrahedron Lett.* 43 (2002) 8169.
[80] A. B. Smith, P. A. Levenberg, P. J. Jerris, R. M. Scarborough, P. M. Wovkulich, Synthesis and reactions of simple 3(2H)-furanones, *J. Am. Chem. Soc.* 103 (1981) 1501.
[81] J.-M. Brégeault, C. Lepetit, F. Ziani-Derdar, O. Mohammedi, L. Salles, A. Deloffre, Epoxidation of tertiary allylic alcohols and subsequent isomerization of tertiary epoxy-alcohols: A comparison of some catalytic systems for demanding ketonization processes, in: R. K. Grasselli, S. T. Oyama, A. M. Gaffney, J. E. Lyons (Eds.), 3rd World Congress on Oxidation Catalysis, Elsevier, Amsterdam, 1997, *Stud. Surf. Sci. Catal.* 110 (1997) 545.
[82] R. A. Johnson, K. B. Sharpless, Catalytic asymmetric epoxidation of allylic alcohols, in: I. Ojima (Ed.), *Catalytic Asymmetric Synthesis*, VCH, New York, 1993, pp. 231–280. Chapter 6A.
[83] T. Katsuki, Asymmetric epoxidation of unfunctionalized olefins and related reactions, in: I. Ojima (Ed.), *Catalytic Asymmetric Synthesis*, VCH, New York,1993, pp. 287–326. Chapter 6B.
[84] A. K. Yudin (Ed.), *Aziridines and Epoxides in Organic Synthesis*, Wiley-VCH, New York, 2006.
[85] H. C. Kolb, M. G. Finn, K. B. Sharpless, Click chemistry: Diverse chemical function from a few good reactions, *Angew. Chem. Int. Ed.* 40 (2001) 2004.
[86] V. V. Fokin, P. Wu, Epoxides and aziridines in click chemistry, in: A. K. Yudin (Ed.), *Aziridines and Epoxides in Organic Synthesis*, Wiley-VCH, New York, 2006. pp. 443–477. Chapter. 12.
[87] H. Takada, M. Shima, Silver catalyst for production of ethylene oxide, method of production thereof, and method of production of ethylene oxide, *US Patent* 6,103,916. Aug. 15, 2000, To Nippon Shokubai Co., Ltd.
[88] J. R. Lockemeyer, Preparation of ethylene oxide and catalyst, *US Patent* 5,929,259, Jul. 27, 1999, To Shell Oil Company.
[89] M. M. Bhasin, Catalyst composition for oxidation of ethylene to ethylene oxide, *US Patent* 5,057,481, Oct. 15, 1991, To Union Carbide Chemical and Plastics Technology Corp.
[90] R. P. Nielsen, J. H. La Rochelle, Ethylene oxide process, *US Patent* 4,012,425, Mar. 15, 1977, To Shell Oil Company.
[91] M. Atkins, J. Couves, M. Hague, B. H. Sakaniki, K. C. Waugh, On the role of Cs, Cl, and subsurface O in promoting selectivity in Ag/α-Al$_2$O$_3$ catalysed oxidation of ethene to ethene epoxide, *J. Catal.* 235 (2005) 103.
[92] S. Linic, M. A. Barteau, On the mechanism of Cs promotion in ethylene epoxidation on Ag, *J. Am. Chem. Soc.* 126 (2004) 8086.
[93] N. Rizkalla, Ethylene oxide catalyst, *US Patent* 6,846,774 B2, Jan. 25, 2005, To Scientific Design Co., Inc.
[94] A. M. Lauritzen, Ethylene oxide catalyst and process for the catalytic production of ethylene oxide, *European Patent EP* 0622015, May 4, 1988, To Shell Int. Research.
[95] C. T. Campbell, M. T. Paffett, The role of chlorine promoters in catalytic ethylene epoxidation over the Ag(110) surface, *Appl. Surf. Sci.* 19 (1984) 28.
[96] S. A. Tan, R. B. Grant, R. M. Lambert, Chlorine-oxygen interactions and the role of chlorine in ethylene oxidation over Ag(111), *J. Catal.* 100 (1986) 383.
[97] K. L. Yeung, A. Gavriilidis, A. Varma, M. M. Bhasin, Effects of 1,2 dichloroethane addition on the optimal silver catalyst distribution in pellets for epoxidation of ethylene, *J. Catal.* 174 (1988) 1.

[98] C.-F. Mao, M. A. Vannice, High surface area α-aluminas III. Oxidation of ethylene over silver dispersed on high surface area α-alumina, *Appl. Catal. A: Gen.* 122 (1995) 61.
[99] W. E. Evans, P. I. Chipman, Process for operating the epoxidation of ethylene, US Patent 6,717,001 B2, Apr. 6, 2004, To Shell Oil Company.
[100] T. A. Nijhuis, M. Makkee, J. A. Moulijn, B. M. Weckhuysen, The production of propylene oxide: Catalytic processes and recent developments, *Ind. Eng. Chem. Res.* 45 (2006) 3447.
[101] J. Kollar, Epoxidation process, US Patent 3,351,635, Nov. 7, 1967, To Halcon International, Inc.
[102] R. A. Sheldon, M. Wallau, I. W. C. E. Arends, U. Schuchardt, Heterogeneous catalysts for liquid-phase oxidations: Philosophers' stones or Trojan horses? *Acc. Chem. Res.* 31 (1998) 485.
[103] H. P. Wulff, F. Wattimena, Olefin epoxidation, US Patent 4,021,454, May 3, 1977, To the Shell Oil Co.
[104] R. A. Sheldon, J. Dakka, Heterogeneous catalytic oxidations in the manufacture of fine chemicals, *Catal. Today* 19 (1994) 215.
[105] M. G. Clerici, G. Bellussi, U. Romano, Synthesis of propylene oxide from propylene and hydrogen peroxide catalyzed by titanium silicalite, *J. Catal.* 129 (1991) 159.
[106] M. G. Clerici, P. Ingallina, Epoxidation of lower olefins with hydrogen peroxide and titanium silicalite, *J. Catal.* 140 (1993) 71.
[107] B. Notari, Titanium silicalites, *Catal. Today* 18 (1993) 163.
[108] G. Thiele, Process for the preparation of epoxides from olefins, US Patent 6,372,924 B2, Apr. 16, 2002, To Degussa-Huls AG.
[109] M. G. Clerici, P. Ingallina, Oxidation reactions with *in situ* generated oxidants, *Catal. Today* 41 (1998) 351.
[110] A. Sato, T. Miyake, Production of propylene oxide, Japan Patent JP 04–352,771, Dec. 7, 1992, To Tosoh Corp.
[111] T. Hayashi, M. Haruta, Selective oxidation of hydrocarbons with gold supported on titania, *Shokubai* 37 (1995) 72.
[112] T. Hayashi, K. Tanaka, M. Haruta, Selective vapor-phase epoxidation of propylene over Au/TiO_2 catalysts in the presence of oxygen and hydrogen, *J. Catal.* 178 (1998) 566.
[113] R. Meiers, U. Dingerdissen, W. F. Hölderich, Synthesis of propylene oxide from propylene, oxygen, and hydrogen catalyzed by palladium–platinum-containing titanium silicalite, *J. Catal.* 176 (1988) 376.
[114] W. F. Hoelderich, F. Kollmer, Oxidation reactions in the synthesis of fine and intermediate chemicals using environmentally benign oxidants and the right reactor system, *Pure Appl. Chem.* 72 (2000) 1273.
[115] W. F. Hoelderich, One-pot reactions: A contribution to environmental protection, *Appl. Catal. A: Gen.* 194 (2000) 487.
[116] W. Laufer, W. F. Hoelderich, Direct oxidation of propylene and other olefins on precious metal containing Ti-catalysts, *Appl. Catal. A: Gen.* 213 (2001) 163.
[117] M. Haruta, S. Tsubota, T. Hayashi, Method for selective oxidation of hydrocarbons by gold-titanium oxide containing catalysts, Japan Patent JP 2,615,432 (Publ. no. 08-127550), May 21, 1996, To Agency for Industrial Science and Technology (AIST).
[118] B. Taylor, J. Lauterbach, W. N. Delgass, Gas phase epoxidation of propylene over small gold ensembles on TS-1, *Appl. Catal. A: Gen.* 291 (2005) 188.
[119] J. Lu, X. Zhang, J. J. Bravo-Suárez, K. K. Bando, S. T. Oyama, Direct propylene epoxidation over barium promoted Au/Ti-TUD catalysts with H_2 and O_2: Effect of Au particle size, *J. Catal.* doi:10.1016/j.jcat.2007.06.006.
[120] J. R. Monnier, P. J. Muehlbauer, Selective monoepoxidation of olefins, US Patent 4,897,498, Jan. 30, 1990, To Eastman Kodak Company.
[121] J. R. Monnier, P. J. Muehlbauer, Selective epoxidation of olefins, US Patent 4,950,773, Aug. 21, 1990, To Eastman Kodak Company.
[122] J. R. Monnier, P. J. Muehlbauer, Selective epoxidation of diolefins and aryl olefins, US Patent 5,138,077, Aug. 11, 1992, To Eastman Kodak Company.
[123] S. D. Barnicki, J. R. Monnier, Use of fluorinated hydrocarbons as reaction media for selective epoxidation of olefins, US Patent 6,011,163, Jan. 4, 2000, To Eastman Chemical Company.

[124] S. D. Barnicki, J. R. Monnier, K. T. Peters, Gas phase process for the epoxidation of non-allylic olefins, US Patent 5,945,550, Aug. 31, 1999, To Eastman Chemical Company.
[125] J. M. Thomas, G. Sankar, M. C. Klunduk, M. P. Attfield, T. Maschmeyer, B. F. G. Johnson, R. G. Bell, The identity in atomic structure and performance of active sites in heterogeneous and homogeneous, titanium-silica epoxidation catalysts, *J. Phys. Chem. B* 103 (1999) 8809.
[126] R. Hutter, T. Mallat, A. Baiker, Titania-silica mixed oxides: II. Catalytic behavior in olefin epoxidation, *J. Catal.* 153 (1995) 177.
[127] M. Dusi, T. Mallat, A. Baiker, Epoxidation of functionalized olefins over solid catalysts, *Catal. Rev.-Sci. Eng.* 42 (2000) 213.
[128] P. Lignier, F. Morfin, S. Mangematin, L. Massin, J.-L. Rousset, V. Caps, Stereoselective stilbene epoxidation over supported gold-based catalysts, *Chem. Commun.* (2007) 186.
[129] M. Klawonn, M. K. Tse, S. Bhor, C. Döbler, M. Beller, A convenient ruthenium-catalyzed alkene epoxidation with hydrogen peroxide as oxidant, *J. Mol. Catal. A: Chem.* 218 (2004) 13.
[130] K. Sato, M. Aoki, M. Ogawa, T. Hashimoto, R. Noyori, A practical method for epoxidation of terminal olefins with 30% hydrogen peroxide under halide-free conditions, *J. Org. Chem.* 61 (1996) 8310.
[131] N. Mizuno, K. Yamaguchi, K. Kamata, Epoxidation of olefins with hydrogen peroxide catalyzed by polyoxometalates, *Coord. Chem. Rev.* 249 (2005) 1944.
[132] J. Rudolph, K. L. Reddy, J. P. Chiang, K. B. Sharpless, Highly efficient epoxidation of olefins using aqueous H_2O_2 and catalytic methyltrioxorhenium/pyridine: Pyridine-mediated ligand acceleration, *J. Am. Chem. Soc.* 119 (1977) 6189.
[133] H. Sugimoto, D. T. Sawyer, Ferric chloride induced activation of hydrogen peroxide for the epoxidation of alkenes and monooxygenation of organic substrates in acetonitrile, *J. Org. Chem.* 50 (1985) 1784.
[134] E. Pizzo, P. Sgarbossa, A. Scarso, R. A. Michelin, G. Strukul, Second-generation electron-poor platinum(II) complexes as efficient epoxidation catalysts for terminal alkenes with hydrogen peroxide, *Organometallics* 25 (2006) 3056.
[135] A. Murphy, G. Dubois, T. D. P. Stack, Ligand and pH influence on manganese-mediated peracetic acid epoxidation of terminal olefins, *Org. Lett.* 6 (2004) 3119.
[136] A. Murphy, G. Dubois, T. D. P. Stack, Efficient epoxidation of electron-deficient olefins with a cationic manganese complex, *J. Am. Chem. Soc.* 125 (2003) 5250.
[137] A. Murphy, T. D. P. Stack, Discovery and optimization of rapid manganese catalysts for the epoxidation of terminal olefins, *J. Mol. Catal. A: Chem.* 251 (2006) 78.
[138] I. W. C. E. Arends, R. A. Sheldon, Recent developments in selective catalytic epoxidations with H_2O_2, *Top. Catal.* 19 (2002) 133.
[139] B. S. Lane, K. Burgess, Metal-catalyzed epoxidations of alkenes with hydrogen peroxide, *Chem. Rev.* 103 (2003) 2457.
[140] G. Grigoropoulou, J. H. Clark, J. A. Elings, Recent developments on the epoxidation of alkenes using hydrogen peroxide as an oxidant, *Green Chem.* 5 (2003) 1.
[141] C. Venturello, R. D'Alolslo, Quaternary ammonium tetrakis(diperoxotungsto)phosphates(3-) as a new class of catalysts for efficient alkene epoxidation with hydrogen peroxide, *J. Org. Chem.* 53 (1988) 1553.
[142] C. Venturello, E. Alneri, M. Ricci, A new, effective catalytic system for epoxidation of olefins by hydrogen peroxide under phase-transfer conditions, *J. Org. Chem.* 48 (1983) 3831.
[143] C. Venturello, R. D'Aloislo, J. C. J. Bart, M. Ricci, A new peroxotungsten heteropoly anion with special oxidizing properties: Synthesis and structure of tetrahexylammonium tetra(diperoxotungsto)phosphate($3-$), *J. Mol. Catal.* 32 (1985) 107.
[144] Y. Mahha, L. Salles, J.-Y. Piquemal, E. Briot, A. Atlamsani, J.-M. Brégeault, Environmentally friendly epoxidation of olefins under phase-transfer catalysis conditions with hydrogen peroxide, *J. Catal.* 249 (2007) 338.
[145] K. Sato, M. Aoki, M. Ogawa, T. Hashimoto, and R. Noyori, A practical method for epoxidation of terminal olefins with 30% hydrogen peroxide under halide-free conditions, *J. Org. Chem.* 61 (1996) 8310.
[146] K. Sato, M. Aoki, M. Ogawa, T. Hashimoto, D. Panyella, R. Noyori, A halide-free method for olefin epoxidation with 30% hydrogen peroxide, *Bull. Chem. Soc. Jpn.* 70 (1997) 905.

[147] Y. Matoba, H. Inoue, J. Akagi, T. Okabayashi, Y. Ishi, M. Ogawa, Epoxidation of allylic alcohols with hydrogen peroxide catalyzed by $[PMo_{12}O_{40}]^{3-}[C_5H_5N\ (CH_2)_{15}CH_3]_3$, *Synth. Commun.* 14 (1984) 865.
[148] S. Sakaguchi, Y. Nishiyama, Y. Ishii, Selective oxidation of monoterpenes with hydrogen peroxide catalyzed by peroxotungstophosphate (PCWP), *J. Org. Chem.* 61 (1996) 5307.
[149] J.-M. Brégeault, Transition-metal complexes for liquid-phase catalytic oxidation: Some aspects of industrial reactions and of emerging technologies, *Dalton Trans.* (2003) 3289.
[150] A. C. Dengel, W. P. Griffith, B. C. Parkin, Studies on polyoxo- and polyperoxo-metalates. Part 1. Tetrameric heteropolyperoxotungstates and heteropolyperoxomolybdates, *J. Chem. Soc. Dalton Trans.* (1993) 2683.
[151] Y. Nakagawa, K. Kamata, M. Kotani, K. Yamaguchi, N. Mizuno, Polyoxovanadometalate-catalyzed selective epoxidation of alkenes with hydrogen peroxide, *Angew. Chem. Int. Ed. Engl.* 44 (2005) 5136.
[152] Y. Nakagawa, N. Mizuno, Mechanism of $[\gamma\text{-}H_2SiV_2W_{10}O_{40}]^{4-}$ -catalyzed epoxidation of alkenes with hydrogen peroxide, *Inorg. Chem.* 46 (2007) 1727.
[153] D. Sloboda-Rozner, P. Witte, P. L. Alsters, R. Neumann, Aqueous biphasic oxidation: A water-soluble polyoxometalate catalyst for selective oxidation of various functional groups with hydrogen peroxide, *Adv. Synth. Catal.* 346 (2004) 339.
[154] J.-M. Brégeault, M. Vennat, L. Salles, J.-Y. Piquemal, Y. Mahha, E. Briot, P. C. Bakala, A. Atlamsani, R. Thouvenot, From polyoxometalates to polyoxoperoxometalates and back again; potential applications, *J. Mol. Catal. A: Chem.* 250 (2006) 177.
[155] Z. Xi, N. Zhou, Y. Sun, K. L. Li, Reaction-controlled phase transfer catalysis for propylene epoxidation to propylene oxide, *Science* 292 (2001) 1139.
[156] W. A. Herrmann, M. Dieter, W. Wagner, J. G. Kuchler, G. Weichselbaumer, R. Fischer, Use of organorhenium compounds for the oxidation of multiple C-C bonds, oxidation based thereon and novel organorhenium compounds, *US Patent* 5,155,247, Oct. 13, 1992, To Hoechst Aktiegesellschaft.
[157] F. E. Kühn, A. M. Santos, W. A. Herrmann, Organorhenium(VII) and organomolybdenum(VI) oxides: Syntheses and application in olefin epoxidation, *Dalton Trans.* (2005) 2483.
[158] J. B. Espenson, M. M. Abu-Omar, Reactions catalyzed by methylrheniumtrioxide, in: *Adv. Chem. Series*,Vol. 253, p. 99, Am. Chem. Society, Washington, DC, 1997.
[159] J. H. Espenson, Atom-transfer reactions catalyzed by methyltrioxorhenium(VII)—mechanisms and applications, *Chem. Commun.* (1999) 479.
[160] W. A. Herrmann, F. E. Kuehn, Organorhenium oxides, *Acc. Chem. Res.* 30 (1997) 169.
[161] W. A. Herrmann, R. M. Kratzer, H. Ding, W. Thiel, H. Glas, Methyltrioxorhenium/pyrazole-A highly efficient catalyst for the epoxidation of olefins, *J. Organomet. Chem.* 555 (1998) 293.
[162] M. C. A. van Vliet, I. W. C. E. Arends, R. A. Sheldon, Methyltrioxorhenium-catalysed epoxidation of alkenes in trifluoroethanol, *Chem. Commun.* (1999) 821.
[163] F. E. Kühn, A. Scherbaum, W. A. Herrmann, Methyltrioxorhenium and its applications in olefin oxidation, metathesis and aldehyde olefination, *J. Organomet. Chem.* 689 (2004) 4149.
[164] D. V. Deubel, The surprising nitrogen-analogue chemistry of the methyltrioxorhenium-catalyzed olefin epoxidation, *J. Am. Chem. Soc.* 125 (2003) 15308.
[165] F. E. Kühn, A. M. Santos, I. S. Gonçalves, C. C. Romão, A. D. Lopes, Organorhenium(VII) and organomolybdenum(VI) oxides: Synthesis and application in oxidation catalysis, *Appl. Organomet. Chem.* 15 (2001) 43.
[166] B. Subramaniam, D. H. Busch, H.-J. Lee, T.-P. Shi, Process for the selective oxidation of propylene to propylene oxide and facile separation of propylene oxide, *U.S. Pat.Appl.*, Oct. 25, 2006, To University of Kansas.
[167] D. E. De Vos, B. F. Sels, M. Reynaers, Y. V. Subba Rao, P. A. Jacobs, Epoxidation of terminal or electron-deficient olefins with H_2O_2, catalysed by Mn-trimethyltriazacyclonane complexes in the presence of an oxalate buffer, *Tetrahedron Lett.* 39 (1998) 3221.
[168] D. E. De Vos, T. Bein, Highly selective epoxidation of alkenes and styrenes with H_2O_2 and manganese complexes of the cyclic triamine 1,4,7-trimethyl-1,4,7-triazacyclononane, *Chem. Commun.* (1996) 917.

[169] R. A. Sheldon, J. A. van Doorn, Metal-catalyzed epoxidation of olefins with organic hydroperoxides. II. The effect of solvent and hydroperoxide structure, *J. Catal.* 31 (1973) 438.
[170] Z. Dawooki, R. L. Kelly, Epoxidation of ethylene catalysed by molybdenum complexes, *Polyhedron* 5 (1986) 271.
[171] H. Mimoun, The role of peroxymetallation in selective oxidative processes, *J. Mol. Catal.* 7 (1980) 1.
[172] J. A. Gnecco, G. Borda, P. Reyes, Catalytic epoxidation of cyclohexene using molybdenum complexes, *J. Child. Chem. Soc.* 49 (2004) 179.
[173] W. F. Brill, N. Indictor, Reactions of *t*-butyl hydroperoxide with olefins, *J. Org. Chem.* 29 (1964) 710.
[174] M. Abrantes, A. M. Santos, J. Mink, F. E. Kühn, C. C. Romào, A simple entry to (5-C_5R_5) chlorodioxomolybdenum(VI) complexes (R = H, CH_3, CH_2Ph) and their use as olefin epoxidation catalysts, *Organometallics* 22 (2003) 2112.
[175] M. Collardon, A. Scarso, P. Sgarbossa, R. A. Michelin, G. Strukul, Asymmetric epoxidation of terminal alkenes with chiral Pt complexes, *J. Am. Chem. Soc.* 128 (2006) 14006.
[176] C. Bassin, A. Gusso, F. Pinna, G. Strukul, Platinum-catalyzed oxidations with hydrogen peroxide: The (enantioselective) epoxidation of α,β-unsaturated ketones, *Organometallics* 14 (1995) 1161.
[177] Y. Gao, J. M. Klunder, R. M. Hanson, H. Masamune, S. Y. Ko, K. B. Sharpless, Catalytic asymmetric epoxidation and kinetic resolution: Modified procedures including *in situ* derivatization, *J. Am. Chem. Soc.* 109 (1987) 5765.
[178] W. Zhang, A. Basak, Y. Kosugi, Y. Hoshino, H. Yamamoto, Enantioselective epoxidation of allylic alcohols by a chiral complex of vanadium: An effective controller system and a rational mechanistic model, *Angew. Chem. Int. Ed. Engl.* 44 (2005) 4389.
[179] M. C. A. van Vliet, I. W. C. E. Arends, R. A. Sheldon, Hexafluoroacetone in hexafluoro-2-propanol: A highly active medium for epoxidation with aqueous hydrogen peroxide, *Synlett* (2001) 1305.
[180] W. A. Adam, H.-G. Degen, C. R. Saha-Möller, Regio- and diastereoselective catalytic epoxidation of chiral allylic alcohols with hexafluoroacetone perhydrate. Hydroxy-group directivity through hydrogen bonding, *J. Org. Chem.* 64 (1999) 1274.
[181] M. C. A. van Vliet, I. W. C. E. Arends, R. A. Sheldon, Hexafluoroacetone in hexafluoro-2-propanol: A highly active medium for epoxidation with aqueous hydrogen peroxide, *Synlett* (2001) 248.
[182] P. F. Wolf, R. K. Barnes, Borate ester-induced decomposition of alkyl hydroperoxides. The epoxidation of olefins by electrophilic oxygen, *J. Org. Chem.* 34 (1969) 3441.
[183] R. A. W. Johnstone, E. Francsicsné-Czinege, Hydrogen peroxide adducts, World Patent WO 17640, April 30, 1998, To Solvay Interox Limited.
[184] J. P. Sankey, R. A. W. Johnstone, Epoxidation of Alkenes, British Patent GB 2,330,358, April 21, 1991, To Contract Chemicals Ltd. and Solvay Interox Ltd.
[185] M. C. A. van Vliet, I. W. C. E. Arends, R. A. Sheldon, Tertiary arsine oxides: Active and selective catalysts for epoxidation with hydrogen peroxide, *Tetrahedron Lett.* 40 (1999) 5239.
[186] A. Berkessel, J. A. Adrio, Kinetic studies of olefin epoxidation with hydrogen peroxide in 1,1,1,3,3,3-hexafluoro-2-propanol reveal a crucial catalytic role for solvent clusters, *Adv. Synth. Catal.* 346 (2004) 275.
[187] S. Y. Tosaki, R. Tsuji, T. Ohshima, M. Shibasaki, Dynamic ligand exchange of the lanthanide complex leading to structural and functional transformation: One-pot sequential catalytic asymmetric epoxidation-regioselective epoxide-opening process, *J. Am. Chem. Soc.* 127 (2005) 2147.
[188] S. E. Jacobsen, F. Mares, P. M. Zambri, Biphase and triphase catalysis. Arsonated polystyrenes as catalysts for epoxidation of olefins by aqueous hydrogen peroxide, *J. Am. Chem. Soc.* 101 (1979) 6946.
[189] T. Hori, K. B. Sharpless, Synthetic applications of arylselenenic and arylseleninic acids. Conversion of olefins to allylic alcohols and epoxides, *J. Org. Chem.* 43 (1978) 1689.
[190] L. Syper, The Baeyer-Villiger oxidation of aromatic aldehydes and ketones with hydrogen peroxide catalyzed by selenium compounds. A convenient method for the preparation of phenols, *Synthesis* (1989) 167.

[191] B. Betzemeier, F. Lhermite, P. Knochel, A selenium catalyzed epoxidation in perfluorinated solvents with hydrogen peroxide, *Synlett* (1999) 489.
[192] G. J. ven Brink, B. C. M. Fernandez, M. C. A. van Vliet, I. W. C. E. Arends, R. A. Sheldon, Selenium catalysed oxidations with aqueous hydrogen peroxide. Part I: Epoxidation reactions in homogeneous solution, *J. Chem. Soc. Perkin Trans. I* (2001) 224.
[193] J. Rebek, R. McCready, New epoxidation reagents derived from alumina and silicon, *Tetrahedron Lett.* 20 (1979) 4337.
[194] M. C. A. van Vliet, D. Mandelli, I. W. C. E. Arends, U. Schuchardt, R. A. Sheldon, Alumina: A cheap, active and selective catalyst for epoxidations with (aqueous) hydrogen peroxide, *Green Chem.* 3 (2001) 243.
[195] D. Mandelli, M. C. A. van Vliet, R. A. Sheldon, U. Schuchardt, Alumina-catalyzed alkene epoxidation with hydrogen peroxide, *Appl. Catal. A: Gen.* 219 (2001) 209.
[196] R. G. Cesquini, J. M. de S. e Silva, C. B. Woitiski, D. Mandelli, R. Rinaldi, U. Schuchardt, Alumina-catalyzed epoxidation with hydrogen peroxide: Recycling experiments and activity of sol-gel alumina, *Adv. Synth. Catal.* 344 (2002) 911.
[197] R. Rinaldi, J. Sepúlveda, U. Schuchardt, Cyclohexene and cyclooctene epoxidation with aqueous hydrogen peroxide using transition metal-free sol-gel alumina as catalyst, *Adv. Synth. Catal.* 346 (2004) 281.
[198] T. K. M. Shing, G. Y. C. Leung, T. Luk, Transition state studies on the dioxirane-mediated asymmetric epoxidation via kinetic resolution and desymmetrization, *J. Org. Chem.* 70 (2005) 7279.
[199] Z. Crane, D. Goeddel, Y. H. Gan, Y. Shi, Highly enantioselective epoxidation of α-β-unsaturated esters by chiral dioxirane, *J. Am. Chem. Soc.* 124 (2002) 8792.
[200] J. C. Lorenz, M. Frohn, X. M. Zhou, J. R. Zhang, Y. Tang, C. Burke, Y. Shi, Transition state studies on the dioxirane-mediated asymmetric epoxidation via kinetic resolution and desymmetrization, *J. Org. Chem.* 70 (2005) 2904.
[201] D. A. Singleton, Z. H. Wang, Isotope effects and the nature of enantioselectivity in the Shi epoxidation. The importance of asynchronicity, *J. Am. Chem. Soc.* 127 (2005) 6679.
[202] A. Picot, P. Millet, X. Lusinchi, Formation d'un sel d'oxaziridinium quaternaire par methylation d'un oxaziranne—mise en evidence de ses proprieties oxydantes, *Tetrahedron Lett.* 17 (1976) 1573.
[203] J. Vachon, C. Perollier, D. Monchaud, C. Marsol, K. Ditrich, J. Lacour, Biphasic enantioselective olefin epoxidation using *Tropos dibenzoazepinium* catalysts, *J. Org. Chem.* 70 (2005) 5903.
[204] P. C. B. Page, D. Barros, B. R. Buckley, A. Ardakani, B. A. Marples, Organocatalysis of asymmetric epoxidation mediated by iminium salts under nonaqueous conditions, *J. Org. Chem.* 69 (2004) 3595.
[205] Y. Ishii, S. Sakaguchi, A new strategy for alkane oxidation with O_2 using N-hydroxyphthalimide (NHPI) as a radical catalyst, *Catal. Surv. Jpn.* 3 (1999) 27.
[206] Y. Ishii, S. Sakaguchi, T. Iwahama, Innovation of hydrocarbon oxidation with molecular oxygen and related reactions, *Adv. Synth. Catal.* 343 (2001) 393.
[207] W. Zhuang, M. Marigo, K. A. Jorgensen, Organocatalytic asymmetric epoxidation reactions in water–alcohol solutions, *Org. Biomol. Chem.* 3 (2005) 3883.
[208] B. Meunier, Ed., *Biomimetic Oxidations Catalyzed by Transition Metal Complexes*, Imperial College Press, London, 2000.
[209] R. van Eldik, J. Reedijk, Homogeneous biomimetic oxidation catalysis, in: *Advances in Inorganic Chemistry*, Vol. 58. Elsevier, Amsterdam, (2006).
[210] D. Mansuy, A brief history of the contribution of metalloporphyrin models to cytochrome P450 chemistry and oxidation catalysis, *Comp. Rend. Chimie* 10 (2007) 392.
[211] J. P. Collman, T. Kodadek, S. A. Raybuck, J. I. Brauman, L. M. Papazian, Mechanism of oxygen atom transfer from high valent iron porphyrins to olefins: Implications to the biological epoxidation of olefins by cytochrome P-450, *J. Am. Chem. Soc.* 107 (1985) 4343.
[212] J. P. Collman, J. I. Brauman, B. Meunier, S. A. Raybuck, T. Kodadek, Epoxidation of olefins by cytochrome P-450 model compounds: Mechanism of oxygen atom transfer, *Proc. Natl. Acad. Sci. USA* 81 (1984) 3245.
[213] S. Banfi, M. Dragoni, F. Montanari, G. Pozzi, S. Quici, Alkene epoxidations catalyzed by chemically robust Mn(III) porphyrins and promoted by HOCl under aqueous-organic 2-phase

conditions in the absence of axial ligand and phase-transfer catalyst - reaction-mechanism and large-scale preparative applications, *Gazz. Chim. Ital.* 123 (1993) 431.

[214] J. P. Collman, L. Zeng, H. J. H. Wang, A. Lei, J. I. Brauman, Kinetics of (porphyrin)manganese (III)-catalyzed olefin epoxidation with a soluble iodosylbenzene derivative, *Eur. J. Org. Chem.* (2006) 2707.

[215] J. R. Lindsay Smith, D. N. Mortimer, The oxidation of organic compounds with iodosylbenzene catalysed by tetra(4-N-methylpyridyl)porphyrinatoiron(III) pentacation: A polar model system for the cytochrome P450 dependent mono-oxygenases, *J. Chem. Soc. Chem. Commun.* (1985) 410.

[216] T. G. Traylor, J. C. Marsters Jr., T. Nakono, B. E. Dunlap, Kinetics of iron(III) porphyrin catalyzed epoxidations, *J. Am. Chem. Soc.* 107 (1985) 5537.

[217] Z. Gross, H. B. Gray, Oxidations catalyzed by metallocorroles, *Adv. Synth. Catal.* 346 (2004) 165.

[218] W. D. Kerber, D. P. Goldberg, High-valent transition metal corrolazines, *J. Inorg. Biochem.* 100 (2006) 838.

[219] M. E. Niño, S. A. Giraldo, E. A. Páez-Mozo, Olefin oxidation with dioxygen catalyzed by porphyrins and phthalocyanines intercalated in α-zirconium phosphate, *J. Mol. Catal. A: Chem.* 175 (2001) 139.

[220] W. Zhang, J. L. Loebach, S. R. Wilson, E. N. Jacobsen, Enantioselective epoxidation of unfunctionalized olefins catalyzed by salen manganese complexes, *J. Am. Chem. Soc.* 112 (1990) 2801.

[221] R. Irie, K. Noda, Y. Ito, N. Matsumoto, T. Katsuki, Catalytic asymmetric epoxidation of unfunctionalized olefin, *Tetrahedron Lett.* 31 (1990) 7345.

[222] H. Sasaki, R. Irie, T. Hamada, K. Suzuki, T. Katsuki, Rational design of Mn-salen catalyst (2): Highly enantioselective epoxidation of conjugated *cis* olefins, *Tetrahedron* 50 (1994) 11827.

[223] W. Zhang, E. N. Jacobsen, Asymmetric olefin epoxidation with sodium hypochlorite catalyzed by easily prepared chiral manganese(III) salen complexes, *J. Org. Chem.* 56 (1991) 2296.

[224] M. Palucki, P. J. Pospisil, W. Zhang, E. N. Jacobsen, Highly enantioselective, low-temperature epoxidation of styrene, *J. Am. Chem. Soc.* 116 (1994) 9333.

[225] M. Palucki, G. J. McCormick, E. N. Jacobsen, Low temperature asymmetric epoxidation of unfunctionalized olefins catalyzed by (salen)Mn(III) complexes, *Tetrahedron Lett.* 36 (1995) 5457.

[226] T. S. Reger, K. D. Janda, Polymer-supported (salen)Mn catalysts for asymmetric epoxidation: A comparison between soluble and insoluble matrices, *J. Am. Chem. Soc.* 122 (2000) 6929.

[227] P. G. Cozzi, Metal–salen Schiff base complexes in catalysis: Practical aspects, *Chem. Soc. Rev.* 33 (2004) 410.

[228] A. Martinez, C. Hemmert, B. Meunier, A macrocyclic chiral manganese(III) Schiff base complex as an efficient catalyst for the asymmetric epoxidation of olefins, *J. Catal.* 235 (2005) 250.

[229] Y. Sawada, K. Matsumoto, S. Kondo, H. Watanabe, T. Ozawa, K. Suzuki, B. Saito, T. Katsuki, Titanium-salan-catalyzed asymmetric epoxidation with aqueous hydrogen peroxide as the oxidant, *Angew. Chem. Int. Ed. Engl.* 45 (2006) 3478.

[230] W. Nam, R. Ho, J. S. Valentine, Iron-cyclam complexes as catalysts for the epoxidation of olefins by 30% aqueous hydrogen peroxide in acetonitrile and methanol, *J. Am. Chem. Soc.* 113 (1991) 7052.

[231] T. G. Traylor, S. Tsuchiya, Y. S. Byun, C. Kim, High-yield epoxidations with hydrogen peroxide and *tert*-butyl hydroperoxide catalyzed by iron(III) porphyrins: Heterolytic cleavage of hydroperoxides, *J. Am. Chem. Soc.* 115 (1993) 2775.

[232] D. Dolphin, T. G. Traylor, L. Y. Xie, Polyhaloporphyrins: Unusual ligands for metals and metal-catalyzed oxidations, *Acc. Chem. Res.* 30 (1997) 251.

[233] G. Anilkumar, B. Bitterlich, F. G. Gelalcha, M. K. Tse, M. Beller, An efficient biomimetic Fe-catalyzed epoxidation of olefins using hydrogen peroxide, *Chem. Commun.* (2007) 289.

[234] K. Chen, M. Costas, J. Kim, A. K. Tipton, L. Que Jr., Olefin *cis*-dihydroxylation versus epoxidation by nonheme iron catalysts: Two faces of an Fe^{III}-OOH coin, *J. Am. Chem. Soc.* 124 (2002) 3026.

[235] M. Costas, L. Que Jr., Ligand topology tuning of iron-catalyzed hydrocarbon oxidations, *Angew. Chem. Int. Ed. Engl.* 41 (2002) 2179.

[236] M. C. White, A. G. Doyle, E. N. Jacobsen, A synthetically useful, self-assembling MMO mimic system for catalytic alkene epoxidation with aqueous H_2O_2, *J. Am. Chem. Soc.* 123 (2001) 7194.

[237] L. P. Wackett, Mechanism and applications of Rieske non-heme iron dioxygenases, *Enzyme Microb. Tech.* 31 (2002) 577.

[238] D. J. Ferraro, L. Gakhar, S. Ramaswamy, Rieske business: Structure-function of Rieske non-heme oxygenases, *Biochem. Biophys. Res. Commun.* 338 (2005) 175.
[239] F. T. de Oliveira, A. Chanda, D. Banerjee, X. Shan, S. Mondal, L. Que Jr., E. L. Bominaar, E. Münck, T. J. Collins, Chemical and spectroscopic evidence for an Fe^V-oxo complex, *Science* 315 (2007) 835.
[240] Y. Mekmouche, S. Ménage, J. Pécaut, C. Lebrun, L. Reilly, V. Schuenemann, A. Trautwein, M. Fontecave, Mechanistic tuning of hydrocarbon oxidations with H_2O_2, catalyzed by hexacoordinate ferrous complexes, *Eur. J. Inorg. Chem.* (2004) 3163.
[241] K. Chen, L. Que Jr., *cis*-Dihydroxylation of olefins by a nonheme iron catalyst. A functional model for Rieske dioxygenases, *Angew. Chem. Int. Ed. Engl.* 38 (1999) 2227.
[242] M. R. Bukowski, P. Comba, A. Lienke, C. Limberg, C. Lopez de Laorden, R. Mas-Ballesté, M. Merz, L. Que Jr., Catalytic epoxidation and 1,2-dihydroxylation of olefins with bispidine-iron(II)/H_2O_2 systems, *Angew. Chem. Int. Ed. Engl.* 45 (2006) 3446.
[243] S. Bhor, M. K. Tse, M. Klawonn, C. Döbler, W. Mägerlein, M. Beller, Ruthenium-catalyzed asymmetric alkene epoxidation with *tert*-butyl hydroperoxide as oxidant, *Adv. Synth. Catal.* 346 (2004) 263.
[244] M. Fujita, L. Que Jr., In situ formation of peracetic acid in iron-catalyzed epoxidations by hydrogen peroxide in the presence of acetic acid, *Adv. Synth. Catal.* 346 (2004) 190.
[245] G. Dubois, A. Murphy, T. D. P. Stack, Simple iron catalyst for terminal alkene epoxidation, *Org. Lett.* 5 (2003) 2469.
[246] M. Ono, I. Okura, On the reaction mechanism of alkene epoxidation with *Methylosinus trichosporium* (OB3b), *J. Mol. Catal.* 61 (1990) 113.
[247] I. J. Higgins, D. J. Best, R. C. Hammond, New findings in methane-utilizing bacteria highlight their importance in the biosphere and their commercial potential, *Nature* 286 (1980) 561.
[248] E. G. Ankudey, H. F. Olivo, T. L. Peeples, Lipase-mediated epoxidation utilizing urea-hydrogen peroxide in ethyl acetate, *Green Chem.* 8 (2006) 923.
[249] M. A. Moreira, T. B. Bitencourt, M. D. G. Nascimento, Optimization of chemo-enzymatic epoxidation of cyclohexene mediated by lipases, *Synth. Commun.* 35 (2005) 2107.
[250] Z. Hu, S. M. Gorun, Methane monooxygenase models, in: B. Meunier (Ed.), *Biomimetic Oxidations Catalyzed by Transition Metal Complexes*, Imperial College Press, London, 2000, pp. 269–307.
[251] K. Faber, (Ed.), *Biotransformations in Organic Chemistry*, Springer-Verlag, Berlin (2000).
[252] S. Panke, M. Held, M. G. Wubbolts, B. Witholt, A. Schmid, Pilot-scale production of (S)-styrene oxide from styrene by recombinant *Escherichia coli* synthesizing styrene monooxygenase, *Biotechnol. Bioeng.* 80 (2002) 33.
[253] S. Panke, M. G. Wubbolts, A. Schmid, B. Wiholt, Production of enantiopure styrene oxide by recombinant *Escherichia coli* synthesizing a two-component styrene monooxygenase, *Biotechnol. Bioeng.* 69 (2000) 91.
[254] T. Kubo, M. W. Peters, P. Meinhold, F. H. Arnold, Enantiomeric epoxidation of terminal alkenes to (R)- and (S)-epoxides by engineered cytochromes P-450 BM-3, *Chem. Eur. J.* 12 (2006) 1216.
[255] V. B. Valodkar, G. L. Tembe, R. N. Ram, H. S. Rama, Catalytic asymmetric epoxidation of unfunctionalized olefins by supported Cu(II)-amino acid complexes, *Catal. Lett.* 90 (2003) 91.
[256] E. Rose, Q. Z. Ren, B. Andrioletti, A unique binaphthyl strapped iron-porphyrin catalyst for the enantioselective epoxidation of terminal olefins, *Chem. Eur. J.* 10 (2004) 224.
[257] C. A. Martinez, J. D. Stewart, Cytochrome P450's: Potential catalysts for asymmetric olefin epoxidations, *Curr. Org. Chem.* 4 (2000) 263.
[258] S. J. Elliot, M. Zhu, L. Tso, H. H. T. Nguyen, J. H. K. Yip, S. I. Chan, Regio- and stereoselectivity of particulate methane monooxygenase from *Methylococcus capsulatus* (Bath), *J. Am. Chem. Soc.* 119 (1997) 9949.
[259] K. McClay, B. G. Fox, R. J. Steffan, Toluene monooxygenase-catalyzed epoxidation of alkenes, *Appl. Environ. Microbiol.* 66 (2000) 1877.
[260] A. Schmid, K. Hofstetter, H. J. Feiten, F. Hollmann, B. Witholt, Integrated biocatalytic synthesis on gram scale: The highly enantioselective preparation of chiral oxiranes with styrene monooxygenase, *Adv. Synth. Catal.* 343 (2001) 732.

[261] A. F. Dexter, F. J. Lakner, R. A. Campbell, L. P. Hager, Highly enantioselective epoxidation of 1,1-disubstituted alkenes catalyzed by chloroperoxidase, *J. Am. Chem. Soc.* 117 (1995) 6412.
[262] K. Furuhashi, (Ed.), *Biological Routes to Optically Active Epoxides*, Wiley, New York, 1992, pp. 167–186.
[263] A. Archelas, R. Furstoss, Synthesis of enantiopure epoxides through biocatalytic approaches, *Annu. Rev. Microbiol.* 51 (1997) 491.
[264] E. T. Farinas, M. Alcalde, F. Arnold, Alkene epoxidation catalyzed by cytochrome P450 BM-3 139–3, *Tetrahedron* 60 (2004) 525.
[265] G. Langhendries, D. E. De Vos, B. F. Sels, I. Vankelecom, P. A. Jacobs, G. V. Baron, Clean catalytic technology for liquid phase hydrocarbon oxidation, *Clean Prod. Proc.* 1 (1998) 21.
[266] R. A. Sheldon, J. A. Van Doorn, Metal-catalyzed epoxidation of olefins with organic hydroperoxides. I. A comparison of various metal catalysts, *J. Catal.* 31 (1973) 427.
[267] L. Sümegi, I. P. Hajdu, I. Nemes, Á. Gedra, On the mechanism of propylene epoxidation catalyzed by molybdenum naphthenate, *React. Kinet. Catal. Lett.* 12 (1979) 57.
[268] R. Neumann, M. Cohen, Solvent-anchored supported liquid phase catalysis: Polyoxometalate-catalyzed oxidations, *Angew. Chem. Int. Ed. Engl.* 36 (1997) 1738.
[269] T. Sakamoto, C. Pac, Selective epoxidation of olefins by hydrogen peroxide in water using a polyoxometalate catalyst supported on chemically modified hydrophobic mesoporous silica gel, *Tetrahedron Lett.* 41 (2000) 10009.
[270] G. Gelbard, T. Gauducheau, E. Vidal, V. I. Parvulescu, A. Crosman, V. M. Pop, Epoxidation with peroxotungstic acid immobilised onto silica-grafted phosphoramides, *J. Mol. Catal. A: Chem.* 182–183 (2002) 257.
[271] T. Luts, H. Papp, Novel ways of Mn-salen complex immobilization on modified silica support and their catalytic activity in cyclooctene epoxidation, *Kinet. Catal.* 48 (2007) 176.
[272] B. M. Choudary, U. Pal, M. L. Kantam, K. V. S. Ranganath, B. Sreedhar, Asymmetric epoxidation of olefins by manganese(III) complexes stabilized on nanocrystalline magnesium oxide, *Adv. Synth. Catal.* 348 (2006) 1038.
[273] D. Chatterjee, A. Mitra, Olefin epoxidation catalysed by Schiff-base complexes of Mn and Ni in heterogenised-homogeneous systems, *J. Mol. Catal. A: Chem.* 144 (1999) 363.
[274] H. Sohrabi, M. Esmaeeli, F. Farzaneh, M. Ghandi, Nickel(macrocycle) complexes immobilized within montmorillonite and MCM-41 as catalysts for epoxidation of olefins, *J. Inclusion Phenom. Macrocyclic. Chem.* 54 (2006) 23.
[275] B. F. Sels, D. E. De Vos, M. Buntinx, F. Pierard, A. Kirsh-De Mesmacker, P. A. Jacobs, Layered double hydroxides exchanged with tungstate as biomimetic catalysts for mild oxidative bromination, *Nature* 400 (1999) 855.
[276] B. F. Sels, D. E. De Vos, P. A. Jacobs, Use of WO_4^{2-} on layered double hydroxides for mild oxidative bromination and bromide-assisted epoxidation with H_2O_2, *J. Am. Chem. Soc.* 123 (2001) 8350.
[277] H. Zhang, S. Xiang, C. Li, Enantioselective epoxidation of unfunctionalised olefins catalyzed by Mn(salen) complexes immobilized in porous materials via phenyl sulfonic group, *Chem. Commun.* (2005) 1209.
[278] A. L. Villa, B. F. Sels, D. E. De Vos, P. A. Jacobs, A heterogeneous tungsten catalyst for epoxidation of terpenes and tungsten-catalyzed synthesis of acid-sensitive terpene epoxides, *J. Org. Chem.* 64 (1999) 7267.
[279] U. Arnold, W. Habicht, M. Doring, Metal-doped epoxy resins—New catalysts for the epoxidation of alkenes with high long-term activities, *Adv. Synth. Catal.* 348 (2006) 142.
[280] B. M. L. Dioos, I. F. J. Vankelecom, P. A. Jacobs, Aspects of immobilisation of catalysts on polymeric supports, *Adv. Synth. Catal.* 348 (2006) 1413.
[281] H. Sellner, J. K. Karjalainen, Preparation of dendritic and non-dendritic styryl-substituted salens for cross-linking suspension copolymerization with styrene and multiple use of the corresponding Mn and Cr complexes in enantioselective epoxidations and Hetero-Diels-Alder reactions, *Chem. Eur. J.* 7 (2001) 2873.
[282] K. Smith, C.-H. Liu, Asymmetric epoxidation using a singly-bound supported Katsuki-type (salen)Mn complex, *Chem. Commun.* (2002) 886.

[283] B. F. Sels, A. L. Villa, D. Hoegaerts, D. E. De Vos, P. A. Jacobs, Application of heterogenized oxidation catalysts to reactions of terpenic and other olefins with H_2O_2, *Top. Catal.* 13 (2000) 223.
[284] S. Bhattcharjee, J. A. Anderson, Novel chiral sulphonato-salen-manganese(III)-pillared hydrotalcite catalysts for the asymmetric epoxidation of styrenes an cyclic alkenes, *Adv. Synth. Catal.* 348 (2006) 151.
[285] W. Adam, C. R. Saha-Möller, O. Weichold, NaY zeolite as host for the selective heterogeneous oxidation of silanes and olefins with hydrogen peroxide catalyzed by methyltrioxorhenium, *J. Org. Chem.* 65 (2000) 2897.
[286] Z. Petrovski, S. S. Braga, A. M. Santos, S. S. Rodrigues, I. S. Gonçalves, M. Pillinger, F. E. Kühn, C. C. Romão, Synthesis and characterization of the inclusion compound of a ferrocenyldiimine dioxomolybdenum complex with heptakis-2,3,6-tri-O-methyl-β-cyclodextrin, *Inorg. Chim. Acta* 358 (2005) 981.
[287] S. Jana, B. Dutta, R. Bera, S. Koner, Anchoring of copper complex in MCM-41 matrix: A highly efficient catalyst for epoxidation of olefins by *tert*-BuOOH, *Langmuir* 23 (2007) 2492.
[288] N. N. Reed, T. J. Dickerson, G. E. Boldt, K. D. Janda, Enantioreversal in the Sharpless asymmetric epoxidation reaction controlled by the molecular weight of a covalently appended achiral polymer, *J. Org. Chem.* 70 (2005) 1728.
[289] S.-H. Cho, B. Ma, S. T. Nguyen, J. T. Hupp, T. E. Albrecht-Smith, A metal-organic framework material that functions as an enantioselective catalyst for olefin epoxidation, *Chem. Commun.* (2006) 2563.
[290] T. Mukaiyama, T. Yamada, Recent advances in aerobic oxygenation, *Bull. Chem. Soc. Jpn.* 68 (1995) 17.
[291] B. B. Wentzel, P. A. Gosling, M. C. Feiters, R. J. M. Nolte, Mechanistic studies on the epoxidation of alkenes with molecular oxygen and aldehydes catalysed by transition metal--diketonate complexes, *J. Chem. Soc. Dalton Trans.* (1998) 2241.
[292] G. Pozzi, F. Cinato, F. Montanari, S. Quici, Efficient aerobic epoxidation of alkenes in perfluorinated solvents catalysed by chiral (salen) Mn complexes, *Chem. Commun.* (1998) 877.
[293] H. Adolfsson, Transition metal-catalyzed epoxidation of alkenes, in: J. E. Bäckvall (Ed.), Modern Oxidation Methods, Wiley-VCH, Weinheim, 2004, p. 1.
[294] R. A. Sheldon, in: R. Ugo (Ed.), Aspects of Homogeneous Catalysis, Vol. 4, Reidel, Dordrecht, 1981, pp. 3–70.
[295] H. Tanaka, K. Hashimoto, K. Suzuki, Y. Kitaichi, M. Sato, T. Ikeno, T. Yamada, Nitrous oxide oxidation catalyzed by ruthenium porphyrin complex, *Bull. Chem. Soc. Jpn.* 77 (2004) 1905.
[296] J. T. Groves, T. E. Nemo, R. S. Myers, Hydroxylation and epoxidation catalyzed by iron-porphyrin complexes. Oxygen transfer from iodosylbenzene, *J. Am. Chem. Soc.* 101 (1979) 1032.
[297] B. C. Schardt, C. L. Hill, Preparation of iodobenzene dimethoxide. A new synthesis of [^{18}O] iodosylbenzene and a reexamination of its infrared spectrum, *Inorg. Chem.* 22 (1983) 1563.
[298] V. V. Zhdankin, P. J. Stang, Recent developments in the chemistry of polyvalent iodine compounds, *Chem. Rev.* 102 (2002) 2523.
[299] D. Macikenas, E. Skrzypczak-Jankun, J. D. Protasiewicz, A new class of iodonium ylides engineered as soluble primary oxo and nitrene sources, *J. Am. Chem. Soc.* 121 (1999) 7164.
[300] P. S. Traylor, D. Dolphin, T. G. Traylor, Sterically protected hemins with electronegative substituents: Efficient catalysts for hydroxylation and epoxidation, *J. Chem. Soc. Chem. Commun.* (1984) 279.
[301] F. Montanari, M. Penso, S. Quici, P. Viganó, Highly efficient sodium hypochlorite olefin epoxidations catalyzed by imidazole or pyridine "tailed" manganese porphyrins under two-phase conditions. Influence of pH and of the anchored ligand, *J. Org. Chem.* 50 (1985) 4888.
[302] I. Tabushi, N. Koga, P-450 type oxygen activation by porphyrin-manganese complex, *J. Am. Chem. Soc.* 101 (1979) 6456.
[303] M. Perrée-Fauvet, A. Gaudemer, Manganese porphyrin-catalysed oxidation of olefins to ketones by molecular oxygen, *J. Chem. Soc. Chem. Commun.* (1981) 874.
[304] R. Wichmann, D. Vasic-Racki, Cofactor regeneration at lab scale, *Adv. Biochem. Eng./Biotechnol.* 92 (2005) 225.
[305] C. Duboc-Toia, S. Menage, C. Lambeaux, M. Fontecave, μ-Oxo diferric complexes as oxidation catalysts with hydrogen peroxide and their potential in asymmetric oxidation, *Tetrahedron Lett.* 38 (1997) 3727.

[306] S. Menage, J. M. Vincent, C. Lambeaux, G. Chottard, A. Grand, M. Fontecave, Alkane oxidation catalyzed by μ-oxo-bridged diferric complexes: A structure/reactivity correlation study, *Inorg. Chem.* 32 (1993) 4766.
[307] A. Stassinopoulos, J. P. Caradonna, A binuclear non-heme iron oxo-transfer analog reaction system: Observations and biological implications, *J. Am. Chem. Soc.* 112 (1990) 7071.
[308] N. Kitajima, H. Fukui, Y. Morooka, A model for methane mono-oxygenase: Dioxygen oxidation of alkanes by use of a μ-oxo binuclear iron complex, *J. Chem. Soc. Chem. Commun.* (1988) 485.
[309] B. P. Much, F. C. Bradley, L. Que Jr., A binuclear iron peroxide complex capable of olefin epoxidation, *J. Am. Chem. Soc.* 108 (1986) 5027.
[310] C. Döbler, G. M. Mehltretter, U. Sundermeier, M. Beller, Dihydroxylation of olefins using air as the terminal oxidant, *J. Organomet. Chem.* 621 (2001) 70.
[311] C. Döbler, G. M. Mehltretter, U. Sundermeier, M. Beller, Osmium-catalyzed dihydroxylation of olefins using dioxygen or air as the terminal oxidant, *J. Am. Chem. Soc.* 122 (2000) 10289.
[312] C. Döbler, G. M. Mehltretter, M. Beller, Atom-efficient oxidation of alkenes with molecular oxygen: Synthesis of diols, *Angew. Chem. Int. Ed. Engl.* 38 (1999) 3026.
[313] T. Punniyamurthy, S. Velusamy, J. Iqbal, Recent advances in transition metal catalyzed oxidation of organic substrates with molecular oxygen, *Chem. Rev.* 105 (2005) 2329.
[314] K. Murata, Y. Kiyozumi, Oxidation of propene by molecular oxygen over Ti-modified silicalite catalysts, *Chem. Commun.* (2001) 1356.
[315] N. Mimura, S. Tsubota, K. Murata, K. K. Bando, J. J. Bravo-Suárez, M. Haruta, S. T. Oyama, Gas-phase radical generation by Ti oxide clusters supported on silica: Application to the direct epoxidation of propylene to propylene oxide using molecular oxygen as an oxidant, *Catal. Lett.* 110 (2006) 47.
[316] G. M. Lobmaiera, H. Trauthweinb, G. D. Freyc, B. Scharbertd, E. Herdtweckc, W. A. Herrmann, Oxovanadium(IV) complexes as molecular catalysts in epoxidation: Simple access to pyridylalk-oxide derivatives, *J. Organomet. Chem.* 691 (2006) 2291.
[317] R. W. Murray, R. Jeyaraman, Dioxiranes: Synthesis and reactions of methyldioxiranes, *J. Org. Chem.* 50 (1985) 2847.
[318] A. L. Baumstark, C. J. McCloskey, Epoxidation of alkenes by dimethyldioxirane: Evidence for a spiro transition state, *Tetrahedron Lett.* 28 (1987) 3311.
[319] P. C. Vasquez, A. L. Baumstark, Epoxidation by dimethyldioxirane. Electronic and steric effects, *J. Org. Chem.* 53 (1988) 3437.
[320] T. E. Lefort, (Ed.), Process for the production of ethylene oxide, *French Patent FR* 729,952, Mar. 27, 1931, To Societe Francaise de Catalyse Generalisee.
[321] T. E. Lefort, (Ed.), Process for the production of ethylene oxide, *US Patent* 1,998,878, Apr. 23, 1935, To Societe Francaise de Catalyse Generalisee.
[322] S. Rojluechai, S. Chavadej, J. Schwank, V. Meeyoo, Activity of ethylene epoxidation over high surface area alumina support Au-Ag catalysts, *J. Chem. Eng. Japan* 39 (2006) 321.
[323] J. T. Jankowiak, M. A. Barteau, Ethylene epoxidation over silver and copper-silver bimetallic catalysts: I. Kinetics and selectivity, *J. Catal.* 236 (2005) 366.
[324] W. M. H. Sachtler, C. Backy, R. A. van Santen, On the mechanism of ethylene epoxidation, *Catal. Rev.-Sci. Eng.* 23 (1981) 127.
[325] V. B. Kazansky, V. A. Shvets, M. Y. Kon, V. V. Nikisha, B. N. Shelimov, Spectroscopic study of the elementary reactions in the coordination sphere of the surface transition metal ions and the mechanism of some related catalytic reactions, in: J. Hightower (Ed.), *Proceedings of the Fifth International Congress on Catalysis*, North-Holland, Amsterdam, 1973, p. 1423.
[326] V. I. Bukhtiyarov, M. Hävecker, V. V. Kaichev, A. Knop-Gericke, R. W. Mayer, R. Schlögl, Atomic oxygen species on silver: Photoelectron spectroscopy and x-ray absorption studies, *Phys. Rev. B* 67 (2003) 235422.
[327] R. M. Lambert, F. J. Williams, R. L. Cropley, A. Palermo, Heterogeneous alkene epoxidation: Past, present and future, *J. Mol. Catal. A: Chem.* 228 (2005) 27.
[328] R. A. van Santen, H. P. C. E. Kuipers, The mechanism of ethylene epoxidation, *Adv. Catal.* 35 (1987) 265.
[329] C. Backx, J. Moolhuysen, P. Geenen, R. A. van Santen, Reactivity of oxygen adsorbed on silver powder in the epoxidation of ethylene, *J. Catal.* 72 (1981) 364.

[330] C. J. Bertole, C. A. Mims, Dynamic isotope tracing: Role of subsurface oxygen in ethylene epoxidation on silver, *J. Catal.* 184 (1999) 224.
[331] C. Stegelmann, P. Stoltze, Microkinetic analysis of transient ethylene oxidation experiments on silver, *J. Catal.* 226 (2004) 129.
[332] Y. Xu, J. Greeley, M. Mavrikakis, Effect of subsurface oxygen on the reactivity of the Ag(111) surface, *J. Am. Chem. Soc.* 127 (2005) 12823.
[333] P. J. van den Hoek, E. J. Baerends, R. A. van Santen, Ethylene epoxidation on silver(110): the role of subsurface oxygen, *J. Phys. Chem.* 93 (1989) 6469.
[334] P. Hayden, R. J. Sampson, C. B. Spencer, H. Pinnegar, Promoted silver catalyst for producing alkylene oxides, *US Patent* 4,007,135, Feb. 8, 1977, To Imperial Chemical Industries, Limited.
[335] R. B. Grant, R. M. Lambert, A single crystal study of the silver-catalyzed selective oxidation and total oxidation of ethylene, *J. Catal.* 92 (1985) 364.
[336] N. W. Cant, W. K. Hall, Catalytic oxidation. VI. Oxidation of labeled olefins over silver, *J. Catal.* 52 (1978) 81.
[337] S. Linic, M. A. Barteau, Formation of a stable surface oxametallacycle that produces ethylene oxide, *J. Am. Chem. Soc.* 124 (2002) 310.
[338] S. Linic, M. A. Barteau, Control of ethylene epoxidation selectivity by surface oxametallacycles, *J. Am. Chem. Soc.* 125 (2003) 4034.
[339] E. M. Cordi, J. L. Falconer, Oxidation of volatile organic compounds on a Ag/Al_2O_3 catalyst, *Appl. Catal. A: Gen.* 151 (1997) 179.
[340] S. Linic, M. A. Barteau, Construction of a reaction coordinate and a microkinetic model for ethylene epoxidation on silver from DFT calculations and surface science experiments, *J. Catal.* 214 (2003) 200.
[341] C. Stegelmann, N. C. Schiødt, C. T. Campbell, P. Stoltze, Microkinetic modeling of ethylene oxidation over silver, *J. Catal.* 221 (2004) 630.
[342] S. Linic, J. T. Jankowiak, M. A. Barteau, Selectivity driven design of bimetallic ethylene epoxidation catalysts from first principles, *J. Catal.* 224 (2004) 489.
[343] J. T. Jankowiak, M. A. Barteau, Ethylene epoxidation over silver and copper-silver bimetallic catalysts: II. Cs and Cl promotion, *J. Catal.* 236 (2005) 379.
[344] J. Lu, M. Luo, H. Lei, C. Li, Epoxidation of propylene on NaCl-modified silver catalysts with air as the oxidant, *Appl. Catal. A: Gen.* 237 (2002) 11.
[345] R. G. Bowman, Silver-based catalyst for vapor phase oxidation of olefins to epoxides, *US Patent* 4,845,253, July 4, 1989, To The Dow Chemical Company.
[346] E. M. Thorsteinson, Carbonate-supported catalytic system for epoxidation of alkenes, *Canadian Patent CP* 1,282,772, April 9, 1991, To the Union Carbide Corporation.
[347] P. V. Geenen, H. J. Boss, G. T. Pott, A study of the vapor phase epoxidation of propylene and ethylene on silver and silver-gold alloy catalysts, *J. Catal.* 77 (1982) 499.
[348] T. Miyazaki, S. Ozturk, I. Onal, S. Senkan, Selective oxidation of propylene to propylene oxide using combinatorial methodologies, *Catal. Today* 81 (2003) 473.
[349] J. Lu, J. J. Bravo-Suárez, M. Haruta, S. T. Oyama, Direct propylene epoxidation over modified $Ag/CaCO_3$ catalysts, *Appl. Catal. A: Gen.* 302 (2006) 283.
[350] R. Pitchai, A. P. Kahn, A. M. Gaffney, Vapor phase oxidation of propylene to propylene oxide, *US Patent* 5,686,380, Nov. 11 1997, To ARCO Chemical Technology.
[351] G. Mul, M. F. Asaro, A. S. Hirschon, R. B. Wilson Jr., Epoxidation of olefins using lanthanide-promoted silver catalysts, *US Patent* 6,392,066, May 21, 2002, To SRI International.
[352] G. Lu, X. Zuo, Epoxidation of propylene by air over modified silver catalyst, *Catal. Lett.* 58 (1999) 67.
[353] W. Mueller-Markgraf, (Ed.), Process for the direct epoxidation of propylene to propylene oxide, *German Patent DE* 0019,529,679A1, Feb. 13, 1997, To Linde AG.
[354] O. P. H. Vaughan, G. Kyuriakou, N. Macleod, M. Tikhov, R. M. Lambert, Copper as a selective catalyst for the epoxidation of propene, *J. Catal.* 236 (2005) 401.
[355] P. Hayden, G. W. Irving, H. Pinnegar, Oxidation of olefins, *British Patent GB* 1,423,399, Feb. 4, 1976, To Imperial Chemical Industries, Ltd.

[356] J. Lu, J. J. Bravo-Suárez, A. Takahashi, M. Haruta, S. T. Oyama, In situ UV-vis studies of the effect of particle size on the epoxidation of ethylene and propylene on supported silver catalysts using molecular oxygen, J. Catal. 232 (2005) 85.
[357] M. A. Barteau, R. J. Madix, Low-pressure oxidation mechanism and reactivity of propylene on silver(110) and relation to gas-phase acidity, J. Am. Chem. Soc. 105 (1983) 344.
[358] J. T. Roberts, R. J. Madix, W. W. Crew, The rate-limiting step for olefin combustion on silver: Experiment compared to theory, J. Catal. 141 (1993) 300.
[359] M. Akimoto, K. Ichikawa, E. Echigoya, Kinetic and adsorption studies on vapor- phase catalytic oxidation of olefins over silver, J. Catal. 76 (1982) 333.
[360] J. T. Roberts, A. J. Capote, R. J. Madix, Surface-mediated cycloaddition: 1,4-addition of atomically adsorbed oxygen to 1,3-butadiene on silver(110), J. Am. Chem. Soc. 113 (1991) 9848.
[361] J. J. Cowell, A. K. Santra, R. M. Lambert, Ultraselective epoxidation of butadiene on Cu{111} and the effects of Cs promotion, J. Am. Chem. Soc. 122 (2000) 2381.
[362] J. T. Roberts, R. J. Madix, Epoxidation of olefins on silver: Conversion of norbornene to norbornene oxide by atomic oxygen on silver(110), J. Am. Chem. Soc. 110 (1980) 8540.
[363] C. Mukoid, S. Hawker, J. P. S. Badyal, R. M. Lambert, Molecular mechanism of alkene epoxidation: A model study with 3,3-dimethyl-1-butene on Ag(111), Catal. Lett. 4 (1990) 57.
[364] A. K. Santra, J. J. Cowell, R. M. Lambert, Ultra-selective epoxidation of styrene on pure Cu{111} and the effects of Cs promotion, Catal. Lett. 67 (2000) 87.
[365] F. J. Williams, D. P. C. Bird, A. Palermo, A. K. Santra, R. M. Lambert, Mechanism, selectivity promotion, and new ultraselective pathways in Ag-catalyzed heterogeneous epoxidation, J. Am. Chem. Soc. 126 (2004) 8509.
[366] A. Klust, R. J. Madix, Selectivity limitations in the heterogeneous epoxidation of olefins: Branching reactions of the oxametallacycle intermediate in the partial oxidation of styrene, J. Am. Chem. Soc. 128 (2006) 1034.
[367] X. Y. Liu, A. Klust, R. J. Madix, C. M. Friend, Structure sensitivity in the partial oxidation of styrene, styrene oxide, and phenylacetaldehyde on silver single crystals, J. Phys. Chem. C 111 (2007) 3675.
[368] D. Torres, N. Lopez, F. Illas, R. M. Lambert, Low-basicity oxygen atoms: A key in the search for propylene epoxidation catalysts, Angew. Chem. Int. Ed. Engl. 46 (2007) 1.
[369] D. Torres, K. M. Neyman, F. Illas, Oxygen atoms on the (1 1 1) surface of coinage metals: On the chemical state of the adsorbate, Chem. Phys. Lett. 429 (2006) 86.
[370] J. W. Medlin, J. R. Monnier, M. A. Barteau, Deuterium kinetic isotope effects in butadiene epoxidation over unpromoted and Cs-promoted silver catalysts, J. Catal. 204 (2001) 71.
[371] J. R. Monnier, J. W. Medlin, M. A. Barteau, Use of oxygen-18 to determine kinetics of butadiene epoxidation over Cs-promoted Ag catalysts" J. Catal. 203 (2001) 362.
[372] R. A. Sheldon, I. W. C. E. Arends, H. E. B. Lempers, Liquid phase oxidation at metal ions and complexes in constrained environments, Catal. Today 41 (1998) 387.
[373] M. G. Clerici, TS-1 and propylene oxide, 20 years later, Presented at the DGMK/SCI Conference "Oxidation and Functionalization: Classical and Alternative Routes and Sources", S.T.O. thanks M.G.C. for the manuscript, October 12–15, 2005, Milan, Italy.
[374] M. Taramasso, G. Perego, B. Notari, Preparation of porous crystalline synthetic material comprised of silicon and titanium oxides, US. Patent 4,410,501, Oct. 18, 1983, To Snamprogetti, S.p.A.
[375] M. Taramasso, G. Manara, V. Fattore, B. Notari, Silica-based synthetic material containing titanium in the crystal lattice and process for its preparation, US Patent 4,666,692, May 19, 1987, To Snamprogetti, S.p.A.
[376] B. Notari, Synthesis and catalytic properties of titanium containing zeolites, in: P. J. Grobet, et al. (Eds.), Innovation in Zeolite Materials Science, Stud. Surf. Sci. Catal. 37 (1987) p. 413.
[377] J. S. Rafelt, J. H. Clark, Recent advances in the partial oxidation of organic molecules using heterogeneous catalysis, Catal. Today 57 (2000) 33.
[378] G. Belussi, M. S. Rigutto, Metal-ions associated to the molecular sieve framework - possible catalytic-oxidation sites, Stud. Surf. Sci. Catal. 85 (1994) 177.
[379] I. W. C. E. Arends, R. A. Sheldon, M. Wallau, U. Schuchardt, Oxidative transformations of organic compounds mediated by redox molecular sieves, Angew. Chem. Int. Ed. Engl. 36 (1997) 1144.

[380] D. E. De Vos, P. L. Buskens, D. L. Vanoppen, P. P. Knops-Gerrits, P. A. Jacobs, *Comprehensive-Supramolecular Chemistry*, Vol. 7, Elsevier, Amsterdam, 1996, p. 647.
[381] A. Corma, M. A. Camblor, P. Esteve, A. Martinez, J. Perez-Pariente, Activity of Ti-beta catalyst for the selective oxidation of alkenes and alkanes, *J. Catal.* 145 (1994) 151.
[382] A. Corma, M. T. Navarro, J. Pérez-Pariente, Synthesis of an ultralarge pore titanium silicate isomorphous to MCM-41 and its application as a catalyst for selective oxidation of hydrocarbons, *Chem. Commun.* (1994) 147.
[383] O. Franke, J. Rathousky, G. Schulz-Ekloff, J. Starek, A. Zukal, New mesoporous titanosilicate molecular sieve," in: J. Weitkamp, H. G. Karge, H. Pfeifer, W. Hölderich (Eds.), Zeolites and Related Microporous Materials: State of the Art 1994, *Stud. Surf. Sci. Catal.* 84 (1994) p. 77.
[384] R. Hutter, D. C. M. Dutoit, T. Mallat, M. Schneider, A. Baiker, Novel mesoporous titania-silica aerogels highly active for the selective epoxidation of cyclic olefins, *Chem. Commun.* (1995) 163.
[385] D. Tantanak, M. A. Vincent, I. H., Hillier, Elucidation of the mechanism of alkene epoxidation by hydrogen peroxide catalyzed by titanosilicates: A computational study, *Chem. Commun.* (1998) 1031.
[386] T. Maschmeyer, M. C. Klunduk, C. M. Martin, D. S. Shephard, J. M. Thomas, B. F. G. Johnson, Modelling the active sites of heterogeneous titanium-centred epoxidation catalysts with soluble silsesquioxane analogues, *Chem. Commun.* (1997) 1847.
[387] C. Neri, B. Anfossi, A. Esposito, F. Buonomo, Process for the epoxidation of olefinic compounds, *Italian Patent IT* 22,608, Dec. 31, 1982, To ANIC, S.p.A.
[388] C. Neri, B. Anfossi, A. Esposito, F. Buonomo, Process for the epoxidation of olefinic compounds, *European Patent EP* 100119, July 13, 1983, To ANIC, S.p.A.
[389] C. Neri, B. Anfossi, A. Esposito, F. Buonomo, Process for the epoxidation of olefinic compounds, *US Patent* 4,833,260, May 23, 1989, To ANIC, S.p.A.
[390] M. G. Clerici, P. Ingallina, Epoxidation of lower olefins with hydrogen peroxide and titanium silicalite, *J. Catal.* 140 (1993) 71.
[391] W. Lin, H. Frei, Photochemical and FT-IR probing of the active site of hydrogen peroxide in Ti silicalite sieve, *J. Am. Chem. Soc.* 124 (2002) 9292.
[392] X. H. Liang, Z. T. Mi, Y. L. Wu, L. Wang, E. H. Xing, Kinetics of epoxidation of propylene over TS-1 in isopropanol, *React. Kinet. Catal. Lett.* 80 (2003) 207.
[393] X. W. Liu, X. S. Wang, X. W. Guo, G. Li, Effect of solvent on propylene epoxidation over TS-1 catalyst, *Catal. Today* 93 (2004) 505.
[394] A. Corma, P. Esteve, A. Martínez, Solvent effects during the oxidation of olefins and alcohols with hydrogen peroxide on Ti-beta catalyst: The influence of the hydrophilicity/hydrophobicity of the zeolite, *J. Catal.* 161 (1996) 11.
[395] N. Jappar, Q. H. Xia, T. Tatsumi, Oxidation activity of Ti-beta synthesized by a dry-gel conversion method, *J. Catal.* 180 (1998) 132.
[396] J. C. van der Waal, H. van Bekkum, Zeolite titanium beta: A versatile epoxidation catalyst. Solvent effects, *J. Mol. Catal. A: Chem.* 124 (1997) 137.
[397] P. Wu, T. Tatsumi, Unique *trans* selectivity of Ti-MWW in epoxidation of *cis/trans* alkenes with hydrogen peroxide, *J. Phys. Chem. B* 106 (2002) 748.
[398] G. Langhendries, D. E. de Vos, G. V. Baron, P. A. Jacobs, Quantitative adsorption measurements on Ti zeolites and relation with α-olefin oxidation with H_2O_2, *J. Catal.* 187 (1999) 453.
[399] M. S. Rigutto, R. de Ruiter, J. P. M. Niederer, H. van Bekkum, Synthesis of aluminum free titanium silicate with the BEA structure using a new and selective template and its use as a catalyst in epoxidations, in: H. Chon, S.-K. Ihm, Y. S. Uh (Eds.), Progress in Zeolite and Microporous Materials, *Stud. Surf. Sci. Catal.* 84 (1994) p. 2245.
[400] G. F. Thiele, E. Roland, Propylene epoxidation with hydrogen peroxide and titanium silicalite catalyst: Activity, deactivation, and regeneration of the catalyst, *J. Mol. Catal. A: Chem.* 117 (1997) 351.
[401] B. Notari, Microporous crystalline titanium silicates, in: D. D. Eley, W. O. Haag, B. C. Gates (Eds.), *Advances in Catalysis*, Vol. 41, Academic Press, San Diego, 1996, p. 253.
[402] P. Ratnasamy, D. Srinivas, H. Knözinger, Active sites and reactive intermediates in titanium silicate molecular sieves, *Adv. Catal.* 48 (2004) 1.

[403] D. R. C. Huybrechts, L. De Bruycker, P. A. Jacobs, Oxyfunctionalization of alkanes with hydrogen peroxide on titanosilicalite, *Nature* 345 (1990) 240.
[404] H. J. Ledon, F. Varescon, Role of peroxo vs. alkylperoxo titanium porphyrin complexes in the epoxidation of olefins, *Inorg. Chem.* 23 (1984) 2735.
[405] G. Belussi, A. Carati, M. G. Clerici, G. Maddinelli, R. Millini, Reactions of titanium silicalite with protic molecules and hydrogen peroxide, *J. Catal.* 133 (1992) 220.
[406] M. G. Clerici, Oxidation of saturated hydrocarbons with hydrogen peroxide, catalyzed by titanium silicalite, *Appl. Catal.* 68 (1991) 249.
[407] E. Karlsen, K. Schöffel, Titanium silicalite catalyzed epoxidation of ethylene with hydrogen peroxide. A theoretical study, *Catal. Today* 32 (1996) 107.
[408] I. V. Yudanov, P. Gisdakis, C. Di Valentin, N. Rösch, Activity of peroxo and hydroperoxo complexes of Ti^{IV} in olefin epoxidation: A density functional theory model study of energetics and mechanism, *Eur. J. Inorg. Chem.* (1999) 2135.
[409] P. E. Sinclair, C. R. A. Catlow, Quantum chemical study of the mechanism of partial oxidation reactivity on titanosilicate catalysts: Active site formation, oxygen transfer, and catalyst deactivation, *J. Phys. Chem. B* 103 (1999) 1084.
[410] C. M. Barker, N. Kaltsoyannis, C. R. A. Catlow, in: A. Galarneau, F. Di Renzo, F. Fajula, J. Vedrine (Eds.), Zeolites and mesoporous materials at the dawn of the 21st century, Stud. Surf. Sci. Catal., Vol. 135. Elsevier, Amsterdam, 2001.
[411] M. Neurock, L. E. Manzer, Theoretical insights on the mechanism of alkene epoxidation by H_2O_2 with titanium silicalite, *Chem. Commun.* (1996) 1133.
[412] M. R. Boccuti, K. M. Rao, A. Zechhina, G. Leofanti, G. Petrini, in: C. Morterra, A. Zecchina, G. Costa (Eds.), Structure and reactivity of surfaces, Stud. Surf. Sci. Catal., Vol. 48, Elsevier, Amsterdam, 1989, p. 133.
[413] S. Bordiga, S. Coluccia, C. Lamberti, L. Marchese, A. Zecchina, F. Boscherini, F. Buffa, F. Genoni, G. Leofanti, G. Petrini, B. Vlaic, XAFS study of Ti-Silicalite: Structure of framework Ti(IV) in the presence and absence of reactive molecules (H_2O, NH_3) and comparison with ultraviolet-visible and IR results, *J. Phys. Chem.* 98 (1994) 4125.
[414] F. Maspero, U. Romano, Oxidation of alcohols with H_2O_2 catalyzed by titanium silicalite-1, *J. Catal.* 146 (1994) 476.
[415] H. X. Gao, G. X. Lu, J. S. Lu, S. Li, Epoxidation of allyl chloride with hydrogen peroxide catalyzed by titanium silicalite 1, *Appl. Catal. A: Gen.* 138 (1996) 27.
[416] A. Corma, P. Esteve, A. Martínez, S. Valencia, Oxidation of olefins with hydrogen peroxide and *tert*-butyl hydroperoxide on Ti-Beta catalyst, *J. Catal.* 152 (1995) 18.
[417] J. C. van der Waal, M. S. Rigutto, H. van Bekkum, Zeolite titanium beta as a selective catalyst in the epoxidation of bulky alkenes, *Appl. Catal. A: Gen.* 167 (1998) 331.
[418] M. G. Clerici, P. Ingallina, Oxidation reactions with *in situ* generated oxidants, *Catal. Today* 41 (1998) 351.
[419] P. Wu, T. Komatsu, T. Yashima, Hydroxylation of aromatics with hydrogen peroxide over titanosilicates with MOR and MFI structures: Effect of Ti peroxo species on the diffusion and hydroxylation activity, *J. Phys. Chem. B* 102 (1998) 9297.
[420] B. K. Min, C. M. Friend, Heterogeneous gold based catalysis for green chemistry: Low-temperature CO oxidation and propene oxidation, *Chem. Rev.* 107 (2007) 2709.
[421] N. Yap, R. P. Andres, W. N. Delgass, Reactivity and stability of Au in and on TS-1 for epoxidation of propylene with H_2 and O_2, *J. Catal.* 226 (2004) 156.
[422] E. E. Stangland, B. Taylor, R. P. Andres, W. N. Delgass, Direct vapor phase propylene epoxidation over deposition-precipitation gold-titania catalysts in the presence of H_2/O_2: Effects of support, neutralizing agent, and pretreatment, *J. Phys. Chem. B* 109 (2005) 2321.
[423] B. Chowdhury, J. J. Bravo-Suárez, M. Daté, S. Tsubota, M. Haruta, Trimethylamine as a Gas-Phase Promoter: Highly efficient epoxidation of propylene over supported gold catalysts, *Angew. Chem. Int. Ed. Engl.* 45 (2006) 412.
[424] A. K. Sinha, S. Seelan, S. Tsubota, M. Haruta, A three-dimensional mesoporous titanosilicate support for gold nanoparticles: Vapor-phase epoxidation of propene with high conversion, *Angew. Chem. Int. Ed. Engl.* 43 (2004) 1546.

[425] A. K. Sinha, S. Seelan, M. Okumura, T. Akita, S. Tsubota, M. Haruta, Three-dimensional mesoporous titanosilicates prepared by modified sol-gel method: Ideal gold catalyst supports for enhanced propene epoxidation, *J. Phys. Chem. B* 109 (2005) 3956.
[426] D. H. Wells Jr., W. N. Delgass, K. T. Thomson, Evidence of defect-promoted reactivity for epoxidation of propylene in titanosilicate (TS-1) catalysts: A DFT study, *J. Am. Chem. Soc.* 126 (2004) 2956.
[427] D. H. Wells Jr., W. N. Delgass, K. T. Thomson, Formation of hydrogen peroxide from H_2 and O_2 over a neutral gold trimer: A DFT study, *J. Catal.* 225 (2004) 69.
[428] D. H. Wells Jr., A. M. Joshi, W. N. Delgass, K. T. Thomson, A quantum chemical study of comparison of various propylene epoxidation mechanisms using H_2O_2 and TS-1 catalyst, *J. Phys. Chem. B* 110 (2006) 14627.
[429] B. Taylor, J. Lauterbach, G. E. Blau, W. N. Delgass, Reaction kinetic analysis of the gas-phase epoxidation of propylene over Au/TS-1, *J. Catal.* 242 (2006) 142.
[430] J. Q. Lu, X. Zhang, J. J. Bravo-Suárez, S. Tsubota, J. Gaudet, S. T. Oyama, Kinetics of propylene epoxidation using H_2 and O_2 over a gold/mesoporous titanosilicate catalyst, *Catal. Today* 123 (2007) 189.
[431] E. E. Stangland, K. B. Stavens, R. P. Andres, W. N. Delgasss, Characterization of gold–titania catalysts via oxidation of propylene to propylene oxide, *J. Catal.* 191 (2000) 332.
[432] C. Sivadinarayana, T. V. Choudhary, L. L. Daemen, J. Eckert, D. W. Goodman, The nature of the surface species formed on AU/TiO_2 during the reaction of H_2 and O_2: An inelastic neutron scattering study, *J. Amer. Chem. Soc.* 126 (2004) 38.
[433] B. Chowdhury, J. J. Bravo-Suárez, N. Mimura, J. Q. Lu, K. K. Bando, S. Tsubota, M. Haruta, *In situ* UV-vis and EPR study on the formation of hydroperoxide species during direct gas phase propylene epoxidation over $Au/Ti-SiO_2$ catalyst, *J. Phys. Chem. B* 110 (2006) 22995.
[434] B. M. Weckhuysen, *In-situ Spectroscopy of Catalysts*; American Scientific Publishers, Stevenson Ranch, CA, 2004.
[435] S. J. Tinnemans, J. G. Mesu, K. Kervinen, T. Visser, T. A. Nijhuis, A. M. Beale, D. E. Keller, A. M. J. van der Eerden, B. M. Weckhuysen, Combining operando techniques in one spectroscopic-reaction cell: New opportunites for elucidating the active site and realted reaction mechanism in catalysis, *Catal. Today* 113 (2006) 3.
[436] K. Tamaru, *Dynamic Heterogeneous Catalysis*, Academic Press, New York (1978).
[437] C. Reed, Y. Xi, S. T. Oyama, Distinguishing between reaction intermediates and spectators: A kinetic study of acetone oxidation using ozone on a silica-supported manganese oxide catalyst, *J. Catal.* 235 (2005) 378.
[438] S. T. Oyama, W. Li, Absolute determination of reaction mechanisms by *in situ* measurements of reaction intermediates, *Top. Catal.* 8 (1999) 75.
[439] J. J. Bravo-Suárez, K. K. Bando, J. Lu, M. Haruta, T. Fujitani, S. T. Oyama, Identification of true reaction intermediates in propylene epoxidation on gold/titanosilicate catalysts by *in situ* UV-vis and XAFS Spectroscopies, *J. Phys. Chem.* 112 (2008) 1115.
[440] C. M. Barker, D. Gleeson, N. Kaltsoyannis, C. R. A. Catlow, G. Sankar, J. M. Thomas, On the structure and coordination of the oxygen-donating species in TiMCM-41/TBHP oxidation catalysts: A density functional theory and EXAFS study, *Phys. Chem. Chem. Phys.* 4 (2002) 1228.
[441] P. E. Sinclair, G. Sankar, C. R. A. Catlow, J. M. Thomas, T. J. Maschmeyer, Computational and EXAFS study of the nature of the Ti(IV) active sites in mesoporous titanosilicate catalysts, *J. Phys. Chem. B* 101 (1997) 4232.
[442] D. Matthey, J. G. Wang, S. Wendt, J. Matthiesen, R. Schaub, E. Lægsgaard, B. Hammer, F. Besenbacher, Enhanced bonding of gold nanoparticles on oxidized $TiO_2(110)$, *Science* 315 (2007) 1692.
[443] A. M. Joshi, W. N. Delgass, K. T. Thompson, Comparison of the catalytic activity of Au_3, Au_4^+, Au_5, and Au_5^- in the gas-phase reaction of H_2 and O_2 to form hydrogen peroxide: A density functional theory investigation, *J. Phys. Chem. B* 109 (2005) 22392.
[444] P. P. Olivera, E. M. Patrito, H. Sellers, Hydrogen peroxide synthesis over metallic catalysts, *Surf. Sci.* 313 (1994) 25.
[445] M. Okumura, J. Coronado, M. J. Soria, M. Haruta, C. Conesay, EPR study of CO and O_2 interaction with supported au catalysts, *J. Catal.* 203 (2001) 168.

[446] C. B. Khouw, C. B. Dartt, J. A. Labinger, M. E. Davis, Studies on the catalytic-oxidation of alkanes and alkenes by titanium silicates, *J. Catal.* 149 (1994) 195.
[447] P. Chaumette, H. Mimoun, L. Saussine, J. Fischer, A. Mitschler, Peroxo and alkylperoxidic molybdenum(VI) complexes as intermediates in the epoxidation of olefins by alkyl hydroperoxides, *J. Organomet. Chem.* 250 (1983) 291.
[448] M. Fujiwara, H. Wessel, H. Park, H. W. Roesky, Formation of titanium tert-butylperoxo intermediate from cubic silicon-titanium complex with tert-butyl hydroperoxide and its reactivity for olefin epoxidation, *Tetrahedron* 58 (2002) 239.
[449] W. R. Thiel, T. Priermeier, The first olefin-substituted peroxomolybdenum complex: Insight into a new mechanism for the molybdenum-catalyzed epoxidation of olefins, *Angew. Chem. Int. Ed. Engl.* 34 (1995) 1737.
[450] W. A. Herrmann, R. M. Kratzer, H. Ding, W. Thiel, H. Glas, Methyltrioxorhenium/pyrazole—A highly efficient catalyst for the epoxidation of olefins, *J. Organomet. Chem.* 555 (1998) 293.
[451] D. V. Deubel, G. Frenking, P. Gisdakis, W. A. Herrmann, N. Rösch, J. Sundermeyer, Olefin epoxidation with inorganic peroxides. solutions to four long-standing controversies on the mechanism of oxygen transfer, *Acc. Chem. Res.* 37 (2004) 645.
[452] S. Dapprich, G. Frenking, Investigation of donor-acceptor interactions: A charge decomposition analysis using fragment molecular orbitals, *J. Phys. Chem.* 99 (1995) 9352.
[453] G. Frenking, N. Fröhlich, The nature of the bonding in transition-metal compounds, *Chem. Rev.* 100 (2000) 717.
[454] D. V. Deubel, Are peroxyformic acid and dioxirane electrophilic or nucleophilic oxidants? *J. Org. Chem.* 66 (2001) 3790.
[455] R. Curci, J. O. Edwards, in: G. Strukul(Ed.), *Catalytic Oxidations with Hydrogen Peroxide as Oxidant*, Kluwer, Dordrecht, The Netherlands, 1992.
[456] R. D. Bach, G. J. Wolber, B. A. Coddens, On the mechanism of metal-catalyzed epoxidation: A model for the bonding in peroxo-metal complexes, *J. Am. Chem. Soc.* 106 (1984) 6098.
[457] C. W. Smith, Oxidation with peroxides, US Patent 2,754,325, Jul. 10, 1956, To Shell Development Company.
[458] M. N. Sheng, J. G. Zajacek, British Patent GB 1,136,923, 1968, To Atlantic Richfield, Co.
[459] P. D. Taylor, M. T. Mocella, Recovery of molybdenum as an aqueous solution from spent catalyst, US Patent 4,315,896, Feb. 16, 1982, To Atlantic Richfield, Co.
[460] N. H. Sweed, Molybdenum epoxidation catalyst recovery, US Patent 4,455,283, June 19, 1984, To Atlantic Richfield Co.
[461] R. B. Poenisch, Process for the recovery of molybdenum from organic solutions, US Patent 4,485,074, Nov. 27, 1984, To Atlantic Richfield Co.
[462] B. H. Isaacs, Regeneration of soluble molybdenum catalysts from spent catalyst streams, US Patent 4,598,057, July 1, 1986, To Atlantic Richfield Co.
[463] T. T. Shih, Lower alkylene oxide purification, US Patent 5,133,839, July 28, 1992, To ARCO Chemical Technology, L.P.
[464] V. S. Sergienko, Structural characteristics of peroxo complexes of group IV and V transition metals, *Review, Crystallogr. Rep.* 49 (2004) 907.
[465] R. A. Sheldon, J. K. Kochi, *Metal-Catalyzed Oxidations of Organic Compounds*, Academic Press, New York, 1981.
[466] C. R. Noller, *Chemistry of Organic Compounds*, 2nd ed., W. B. Saunders, Philadelphia 1957.
[467] G. L. Linden, M. F. Farona, A resin-bound vanadyl catalyst for the epoxidation of olefins, *Inorg. Chem.* 16 (1970) 3170.
[468] M. B. Ward, K. Mizuno, J. H. Lunsford, Epoxidation of propylene over molybdenum-Y zeolites, *J. Molec. Catal.* 27 (1984) 1.
[469] J. Sobczak, J. J. Ziòłkowski, The molybdenum(V) complexes as the homogeneous and heterogenized catalysts in epoxidation reactions of olefins with the organic hydroperoxides, *J. Mol. Catal.* 3 (1977/1978) 165.
[470] S. Ivanov, R. Boeva, S. Tanielyan, Catalytic epoxidation of propylene with *tert*-butyl hydroperoxide in the presence of modified carboxy cation-exchange resin "Amberlite" IRC-50, *J. Catal.* 56 (1979) 150.

[471] S. Bhaduri, H. Khwaja, Polymer-supported complexes. Part 3. Synthesis of a polystyrene-anchored molybdenum(V) dithiocarbamato-derivative and its applications in reactions involving *t*-butyl hydroperoxide, *J. Chem. Soc., Dalton Trans.* (1983) 415.
[472] D. C. Sherrington, Polymer-supported metal complex alkene epoxidation catalysts, *Catal. Today* 57 (2000) 87.
[473] A. O. Chong, K. B. Sharpless, Mechanism of the molybdenum and vanadium catalyzed epoxidation of olefins by alkyl hydroperoxides, *J. Org. Chem.* 42 (1977) 1587.
[474] M. N. Sheng, J. G. Zajacek, Hydroperoxide Oxidations Catalyzed by Metals-I: The Epoxidation of Olefins, in: *Oxidation of Organic Compounds-II*, Adv. Chem. Series 76, Am. Chem. Society, Washington, DC, 1968, p. 418.
[475] H. Mimoun, Oxygen transfer from inorganic and organic peroxides to organic substrates: A common mechanism? *Angew. Chem. Int. Ed. Engl.* 21 (1982) 734.
[476] C. Y. Wu, H. E. Swift, Selective olefin epoxidation at high hydroperoxide-to-olefin ratios, *J. Catal.* 43 (1976) 380.
[477] R. A. Sheldon, Molybdenum-catalyzed epoxidation of olefins with alkyl hydroperoxides. 1. Kinetic and product studies, *Rec. Trav. Chim. Pays-Bas* 92 (1973) 253.
[478] J. E. McKeon, D. W. Connell, Mechanisms of the borate ester induced decomposition of alkyl hydroperoxides, *J. Org. Chem.* 40 (1975) 1875.
[479] J. Sundermeyer, Metal-mediated oxyfunctionalization of organic substrates by organometallic intermediates—More recent developments and perspectives, *Angew. Chem. Int. Ed. Engl.* 32 (1993) 1144.
[480] P. Gisdakis, I. V. Yudanov, N. Rösch, Olefin epoxidation by molybdenum and rhenium peroxo and hydroperoxo compounds: A density functional study of energetics and mechanisms, *Inorg. Chem.* 40 (2001) 3755.
[481] K. B. Sharpless, J. M. Townsend, D. R. Williams, Mechanism of epoxidation of olefins by covalent peroxides of molybdenum(VI), *J. Am. Chem. Soc.* 94 (1972) 295.
[482] K. B. Sharpless, T. C. Flood, Oxotransition metal oxidants as mimics for the action of mixed-function oxygenases. "NIH shift" with chromyl reagents, *J. Am. Chem. Soc.* 93 (1971) 2316.
[483] H. Shi, X. Wang, R. M. Hua, Epoxidation of alpha, beta-unsaturated acids catalyzed by tungstate (VI) or molybdate (VI) in aqueous solvents: a specific direct oxygen transfer mechanism, *Tetrahedron* 61 (2005) 1297.
[484] D. V. Deubel, J. Sundermeyer, G. Frenking, Mechanism of the olefin epoxidation catalyzed by molybdenum diperoxo complexes: Quantum-chemical calculations give an answer to a long-standing question, *J. Am. Chem. Soc.* 122 (2000) 10101.
[485] D. V. Deubel, J. Sundermeyer, G. Frenking, Olefin epoxidation with transition metal 2-peroxo complexes: The control of reactivity, *Eur. J. Inorg. Chem.* (2001) 1819.
[486] A. Hroch, G. Gemmecker, W. R. Thiel, Metal-catalyzed oxidations, 10 New insights into the mechanism of hydroperoxide activation by investigation of dynamic processes in the coordination sphere of seven-coordinated molybdenum peroxo complexes, *Eur. J. Inorg. Chem.* (2000) 1107.
[487] C. Di Valentin, P. Gisdakis, I. V. Yudanov, N. Rösch, Olefin epoxidation by peroxo complexes of Cr, Mo, and W. A comparative density functional study, *J. Org. Chem.* 65 (2000) 2996.
[488] H. Mimoun, M. Mignard, P. Brechot, L. Saussine, Selective epoxidation of olefins by oxo[N-(2-oxidophenyl)salicylidenaminato]vanadium(V) alkylperoxides. On the mechanism of the Halcon epoxidation process, *J. Am. Chem. Soc.* 108 (1986) 3711.
[489] S. S. Woodard, M. G. Finn, K. B. Sharpless, Mechanism of asymmetric epoxidation. 1. Kinetics, *J. Am. Chem. Soc.* 113 (1991) 106.
[490] M. G. Finn, K. B. Sharpless, Mechanism of asymmetric epoxidation. 2. Catalyst structure, *J. Am. Chem. Soc.* 113 (1991) 113.
[491] E. J. Corey, On the origin of the enantioselectivity in the Katsuki-Sharpless epoxidation procedure, *J. Org. Chem.* 55 (1990) 1693.
[492] M. G. Finn, K. B. Sharpless, On the mechanism of asymmetric epoxidation with titanium tartrate catalysts, in: J. D. Morrison(Ed.), *Asymmetric Synthesis*, Vol. 5, Academic Press, New York, 1984, pp. 247–301, Chapt. 8.

[493] Y.-D. Wu, D. K. W. Lai, A density functional study on the stereocontrol of the sharpless epoxidation, *J. Am. Chem. Soc.* 111 (1995) 11327.
[494] M. Cui, W. Adam, J. H. Shen, Y. M. Luo, X. J. Tan, K. X. Chen, R. Y. Ji, H. L. Jiang, A density-functional study of the mechanism for the diastereoselective epoxidation of chiral allylic alcohols by the titanium peroxy complexes, *J. Org. Chem.* 67 (2002) 1427.
[495] D. Wei, W. Chuei, G. L. Haller, Catalytic behavior of vanadium substituted mesoporous molecular sieves, *Catal. Today* 51 (1999) 501.
[496] W. A. Herrmann, R. W. Fischer, W. Scherer, M. U. Rauch, Methyltrioxorhenium(VII) as catalyst for epoxidations: Structure of the active species and mechanism of catalysis, *Angew. Chem. Int. Ed. Engl.* 32 (1993) 1157.
[497] A. Al-Ajlouni, H. Espenson, Epoxidation of styrenes by hydrogen peroxide as catalyzed by methylrhenium trioxide, *J. Am. Chem. Soc.* 117 (1995) 9243.
[498] W. A. Herrmann, J. D. G. Correia, G. R. J. Artus, R. W. Fischer, C. C. Romão, Multiple bonds between main group elements and transition metals, 155. (Hexamethylphosphoramide) methyl (oxo) bis(η^2-peroxo)rhenium(VII), the first example of an anhydrous rhenium peroxo complex: Crystal structure and catalytic properties, *J. Organomet. Chem.* 520 (1996) 139.
[499] W. Adam, C. R. Saha-Möller, O. Weichold, Epoxidation of *trans*-cyclooctene by methyltrioxorhenium/H_2O_2: Reaction of *trans*-epoxide with the monoperoxo complex, *J. Org. Chem.* 65 (2000) 5001.
[500] P. Gisdakis, N. Rösch, Solvent effects on the activation barriers of olefin epoxidation - A density functional study, *Eur. J. Org. Chem.* (2001) 719.
[501] C. di Valentin, R. Gandolfi, P. Gisdakis, N. Rösch, Allylic alcohol epoxidation by methyltrioxorhenium: A density functional study on the mechanism and the role of hydrogen bonding, *J. Am. Chem. Soc.* 123 (2001) 2365.
[502] O. Pestovski, R.v. Eldik, P. Huston, J. H. Espenson, Mechanistic study of the co-ordination of hydrogen peroxide to methylrhenium trioxide, *J. Chem. Soc. Dalton Trans.* (1995) 133.
[503] P. Gisdakis, W. Antonczak, S. Köstlmeier, W. A. Herrmann, N. Rösch, Olefin epoxidation by methyltrioxorhenium: A density functional study on energetics and mechanisms, *Angew. Chem. Int. Ed. Engl.* 37 (1998) 2211.
[504] F. E. Kühn, J. Zhao, W. A. Herrmann, Chiral monomeric organorhenium(VII) and organomolybdenum(VI) compounds as catalysts for chiral olefin epoxidation reactions, *Tetrahedron Asymm.* 16 (2005) 3469.
[505] R. W. Alder, A. P. Davis, The design of organic catalysis for epoxidation by hydrogen peroxide, *J. Mol. Model.* 12 (2006) 649.
[506] M. Freccero, R. Gandolfi, M. Sarzi, A. Rastelli, Competition between peroxy acid oxygens as hydrogen bond acceptors in B3LYP transition structures for epoxidations of allylic alcohols with peroxyformic acid, *J. Org. Chem.* 64 (1999) 3853.
[507] W. Adam, A. K. Smerz, Solvent effects in the regio- and diastereoselective epoxidations of acyclic allylic alcohols by dimethyldioxirane: Hydrogen bonding as evidence for a dipolar transition state, *J. Org. Chem.* 61 (1996) 3506.
[508] R. D. Bach, O. Dmitrenko, W. Adam, S. Schambony, Relative reactivity of peracids versus dioxiranes (DMDO and TFDO) in the epoxidation of alkenes. A combined experimental and theoretical analysis, *J. Am. Chem. Soc.* 125 (2003) 924.
[509] M. Fioroni, K. Burger, A. E. Mark, D. Roccatano, Model of 1,1,1,3,3,3-hexafluoro-propan-2-ol for molecular dynamics simulations, *J. Phys. Chem. B* 105 (2001) 10967.
[510] P. C. B. Page, D. Barros, B. R. Buckley, B. A. Marples, Organocatalysis of asymmetric epoxidation mediated by iminium salts: Comments on the mechanism, Tetrahedron, *Asymmetry* 16 (2005) 3488.
[511] M. R. Biscoe, R. Breslow, Oxaziridinium salts as hydrophobic epoxidation reagents: Remarkable hydrophobically-directed selectivity in olefin epoxidation, *J. Am. Chem. Soc.* 127 (2005) 10812.
[512] J. T. Groves, Key elements of the chemistry of cytochrome P-450: The oxygen rebound mechanism, *J. Chem. Educ.* 65 (1985) 928.
[513] J. M. Garrison, T. C. Bruice, Intermediates in the epoxidation of alkenes by cytochrome P-450 models. 3. Mechanism of oxygen transfer from substituted oxochromium(V) porphyrins to olefinic substrates, *J. Am. Chem. Soc.* 111 (1989) 191.

[514] W. Adam, K. J. Roschmann, C. R. Saha-Möller, D. Seebach, cis-Stilbene and (1,2,3)-(2-ethenyl-3-methoxycyclopropyl)benzene as mechanistic probes in the MnIII(salen)-catalyzed epoxidation: Influence of the oxygen source and the counterion on the diastereoselectivity of the competitive concerted and radical-type oxygen transfer, *J. Am. Chem. Soc.* 124 (2002) 5068.
[515] N. S. Finney, P. J. Pospisil, S. Chang, M. Palucki, R. G. Konsler, K. B. Hansen, E. N. Jacobsen, On the viability of oxametallacyclic intermediates in the (salen)Mn-catalyzed asymmetric epoxidation, *Angew. Chem., Int. Ed. Engl.* 36 (1997) 1720.
[516] J. P. Collman, A. S. Chien, T. A. Eberspacher, J. I. Brauman, Multiple active oxidants in cytochrome P-450 model oxidations, *J. Am. Chem. Soc.* 122 (2000) 11098.
[517] D. Mohajer, S. Tangestaninejad, Efficient olefin epoxidation with tetrabutylammonium periodate catalyzed by manganese porphyrin in the presence of imidazole, *Tetrahedron Lett.* 35 (1994) 945.
[518] A. M. Daly, M. F. Renehan, D. G. Gilheany, High enantioselectivities in an (E)-alkene epoxidation by catalytically active chromium salen complexes. Insight into the catalytic cycle, *Org. Lett.* 3 (2001) 663.
[519] J. P. Collman, V. J. Lee, C. J. Kellen-Yuen, X. Zhang, J. A. Ibers, J. I. Brauman, Threitol-strapped manganese porphyrins as enantioselective epoxidation catalysts of unfunctionalized olefins, *J. Am. Chem. Soc.* 117 (1995) 692.
[520] S. J. Yang, W. Nam, Water-soluble iron porphyrin complex-catalyzed epoxidation of olefins with hydrogen peroxide and tert-butyl hydroperoxide in aqueous solution, *Inorg. Chem.* 37 (1998) 606.
[521] B. Meunier, Metalloporphyrins as versatile catalysts for oxidation reactions and oxidative DNA cleavage, *Chem. Rev.* 92 (1992) 1411.
[522] P. R. Ortiz de Montellano (Ed.), *Cytochrome P450 Structure, Mechanism, and Biochemistry*, 2nd ed, Plenum, New York, 1995.
[523] K. M. Kadish, K. M. Smith, R. Guilard (Eds.), *The Porphyrin Handbook*, Academic Press, New York, 2000.
[524] E. Rose, B. Andrioletti, S. Zrig, M. Q. Ethéve, Enantioselective epoxidation of olefins with chiral metalloporphyrin catalysts, *Chem. Soc. Rev.* 34 (2005) 573.
[525] S. Franzen, M. P. Roach, Y.-P. Chen, R. B. Dyer, W. H. Woodruff, J. H. Dawson, The unusual reactivities of *Amphitrite ornata* dehaloperoxidase and *Notomastus lobatus* chloroperoxidase do not arise from a histidine imidazolate proximal heme iron ligand, *J. Am. Chem. Soc.* 120 (1998) 4658.
[526] J. H. Dawson, Probing structure-function relations in heme-containing oxygenases and peroxidases, *Science* 240 (1988) 433.
[527] R. W. Lee, P. C. Nakagaki, T. C. Bruice, The kinetics for the reaction of hypochlorite with a manganese(III) porphyrin and subsequent epoxidation of alkenes in a homogeneous solution, *J. Am. Chem. Soc.* 111 (1989) 1368.
[528] J. Razenberg, A. W. Vandermade, J. W. Smeets, R. J. M. Nolte, Cyclohexene epoxidation by the mono-oxygenase model (tetraphenylporphyrinato)manganese(III) acetate-sodium hypochlorite, *J. Mol. Catal.* 31 (1985) 271.
[529] M. C. Feiters, A. E. Rowan, R. J. M. Nolte, From simple to supramolecular cytochrome P450 mimics, *Chem. Soc. Rev.* 29 (2000) 375.
[530] J. P. Collman, X. M. Zhang, V. J. Lee, E. S. Uffelman, J. I. Brauman, Regioselective and enantioselective epoxidation catalyzed by metalloporphyrins, *Science* 261 (1993) 1404.
[531] F. G. Doro, J. R. L. Smith, A. G. Ferreira, M. D. Assis, Oxidation of alkanes and alkenes by iodosylbenzene and hydrogen peroxide catalysed by halogenated manganese porphyrins in homogeneous solution and covalently bound to silica, *J. Mol. Catal. A: Chem.* 164 (2000) 97.
[532] M. L. Merlau, M. D. P. Mejia, S. T. Nguyen, J. T. Hupp, Enhanced activity of manganese(III) porphyrin epoxidation catalysts through supramolecular complexation, *J. Mol. Catal. A: Chem.* 156 (2000) 79.
[533] T. G. Groves, W. Watanabe, The mechanism of olefin epoxidation by oxo-iron porphyrins. Direct observation of an intermediate, *J. Am. Chem. Soc.* 108 (1986) 507.
[534] T. G. Traylor, C. Kim, J. L. Richards, F. Xu, C. L. Perrin, Reactions of iron(III) porphyrins with oxidants. Structure-reactivity studies, *J. Am. Chem. Soc.* 117 (1995) 3468.
[535] Y. M. Goh, W. Nam, Significant electronic effect of porphyrin ligand on the reactivities of high-valent iron(IV) oxo porphyrin cation radical complexes, *Inorg. Chem.* 38 (1999) 914.

[536] J. T. Groves, W. J. Kruper Jr., R. C. Haushater, Hydrocarbon oxidations with oxometalloporphinates. Isolation and reactions of a (porphinato)manganese(V) complex, *J. Am. Chem. Soc.* 102 (1980) 6375.
[537] M. F. Powell, E. F. Pai, T. C. Bruice, Study of (tetraphenylporphinato)manganese(III)-catalyzed epoxidation and demethylation using p-cyano-N,N-dimethylaniline N-oxide as oxygen donor in a homogeneous system. Kinetics, radiochemical ligation studies, and reaction mechanism for a model of cytochrome P-450, *J. Am. Chem. Soc.* 106 (1984) 3277.
[538] P. Shannon, T. C. Bruice, A novel P-450 model system for the N-dealkylation reaction, *J. Am. Chem. Soc.* 103 (1981) 4580.
[539] E. Guilmet, B. Meunier, A new catalytic route for the epoxidation of styrene with sodium hypochlorite activated by transition metal complexes, *Tetrahedron Lett.* 21 (1980) 4449.
[540] J. T. Groves, W. J. Kruper Jr., Preparation and characterization of an oxoporphinatochromium(V) complex, *J. Am. Chem. Soc.* 101 (1979) 7613.
[541] J. T. Groves, R. C. Haushalter, M. Nakamura, T. E. Nemo, B. J. Evans, High-valent iron-porphyrin complexes related to peroxidase and cytochrome P-450, *J. Am. Chem. Soc.* 103 (1981) 2884.
[542] T. G. Traylor, W.-P. Fann, D. Bandyopadhyay, A common heterolytic mechanism for reactions of iodosobenzenes, peracids, hydroperoxides, and hydrogen peroxide with iron(III) porphyrins, *J. Am. Chem. Soc.* 111 (1989) 8009.
[543] P. Battioni, J. P. Renaud, J. F. Bartoli, M. Reina-Artiles, M. Fort, D. Mansuy, Monooxygenase-like oxidation of hydrocarbons by hydrogen peroxide catalyzed by manganese porphyrins and imidazole: selection of the best catalytic system and nature of the active oxygen species, *J. Am. Chem. Soc.* 110 (1988) 8462.
[544] B. Meunier, Potassium monopersulfate-Just another primary oxidant or a highly versatile oxygen atom donor in metalloporphyrin-mediated oxygenation and oxidation reactions, *New J. Chem.* 16 (1992) 203.
[545] J. T. Groves, Y. Watanabe, Oxygen activation by metalloporphyrins related to peroxidase and cytochrome P-450. Direct observation of the oxygen-oxygen bond cleavage step, *J. Am. Chem. Soc.* 108 (1986) 7834.
[546] J. T. Groves, Y. Watanabe, Reactive iron porphyrin derivatives related to the catalytic cycles of cytochrome P-450 and peroxidase. Studies of the mechanism of oxygen activation, *J. Am. Chem. Soc.* 110 (1988) 8443.
[547] T. G. Traylor, W. A. Lee, D. V. Stynes, Model compound studies related to peroxidases. Mechanisms of reactions of hemins with peracids, *J. Am. Chem. Soc.* 106 (1984) 755.
[548] W. A. Lee, T. C. Bruice, Homolytic and heterolytic oxygen-oxygen bond scissions accompanying oxygen transfer to iron(III) porphyrins by percarboxylic acids and hydroperoxides. A mechanistic criterion for peroxidase and cytochrome P-450, *J. Am. Chem. Soc.* 107 (1985) 513.
[549] T. G. Traylor, C. Kim, W.-P. Fann, C. L. Perrin, Reactions of hydroperoxides with iron(III) porphyrins: Heterolytic cleavage followed by hydroperoxide oxidation, *Tetrahedron* 54 (1998) 7977.
[550] D. Mansuy, J. F. Bartoli, J. C. Chottard, M. Lange, Metalloporphyrin-catalyzed hydroxylation of cyclohexane by alkyl hydroperoxides: pronounced efficiency of iron-porphyrins, *Angew. Chem. Int. Ed. Engl.* 19 (1980) 909.
[551] O. Almarsson, T. C. Bruice, A homolytic mechanism of O-O bond scission prevails in the reactions of alkyl hydroperoxides with an octacationic tetraphenylporphinato-iron(III) complex in aqueous solution, *J. Am. Chem. Soc.* 117 (1995) 4533.
[552] C. Walling, Fenton's reagent revisited, *Acc. Chem. Res.* 8 (1975) 125.
[553] R. Panicucci, T. C. Bruice, Dynamics of the reaction of hydrogen peroxide with a water soluble non.mu.-oxo dimer forming iron(III) tetraphenylporphyrin. 2. The reaction of hydrogen peroxide with 5,10,15,20-tetrakis(2,6-dichloro-3-sulfonatophenyl)porphinato iron(III) in aqueous solution, *J. Am. Chem. Soc.* 112 (1990) 6063.
[554] T. G. Traylor, F. Xu, Mechanisms of reactions of iron(III) porphyrins with hydrogen peroxide and hydroperoxides: Solvent and solvent isotope effects, *J. Am. Chem. Soc.* 112 (1990) 178.
[555] W. Nam, H. J. Han, S.-Y. Oh, Y. J. Lee, M.-H. Choi, S.-Y. Han, C. Kim, S. K. Woo, W. Shin, New insights into the mechanisms of O-O bond cleavage of hydrogen peroxide and *tert*-alkyl hydroperoxides by iron(III) porphyrin complexes, *J. Am. Chem. Soc.* 122 (2000) 8677.

[556] J. P. Collman, J. I. Brauman, B. Meunier, T. Hayashi, T. Kodadek, S. A. Raybuck, Epoxidation of olefins by cytochrome P-450 model compounds: Kinetics and stereochemistry of oxygen atom transfer and origin of shape selectivity, *J. Am. Chem. Soc.* 107 (1985) 2000.
[557] L. Michaelis, M. L. Menten, Die kinetic der invertinwirkung, *Biochem. Z.* 49 (1913) 333.
[558] J. S. Lindsey, I. C. Schreiman, H. S. Hsu, P. C. Kearney, A. M. Marguerettaz, Rothemund and Adler-Longo reactions revisited: Synthesis of tetraphenylporphyrins under equilibrium conditions, *J. Org. Chem.* 52 (1987) 827.
[559] D. J. Liston, B. O. West, Oxochromium compounds. 2. Reaction of oxygen with chromium(II) and chromium(III) porphyrins and synthesis of a μ-oxo chromium porphyrin derivative, *Inorg. Chem.* 24 (1985) 1568.
[560] T. G. Traylor, A. R. Miksztal, Mechanisms of hemin-catalyzed epoxidations: Electron transfer from alkenes, *J. Am. Chem. Soc.* 109 (1987) 2770.
[561] A. J. Castellino, T. C. Bruice, Intermediates in the epoxidation of alkenes by cytochrome P-450 models. 1. *cis*-Stilbene as a mechanistic probe, *J. Am. Chem. Soc.* 110 (1998) 158.
[562] A. J. Castellino, T. C. Bruice, Radical intermediates in the epoxidation of alkenes by cytochrome P-450 model systems. The design of a hypersensitive radical probe, *J. Am. Chem. Soc.* 110 (1988) 1313.
[563] M. C. Curet-Arana, G. A. Emberger, L. J. Broadbelt, R. Q. Snurr, Quantum chemical determination of stable intermedicates for alkene epoxidation with Mn-prophyrin catalysts, *J. Molec. Catal. A*, In press.
[564] J. P. Collman, T. Kodadek, S. A. Raybuck, B. Meunier, Oxygenation of hydrocarbons by cytochrome P-450 model compounds: Modification of reactivity by axial ligands, *Proc. Natl. Acad. Sci. USA* 80 (1983) 7039.
[565] K. B. Sharpless, A. Y. Teranishi, J.-E. Backvall, Chromyl chloride oxidations of olefins. Possible role of organometallic intermediates in the oxidations of olefins by oxo transition metal species, *J. Am. Chem. Soc.* 99 (1977) 3120.
[566] D. Ostovic, T. C. Bruice, Intermediates in the epoxidation of alkenes by cytochrome P-450 models. 5. Epoxidation of alkenes catalyzed by a sterically hindered (meso-tetrakis(2,6-dibromophenyl) porphinato)iron(III) chloride, *J. Am. Chem. Soc.* 111 (1989) 6511.
[567] T. G. Traylor, A. R. Miksztal, Alkene epoxidations catalyzed by iron(III), manganese(III), and chromium(III) porphyrins. Effects of metal and porphyrin substituents on selectivity and regiochemistry of epoxidation, *J. Am. Chem. Soc.* 11 (1989) 7443.
[568] D. Rutkowska-Zbik, R. Tokarz-Sobieraj, M. Witko, Quantum chemical description of oxygen activation process on Co, Mn, and Mo porphyrins, *J. Chem. Theory Comput.* 3 (2007) 914.
[569] R. Zhang, M. Newcomb, Laser flash photolysis formation and direct kinetic studies of manganese (V)-oxo porphyrin intermediates, *J. Am. Chem. Soc.* 125 (2003) 12418.
[570] G. Yin, M. Buchalova, A. M. Danby, C. M. Perkins, D. Kitko, J. D. Carter, W. M. Scheper, D. H. Busch, Olefin epoxidation by the hydrogen peroxide adduct of a novel non-heme manganese(IV) complex: Demonstration of oxygen transfer by multiple mechanisms, *Inorg. Chem.* 45 (2006) 3467.
[571] R. D. Bach, M. D. Su, J. L. Andres, H. B. Schlegel, Structure and reactivity of diamidoiron(III) hydroperoxide. The mechanism of oxygen-atom transfer to ammonia, *J. Am. Chem. Soc.* 115 (1993) 8763.
[572] J. W. Sam, X. J. Tang, J. Peisach, Electrospray mass spectrometry of iron bleomycin: Demonstration that activated Bleomycin is a ferric peroxide complex, *J. Am. Chem. Soc.* 116 (1994) 5250.
[573] R. Y. N. Ho, G. Roelfes, B. L. Feringa, L. Que Jr., Raman evidence for a weakened O-O bond in mononuclear low-spin iron(III)-hydroperoxides, *J. Am. Chem. Soc.* 121 (1999) 264.
[574] P. Wadhwani, M. Mukherjee, D. Bandyopadhyay, The prime reactive intermediate in the iron(III) porphyrin complex catalyzed oxidation reactions by *tert*-butyl hydroperoxide, *J. Am. Chem. Soc.* 123 (2001) 12430.
[575] C. Kim, K. Chen, J. Kim, L. Que Jr., Stereospecific alkane hydroxylation with H_2O_2 catalyzed by an iron(II)-tris(2-pyridylmethyl)amine complex, *J. Am. Chem. Soc.* 119 (1997) 5964.
[576] M. Newcomb, D. Aebisher, R. Shen, R. E. P. Chandrasena, P. F. Hollenberg, M. J. Coon, Kinetic isotope effects implicate two electrophilic oxidants in cytochrome P450-catalyzed hydroxylations, *J. Am. Chem. Soc.* 125 (2003) 6064.

[577] W. Nam, M. H. Lim, H. J. Lee, C. Kim, Evidence for the participation of two distinct reaction intermediates in iron (III) porphyrin complex-catalyzed epoxidation reactions, *J. Am. Chem. Soc.* 122 (2000) 6641.

[578] W. Nam, M. H. Lim, S. K. Moon, C. Kim, Participation of two distinct hydroxylating intermediates in iron(III) porphyrin complex-catalyzed hydroxylation of alkanes, *J. Am. Chem. Soc.* 122 (2000) 10805.

[579] N. A. Stephenson, A. T. Bell, A study of the mechanism and kinetics of cyclooctene epoxidation catalyzed by iron(III) tetrakispentafluorophenyl porphyrin, *J. Am. Chem. Soc.* 127 (2005) 8635. (2). (3). (4).

[580] N. A. Stephenson, A. T. Bell, Influence of solvent composition on the kinetics of cyclooctene epoxidation by hydrogen peroxide catalyzed by iron(III) [tetrakis(pentafluorophenyl)] porphyrin chloride [(F_{20}TPP)FeCl], *Inorg. Chem.* 45 (2006) 2758.

[581] N. A. Stephenson, A. T. Bell, Effects of methanol on the thermodynamics of iron(III) [tetrakis (pentafluorophenyl)]porphyrin chloride dissociation and the creation of catalytically active species for the epoxidation of cyclooctene, *Inorg. Chem.* 45 (2006) 5591.

[582] N. A. Stephenson, A. T. Bell, The influence of substrate composition on the kinetics of olefin epoxidation by hydrogen peroxide catalyzed by iron(III) [tetrakis(pentafluorophenyl)] porphyrin, *J. Mol. Catal. A.* 258 (2006) 231.

[583] W. Nam, M. H. Lim, S.-Y. Oh, J. H. Lee, S. K. Woo, C. Kim, W. Shin, Remarkable anionic axial ligand effects of iron(III) porphyrin complexes on the catalytic oxygenations of hydrocarbons by H_2O_2 and the formation of oxoiron(IV) porphyrin intermediates by *m*-chloroperoxybenzoic acid, *Angew. Chem. Int. Ed. Engl.* 39 (2000) 3646.

[584] H. Schiff, Aldehyd derivative einiger amide, *Annal. Chemie Pharmacie (Liebigs Annal. Chem.)* 148 (1868) 330. Also *Annal. Chemie Phar.* Suppl. 3 (1864) 343.

[585] K. Srinivasan, P. Michaud, J. K. Kochi, Epoxidation of olefins with cationic (salen)manganese(III) complexes. The modulation of catalytic activity by substituents, *J. Am. Chem. Soc.* 108 (1986) 2309.

[586] M. Palucki, N. S. Finney, P. J. Pospisil, M. L. Güler, T. Ishida, E. N. Jacobsen, The mechanistic basis for electronic effects on enantioselectivity in the (salen)Mn(III)-catalyzed epoxidation reaction, *J. Am. Chem. Soc.* 120 (1998) 948.

[587] T. Katsuki, Catalytic asymmetric oxidations using optically active (salen)manganese(III) complexes as catalysts, *Coord. Chem. Rev.* 140 (1995) 189.

[588] T. Katsuki, Mn-salen catalyst, competitor of enzymes, for asymmetric epoxidation, *J. Mol. Catal. A: Chem.* 113 (1996) 87.

[589] Y. N. Ito, T. Katsuki, Asymmetric catalysis of new generation chiral metallosalen complexes, *Bull. Chem. Soc. Jpn.* 72 (1999) 603.

[590] C. T. Dalton, K. M. Ryan, V. M. Wall, C. Bousquet, D. G. Gilheany, Recent progress towards the understanding of metal–salen catalysed asymmetric alkene epoxidation, *Top. Catal.* 5 (1998) 75.

[591] T. Linker, The Jacobsen-Katsuki epoxidation and its controversial mechanism, *Angew. Chem. Int. Ed. Engl.* 36 (1997) 2060.

[592] D. Feichtinger, D. A. Plattner, Direct proof for O=Mn^V(salen) complexes, *Angew. Chem. Int. Ed. Engl.* 36 (1997) 1718.

[593] D. Feichtinger, D. A. Plattner, Oxygen transfer to manganese–salen complexes: An electrospray tandem mass spectrometric study, *J. Chem. Soc. Perkin Trans.* 2 (2000) 1023.

[594] D. Feichtinger, D. A. Plattner, Probing the reactivity of oxomanganese-salen complexes: An electrospray tandem mass spectrometric study of highly reactive intermediates, *Chem. Eur. J.* 7 (2001) 591.

[595] E. M. McGarrigle, D. G. Gilheany, Chromium- and manganese-salen promoted epoxidation of alkenes, *Chem. Rev.* 105 (2005) 1563.

[596] N. J. Kerrigan, H. Muller-Bunz, D. G. Gilheany, Salen ligands derived from *trans*-1,2-dimethyl-1,2-cyclohexanediamine: preparation and application in oxo-chromium salen mediated asymmetric epoxidation of alkenes, *J. Mol. Catal. A: Chem.* 227 (2005) 163.

[597] A. Scheurer, H. Maid, F. Hampel, R. W. Saalfrank, L. Toupet, P. Mosset, R. Puchta, N. J. R. van Eikema Hommes, Influence of the conformation of salen complexes on the stereochemistry of the asymmetric epoxidation of olefins, *Eur. J. Org. Chem.* (2005) 2566.

[598] T. P. Yoon, E. N. Jacobsen, Privileged chiral catalysts, *Science* 299 (2003) 1691.

[599] M. J. Patel, B. M. Trivedi, Synthesis and catalytic activity of binuclear Mn(III,III)-BINOL complexes for epoxidation of olefins, *Appl. Organomet. Chem.* 20 (2006) 521.
[600] J. P. Collman, L. Zeng, J. I. Brauman, Donor ligand effect on the nature of the oxygenating species in Mn^{III}(salen)-catalyzed epoxidation of olefins: Experimental evidence for multiple active oxidants, *Inorg. Chem.* 43 (2004) 2672.
[601] S.-E. Park, W. J. Song, Y. O. Ryu, M. H. Lim, R. Song, K. M. Kim, W. Nam, Parallel mechanistic studies on the counterion effect of manganese salen and porphyrin complexes on olefin epoxidation by iodosylarenes, *J. Inorg. Biochem.* 99 (2005) 424.
[602] W. Adam, K. J. Roschmann, C. R. Saha-Möller, A novel counterion effect on the diastereoselectivity in the Mn^{III}(salen)-catalyzed epoxidation of phenyl-substituted *cis*-alkenes, *Eur. J. Org. Chem.* (2000) 3519.
[603] D. Schröder, S. Shaik, H. Schwarz, Two-state reactivity as a new concept in organometallic chemistry, *Acc. Chem. Res.* 33 (2000) 139.
[604] Y. G. Abashkin, S. K. Burt, (Salen)Mn-catalyzed epoxidation of alkenes: A two-zone process with different spin-state channels as suggested by DFT study, *Org. Lett.* 6 (2004) 59.
[605] L. Cavallo, H. Jacobsen, Manganese-salen complexes as oxygen-transfer agents in catalytic epoxidations—A density functional study of mechanistic aspects, *Eur. J. Inorg. Chem.* (2003) 892.
[606] C. Linde, M. Arnold, B. Åkermark, P.-O. Norrby, Is there a radical intermediate in the (salen) Mn-catalyzed epoxidation of alkenes?, *Angew. Chem. Int. Ed. Engl.* 36 (1997) 1723.
[607] K. A. Jørgensen, B. Schiøtt, Metallaoxetanes as intermediate in oxygen-transfer reactions—reality or fiction?, *Chem. Rev.* 90 (1990) 1483.
[608] S. Fritzche, P. Lönnecke, T. Höcher, E. H. Hawkins, Soluble monometallic salen complexes derived from O-functionalized salicylaldehydes as metalloligands for synthesis of heterobimetallic complexes, *Z. Anorg. Allg. Chem.* 632 (2006) 2256.
[609] H. Wang, B. Mandimutsira, R. C. Todd, B. Ramdhanie, J. P. Fox, D. P. Goldberg, Catalytic sulfoxidation and epoxidation with a Mn(III) triazacorrole: Evidence for a "third oxidant" in high-valent porphyrinoid oxidations, *J. Am. Chem. Soc.* 126 (2004) 18.
[610] W. Nam, S. J. Baek, K. I. Liao, J. S. Valentine, Epoxidation of olefins by iodosylbenzene catalyzed by non-porphyrin metal complexes, *Bull. Kor. Chem. Soc.* 15 (1994) 1112.
[611] D. C. Haines, D. R. Tomchick, M. Machius, J. A. Peterson, Pivotal role of water in the mechanism of P450BM-3, *Biochemistry* 40 (2001) 13456.
[612] K. G. Ravichandran, S. S. Boddupalli, C. A. Hasermann, J. A. Peterson, J. Deisenhofer, Crystal structure of hemoprotein domain of P450BM-3, a prototype for microsomal P450's, *Science* 261 (1993) 731.
[613] H. Li, T. L. Poulos, The structure of the cytochrome P450BM-3 haem domain complexed with the fatty acid substrate, palmitoleic acid, *Nat. Struct. Biol.* 4 (1997) 140.
[614] T. L. Poulos, The role of the proximal ligand in heme enzymes, *J. Biol. Inorg. Chem.* 1 (1996) 356.
[615] D. B. Goodin, When an amide is more like histidine than imidazole: The role of axial ligands in heme catalysis, *J. Biol. Inorg. Chem.* 1 (1996) 360.
[616] I. M. C. M. Rietjens, A. M. Osman, C. Veeger, O. Zakharieva, J. Antony, M. Grodzicki, A. X. Trautwein, On the role of the axial ligand in heme-based catalysis of the peroxidase and P450 type, *J. Biol. Inorg. Chem.* 1 (1996) 372.
[617] S. Nagano, J. R. Cupp-Vickery, T. L. Poulos, Crystal structures of the ferrous dioxygen complex of wild-type cytochrome P450eryF and its mutants, A245S and A245T: Investigation of the proton transfer system in P450eryF, *J. Biol. Chem.* 280 (2005) 22102.
[618] V. Y. Kuznetsov, E. Blair, P. J. Farmer, T. L. Poulos, A. Pifferitti, I. F. Sevrioukova, The Putidaredoxin reductase-putidaredoxin electron transfer complex: Theoretical and experimental studies, *J. Biol. Chem.* 280 (2005) 16135.

SECTION 2
Homogeneous Catalysis

CHAPTER 2

Unprecedented Selectivity in the H_2O_2 Epoxidation of Simple Alkenes Imparted by Soft Pt(II) Lewis Acid Catalysts

Giorgio Strukul and **Alessandro Scarso**

Contents		
	1. Introduction	104
	2. Catalyst Synthesis and Lewis Acid Properties	105
	3. General Epoxidation Activity	106
	4. Regioselectivity	107
	5. Diastereoselectivity	108
	6. Enantioselectivity	109
	7. Reactions in Micellar Media	110
	8. Reaction Mechanism	113
	9. Conclusions	113
	References	115

Abstract The use of a class of pentafluorophenyl Pt(II) complexes as catalysts allows the efficient epoxidation of simple terminal alkenes with environmentally benign hydrogen peroxide as the oxidant. Key features of this system are very high substrate selectivity, regioselectivity, and enantioselectivity, at least for this class of substrates. These properties are related to the soft Lewis acid character of the metal center that makes it relatively insensitive to water but, at the same time, capable of increasing the electrophilicity of the substrate by coordination. The reversal of the traditional electrophile/nucleophile roles in epoxidation helps explain the unprecedented reactivity observed.

Key Words: Platinum complexes, Epoxidation, Pentafluorophenyl, Hydrogen peroxide, Terminal alkenes, Regioselectivity, Enantioselectivity.   © 2008 Elsevier B.V.

Dipartimento di Chimica, Università Ca' Foscari di Venezia, 30123 Venice, Italy

Mechanisms in Homogeneous and Heterogeneous Epoxidation Catalysis   © 2008 Elsevier B.V.
DOI: 10.1016/B978-0-444-53188-9.00002-X   All rights reserved.

1. INTRODUCTION

The oxidation of alkenes to the corresponding epoxides is a well-documented reaction that has been investigated for decades because epoxides are important commodity products and, at the same time, pivotal building blocks for organic synthesis, both from industrial and academic standpoints [1]. Although heterogeneous methods for the epoxidation of alkenes have been developed [2,3], the highest selectivities have been observed under homogeneous conditions [4] with metal-containing or purely organic catalysts.

Several oxidants have been tested over the years for epoxidation; nevertheless, in recent years, hydrogen peroxide has emerged as the oxidant of choice for many transformations because it is environmentally benign, having high atom efficiency [5], and because it can be handled and stored safely, and because it produces water as the only by-product [6,7].

The number of transition metal complexes able to efficiently activate hydrogen peroxide toward different alkenes is relatively large [4]. Nevertheless, most of them are generally active toward a limited class of substrates such as allylic alcohols, where the presence of the hydroxyl group allows easy coordination to the metal active site [8], or unfunctionalized alkenes, where good performance could be observed only for electron-rich C=C double bonds or styrene derivatives where peculiar reactivity is imparted by the presence of the conjugated aromatic ring. In this respect, Ti-silicalites are an exception [9] because terminal alkenes can be epoxidized with high yields and selectivities. However, this is due more to their peculiar reactivity rather than to the intrinsic electronic properties of the Ti centers dispersed within the zeolite matrix. In this framework, a lack of methods is evident for the efficient and selective epoxidation of terminal, unfunctionalized alkenes that are intrinsically poorly reactive substrates toward traditional electrophilic oxidation. In this respect, worthy of mention are complexes of Ru(III) [10], W(VI) [11–14], Mn(II) [15–19], Re(V) [20,21], and Fe(III) [22,23], although relatively high metal loading, presence of additives, moderately high temperatures, and, in many cases, use of overstoichiometric amounts of hydrogen peroxide are generally required because of parallel partial decomposition of the oxidant catalyzed by the complex itself. Very recently, bis-(μ-hydroxo)-bridged di-vanadium species contained in a peroxotungstate framework have been reported to be highly selective and efficient for H_2O_2 epoxidation of alkenes, in particular toward terminal alkenes with very good regioselectivity in diene epoxidation [24–26].

So far, catalyst design has aimed mainly at oxidant activation and little attention has been paid to the interaction between the metal center and the alkene. A requirement for a successful epoxidation system of wide scope for simple terminal alkenes would seem to be a new catalyst design focusing on activation of the substrate instead of the oxidant. This suggests noble metals as applicable catalytic centers, because of their affinity for terminal alkenes versus internal ones. This results in a change of role for the catalyst from electrophile to nucleophile in the system.

Very recently, we reported the synthesis and peculiar activity of second-generation [27], electron-poor Pt(II) complexes containing a pentafluorophenyl residue with general formula $[(P-P)Pt(C_6F_5)(H_2O)]^+$ (P-P = diphosphine) and their application as selective epoxidation catalysts toward terminal unfunctionalized alkenes (for first-generation epoxidation catalysts with general formula [(PP)Pt(CF$_3$)(H$_2$O)]$^+$, see refs. [28–32]). These complexes produced with only 2 mol% catalyst loading, yields up to 89%, at mild conditions, and with the use of only one equivalent of hydrogen peroxide [27].

Herein we provide insight into their extremely high substrate selectivity and explore in detail aspects like regioselectivity, diasteroselectivity, and enantioselectivity in the oxidation of a wide range of substrates. Further details into the selectivity of this catalyst are provided by a mechanistic investigation performed employing the so-called "reaction progress kinetic analysis" approach recently developed by Blackmond [33].

2. CATALYST SYNTHESIS AND LEWIS ACID PROPERTIES

New complexes of the type [(P-P)Pt(C$_6$F$_5$)(H$_2$O)]$^+$ (P-P = diphosphine) were synthesized according to the route indicated in Fig. 2.1 and were characterized by elemental analysis, multinuclear ^{1}H, ^{31}P{^{1}H}, and ^{19}F{^{1}H} nuclear magnetic resonance (NMR) spectroscopy [27]. The synthetic pathway is very flexible, allowing the preparation of homologous complexes with a wide variety of diphosphine ligands. These are all commercially available except **2g** which was prepared following a procedure reported in the literature [34].

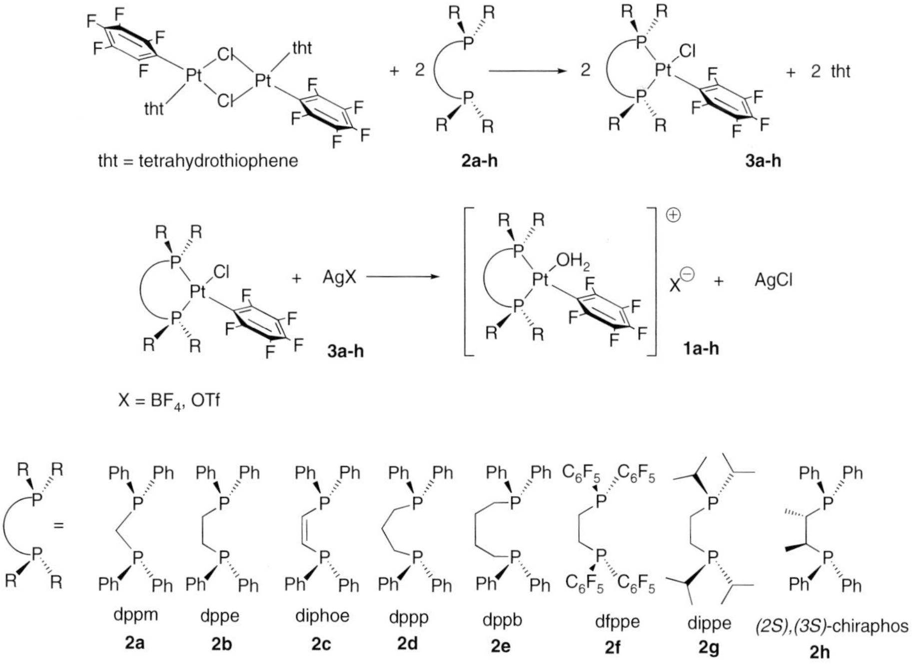

FIGURE 2.1 Synthesis of the Pt(II) monocationic catalysts **1a–h** bearing a perfluorophenyl residue.

The Lewis acid character of metal complexes is a key issue in the activation of oxidants for catalytic oxygen transfer reactions [35–37], and in our studies on the Baeyer–Villiger oxidation of ketones [38,39], we observed several times that high activity correlates well with high Lewis acidity of metal catalysts. Evidence of the Lewis acid character of complexes **1a–h** as the result of the electron withdrawing ability of the pentafluorophenyl ligand and the concomitant effect of the diphosphine ligands was assessed using 2,6-dimethyl-phenyl isocyanide as a molecular probe. In fact, the value of the wavenumber shift ($\Delta v = v(C\equiv N)_{coord} - v(C\equiv N)_{free}$) for the $C\equiv N$ stretching of 2,6-dimethyl-phenyl isocyanide provides valuable information about the electrophilicity of the isocyanide carbon atom which is known to correlate well with the Lewis acidity of the metal complex [40,41]. Substitution of the coordinating water molecule with the isocyanide moiety in complexes **1a–h** provided a series of homologous complexes for which a comparison of the $C\equiv N$ stretching frequencies between the free ligand and the coordinated ligand (Δv) is reported in Table 2.1.

A correlation between the acidity (Table 2.1) and the catalytic activity of complexes in the epoxidation of 1-octene, taken as reference reaction, indicated that the maximum activity was obtained with complexes such as **1b, 1c,** and **1h** characterized by an intermediate acidity [27].

3. GENERAL EPOXIDATION ACTIVITY

The scope of the reaction was thoroughly investigated with catalyst **1b** exploring the reactivity toward different substrates bearing alkyl substitution as well as various functional groups on the alkyl chain. Experimental conditions adopted were 2 mol% catalyst, substrate/H_2O_2 = 1/1, solvent dichloroethane (DCE) at room temperature (RT). We found that disubstituted alkenes (e.g., cyclohexene, methylene cyclohexane) as well as styrene are not suitable substrates, but terminal double bonds can be efficiently epoxidized [27]. As shown in Table 2.2, **1b** shows high activity toward monofunctionalized linear terminal alkenes with a slight decrease in productivity with increase in the length of the alkyl chain (entries 2–4).

TABLE 2.1 Δv of 2,6-dimethyl-phenyl Isocyanide Pt(II) derivatives as a function of the different ligands **2a–h**

Ligand	Δv (cm^{-1})
2a	82.72
2b	78.91
2c	80.95
2d	78.20
2e	77.10
2f	93.22
2g	71.68
2h	78.67

TABLE 2.2 Catalytic epoxidation of various alkenes with hydrogen peroxide mediated by **1b**

Entry	Substrate	Time (h)	Yield (%)[a]
1	(alkene)	20	78[b]
2	(alkene)	3.5	96
3	(alkene)	4	81
4	(alkene)	4	81
5	(alkene)	24	0
6	(alkene)	4	82
7	(alkene)	24	4
8	(alkene)	6	55
9	(alkene)	6	38

Experimental conditions: substrate 0.83 mmol, H_2O_2 0.83 mmol, [**1b**] 2 mol%, solvent 1 ml dichloroethane (DCE) at room temperature (RT).
[a] Yield (conversion × selectivity) determined by GC analysis.
[b] Reaction performed at 0 °C.

Methyl substitution in the alkyl chain of the substrate resulted in a decrease of activity, the extent of which was strongly dependent on the methyl position. It clearly appears that the present catalytic system is specific for terminal alkenes and is very sensitive to the steric properties of the substrate.

Suitable alkenes were also allyl benzene derivatives, which can be considered as β-substituted terminal double bonds (entries 8–9). Substitution with methoxy residues decreased the yield of the epoxide probably because of increased steric bulkiness or competition of the oxygen donor with the C=C double bond for coordination at Pt. At the same time, the system did not withstand the presence of coordinating heteroatoms in the side chain; in fact, 3-butenol, 3-cyano-propene, allyl chloride, 5-hexenoic acid, and allyl imidazole were all nonreactive substrates.

4. REGIOSELECTIVITY

The high selectivity of catalyst **1b** toward terminal unfunctionalized alkenes is highlighted by the experiments reported in Fig. 2.2. As a typical example, *cis*-1,4-hexadiene bears both terminal and *cis* C=C bonds and in the presence of a stoichiometric amount of *m*-chloroperbenzoic acid (*m*-CPBA) leads mainly to *cis*-4,5-epoxy-1-hexene due to the electrophilic epoxidation of the more electron-rich internal double bond. On the contrary, when the epoxidation is performed with catalyst **1b** and one equivalent of H_2O_2, the regioselectivity of the reaction is completely inverted, favoring the product with the terminal oxirane ring. The same applies to *trans*-1,4-hexadiene or dienes bearing substituents in the

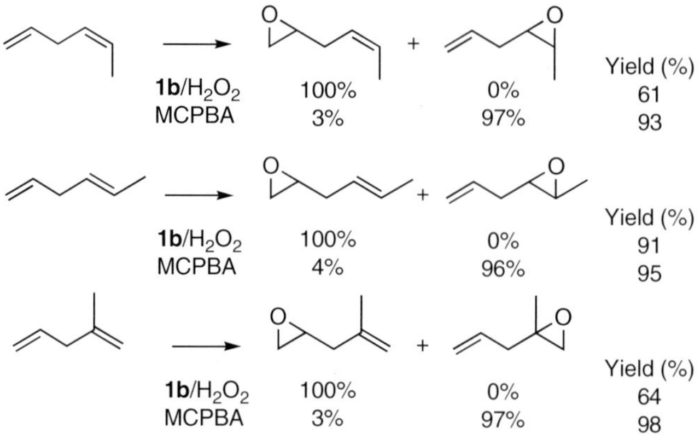

FIGURE 2.2 Regioselectivity in the epoxidation of dienes. Reactions performed at room temperature (RT) with either 2 mol% catalyst **1b** in 24 h or stoichiometric amount of m-chloroperbenzoic acid (m-CPBA) in 0.5 h.

2 position. The complete regioselectivity toward unsubstituted terminal olefins observed with the Pt system is, to the best of our knowledge, for some substrates comparable and for others better than those observed with the best catalyst reported so far, that is, V(III)-containing polyoxometalates [23–26]. Similarly to the latter system, the exceptional regioselectivity observed with **1b** supports the existence of stringent steric requirements, although the high substrate recognition ability suggests that electronic effects should also be carefully analyzed.

5. DIASTEREOSELECTIVITY

Spurred by the high selectivity showed by catalyst **1b**, we investigated also the diastereoselectivity (d.e.) of this complex toward the epoxidation of a racemic chiral terminal alkene such as 4-methylhexene. This substrate differs from 4-methylpentene only in having a longer alkyl chain; however, the former reacts more slowly (25% yield after 24 h vs 59% yield in 3 h at RT) compared to the shorter analog, with 25% d.e. in favor of the epoxide product with the oxirane ring *anti* to the methyl group in the β position. The d.e. observed at RT for 4-methylhexene epoxidation (Fig. 2.3) increases up to 32% with 21% yield when the reaction was performed at 0 °C. The d.e. observed is low, but this is no surprise because of the small steric difference between methyl and ethyl groups in the chiral alkene substrate, as well as the absence of functional groups that are often responsible for substrate orientation. A well-known example of such behavior is the oxidation of chiral allylic alcohols that often show a high degree of d.e. with a product distribution that is dependent on the catalyst/oxidant combination (some examples are given in refs. [42–48]).

In isolated terminal dienes, the epoxidation of the first C=C double bond creates a stereo center (racemic monoepoxide) and the subsequent epoxidation of the remaining alkene moiety occurs in a diastereoselective fashion. In Fig. 2.4

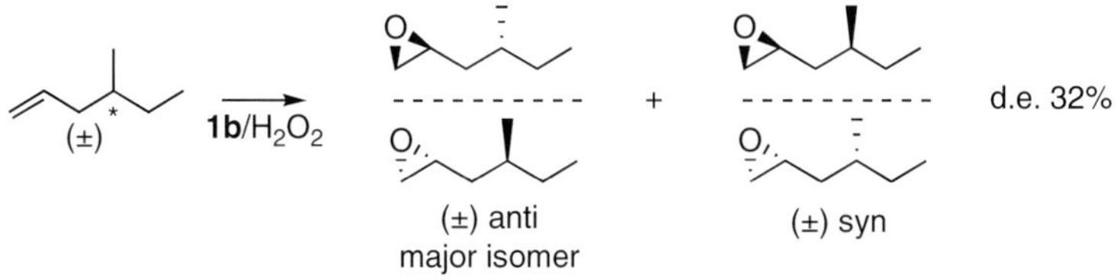

FIGURE 2.3 Diastereoselective epoxidation of a chiral terminal alkene with hydrogen peroxide catalyzed by **1b** (2 mol%).

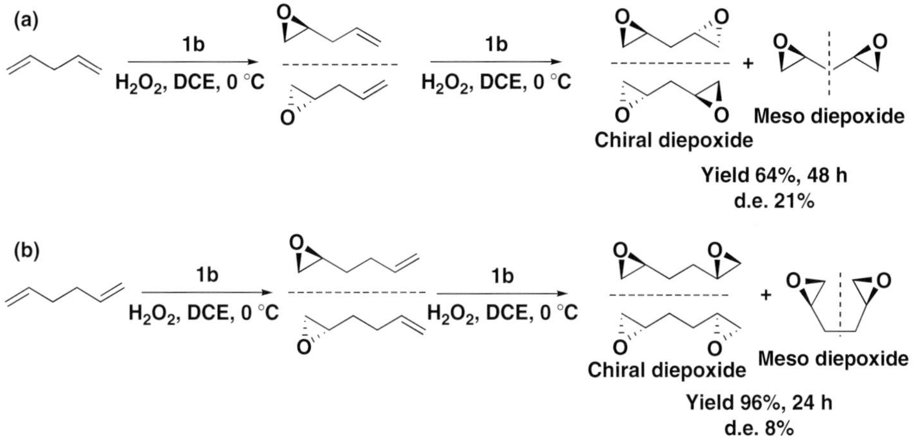

FIGURE 2.4 Diastereoselective oxidation of isolated terminal dienes (a) 1,4-pentadiene and (b) 1,5-hexadiene with excess of hydrogen peroxide catalyzed by **1b** (2 mol%).

are shown the results of the diastereoselective epoxidation of 1,4-pentadiene and 1,5-hexadiene with catalyst **1b** and an excess of hydrogen peroxide.

At 0 °C, the reaction with 1,4-pentadiene led to lower yield in diepoxides but with a higher d.e. compared to the reaction performed with the longer substrate for which higher yields but lower d.e. were observed. Both effects can be probably ascribed to the strong steric sensitivity characteristic of catalyst **1b**. In 1,5-hexadiene, the two double bonds are remote from each other and they react almost independently, behaving similarly to isolated double bonds with high yield in diepoxide and low mutual sensing as confirmed by the low d.e. On the contrary, with the smaller 1,4-pentadiene, the two double bonds are closer and this decreases the yield in diepoxide while increasing the d.e.

6. ENANTIOSELECTIVITY

The general synthetic scheme outlined in Fig. 2.1 allows the facile preparation of a series of chiral catalysts using commercial chiral diphosphines as ligands. The use of complexes of the type [(P-P)Pt(C_6F_5)(H_2O)]OTf represents the most versatile, active, and enantioselective system so far developed for the asymmetric epoxidation of terminal unfunctionalized alkenes (Fig. 2.5 and Table 2.3) [49].

Catalyst **1h** proved to be the best for the enantioselective epoxidation of terminal alkenes. Table 2.4 reports some representative data. Excellent enantioselectivities can be observed in many cases. Dienes were also investigated (Table 2.5). In this case, the epoxidation occurred exclusively at the terminal double bond with complete regioselectivity and ee up to 98%. To the best of our knowledge, any other chiral metal catalyst reported in the literature would lead to the electrophilic asymmetric epoxidation of the more electron-rich double bond [50,51].

7. REACTIONS IN MICELLAR MEDIA

The enantioselective epoxidation reported above was also studied in water/surfactant (micellar) media in order to find a greener version that avoided the use of chlorinated solvents [52]. Initially a surfactant screening was performed testing anionic, cationic, and nonionic surfactants with best performance being observed with surfactants related to the Triton family (Table 2.6). It was observed that the right balance of polarity, presence, and size of functional groups in the micellar aggregate are all critical parameters to ensure good activity and selectivity.

The use of biphasic catalysis opens the way for the possible recycling of the chiral catalyst by extracting the aqueous phase with a solvent in which **1h** and the surfactant are both insoluble. This is an important issue when soluble catalysts

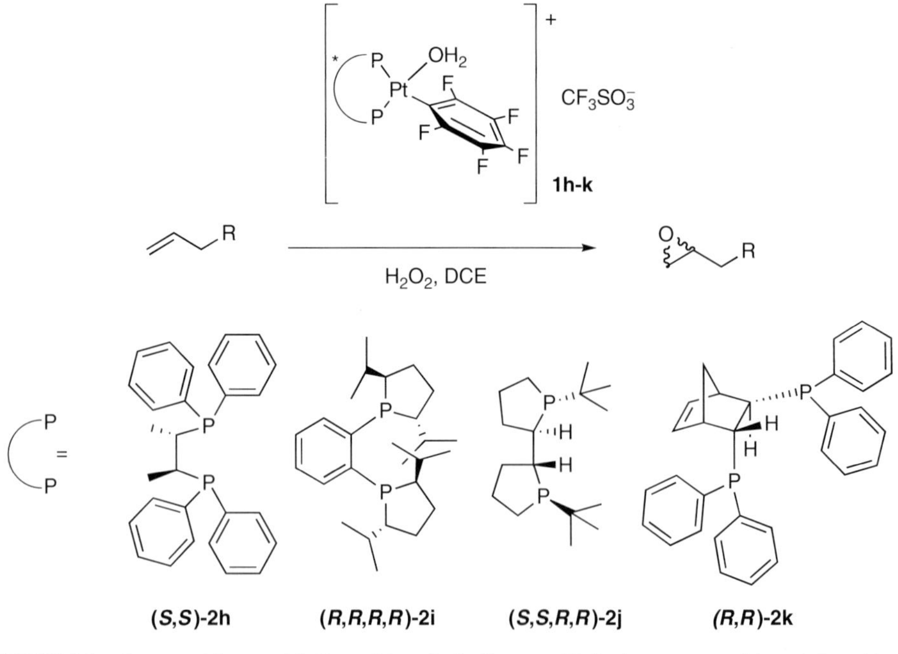

FIGURE 2.5 Asymmetric epoxidation of terminal alkenes with hydrogen peroxide catalyzed by Pt(II) chiral complexes **1h–k**.

TABLE 2.3 Catalytic asymmetric epoxidation of 4-methyl-1-pentene with hydrogen peroxide mediated by chiral Pt(II) catalysts **1h–k**

Entry	Catalyst	T (°C)	Time (h)	Yield (%)	ee (%)	Abs. config.
1	(S,S)-**1h**	20	4	56	58	(R)
2	(S,S)-**1h**	−10	48	60	75	(R)
3	(R,R,R,R)-**1i**	−10	48	5	49	(S)
4	(S,S,R,R)-**1j**	−10	48	4	36	(S)
5	(R,R)-**1k**	20	6	58	59	(S)
6	(R,R)-**1k**	−10	48	71	72	(S)

Experimental conditions as in Table 2.2.

TABLE 2.4 Catalytic asymmetric epoxidation of terminal alkenes with hydrogen peroxide mediated by **1h**

Entry	Substrate	T (°C)	Time (h)	Yield (%)	ee (%)
1		−25	24	78[a]	64
2		−10	48	63	78
3		−10	20	77	68
4		−10	20	48	83
5		−10	48	88	79
6		−10	48	81	71
7		20	24	75	66
8		−10	48	45	82
9		−10	48	27	84
10		−10	24	64	87

Experimental conditions as in Table 2.2.
[a] Yield determined by ^{1}H NMR integration.

are used, because in homogeneous catalysis catalyst separation has always been the major hurdle preventing widespread applications. We found that operating with Triton-X114 in a 2:1 substrate/oxidant excess, extracting the water phase with hexane and simply adding fresh substrate and 35% H_2O_2 for the next runs,

TABLE 2.5 Catalytic asymmetric epoxidation of dienes with hydrogen peroxide mediated by **1h**

Entry	Substrate	T (°C)	Time (h)	Yield (%)	Terminal epoxidation/ internal epoxidation	ee (%)
1		−10	48	93	100/0	63
2		−10	48	96	100/0	86
3		−10	48	66	100/0[a]	98

Experimental conditions as in Table 2.2.
[a] Monosubstituted/disubstituted epoxide.

TABLE 2.6 Catalytic enantioselective epoxidation of terminal alkenes with H_2O_2 mediated by **1h** in micellar media

Entry	Substrate	Time (h)	Yield (%)	ee. (%)	Solvent/additive
1		4	56	58	DCE
2		24	0	–	H_2O
3		24	0	–	H_2O/SDS[a]
4		24	0	–	H_2O/CTABr[b]
5		6	28	84	H_2O/Zwitterionic[c]
6		6	41	84	H_2O/POA[d]
7		6	61	82	H_2O/Triton-X114[e]
8		6	51	82	H_2O/Triton-X100[f]
9		6	84	74	H_2O/Triton-X100[f]
10		6	78	57	H_2O/Triton-X100[f]
11		6	81	61	H_2O/Triton-X100[f]
12		20	23	74[d]	H_2O/Triton-X100[f]

Experimental conditions as in Table 2.2.
[a] Sodium dodecylsulfate.
[b] Cetyltrimethylammonium bromide.
[c] N-Dodecyl-N,N-dimethyl-3-ammonio-1-propanesulfonate.
[d] Polyoxyethylene alcohol ($C_{12}H_{25}$-$C_{18}H_{37}$)-$(OCH_2CH_2)_5$.
[e] Polyoxyethylene(8)isooctylcyclohexyl ether.
[f] Polyoxyethylene(10)isooctyl phenyl ether.

makes it possible to perform up to three enantioselective epoxidation cycles with both constant yield and enantioselectivity. The major limitation to further recycling was not due to catalyst decomposition, but rather due to the reaction medium losses that are unavoidably involved in carrying out the different operations manually on a small lab scale.

8. REACTION MECHANISM

Insight into the mechanism of epoxidation catalyzed by **1b** was obtained by performing a series of kinetic experiments following the approach elegantly described by Blackmond in a recent review article [33]. Starting from a small number of kinetic experiments following the reaction progress from the beginning to completion, this kinetic method allows the determination of catalyst and substrates order, as well as possible inactivation phenomena like catalyst decomposition or product inhibition. The kinetic analysis leads to the following rate expression:

$$\text{Rate} = \frac{d[\text{epox}]}{dt} = k[\text{Pt}_0][\text{alkene}]$$

Rate data were integrated with NMR studies under identical experimental conditions: concentrations, solvent, temperature, and so on. Essential findings were: (a) the existence of a fast equilibrium in metal-alkene formation and (b) the full interaction of the starting complex with hydrogen peroxide to form a hydrogen bonded adduct with the fluorine atoms in the $-C_6F_5$ ligand. The scheme reported in Fig. 2.6 is suggested as the possible mechanism for the present catalytic epoxidation system.

This scheme takes into account all the spectroscopic evidence and is in agreement with the kinetic data under the assumption (confirmed by NMR evidence) that $[\text{Pt}_0]$ is essentially given by the solvento and aquo complexes carrying H_2O_2 hydrogen bound to the C_6F_5 ligand. It also accounts for the steric effects observed in Table 2.2 as metal-alkene formation precedes the rate determining step, as well as for the nucleophilic character of the oxygen transfer step as demonstrated by the unusual regioselectivity observed in dienes epoxidation. This is, to the best of our knowledge, a rare example of epoxidation of unfunctionalized alkenes by means of activation of electrophilic alkene and nucleophilic oxidant by metal catalysts. In fact, this type of oxidation, characterized by concomitant substrate and oxidant activation, is much more common for electron-poor substrates, as demonstrated recently for the asymmetric nucleophilic epoxidation of chalcone with hydrogen peroxide catalyzed by polyleucine catalysts [53].

The reaction pathway reported here is profoundly different from traditional electrophilic oxidation in which the role of the metal is essentially devoted to oxidant activation and reactivity is favored by electron-rich olefins. Here, a new paradigm is revealed specific for electron-poor substrates.

9. CONCLUSIONS

The use of the Pt(II) complexes of the class $[(\text{P-P})\text{Pt}(C_6F_5)(H_2O)]^+$ (P-P = diphosphine) was demonstrated to be highly active and unusually selective compared to other metal catalysts for the epoxidation of terminal monofunctionalized

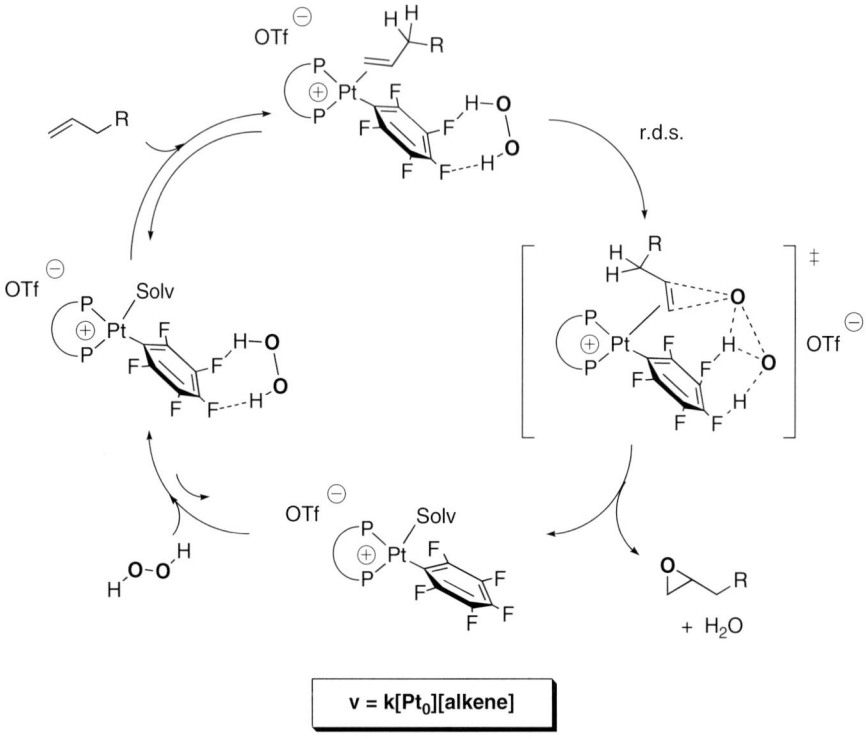

FIGURE 2.6 Mechanistic hypothesis for terminal alkene epoxidation with hydrogen peroxide mediated by **1b** on the basis of the results of reaction-progress kinetic analysis and ^{31}P and ^{19}F NMR studies.

alkenes with hydrogen peroxide as oxidant. In addition to (a) high activity under mild experimental conditions, (b) catalyst loading as low as 2%, and (c) no need for overstoichiometric amounts of oxidant, key features of the catalysts are (d) extreme substrate selectivity and (e) exceptional regioselectivity [24–26], which, in particular with some dienes, is the highest so far reported for metal-mediated epoxidation. Such properties, along with the straightforward synthesis of the catalyst, allowed the development of an asymmetric and highly regioselective version of the reaction system whose results are well interpreted on the basis of the mechanism discussed. The soft character of these complexes makes them insensitive to the presence of large amounts of water, a problem generally affecting traditional early transition metal Lewis acid catalysts that generally prevents (with some notable exceptions) the use of hydrogen peroxide as the terminal oxidant. On the other hand, their remarkable Lewis acidity allows coordinated alkene substrates to be susceptible to nucleophilic attack, thus inverting the traditional reactivity of alkenes in epoxidation.

REFERENCES

[1] P. W. N. M. van Leeuwen, Epoxidation: Large scale commodities and small scale beauties, *Homogeneous Catalysis*, Kluwer Academic Publishers, Dordrecht, 2004, p. 299.
[2] D. E. De Vos, B. F. Sels, P. A. Jacobs, Practical heterogenous catalysts for epoxide production, *Adv. Synth. Catal.* 345 (2003) 457.
[3] R. M. Lambert, F. J. Williams, R. L. Cropley, A. Palermo, Heterogeneous alkene epoxidation: Past, Present and Future, *J. Mol. Cat. A.* 228 (2005) 27.
[4] B. S. Lane, K. Burgees, Metal-catalyzed epoxidations of alkenes with hydrogen peroxide, *Chem. Rev.* 103 (2003) 2457.
[5] B. M. Trost, Atom economy-A challenge for organic symthesis: Homogeneous catalysis leads the way, *Angew. Chem. Int. Ed.* 34 (1995) 259.
[6] G. Strukul (Ed.), *Catalytic Oxidations with Hydrogen Peroxide as Oxidant*, Kluwer Academic, Dordrecht, 1992, p. 13.
[7] G. Grigoropoulou, J. H. Clark, J. A. Elings, Recent developments on the epoxidation of alkenes using hydrogen peroxide as an oxidant, *Green Chem.* 5 (2003) 1.
[8] W. Adam, T. Wirth, Hydroxy group directivity in the epoxidation of chiral allylic alcohols: Control of diastereoselectivity through allylic strain and hydrogen bonding, *Acc. Chem. Res.* 32 (1999) 703.
[9] B. Notari, R. J. Willey, M. Panizza, G. Busca, Which sites are the active sites in TiO_2–SiO_2 mixed oxides? *Catal. Today* 18 (1993) 163.
[10] M. Klawonn, M. K. Tse, S. Bhor, C. Döbler, M. Beller, A convenient Ruthenium-catalyzed alkene epoxidation with hydrogen peroxide as oxidant, *J. Mol. Catal. A.* 218 (2004) 13.
[11] K. Sato, M. Aoki, M. Ogawa, T. Hashimoto, R. Noyori, A practical method for epoxidation of terminal olefins with 30% hydrogen peroxide under halide-free conditions, *J. Org. Chem.* 61 (1996) 8310.
[12] C. Venturello, E. Alneri, M. Ricci, A new, effective catalytic system for epoxidation of olefins by hydrogen peroxide under phase-transfer conditions, *J. Org. Chem.* 48 (1983) 3831.
[13] C. Venturello, R. D'Aloisio, Quaternary ammonium tetrakis(diperoxotungsto)phosphates$^{(3-)}$ as a new class of catalysts for efficient alkene epoxidation with hydrogen peroxide, *J. Org. Chem.* 53 (1988) 1553.
[14] N. Mizuno, K. Yamaguchi, K. Kamala, Epoxidation of olefins with hydrogen peroxide catalyzed by polyoxometalates, *Coord. Chem. Rev.* 249 (2005) 1944.
[15] S. Banfi, F. Montanari, S. Quici, S. V. Barkanova, O. L. Lakiya, V. N. Kopranenkov, E. A. Luk'yanets, Porphyrins and azaporphines as catalysts in alkene epoxidations with peracetic acid, *Tetrahedron Lett.* 36 (1995) 2317.
[16] A. Murphy, A. Pace, T. D. P. Stack, Legand and pH influence on manganese-mediated peracetic acid epoxidation of terminal olefins, *Org. Lett.* 6 (2004) 3119.
[17] A. Murphy, G. Dubois, T. D. P. Stack, Efficient epoxidation of electron-deficient olefins with a cationic manganese complex, *J. Am. Chem. Soc.* 125 (2003) 5250.
[18] D. E. De Vos, B. F. Sels, M. Reynaers, Y. V. Subba Rao, P. A. Jacobs, Epoxidation of terminal or electron-deficient olefins with H_2O_2, catalysed by Mn-trimethyltriazacyclonane complexes in the presence of an oxalate buffer, *Tetrahedron Lett.* 39 (1998) 3221.
[19] A. Murphy, T. D. Stack, Discovery and optimization of rapid manganese catalysts for the epoxidation of terminal olefins, *J. Mol. Catal. A* 251 (2006) 78.
[20] J. Rudolph, K. L. Reddy, J. P. Chiang, K. B. Sharpless, Highly efficient epoxidation of olefins using aqueous H_2O_2 and catalytic Methyltrioxorhenium/Pyridine: Pyridine-mediated ligand acceleration, *J. Am. Chem. Soc.* 119 (1997) 6189.
[21] C. Copéret, H. Adolfson, K. B. Sharpless, A simple and efficent method for epoxidation of terminal alkenes, *Chem. Commun.* (1997) 1565.
[22] G. Dubois, A. Murphy, T. D. Stack, Simple iron catalyst for terminal alkene epoxidation, *Org. Lett.* 5 (2003) 2469.

[23] M. C. White, A. G. Doyle, E. N. Jacobsen, A synthetically useful, self-assembling MNO minic system for catalytic alkene epoxidaion with aqueous H_2O_2, *J. Am. Chem Soc.* 123 (2001) 7194.
[24] Y. Nakagawa, K. Kamata, M. Kotani, K. Yamaguchi, N. Mizuno, Polyoxovanadometalate-catalyzed selective epoxidation of alkenes with hydrogen peroxide, *Angew. Chem. Int. Ed.* 44 (2005) 5136.
[25] K. Kamata, Y. Nakagawa, K. Yamaguchi, N. Mizuno, Efficient, regioselective epoxidation of dienes with hydrogen peroxide catalyzed by $[\gamma\text{-SiW}_{10}O_{34}(H_2O)_2]^{4-}$, *J. Catal.* 224 (2004) 224.
[26] N. Mizuno, Y. Nakagawa, K. Yamaguchi, Bis(μ-hydroxo) bridged di-Vanadium-catalyzed selective epoxidation of alkenes with H_2O_2, *J. Mol. Catal. A* 251 (2006) 286.
[27] E. Pizzo, P. Sgarbossa, A. Scarso, R. A. Michelin, G. Strukul, Second-generation electron-poor platinum(II) complexes as efficient epoxidation catalysts for terminal alkenes with hydrogen peroxide, *Organometallics* 25 (2006) 3056.
[28] C. Baccin, A. Gusso, F. Pinna, G. Strukul, Platinum-catalyzed oxidations with hydrogen peroxide: The (Enantioselective) epoxidation of α,β-unsaturated ketones, *Organometallics* 14 (1995) 1161.
[29] R. Sinigaglia, R. A. Michelin, F. Pinna, G. Strukul, Asymmetric epoxidation of simple olefins catalyzaed by chiral diphosphine-modified Platinum(II) complexes", *Organometallics* 6 (1987) 728.
[30] A. Zanardo, R. A. Michelin, F. Pinna, G. Strukul, Epoxidation of olefins catalyzed by chelating diphosphine-platinum(II) complexes. Ring-size and ring-shape effects on the catalytic activity, *Inorg. Chem.* 28 (1989) 1648.
[31] G. Strukul, R. A. Michelin, Catalytic epoxidation of 1-Octene with diluted hydrogen peroxide. On the basic role of hydroxo complexes of platinum(II) and related species, *J. Am. Chem. Soc.* 107 (1985) 7563.
[32] A. Zanardo, F. Pinna, R. A. Michelin, G. Strukul, Kinetic study of the epoxidation of 1-octene with hydrogen peroxide catalyzed by platinum(II) complexes. Evidence of the involvement of two metal species in the oxygen-transfer step, *Inorg. Chem.* 27 (1988) 1966.
[33] D. G. Blackmond, Reaction progress kinetic analysis: A powerful methodology for mechanistic studies of complex catalytic reactions, *Angew. Chem. Int. Ed.* 44 (2005) 4302.
[34] M. D. Fryzuk, T. Jones, F. W.B. Einstein, Reactivity of electron-rich binuclear rhodium hydrides. Synthesis of bridging alkenyl hydrides and X-ray crystal structure of $[[(Me_2CH)_2PCH_2CH_2P(CHMe_2)_2]Rh]_2(\mu\text{-H})(\mu\text{-}\eta^2\text{-CH=CH}_2)$, *Organometallics* 3 (1984) 185.
[35] A. Corma, H. Garcia, Lewis acids: From conventional homogeneous to green homogeneous and heterogeneous catalysis, *Chem. Rev.* 103 (2003) 4307.
[36] A. Corma, H. Garcia, Lewis acids as catalysts in oxidation reactions: From homogeneous to heterogeneous systems, *Chem. Rev.* 102 (2002) 3837.
[37] G. Strukul, Lewis acid behavior of cationic complexes of palladium(II) and platinum (II): Some examples of catalytic applications, *Top. Catal.* 19 (2002) 33.
[38] R. A. Michelin, E. Pizzo, A. Scarso, P. Sgarbossa, G. Strukul, A. Tassan, Baeyer-Villiger oxidation of ketones catalyzed by platinum(II) Lewis acid complexes containing coordinated electron-poor fluorinated diphosphines, *Organometallics* 24 (2005) 1012.
[39] A. Brunetta, G. Strukul, Epoxidation versus Baeyer-Villiger oxidation: The possible role of Lewis acidity in the conrol of selectivity in catalysis by transition metal complexes, *Eur. J. Inorg. Chem.* 5 (2004) 1030.
[40] R. A. Michelin, M. F. C. G. da Silva, A. J. L. Pombeiro, Aminocarbene complexes derived from nucleophilic addition to isocyanide ligands, *Coord. Chem. Rev.* 218 (2001) 75.
[41] U. Belluco, R. A. Michelin, P. Uguagliati, B. Crociani, Mechanisms of nucleophilic and electrophilic attack on carbon bonded palladium(II) and platinum(II) complexes, *J. Organomet. Chem.* 250 (1983) 565.

[42] W. Adam, H.-G. Degen, C. R. Saha-Möller, Regio- and Diastereoselective catalytic epoxidation of chiral allylic alcohols with hexafluoroacetone perhydrate. Hydroxy-group directivity through hydrogen bonding, *J. Org. chem.* 64 (1999) 1274.
[43] W. Adam, V. R. Stegmann, C. R. Saha-Möller, Regio- and Diastereoselective epoxidation of chiral allylic alcohols catalyzed by manganese(salen) and iron (porphyrin) complexes, *J. Am. Chem. Soc.* 121 (1999) 1879.
[44] W. Adam, C. M. Mitchell, C. R. Saha-Möller, Regio- and Diastereoselective Catalytic epoxidation of acyclic allylic alcohols with methyltrioxorhenium: A mechanistic comparison with metal (peroxy and peroxo complexes) and nonmetal (peracids and dioxirane) oxidants, *J. Org. Chem.* 64 (1999) 3699.
[45] G. Della Sala, L. Giordano, A. Lattanzi, A. Proto, A. Scettri, Metallocene-catalyzed diastereoselective epoxidation of allylic alcohols, *Tetrahedron* 56 (2000) 3567.
[46] W. Adam, P. L. Alsters, R. Neumann, C. R. Saha-Möller, D. Sloboda-Rozner, R. Zhang, A highly chemoselective, diastereoselective, and regioselective epoxidation of chiral allylic alcohols with hydrogen peroxide, catalyzed by sandwich-type polyoxometalates: Enhancement of reactivity and control of selectivity by the hydroxy group through metal-alcoholate bonding, *J. Org. Chem.* 68 (2003) 1721.
[47] M. Dusi, T. Mallat, A. Baiker, Epoxidation of functionalized olefins over solid catalysts, *Catal. Rev. Sci. Eng.* 42 (2000) 213.
[48] W. Adam, W. Malisch, K. J. Roschmann, C. R. Saha-Möller, W. A. Shenk, Catalytic oxidations by peroxy, peroxo and oxo metal complexes: An interdisciplinary account with a personal view, *J. Organom. Chem.* 661 (2002) 3.
[49] V. B. Valodka, G. L. Tembe, R. N. Am, H. S. Rama, Catalytic asymmetric epoxidation of unfunctionalized olefins by supported Cu(II)-Amino acid complexes, *Catal. Lett.* 90 (2003) 91.
[50] C. Bonini, G. Righi, A critical outlook and comparison of enantioselective oxidation methodologies of olefins, *Tetrahedrom* 58 (2002) 4981.
[51] Q.-H. Xia, H.-Q. Ge, C.-P. Ye, Z.-M. Liu, K.-X. Su, Advances in homogeneous and heterogeneous catalytic asymmetric epoxidation, *Chem. Rev.* 105 (2005) 1603.
[52] M. Colladon, A. Scarso, G. Strukul, Towards a greener epoxidation method. Use of water-surfactant media and catalyst recycling in the platinum-catalyzed asymmetric epoxidation of terminal alkenes with hydrogen peroxide, *Adv. Synth. Catal.* 349 (2007) 797.
[53] S. P. Mathew, S. Gunathilagan, S. M. Roberts, D. G. Blackmond, Mechanistic insights from reaction progress kinetic analysis of the polypeptide-catalyzed epoxidation of chalcone, *Org. Lett.* 7 (2005) 4847.

CHAPTER 3

Lewis Acid Catalyzed Epoxidation of Olefins Using Hydrogen Peroxide: Growing Prominence and Expanding Range

Daryle H. Busch, Guochuan Yin, and **Hyun-Jin Lee**

Contents

1. Introduction — 120
2. Industrial Processes for Propylene Oxide Manufacture: Present and Future — 123
3. A New Pressure Intensified Epoxidation Process for Light Olefins — 126
4. Expanding Range of Lewis Acid Catalysis Chemistry — 130
5. Mid- to Late Transition Metal Catalysts also Perform Epoxidation Reactions by the Lewis Acid Mechanism — 131
   - 5.1. Iron compounds that catalyze lewis acid epoxidation reactions — 131
   - 5.2. Early examples of catalysis by lewis acid adducts involving manganese — 133
   - 5.3. A highly selective Mn(IV) catalyst for epoxidation reactions — 134
- Acknowledgments — 148
- References — 149

Abstract

Among the many methods for synthesizing epoxides, Lewis acid catalyzed olefin epoxidation provides a special combination of capabilities including gentle and green processing and selectivity of multiple kinds. The subject has contributed both to the understanding of important basic science and a large and growing inventory of impressive oxidation reactions. Some big challenges have been met by Lewis acid catalysts, but it has yet to provide the

Department of Chemistry and Center for Environmentally Beneficial Catalysis, University of Kansas, Lawrence, Kansas 66047

underlying science in a major industrial process. Historically, the cost of hydrogen peroxide has been a limitation, but recent developments, especially the on-site and/or *in situ* generation of H_2O_2 has been demonstrated for promising alternative processes, especially in the light olefin epoxidation industry. For commodity chemicals, costs and environmental issues are accompanied by other related challenges such as catalyst durability and productivity that thwart otherwise elegant processes. The case of the long known and arguably unique catalyst methyltrioxorhenium (MTO) for light olefin oxidation is explored in this context. Attention is also directed to the growing realization that, in combination with the right ligands, late transition metal elements, including manganese and iron also catalyze olefin epoxidation reactions. Ordinarily manganese and iron perform epoxidations by the oxygen rebound mechanism of Groves [1], and engage in one-electron redox chemistry. To produce only epoxides in reacting with many common olefins, the activated catalyst must not abstract hydrogen atoms which would open radical routes to other products. Ligand design has produced highly selective Mn(IV) catalysts capable of converting many olefins into their oxides but limited in their ability to initiate radical processes by hydrogen abstraction. Earlier studies both in laboratory chemistry and in biochemistry have identified processes as Lewis acid catalyzed by late transition metal complexes, especially for iron where such H_2O_2 oxidations have been broadly accepted. Iodosylbenzene oxidations using manganese catalysts have also been suspected of operating by this mechanism.

Key Words: Lewis acid adducts, Radical oxidations, Epoxidation, Hydrogen peroxide, Bond dissociation energy, Catalyst durability, Methyltrioxorhenium, Cross-bridged cyclam, Mn(IV), Late transition metal, Propylene oxide, Titanium silicalite (TS-1) catalyst, Ethylanthrahydroquinone/H_2 process, Polyoxometallates, Mn(IV) catalyst, Hydrogen abstraction, Rebound mechanism, Isotopic label, *t*-BuOOH, Peroxide adduct.   © 2008 Elsevier B.V.

1. INTRODUCTION

This chapter is concerned with the use of transition metal complexes as catalysts for the H_2O_2 epoxidation of olefins in industrial processes and how the mechanisms available to a given transition metal catalyst affect the chemical, economical, and environmental viability of the processes. Central to this subject is how two-electron oxygen transfer of the kind that converts olefins into epoxides can occur by multiple mechanisms and, in fact, how a single kind of activated catalyst system might actually carry out epoxidations by more than one mechanism, depending on details of catalyst structure and reaction conditions. Many compounds of so-called early transition metal atoms in high oxidation states like titanium(IV), zirconium(IV), molybdenum(VI), tungsten(VI), and rhenium(VII) have been much studied for H_2O_2 oxidation reactions in which the H_2O_2 first binds to the early transition metal atom, which polarizes it, and then transfers an oxygen atom directly to a nucleophilic substrate, like an olefin, in a Lewis acid promoted process. It is also a property of these metal ions that they perform much chemistry without changing oxidation states. In contrast, elements occurring later in the periodic table, such as manganese, iron, cobalt, and copper, are best known for changing their oxidation states in the course of oxidation catalysis. In familiar processes, when iron and manganese are involved in oxidation catalysis they first bind oxygen atoms (often from the oxidant) when they are oxidized and then undergo reduction as they transfer these oxygen atoms to substrates. These late transition elements are said to perform oxidations by the rebound mechanism, a term coined by Groves [1]. Growing evidence indicates that these familiar expectations are merely convenient simplifications of the catalytic capabilities of the metallic elements.

Epoxides are formed by a variety of reactions, including dehydration of glycols, and oxygen insertion by peroxo radicals, peroxy acids, high valent late transition metal "oxo" complexes, and Lewis acid adducts of H_2O_2. In catalytic applications, radical oxidations are attractive because they are powerful and can use freely available molecular oxygen, but their selectivity is often limited. In contrast, Lewis acid catalyzed epoxidations often proceed with high selectivity, but require a manufactured oxidant, such as hydrogen peroxide or an organic hydroperoxide. In an era of earth literacy, response to the demands of sustainability turns attention to processes that are highly selective and use green reactants. These considerations direct this chapter to catalytic epoxidations performed by H_2O_2 and Lewis acid catalysts. Lewis acid catalysis with H_2O_2 is chosen over stoichiometric peroxy acid oxidation because the latter produces a molar equivalent of organic acid as a by-product while water is the only by-product of H_2O_2 oxidation.

As will be discussed, substantial challenges to developing Lewis acid catalysts for the epoxidation of olefins include catalyst reactivity, selectivity, durability, and recycle. In their search for total enantiomeric selectivity in oxidation reactions, Sharpless and others remark on the marvel of enzyme selectivities [2–4]. In that context, it is clear that the selectivity of enzymes results from control of the catalyst/substrate encounter, which may involve binding between the two. The cytochromes P450 exemplify substrate selectivity by enzymes [5,6]. The activated site, characterized long ago in horseradish peroxidase as compound I, is a model of molecular oxidizing power and its selectivity is derived from substrate binding [7]. This leads to the concept that any substrate that a P450 can bind will be oxidized by the P450 equivalent of compound I. The enormous commitment of talent, research time and resources to chiral catalysis [2–4], and the highly empirical nature of the fabulous advances in the field emphasize the fact that selective interactions of the kinds that determine fleeting catalyst/substrate binding are among the most difficult targets of molecular design. For these reasons and because there are less challenging approaches, oxidation product selectivity is usually achieved on the basis of mechanistic considerations and precise control of process parameters, rather than by selective substrate binding to the catalyst. This is consistent with the fact that chemoselectivity is the more common issue in large-scale industrial oxidation reactions.

The hydrogen abstraction reactions that accompany the radical process for epoxidation often produce waste products in addition to the desired product. Similarly, the complicated discussions of mechanistic issues associated with the very successful chiral epoxidations by manganese salen catalysts, including Jacobsen's catalyst, reflect the versatility of high valent manganese oxidation–reduction chemistry [8]. Mn(V) is a powerful oxidant when coordinated to salen and other similar ligands, but has the ability to react by either one- or two-electron processes. Indeed, high valent late transition metal complexes are capable of engaging in both hydrogen abstraction reactions and oxygen atom insertion reactions. In view of this mechanistic versatility, catalysts in this class should be expected to function with some lack of selectivity, depending on the substrates and reaction conditions. The parallel case of catalysis by certain iron-centered enzymes is clear from the duality of catalyst families derived from that late transition metal. Peroxidases serve as one-electron oxidants while peroxygenases transfer oxygen atoms to substrates. In contrast to the expectation for multiple products in systems making use of O_2-based radical reactions and those applying high selectivity valent late transition metal catalysts, Lewis acid catalyzed hydrogen peroxide epoxidations have gained the reputation for high [9,10].

In the history of epoxidation reactions, landmarks are found in regio- and enantioselective reactions. Scheme 3.1 shows both kinds of selectivity for the classic case of geraniol. Organic peracids selectively epoxide the 6,7-double bond because the OH group deactivates the 2–3 bond [11], and Sharpless and Michaelson found that early transition metals are highly selective for the 2,3 position [12], resolving a classic selectivity issue. In addition to the solution of challenging regio- and enantioselective problems, Sharpless and coworkers explored broad capabilities in the area of Lewis acid catalyzed epoxidations, using alkylhydroperoxides as oxidants [2,13,14]. They identify compounds of periodic groups 3–6, all of the available lanthanide elements, and uranium as capable of epoxidation of allylic alcohols, using *t*-BuOOH as the oxidant [2].

This listing illustrates the common perspective that "early transition metal ions" perform oxidations by the Lewis acid pathway while "late transition metal ions" produce epoxides by the rebound mechanism. That generality has

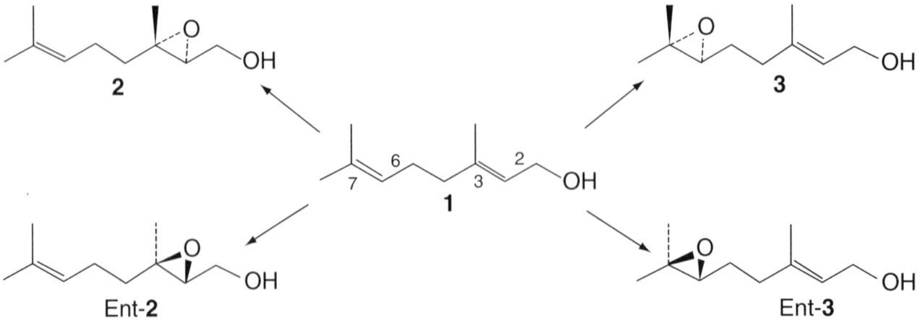

SCHEME 3.1 Reprinted with permission from *Angew. Chem. Int. Ed.* 41 (2002) 2027, Scheme 1. Copyright (2002) Wiley-VCH.

been challenged in recent years and is discussed below. Finally, the limitations of the Lewis acid pathway found by these and other investigators include, among others, the reactivity of the catalyst system (often too low), and catalyst durability (too few turnovers), factors that are terminal if industrial processes are sought. As discussion proceeds, reasons are given for an optimistic view of the possible utility of this family of catalysts for certain applications. Large-scale propylene oxide synthesis is discussed as one possible target.

2. INDUSTRIAL PROCESSES FOR PROPYLENE OXIDE MANUFACTURE: PRESENT AND FUTURE

Industry produces over six million metric tons of propylene oxide each year and new production facilities are planned or are under construction. The expansion of the industry continues to provide the opportunity for implementation of new processes, a change that is in the interest of both economical and environmental sustainability. The most practiced processes (Scheme 3.2) are aged and bothered by burdens that call for their replacement by new technologies [15,16]. The chlorohydrin process uses chlorine and lime. First, the chlorine forms the chlorohydrin and then the lime extracts HCl to form the epoxide, resulting in large amounts of calcium chloride in huge waste streams. The peroxidation process begins with isobutane and O_2, forming t-BuOOH, which is then used to produce the epoxide. The process operates in the gas phase at elevated temperature and produces two molar equivalents of t-butanol for every mole of propylene oxide formed.

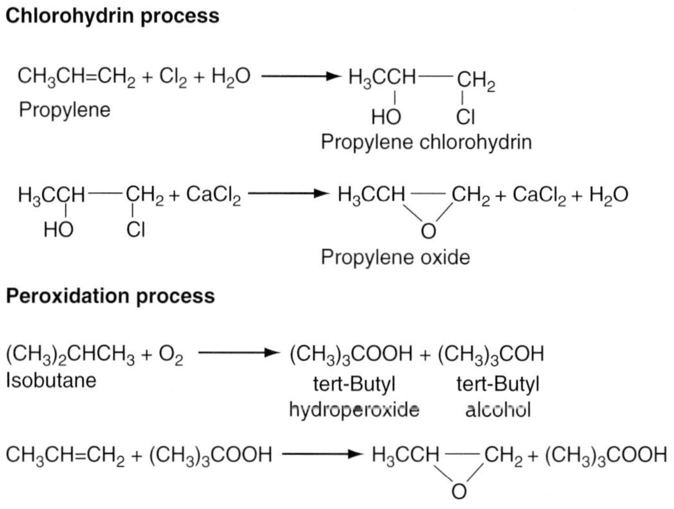

Chlorohydrin process

$$CH_3CH=CH_2 \text{ (Propylene)} + Cl_2 + H_2O \longrightarrow \underset{\text{Propylene chlorohydrin}}{H_3C\underset{OH}{C}H-\underset{Cl}{C}H_2}$$

$$H_3C\underset{OH}{C}H-\underset{Cl}{C}H_2 + CaCl_2 \longrightarrow \underset{\text{Propylene oxide}}{H_3CCH\overset{O}{\diagdown\diagup}CH_2} + CaCl_2 + H_2O$$

Peroxidation process

$$\underset{\text{Isobutane}}{(CH_3)_2CHCH_3} + O_2 \longrightarrow \underset{\substack{\text{tert-Butyl} \\ \text{hydroperoxide}}}{(CH_3)_3COOH} + \underset{\substack{\text{tert-Butyl} \\ \text{alcohol}}}{(CH_3)_3COH}$$

$$CH_3CH=CH_2 + (CH_3)_3COOH \longrightarrow H_3CCH\overset{O}{\diagdown\diagup}CH_2 + (CH_3)_3COH$$

SCHEME 3.2

Two early alternative processes for PO production used the potent organometallic catalyst, methyltrioxorhenium (MTO), in media from which water was diligently removed from aqueous solutions of hydrogen peroxide. The first process, from the laboratories of MTO pioneer Herrmann [17], also operated at low temperatures, for example, $-10\,°C$ in order to minimize hydrolysis of the PO product to propylene glycol. Such low temperatures are a burden in industrial operation of strongly exothermic reactions. Crocco's improved MTO process still worked at minimizing the concentration of water and used O_2 reduction to H_2O_2 by oxidation of a solvent component, a ternary benzylic alcohol, but their operating temperature was favorably some $20\,°C$ higher. This MTO process also benefited from the use of an accelerating ligand, preferably pyridine [18].

Venturello opened a field, now much explored, in which Na_2WO_4 and H_2WO_4 are partnered with acids and phase-transfer reagents (PTRs) to catalyze H_2O_2 epoxidations of olefins [19–21]. A heteropoly anion, $[PO_4\{W(O)(O_2)_2\}_4]^{3-}$, having the structure shown in Fig. 3.1 has been isolated, characterized, and a number of its salts prepared with cations of varying phase-transfer capabilities. Productive catalytic behavior is displayed even for substrates regarded as difficult. Methodologies were developed to minimize the hydrolysis of the epoxide produced in reactions involving aqueous H_2O_2 as the oxidant. Today, their preferred solvents, such as dichloroethylene, are not considered green.

An efficient related process by Noyori used no organic solvent, did not use a catalyst prepared in advance, and performed the epoxidations with a biphasic system involving 30% water/H_2O_2 and the neat substrate. This process used aminomethylphosphonic acid instead of phosphate and a long-chain quaternary ammonium hydrosulfate as PTR [22,23]. Rapid conversion and high selectivity were achieved with terminal olefins, 1-octene, 1-decene, and 1-dodecane, which are difficult to epoxidize. Typically using terminal olefins, this process operates at $90\,°C$ with stirring at 1,000 rpm for 2 h with yields of 86–97%. Reedijk et al. report a simplified system that converts a range of terminal and cyclic olefins to epoxides at lower temperatures. The preferred catalyst system uses equimolar concentrations of Na_2WO_4 and H_2WO_4 (2 mol%), partnered with $ClCH_2COOH$ (1.6 mol%). Conversions of terminal olefins at $60\,°C$ were 54% (1-octene) and 60% (1-hexene) for 4 h reactions. The more reactive cyclic olefins gave high conversions and yields in minutes [24].

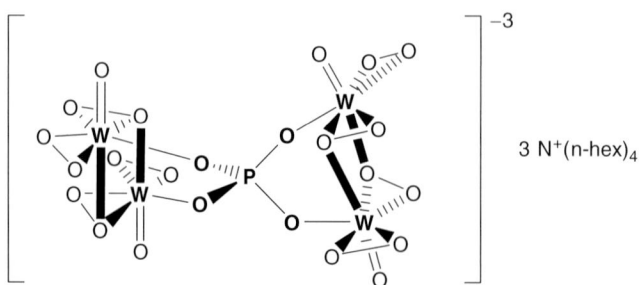

FIGURE 3.1 Structure of Venturello's catalyst.

Using catalysts they attribute to Venturello [21], Xi and coworkers offer an alternative propylene oxide process based on a phenomenon that they label "reaction-controlled, phase-transfer catalysis" [25]. Their catalysts I, [π-$C_5H_5NC_{16}H_{33}$]$_3$[$PO_4(WO_3)_4$], which is a "Venturello compound," forms a complex with the dianion of H_2O_2 that is soluble in the organic solvent, 4:3 xylene: tributylphosphate. However, the peroxide-free catalyst is insoluble in that medium. Consequently, reaction is initiated by the addition of the aqueous H_2O_2 solution to the previously prepared reaction system. As the catalyst forms the peroxo complex, it dissolves and is available to oxidize the substrate. A reaction time of 6 h at 65 °C converted 89% of the propylene to propylene oxide with 95% selectivity, and formation of a small amount of by-product propylene glycol. A feature emphasized by the authors is the ease of recycle of the catalyst in this unique system. As the catalyst is soluble only in the presence of peroxide, exhaustion of the H_2O_2 results in spontaneous precipitation of the catalyst which can be recovered and used again. Repeated cycles showed that 90% catalyst recovery can be routine and that the recovered catalyst retains full activity. These researchers also demonstrated the ancillary generation of H_2O_2 from O_2 by the well-known 2-ethylanthrahydroquinone/H_2 process. It should be noted that the choice of the H_2 plus O_2 to H_2O_2 process appears to influence the solvents available for the epoxidation process.

Mizuno [26,27] and Neumann [28,29] studied tungsten-based polyoxometallates, or heteropolyanions, as catalysts for epoxidations and found much variation in catalytic effectiveness and durability. Mizuno et al. [30] found very promising behavior by [$SiW_{10}O_{34}(H_2O)_2$]$^{4-}$, which differs from a silicon-centered, 12-acid anion of the textbook Keggin structure [31] by removal of two WO_3 units. Under their reaction conditions, which were not optimized for the competition, the "SiW_{10}" catalyst outperformed Venturello's catalyst (see above) and simpler tungstate catalysts. More significantly, SiW_{10} showed excellent durability in experiments that made repetitive use of the catalyst, showing no structural change, under conditions where the 12-acid of phosphorus, $H_3PW_{12}O_{40}$, is decomposed into Venturello's catalyst, [{W(O)(O_2)$_2$}$_4$(PO_4)]$^{3-}$. This latter observation bodes well for both of these heteropolytungstate catalysts, $H_3PW_{12}O_{40}$ and [{W(O)(O_2)$_2$}$_4$(PO_4)]$^{3-}$. Mizuno's catalyst is also impressive in its ready application to the oxidation to both terminal olefins and light olefins, each of which adds a challenge to the epoxidation process.

Under the heading "Propylene oxide routes takeoff," *Chemical and Engineering News* highlighted the plans to introduce H_2O_2 into large-scale production of the commodity, propylene oxide [32]. BASF and Dow announced plans to commercialize an innovative hydrogen peroxide/propylene oxide (HPPO) technology that will produce 300,000 metric tons of PO annually, using a new 230,000 ton H_2O_2 plant to be built on-site. While not revealing details of the process they claim that it has a fine footprint, will produce no by-products except water, and will reduce energy consumption by 35% and waste water generation by 70–80%. A Korean Company, SKC, has licensed a different HPPO process technology for PO manufacture from two German companies, Degussa, a specialty chemicals company, and Uhde, a high-tech engineering company. Plans are to launch a 100,000 ton/annum production facility at the beginning of 2008. A joint venture between Degussa and Headwaters is planned to supply the necessary oxidant by acquiring an H_2O_2 plant from the Finnish company, Kemira Oyj, and expanding its production. This growth in the world supply of PO reflects both the expanding markets for propylene oxide and the great need for much simplified technologies that make use of a truly green oxidant, in this case, H_2O_2. Costs are reduced by making hydrogen peroxide on-site and using it without purification.

3. A NEW PRESSURE INTENSIFIED EPOXIDATION PROCESS FOR LIGHT OLEFINS

To be considered green, a chemical process needs to operate at high atom economy, involve minimum hazardous substances and procedures, consume minimum energy, and produce little waste [33,34]. But being green is not enough. A viable chemical process must also be economically sustainable. Figure 3.2 summarizes a process under development in these laboratories that is focused on propylene oxide [35, 102]. It uses long-known MTO [36,37] as the homogeneous catalyst, under moderate conditions (30 °C, 20 bar of N_2 pressure), and safe materials [propylene, excess aqueous H_2O_2 as oxidant, methanol as solvent, and pyridine-N-oxide (PyNO) as accelerating ligand], with a simple separation scheme for the product, solvent and water, the only significant by-product.

The performance of this catalyst system is compared with the traditional industrial processes in Table 3.1. In view of the advances made in PO technology

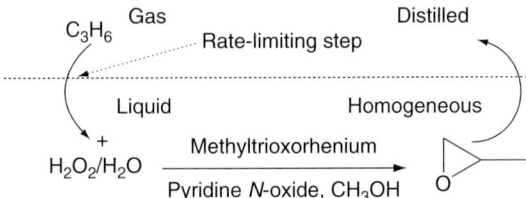

FIGURE 3.2 Pressure intensified, liquid/gas biphasic catalytic propylene oxide process. Reprinted with permission from reference [102].

TABLE 3.1 Comparison of CEBC and industrial processes

System		Conditions	Yield PO
Chlorohydrin	Cl_2, caustic, lime	45–90 °C, 1.1–1.9 bar	$S = 88$–95%
Hydroperoxide	W, V, Mo	100–130 °C, 15–35.2 bar	95% ($S = 97$–98%)/2 h
This work	MTO	30 °C, 17 bar N_2, PyNO	+98% ($S = 99$%)/<2 h

MTO, methyltrioxorhenium; PyNO, pyridine-N-oxide; S, selectivity.

that facilitate the use of H_2O_2, as mentioned above, we consider the selectivity, conversion, and green signature of the process to be particularly promising. Methanol was chosen as the solvent because of performance and good oxidant miscibility with aqueous H_2O_2, and the added ligand, PyNO. As Fig. 3.2 shows, the reaction system is biphasic, involving a gaseous substrate while the catalyst and oxidant are in the liquid phase.

Pressure intensification is an innovative feature of this process and is displayed in Fig. 3.3. During the study of reactions in solvents composed on an organic component plus comparable volumes of co-solvent CO_2, it was found that adding substantial CO_2 pressure (45 bar) provided no improvement in conversion and yield. However, use of ~45 bar of N_2 gas resulted in an increase from about 80% conversion, to essentially total substrate conversion, with excellent selectivity, in less than 2 h. The pressure of the low solubility gas N_2 improved the transport of propylene gas into the liquid phase and resulted in higher productivity with small waste and energy debits.

Although the gentle reaction conditions and apparent sustainability of this oxidation process provide encouraging indications of likely success, the long-known cleanly reactive MTO catalyst has not previously been exploited. Since its discovery in 1978 by Beattie and Jones [38], the unique nature of MTO has stimulated enormous amounts of research, with contributions from many laboratories. Most notable was the discovery by Herrmann and his coworkers that MTO is a superb oxidation catalyst when used with H_2O_2 [39,40]. MTO is soluble in most solvents and stable. Although many compounds of the formula $RReO_3$ exist, MTO is unique in being the most robust and active [41–43].

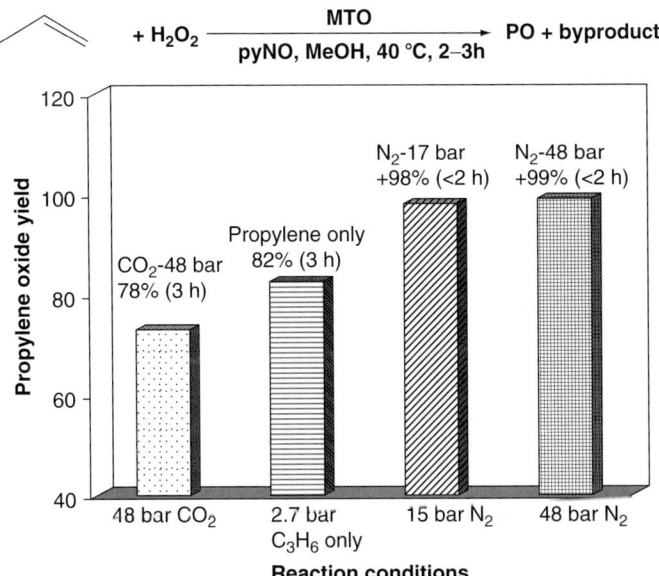

FIGURE 3.3 Pressure intensification and productivity of the new PO process. Reprinted with permission from reference [102].

A fundamental issue for the practical use of MTO is its stability. The structure of MTO, with a methyl group forming an organometallic bond to a rhenium(VII) atom, suggests fragility. The excellent research on MTO by Espenson [44,45], Herrmann [36,37], Sharpless [46], and others, teaches some pertinent lessons on the durability of both the MTO oxidation catalyst and the PO product, which have many facets. A relevant quote from Espenson is "MTO is a remarkably stable substance both as the pure solid and in dilute solutions, in which neither oxygen nor acid nor water has deleterious effects" [44]. In contrast, both acid and water affect the function of the catalyst, a situation which introduces H_2O_2 and cofactors into the catalyst stability issue.

There are two catalytically active intermediates of MTO in H_2O_2 oxidations which exist as the 1:1 and 2:1 H_2O_2:MTO adducts, and are known in the literature as **A** and **B**. The 1:1 adduct **A** has a limited catalyst life but it is not clear whether compound **A** is terminated by nucleophilic removal of its methyl group by hydroxide attack to form MeOH and ReO_4^- (Eq. 3.1), or if, during formation of **A** from MTO and HO_2^-, some fraction of the events involve destruction of MTO, forming methanol and rhenate ion (Eq. 3.2). According to the literature [47], the two pathways described by Eqs. 3.1 and 3.2 are kinetically indistinguishable in this system.

$$CH_3Re(O)_2(O_2) + OH^- \rightarrow CH_3OH + ReO_4^- \qquad (3.1)$$

$$CH_3ReO_3 + HO_2^- \rightarrow CH_3OH + ReO_4^- \qquad (3.2)$$

Remarkably, the 2:1 adduct **B** is itself indefinitely stable in the absence of base, although it does disproportionate H_2O_2 into O_2 and water, simultaneously regenerating MTO, at a very slow rate. Reactions like those described here in aqueous solution are invariably conducted in the presence of added acid to retard the base hydrolysis of MTO. Of course, this creates a problem for epoxidation reactions since epoxides hydrolyze to glycols in acidic media. Also, the 2:1 adduct **B** is inseparably linked to the catalyst instability associated with the 1:1 adduct **A** by two chemical reactions, the reversible dissociation of H_2O_2 molecules from **B** (Eq. 3.3), and substrate oxidation which converts a bound peroxo group into an oxide, forming compound **A** (Eq. 3.4).

$$CH_3Re(O)(O_2)_2 + H_2O \rightarrow CH_3Re(O)_2(O_2) + H_2O_2 \qquad (3.3)$$

$$CH_3Re(O)(O_2)_2 + X \rightarrow CH_3Re(O)_2(O_2) + XO \qquad (3.4)$$

It follows that a key property is the stability of the catalyst. Figure 3.4 shows the cumulative yields in two sets of five successive measurements using a single catalyst sample, with fresh substrate and aqueous H_2O_2 added before each run. Two alcohols, MeOH and *t*-BuOH, were used as solvents and 2-fluoropyridine as an added axial ligand. The linearity of the data indicates the lack of deactivation.

The reactions of MTO with H_2O_2 to form **A** and **B** have been studied by Espenson and coworkers in methanol [48], acetonitrile [49], 1:1 aqueous acetonitrile [50], nitromethane [51], and aqueous solution [52,53]. Table 3.2 summarizes the forward rates and equilibrium constants for formation of compounds **A** and **B** in these media (Eqs. 3.5 and 3.6) [51].

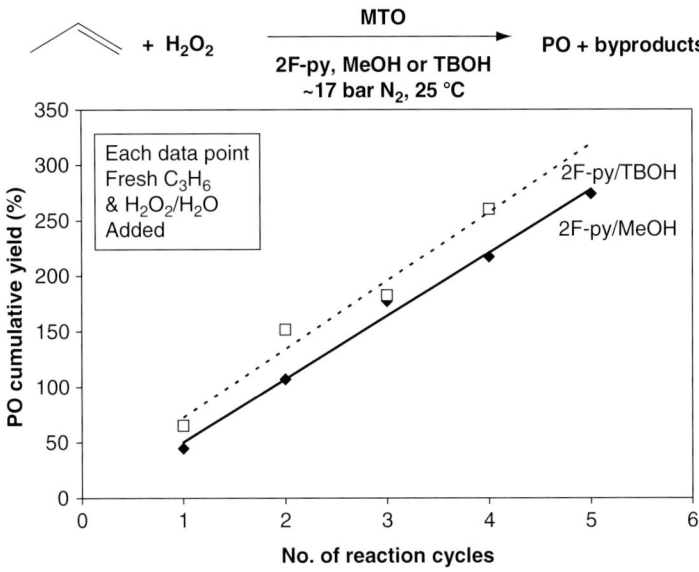

FIGURE 3.4 Catalyst stability in five successive runs using 2F-pyridine as axial ligand at 25 °C. Reprinted with permission from reference [102].

TABLE 3.2 Equilibrium constants, K_1 and K_2, and rate constants, k_1 and k_2, for formation of **A** and **B** in Various Media[a]

Solvent	K_1 (M^{-1})	K_2 (M^{-1})	k_1 (10^{-2} l/mol/s)	k_2 (10^{-2} l/mol/s)
CD$_3$NO$_2$	NA	1,300	7.4	1.4
CH$_3$CN	210	700	10	1.0
CH$_3$OH	261	814	68	1.8
H$_2$O	7.7	145	8000	520
1:1 H$_2$O/CH$_3$CN	13	136	3250	105

[a] Reprinted with permission from Ref. [51]. Copyright (1998) American Chemical Society.

$$CH_3ReO_3 + H_2O_2 \rightarrow CH_3Re(O)_2(O_2) + H_2O; K_1 \quad (3.5)$$

$$CH_3Re(O)_2(O_2) + H_2O_2 \rightarrow CH_3Re(O)(O_2)_2 + H_2O; K_2 \quad (3.6)$$

Most striking is the fact that the equilibrium constant for formation of **B** is greater than that for **A**, the reverse of the usual relative stabilities of complexes formed by successively adding ligands. This "cooperativity" suggests that a structural advantage is associated with the second step. From the practical point of view, this fact, $K_2 > K_1$, should be beneficial to the in-service lifetime of derived catalyst systems since **B** is stable while **A** is associated with MTO destruction. The rate constants for formation of **A** and **B** work in the opposite direction since **B** forms more slowly than **A**. The data in Table 3.2 suggest that the H$_2$O$_2$ complex formation rates are much greater in water than in organic solvents, but that the equilibrium constants are much smaller. Thus, mixed solvents show some possibility of providing intermediate values at least for the rates of complex formation.

The discovery that ligands, such as pyridine, its derivatives, and other nitrogen bases improve the catalyst [46,54,55] was a major advance for MTO systems. The presence of pyridine (or some other ligand) in the reaction system slows the hydrolysis of PO to propylene glycol [42,54] and it also stabilizes the MTO against decomposition, despite the fact that MTO is unstable in basic media. Pyridine also accelerates both the H_2O_2 binding process and the epoxidation process [46,51]. The recently determined binding constants between MTO and substituted pyridines [51] show that pyridines with electron-withdrawing substituents bind most weakly (4-cyanopyridine, $K = 7.21$ l/mol), and those with electron-donating substituents bind most strongly (4-methylpyridine, $K = 732$ l/mol), while the value for pyridine itself is 200 l/mol.

Insight into the nature of MTO as a Lewis acid is found in its affinities for these monodentate ligands. Surprisingly, PyNO gives essentially the same value for its MTO binding constant as does pyridine itself, despite the fact that, based on equilibrium constants [51], pyridine is almost five orders of magnitude stronger as a Brønsted base. The result is the availability of a strong ligand (PyNO) for MTO that exerts little effect on the pH of the medium.

While MTO cannot be modified, considerable work is necessary to optimize an MTO catalyst system. There is much to be examined: the nature of the medium is critical, perhaps methanol, its water content and the activity of hydrogen ion therein, the cofactors including "accelerating ligands," perhaps PyNO, and, perhaps, a noncoordinating buffer, concentrations and relative concentrations of all reactants, temperature, pressure, and how the pressure is applied.

4. EXPANDING RANGE OF LEWIS ACID CATALYSIS CHEMISTRY

There are good reasons for the perception, described earlier, that it is the "early transition metals" whose high valent compounds perform clean Lewis acid epoxidations. Many of those high oxidation states are indeed stable and, for most of these elements, changes in the oxidation state of the metallic element are not a common feature of their chemistries. Sharpless's list of elements producing Lewis acid epoxidation catalysis is best viewed as an approximation which only marginally includes chromium and vanadium, elements whose chemistries also resemble those of middle to late transition metals. Certainly the chemistries of manganese, iron, cobalt, ruthenium, rhodium, palladium, and osmium are characterized by frequent oxidation state changes by the metal ion; that is, redox in which the substrate is most often involved. The *current wisdom* is that early transition metal ions, that is, those of groups 3–6, commonly catalyze peroxide oxidations without undergoing changes in oxidation state. In contrast, middle to late transition metal ions, those of groups 7–11, perform such oxidations by processes that reduce their own oxidation states. For this crude classification to be functional, "late" has to begin pretty early. It must include manganese and iron, since these elements so often transfer oxygen atoms from their coordination spheres to substrates by pathways often given the *rebound mechanism* label [1].

5. MID- TO LATE TRANSITION METAL CATALYSTS ALSO PERFORM EPOXIDATION REACTIONS BY THE LEWIS ACID MECHANISM

5.1. Iron compounds that catalyze lewis acid epoxidation reactions

Recent developments have indicated that certain peroxide adducts of iron may participate in oxygen transfer processes in biological and biomimetic oxidations. In a landmark study, Valentine *et al.* provided compelling evidence for Lewis acid catalysis of olefin epoxidation by a complex of iron, in a system using H_2O_2 as the oxidant and a cyclam complex as the source of iron [56]. Since Lewis acid epoxidations usually retain the diastereomeric structure of *cis*-stilbene, it is particularly significant that the oxidation of *cis*-stilbene in Valentine's system gave a high yield (26%) of *cis*-stilbene oxide and less than 2% of the *trans*-oxide. In contrast, both H_2O_2 and ROOH oxidized 2,4,6-tri-*tert*-butylphenol (TBPH) and methanol in the presence of the iron–cyclam complexes, but only H_2O_2, and not ROOH, performed the epoxidation of cyclohexene under the same conditions. These results led the authors to conclude that these two types of reactions do not occur by way of a common intermediate. The inference that two intermediates function in parallel in this system supports the suggestion that the epoxidation reaction occurs by direct reaction of the olefin with the hydroperoxide adduct of the iron catalyst, whereas the intermediate responsible for the TBPH and methanol oxidations is formed by cleavage of the O–O bond.

Coon and his collaborators [57] found that, in a certain P450 epoxidation enzyme, mutating a protein residue, which apparently would only perturb the delivery of protons, changed the ratio of epoxidation to hydroxylation products. In their opinion, this kind of product ratio change could not be explained in terms of a single oxenoid-iron intermediate, (porphyrin)$Fe^{IV} = O^+$, serving both the epoxidation and the hydroxylation processes. Therefore, they suggested that a hydroperoxo-iron intermediate could be responsible for the olefin epoxidation mediated by cytochrome P450 (Scheme 3.3). Significantly, such a hydroperoxo-iron species is the obvious intermediate that must form first and undergo transformation in order to form the oxenoid-iron intermediate. The authors also pointed out that the versatility in oxidation reactions may, in part, be attributed to the ability of P450 enzymes to use the peroxo **1**, hydroperoxo **2**, or oxenoid-iron species **3** as the active oxidant depending on the substrate and the type of reaction involved. In biological heme catabolism, it has also been concluded [58] that, in the key step of heme oxygenation, a (porphyrin)Fe^{III}-OOH intermediate is responsible for adding a OH group to the α-*meso*-carbon to give α-*meso*-hydroxy heme, then proceeding stepwise to biliverdin. Also, alkyl peroxide adducts of iron were later suggested by Nam *et al.* as parallel intermediates for epoxidation in addition to oxenoid-iron(V) porphyrin (Scheme 3.4) [59]. The authors suggest that protic solvents such as methanol may function as general acid catalysts to increase

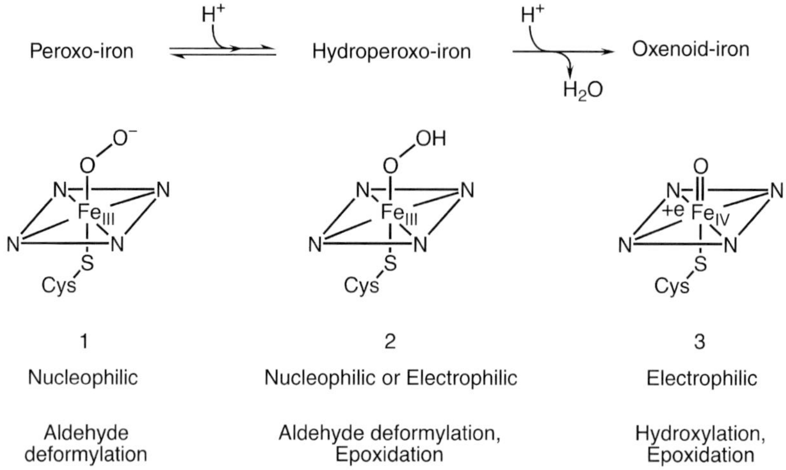

SCHEME 3.3 Proposed versatility of P450 oxygenating species. The nucleophilic or electrophilic properties and typical reactions catalyzed are indicated under the structures of the peroxo, hydroperoxo, and oxenoid species. [Reprinted with permission from *Proc. Natl. Acad. Sci. USA* 95 (1998) 3559, Figure 4. Copyright (1998) National Academy of Sciences, USA.]

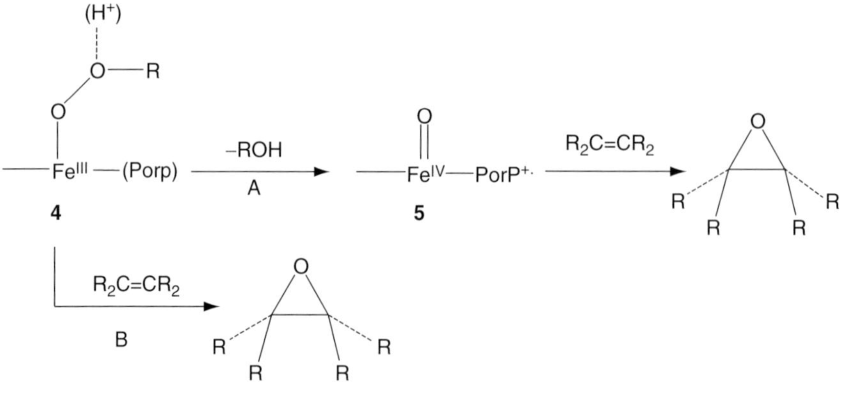

SCHEME 3.4 Proposed mechanism for olefin epoxidation in protic and aprotic solvents. [Reprinted with permission from *J. Am. Chem. Soc.* 122 (2000) 6645, Scheme 2. Copyright (2000) American Chemical Society.]

the rate of the O–O bond cleavage in the hydroperoxy heme intermediate, **4**, to form the oxenoid-iron intermediate, **5**, which epoxidizes olefins by the usual oxygen rebound mechanism. However, in the absence of general acid catalysis, such as in an *aprotic* solvent, the rate of the O–O bond cleavage of **4** is relatively slow, so that it transfers its oxygen atom to the olefin substrate more rapidly than it transforms into **5**.

Newcomb's studies of intramolecular and intermolecular kinetic isotope effects (KIEs) provided additional support for the suggestion that the hydroperoxide adduct of an iron porphyrin can serve as a second reactive intermediate in cytochrome P450 hydroxylation reactions. This was proposed as an alternative to the general belief that an oxenoid intermediate, the classic "Compound I," is the reactive intermediate and that the reaction proceeds by the often invoked oxygen rebound mechanism [60,61]. Studies of the hydroxylation of the enantiomers of trans-2-(p-trifluoromethyl phenyl)cyclopropylmethane by hepatic cytochrome P450 provided strong evidence for two oxidant intermediates. It was proposed that Fe–OOH and (porphyrin)Fe^{IV}=O^{+} are involved in the hydroxylation process rather than two different spin states, that is, low spin and high spin iron in the common oxenoid moiety [61]. A hydroperoxide adduct of iron was also suggested by Sam et al. as an intermediate in DNA cleavage by bleomycin, as a result of electrospray mass spectrometric studies of activated bleomycin [62]. Furthermore, Que et al. also provided evidence for the existence of [LFe–OOH] by EMI-MS and Raman spectroscopy of model compounds [63,64].

On the basis of these varied studies on the intermediates responsible for the versatility of oxidations catalyzed by compounds of iron, the existence and function of the hydroperoxide complex of iron in intermediate and high oxidation states is widely accepted. This perspective is consistent with the broader view that, like water flowing from high ground, exothermic oxidation reactions have, by nature, multiple pathways to products. The barriers and conditions determine the specific pathways that dominate.

5.2. Early examples of catalysis by lewis acid adducts involving manganese

In view of its wealth of high oxidation states, manganese should be expected to generate multiple activated intermediates. Accordingly, a number of studies indicate that iodosylbenzene adducts of Mn(III) or Mn(V) with salen or corrole ligands are possibly involved in oxygen transfer processes, such as olefin epoxidation [65–67]. Goldberg et al. reported that the isolated and well characterized Mn(V) oxenoid complex, (corrolazine)Mn^{V}=O, is incapable of epoxidation in the absence of additional oxidant, whereas in the presence of iodosylbenzene, it catalyzes cis-stilbene epoxidation. This suggests that the high oxidation state Mn(V) complex functions as a Lewis acid catalyst for olefin epoxidation (Scheme 3.5) [67]. At about the same time, Collman et al. investigated the stereoselectivity of olefin epoxidation catalyzed by Mn^{III}(salen) complexes with various iodosylbenzene derivatives (PhIO, C$_6$F$_5$IO, and MesIO), and found that the cis/trans ratios of stilbene oxides were strongly dependent on the oxidant and the reaction conditions. The authors suggested that, in this case, an oxenoid intermediate, O=Mn^{V}(salen), and a complex between the catalyst and the terminal oxidant affect the epoxidation, competitively, along parallel pathways. Otherwise, a single active intermediate, O=Mn^{V}(salen), would consistently produce similar cis/trans ratios of stilbene oxides with different iodosylbenzenes [65].

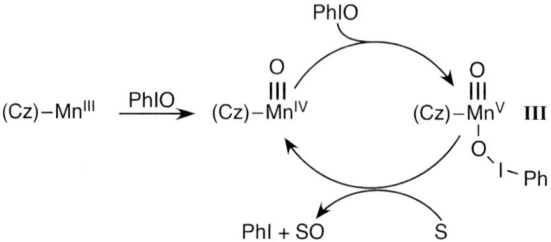

SCHEME 3.5 Reprinted with permission from *J. Am. Chem. Soc.* 126 (2004) 19, Scheme 1. Copyright (2004) American Chemical Society.

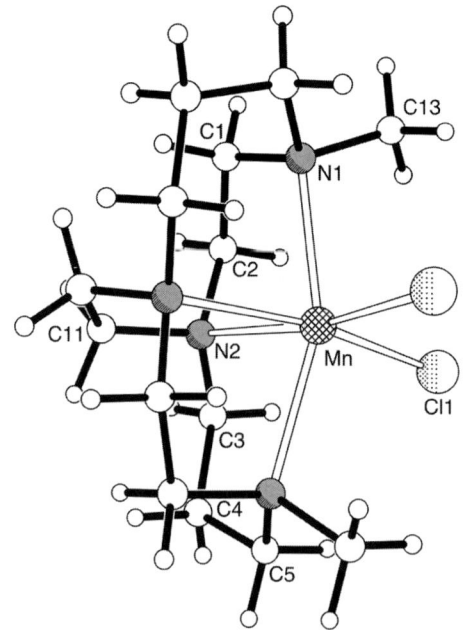

FIGURE 3.5 Structure of $Mn^{II}(Me_2EBC)Cl_2$.

5.3. A highly selective Mn(IV) catalyst for epoxidation reactions

The first established examples of manganese catalyzed epoxidations by a Lewis acid pathway using hydrogen peroxide and alkyl hydroperoxides used the novel oxidation catalyst, $Mn(Me_2EBC)Cl_2$, which is based on an ethylene cross-bridged cyclam ligand (Me_2EBC is 4,11-dimethyl-1,4,8,11-tetraazabicyclo[6.6.2]hexadecane) (Fig. 3.5) [68]. This catalyst performs highly selective oxidations with excellent efficiency, a capability well tested in its application in laundry detergents [69,70]. For this purpose, selectivity means the removal of easily oxidized stains without damage to the relatively resistant dyes and the cotton fabric. Because oxenoid Mn(V) species are well known to be highly reactive for both the epoxidation of olefinic linkages and the hydroxylation of C–H bonds in aliphatic hydrocarbons [1,71–73], a ligand very different from those stabilizing transition metals in higher oxidation states was needed. The iron porphyrin complexes of such enzymes as cytochromes P450 are powerful when oxidized and were viewed as models to be avoided. The ligands should not be negatively charged and should not have delocalized π-electron systems that include the donor atoms. The ligands should also be redox inactive and, to the extent possible, should not contain lone electron pairs of π-symmetry. The ternary nitrogen atoms of the bridged macrocycle satisfied this requirement. These features, which suggested polydentate ligands containing only nitrogen donors and the need for a complex of exceptional kinetic stability led to the constraining topology of a tetradentate cross-bridged macrocycle. Expectations were partially confirmed when detailed investigations revealed Mn(IV) as the highest oxidation state achieved by manganese in this catalyst system, combined with results from kinetic studies that showed, in comparison with macrocyles and branched ligands, many orders of magnitude stabilization of the complexes with respect to ligand dissociation in both acidic and basic media [68,74].

5.3.1. Hydrogen abstracting ability of the Mn(Me$_2$EBC)Cl$_2$ catalyst system

The redox potential for the Mn^{4+}/Mn^{3+} couple for the Mn(Me$_2$EBC)Cl$_2$ catalyst is +0.756 V versus SHE, and the Mn(IV) complex has only weak hydrogen abstracting ability [74]. Using the method of Bordwell [75] and Mayer [76], the bond dissociation energies (BDEs) of OH and water ligated to the Mn(III) complexes were calculated. These values indicate the hydrogen atom abstracting abilities of the corresponding LMn^{IV}-OH and LMn^{IV}=O moieties, and they are surprisingly similar, that is, 83.0 and 84.3 kcal/mol (Scheme 3.6). Further, experimental

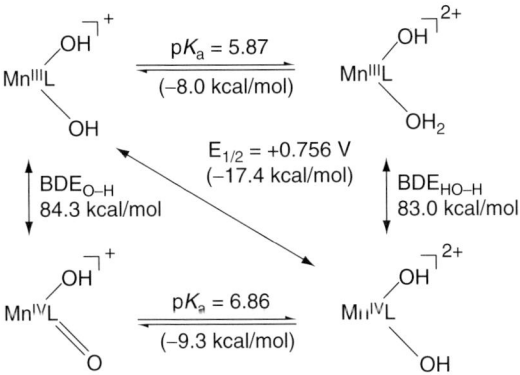

SCHEME 3.6 Theoretical calculation of the bond dissociation energies of OH$^-$ and H$_2$O ligated on the Mn(Me$_2$EBC) complex. [Reprinted with permission from [77], Scheme 1. Copyright (2007) American Chemical Society.]

determination of hydrogen abstracting ability was carried out by treating freshly synthesized [Mn^{IV}(Me$_2$EBC)(OH)$_2$](PF$_6$)$_2$ with selected substrates having known C–H BDEs. The results confirmed the prediction that hydrogen atom abstraction should not be a problem because these low BDE$_{OH}$ values for the OH bonds and supporting experiments indicated that only substantially activated C–H bonds such as substrates forming aromatics, activated allylics, and cyclic olefins will give up their hydrogens to this oxidant. Given the similarities in the BDE values for the two functional groups and, consequently, their almost identical thermodynamic hydrogen abstracting abilities, it was surprising that the LMn^{IV}(O)(OH)$^+$ moiety performs hydrogen atom abstractions 10–20 times faster than LMn^{IV}(OH)$_2^{2+}$ [77]. These hydrogen abstraction studies explain one aspect of the selectivity of this novel catalyst: conjugated color sources might be subject to attack by hydrogen abstraction, but more robust structural features should be safe. Two issues remained to be clarified: (1) does this catalyst perform epoxidations by the usual rebound mechanism and (2) to what extent does this system generate radical species?

5.3.2. Olefin epoxidation by Mn(Me$_2$EBC)Cl$_2$ using H$_2$O$_2$

The mechanisms by which epoxidation are conducted by the Mn(Me$_2$EBC)Cl$_2$ catalyst using H$_2$O$_2$ and *t*-BuOOH give insight on the high selectivity of this catalyst system [78–80]. To investigate the oxygen transfer process from H$_2$O$_2$ mediated by Mn(Me$_2$EBC)Cl$_2$, selected olefins were subjected to epoxidation and substantial yields of epoxide were obtained in each case (Table 3.3). Epoxidation of cyclohexene provided an 18% yield of cyclohexene oxide and a 13.3% yield of cyclohexene-1-one. Styrene gave a 45.5% yield of oxide and 2.8% yield of benzaldehyde. Norbornylene provided a 32% yield of norbornylene oxide. Significantly, epoxidation of *cis*-stilbene provided a 17.5% yield of *cis*-stilbene oxide, accompanied by a 2% yield of *trans*-stilbene oxide and a 2.6% yield of benzaldehyde. The latter result is very similar to that reported for the iron-cyclam/H$_2$O$_2$ system in which the HOO$^-$ adduct was proposed as the intermediate for epoxidation [56].

TABLE 3.3 Epoxidation of various olefins by Mn(Me$_2$EBC)Cl$_2$ and H$_2$O$_2$ in acetone/water media[a]

Substrate	Product	Yield (%)
Cyclohexene	Cyclohexene oxide	18.0
	Cyclohexen-1-one	13.3
Styrene	Styrene oxide	45.5
	Benzaldehyde	2.8
Norbornylene	Norbornylene oxide	32.0
cis-Stilbene	*cis*-Stilbene oxide	17.5
	trans-Stilbene oxide	2.0
	Benzaldehyde	2.6

Source: Reprinted with permission from *J. Am. Chem. Soc.* 127 (2005) 17171, Table 1. Copyright (2005) American Chemical Society.
[a] Reaction conditions: solvent, acetone/water (4:1), cat. 1 mM, olefin 0.1 M, 50% H$_2$O$_2$ 1 ml, added stepwise by 0.2 ml/0.5 h, room temperature, yield determined by gas chromatography with internal standard.

Product distributions from various olefins clearly show that at least two mechanisms are involved. Cyclohexene provided a higher yield of oxide than that of cyclohexene-1-one. The formation of cyclohexene-1-one suggests that the ROO• radical serves as a reactive intermediate in this system, but a radical process, operating alone, would produce mostly the ketone and alcohol products with very little epoxide. Most significantly, cis-stilbene gave cis-stilbene oxide as the dominant product with only a minor yield of the trans-stilbene oxide, whereas a radical process would give mostly trans-product [81].

Since the literature reports that the moieties, Mn^V=O moieties [82–85], and, in some cases Mn^{IV}=O [86,87], in their complexes with porphyrins, salens, or other ligands, react directly with olefins producing epoxides, it might be expected that the manganese oxo species, in this newly developed $Mn(Me_2EBC)Cl_2$ system, whether it contains Mn(IV) or Mn(V), would perform epoxidation reactions. Routine synthesis of the pure crystalline Mn(IV) complex (Fig. 3.6) greatly facilitated the investigation of mechanistic issues in this system. The first question, whether this catalytic reaction proceeds by the oxygen rebound mechanism, was easily addressed because the first pK_a value of $Mn^{IV}(Me_2EBC)(OH)_2^{2+}$, 6.86, indicated that substantial amounts of the compound would naturally exist as the oxo complex, $Mn^{IV}(Me_2EBC)(O)(OH)^+$, in neutral media [74]. Attempts to produce direct epoxidation reactions by $[Mn^{IV}(Me_2EBC)(OH)_2](PF_6)_2$ using various olefins, including norbornylene, styrene, and cis-stilbene, failed uniformly. Even with reaction times of several days, this Mn^{IV}=O species is not capable of epoxidation of olefins.

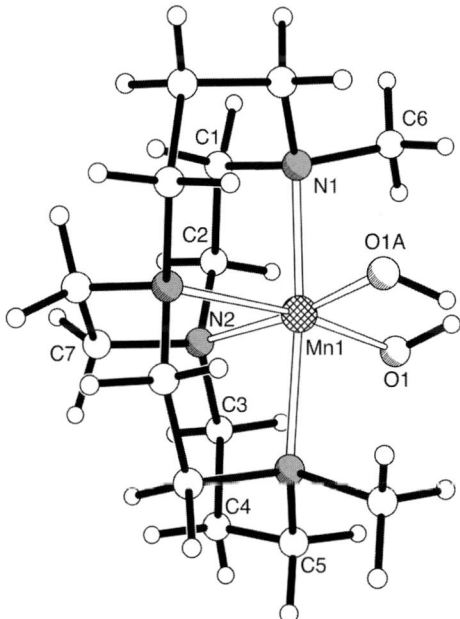

FIGURE 3.6 X-ray crystal structure of $[Mn^{IV}(Me_2EBC)(OH)_2]^{2+}$. Reprinted with permission from reference [74].

As reported in early studies of these compounds, the Mn(IV) derivative is unstable in base, decomposing to the corresponding Mn(III) complex, which is reliably recoverable in 88 ± 1% yield [74]. A rational process for this decomposition could involve the disproportionation of Mn(IV) into equal amounts of Mn(V) and Mn(III) (Eq. 3.7), followed by rapid self-destruction of the highly reactive Mn(V) complex. However, this possibility was ruled out by trapping experiments that failed to detect the putative Mn(V) intermediate.

$$2Mn(IV) \rightarrow Mn(III) + Mn(V) \quad (3.7)$$

Further, when treating the Mn(II), Mn(III), or Mn(IV) complex with various oxidants, including H_2O_2 and t-BuOOH, in aqueous solution, no evidence could be found for the existence of any species containing the $Mn^V=O$ group, and $[Mn^{IV}(Me_2EBC)(OH)_2]^{2+}$ has consistently been confirmed by ultraviolet-visible spectroscopy to be the dominant species in neutral or acidic solutions. Also, it should be recalled that the presence in the main ligand of π-systems that transmit electron density to the $Mn^V=O$ moiety, as in the cases of porphyrins and salen ligands, may be necessary to stabilize those $Mn^V=O$ species. In that context it is not surprising that stable $Mn^V=O$ derivatives do not exist with these bridged cyclam ligands since their complexes contain only tertiary nitrogen donors and no π-electron systems except those of the two monodentate oxo or hydroxo ligands derived from water.

Because of their reactivities, Mn(V) derivatives of porphyrins that were the subjects of early oxidation studies did not include isolated, well-characterized oxenoid complexes to demonstrate their epoxidizing abilities. The elegant experiments that provided clear proof for the oxygen rebound mechanism used isotopically labeled water, $H_2^{18}O$, as the solvent for olefin epoxidation, and involved two sequential reactions [88,89]. In the first reaction, the ^{18}O label confirmed the equilibration of the metal oxenoid species with the labeled water, $H_2^{18}O$, to form the labeled catalyst intermediate LM = ^{18}O (L = porphyrin). In the second step, the labeled metal complex LM = ^{18}O was identified as the source of the oxygen atom that is transferred to the olefin, thereby supporting the oxygen rebound mechanism. The basis of this work is the rapid ligand exchange of ^{18}O between M=O and ^{18}O-water.

In the highly selective manganese catalyst derived from the cross-bridged cyclam, $Mn(Me_2EBC)Cl_2$, the exchange of ^{18}O was observed between the species containing the $Mn^{IV}=O$ unit and $H_2^{18}O$ (Scheme 3.7). When $[Mn^{IV}(Me_2EBC)(OH)_2]^{2+}$ was dissolved in $H_2^{18}O$ (95% ^{18}O), the original ms peak for $[Mn^{IV}(Me_2EBC)(O)(OH)]^+$ at $m/z = 342$, observed for normal aqueous solutions, disappeared and a new ms peak for $[Mn^{IV}(Me_2EBC)(^{18}O)(^{18}OH)]^+$ appeared immediately at $m/z = 346$. This confirmed the ^{18}O exchange between the bridged cyclam manganese oxo complex and water. Furthermore, this kind of rapid ^{18}O exchange has also been demonstrated by low-frequency resonance Raman (rR) [74].

Subsequently, the isotope labeling experiment using water, $H_2^{18}O$ (95% ^{18}O), was performed during olefin epoxidation using cis-stilbene. cis-Stilbene is a useful substrate when seeking mechanistic information from isotope labeling experiments, since unlike many olefins, cis-stilbene can form two epoxides, the cis and trans isomers. The product analysis by gas chromatography-mass spectrometry shows that the amount of ^{18}O introduced into the epoxide product from experiments conducted in $H_2^{18}O$ (95% ^{18}O), is identical within experimental error to that obtained in the control experiments using normal water [1.7 ± 0.4% incorporation of ^{18}O from $H_2^{18}O$ vs 1.7 ± 0.3% incorporation of ^{18}O from normal water in cis-stilbene oxide and 3.8 ± 0.4% incorporation of ^{18}O from $H_2^{18}O$ vs 2.6 ± 1% incorporation of ^{18}O from normal water in trans-stilbene oxide (The ^{18}O signals in control experiments with normal water are due to the background signal of the mass spectrometer.); [see Schemes 3.8 and 3.9]. The results of these ^{18}O labeling experiments strongly support the conclusion that manganese oxo species, such as $Mn^{IV}(O)(OH)^+$, $Mn^{IV}(O)_2$, and $Mn^{V}(O)_2^+$, are not responsible for the oxygen transfer process. Since the Mn-oxenoid species rapidly equilibrates with $H_2^{18}O$, if it is the oxidant, it must transfer ^{18}O to the substrate. To the contrary, only baseline ^{18}O is found in the substrate. Therefore, any mechanism requiring the Mn=O group to transfer its oxygen directly to the substrate, including the rebound mechanism, is eliminated.

Building on an earlier conclusion, epoxidations of various olefins by this manganese catalyst revealed at least two mechanisms but not the oxygen rebound mechanism. To extend these studies, the epoxidation of cis-stilbene was performed with 2% H_2O_2 under an atmosphere of $^{18}O_2$ (Schemes 3.8 and 3.10). The deviation in the ^{16}O content from 100% in case of cis-stilbene oxide approximates

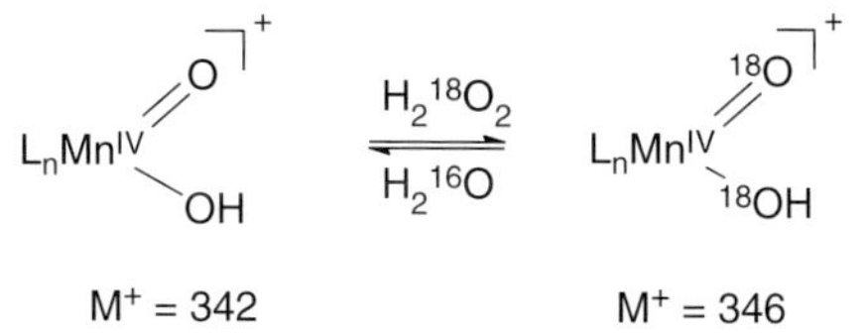

SCHEME 3.7

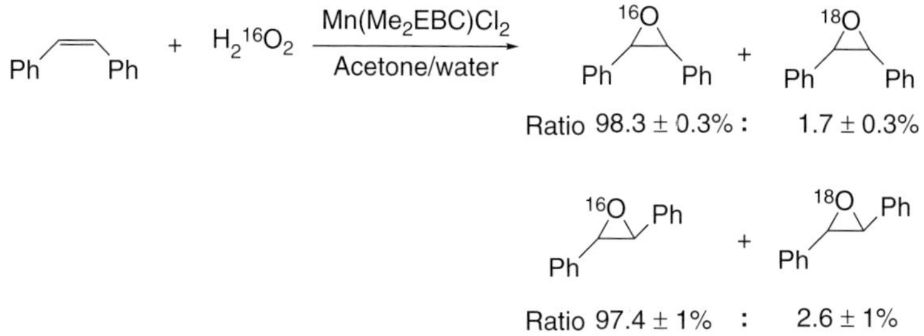

SCHEME 3.8

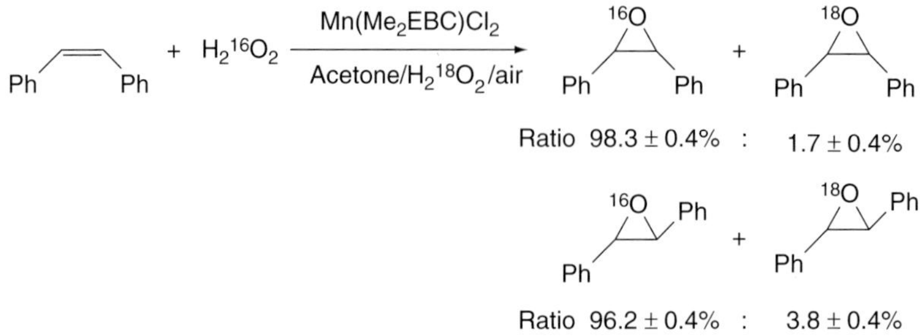

SCHEME 3.9

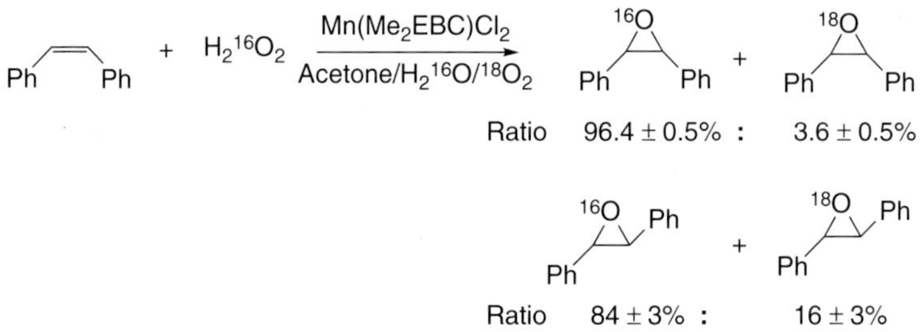

Ph−CH=CH−Ph + H$_2$^{16}O$_2$ $\xrightarrow{\text{Mn(Me}_2\text{EBC)Cl}_2}{\text{Acetone/H}_2{}^{18}\text{O}_2\text{/air}}$ Ph−(^{16}O)−Ph (trans) + Ph−(^{18}O)−Ph (trans)

Ratio 98.3 ± 0.4% : 1.7 ± 0.4%

Ph−(^{16}O)−Ph (cis) + Ph−(^{18}O)−Ph (cis)

Ratio 96.2 ± 0.4% : 3.8 ± 0.4%

SCHEME 3.10

Ph−CH=CH−Ph + H$_2$^{16}O$_2$ $\xrightarrow{\text{Mn(Me}_2\text{EBC)Cl}_2}{\text{Acetone/H}_2{}^{16}\text{O}/{}^{18}\text{O}_2}$ Ph−(^{16}O)−Ph (trans) + Ph−(^{18}O)−Ph (trans)

Ratio 96.4 ± 0.5% : 3.6 ± 0.5%

Ph−(^{16}O)−Ph (cis) + Ph−(^{18}O)−Ph (cis)

Ratio 84 ± 3% : 16 ± 3%

experimental error (3.6 ± 0.5% vs 1.7 ± 0.3% in the blank), so it is concluded from this experiment that the *cis* product derives its oxygen from sources other than O$_2$. This rules out a peroxy radical pathway for formation of this isomer of the epoxide. The small amount of *trans*-stilbene oxide (~2%) contained a substantial fraction (16 ± 3%) of oxygen from ^{18}O$_2$, implicating the expected radical pathway of the sort described in literature [90]. The remaining oxygen (84 ± 3%) may either come from a nonradical mechanism or may be the result of the abundance of unlabeled hydrogen peroxide, its liberation of O$_2$, and/or its involvement in a radical process. The results clearly show that the dominant pathway for epoxidation is not a radical pathway. This conclusion is also consistent with the catalytic results described above.

The results of studies described above have eliminated water from the solvent and O$_2$ from the air as the source of the oxygen in the product of the main epoxidation reaction, and strongly indicated that the oxygen is derived from the oxidant directly. Closure is found in the complementary experiment in which the label resides on the peroxide. Epoxidation of *cis*-stilbene with 2% H$_2$^{18}O$_2$ (90% enrichment of ^{18}O; used as received due to cost) and open to the air resulted in 89.9 ± 0.8% incorporation of ^{18}O in *cis*-stilbene oxide and 72.5 ± 2.4% incorporation in *trans*-stilbene oxide (Scheme 3.11). The reproducibility is encouraging,

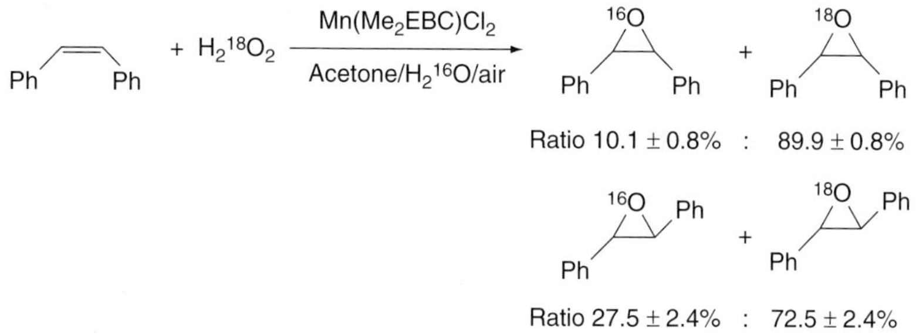

SCHEME 3.11

Ph–CH=CH–Ph + $H_2^{18}O_2$ →[Mn(Me$_2$EBC)Cl$_2$, Acetone/H_2^{16}O/air] ^{16}O-epoxide(Ph,Ph) + ^{18}O-epoxide(Ph,Ph)

Ratio 10.1 ± 0.8% : 89.9 ± 0.8%

^{16}O-epoxide(Ph,Ph) + ^{18}O-epoxide(Ph,Ph)

Ratio 27.5 ± 2.4% : 72.5 ± 2.4%

but these experiments use very expensive labeled reactant and were run with small samples so that their accuracy is limited. Even assuming an optimistic ±5%, the observed values would be the same as those from the use of labeled O_2. That is, for labeling experiments with $^{18}O_2$ and with $H_2^{18}O_2$, the oxygen in cis-stilbene oxide comes exclusively from the hydrogen peroxide (95 ± 5%), and that in trans-stilbene oxide has two sources, 80 ± 5% from hydrogen peroxide and 20 ± 5% from dioxygen. These results and the preceding discussions lead to the conclusion that, in the dominant reaction, cis-stilbene is converted to the corresponding epoxide by a nonradical pathway in which the oxygen comes directly from hydrogen peroxide, not via a rebound mechanism involving oxygenation of the manganese atom [recall that the Mn(IV) complex will not produce the epoxide].

The results summarized for the epoxidation of cis-stilbene by Mn(Me$_2$EBC)Cl$_2$ and H_2O_2 are reminiscent of epoxidations mediated by such early transition metal species as tungstates, molybdates, MTO, and titanium compounds; that is, early transition metal species in high oxidation states. In these systems, it is generally understood that H_2O_2 forms a complex with the early transition metal species and that the adduct is responsible for direct oxygen transfer to the double bond by a concerted pathway, and the oxidation state of metal ion remains unchanged throughout the catalytic cycle [91]. In the case of Mn(Me$_2$EBC)Cl$_2$, the chlorides are quickly replaced by water molecules in solution and, in the presence of hydrogen peroxide, the manganese is easily oxidized to the tetravalent state. During the catalytic process, the same species is dominant whether the manganese is added as Mn(II), Mn(III), or Mn(IV), and no evidence supports the existence of Mn(V). This activated catalyst is not like the well-known oxenoid moieties, Mn^V=O or, figuratively, Fe^V=O, which possess high oxidizing power and can transfer oxygen atoms directly to olefins forming epoxides or can abstract hydrogen atoms from saturated hydrocarbons, resulting in hydroxylation of saturated C–H bonds [92,93]. Mn^{IV}(Me$_2$EBC)(OH)$_2^{2+}$ demonstrates a modest oxidizing power with a Mn^{4+}/Mn^{3+} redox potential of +0.756 V, limited hydrogen abstracting ability (BDE values of 83–84 kcal/mol), and is incapable of epoxidizing olefins by itself. The demonstrated chemical nature on this Mn^{IV}=O moiety led to the proposal that, in the process of catalyzing olefin epoxidation, this

Mn(IV) ion functions as a Lewis acid, in the same manner as the early transition metal ions, to activate the O–O bond in hydrogen peroxide through ligation but with cleavage of the O–O bond only during the transfer of an oxygen atom to the olefin double bond [25,94–96].

The suggested mechanism is summarized in Scheme 3.12. The Mn(II) ion is first oxidized to the Mn(IV) by H_2O_2; then, through ligand exchange, HO_2^- is ligated to the Mn(IV) ion to form a peroxymanganic cation, $[Mn^{IV}(Me_2EBC)(O)(OOH)]^+$. The highly electropositive Mn^{IV} polarizes the O–O bond, making it more electrophilic, and this facilitates its two-electron oxygen atom transfer directly from the HO_2^- ligand of Mn(IV) complex to the double bond of the substrate. During the catalytic cycle, the plus four oxidation state of the manganese ion remains unchanged. In order to capture the inorganic peracid that is the expected main reactive intermediate, $LMn^{IV}(O)(OOH)^+$, the mass spectrum was determined for solutions in which $[Mn^{II}(Me_2EBC)(OH_2)_2]^{2+}$ was added to catalyze the epoxidation reaction by hydrogen peroxide. Indeed, the mass spectra of such solutions show a moderate ms peak at $m/z = 358$, establishing the presence of the HO_2^- complex, $[Mn^{IV}(Me_2EBC)(O)(OOH)]^+$, as a prominent species in these catalytic systems (Fig. 3.7), a conclusion confirmed by accurate mass measurement (M^+: calculated 358.1777; found 358.1761). Although the hydroperoxide adducts of manganese have been frequently proposed in biological oxidations such as O_2 evolution in PSII [97,98], prior to these studies, a hydroperoxide adduct of a manganese catalyst had not been detected and identified.

5.3.3. Olefin epoxidation with Mn(Me$_2$EBC)Cl$_2$ using *t*-BuOOH

The results of catalytic epoxidation of various olefins, using *t*-BuOOH as the terminal oxidant and Mn(Me$_2$EBC)Cl$_2$ as the catalyst, are summarized in Table 3.4. The color of the reaction mixture turns to purple upon addition of *t*-BuOOH and the ultraviolet-visible spectrophotometry shows that the manganese is present predominantly in the tetravalent state, and no dioxygen evolution was observed.

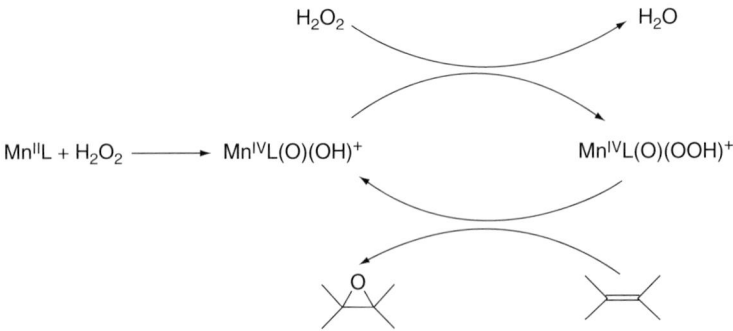

SCHEME 3.12 Mechanism for olefin epoxidation by $Mn^{II}(Me_2EBC)Cl_2$ and H_2O_2. [Reprinted with permission from [79]. Scheme 3. Copyright (2006) American Chemical Society.]

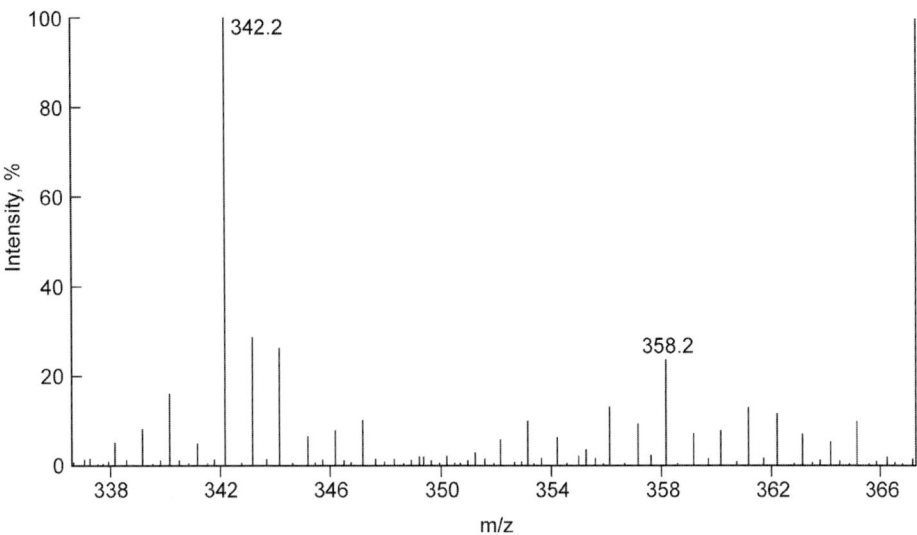

FIGURE 3.7 Electrospray ionization mass spectrum identifying $Mn^{IV}(Me_2EBC)(O)(OOH)^+$ at $m/z = 358$ (acetone/water solution containing aqueous H_2O_2). [Reprinted with permission from [79]. Figure 2. Copyright (2006) American Chemical Society.]

TABLE 3.4 Epoxidation of Various Olefins by $Mn(Me_2EBC)Cl_2$ with t-BuOOH[a]

Substrate	Product	Yield (%)
Cyclohexene	Cyclohexene oxide	1.4
	Cyclohexen-1-one	32.3
Styrene	Styrene oxide	4.8
	Benzaldehyde	24
Norbornylene	Norbornylene oxide	36.0
cis-Stilbene	cis-Stilbene oxide	0.8
	trans-Stilbene oxide	13.5
	Benzaldehyde	6.1

Source: Reprinted with permission from *Inorg. Chem.* 46 (2007) 2177, Table 1. Copyright (2007) American Chemical Society.
[a] Reaction conditions: solvent, acetone/water (4:1), cat. 1 mM, olefin 0.1 M, 0.53 M t-BuOOH, room temperature, 14 h, yield determined by gas chromatography with internal standard.

The reaction of cyclohexene produces mainly the allylic hydrogen abstraction product, cyclohexen-1-one (yield 32.3%), along with a small amount of cyclohexene oxide (yield 1.4%). With styrene, the dominant reaction involves C C bond breaking, yielding benzaldehyde (24%) with only small amounts of styrene oxide (yield 4.8%). cis-Stilbene produces mostly trans-stilbene oxide (yield 13.5%) and minor amounts of cis-stilbene oxide (yield 0.8%) plus a 6.1% yield of benzaldehyde. These results are distinctly different from those obtained from usage of hydrogen peroxide as the terminal oxidant (see above), and strongly suggest a dominant radical pathway for this oxidation reaction, as should be expected on the basis of earlier literature reports [90,99].

Significantly, the reaction of norbornylene with *t*-BuOOH in the presence of Mn(Me$_2$EBC)Cl$_2$, provides a relatively high yield of norbornylene oxide (36%), and the results of detailed studies on the effect of oxidant concentration on product distribution are in disagreement with a single radical pathway. As described by Pecoraro and coworkers [90], ROO• radicals typically contribute to both hydrogen abstraction and oxygen atom transfer, while RO• radicals perform only hydrogen abstractions. If radical processes alone were responsible, the yield of hydrogen abstraction products should increase in proportion to the increase in yield of the epoxidation products. Figure 3.8 shows that these expected parallel yield increases are not observed. Norbornylene conversion and epoxide yield both increase as the initial concentration of *t*-BuOOH is increased, but the total percentage yield of hydrogen abstraction products actually shows a small decrease with time. This behavior strongly suggests that more than one oxygen transfer mechanism must be involved in this reaction system. The peroxy radical pathway, initiated by hydrogen abstraction, must be accompanied by an additional process that contributes to the epoxidation of the olefin.

Labeling experiments with H$_2^{18}$O and ^{18}O$_2$ helped determine whether a radical process is involved in the reaction [90]. The product analysis by gas chromatography–mass spectrometry for the ^{18}O-water epoxidation of norbornylene, using *t*-BuOOH in decane as the terminal oxidant with [Mn^{II}(Me$_2$EBC)(OH$_2$)$_2$]$^{2+}$ as catalyst in 4:1 acetone/H$_2^{18}$O (95% ^{18}O), showed that no ^{18}O from H$_2^{18}$O is introduced into the epoxide product. This strongly supports the conclusion that the oxygen in the epoxide does not come directly from Mn^{IV}(Me$_2$EBC)(OH)$_2^{2+}$, Mn^{IV}(Me$_2$EBC)(O)

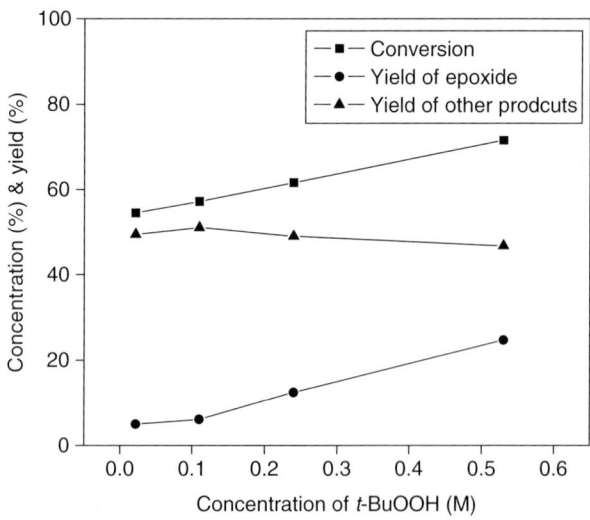

FIGURE 3.8 Influence of concentration of oxidant on norbornylene epoxidation. [Reprinted with permission from *Inorg. Chem.* 46 (2007) 2177, Figure 2. Copyright (2007) American Chemical Society.]

(OH)$^+$, or Mn^{IV}(Me$_2$EBC)(O)$_2$, and is consistent with the earlier conclusion that the Mn(IV) complex is not capable of direct epoxidation of norbornylene. However, a distinct difference exists between the two terminal oxidants for these systems, in that t-BuOOH reacts predominantly via a radical pathway whereas such a reaction path is very clearly minor in the H$_2$O$_2$ oxidation of olefins in the same catalyst system.

To address whether O$_2$ is involved in t-BuOOH epoxidation by a radical pathway, the corresponding ^{18}O$_2$ labeling experiments were conducted. The product analysis by gas chromatography–mass spectrometry shows substantial incorporation of ^{18}O from ^{18}O$_2$ into norbornylene oxide (74.6 ± 2% ^{18}O incorporation vs 4.4 ± 0.1% ^{18}O incorporation in the control experiment, see Schemes 3.13 and 3.14), strongly indicating that the dioxygen oxidation reaction involves a radical process. However, it also reveals that, there are two oxygen sources in the epoxide: 74.6 ± 2% of the oxygen comes from the labeled gaseous dioxygen by a radical process and the rest of the oxygen, 25.4 ± 2%, from the added oxidant, t-BuOOH, either by a radical or a nonradical process. One might argue that this 25.4 ± 2% oxygen could be derived from the t-BuOO• radical, either through its direct reaction or through the unlabeled dioxygen, ^{16}O$_2$, produced from the t-BuOO•. The latter would simply dilute the labeled dioxygen. In this way, both the t-BuOO• radical and its derived dioxygen, ^{16}O$_2$, could serve as the oxygen sources for the ^{16}O-epoxide.

A typical radical reaction has been demonstrated by Pecoraro and coworkers [90], in which cyclohexene oxidation was performed using a small amount of t-BuOOH as the initiator, with Mn$_2$(2-OHsalphen)$_2$ as the catalyst, under an ^{18}O$_2$ atmosphere. The product analysis showed that the oxygen sources in the cyclohexene oxidation products were derived entirely from ^{18}O$_2$, in general agreement with an exclusive radical oxidation reaction. We have also carried out isotopically labeled ^{18}O$_2$ experiments for norbornylene oxidation, using t-BuOOH as the initiator, with Mn(Me$_2$EBC)Cl$_2$ as the catalyst, under conditions identical to those described by Pecoraro. The results reveal that, even though the concentration of t-BuOOH is low, the norbornylene oxide product still contains only 71.8 ± 1.8% ^{18}O incorporation from isotopic dioxygen, ^{18}O$_2$, with 28.2 ± 1.8%

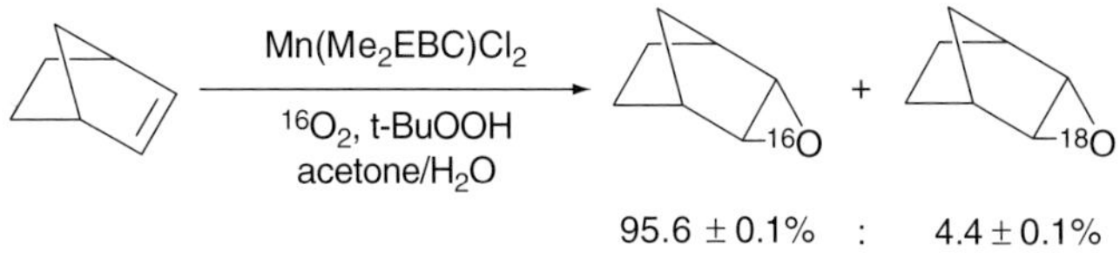

SCHEME 3.13

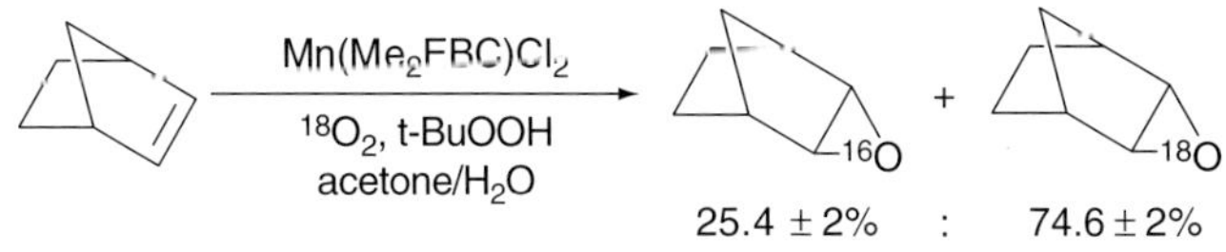

SCHEME 3.14

^{16}O traceable to the initiator, *t*-BuOOH. This suggests a mechanistic difference between this system and the catalyst system described by Pecoraro. It is concluded that the data both from the isotopic labeling of dioxygen and from the previously discussed kinetic analysis require at least two pathways for the olefin. One of these pathways proceeds to the olefin epoxide by the expected peroxy radical mechanism. The other pathway appears to require direct transfer of an oxygen atom from the oxidant, *t*-BuOOH, to the olefin but not by the familiar rebound mechanism, since the oxygen has not equilibrated with the water in the solvent.

As stated above, *cis*-stilbene is a frequently used substrate for the study of olefin epoxidation mechanisms [100] because of mechanistic information associated with the ratio of the *cis*- and *trans*-isomers in the stilbene oxide product. Catalytic *t*-BuOOH epoxidation of *cis*-stilbene was performed, with our Mn(Me$_2$EBC)Cl$_2$ catalyst, under ^{18}O$_2$ and product analysis shows that *cis*-stilbene oxide contains 25 ± 0.3% incorporation of ^{18}O from atmospheric ^{18}O$_2$ versus 1.7 ± 0.3% ^{18}O in a control experiment using ordinary air, whereas *trans*-stilbene oxide contains 55.1 ± 1% incorporation of ^{18}O from ^{18}O$_2$ versus 2.6 ± 1% ^{18}O incorporation in a control experiment. Similar ^{18}O incorporation ratios are expected for the *cis*- and *trans*-stilbene oxides if all epoxidation products result from only one reaction pathway and a single reactive intermediate. However, the incorporation of ^{18}O in *cis*-stilbene oxide (25 ± 0.3%) is definitely different from that in *trans*-stilbene oxide (55.1 ± 1%). This result leads to the conclusion that, at least two distinct reactive intermediates occur in these epoxidation reactions. This is also consistent with the results described above for norbornylene epoxidation.

As discussed, epoxidation of *cis*-stilbene provides mostly *trans*-stilbene oxide (yield of 13.5%) and that isomer contains excess ^{18}O derived from ^{18}O$_2$ (55.1 ± 1%), and this is consistent with a radical pathway in which ROO$^\bullet$ is the reactive intermediate (Scheme 3.15, pathway A) [87,96]. The substantially greater yield of cyclohexen-1-one over cyclohexene oxide in cyclohexene oxidation, and the dominance of products resulting from C–C bond cleavage in styrene oxidation suggest the same radical mechanism. The remaining viable model for the second reactive intermediate leading to the minor epoxidation product, that is, *cis*-stilbene epoxide, from *cis*-stilbene, which displays a lower ^{18}O incorporation (25 ± 0.3%), is the Lewis acid pathway in which an oxygen atom is transferred directly from the alkyl hydroperoxide adduct of Mn(IV) complex to the olefinic double bond. As described above, earlier studies provided strong support for the corresponding mechanism for olefin epoxidation by the hydrogen peroxide adduct of the Mn(IV) complex, Mn^{IV}(Me$_2$EBC)(O)(OOH)$^+$ (Scheme 3.15, pathway B). The Mn(IV) complex is the dominant manganese species in the solution during the oxidation reaction (vide supra).

The alkyl hydroperoxide adduct of that Mn(IV) complex, Mn^{IV}(Me$_2$EBC)(O)(OOR)$^+$, should be formed by simple ligand exchange in which *t*-BuOO$^-$ replaces the hydroxo ligand in Mn^{IV}(Me$_2$EBC)(O)(OH)$^+$ or by oxidation of Mn^{III}(Me$_2$EBC)(O)(OH) by the ROO$^\bullet$ radical, again replacing the hydroxo ligand. Within the adduct, the high charge of Mn(IV) and its potent Lewis acid character polarizes the O–O bond of the bound alkyl peroxide, making it susceptible to nucleophilic

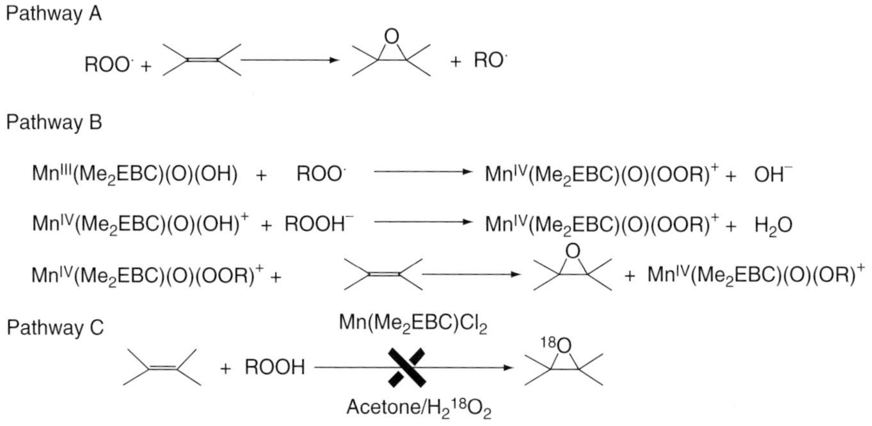

SCHEME 3.15 Proposed mechanisms for olefin epoxidation. Pathways A and B have been implicated for the system using Mn(Me$_2$EBC)Cl$_2$ and *t*-BuOOH. [Reprinted with permission from *Inorg. Chem.* 46 (2007) 2178, Scheme 3. Copyright (2007) American Chemical Society.]

attack by the olefinic double bond (Scheme 3.15, pathway B), an established process for high oxidation state early transition element species, including titanium(IV), vanadium(V), molybdenum(VI), tungsten(VI), and rhenium(VII) [25,94–96]. For completeness, Scheme 3.15 includes pathway C, the oxygen rebound mechanism. However, as shown earlier, pathway C does not occur in the systems under study here.

In order to detect the expected second reactive intermediate, mass spectra were determined for solutions in which [Mn^{II}(Me$_2$EBC)(OH$_2$)$_2$]$^{2+}$ and its oxidized derivatives were catalyzing the *t*-BuOOH epoxidation reaction, Mn^{IV}(Me$_2$EBC)(O)(OOR)$^+$ (Fig. 3.9). A strong peak as $m/z = 342$ is assigned to the singly charged complex of the oxidized catalyst, Mn^{IV}(Me$_2$EBC)(O)(OH)$^+$. A weak ms peak at $m/z = 414$ can be attributed to [Mn^{IV}(Me$_2$EBC)(O)(*t*-OOBu)]$^+$ because accurate mass measurement provides precise support for that assignment (M$^+$: calculated 414.2403; found 414.2402). This provided the first direct evidence for the existence of the alkyl hydroperoxide adduct of a high oxidation state manganese species. In heme and non-heme iron monoxygenase models, Fe–OOR species (R=H, alkyl) have been proposed as the reactive intermediates in certain oxygenation processes, and ms and Raman spectra have also provided evidence for the existence of Fe–OOH species [56,59,63,64]. In retrospect, one should expect the alkyl peroxide adduct of Mn(IV) to be an active intermediate for epoxidation in this and other manganese systems in which the Mn(IV)/Mn(III) couple has a modest potential, because Mn(IV) ion is reminiscent of Ti(IV). As stated above, the early transition metals, Ti, V, W, Mo, and Re, are well known to form metallo-peracid species when they function as epoxidation catalysts [25,94–96]. Although the alkyl peroxide adduct of this Mn(IV) complex plays only a minor role in the chemically catalyzed oxygen transfer process studied here, we suggest that the same process

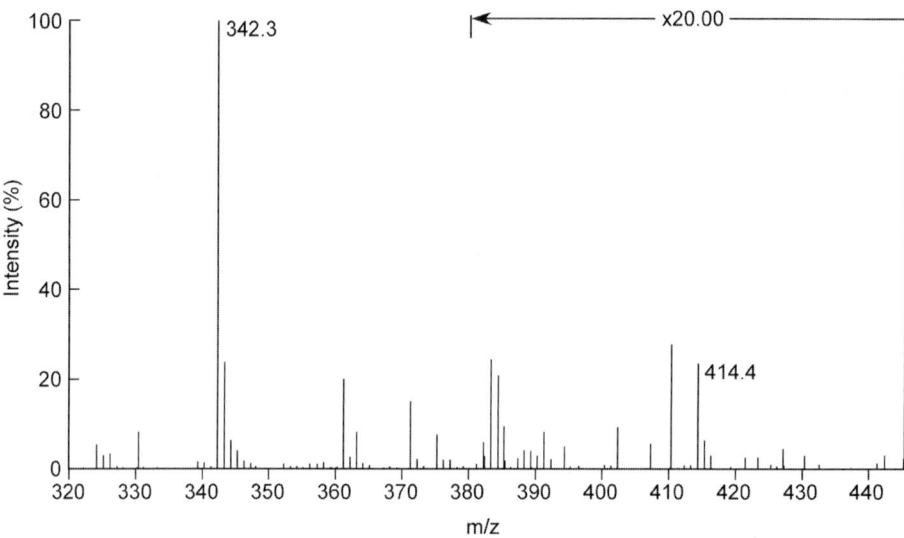

FIGURE 3.9 Electrospray ionization mass spectrum identifying $Mn^{IV}(Me_2EBC)(O)(t\text{-}OOBu)^+$ at $m/z = 414$ (acetone/water solution containing aqueous t-BuOOH). [Reprinted with permission from *Inorg. Chem.* 46 (2007) 2179, Figure 3. Copyright (2007), American Chemical Society].

could be much more important in such highly selective systems as those of biological origin.

The difference in behavior of the two terminal oxidants, H_2O_2 and t-BuOOH, in these systems deserves brief comment. In the case of H_2O_2, the Lewis acid epoxidation reaction, pathway B in Scheme 3.15, dominates the process, whereas the radical process of pathway A is dominant in the t-BuOOH case. This contrasting mechanistic behavior can be rationalized on the basis of the physical parameters of the O–O bonds in these two oxidants. The BDE of the O–O bond in H_2O_2 is 213.8 kJ/mol which is 21.8 kJ/mol higher than that of O–O bond in t-BuOOH (192 kJ/mol [101]). The Mn–O–O–t-Bu moiety more easily undergoes homolytic O–O bond cleavage than does Mn–O–O–H group, and this favors a radical pathway for the t-BuOOH derivative. Therefore, it is suggested that the hydrogen peroxide adduct of the manganese complex is the main intermediate in H_2O_2 systems, leading to epoxidation, accompanied by only a minor radical process, while, in contrast, with t-BuOOH as the oxidant, a radical process dominates with minor participation by the alkyl peroxide/manganese adduct and relatively little epoxidation by that pathway.

ACKNOWLEDGMENTS

The authors sincerely thank our coworkers and collaborators whose names are listed in the references to our joint work. Support by the Procter and Gamble Company of the mechanistic studies of the bridged-cyclam/manganese catalysts is gratefully acknowledged. PO work was supported by NSF

Grant (CHE-0328185). We also acknowledge the National Science Foundation Engineering Research Center Grant (EEC-0310689) for partial support.

REFERENCES

[1] J. T. Groves, Key elements of the chemistry of cytochrome P-450. The oxygen rebound mechanism, *J. Chem. Edu.* 65 (1985) 928.
[2] K. B. Sharpless, Searching for new reactivity (Nobel Lecture), *Angew. Chem. Int. Ed.* 41 (2002) 2024.
[3] W. S. Knowles, Asymmetric hydrogenations (Nobel Lecture), *Angew. Chem. Int. Ed.* 41 (2002) 1998.
[4] R. Noyori, Asymmetric catalysis: Science and opportunities (Nobel Lecture), *Angew. Chem. Int. Ed.* 41 (2002) 2008.
[5] P. R. Ortiz de Montellano, *Cytochrome P-450: Structure, Mechanism, and Biochemistry*. Plenum Press, New York, NY, 1986.
[6] T. D. Porter, M. J. Coon, Cytochrome P-450. Multiplicity of isoforms, substrates, and catalytic and regulatory mechanisms, *J. Biol. Chem.* 266 (1991) 13469.
[7] B. Chance, Enzyme-substrate compounds of horseradish peroxidase and peroxides. II. Kinetics of formation and decomposition of the primary and secondary complexes, *Arch. Biochem. Biophys.* 22 (1949) 224.
[8] E. M. McGarrigle, D. G. Gilheany, Chromium- and Manganese-salen Promoted Epoxidation of Alkenes, *Chem. Rev.* 105 (2005) 1564.
[9] A. Molnar, G. A. Olah, Oxidation of Alkenes, in: *Hydrocarbon Chemistry*, Wiley-Interscience, Hoboken, NJ, 2003, pp. 452–461.
[10] V. Conte, F. Di Furia, Catalytic oxidations with hydrogen peroxide as oxidant. Peroxometal complexes derived from hydrogen peroxide. Some applications in organic synthesis. *Catalysis by Metal Complexes* 9 (1992) 223–252.
[11] H. B. Henbest, R. A. I. Wilson, Aspects of stereochemistry. I. Stereospecificity in formation of epoxides from cyclic allylic alcohols, *J. Chem. Soc.* (1957) 1958.
[12] K. B. Sharpless, R. C. Michaelson, High stereo- and regioselectivities in the transition metal catalyzed epoxidations of olefinic alcohols by tert-butyl hydroperoxide, *J. Am. Chem. Soc.* 95 (1973) 6136.
[13] S. S. Woodard, M. G. Finn, K. B. Sharpless, Mechanism of asymmetric epoxidation. I. Kinetics, *J. Am. Chem. Soc.* 113 (1991) 106.
[14] M. G. Finn, K. B. Sharpless, Mechanism of asymmetric epoxidation. I. Kinetics, *J. Am. Chem. Soc.* 113 (1991) 113.
[15] D. Trent, Propylene Oxide, in: *Encyclopedia of Chemical Technology*, Vol. 30, 4th ed., John Wiley & Sons, New York, NY, 1996, pp. 271–302.
[16] W. Gerhartz, Y. S. Yamamoto, L. Kaudy, J. F. Rounsaville, and J. Schulz (Eds.), *Ullmann's Encyclopedia of Industrial Chemistry*, Vol. A9, 5th ed., Verlag Chemie, Weinheim, 1987, pp. 531–564.
[17] W. A. Herrmann, M. Dieter, W. Wagner, J. G. Kuchler, G. Weichselbaumer, R. Fischer, Use of organorhenium compounds for the oxidation of multiple C–C bonds, oxidation based thereon and novel organorhenium compounds, U.S. Patent 5155247, Oct. 13, 1992, Assigned to Hoechst Aktiegesellschaft.
[18] G. L. Crocco, W. F. Shum, J. G. Zajacek, S. Kesling Jr., Epoxidation process, U.S. Patent, 5166372, Assigned to Arco Chemical Technology.
[19] C. Venturello, R. D'Alolslo, Quaternary ammonium tetrakis(diperoxotungsto)phosphates(3-) as a new class of catalysts for efficient alkene epoxidation with hydrogen peroxide, *J. Org. Chem.* 53 (1988) 1553.
[20] C. Venturello, E. Alneri, M. Ricci, A new, effective catalytic system for epoxidation of olefins by hydrogen peroxide under phase-transfer conditions, *J. Org. Chem.* 48 (1983) 3831.
[21] C. Venturello, R. D'Aloislo, J. C. J. Bart, M. Ricci, A new peroxotungsten heteropoly anion with special oxidizing properties: Synthesis and structure of tetrahexylammonium tetra(diperoxotungsto)phosphate(3-), *J. Mol. Catal.* 32 (1985) 107.

[22] K. Sato, M. Aki, M. Ogawa, T. Hashimoto, R. Noyori, A practical method for epoxidation of terminal olefins with 30% hydrogen peroxide under halide-free conditions, *J. Org. Chem.* 61 (1996) 8310.
[23] R. Noyori, M. Aoki, K. Sato, Green oxidation with aqueous hydrogen peroxide, *Chem. Comm.* (2003) 1977.
[24] P. U. Maheswari, P. de Hoog, R. Hage, P. Gamez, J. Reedijk, A Na2WO4/H2WO4-based highly efficient biphasic catalyst towards alkene epoxidation, using dihydrogen peroxide as oxidant, *Adv. Synth. Catal.* 347 (2005) 1759.
[25] Z. Xi, N. Zou, Y. Sun, K. Li, Reaction-controlled phase-transfer catalysis for propylene epoxidation to propylene oxide, *Science* 292 (2001) 1139.
[26] T. Okuhara, N. Mizuno, M. Misono, Catalytic chemistry of heteropoly compounds, *Adv. Catal.* 41 (1996) 113.
[27] N. Mizumo, C. Nozaki, I. Kiyoto, M. Misono, Highly efficient utilization of hydrogen peroxide for selective oxygenation of alkanes catalyzed by Diiron-substituted polyoxometalate precursor, *J. Am. Chem. Soc.* 120 (1998) 9267.
[28] R. Neumann, Polyoxometalate complexes in organic oxidation chemistry, *Prog. Inorg. Chem.* 47 (1998) 317.
[29] R. Neumann, M. Gara, The Manganese-containing polyoxometalate, [WZnMnII2(ZnW9O34)2]12-, as a remarkably effective catalyst for hydrogen peroxide mediated oxidations, *J. Am. Chem. Soc.* 117 (1995) 5066.
[30] F. A. Cotton, G. Wilkinson, *Advanced Inorganic Chemistry*, 5th ed., John Wiley, New York, NY, 1988, p. 816.
[31] K. Kamata, K. Yonehara, Y. Sumida, K. Yamaguchi, S. Hikichi, N. Mizuno, Efficient epoxidation of olefins with ^{3}99% selectivity and use of hydrogen mediated peroxide, *Science* 300 (2003) 964.
[32] H. Tullo, P. L. Short, Prepylene oxide routes take off, *Chem. Eng. News* 84(41), (2006) 22.
[33] P. T. Anastas, J. C. Warner, *Green Chemistry. Theory and Practice*, Oxford University Press, Oxford, UK, 1998.
[34] M. Lancaster, *Green Chemistry: An Introductory Text*. Royal Society of Chemistry, Cambridge, UK, 2002.
[35] B. Subramaniam, D. H. Busch, H.-J. Lee, T.-P. Shi, Process for the selective oxidation of propylene to propylene oxide and facile separation of propylene oxide, U.S. Provisional Patent filled 10/25/05; converted to U.S. Patent Application, 10/25/06, Assigned to University of Kansas.
[36] F. E. Kuehn, A. M. Santos, W. A. Herrmann, Organorhenium (VII) and organomolybdenum (VI) oxides: Syntheses and application in olefin epoxidation, *Dalton Trans.* (2005) 2483.
[37] W. A. Herrmann, F. E. Kuehn, Organorhenium Oxides, *Acc. Chem. Res.* 30 (1997) 169.
[38] R. Beattie, P. J. Jones, Methyltrioxorhenium. An air-stable compound containing a carbon-rhenium bond, *Inorg. Chem.* 18 (1979) 2318.
[39] W. A. Herrmann, R. W. Fischer, M. U. Rauch, W. Scherer, Multiple bonds between main-group elements transition metals. 125. Alkylrhenium oxides as homogeneous epoxidation catalysts: Activity, selectivity, stability, deactivation, *J. Mol. Catal.* 86 (1994) 243.
[40] W. A. Herrmann, R. W. Fischer, D. W. Marz, Multiple bonding between main-group elements transition metals. 100. Part 2, Methyltrioxorhenium as catalyst for olefin oxidation, *Angew. Chem. Int. Ed. Engl.* 30 (1991) 1638.
[41] F. E. Kühn, A. M. Santos, W. A. Herrmann, Organorhenium(VII) and organomolybdenum(VI) oxides: Syntheses and application in olefin epoxidation, *J. Chem. Soc., Dalton Trans.* (2005) 2483.
[42] F. E. Kühn, W. A. Herrmann, Methyltrioxorhenium(VII) as an oxidation catalyst, *Aqueous Phase Organometallic Catalysis*, in: B. Cornils, W.A. Herrmann, Eds., 2nd ed., Wiley-VCH Verlag GmbH & Co., KGaA, Weinheim, 2004, p. 488.
[43] F. E. Kühn, W. A. Herrmann, Organorhenium Oxides *Acc. Chem. Res.* 30 (1997) 169.
[44] J. B. Espenson, M. M. Abu-Omar, Reactions catalyzed by methylrheniumtrioxide, *Adv. Chem. Ser.* 253, American Chemistry Society, Washington, DC, 1997, p. 99.
[45] J. H. Espenson, Atom-transfer reactions catalyzed by methyltrioxorhenium(VII) – mechanisms and applications, *Chem. Commun.* (1999) 479.

[46] J. Rudolph, K. L. Reddy, J. P. Chiang, K. B. Sharpless, Highly efficient epoxidation of olefins using aqueous H2O2 and catalytic Methyltrioxorhenium/pyridine; Pyridine-mediated ligand acceleration, *J. Am. Chem. Soc.* 119 (1997) 6189.
[47] M. M. Abu-Omar, P. J. Hansen, J. H. Espenson, Deactivation of Methylrhenium trioxide-peroxide catalysts by diverse and competing pathways, *J. Am. Chem. Soc.* 118 (1996) 4966.
[48] Z. Shu, J. H. Espenson, Kinetics and mechanism of oxidation of anilines by hydrogen peroxide as catalyzed by Methylrhenium trioxide, *J. Org. Chem.* 60 (1995) 1326.
[49] W. D. Wang, J. H. Espenson, Thermal and photochemical reactions of methylrhenium diperoxide: Formation of methyl hydroperoxide in acetonitrile, *Inorg. Chem.* 36 (1997) 5069.
[50] M. M. Abu-Omar, J. H. Espenson, Oxidations of ER3 (E = P, As, or Sb) by hydorgen peroxide: Methylrhenium trioxide as catalyst, *J. Am. Chem. Soc.* 117 (1995) 272.
[51] W. D. Wang, J. H. Espenson, Effects of pyridine and its derivatives on the equilibria and kinetics pertaining to epoxidation reactions catalyzed by methyltrioxorhenium, *J. Am. Chem. Soc.* 120 (1998) 11335.
[52] S. Yamazaki, J. H. Espenson, P. Huston, Equilibria and kinetics of the reactions between hydrogen peroxide and methyltrioxorhenium in aqueous perchloric acid solutions, *Inorg. Chem.* 32 (1993) 4683.
[53] P. J. Hansen, J. H. Espenson, Oxidation of chloride ions by hydrogen peroxide, catalyzed by methylrhenium trioxide, *Inorg. Chem.* 34 (1995) 5839.
[54] C. Coperet, H. Adolfsson, K. B. Sharpless, A simple and efficient method for epoxidation of terminal alkenes, *J. Chem. Soc. Chem. Commun.* (1997) 1565.
[55] W. A. Herrmann, R. M. Kratzer, F. E. Kuhn, J. J. Haider, R. W. Fischer, Multiple bonds between transition metals and main-group elements. Part 168. Methyltrioxorhenium/Lewis base catalysts in olefin epoxidation, *J. Organomet. Chem.* 549 (1997) 319.
[56] W. Nam, R. Ho, J. S. Valentine, Iron-cyclam complexes as catalysts for the epoxidation of olefins by 30% aqueous hydrogen peroxide in acetonitrile and methanol, *J. Am. Chem. Soc.* 113 (1991) 7052.
[57] D. N. Vaz, D. F. McGinnity, M. J. Coon, Epoxidation of olefins by cytochrome P450: Evidence from site-specific mutagenesis for hydroperoxo-iron as an electrophilic oxidant, *Proc. Natl. Acad. Sci. USA* 95 (1998) 3555.
[58] L. Avila, H. Huang, C. O. Damaso, S. Lu, P. Moënne-Loccoz, M. Rivera, Coupied oxidation vs heme oxygenation: Insights from axial ligand mutants of mitochondrial cytochrome b5, *J. Am. Chem. Soc.* 125 (2003) 4103.
[59] W. Nam, M. H. Lim, H. J. Lee, C. Kim, Evidence for the participation of two distinct reactive intermediates in Iron(III) porphyrin complex-catalyzed epoxidation reactions, *J. Am. Chem. Soc.* 122 (2000) 6641.
[60] M. Newcomb, D. Aebisher, R. Shen, R. E. P. Chandrasena, P. F. Hollenberg, M. J. Coon, Kinetic isotope effects implicate two electrophilic oxidants in cytochrome P450-Catalyzed hydroxylations, *J. Am. Chem. Soc.* 125 (2003) 6064.
[61] P. H. Toy, M. Newcomb, M. J. Coon, A. D. N. Vaz, Two distinct electrophilic oxidants effect hydroxylation in cytochrome P-450-catalyzed reactions, *J. Am. Chem. Soc.* 120 (1998) 9718.
[62] J. W. Sam, X. J. Tang, J. Peisach, Electrospray mass spectrometry of iron Bleomycin: Demonstration that activated Bleomycin is ferric peroxide complex, *J. Am. Chem. Soc.* 116 (1994) 5250.
[63] R. Y. N. Ho, G. Roelfes, B. L. Feringa, L. Que Jr., Raman evidence for a weakened O-O bond in mononuclear low-spin Iron(III)-hydroperoxides, *J. Am. Chem. Soc.* 121 (1999) 264.
[64] M. Lubben, A. Meetsma, E. C. Wilkinson, B. Feringa, L. Que Jr., Nonheme iron centers in oxygen activation: Characterization of an iron(III) hydroperoxide intermediate, *Angew. Chem. Int. Ed. Engl.* 34 (1995) 1512.
[65] J. P. Collman, L. Zeng, J. I. Brauman, Donor ligand effect on the nature of the oxygenating species in MnIII(salen)-catalyzed epoxidation of olefins: Experimental evidence for multiple active oxidants, *Inorg. Chem.* 43 (2004) 2672.
[66] R. B. VanAtta, C. C. Franklin, J. S. Valentine, Oxygenation of organic substrates by iodosylbenzene catalyzed by soluble manganese, iron, cobalt, or copper salts in acetonitrile, *Inorg. Chem.* 23 (1984) 4121.
[67] S. H. Wang, B. S. Mandimutsira, R. Todd, B. Ramdhanie, J. P. Fox, D. P. Goldberg, Catalytic Sulfoxidation and Epoxidation with a Mn(III) Triazacorrole: Evidence for a "Third Oxidant" in high-valent porphyrinoid oxidations, *J. Am. Chem. Soc.* 126 (2004) 18.

[68] T. J. Hubin, J. M. McCormick, S. R. Collinson, M. Buchalova, C. M. Perkins, N. W. Alcock, P. K. Kahol, A. Raghunathan, D. H. Busch, New Iron(II) and Manganese(II) complexes of two ultra-rigid, cross-bridged tetraazamacrocycles for catalysis and biomimicry, *J. Am. Chem. Soc.* 122 (2000) 2512.

[69] D. H. Busch, S. R. Collinson, T. J. Hubin, Catalysts and methods for catalytic oxidation. (Sept. 11, 1998) WO 98/39098, Assigned to University of Kansas.

[70] D. H. Busch, S. R. Collinson, T. J. Hubin, R. Labeque, B. K. Williams, J. P. Johnston, D. Kitko, J. Burckett-St. Laurent, C. M. Perkins, Bleach compositions. (Sept. 11, 1998) WO 98/39406, Assigned to Procter and Gamble Company.

[71] J. M. Garrison, T. C. Bruice, Intermediates in the epoxidation of alkenes by cytochrome P-450 models. 3. Mechanism of oxygen transfer from substituted oxochromium(V) porphyrins to olefinic substrates, *J. Am. Chem. Soc.* 111 (1989) 191.

[72] J. P. Collman, A. S. Chien, T. A. Eberspacher, J. I. Brauman, Multiple active oxidants in cytochrome P-450 model oxidations, *J. Am. Chem. Soc.* 122 (2000) 11098.

[73] W. Zhang, E. N. Jacobsen, Asymmetric olefin epoxidation with sodium hypochlorite catalyzed by easily prepared chiral manganese(III) salen complexes, *J. Org. Chem.* 56 (1991) 2296.

[74] G. Yin, J. M. McCormick, M. Buchalova, A. M. Danby, K. Rodgers, K. Smith, C. Perkins, D. Kitko, J. Carter, W. M. Scheper, D. H. Busch, Synthesis, characterization, and solution properties of a novel cross-bridged cyclam Manganese(IV) complex having two terminal hydroxo ligands, *Inorg. Chem.* 45 (2006) 8052.

[75] F. G. Bordwell, J. P. Cheng, J. A. Harrelson, Homolytic bond dissociation energies in solution from equilibrium acidity and electrochemical data, *J. Am. Chem. Soc.* 110 (1988) 1229.

[76] J. M. Mayer, Hydrogen atom abstraction by metal-oxo complex: Understanding the analogy with organic radical reations, *Acc. Chem. Res.* 37 (1998) 441.

[77] G. Yin, A. M. Danby, D. Kitko, J. D. Carter, W. M. Scheper, D. H. Busch, Understanding the selectivity of a moderate oxidation catalyst. Hydrogen abstraction by a fully characterized activated catalyst, the robust dihydroxo Manganese(IV) complex of a bridged cyclam, *J. Am. Chem. Soc.* 129 (2007) 1512.

[78] G. Yin, M. Buchalova, A. M. Danby, C. M. Perkins, D. Kitko, J. D. Carter, W. M. Scheper, D. H. Busch, Olefin oxygenation by the hydroperoxide adduct of a nonheme Manganese(IV) complex: Epoxidations by a metallo-peracid produces gentle selective oxidations, *J. Am. Chem. Soc.* 127 (2005) 17170.

[79] G. Yin, M. Buchalova, A. M. Danby, C. M. Perkins, D. Kitko, J. D. Carter, W. M. Scheper, D. H. Busch, Olefin epoxidation by the hydrogen peroxide adduct of a novel non-heme Manganese(IV) complex: Demonstration of oxygen transfer by multiple mechanisms, *Inorg. Chem.* 45 (2006) 3467.

[80] G. Yin, A. M. Danby, D. Kitko, J. D. Carter, W. M. Scheper, D. H. Busch, Olefin epoxidation by alkyl hydroperoxide with a novel cross-bridged cyclam manganese complex: Demonstration of oxygenation by two distinct reactive intermediates. [Erratum to document cited in CA146:316400], *Inorg. Chem.* 46 (2007) 2173.

[81] G. He, T. C. Bruice, The rate-limiting step in the one-electron oxidation of an alkene by oxo[mesotetrakis(2,6-dibromophenyl)porphinato]chromium(V) is the formation of a charge-transfer complex, *J. Am. Chem. Soc.* 113 (1991) 2747.

[82] B. Meunier, E. Guilmet, M. De Carvalho, R. Poilblanc, Sodium hypochlorite: A convenient oxygen source for olefin epoxidation catalyzed by (porphyrinato)manganese complexes, *J. Am. Chem. Soc.* 106 (1984) 6668.

[83] P. Battioni, J. P. Renaud, J. F. M. Bartoli, M. Reina-Artiles, M. Fort, D. Mansuy, Monooxygenase-like oxidation of hydrocarbons by hydrogen peroxide catalyzed by manganese porphyrins and imidazole: Selection of the best catalytic system and nature of the active oxygen species, *J. Am. Chem. Soc.* 110 (1988) 8462.

[84] J. P. Collman, V. J. Lee, C. J. Kellen-Yuen, X. Zhang, J. A. Ibers, J. I. Brauman, Threitol-strapped manganese porphyrins as enantioselective epoxidation catalysts of unfunctionalized olefins, *J. Am. Chem. Soc.* 117 (1995) 692.

[85] A. Murphy, G. Bubois, T. D. P. Stack, Efficient epoxidation of electron-deficient olefins with a cationic manganese complex, *J. Am. Chem. Soc.* 125 (2003) 5250.

[86] J. T. Groves, M. K. Stern, Synthesis, characterization, and reactivity of oxomanganese(IV) porphyrin complexes, *J. Am. Chem. Soc.* 110 (1988) 8628.

[87] R. Arasasingham, G. He, T. C. Bruice, Mechanism of manganese porphyrin-catalyzed oxidation of alkenes. Role of manganese(IV)-oxo species, *J. Am. Chem. Soc.* 115 (1993) 7985.

[88] J. Bernadou, A. Fabiano, A. Robert, B. Meunier, Redox tautomerism in high-valent metal-oxo-aquo complexes.Origin of the oxygen atom in epoxidation reactions catalyzed by water-soluble metalloporphyrins, *J. Am. Chem. Soc.* 116 (1994) 9375.

[89] J. T. Groves, J. Lee, S. S. Marla, Detection and characterization of an oxomanganese(V) porphyrin complex by rapid-mixing Stopped-flow spectrophotometry, *J. Am. Chem. Soc.* 119 (1997) 6269.

[90] M. T. Caudle, P. Riggs-Gelasco, A. K. Gelasco, J. E. Penner-Hahn, V. L. Pecoraro, Mechanism for the homolytic cleavage of alkyl hydroperoxides by the Manganese(III) dimer MnIII2(2-OHsalph)2, *Inorg. Chem.* 35 (1996) 3577.

[91] D. V. Deubel, G. Frenking, P. J. Gisdakis, W. A. Herrmann, N. Rösch, J. Sundermeyer, Olefin epoxidation with inorganic peroxides. Solutions to four long-standing controversies on the mechanism of oxygen transfer, *Acc. Chem. Res.* 37 (2004) 645.

[92] J. T. Groves, J. Lee, S. S. Marla, Detection and characterization of an Oxomanganese(V) Porphyrin complex by rapid mixing stopped-flow spectrophotometry, *J. Am. Chem. Soc.* 119 (1997) 6269.

[93] D. Ostovic, T. C. Bruice, Mechanism of alkene epoxidation by iron. chromium and manganese higher valent oxo-metalloporphyrins, *Acc. Chem. Res.* 25 (1992) 314.

[94] T. Katsuki, K. B. Sharpless, The first practical method for asymmetric epoxidation, *J. Am. Chem. Soc.* 102 (1980) 5974.

[95] P. Chaumette, H. Mimoun, L. Saussine, J. Fischer, A. Mitachler, Peroxo and alkylperoxidic molybdinum(V), complexes as intermediates in the epoxidation of olefins by alkyl hydroperoxides, *J. Organomet. Chem.* 250 (1983) 291.

[96] W. A. Herrmann, R. M. Kratzer, H. Ding, W. Thiel, H. Glas, Methyltrioxorhenium/pyrazole a highly efficient calayst in the epoxidation of olefins, *J. Organomet. Chem.* 555 (1998) 293.

[97] J. S. Vrettos, G. W. Brudvig, Oxygen evolution, In: Que Jr., W. B. Tolman, (Eds.), *Comprehensive Coordination Chemistry II* Vol. 8, Elsevier, 2004, pp. 507–547.

[98] C. Tommos, G. T. Babcock, Oxygen production in nature: A light-driven metalloradical enzyme process, *Acc. Chem. Res.* 31 (1998) 18.

[99] W. F. Brill, The origin of epoxides in the liquid phase oxidation of olefins with molecular oxygen, *J. Am. Chem. Soc.* 85 (1984) 141.

[100] W. Adam, K. J. Roschmann, C. R. Saha-Möller, D. Seebach, cis-Stilbene and (1a,2b,3a)-(2-Ethenyl-3-methoxycyclopropyl)benzene as mechanistic probes in the MnIII(salen)-catalyzed epoxidation: Influence of the oxygen source and the counterion on the diastereoselectivity of the competitive concerted and radical-type oxygen transfer, *J. Am. Chem. Soc.* 124 (2002) 5068.

[101] J. A. Dean, *Lange's Handbook of Chemistry*, 4th ed., McGraw-Hill, New York, NY, 1973, pp. 4–32.

[102] H.-J. Lee, T. P. Shi, D. H. Busch, B. Subramaniam, A greener pressure intensified propylene epoxidation process with facile product separation, *Chem. Eng. Sci.* 62 (2007) 7282.

CHAPTER 4

Activation of Hydrogen Peroxide by Polyoxometalates

Noritaka Mizuno, Kazuya Yamaguchi, Keigo Kamata, and Yoshinao Nakagawa

Contents		
	1. Introduction	156
	2. Activation of Hydrogen Peroxide by Polyoxometalates	157
	2.1. Peroxotungstates	158
	2.2. Lacunary polyoxotungstates	159
	2.3. Transition-metal-substituted polyoxometalates	164
	3. Catalytic Oxidation by Polyoxometalates	166
	3.1. Peroxotungstates	166
	3.2. Lacunary polyoxotungstates	166
	3.3. Transition-metal-substituted polyoxometalates	168
	4. Conclusions	170
	5. Future View	170
	Acknowledgments	171
	References	171

Abstract	Numerous catalytic H_2O_2-based oxidation reactions by polyoxometalates (POMs) have been developed. POMs for H_2O_2-based oxidations can be classified into three groups from the standpoints of their structures and H_2O_2 activation properties: (a) di- and tetranuclear small peroxotungstates, (b) lacunary polyoxotungstates, and (c) transition-metal-substituted POMs. This chapter focuses on the activation of H_2O_2 on the basis of groups (a)–(c) including reaction mechanisms. In addition, recent progress in the selective oxidation with H_2O_2 catalyzed by POMs is comprehensively summarized.

Department of Applied Chemistry, School of Engineering, The University of Tokyo, 7-3-1 Hongo, Bunkyo-ku, Tokyo 113-8656, Japan

Mechanisms in Homogeneous and Heterogeneous Epoxidation Catalysis
DOI: 10.1016/B978-0-444-53188-9.00004-3

© 2008 Elsevier B.V.
All rights reserved.

Key Words: Polyoxometalate, Peroxotungstate, Lacunary polyoxotungstate, Transition-metal-substituted polyoxometalate, Hydrogen peroxide, Tungsten, Vanadium, Titanium, Dinuclear, Peroxo, Hydroperoxo, Sulfoxidation, Baeyer–Villiger oxidation. © 2008 Elsevier B.V.

1. INTRODUCTION

Catalytic oxidation is a key technology for converting petroleum-based feedstocks to useful chemicals [1–4]. Epoxidation of olefins is an important reaction in the laboratory as well as in the chemical industry, because epoxides are widely used as raw materials for epoxy resins, paints, surfactants, and intermediates in organic synthesis. However, a noncatalytic process based on chlorine (chlorohydrin process) and catalytic processes based on organic peroxides and peracids are still used extensively. These processes have disadvantages from economical and environmental standpoints because they are capital intensive, produce large outputs of chloride-laden sewage (chlorohydrin process), and result in the formation of two coupled products (organic peroxide process). In contrast with such antiquated processes, the catalytic epoxidation with H_2O_2 as an oxidant offers advantages because the only by-product is water and H_2O_2 has a high content of active oxygen species.

For H_2O_2-based epoxidation, various sophisticated catalysts such as well-defined molecular catalysts, biomimetic catalysts related to the heme enzyme of cytochrome P-450 and the nonheme enzyme of methane monooxygenase, and isolated single-site heterogeneous catalysts can effectively activate oxidants and/or substrates, resulting in high specific reactivity and high chemo-, regio-, and stereoselectivity [5–10]. The activation of H_2O_2 by transition metals leads to the generation of a large number of metal peroxo complexes with various coordination modes (η^2-O_2, μ-η^1:η^1-O_2, μ-η^1:η^2-O_2, μ-η^2:η^2-O_2, OOH, etc.). Such structures of metal-coordinated active oxygen species play an important role in various oxidative transformations of organic substances [11–14]. Recently, multinuclear metal complexes with two or more metal atoms that contain bridging peroxo groups have attracted much attention due to their remarkable oxygenation activity. For example, polynuclear d^0-transition metal complexes containing μ-η^1:η^2-peroxo moieties, bioinspired Cu-based dinuclear μ-η^2:η^2-peroxo complexes, and related bis(μ-oxo) species show high activity for oxidations, such as oxygen transfer to C=C double bonds and activation of aromatic and aliphatic C–H bonds [15–25]. However, degradation of most organometallic complexes is almost inevitable under oxidative conditions, and catalytic lifetimes are limited. Therefore, the development of selective oxidation systems with H_2O_2 using robust inorganic catalysts containing structurally well-defined polynuclear active sites can contribute to the development of "green" or "environmentally conscious" chemical processes.

Polyoxometalates (POMs) are early transition metal (V, Nb, Ta, W, Mo, etc.) oxygen anion clusters that have been applied in various fields, such as structural chemistry, analytical chemistry, surface science, medicine, electrochemistry, photochemistry, and catalysis. POMs have the following advantages as oxidation catalysts: (a) their redox and acid–base properties can be controlled by changing their chemical composition, (b) they are not susceptible to oxidative and thermal degradation compared with organometallic complexes, and (c) their catalytically active sites can be designed, for example, by inserting transition metals and inorganic ligands in lacunary POMs (POMs with metal vacancies). Because metal atoms in their structure can be removed and replaced, POMs-based oxidation catalysts can be designed at the atomic and/or molecular levels [19–25].

Numerous catalytic H_2O_2-based oxidations by POMs have been developed over the years. POMs for H_2O_2-based oxidations can be classified into three groups from the standpoints of their structures and the mode of H_2O_2 activation: (a) di- and tetranuclear small peroxotungstates, (b) lacunary polyoxotungstates, and (c) transition-metal-substituted POMs. In Section 2, the activation of H_2O_2 in relation to the oxidation mechanism by the above three groups of POMs are described based on our recent studies. In Section 3, examples of recent developments in selective oxidation with H_2O_2 homogeneously catalyzed by POM compounds are comprehensively summarized. Other recent review articles describe homogeneous and heterogeneous transition–metal-catalyzed oxidations in more detail [5–10,19–25].

2. ACTIVATION OF HYDROGEN PEROXIDE BY POLYOXOMETALATES

Various kinds of POMs are effective catalysts for H_2O_2-based environmentally friendly oxidations. The most frequently used POMs for the oxidations are Keggin-type POMs. Phosphotungstic acid reacts with H_2O_2 to give various small peroxotungstates such as $[PO_4\{WO(O_2)_2\}_4]^{3-}$ (1, Fig. 4.1) and $[\{WO(O_2)_2(H_2O)_2\}(\mu\text{-}O)]^{2-}$, which can catalyze the oxidation of various organic substances [26–34]. Lacunary POMs can be the catalyst precursors of polynuclear peroxo species because their vacant sites can activate H_2O_2. The tetraperoxo species, $[\beta_3\text{-}Co^{II}W_{11}O_{35}(O_2)_4]^{10-}$, is prepared by the reaction of monovacant $[\alpha\text{-}Co^{III}W_{11}O_{39}]^{9-}$ with H_2O_2 [35]. The diperoxo species is formed on a divacant site of the γ-isomer of Keggin-type silicodecatungstate, which can efficiently catalyze the oxidation of various organic substances [36–39]. In contrast with small monomeric and dimeric peroxometalates, peroxo species formed on polynuclear sites are expected to show specific reactivity and selectivity due to their unique electronic and structural characters. Lacunary POMs are most often used as precursors of transition-metal-substituted POMs. The oxo ligands at the vacant sites readily react with various kinds of transition metal cations. To date, various kinds of transition-metal-substituted POMs have been synthesized and some of them can activate H_2O_2 efficiently.

There are only a few reports on the structures of active oxygen species formed on POMs. Recently, we have clarified some of the structures of the active oxygen species formed on POMs with nuclear magnetic resonance (NMR) spectroscopy, coldspray ionization mass spectrometry (CSI-MS), and single crystal X-ray structural analysis. In this section, we focus on the activation of H_2O_2 by the species (a)–(c) and present results from our recent studies. Some reaction mechanisms for H_2O_2-based oxidations by POMs are also described.

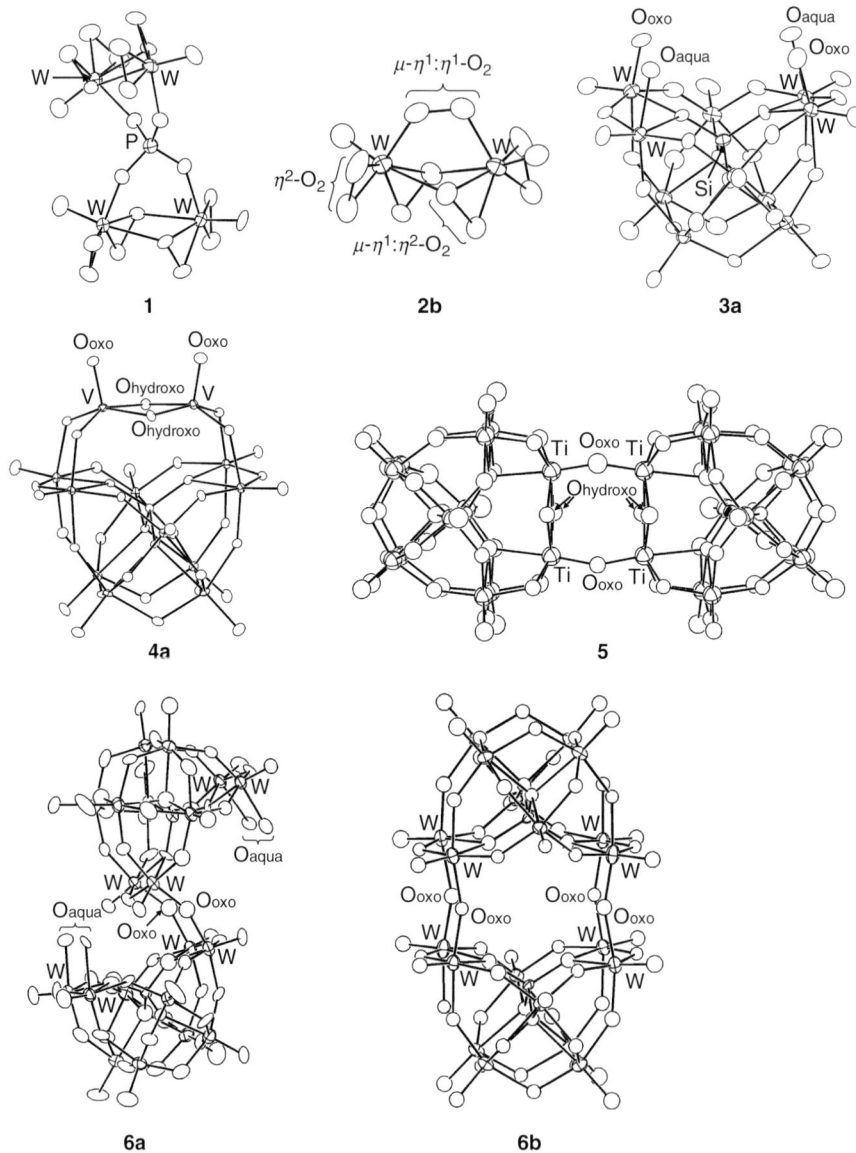

FIGURE 4.1 Molecular structures of various tungstates.

2.1. Peroxotungstates

Tungsten-based epoxidation systems with H_2O_2 have attracted much attention because of their high reactivity compared with molybdenum analogues and inherently poor activity for the decomposition of H_2O_2. The tetranuclear peroxotungstate **1** was isolated and characterized crystallographically by Venturello and coworkers [26,27]. The anion **1** consists of the PO_4^{3-} anion and two

$[W_2O_2(O_2)_4]$ species. Spectroscopic and kinetic investigations showed that **1** was a catalytically important species for the oxidation of various substances with H_2O_2 [31–34]. Peroxotungstates containing phosphorus or arsenic ligands are generally much more active than the dinuclear isopolyperoxotungstate $[\{WO(O_2)_2(H_2O)_2\}(\mu\text{-}O)]^{2-}$ for the catalytic epoxidation of olefins [32,33,40–45], but the effects of these ligands on the reactivity for epoxidation have not yet been clarified.

We have recently demonstrated novel oxidation systems catalyzed by various tungstates. The isolated potassium salt of $[\{WO(O_2)_2(H_2O)_2\}(\mu\text{-}O)]^{2-}$ catalyzed the highly chemo-, regio-, and diastereoselective and stereospecific epoxidation of various allylic alcohols with only 1 equiv of H_2O_2 in water solvent [46,47]. It is noted that the reactivity of the dinuclear peroxotungstate with H_2O_2 in organic solvents was quite different from that in water. The reaction of the tetra-n-butylammonium (TBA) salt of dinuclear peroxotungstate, $[\{WO(O_2)_2\}_2(\mu\text{-}O)]^{2-}$ (**2a**, Eq. 4.1), with H_2O_2 in CH_3CN gave the novel peroxo-bridging dinuclear tungsten species **2b** (Fig. 4.1) (Eq. 4.1) [48]. The ^{183}W NMR spectrum of **2a** in CD_3CN showed one resonance at -587.5 ppm ($\Delta v_{1/2} = 24.3$ Hz), suggesting that **2a** is a single species. Upon addition of 4 equiv of H_2O_2 relative to **2a**, one new ^{183}W NMR signal at -462.4 ppm ($\Delta v_{1/2} = 3.8$ Hz) was observed, which is in agreement with C_2 symmetry. Single crystal X-ray structural analysis showed that **2b** consisted of a $\mu\text{-}\eta^1:\eta^1\text{-}O_2^{2-}$ bridging group and two neutral $\{W(=O)(O_2)_2\}$ units, in which seven oxygen atoms were coordinated to a tungsten atom in a distorted pentagonal bipyramidal arrangement. There were three kinds of peroxo ligands in **2b**; η^2-peroxo, $\mu\text{-}\eta^1:\eta^2$-peroxo, and $\mu\text{-}\eta^1:\eta^1$-peroxo groups. The O–O bond lengths (1.46-1.48 Å) determined from the X-ray structural analysis are typical for peroxo ligands and comparable to those for previously reported peroxotungstates. Although several molecular structures of the $(\mu\text{-}\eta^1:\eta^1$-peroxo) dimetal complexes for Co, Mn, Fe, Ge, Rh, Cu, Ir, Pt, Pd, and Sn have been reported, nothing is known of the $(\mu\text{-}\eta^1:\eta^1$-peroxo)$d^0$-dimetal complexes. Therefore, compound **2b** was the first structurally determined $(\mu\text{-}\eta^1:\eta^1$-peroxo)$d^0$-dimetal complex.

$$[\{WO(O_2)_2\}(\mu\text{-}O)]^{2-} (\mathbf{2a}) + H_2O_2 \rightarrow$$
$$[\{WO(O_2)_2\}(\mu\text{-}\eta^1:\eta^1\text{-}O_2)]^{2-} (\mathbf{2b}) + H_2O \quad (4.1)$$

The stoichiometric epoxidation of cyclo-, *cis*-2-, *cis*-3-, *trans*-2-, and 1-octenes with **2b** produced the corresponding epoxides in 97, 100, 94, 93, and 91% yields, respectively, suggesting that **2b** has 1 equiv of active oxygen species for the epoxidation. On the other hand, the stoichiometric epoxidation with **2a** hardly proceeded. The catalytic epoxidation proceeded by the reaction of **2b** with an olefin to form **2a** and the corresponding epoxide followed by the regeneration of **2b** by the reaction of **2a** with H_2O_2.

2.2. Lacunary polyoxotungstates

The γ-isomer of a divacant Keggin-type silicodecatungstate $[\gamma\text{-}SiW_{10}O_{36}]^{8-}$ [49] has been used as a precursor for dimetal-substituted POMs involving Ti, V, Cr, Mn, Fe, Mo, and W metals [50–59] and organic–inorganic hybrids [60–63]. The oxo ligands at the vacant sites are basic enough to react with not only metal cations but also H^+. A protonated divacant silicodecatungstate $[\gamma\text{-SiW}_{10}O_{34}(H_2O)_2]^{4-}$ (**3a**, Fig. 4.1), which efficiently catalyzed olefin epoxidation with H_2O_2, has been synthesized [36–39]. The single crystal X-ray structural analysis of **3a** revealed that two of the four oxo ligands located at the vacant sites were selectively protonated to give aquo ligands. The TBA salt of **3a** catalyzed oxygen-transfer reactions of various substances such as olefins, allylic alcohols, and sulfides with 30% aqueous H_2O_2 [36–39].

The time course for the epoxidation of cyclooctene with H_2O_2 catalyzed by **3a** showed an induction period (~40 min, that disappeared upon pretreatment of **3a** with H_2O_2 for ~60 min) (Fig. 4.2). The induction period was a result of the reaction of **3a** with H_2O_2 to form catalytically active species. The ^{29}Si NMR spectrum of **3a** showed one resonance at −83.5 ppm, showing a single species. The ^{183}W NMR spectrum of **3a** with C_2 symmetry showed five signals at −95.7, −98.9, −118.2, −119.6, and −195.7 ppm with an intensity ratio of 1:1:1:1:1, respectively. Upon addition of 10 equiv of H_2O_2 with respect to **3a**, one new ^{29}Si NMR signal appeared at −84.1 ppm and five new ^{183}W NMR signals appeared at −125.9, −135.4, −155.3, −219.7, and −554.6 ppm with an intensity ratio of 1:1:1:1:1, respectively (Fig. 4.3). Since the coordination of strong σ donors such as a peroxo groups causes upfield shifts of ^{95}Mo and ^{183}W NMR signals [64–67], the signal at −554.6 ppm can be assigned to tungsten atoms with peroxo ligands. Thus, the NMR results show the formation of a single species (**3b**, Eq. 4.2) with the C_2 symmetry.

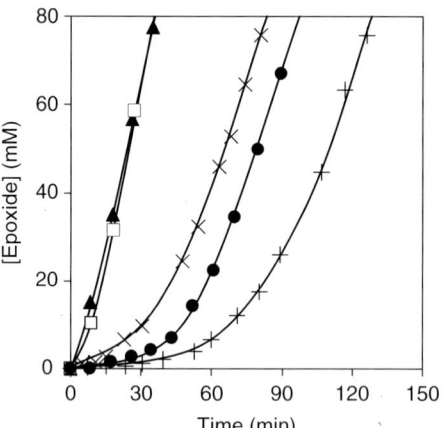

FIGURE 4.2 Reaction profiles of the epoxidation of cyclooctene with 30% aqueous H_2O_2 catalyzed by **3a**: [**3a**] (1.18 mM), [cyclooctene] (0.74 M), [H_2O_2] (0.15 M), [H_2O] (0.65 M), acetonitrile (6 ml), reaction temperature (305 K). (●) Without pretreatment. (▲) With pretreatment of 30% aqueous H_2O_2 before addition of cyclooctene for 1 h at 305 K. (+) With pretreatment of cyclooctene before addition of 30% aqueous H_2O_2 for 60 min at 305 K. (□) With addition of $HClO_4$ (0.30 mM) without pretreatment. (×) With use of **3b** (1.18 mM) instead of **3a** without pretreatment.

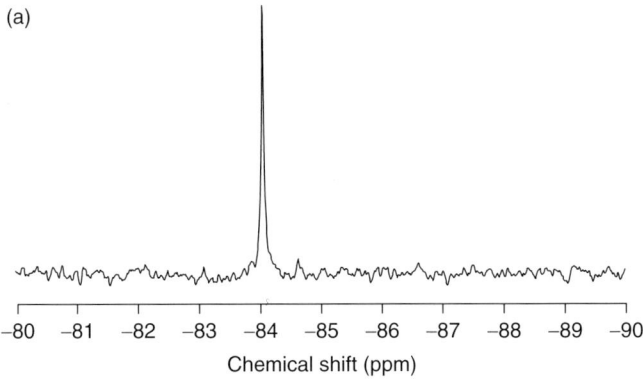

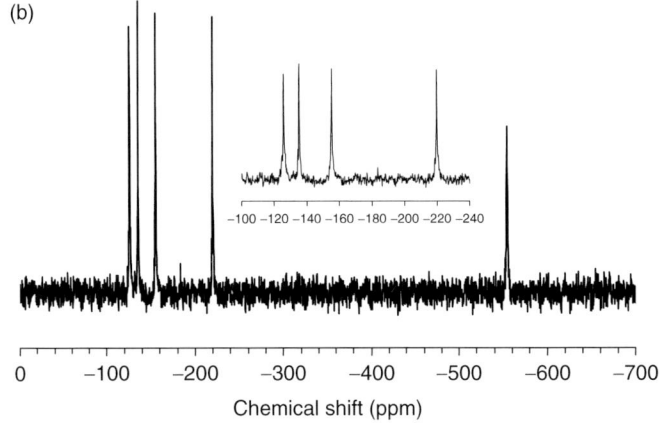

FIGURE 4.3 (a) ^{29}Si and (b) ^{183}W NMR of **3a** in CD$_3$CN treated with 10 equiv 30% aqueous H$_2$O$_2$ with respect to **3a** at 263 K for 3 h.

After the treatment of **3a** with H$_2$O$_2$ followed by the addition of an excess of diethyl ether, **3b** could successfully be isolated. The positive-ion coldspray ionization mass spectrum of the isolated **3b** showed a most intense parent ion peak centered at $m/z = 3654.8$ with an isotopic distribution that agreed with the calculated pattern of $[(TBA)_5SiW_{10}O_{32}(O_2)_2]^+$. All these results show that the two oxo groups (O^{2-}) are replaced by two peroxo groups (O$_2^{2-}$) on the divacant lacunary site with retention of the γ-Keggin framework (Eq. 4.2).

$$[\gamma\text{-SiW}_{10}O_{34}(H_2O)_2]^{4-} \text{ (3a)} + 2H_2O_2 \rightarrow [\gamma\text{-SiW}_{10}O_{32}(O_2)_2]^{4-} \text{ (3b)} + 4H_2O \quad (4.2)$$

No stoichiometric epoxidation of cyclooctene with the isolated compound **3b** proceeded, showing that **3b** is inactive for the present epoxidation. An induction period (30 min) was still observed for the epoxidation catalyzed by **3b** with H$_2$O$_2$

(Fig. 4.2). The reaction of **3a** with excess H_2O_2 (20 equiv relative to **3a**) was traced with *in situ* infrared (IR), ^{29}Si and ^{183}W NMR, and CSI-MS and the formation of **3b** was completed within 10 min, supporting that an induction period (30 min) for the catalytic epoxidation by **3b** was shorter than that by **3a** (40 min). It is therefore probable that the reaction of **3b** with H_2O_2 leads to the formation of a reactive species (**3c**) and that the induction period observed in the catalytic epoxidation corresponds to the slow formation of **3c** from **3b**. Detection of **3c** with ^{29}Si and ^{183}W NMR spectroscopy was unsuccessful, probably because the quantity of **3c** was very small and below the detection limits of ^{29}Si and ^{183}W nuclei (S/N ratio of ^{29}Si and ^{183}W NMR spectra = 7–10/1). It has been reported that the epoxidation by molybdenum peroxo complexes is initiated by the addition of hydroperoxides, and that the proton-transfer reaction from the hydroperoxides to a peroxo ligand is a key step determining the catalytic activity [68–71]. Similarly, H_2O_2 could function as a proton donor to form a hydroperoxo species (**3c**) such as $[\gamma\text{-SiW}_{10}O_{32}(O_2)(OOH)]^{3-}$ through protonation of **3b**. The induction period almost disappeared with addition of a small amount of $HClO_4$ (0.125–0.250 equiv relative to **3a**). In addition, a good linear correlation for the epoxidation of cyclooctene in the presence of $HClO_4$ was observed between the reaction rates and the added amounts of $HClO_4$. All these results support the notion that the hydroperoxo species **3c** is generated and acts as an active species for the present oxidation.

After the first report on **3a**-catalyzed olefin epoxidation with H_2O_2, experimental and theoretical studies on **3a** were reported by several research groups [72–76]. While it was suggested that the formula of **3a** was $[\gamma\text{-SiW}_{10}O_{34}(H_2O)_2]^{4-}$ with two aqua ligands, the structural assignment of the four protons on the lacunary sites still remains unclear. Protons on the vacant sites are key features of the catalytic epoxidation mechanism, which may affect the formation and reactivity of the tungsten peroxide species. Density functional theory (DFT) calculations on the structure of tetraprotonated form of $[\gamma\text{-H}_4\text{SiW}_{10}O_{36}]^{4-}$ were presented by Musaev *et al.* [72]. Based on DFT calculations at the B3LYP/Lanl2dz + d(Si) level, the structure of $[\gamma\text{-SiW}_{10}O_{32}(OH)_4]^{4-}$ with four hydroxo ligands was calculated to be more stable by 8.0 kcal/mol than that of $[\gamma\text{-SiW}_{10}O_{34}(H_2O)_2]^{4-}$ with two aqua and two oxo(terminal) ligands. Therefore, it was concluded that the structure of the tetraprotonated form of $[\gamma\text{-H}_4\text{SiW}_{10}O_{36}]^{4-}$ should be formulated as $[\gamma\text{-SiW}_{10}O_{32}(OH)_4]^{4-}$ and that the asymmetry in the W–OH bond lengths was a result of the existence of hydrogen bonding interactions (Fig. 4.4). In addition, the theoretical mechanism for the epoxidation of ethylene with H_2O_2 by the calculated model of $[\gamma\text{-SiW}_{10}O_{32}(OH)_4]^{4-}$ was taken to consist of two steps [73]. In the first step, the hydroperoxo species (W–OOH) was produced by the reaction of W–OH and H_2O_2. In the second step, the O–O bond of the W–OOH was cleaved resulting in the formation of an epoxide and this step was the rate-determining step. It was suggested that the presence of Me_4N^+ countercations significantly reduced the activation barrier due to ion-pairing effects.

On the other hand, Bonchio and coworkers suggested that $[\gamma\text{-SiW}_{10}O_{34}(H_2O)_2]^{4-}$ was the active epoxidation catalyst [74]. The titration of **3a** by tetra-*n*-butylammonium hydroxide (TBAOH) indicated that only two out of

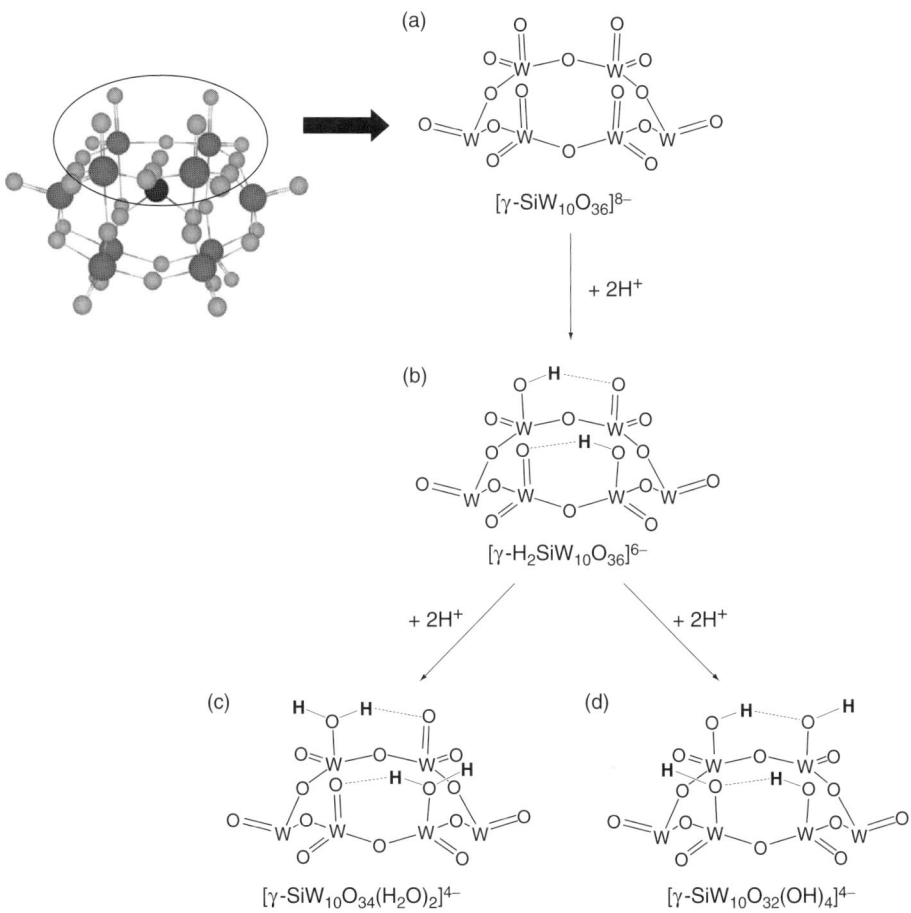

FIGURE 4.4 Schematic representations of the lacunary sites on the calculated (a) $[\gamma\text{-SiW}_{10}\text{O}_{36}]^{8-}$, (b) $[\gamma\text{-H}_2\text{SiW}_{10}\text{O}_{36}]^{6-}$, (c) $[\gamma\text{-SiW}_{10}\text{O}_{34}(\text{H}_2\text{O})_2]^{4-}$, and (d) $[\gamma\text{-SiW}_{10}\text{O}_{32}(\text{OH})_4]^{4-}$.

four acidic protons on the POM surface played a major role in promoting oxygen transfer. Upon addition of 2 equiv of TBAOH relative to **3a**, the ^{183}W NMR spectrum showed a C_{2v} structure, which necessarily implies fast exchange between the remaining protons. Using relativistic DFT calculations using ZORA-BP86/TZP level theory including solvent effects with the COSMO method, inspection of the molecular electrostatic potentials for the calculated structure of $[\gamma\text{-H}_2\text{SiW}_{10}\text{O}_{36}]^{6-}$ showed that the mono-protonated lacunary oxygens still retained a significant electron density that did not prevent a regioselective double protonation of each site to form $[\gamma\text{-SiW}_{10}\text{O}_{34}(\text{H}_2\text{O})_2]^{4-}$ (Fig. 4.4). The energies of $[\gamma\text{-SiW}_{10}\text{O}_{34}(\text{H}_2\text{O})_2]^{4-}$ and $[\gamma\text{-SiW}_{10}\text{O}_{32}(\text{OH})_4]^{4-}$ were so close as to be very sensitive to the method/basis set combination adopted. However, the optimized geometry of $[\gamma\text{-SiW}_{10}\text{O}_{34}(\text{H}_2\text{O})_2]^{4-}$ fitted better with the X-ray structure of **3a**.

2.3. Transition-metal-substituted polyoxometalates

It was reported for the first time by Hill and Brown that transition-metal-substituted POMs $[PW_{11}O_{39}M'(OH_2)]^{5-}$ ($M' = Mn^{II}, Co^{II}$) catalyzed the oxidation of olefins with PhIO [77]. The M' atom easily becomes coordinatively unsaturated by elimination of the aquo ligand, and the resulting polyanion is regarded as an inorganic metalloporphyrin analogue [78]. Transition-metal-substituted POMs are oxidatively and hydrolytically stable compared with organometallic complexes, and the geometry and composition of their active sites can be controlled. These advantages have been applied to the development of biomimetic catalysis related to the heme enzyme, cytochrome P-450, and the nonheme enzyme in methane monooxygenase analogues. Until now, numerous catalytic oxidations by transition-metal-substituted POMs have been developed and some transition-metal-substituted POMs with "controlled" active sites can activate H_2O_2 efficiently. For example, we have reported that the Keggin-type di-vanadium-substituted silicotungstate $[SiW_{10}O_{38}V_2(\mu\text{-}OH)_2]^{4-}$ (**4a**, Fig. 4.1) with a {VO-$(\mu\text{-}OH)_2$-VO} core [50,51] and the dimeric di-titanium-substituted silicotungstate $[\{\gamma\text{-}SiW_{10}O_{36}Ti_2(\mu\text{-}OH)_2\}_2(\mu\text{-}O)_2]^{8-}$ (**5**, Fig. 4.1) with a {Ti-$(\mu\text{-}OH)_2$-Ti} core [52] can catalyze the epoxidation of various olefins using H_2O_2 with a high epoxide yield and a high efficiency of H_2O_2 utilization under very mild reaction conditions. In Section 3, the activation of H_2O_2 by the di-vanadium-substituted silicotungstate **4a** is described.

The IR, ultraviolet visible (UV–vis), and NMR results strongly indicated that the {VO-$(\mu\text{-}OH)_2$-VO} core is preserved in the cluster framework of **4a** during the catalytic process and that tungstate compounds such as **3a** are not the true catalytically active species when the vanadium compound is present. The reaction of **4a** with H_2O_2 in 1,2-dichloroethane gave $[\gamma\text{-}SiW_{10}O_{38}V_2(\mu\text{-}OH)(\mu\text{-}OOH)]^{4-}$ (**4b**, Eq. 4.3). The *in situ* formation of **4b** by the treatment of **4a** with H_2O_2 (96% aqueous solution) was confirmed on the basis of the following results. The negative-ion CSI-MS of **4b** in 1,2-$C_2H_4Cl_2$ showed a most intense parent-ion peak centered at m/z 3354, with an isotopic distribution that agreed with the pattern calculated for $[(TBA)_3SiV_2W_{10}O_{38}(\mu\text{-}OH)(\mu\text{-}OOH)]^-$. One new ^{51}V NMR signal of **4b** appeared at $\sim$–530 ppm in 1,2-$C_2D_4Cl_2$ upon the addition of H_2O_2, and the intensity of the signal of **4a** at –562 ppm decreased by 93 ± 1% (Fig. 4.5). Six new ^{183}W NMR signals of **4b** appeared at –79, –83, –92, –104, –127, and –132 ppm with an intensity ratio of 2:2:1:1:2:2, respectively, in 1,2-$C_2D_4Cl_2$ upon the addition of 15 equiv of H_2O_2, and the signals at δ = –81, –95, and –128 ppm due to **4a** almost completely disappeared (Fig. 4.6). The POM **4b** showed two ^1H NMR signals at 5.10 and 9.45 ppm with an intensity ratio of 1:1, respectively. All these NMR results were consistent with the C_s symmetry of **4b** with a symmetry plane containing two V–O–V bridging oxygen atoms.

$$[\gamma\text{-}SiW_{10}O_{38}V_2(\mu\text{-}OH)_2]^{4-} (\textbf{4a}) + H_2O_2 \rightarrow$$
$$[\gamma\text{-}SiW_{10}O_{38}V_2(\mu\text{-}OH)(\mu\text{-}OOH)]^{4-} (\textbf{4b}) + H_2O \quad (4.3)$$

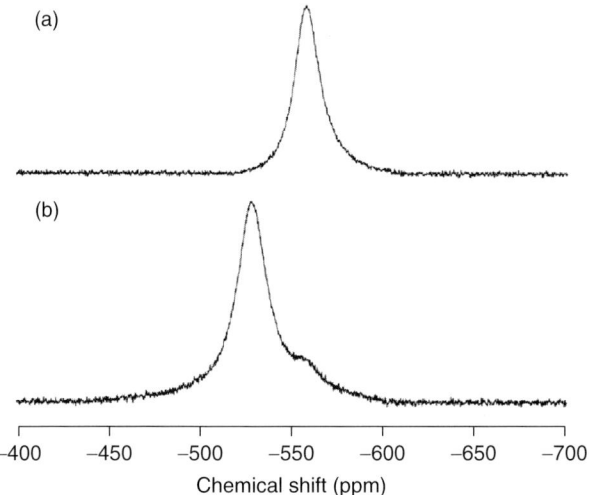

FIGURE 4.5 ^{51}V NMR spectra of (a) **4a** and (b) **4a** treated with 20 equiv 95% aqueous H_2O_2 with respect to **4a** for 2 h. Conditions: [**4a**] = 1.67 mM, 1,2-$C_2D_4Cl_2$, and 253 K.

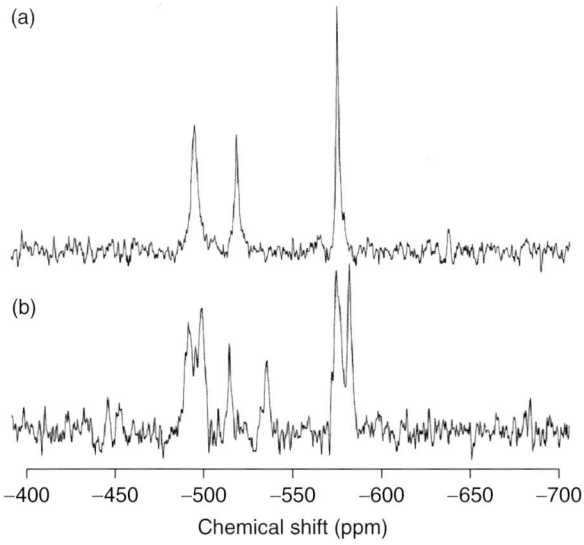

FIGURE 4.6 ^{183}W NMR spectra of (a) **4a** and (b) **4a** treated with 15 equiv 95% aqueous H_2O_2 with respect to **4a** for 2 h. Conditions: [**4a**] = 0.1 M, 1,2-$C_2D_4Cl_2$, and 253 K.

The reactivity of **4a** with H_2O_2 as well as the catalytic activity of **4a** depended considerably on the solvents used. The catalyst **4a** was almost inactive for the epoxidation of 1-octene in CH_3CN, 1,2-dichloroethane, and acetone, but was active in mixed solvents containing secondary or tertiary alcohols. Among the solvents used, the 1:1 (v/v) mixed solvent of CH_3CN and t-BuOH gave the highest yield. For the ^{51}V NMR spectra of **4a** treated with 10 equiv of H_2O_2 (30% aqueous) at 293 K in the mixed solvent, in which **4a** showed the highest activity, a new signal at −595 ppm (**4c**) was observed in addition to the two signals of **4a** and **4b**. Upon the addition of 10 equiv of H_2O_2 relative to **4a**, each signal intensity reached a constant value within 1 min, and the sum of the signal intensities was unchanged. The ratio of the concentration of **4b** to that of **4a** was proportional to $[H_2O_2]/[H_2O]$, supporting the reversible formation of **4b**. The ratio of the concentration of **4c** to that of **4a** was proportional to $[H_2O_2]/[H_2O]^2$, suggesting the successive dehydration of **4b** to form **4c**. The two vanadium atoms in **4c** were equivalent because only one ^{51}V NMR signal was observed for **4c**. In addition, **4c** was formed by the dehydration of **4b**. Therefore, a μ-η^2:η^2-peroxo species is a possible active oxygen species for the **4a**-catalyzed oxidation.

3. CATALYTIC OXIDATION BY POLYOXOMETALATES

In this section, examples of recent developments in the selective oxidation with H_2O_2 homogeneously catalyzed by POMs are comprehensively summarized according to the above mentioned classification.

3.1. Peroxotungstates

Tungsten-based compounds are effective catalysts because they use H_2O_2 efficiently and give high selectivity to epoxides. The peroxotungstate **1** can catalyze the H_2O_2-based epoxidation of allylic alcohols, monoterpenes, and α,β-unsaturated carboxylic acids, as well as the oxidation of alcohols, amines, and alkynes, and the oxidative transformation of diols (Ishii–Venturello system) [26–30]. Peroxotungstates containing phosphorous or arsenic ligands are generally much more active than $[\{WO(O_2)_2(H_2O)_2\}(\mu\text{-}O)]^{2-}$ for the catalytic epoxidation of terminal olefins [32,33,40–45], while the sulfate species $[SO_4\{WO(O_2)_2\}_2]^{2-}$ was the most active for the stoichiometric epoxidation of (R)-(+)-limonene among $[XO_4\{WO(O_2)_2\}_2]^{2-}$ (X = HAs, HP, and S) peroxotungstates [79,80].

Recently, we have showed that the highly chemo-, regio-, and diastereoselective and stereospecific epoxidation of various allylic alcohols with only 1 equiv of H_2O_2 in water solvent could be efficiently catalyzed by the isolated potassium salt of the dinuclear peroxotungstate $[\{WO(O_2)_2(H_2O)_2\}(\mu\text{-}O)]^{2-}$ [46,47]. Noyori and coworkers reported an H_2O_2-based green route to epoxides without halides and organic solvents using a catalytic system consisting of tungstate, (aminomethyl) phosphonic acid, and methyltrioctylammonium hydrogensulfate [81]. Although (aminomethyl)phosphonic acid decomposed into phosphoric acid under the reaction conditions, it facilitated the epoxidation more than phosphoric acid.

3.2. Lacunary polyoxotungstates

The TBA salt of **3a** exhibited the highest activity for the epoxidation of 1-octene with H_2O_2 among various POMs ($[\alpha\text{-}SiW_{12}O_{40}]^{4-}$, $[\gamma\text{-}SiW_{12}O_{40}]^{4-}$, $[\alpha\text{-}SiW_{11}O_{39}]^{8-}$, $[\alpha\text{-}SiW_9O_{34}]^{10-}$, and peroxotungstates (**1** and $[\{WO(O_2)_2(H_2O)_2\}(\mu\text{-}O)]^{2-}$)).

High yields and efficiency for H$_2$O$_2$ utilization were achieved for **3a**-catalyzed epoxidation of various terminal, internal, and cyclic olefins (Eq. 4.4). For epoxidation of primary allylic alcohols, the corresponding epoxy alcohols were obtained selectively with small amounts of α,β-unsaturated aldehydes.

$$(4.4)$$

For the catalytic oxidation of thianthrene 5-oxide (SSO), which is a probe for the electronic character of the oxidant [82,83], the X$_{SO}$ (X$_{SO}$ = (nucleophilic oxidation)/(total oxidation)) value obtained with **3a** (0.04) was lower than those of **1** (0.18) and [{WO(O$_2$)$_2$(H$_2$O)$_2$}(μ-O)]$^{2-}$ (0.35), suggesting that the active oxygen species in **3a** is the most electrophilic among these tungstates under the present conditions. The *trans* diastereoselectivity (*trans/cis* = 81:19) for the epoxidation of 3-methyl-1-cyclohexene by **3a** was higher than those by **1** (56:44) and [{WO(O$_2$)$_2$(H$_2$O)$_2$}(μ-O)]$^{2-}$ (55:45) [84,85]. These results for the oxidation of SSO and epoxidation of 3-methyl-1-cyclohexene suggest that a strong electrophilic oxidant species with steric hindrance is formed by the reaction of **3a** with H$_2$O$_2$.

Ren and coworkers reported that the addition of imidazole, phosphate, or carboxylate significantly enhanced the reaction rate and the efficiency of H$_2$O$_2$ utilization for organic **3a**-catalyzed oxygenation of organic sulfides [76]. For example, use of **3a** and imidazole, both at 1% molar concentration, resulted in the quantitative conversion of phenylsulfide to sulfoxide with 1 equiv of H$_2$O$_2$, and to sulfone with 2 equiv of H$_2$O$_2$, respectively. The introduction of organic ligands on the lacunary sites of **3a** affected reactivity and selectivity. An organophosphoryl polyoxotungstate derivative, [γ-SiW$_{10}$O$_{36}$(PhPO)$_2$]$^{4-}$, synthesized by the reaction of [γ-SiW$_{10}$O$_{36}$]$^{8-}$ with PhPO(OH)$_2$ [61], showed high catalytic activity for the epoxidation of olefins, alcohol oxidation, and sulfoxidation with H$_2$O$_2$ under microwave irradiation [75]. Such a functionalization of lacunary POMs could stabilize the vacant structure under microwave irradiation and control the reactivity by tuning the steric and electronic properties of organophosphoic acids.

It was reported that the dehydrative condensation of **3a** results in the formation of two structurally different disilicoicosatungstates, [{γ-SiW$_{10}$O$_{32}$(H$_2$O)$_2$}$_2$(μ-O)$_2$]$^{4-}$ (**6a**, Fig. 4.1) and [(γ-SiW$_{10}$O$_{32}$)$_2$(μ-O)$_4$]$^{8-}$ (**6b**, Fig. 4.1) [86,87]. The protonation of **3a** gave **6a**, which has an "S-shaped" structure and involves the aquo ligands on the terminal tungsten atoms. By contrast, the dehydration of **3a** resulted in the formation of the "closed-shape" cluster **6b**, which contains neither vacant sites nor terminal aquo ligands. The "S-shaped" cluster **6a** could catalyze the Baeyer–Villiger oxidation of cycloalkanones with H_2O_2 and showed high selectivity ($\geq$90%) for the corresponding lactones, while **3a** and **6b** with the common [γ-SiW$_{10}$O$_{32}$] fragment were almost inactive [86]. Larger scale (10-fold scale-up) reactions with cyclobutanone, cyclopentanone, and 2-adamantanone exhibited high turnover numbers (based on **6a**) of 2000, 1900, and 2000, respectively (Eq. 4.5). These values are much higher than those reported for Baeyer–Villiger oxidations with H_2O_2 and other catalysts; for cyclobutanone: [Pt^{II}(CF$_3$)(dppe)-(CH$_2$Cl$_2$)](BF$_4$) (dppe = 1,4-bis(diphenylphosphanyl)ethane), TON = 333) [88], bis[3,5-bis(trifluoromethyl)phenyl] diselenide (TON=178) [89]; for 2-adamantanone: Sn-beta zeolite (TON=140) [90], bis[3,5-bis(trifluoromethyl)phenyl] diselenide (TON=198) [89]. In addition, **6a** showed acid-catalyzed Mukaiyama-aldol, carbonyl-ene, and Diels–Alder reactions, while **3a** and **6b** showed base-catalyzed Knoevenagel and cyanosilylation reactions [87]. Such different catalytic performances of **3a**, **6a**, and **6b** may arise from their acid or base properties.

$$\text{(4.5)}$$

3.3. Transition-metal-substituted polyoxometalates

Some transition-metal-substituted POMs involving Ti, V, Mn, Zn, Fe, and Ni can act as effective catalysts for H_2O_2-based oxidation without promotion of the nonproductive decomposition of H_2O_2 [50–52,91–102]. Neumann and Gara reported an efficient epoxidation of olefins and the oxidation of secondary alcohols to ketones in biphasic (water–organic) reaction media with 30% aqueous H_2O_2 catalyzed by [WZnMn$^{II}_2$(ZnW$_9$O$_{34}$)$_2$]$^{12-}$ [91]. Manganese containing [WZnMn$^{II}_2$(ZnW$_9$O$_{34}$)$_2$]$^{12-}$ was oxidatively and hydrolytically stable over a range of $\geq$12,500 turnovers for the epoxidation of cyclooctene. A tungsten-peroxo intermediate activated by an adjacent manganese atom was proposed as an active species. Other manganese-containing POMs such as [{Mn^{II}(H$_2$O)$_3$}$_2$(WO$_2$)$_2$ (BiW$_9$O$_{33}$)$_2$]$^{10-}$ [92], [{Mn^{II}(H$_2$O)}$_3$(SbW$_9$O$_{33}$)$_2$]$^{12-}$ [92], [{Mn^{II}(H$_2$O)$_3$}$_2$ {Mn^{II}(H$_2$O)$_2$}$_2$ (TeW$_9$O$_{33}$)$_2$]$^{8-}$ [92], and [{Mn^{II}(OH$_2$)Mn$^{II}_2$PW$_9$O$_{34}$}$_2$(PW$_6$O$_{26}$)]$^{17-}$ [93] were also active for the epoxidation of olefins.

A self-assembled POM $[ZnWZn_2(ZnW_9O_{34})_2]^{12-}$ prepared *in situ* in water by mixing zinc nitrate, sodium tungstates, and nitric acid was active for the oxidation of alcohols, diols, amines, and pyridines with H_2O_2 [94,95]. The catalyst was shown by ^{183}W NMR to be stable in aqueous solutions in the presence of H_2O_2 and showed only minimal nonproductive decomposition of the oxidant.

In contrast with a tetranuclear ferric Wells–Dawson-type sandwich POM, $[Fe^{III}_4(OH_2)_2(P_2W_{15}O_{56})_2]^{12-}$ [96], a triferric sandwich-type POM, $[\{Na(OH_2)\}\{Fe^{III}(OH_2)\}Fe^{III}_2\{P_2W_{15}O_{56}\}_2]^{14-}$ [97], di-iron containing sandwich-type POM, $[Fe^{III}_2\{Na(OH_2)\}_2\{P_2W_{15}O_{56}\}_2]^{16-}$ [98], and nickel-containing sandwich-type POM $[Fe^{III}_2\{Ni(OH_2)\}_2\{P_2W_{15}O_{56}\}_2]^{16-}$ [97] showed catalytic activity for epoxidation. The di-iron-substituted silicotungstate, $[\gamma\text{-}SiW_{10}\{Fe(OH_2)\}_2O_{38}]^{6-}$, could catalyze the selective oxidation of olefins as well as paraffins with highly efficient utilization of H_2O_2 [99–101].

Nickel-substituted quasi-Wells–Dawson-type polyfluorooxometalate, $[Ni^{II}(H_2O)H_2F_6NaW_{17}O_{55}]^{9-}$, was the most active for epoxidation of olefins and allylic alcohols with H_2O_2 among various polyfluorooxometalates, $[M(L)H_2F_6NaW_{17}O_{55}]^{q-}$ (M = Zn^{II}, Co^{II}, Mn^{II}, Fe^{II}, Ru^{II}, Ni^{II}, and V^V) [102]. Partial substitution of an oxide position by fluoride had a significant electronic effect on a neighboring transition metal center.

As mentioned above, **3a** and various transition-metal-substituted POMs are effective homogeneous catalysts for the oxidation of various substrates. However, these require either an excess of H_2O_2 with respect to the olefin or an excess of olefin with respect to H_2O_2 to attain a high yield of epoxide or high efficiency of H_2O_2 utilization. The **4a**-catalyzed epoxidation system required only 1 equiv of H_2O_2 relative to the olefin and produced the corresponding epoxide with high yield. Nonactivated aliphatic terminal C_3–C_{10} olefins including propylene could be epoxidized with $\geq 99\%$ selectivity and $\geq 87\%$ efficiency of H_2O_2 utilization. Notably, the **4a**-catalyzed system showed unique stereospecificity, diastereoselectivity, and regioselectivity that are quite different from those of **1**, **2a**, and **3a** [50,51].

For the competitive epoxidation of *cis*- and *trans*-2-octenes with **4a**, the ratio of the formation rate of *cis*-2,3-epoxyoctane to that of the *trans* isomer is $\geq 3 \times 10^2$, which is much larger than the ratios (1.3–11.5) reported for other stereospecific epoxidation systems. The epoxidation of 3-substituted cyclohexenes, such as 3-methyl-1-cyclohexene and 2-cyclohexen-1-ol, showed an unusual diastereoselectivity: the corresponding epoxides were formed highly diastereoselectively with the oxirane ring *trans* to the substituents (*anti* configuration) (Eq. 4.6). In addition, the more accessible but less nucleophilic double bonds in nonconjugated dienes, such as *trans*-1,4-hexadiene, (R)-(+)-limonene, 7-methyl-1,6-octadiene, and 1-methyl-1,4-cyclohexadiene, were epoxidized in high yields in a highly regioselective manner. The values of [less-substituted epoxide]/[total epoxides] (≥ 0.88) are much higher than those reported for the other epoxidation systems. Such unique stereospecificity, diastereoselectivity, and regioselectivity have never been reported. The possible intermediate $\mu\text{-}\eta^2\text{:}\eta^2$-peroxo species **4c** may be sterically hindered, leading to the unique stereospecificity, diastereoselectivity, and regioselectivity in the epoxidation catalysis.

$$\begin{array}{c}\text{(scheme)}\end{array} \tag{4.6}$$

Reaction conditions: substrate 100 μmol; 4a (1.5 μmol); 30% H₂O₂ (100 μmol); CH₃CN/t-BuOH (1.5/1.5 mL); 293 K, 24 h.

- X = Me: 91% yield (syn/anti = 5/95)
- X = OH: 87% yield (syn/anti = 12/88)
- 1,5-hexadiene monoepoxide: 91% yield
- (R)-(+)-limonene epoxide: 89% yield

4. CONCLUSIONS

Various lacunary and transition-metal-substituted POMs are efficient homogeneous catalysts for liquid-phase oxidations with environmentally friendly H_2O_2 as a sole oxidant. Our recent studies on the catalyst design of POMs showed that various POMs and peroxotungstate could be used as effective catalysts for selective oxygen transfer reactions, such as epoxidation, sulfoxidation, and Baeyer–Villiger oxidation. The high efficiency of H_2O_2 utilization, which means that negligible decomposition of H_2O_2 occurs to form molecular oxygen, reduces by-product formation and the risk of building an explosive atmosphere, and leads to the simple, efficient, and safe oxidation processes. For the present epoxidation and sulfoxidation with H_2O_2, the catalytically active sites play an important role in the activation of H_2O_2. The reaction of the catalyst with H_2O_2 results in the generation of the active species which may be hydroperoxo or bridging peroxo species. These electrophilic active oxygen species attack the C=C double bond of olefins to give the corresponding epoxide with high selectivity and efficient H_2O_2 utilization. In addition, unique stereospecificity, diastereoselectivity, and regioselectivity are observed due to the sterically hindered active oxygen species embedded in the POM frameworks.

5. FUTURE VIEW

These systems are homogeneous, and heterogeneous catalysts are more desirable from standpoints of the catalyst/product separation and catalyst recycling. Recently, we have synthesized an organic–inorganic hybrid support synthesized by covalently anchoring an N-octyldihydroimidazolium cation fragment onto SiO_2 [103,104]. Various POMs were immobilized onto the support. The heterogenized catalysts were capable of epoxidizing a broad range of olefins in heterogeneous ways with maintenance of the catalytic performance of the corresponding homogeneous analogues and the recovered catalyst could be reused. In addition, we have synthesized various kinds of porous POMs [105–113]. They have pores between particles or within their crystal lattices, and show unique sorption and catalytic properties. Such unique properties of porous POMs are promising as heterogeneous catalysts for stereo- and shape-selective H_2O_2-based oxidation.

ACKNOWLEDGMENTS

This work was accomplished through the tremendous efforts of the coworkers in our laboratory and was financially supported by the Core Research for Evolutional Science and Technology (CREST) program of the Japan Science and Technology Agency (JST), Development in a New Interdisciplinary Field Based on Nanotechnology and Materials Science programs, and the Grants-in-Aid for Scientific Researches of the Ministry of Education, Culture, Sports, Science, and Technology.

REFERENCES

[1] R. A. Sheldon, J. K. Kochi, *Metal Catalyzed Oxidations of Organic Compounds*, Academic Press, New York, 1981.
[2] C. L. Hill, Advances in Oxygenated Processes, in: A.L. Baumstark (Ed.), Vol. 1, JAI Press, London, 1988, p. 1.
[3] M. Hudlicky, *Oxidations in Organic Chemistry*, ACS Monograph Series, American Chemical Society, Washington, DC, 1990.
[4] J.-E. Bäckvall, *Modern Oxidation Methods*, Wiley-VCH, Weinheim, 2004.
[5] K. Chen, M. Costas, L. Que Jr., Spin state tuning of non-heme iron-catalyzed hydrocarbon oxidations: Participation of Fe^{III}—OOH and Fe^V=O intermediates, *J. Chem. Soc., Dalton Trans.* 5 (2002) 672.
[6] B. S. Lane, K. Burgess, Metal-catalyzed epoxidations of alkenes with hydrogen peroxide, *Chem. Rev.* 103 (2003) 2457.
[7] R. Noyori, M. Aoki, K. Sato, Green oxidation with aqueous hydrogen peroxide, *Chem. Commun.* (2003) 1977.
[8] D. E. De Vos, B. F. Sels, P. A. Jacobs, Practical heterogeneous catalysts for epoxide production, *Adv. Synth. Catal.* 345 (2003) 457.
[9] J.-M. Brégeault, Transition-metal complexes for liquid-phase catalytic oxidation: Some aspects of industrial reactions and of emerging technologies, *Dalton Trans.* 17 (2003) 3289.
[10] G. Grigoropoulou, J. H. Clark, J. A. Elingsb, Recent developments on the epoxidation of alkenes using hydrogen peroxide as an oxidant, *Green Chem.* 5 (2003) 1.
[11] E. C. Niederhoffer, J. H. Timmons, A. E. Martell, Thermodynamics of oxygen binding in natural and synthetic dioxygen complexes, *Chem. Rev.* 84 (1984) 137.
[12] Thematic issue on "Metal-Dioxygen Complexes."*Chem. Rev.* 94 (1994) 567–856.
[13] Thematic issue on "Bioinorganic Enzymology." *Chem. Rev.* 96 (1996) 2237–3042.
[14] Thematic issue on "Biomimetic Inorganic Chemistry,"*Chem. Rev.* 104 (2004) 347–1200.
[15] E. I. Solomon, P. Chen, M. Metz, S.-K. Lee, A. E. Palmer, Oxygen binding, activation, and reduction to water by copper proteins, *Angew. Chem. Int. Ed.* 40 (2001) 4570.
[16] L. Que Jr., W. B. Tolman, Bis(μ-oxo)dimetal diamond cores in copper and iron complexes relevant to biocatalysis, *Angew. Chem. Int. Ed.* 41 (2002) 1114.
[17] L. Q. Hatcher, K. D. Karlin, Oxidant types in copper-dioxygen chemistry: The ligand coordination defines the Cu_n-O_2 structure and subsequent reactivity, *J. Biol. Inorg. Chem.* 9 (2004) 669.
[18] E. Y. Tshuva, S. J. Lippard, Synthetic models for non-heme carboxylate-bridged diiron metalloproteins: Strategies and tactics, *Chem. Rev.* 104 (2004) 987.
[19] Thematic issue on "Polyoxometalates."*Chem. Rev.* 98 (1998) 1–390.

[20] C. L. Hill, C. Chrisina, M. P. McCartha, Homogeneous catalysis by transition metal oxygen anion clusters, *Coord. Chem. Rev.* 143 (1995) 407.
[21] T. Okuhara, N. Mizuno, M. Misono, Catalytic chemistry of heteropoly compounds, *Adv. Catal.* 41 (1996) 113.
[22] R. Neumann, Polyoxometalate complexes in organic oxidation chemistry, *Prog. Inorg. Chem.* 47 (1998) 317.
[23] I. V. Kozhevnikov, *Catalysis by Polyoxometalates*. John Wiley & Sons, Ltd., Chichester, UK (2002).
[24] C. L. Hill, Polyoxometalates: Reactivity, In *Comprehensive Coordination Chemistry II: Transition Metal Groups 3–6*, Vol. 4, Elsevier Science, New York, 2004, p. 679.
[25] N. Mizuno, K. Kamata, K. Yamaguchi, in: R. Richards (Ed.), *Surface and Nanomolecular Catalysis*, Taylor and Francis Group, Liquid-phase oxidations catalyzed by polyoxometalates, New York, (2006), p. 463.
[26] C. Venturello, E. Alneri, M. Ricci, A new, effective catalytic system for epoxidation of olefins by hydrogen peroxide under phase-transfer conditions, *J. Org. Chem.* 48 (1983) 3831.
[27] C. Venturello, R. D'Aloisio, J. C. J. Bart, M. Ricci, A new peroxotungsten heteropoly anion with special oxidizing properties: Synthesis and structure of tetrahexylammonium tetra(diperoxotungsto)phosphate(3–), *J. Mol. Catal.* 32 (1985) 107.
[28] Y. Ishii, K. Yamawaki, T. Ura, H. Yamada, T. Yoshida, M. Ogawa, Hydrogen peroxide oxidation catalyzed by heteropoly acids combined with cetylpyridinium chloride. Epoxidation of olefins and allylic alcohols, ketonization of alcohols and diols, and oxidative cleavage of 1,2-Diols and olefins, *J. Org. Chem.* 53 (1988) 3587.
[29] Y. Ishii, K. Yamawaki, T. Yoshida, T. Ura, M. Ogawa, Oxidation of olefins and alcohols by peroxomolybdenum complex derived from tris(cetylpyridinium) 12-molybdophosphate and hydrogen peroxide, *J. Org. Chem.* 52 (1987) 1868.
[30] S. Sakaguchi, Y. Nishiyama, Y. Ishii, Selective oxidation of monoterpenes with hydrogen peroxide catalyzed by peroxotungstophosphate (pcwp), *J. Org. Chem.* 61 (1996) 5307.
[31] C. Aubry, G. Chottard, N. Platzer, J.-M. Brégeault, R. Thouvenot, F. Chauveau, C. Huet, H. Ledon, Reinvestigation of epoxidation using tungsten-based precursors and hydrogen peroxide in a biphase medium, *Inorg. Chem.* 30 (1991) 4409.
[32] A. C. Dengel, W. P. Griffith, B. C. Parkin, Studies on polyoxo- and polyperoxo-metalates. Part I. Tetrameric heteropolyperoxotungstates and heteropolyperoxomolybdates, *J. Chem. Soc., Dalton Trans.* 18 (1993) 2683.
[33] L. Salles, C. Aubry, R. Thouvenot, F. Robert, C. Dorémieux-Morin, G. Chottard, H. Ledon, Y. Jeannin, J.-M. Brégeault, ^{31}P and ^{183}W NMR spectroscopic evidence for novel peroxo species in the "$H_3[PW_{12}O_{40}]\cdot yH_2O/H_2O_2$" system. Synthesis and X-ray structure of tetrabutylammonium (µ-Hydrogen phosphato)bis(µ-peroxo)bis(oxoperoxotungstate) (2–): A catalyst of olefin epoxidation in a biphase medium, *Inorg. Chem.* 33 (1994) 871.
[34] D. C. Duncan, R. C. Chambers, E. Hecht, C. L. Hill, Mechanism and dynamics in the $H_3[PW_{12}O_{40}]$-Catalyzed selective epoxidation of terminal olefins by H_2O_2. Formation, reactivity, and stability of $\{PO_4[WO(O_2)_2]_4\}^{3-}$, *J. Am. Chem. Soc.* 117 (1995) 681.
[35] J. Server-Carrió, J. Bas-Serra, M. E. González-Núñez, A. García-Gastaldi, G. B. Jameson, L. C. W. Baker, R. Acerete, Synthesis, characterization, and catalysis of β_3-$[(Co^{II}O_4)W_{11}O_{31}(O_2)_4]^{10-}$, the first keggin-based true heteropoly dioxygen (peroxo) anion. Spectroscopic (ESR, IR) evidence for the formation of superoxo polytungstates, *J. Am. Chem. Soc.* 121 (1999) 977.
[36] K. Kamata, K. Yonehara, Y. Sumida, K. Yamaguchi, S. Hikichi, N. Mizuno, Efficient epoxidation of olefins with ≥99% selectivity and use of hydrogen peroxide, *Science* 300 (2003) 964.
[37] K. Kamata, Y. Nakagawa, K. Yamaguchi, N. Mizuno, Efficient, regioselective epodixation of dienes with hydrogen peroxide catalyzed by $[\gamma\text{-}SiW_{10}O_{34}(H_2O)_2]^{4-}$, *J. Catal.* 224 (2004) 224.
[38] N. Mizuno, K. Yamaguchi, K. Kamata, Epoxidation of olefins with hydrogen peroxide catalyzed by polyoxometalates, *Coord. Chem. Rev.* 249 (2005) 1944.
[39] K. Kamata, M. Kotani, K. Yamaguchi, S. Hikichi, N. Mizuno, Olefin epoxidation with hydrogen peroxide catalyzed by lacunary polyoxometalate $[\gamma\text{-}SiW_{10}O_{34}(H_2O)_2]^{4-}$, *Chem. Eur. J.* 13 (2007) 639.
[40] N. M. Gresley, W. P. Griffith, B. C. Parkin, J. P. White, D. J. Williams, The crystal structures of $[NMe_4][(Me_2AsO_2)\{MoO(O_2)_2\}_2]$, $[NMe_4][(Ph_2PO_2)\{MoO(O_2)_2\}_2]$, $[NBu^n_4][(Ph_2PO_2)\{WO(O_2)_2\}_2]$

and [NH$_4$][(Ph$_2$PO$_2$){MoO(O$_2$)$_2$(H$_2$O)}] and their use as catalytic oxidants, *J. Chem. Soc., Dalton Trans.* 10 (1996) 2039.

[41] W. P. Griffith, B. C. Parkin, A. J. P. White, D. J. Williams, The crystal structures of [NMe$_4$]$_2$[(PhPO$_3$){MoO(O$_2$)$_2$}$_2$–{MoO(O$_2$)$_2$(H$_2$O)}] and [NBu$^n_4$]$_2$[W$_4$O$_6$(O$_2$)$_6$(OH$_2$)(H$_2$O)$_2$] and their use as catalytic oxidants, *J. Chem. Soc., Dalton Trans.* 19 (1995) 3131.

[42] N. M. Gresley, W. P. Griffith, A. J. P. White, D. J. Williams, Crystal structures of [(Me$_2$AsO$_2$)$_2$ {Me$_2$AsO(OH)}{WO(O$_2$)}$_2$O]·H$_2$O and [NMe$_4$][(Ph$_2$PO$_2$)$_2${WO$_2$(O$_2$)}$_2${WO(O$_2$)(OH)(OH$_2$)}]·EtOH ·H$_2$O· 0.5MeOH, *J. Chem. Soc., Dalton Trans.* 1 (1997) 89.

[43] W. P. Griffith, B. C. Parkin, A. J. P. White, D. J. Williams, A novel hexanuclear heteropolyperoxo oxidation catalyst: Preparation, X-ray Crystal Structure and Reactions of [NMe$_4$]$_3$[(MePO$_3$){MePO$_2$(OH)} W$_6$O$_{13}$(O$_2$)$_4$(OH)$_2$(OH$_2$)]·4H$_2$O, *J. Chem. Soc., Chem. Commun.* 21 (1995) 2183.

[44] J.-Y. Piquemal, L. Salles, C. Bois, F. Robert, J.-M. Brégeault, Synthesis and X-ray structure of tetrabutylammonium (μ-hydrogenoarsenato)bis(μ-peroxo)bis(oxoperoxotungstate)2–) and of a methylarsenato-analog: New active oxygen-to-olefin transfer agents, *C. R. Acad. Sci. Paris, Série II* 319 (1994) 1481.

[45] A. J. Bailey, W. P. Griffith, B. C. Parkin, Heteropolyperoxo- and isopolyperoxo-tungstates and -molybdates as catalysts for the oxidation of tertiary amines, alkenes and alcohols, *J. Chem. Soc., Dalton Trans.* 11 (1995) 1833.

[46] K. Kamata, K. Yamaguchi, S. Hikichi, N. Mizuno, [{W=(O)(O$_2$)$_2$(H$_2$O)}$_2$(μ-O)]$^{2-}$-Catalyzed epoxidation of allylic alcohols in water with high selectivity and utilization of hydrogen peroxide, *Adv. Synth. Catal.* 345 (2003) 1193.

[47] K. Kamata, K. Yamaguchi, N. Mizuno, Highly selective, recyclable epoxidation of allylic alcohols with hydrogen peroxide in water catalyzed by dinuclear peroxotungstate, *Chem. Eur. J.* 10 (2004) 4728.

[48] K. Kamata, S. Kuzuya, K. Uehara, S. Yamaguchi, N. Mizuno, μ-η^1:η^1-Peroxo-bridged dinuclear peroxotungstate catalytically active for epoxidation of olefins, *Inorg. Chem.* 46 (2007) 3768.

[49] A. Tézé, G. Hervé, α-, β-, and γ-Dodecatungstosilicic acids: Isomers and related lacunary compounds, *Inorg. Synth.* 27 (1990) 85.

[50] Y. Nakagawa, K. Kamata, M. Kotani, K. Yamaguchi, N. Mizuno, Polyoxovanadometalate-catalyzed selective epoxidation of alkenes with hydrogen peroxide, *Angew. Chem. Int. Ed.* 44 (2005) 5136.

[51] Y. Nakagawa, N. Mizuno, Mechanism of [γ-H$_2$SiV$_2$W$_{10}$O$_{40}$]$^{4-}$-catalyzed epoxidation of alkenes with hydrogen peroxide, *Inorg. Chem.* 46 (2007) 1727.

[52] Y. Goto, K. Kamata, K. Yamaguchi, K. Uehara, S. Hikichi, N. Mizuno, Synthesis, structural characterization, and catalytic performance of dititanium-substituted γ-Keggin silicotungstate, *Inorg. Chem.* 45 (2006) 2347.

[53] J. Canny, R. Thouvenot, A. Tézé, G. Hervé, M. Leparulo-Loftus, M. T. Pope, Disubstituted tungstosilicates. 2. γ- and β-Isomers of tungstovanadosilicate, [SiV$_2$W$_{10}$O$_{40}$]$^{6-}$: Syntheses and structure determinations by tungsten-183, vanadium-51 and silicon-29 NMR spectroscopy, *Inorg. Chem.* 30 (1991) 976.

[54] K. Wassermann, H.-J. Lunk, R. Palm, J. Fuchs, N. Steinfeldt, R. Stösser, M. T. Pope, Polyoxoanions derived from [γ-SiO$_4$W$_{10}$O$_{32}$]$^{8-}$-containing oxo-centered dinuclear chromium(III) carboxylato complexes: Synthesis and single-crystal structural determination of [γ-SiO$_4$W$_{10}$O$_{32}$(OH) Cr$_2$(OOCCH$_3$)$_2$(OH$_2$)$_2$]$^{5-}$, *Inorg. Chem.* 35 (1996) 3273.

[55] X.-Y. Zhang, C. J. O'Connor, G. B. Jameson, M. T. Pope, High-valent manganese in polyoxotungstates. 3. Dimanganese complexes of γ-Keggin anions, *Inorg. Chem.* 35 (1996) 30.

[56] C. Nozaki, I. Kiyoto, Y. Minai, M. Misono, N. Mizuno, Synthesis and characterization of diiron(III) -substituted silicotungstate, [γ(1,2)-SiW$_{10}${Fe(OH$_2$)}$_2$O$_{38}$]$^{6-}$, *Inorg. Chem.* 38 (1999) 5724.

[57] B. Botar, Y. V. Geletii, P. Kögerler, D. G. Musaev, K. Morokuma, I. A. Weinstock, C. L. Hill, The true nature of the di-iron(III) γ-Keggin structure in water, catalytic aerobic oxidation and chemistry of an unsymmetrical trimer, *J. Am. Chem. Soc.* 128 (2006) 11268.

[58] E. Cadot, V. Béreau, B. Marg, S. Halut, F. Sécheresse, Syntheses and characterization of γ-[SiW$_{10}$M$_2$S$_2$O$_{38}$]$^{6-}$ (M = Mo^v, W^v). Two Keggin oxothio heteropolyanions with a metal-metal bond, *Inorg. Chem.* 35 (1996) 3099.

[59] A. Tézé, E. Cadot, V. Béreau, G. Hervé, About the Keggin isomers: Crystal structure of [N(C$_4$H$_9$)$_4$]$_4$-γ-[SiW$_{12}$O$_{40}$], the γ-isomer of the Keggin ion. Synthesis and ^{183}W NMR characterization of the mixed γ-[SiMo$_2$W$_{10}$O$_{40}$]$^{n-}$ ($n = 4$ or 6), *Inorg. Chem.* 40 (2001) 2000.

[60] F. Xin, M. T. Pope, Polyxometalate derivates with multiple organic groups, 3. Synthesis and structure of bis(phenyltin) bis(decatungsto silicate), [(PhSnOH$_2$)$_2$ (γ-SiW$_{10}$O$_{36}$)$_2$]$^{10-}$, *Inorg. Chem.* 35 (1996) 5693.

[61] C. R. Mayer, P. K. Herson, R. Thouvenot, Organic-inorganic hybrids based on polyoxometalates. 5. synthesis and structural characterization of bis(organophosphoryl)decatungstosilicates [γ-SiW$_{10}$O$_{36}$((RPO)$_2$]$^{4-}$, *Inorg. Chem.* 38 (1999) 6152.

[62] C. R. Mayer, I. Fournier, R. Thouvenot, Bis- and tetrakis(organosilyl) decatungstosilicate, [γ-SiW$_{10}$O$_{36}$(RSi)$_2$O]$^{4-}$ and [γ-SiW$_{10}$O$_{36}$(RSiO)$_4$]$^{4-}$: Synthesis and structural determination by multinuclear NMR spectroscopy and matrix-assisted laser desorption/desorption time-of-flight mass spectrometry, *Chem. Eur. J.* 6 (2000) 105.

[63] C. R. Mayer, M. Hervé, H. Lavanant, J.-C. Blais, F. Sécheresse, Hybrid cyclic dimers of divacant heteropolyanions: Synthesis, mass spectrometry (MALDI-TOF and ESI-MS) and NMR multinuclear characterization, *Eur. J. Inorg. Chem.* 5 (2004) 973.

[64] O. W. Howarth, Vanadium-51 NMR *Prog. Nucl. Magn. Reson. Spectrosc.* 22 (1990) 453.

[65] B. Piggott, S. F. Wong, D. Williams, ^{95}Mo NMR Studies of complexes containing the Mo$_2$O$_5^{2+}$ core and crystal structure of Mo$_2$O$_5$[SC$_6$H$_4$NHCH$_2$C$_5$H$_4$N]$_2$(C$_3$H$_7$NO)$_3$, *Inorg. Chim. Acta.* 141 (1988) 275.

[66] H. Nakajima, T. Kudo, N. Mizuno, Reaction of metal, carbide, and nitride of tungsten with hydrogen peroxide characterized by ^{183}W nuclear magnetic resonance and raman spectroscopy, *Chem. Mater.* 11 (1999) 691.

[67] N. J. Cambell, A. C. Dengel, C. J. Edwards, W. P. Griffith, Studies on transition metal peroxo complexes. Part 8. The nature of peroxomolybdates and peroxotungstates in aqueous solution. *J. Chem. Soc., Dalton Trans.* 6 (1989) 1203.

[68] W. R. Thiel, T. Priermeier, The first olefin-substituted peroxomolybdenum complex: Insight into a new mechanism for the molybdenum-catalyzed epoxidation of olefins, *Angew. Chem. Int. Ed. Engl.* 34 (1995) 1737.

[69] W. R. Thiel, Metal-catalyzed oxidations, 4. The reaction of molybdenumperoxo complexes with brønsted and lewis acids, *Chem. Ber* 129 (1996) 575.

[70] A. Hroch, G. Gemmecker, W. R. Thiel, Metal-catalyzed oxidations, 10 new insights into the mechanism of hydroperoxide activation by investigation of dynamic processes in the coordination sphere of seven-coordinated molybdenum peroxo complexes, *Eur. J. Inorg. Chem.* 5 (2000) 1107.

[71] G. Wahl, D. Kleinhenz, A. Schorm, J. Sundermeyer, R. Stowasser, C. Rummy, G. Bringmann, C. Fickert, W. Kiefer, Peroxomolybdenum complexes as epoxidation catalysts in biphasic hydrogen peroxide activation: Raman spectroscopic studies and density functional calculations, *Chem. Eur. J.* 5 (1999) 3237.

[72] D. G. Musaev, K. Morokuma, Y. V. Geletii, C. L. Hill, Computational modeling of di-transition-metal-substituted γ-Keggin polyoxometalate anions. Structural refinement of the protonated divacant lacunary silicodecatungstate, *Inorg. Chem.* 43 (2004) 7702.

[73] R. Prabhakar, K. Morokuma, C. L. Hill, D. G. Musaev, Insights into the mechanism of selective olefin epoxidation catalyzed by [γ-(SiO$_4$)W$_{10}$O$_{32}$H$_4$]$^{4-}$. A computational study, *Inorg. Chem.* 45 (2006) 5703.

[74] A. Sartorel, M. Carraro, A. Bagno, G. Scorrano, M. Bonchio, Asymmetric tetraprotonation of γ-[(SiO$_4$)W$_{10}$O$_{32}$]$^{8-}$ triggers a catalytic epoxidation reaction: Perspectives in the assignment of the active catalyst, *Angew. Chem. Int. Ed.* 46 (2007) 3255.

[75] M. Carraro, L. Sandei, A. Sartorel, G. Scorrano, M. Bonchio, Hybrid polyoxotungstates as second-generation POM-based catalysts for microwave-assisted H$_2$O$_2$ activation, *Org. Lett.* 8 (2006) 3671.

[76] T. D. Phan, M. A. Kinch, J. E. Barker, T. Ren, Highly efficient utilization of H$_2$O$_2$ for oxygenation of organic sulfides catalyzed by [γ-SiW$_{10}$O$_{34}$(H$_2$O)$_2$]$^{4-}$, *Tetrahedron Lett.* 46 (2005) 397.

[77] C. L. Hill, R. B. Brown Jr., Sustained epoxidation of olefins by oxygen donors catalyzed by transition metal-substituted polyoxometalates, oxidatively resistant inorganic analogs of metalloporphyrins, *J. Am. Chem. Soc.* 108 (1986) 536.

[78] C. L. Hill, in: C. L. Hill (Ed.), *Activation and Functionalization of Alkanes, Catalytic Oxygenation of unactivated carbon-hydrogen bonds*: Superior oxo transfer catalysts and the inorganic metalloporphyrin, Wiley, New York, 1989, p. 243.
[79] L. Salles, F. Robert, V. Semmer, Y. Jeannin, J.-M. Brégeault, Novel di- and trinuclear oxoperoxosulfato species in molybdenum(VI) and tungsten(VI) chemistry: The key role of pairs of bridging peroxo groups. *Bull. Soc. Chim. Fr.* 133 (1996) 319.
[80] L. Salles, J.-Y. Piquemal, R. Thouvenot, C. Minot, J.-M. Brégeault, Catalytic epoxidation by heteropolyoxoperoxo complexes: From novel precursors or catalysts to a mechanistic approach, *J. Mol. Catal. A: Chem.* 117 (1997) 375.
[81] K. Sato, M. Aoki, M. Ogawa, T. Hashimoto, D. Panyella, R. Noyori, A halide-free method for olefin epoxidation with 30% hydrogen peroxide, *Bull. Chem. Soc. Jpn.* 70 (1997) 905.
[82] W. Adam, D. Golsch, Thianthrene 5-oxide (SSO) as a mechanistic probe of the electrophilic character in the oxygen transfer by dioxiranes, *Chem. Ber.* 127 (1994) 1111.
[83] W. Adam, D. Golsch, Probing for electronic and steric effects in the peracid oxidation of thianthrene 5-oxide, *J. Org. Chem.* 62 (1997) 115.
[84] W. Adam, A. Corma, H. García, O. Weichold, Titanium-catalyzed heterogeneous oxidations of silanes, chiral allylic alcohols, 3-alkylcyclohexanes, and thianthrene 5-oxide: A comparison of the reactivities and selectivites for the large-pore zeolite Ti-β, the mesoporous Ti-MCM-41, and the layered alumosilicate Ti-ITQ-2, *J. Catal.* 196 (2000) 339.
[85] W. Adam, C. M. Mitchell, C. R. Saha-Mçller, Steric and electronic efforts in the diastereoselective catalytic epoxidation of cyclic allylic alcohols with methyltrioxorhenium (MTO), *Eur. J. Org. Chem.* 4 (1999) 785.
[86] A. Yoshida, M. Yoshimura, K. Uehara, S. Hikichi, N. Mizuno, Formation of S-shaped disilicoicosatungstate and efficient Baeyer-Villiger oxidation with hydrogen peroxide, *Angew. Chem. Int. Ed.* 45 (2006) 1956.
[87] A. Yoshida, S. Hikichi, N. Mizuno, Acid-base catalyses by dimeric disilicoicosatungstates and divacant γ-Keggin-type silicodecatungstate parent: Reactivity of the polyoxometalate compounds controlled by step-by-step protonation of lacunary W=O sites, *J. Organomet. Chem.* 692 (2007) 455.
[88] G. Strukul, Transition metal catalysis in the Baeyer-Villiger oxidation of ketones, *Angew. Chem. Int. Ed.* 37 (1998) 1198.
[89] G.-J. ten Brink, J.-M. Vis, I. W. C. E. Arends, R. A. Sheldon, Selenium-catalyzed oxidations with aqueous hydrogen peroxide. 2. Baeyer-Villiger reactions in homogeneous solution, *J. Org. Chem.* 66 (2001) 2429.
[90] M. Renz, T. Blasco, A. Corma, V. Fornes, R. Jensen, L. Nemeth, Selective and shape-selective Baeyer-Villiger oxidations of aromatic aldehydes and cyclic ketones with Sn-beta zeolites and H_2O_2, *Chem. Eur. J.* 8 (2002) 4708.
[91] R. Neumann, M. Gara, The manganese-containing polyoxometalate, $[WZnMn^{II}_2(ZnW_9O_{34})_2]^{12-}$, as a remarkable effective catalyst for hydrogen peroxide mediated oxidations, *J. Am. Chem. Soc.* 117 (1995) 5066.
[92] M. Bösing, A. Nöh, I. Loose, B. Krebs, Highly efficient catalysts in directed oxygen-transfer processes: Synthesis, structures of novel manganese-containing heteropolyanions, and applications in regioselective epoxidation of dienes with hydrogen peroxide, *J. Am. Chem. Soc.* 120 (1998) 7252.
[93] M. D. Ritorto, T. M. Anderson, W. A. Neiwert, C. L. Hill, Decomposition of A-type sandwiches. Synthesis and characterization of new polyoxometalates incorporating multiple d-electron-centered units, *Inorg. Chem.* 43 (2004) 44.
[94] D. Sloboda-Rozner, P. L. Alsters, R. Neumann, A water-soluble and "self-assembled" polyoxometalate as a recyclable catalyst for oxidation of alcohols in water with hydrogen peroxide, *J. Am. Chem. Soc.* 125 (2003) 5280.
[95] D. Sloboda-Rozner, P. Witte, P. L. Alsters, R Neumann, Aqueous biphasic oxidation: A water-soluble polyoxometalate catalyst for selective oxidation of various functional groups with hydrogen peroxide, *Adv. Synth. Catal.* 346 (2004) 339.
[96] X. Zhang, Q. Chen, D. C. Duncan, C. F. Campana, C. L. Hill, Multiiron polyoxoanions. Syntheses, characterization, X-ray crystal structures, and catalysis of H_2O_2-based hydrocarbon oxidation by $[Fe^{III}_4(H_2O)_2(P_2W_{15}O_{56})_2]^{12-}$, *Inorg. Chem.* 36 (1997) 4208.

[97] T. M. Anderson, X. Zhang, K. I. Hardcastle, C. L. Hill, Reactions of trivacant Wells-Dawson heteropolytungstates. Ionic strength and Jahn-Teller effects on formation in multi-iron complexes, *Inorg. Chem.* 41 (2002) 2477.
[98] X. Zhang, T. M. Anderson, Q. Chen, C. L. Hill, A Baker-Figgis isomer of conventional sandwich polyoxometalates. $H_2Na_{14}[Fe^{III}_2(NaOH_2)_2(P_2W_{15}O_{56})_2]$, a diiron catalyst for catalytic H_2O_2-based epoxidation, *Inorg. Chem.* 40 (2001) 418.
[99] N. Mizuno, C. Nozaki, I. Kiyoto, M. Misono, Highly efficient utilization of hydrogen peroxide for selective oxygenation of alkanes catalyzed by diiron-substituted polyoxometalate precursor, *J. Am. Chem. Soc.* 120 (1998) 9267.
[100] N. Mizuno, C. Nozaki, I. Kiyoto, M. Misono, Selective oxidation of alkenes catalyzed by di-iron-substituted silicotungstate with highly efficient utilization of hydrogen peroxide, *J. Catal.* 182 (1999) 285.
[101] N. Mizuno, I. Kiyoto, C. Nozaki, M. Misono, Remarkable structure dependence of intrinsic catalytic activity for selective oxidation of hydrocarbons with hydrogen peroxide catalyzed by iron-substituted silicotungstates, *J. Catal.* 181 (1999) 171.
[102] R. Ben-Daniel, A. M. Khenkin, R. Neumann, The nickel-substituted quasi-Wells-Dawson-type polyfluoroxometalate, $[Ni^{II}(H_2O)H_2F_6NaW_{17}O_{55}]^{9-}$, as a uniquely active nickel-based catalyst for the activation of hydrogen peroxide and the epoxidation of alkenes and alkenols, *Chem. Eur. J.* 6 (2000) 3722.
[103] K. Yamaguchi, C. Yoshida, S. Uchida, N. Mizuno, Peroxotungstate immobilized on ionic liquid-modified silica as a heterogeneous epoxidation catalyst with hydrogen peroxide, *J. Am. Chem. Soc.* 127 (2005) 530.
[104] J. Kasai, Y. Nakagawa, S. Uchida, K. Yamaguchi, N. Mizuno, $[\gamma\text{-}1,2\text{-}H_2SiV_2W_{10}O_{40}]$ Immobilized on surface-modified SiO_2 as a heterogeneous catalyst for liquid-phase oxidation with H_2O_2, *Chem. Eur. J.* 12 (2006) 4176.
[105] S. Uchida, M. Hashimoto, N. Mizuno, A breathing ionic crystal displaying selective binding of small alcohols and nitriles: $K_3[Cr_3O(OOCH)_6(H_2O)_3][\alpha\text{-}SiW_{12}O_{40}]\cdot 16I\,I_2O$, *Angew. Chem. Int. Ed.* 41 (2002) 2814.
[106] S. Uchida, N. Mizuno, Unique guest-inclusion properties of a breathing ionic crystal of $K_3[Cr_3O(OOCH)_6(H_2O)_3][\alpha\text{-}SiW_{12}O_{40}]\cdot 16H_2O$, *Chem. Eur. J.* 9 (2003) 5850.
[107] S. Uchida, N. Mizuno, Zeotype ionic crystal of $Cs_5[Cr_3O(OOCH)_6(H_2O)_3][\alpha\text{-}CoW_{12}O_{40}]\cdot 7.5H_2O$ with shape-selective adsorption of water, *J. Am. Chem. Soc.* 126 (2004) 1602.
[108] S. Uchida, R. Kawamoto, T. Akatsuka, S. Hikichi, N. Mizuno, Structures and sorption properties of ionic crystals of macrocation-Dawson-type polyoxometalates with different charges, *Chem. Mater.* 17 (2005) 1367.
[109] R. Kawamoto, S. Uchida, N. Mizuno, Amphiphilic guest sorption of $K_2[Cr_3O(OOCC_2H_5)_6(H_2O)_3]_2[\alpha\text{-}SiW_{12}O_{40}]$ ionic crystal, *J. Am. Chem. Soc.* 127 (2005) 10560.
[110] N. Mizuno, S. Uchida, Structures and sorption properties of ionic crystals of polyoxometalates with macrocation, *Chem. Lett.* 35 (2006) 688.
[111] C. Jiang, A. Lesbani, R. Kawamoto, S. Uchida, N. Mizuno, Channel-selective independent sorption and collection of hydrophilic and hydrophobic molecules by $Cs_2[Cr_3O(OOCC_2H_5)_6(H_2O)_3]_2[\alpha\text{-}SiW_{12}O_{40}]$ ionic crystal, *J. Am. Chem. Soc.* 128 (2006) 14240.
[112] K. Uehara, H. Nakao, R. Kawamoto, S. Hikichi, N. Mizuno, 2D-grid layered Pd-based cationic infinite coordination polymer/polyoxometalate crystal with hydrophilic sorption, *Inorg. Chem.* 45 (2006) 9448.
[113] K. Okamoto, S. Uchida, T. Ito, N. Mizuno, Self-organization of all-inorganic dodecatungstophosphate nanocrystallites, *J. Am. Chem. Soc.* 129 (2007) 7378.

CHAPTER 5

Oxaziridinium Salt-Mediated Catalytic Asymmetric Epoxidation

Philip C. Bulman Page and **Benjamin R. Buckley**

Contents		
	1. Introduction	178
	2. Page Group Findings	184
	2.1. Catalyst structure	185
	2.2. The reaction parameters	186
	2.3. Initial findings	189
	2.4. Catalysts based on dibenzo[c,e]azepinium salts	194
	2.5. Catalysts based on a binaphthalene structure	196
	2.6. Catalytic asymmetric anhydrous epoxidation mediated by tetraphenylphosphonium monoperoxysulphate and iminium salts	199
	2.7. Comments on the mechanism	212
	Acknowledgements	214
	References	214

Abstract	The development of new systems for catalytic asymmetric epoxidation is of practical and fundamental importance in chemistry today. We report herein our endeavours in this challenging area and describe several catalyst systems based on iminium salts. This review illustrates the effects of reaction parameters, catalyst structure development, and the formation of a new non-aqueous oxidation system. Enantiomeric excesses of up to 97% and catalyst loadings as low as 0.1 mol% have been achieved, and spectroscopic evidence for an oxaziridinium ion intermediate has been obtained.
	Key Words: Iminium, Oxaziridinium, Oxaziridine, Ketiminium, Oxone, Tetraphenylphosphonium monoperoxysulphate, Isopinocampheylamine, Alkene, Epoxide, Enantiomeric excess, Asymmetric synthesis, Organocatalysis, 2-(2-Bromoethyl)benzaldehyde, Levcromakalim, Dihydroisoquinolinium, Spiro, Azepinium, Benzopyran, Dielectric constant, Binol © 2008 Elsevier B.V.

Department of Chemistry, Loughborough University, Loughborough LE11 3TU, United Kingdom

Mechanisms in Homogeneous and Heterogeneous Epoxidation Catalysis
DOI: 10.1016/B978-0-444-53188-9.00005-5

© 2008 Elsevier B.V.
All rights reserved.

1. INTRODUCTION

Oxaziridinium salts are the quarternized analogues of oxaziridines, and as a result of being more electrophilic, transfer oxygen more efficiently to nucleophilic substrates. The first oxaziridinium salt, described by Lusinchi in 1976 [1–3], was based on a steroidal pyrrolinic skeleton. Through peracid oxidation of the steroidal imine and quaternization using methylfluorosulphonate, it was shown that an oxaziridinium species could be formed (Scheme 5.1). This new species was rather unstable, and upon decomposition reverted to an iminium salt, which could be directly prepared from the imine. However, it was not until some 11 years later that the potential of this type of system to transfer oxygen was realized [4,5].

Using an oxaziridinium salt derived from dihydroisoquinoline, Lusinchi was able to transfer the oxygen atom to several simple alkenes in good yield (Scheme 5.2) [5,6]. Following this work, the first enantiomerically pure oxaziridinium salt was prepared [7]. Quaternization of chiral oxaziridine (1), derived from (1S,2R)-(+)-norephedrine, produced the oxaziridinium salt (2) (Scheme 5.3).

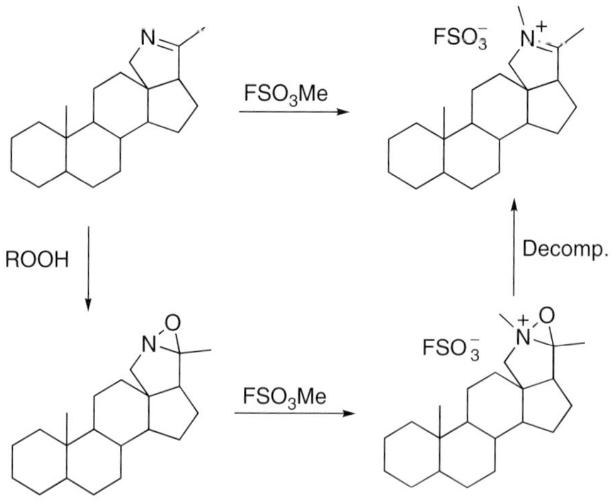

SCHEME 5.1 The first example of an oxaziridinium salt developed by Lusinchi.

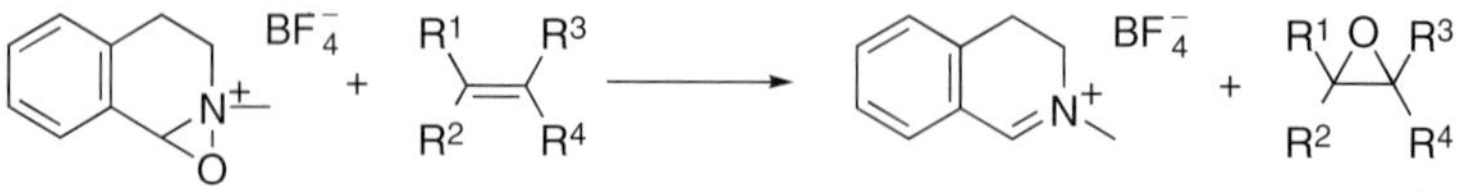

SCHEME 5.2 Oxygen transfer to alkenes mediated by a oxaziridinium salt derived from tetrahydroisoquinoline.

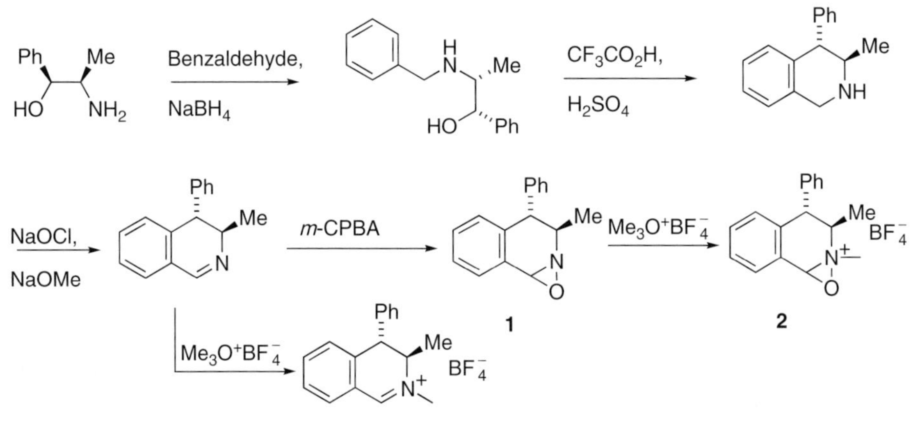

SCHEME 5.3 The first enantiomerically pure oxaziridinium salt derived from (1S,2R)-(+)-norephidrine.

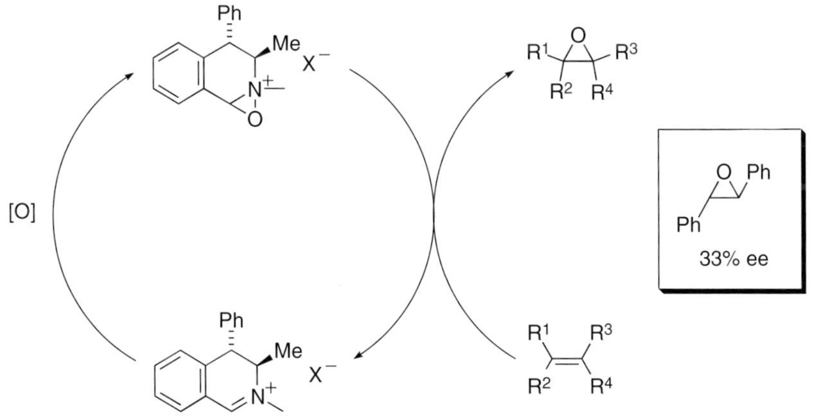

SCHEME 5.4 Catalytic cycle for epoxidations mediated by oxaziridinium salts.

This oxaziridinium salt was also able to transfer oxygen to olefins, and induced moderate enantiocontrol, epoxidizing *trans*-stilbene with 33% ee. With the side product of the reaction being an iminium salt, there was potential to develop this chemistry catalytically. If this iminium salt could be re-oxidized to the oxaziridinium *in situ*, catalytic transfer of oxygen to alkenes could be achieved (Scheme 5.4).

Lusinchi was able to develop such a catalytic system using the triple salt Oxone ($KHSO_5 \cdot KHSO_4 \cdot K_2SO_4$), similar to that developed for dioxiranes; however, less pH control is required as there is no competing Baeyer–Villiger oxidation. The enantiomerically pure iminium salt (3) is thus able to epoxidize *trans*-stilbene catalytically (20 mol%) with the same degree of selectivity as does the stoichiometric oxaziridinium salt (33% ee). Lusinchi has also shown that oxaziridinium salts are capable of transferring oxygen to other nucleophilic substrates such as sulphides to form sulphoxides [8], amines to form nitrones, and imines to form oxaziridines [9].

Lusinchi's group has reported the only X-ray determination of an oxaziridinium salt, of compound 2 [7,10]. Its geometry is similar to that of the parent oxaziridine, and the N–O bond length of 1.468 Å in the oxaziridinium salt is shortened compared with the mean bond length of 1.508 Å observed for oxaziridines. It is also interesting to note that the oxaziridine ring is perpendicular to the isoquinoline ring.

Since Lusinchi's early work, several groups have identified an interest in oxaziridinium/iminium salt chemistry. Aggarwal and Wang produced a cyclic binaphthalene-derived iminium salt, which was shown to be effective in the epoxidation of 1-phenylcyclohexene, giving 71% ee [11]. This catalyst exhibited high substrate dependency, the best ee for other olefins tested being only 45% (Fig. 5.1).

Armstrong has shown that even acyclic iminium salts can mediate epoxidation by Oxone [12], but enantiomeric excesses are low [13]. By condensing *N*-trimethylsilylpyrrolidine (4) with a range of aromatic aldehydes in the presence of trimethylsilyltriflate, Armstrong was able to produce a range of substituted exocyclic iminium salts (Scheme 5.5). Only those compounds with electron-withdrawing groups present on the aromatic ring were active mediators. Catalytic reactions were carried out with the *ortho*-Cl (5) derivative, giving good conversion to epoxide (Scheme 5.6).

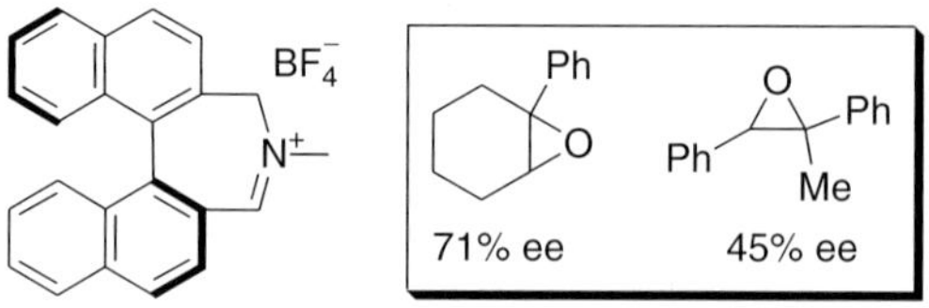

FIGURE 5.1 Aggarwal's binaphthalene-based iminium salt catalyst.

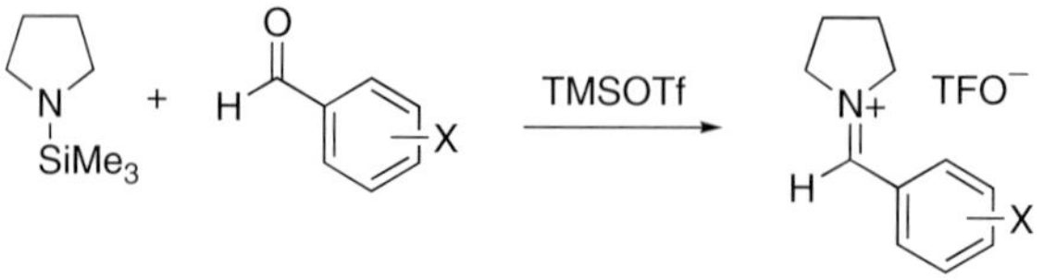

SCHEME 5.5 Armstrong's synthesis of exocyclic iminium salts.

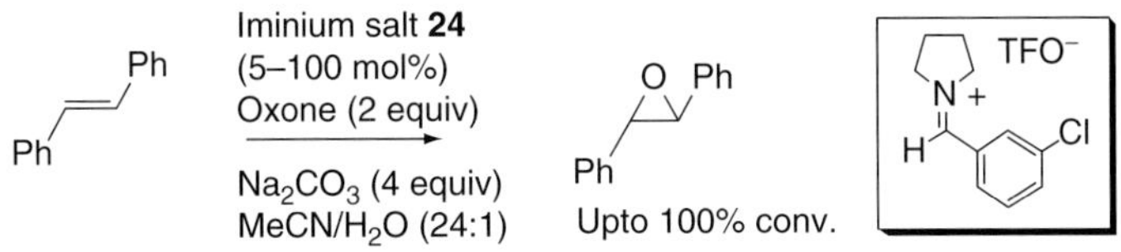

SCHEME 5.6 Epoxidation of *trans*-stilbene with Armstrong's catalyst (**5**).

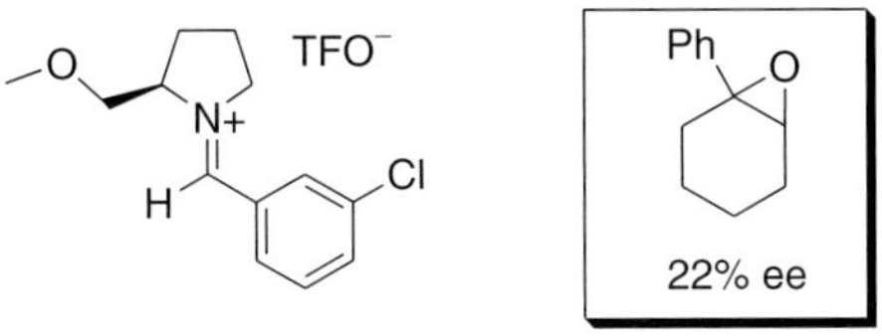

FIGURE 5.2 Armstrong's most successful chiral iminium salt catalyst.

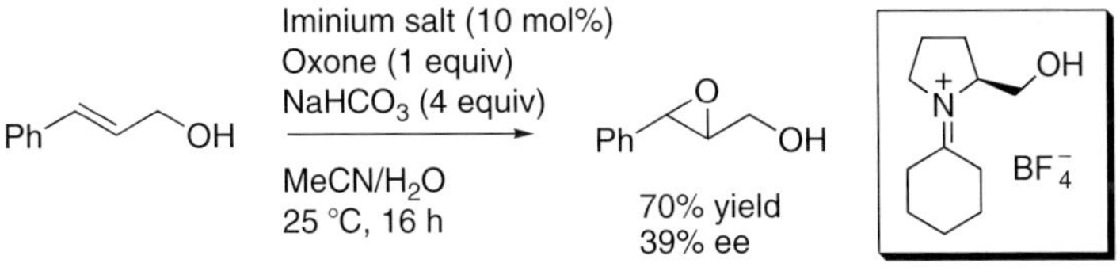

SCHEME 5.7 Komatsu's ketiminium salt-mediated epoxidation.

Despite many attempts to produce chiral variants of this catalyst system, Armstrong was unable to obtain significant ees, catalyst (**6**) giving only 22% ee for 1-phenylcyclohexene (Fig. 5.2).

Recently, two other groups have shown that exocyclic iminium salts can be useful mediators in asymmetric epoxidation. Komatsu has developed a system based on ketiminium salts [14], prepared through the condensation of aliphatic cyclic amines with ketones. A chiral variant was also produced, derived from prolinol and cyclohexanone, which gave 70% yield and 39% ee for cinnamyl alcohol (Scheme 5.7).

Moderate ees have been achieved by Yang using another exocyclic iminium salt system [15]. These salts are not isolated, but are generated *in situ*, thus circumventing the difficulties inherent in the preparation and isolation of unstable acyclic iminium salts. A major drawback to this type of system is the necessary high catalyst loadings; for an efficient rate of reaction, up to 50 mol% of iminium salt is generally required. Nevertheless, ees of up to 65% have been achieved. A range of amines and aldehydes were screened, and a novel proline-based amine and a branched hexanal were found to be the best precursors (Scheme 5.8).

Armstrong has also reported an *in situ* epoxidation, mediated by an intramolecular oxaziridinium salt, which gave good regioselectivity (Scheme 5.9) [16].

With a modification of this procedure, using an oxaziridine, Armstrong was able to demonstrate the synthetic utility of this method by introducing a chiral amine to afford enantiomerically enriched products, and greater than 98% ee was obtained for **(7)**, which is a terminal epoxide, typically very testing substrates for asymmetric epoxidation (Scheme 5.10) [17]. A loss of selectivity was, however, observed when the chain length between the aldehyde and the alkene exceeded three atoms.

More recently Bohé, a former co-worker with Lusinchi, has reported an improved achiral catalyst that prevents some of the common side reactions observed in iminium salt-mediated epoxidation [18]. Two factors are known to reduce the catalytic efficiency of the epoxidation process: hydrolysis of the iminium salt directly by the reaction medium, which generally only affects the acyclic systems; and loss of active oxygen from the intermediate oxaziridinium species, through a reaction that does not regenerate the iminium species, which

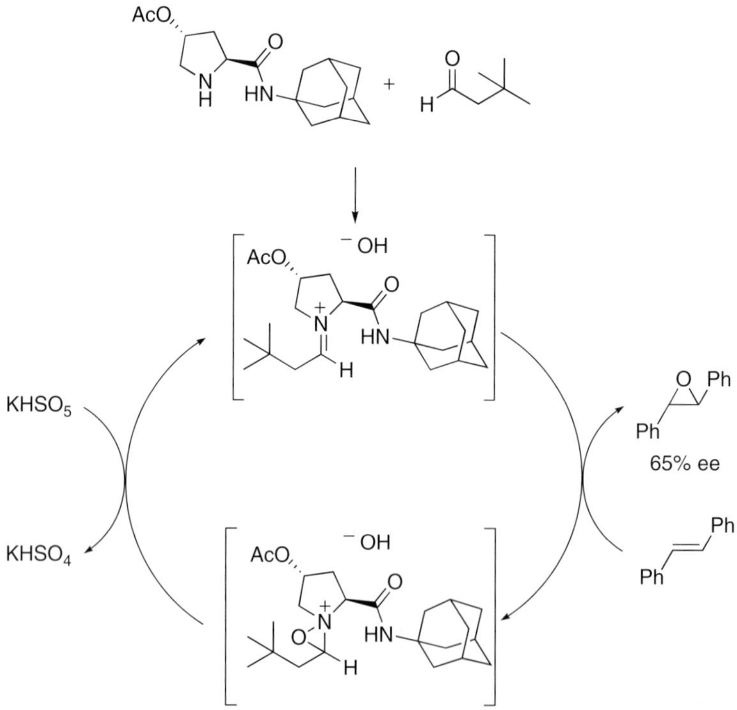

SCHEME 5.8 Yang's *in situ* iminium salt epoxidation system.

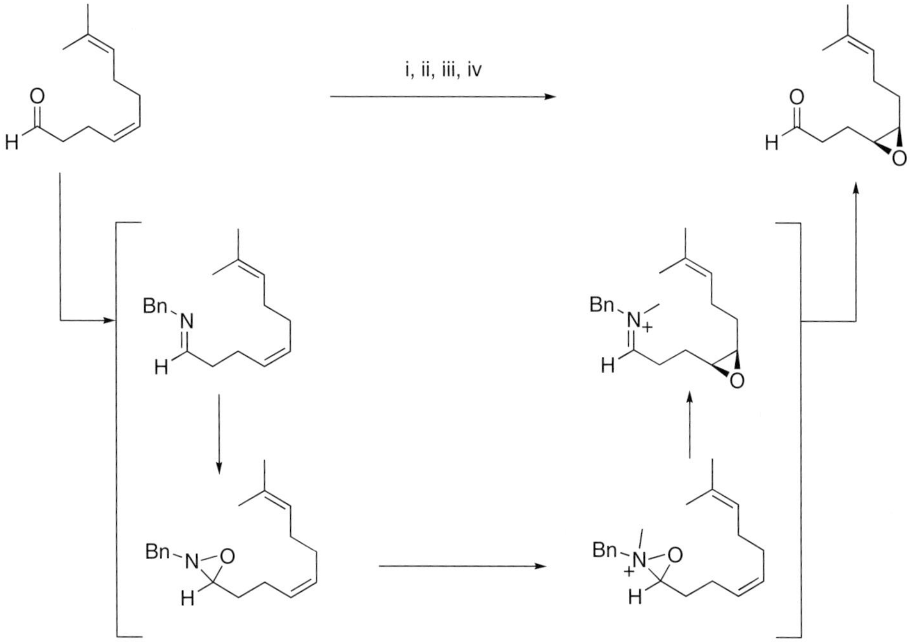

Reagents and conditions: i: BnNH$_2$, 4 Å mol. sieves, DCM; ii: Oxone, NaHCO$_3$, MeCN/H$_2$O; iii: MeOTf, DCM; iv: NaHCO$_3$ (aq.).

SCHEME 5.9 Armstrong's intramolecular epoxidation.

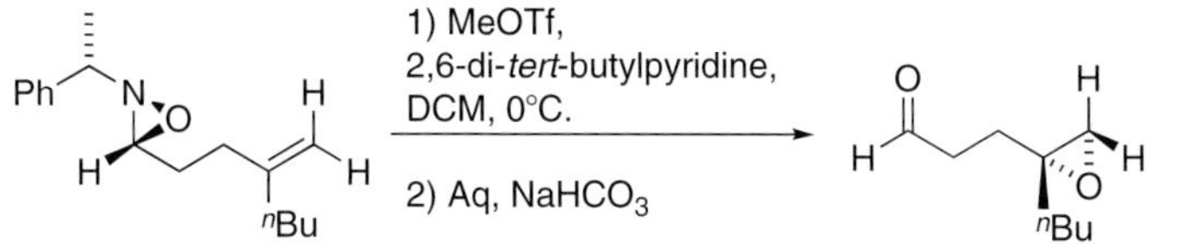

SCHEME 5.10 Armstrong's chiral version of the intramolecular epoxidation reaction.

Yield: 55%; ee: >98%

can occur in all systems containing protons α to the nitrogen atom. This latter process is an irreversible base-catalysed isomerization (Scheme 5.11).

A dramatic increase in catalyst efficiency is observed when the 3,3-disubstituted dihydroisoquinolinium salt **(8)** is used in place of catalyst **(9)**, thus eliminating the base-catalysed isomerization (Scheme 5.12).

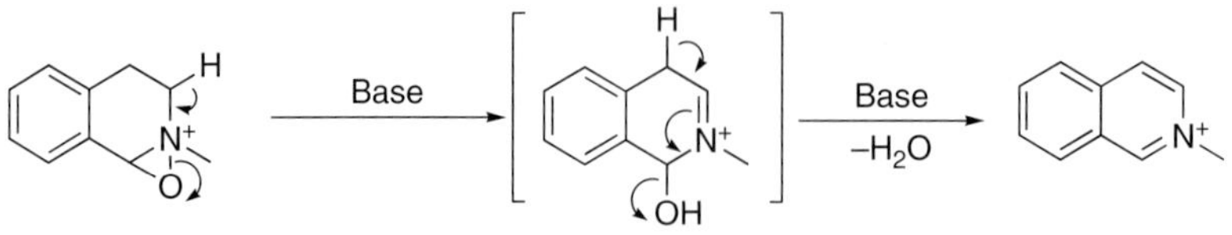

SCHEME 5.11 Bohé's proposed irreversible base-catalysed isomerization.

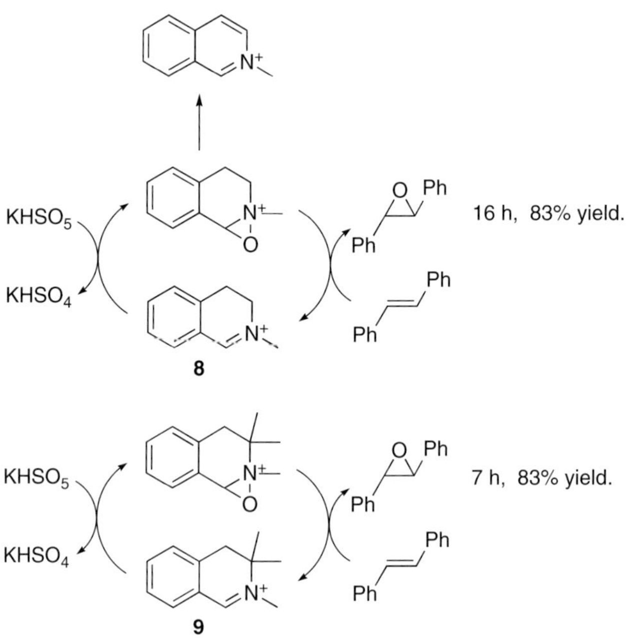

SCHEME 5.12 Bohé's improved achiral catalyst for catalytic epoxidation.

2. PAGE GROUP FINDINGS

In the search for a new and highly enantioselective system for iminium salt-mediated catalytic asymmetric epoxidation, several parameters and catalyst substructures were examined. The optimum method of oxidation, using Oxone, was established after some experimentation, and a possible catalytic cycle for an oxaziridinium ion as the oxidative intermediate and Oxone as the oxidant is depicted in Scheme 5.13 [19–21]. The first stage is the formation of an initial adduct **(10)**, uncharged at nitrogen, formed by (probably reversible) nucleophilic attack of the oxidant on the iminium salt. This is followed by irreversible expulsion of sulphate to give the oxaziridinium ion, which may be the rate-determining step under the reaction conditions. Oxygen may then be transferred to a substrate

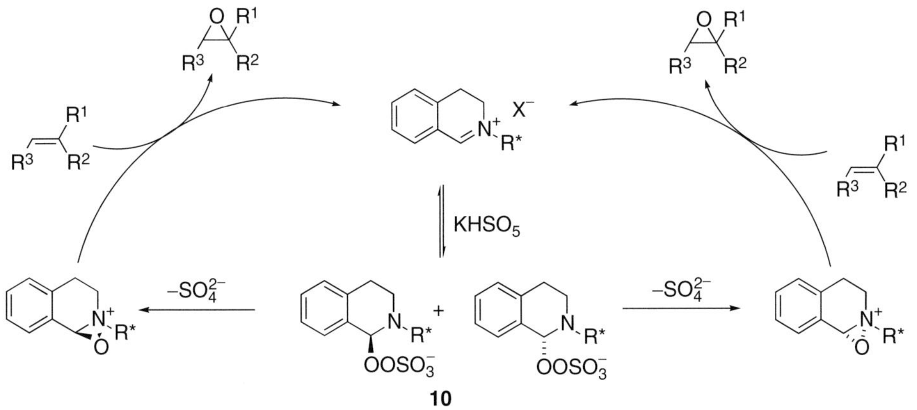

SCHEME 5.13 Catalytic cycle for the oxaziridinium ion as an oxidative intermediate in the epoxidation reaction.

in a subsequent step, the rate of which would not be expected to have any great solvent dependence. An interesting but complicating feature of these processes is that it is not one but two diastereoisomeric oxaziridinium salts which may be formed, by attack of oxidant at the *si* or *re* face of the iminium species. Each may deliver the oxygen atom to either of the prochiral faces of the alkene substrate with a different degree of enantiocontrol, and the resulting oxaziridinium species may be in competition for the alkene substrate.

2.1. Catalyst structure

An ideal method for testing a wide variety of substructures was developed through the condensation of enantiomerically pure chiral primary amines with 2-(2-bromoethyl)benzaldehyde (11) as shown in Scheme 5.14 [19,21].

The iminium salts prepared by this method have the advantage that they are extremely easy to prepare on any scale and that the structural variation available is large, because the chirality is resident in the amine component. Treatment of isochroman (12) with bromine in carbon tetrachloride under reflux for 1 h followed by exposure to concentrated hydrobromic acid provides 2-(2-bromoethyl)benzaldehyde (11) in 65% yield [22]. Primary amines condense smoothly with this material to furnish the corresponding dihydroisoquinolinium bromides. These organic salts are generally oils, and the inherent difficulties in purification by conventional methods necessitated a change in counterion. Addition of a solution of sodium tetraphenylborate at the end of the reaction, in the minimum amount of acetonitrile, induces rapid formation of the corresponding tetraphenylborate salts as crystalline solids [21]. Overall yields of catalyst are generally between 30 and 80%, limited in part as a consequence of a side reaction, elimination of hydrogen

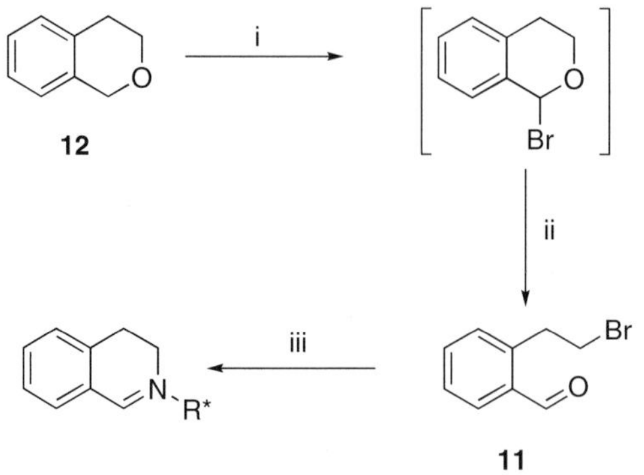

Reagents and conditions: i: Br_2, CCl_4, 1 h; ii: HBr(conc), Δ, 10min;
iii: a) R*NH_2, EtOH, 0 °C-r.t.,12 h b) $NaBPh_4$, MeCN, 5 min.

SCHEME 5.14 The 2-(2-bromoethyl)benzaldehyde method for forming dihydroisoquinolinium salts.

bromide from the bromoethyl moiety of the precursor. No chromatography is necessary at any point in this sequence.

Very hindered amines give inferior yields of iminium salt, typically 25–30%, presumably due to an increased tendency to act as bases rather than nucleophiles, as evidenced by the increased levels of 2-vinylbenzaldehyde [19].

A range of structurally different chiral primary amines was converted into the corresponding iminium tetraphenylborate salts (Fig. 5.3) and tested in the asymmetric epoxidation of a standard test substrate, 1-phenylcyclohexene, using Oxone (4 equiv) as the stoichiometric oxidant, sodium carbonate (8 equiv) as base, in acetonitrile/water (2:1) at 0 °C (Table 5.1) [19,21].

With the first two entries in Table 5.1, using the structurally simplest amines **13** and **14**, no asymmetric induction was observed, and it became clear that a conformationally more defined and rigid system was required to impart reasonable enantioselectivities. Both the camphor **20**-, **21**- and methyl **16**-based systems gave low ees, although these are two of the more common systems upon which chiral auxiliaries have been based. The fenchyl derivative **19** is the most selective under these reaction conditions. However, the *N*-(isopinocampheyl) dihydroisoquinolinium salt **17**, which is considerably less sterically hindered than the fenchyl, is almost as selective, giving a better yield and increased rate of reaction.

2.2. The reaction parameters

Having found a catalyst which exhibited a good reaction profile in terms of ees and rates, an examination of some of the reaction parameters was carried out in order to optimize the reaction conditions with respect to the enantioselectivity of

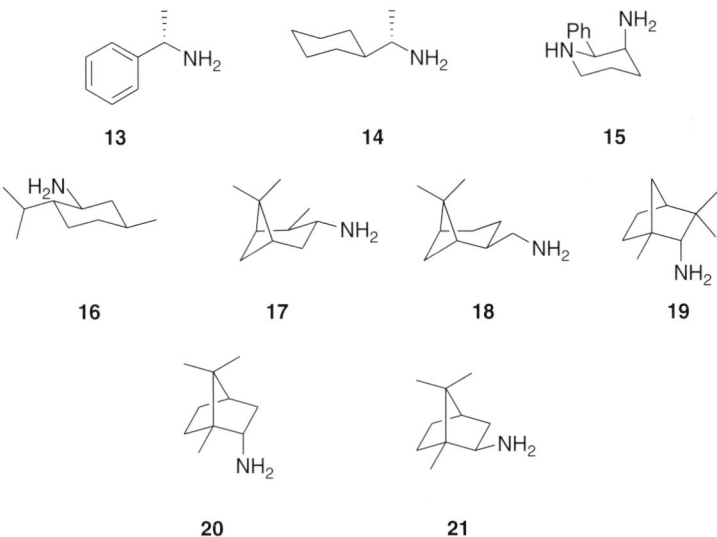

FIGURE 5.3 Initial primary amines condensed with 2-(2-bromoethyl)benzaldehyde **(11)** to form iminium salts.

TABLE 5.1 Epoxidation of 1-Phenylcyclohexene with dihydroisoquinolinium tetraphenylborate salts derived from chiral primary amines[a]

Entry	Amine precursor	Catalyst load (mol%)	Yield (%)	ee (%)	Configuration
1	13	5.0	54	0	—
2	14	5.0	70	0	—
3	15	1.0	39	25	(−)-(1S,2S)
4	16	0.5	63	19	(+)-(1R,2R)
5	17	0.5	68	27	(+)-(1R,2R)
6	18	0.5	66	12	(−)-(1S,2S)
7	19	0.5	45	32	(+)-(1R,2R)
8	20	0.5	60	18	(+)-(1R,2R)
9	21	0.5	58	8	(+)-(1R,2R)
10	22	0.5	47	14	(−)-(1S,2S)

[a] Conditions: Oxone (4 equiv), Na_2CO_3 (8 equiv), 0 °C, H_2O/MeCN 1:2, reactions monitored by TLC.

the oxygen transfer process. The N-(isopinocampheyl) dihydroisoquinolinium salt **(22)** was chosen as the model catalyst for optimization studies[1].

22

[1] The IUPAC name for (_)-isopinocampheylamine is (_)-6-(1R,2R,3R,5S)-2,2,6-trimethylbicyclo[3.1.1]hept-3-ylamine.

2.2.1. Effect of counterion

In addition to the original tetraphenylborate, the corresponding tetrafluoroborate, hexafluorophosphate, perchlorate and periodate salts were prepared. All of these were tested in the asymmetric catalytic epoxidation of 1-phenylcyclohexene. A catalyst loading of 5 mol% was used in a 1:1 water/acetonitrile solvent in the presence of 2 equiv of Oxone and 4 equiv of sodium carbonate at 0 °C. The enantioselectivities obtained exhibited an interesting trend. The periodate salt gave a similar ee (35%) to that of the tetraphenylborate species (40% ee), while the fluoride-containing counterions afforded lower ees (28%). The perchlorate salt also furnished inferior enantioselectivities (20% ee). All of the salts however invariably produced the same major enantiomer of the epoxide product (R,R), and all of the reactions were complete in a similar timescale (~45 min).

2.2.2. Effect of the solvent system

The standard conditions employed within the Page group consist of the solvent system, composed of a 1:1 mixture of acetonitrile and water, 2 equiv of Oxone and 4 equiv of sodium carbonate at 0 °C. An increase in water to acetonitrile ratio is accompanied by an increase in the reaction rate. For example, the yield of 1-phenylcyclohexene oxide after 1 h using catalyst **(17)** was 30% at 0 °C when a 1:1 ratio of the two solvents was used, but the yield was essentially quantitative using a 2:1 (water/acetonitrile) ratio. Reducing the amount of Oxone and base by a factor of 2 (i.e., using 1 equiv of Oxone and 2 equiv of sodium carbonate), resulted in incomplete conversion after 1 h in the improved (2:1) solvent system. Higher catalyst loadings, however, accelerate the rate of reaction to an extent that outweighs the effect of water content.

An investigation into the co-solvent employed was also carried out [21]. The solvents were selected so that they differed significantly in dielectric constant (ε, indicated by the values in parentheses): dichloromethane (8.9), trifluoroethanol (26.7), acetonitrile (37.5), water (78.4) and formamide (111). The epoxidation of 1-phenylcyclohexene with catalyst **(17)** was tested using these co-solvents with water in a 1:1 ratio. Epoxidation did not occur in dichloromethane; this is perhaps due to the poor miscibility of the two solvents, thus limiting the availability of the inorganic oxidant in the organic phase. No reaction was also observed in formamide. This could be due to the iminium species being too well stabilized/solvated, and the possibility of an irreversible attack by the formamide cannot be dismissed. In trifluoroethanol, the reaction had a similar profile to that in acetonitrile; both reactions were complete in 30 min, but the ee was somewhat lower (26% ee in trifluoroethanol and 40% ee in acetonitrile).

2.2.3. Effect of temperature

Variation in reaction temperature is severely limited by the stability and solubility of Oxone. When the reaction was carried out at -10 °C it was sluggish, perhaps because the solubility of the inorganic oxidant and base in water was dramatically reduced [19,21]. When an increased volume of water was employed (3:1 ratio with acetonitrile), oxidation of 1-phenylcyclohexene mediated by catalyst **(17)** (5 mol%) resulted in complete consumption of the starting material within 45 min, and afforded the corresponding epoxide in slightly improved yield. The enantioselectivity was slightly reduced (35%) from that of the reaction carried out at 0 °C (40%).

When the oxidation was carried out at ambient temperature, negligible conversion to epoxide was observed; this is believed to be due to the instability of Oxone in the basic medium at this temperature [23].

2.2.4. Effect of catalyst loading

Catalyst loading was expected to have a large effect on the enantioselectivity of the process. The effect, however, was negligible when using catalyst **22**: decreased loadings resulted in longer reaction times, but the enantioselectivity of the system remained fairly consistent, reaching a maximum at 2 mol% (Fig. 5.4) [19]. Enantioselectivities below this loading decreased from 33% ee at 2 mol% loading, to 18% ee at 0.3 mol% catalyst loading.

2.3. Initial findings

After this early work in setting up the optimum reaction conditions, a further, more detailed, study of catalyst structure was instigated. A range of dihydroisoquinolinium salts containing alcohol, ether and acetal functionalities was tested [24].

2.3.1. Alcohol-containing iminium salts

We hoped that catalysts containing alcohol moieties, which were readily available from 1,2-aminoalcohols, might offer increased ees. Epoxidation reactions using this type of functionality, however, resulted in a sluggish reactivity and low enantioselectivity [24]. This inhibited reactivity is thought to stem from the existence of an equilibrium between the ring-open iminium salt (active) and the ring-closed oxazolidine (inactive) forms of the catalysts under the slightly alkaline reaction conditions (Scheme 5.15). It is known that such dihydroisoquinolinium

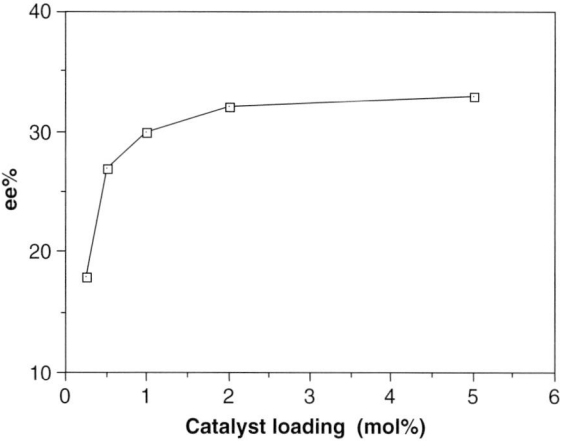

FIGURE 5.4 Effect of catalyst loading.

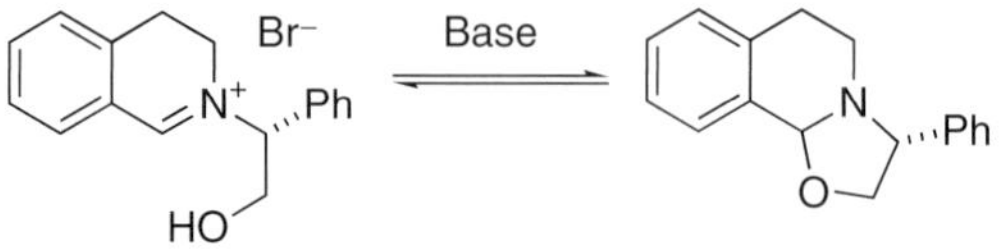

SCHEME 5.15 Base-induced ring closure of hydroxy dihydroisoquinolinium salts to oxazolidines.

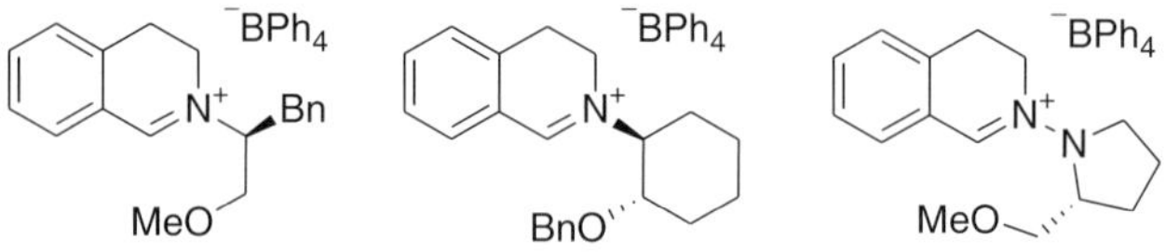

FIGURE 5.5 Some of the aminoether-based catalysts tested in the epoxidation of 1-phenylcyclohexene.

salts can undergo base-induced ring closure to form the corresponding oxazolidines with high diastereoselectivity and yields [25–27].

2.3.2. Catalysts from aminoether precursors

Several aminoether-based dihydroisoquinolinium salts were produced, and they proved to be much more active than the related derivatives of the parent aminoalcohols, but again poor enantioselectivity was observed in the epoxidation of 1-phenylcyclohexene (Fig. 5.5) [24]. This suggested that the size of the ether substituent in such catalysts is not particularly important for asymmetric induction during oxygen transfer to the alkene.

2.3.3. A Catalyst from an aminoacetal precursor

(1S,2S)-5-Amino-2,2-dimethyl-4-phenyl-1,3-dioxane (23) reacted smoothly with 2-(2-bromoethyl)benzaldehyde (11) under the usual conditions to furnish the corresponding dihydroisoquinolinium tetraphenylborate salt in greater than 75% yield (24) (Scheme 5.16) [24].

This iminium salt was tested in the catalytic asymmetric epoxidation of several alkenes at 0 °C, and a comparison of the results with those obtained using catalyst (17) is presented in Table 5.2 [24]. These results indicated that catalyst (24) in general induces much higher enantioselectivity in asymmetric epoxidation than others that were screened, providing in some cases dramatic improvements in ee over catalyst (17).

A feature of catalyst (24) is the *cis* relationship between the nitrogen heterocycle and the phenyl group. This implies that either the phenyl or the dihydroisoquinolinium group must be axial if the dioxane retains a chair conformation, as in (25) or (26) (Scheme 5.17).

Despite the similar size of the two substituents, ^{1}H NMR spectroscopy suggests the presence of only one conformer at ambient temperature [19,24]; first, all the proton signals are sharp and the coupling constants corresponding

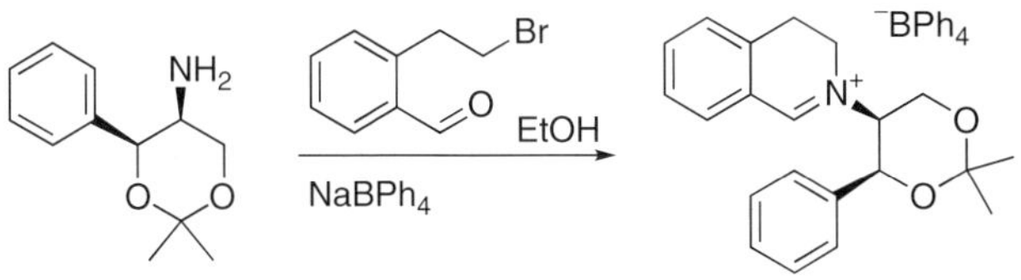

SCHEME 5.16 The amino acetal catalyst derived from (1S,2S)-5-amino-2,2-dimethyl-4-phenyl-1,3-dioxane **(23)**.

TABLE 5.2 Catalytic asymmetric epoxidation using catalysts **(17)** and **(24)**[a]

Epoxide	Catalyst 17			Catalyst 24		
	Yield (%)	ee (%)	Configuration	Yield (%)	ee (%)	Configuration
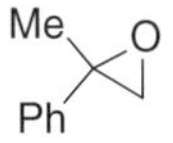 Me, Ph epoxide	68	8	(+)-(R)	64	20	(+)-(R)
Ph, Ph, Me epoxide	72	15	(+)-(1R,2R)	52	52	(−)-(1S,2S)
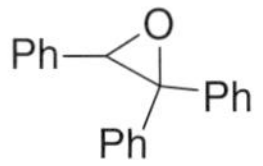 Ph, Ph, Ph epoxide	43	5	(+)-(S)	54	59	(+)-(S)
Cyclohexyl-Ph epoxide	68	40	(+)-(1R,2R)	55	41	(−)-(1S,2S)
Indene oxide	34	3	(+)-(1S,2R)	52	17	(+)-(1S,2R)

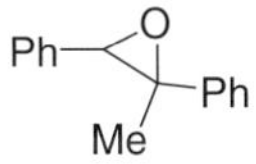

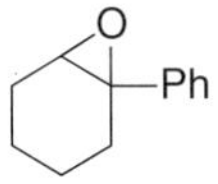

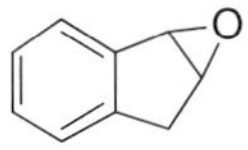

[a] Conditions: Oxone (2 equiv), sodium carbonate (4 equiv), water/acetonitrile (1:1), 0 °C, 5 mol% catalyst.

to each of the protons of the 1,3-dioxane ring are consistent with a chair conformation, in accord with previous reports of substituted 2,2-dimethyl-1,3-dioxane rings [28]. Further, in the ^{13}C NMR spectrum, the geminal methyl groups appear at 17.98 and 28.68 ppm (axial and equatorial, respectively); this is also consistent with a chair conformation [29,30]. Conformer **(26)** would be expected to be the thermodynamically favoured one as a result of reduced 1,3-diaxial interactions and the operation of the gauche effect.

It is also tempting to propose a stabilizing interaction between the electron cloud associated with the oxygen atom lone pairs and the electron-depleted carbon atom of the iminium unit. This suggestion in this case is supported by single-crystal X-ray analysis (Fig. 5.6), although other catalysts of this general

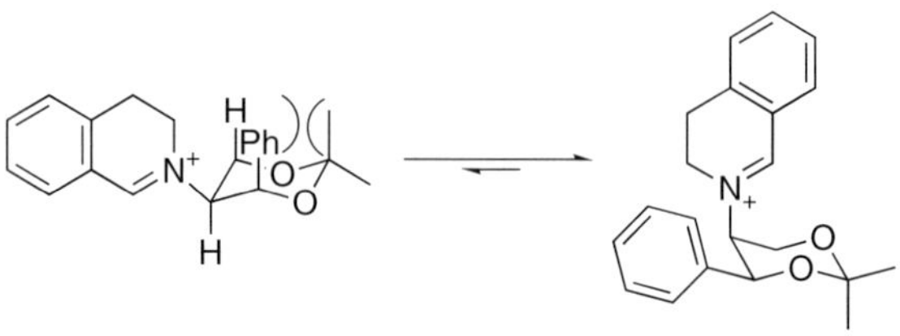

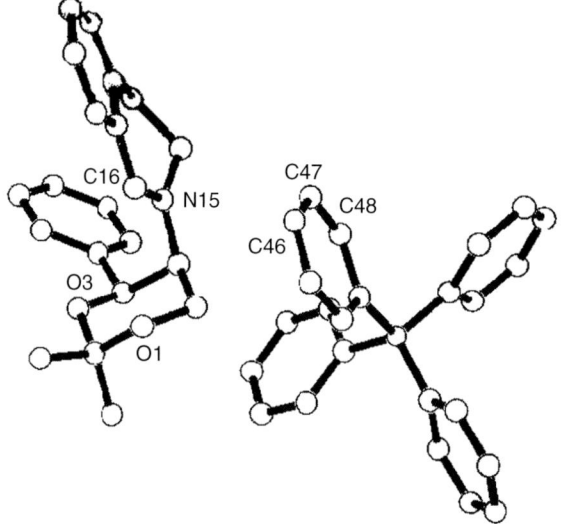

SCHEME 5.17 For a chair conformation one of the substituents must be axial.

FIGURE 5.6 X-ray crystal structure of catalyst **(24)**.

structure do not all show the same orientation of the dihydronaphthalene unit in the solid state. It is interesting that the X-ray analysis does not indicate a twist-boat conformation in any example. The relative success of the dioxane-derived catalyst may stem partly from high conformational rigidity, perhaps a result of the stereoelectronic effects discussed above.

In conformer **(26)**, the phenyl substituent may hinder the attack of the oxidant at that side of the iminium bond, rendering the opposite side more accessible. This arrangement would then produce a high preponderance of one of the two possible diastereoisomeric oxaziridinium intermediates (Scheme 5.18), and enantiocontrol might then result from the process of oxygen transfer from just one diastereoisomer to the substrate.

This high degree of conformational rigidity may be absent from the dihydroisoquinolinium salt **(17)**, derived from (−)-isopinocampheylamine (Fig. 5.7). In that case, rotation around the bond between the nitrogen atom and the chiral unit would then result in both diastereotopic faces of the iminium moiety

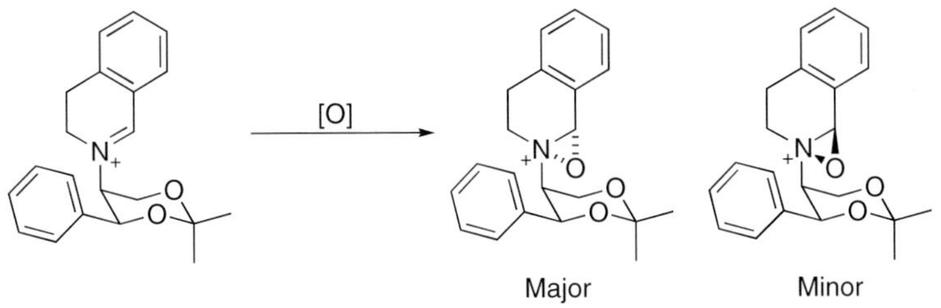

SCHEME 5.18 The two possible diastereoisomers produced in oxaziridinium formation.

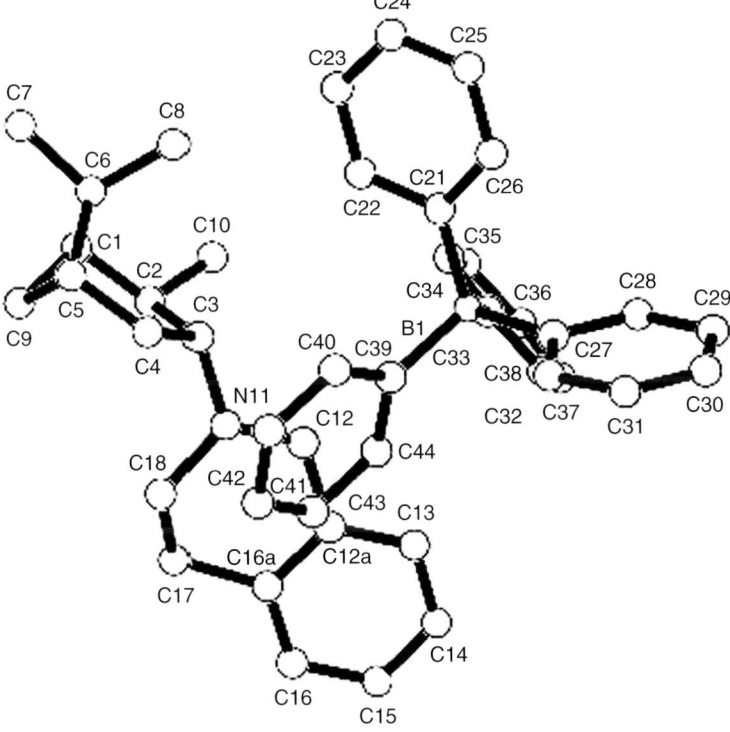

FIGURE 5.7 X-ray crystal structure of catalyst **(17)**.

becoming susceptible to attack by the oxidant, and the two diastereoisomeric oxaziridinium salts so formed may be very different in their potential for asymmetric induction in the epoxidation process.

Two transition states have been proposed for the epoxidation of alkenes by dioxiranes and oxaziridines, the spiro and the planar (Fig. 5.8). In the spiro transition state, the alkene approaches the oxaziridinium moiety in such a way that the axis of the carbon–carbon double bond is perpendicular to the carbon–nitrogen bond axis. In the planar transition state, the two components approach one another so that their axes are parallel to one another, and they and the oxygen atom are in the same plane.

Planar   Spiro

FIGURE 5.8 Geometrical approaches between the substrate and the oxaziridinium species.

The spiro transition state is now generally accepted as the mechanism in operation during both dioxirane- and oxaziridine-mediated epoxidation. This conclusion is supported by theoretical and computational studies [31–33].

2.4. Catalysts based on dibenzo[c,e]azepinium salts

In our ongoing efforts to develop new and more selective catalysts based on iminium salts, a new family of catalyst was produced, in which the dihydroisoquinolinium moiety has been replaced by a biphenyl structure fused to a seven-membered azepinium salt [34]. A similar system was developed some years ago by Aggarwal but with axial chirality, achiral at the nitrogen [11]; the system gave some good results, although the enantioselectivity of the catalyst was dependent upon the substitution pattern of the alkene.

The preparation of this new family of catalysts was achieved by starting from an enantiomerically pure primary amine and 2-[2-(bromomethyl)phenyl]-benzaldehyde (**27**) (Scheme 5.19). 2-[2-(Bromomethyl)phenyl]benzaldehyde was prepared from the corresponding dibenzoxepine (**28**), by treatment with molecular bromine in carbon tetrachloride under reflux, following a similar procedure already proven in the dihydroisoquinolinium salt series. The catalysts were synthesized in three steps starting from commercially available 2,2′-biphenyl dimethanol (**29**).

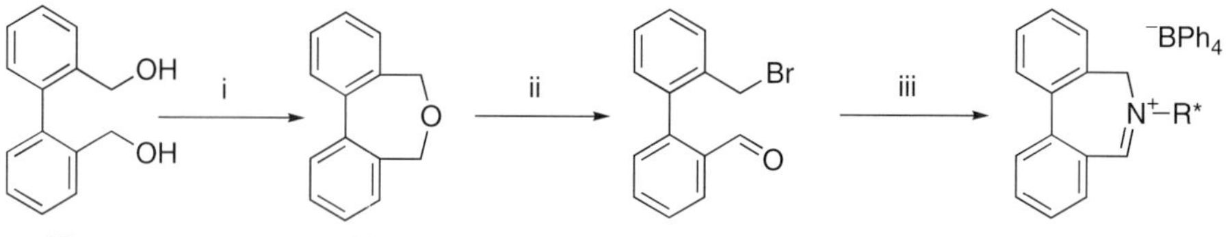

Reagents and Conditions: i: HBr (24%, in water), 100 °C, 40 min, 85%; ii: Br$_2$, CCl4, Δ, 1 h, 59%; iii: a) R*NH2, EtOH, 0 °C-r.t., 12 h, b) NaBPh$_4$, MeCN, 5 min.

SCHEME 5.19 Formation of the dibenzo[c,e]azepinium salts.

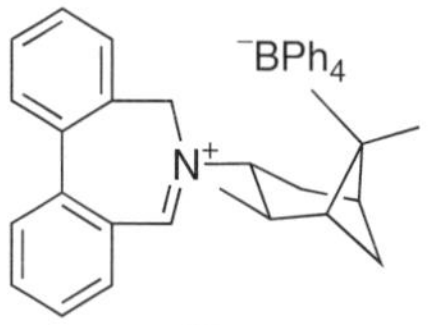

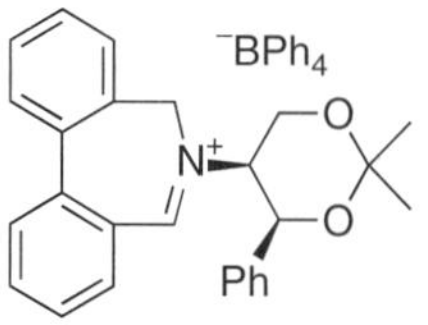

Initially, two new iminium salt catalysts were prepared following this procedure: catalyst (**30**), derived from the (−)-IPC amine (**17**), in 60% yield, and the 1,3-dioxane catalyst (**31**), derived from amine (**23**), in 68% yield. These two amine precursors were selected as they are the parent compounds of our earlier most effective dihydroisoquinolinium catalysts (**17** and **24**, respectively).

Epoxidation reactions were carried out using these new iminium salts, and a comparison of the results obtained with the original dihydroisoquinolinium catalysts (**17**) and (**24**) is displayed in Table 5.3. It is clear that both of the new

TABLE 5.3 Catalytic asymmetric epoxidation mediated by the new dibenzo[c,e]azepinium salts (**30** and **31**); a comparison to the corresponding dihydroisoquinolinium salts (**17** and **24**)[a]

	Catalyst			
	17	30	24	31
Epoxide	In each case: ee (%)[b], Conversion (%)[c], major enantiomer[d]			
Me,Ph epoxide	8, 68[e]	3, 90	20, 64[e]	24, 100
	(+)-R	(+)-R	(+)-R	(+)-R
O,Ph,Ph epoxide	−	0, 95	15, 56[e]	15, 90
			(−)-S,S	(−)-S,S
Ph,Ph,Me epoxide	15, 72[e]	14, 93	52, 52[e]	37, 95
	(+)-1R,2R	(+)-1R,2R	(−)-1S,2S	(−)-1S,2S
Ph,Ph epoxide	5, 43[e]	17, 100	59, 54[e]	59, 90
	(+)-S	(+)-S	(+)-S	(+)-S
cyclohexene-Ph epoxide	40, 68[e]	29, 100	41, 55[e]	60, 100
	(+)-1R,2R	(+)-1R,2R	(−)-1S,2S	(−)-1S,2S
indene oxide	3, 34[e]	8, 95	17, 52[e]	10, 100
	(+)-1S,2R	(+)-1S,2R	(+)-1S,2R	(+)-1S,2R

[a] Epoxidation conditions: Iminium salt (5 mol%), Oxone (2 equiv), Na_2CO_3 (4 equiv), $MeCN:H_2O$ (1:1), 0 °C, 2 h.
[b] Enantiomeric excess determined by ^1H NMR with Eu(hfc)$_3$ (0.1 mol equiv) as chiral shift reagent or by Chiral HPLC on a Chiracel OD column.
[c] Conversion evaluated from the ^1H NMR by integration alkene versus epoxide.
[d] The absolute configuration of the major enantiomer was determined by comparison to those reported in the literature.
[e] Isolated yield.

seven-membered ring catalysts are dramatically more reactive in the epoxidation reaction than are the six-membered ring equivalents. Catalyst (**30**) gives in general poorer ees than the catalyst (**80**), but in some cases provides superior ees to that of its six-membered ring counterpart (**17**). For example, 1-phenyl-3,4-dihydronaphthalene oxide is formed with 20% ee when catalyst (**17**) is employed, but when catalyst (**30**) is used an ee of 38% is observed. Catalyst (**31**) provides ees of up to 60%, although with a somewhat different pattern of selectivity from catalyst (**24**), for example giving an improved 60% ee for 1-phenylcyclohexene oxide (catalyst **24** gives 41% ee), but an identical 59% ee for triphenylethylene oxide. It is also worth noting that all the reactions with catalyst (**31**) are complete within 10 min or less at 0 °C, making it one of our most reactive iminium salt catalysts discovered to date. Lacour has shown that dichloromethane may be used instead of acetonitrile, using catalyst (**31**) with a TRISPHAT counterion, if a crown ether is added to the mixture [35–38].

2.5. Catalysts based on a binaphthalene structure

Because of the encouraging findings described in Section 2.4, we felt that a logical continuation of this work would be the synthesis of the axially chiral binaphthalene analogues. As indicated above, Aggarwal has described an iminium salt catalyst based upon a binaphthalene unit, but this has a methyl group substitution at the nitrogen atom. We believed that addition of a chiral group at the nitrogen atom, as we had already carried out for the simpler systems, would lead to greater substrate control and higher enantioselectivity. One of the inherent problems associated with the presence of additional chiral centres introduced in this way is the advent of matched and mismatched pairs of diastereoisomers, something which had to be considered during the catalyst synthesis. We predicted that if the binaphthalene component were of the *R*-configuration and the 1,3-dioxane component were of the 4*S*,5*S* configuration, then this would be a matched pair, as Aggarwal's *S*-iminium salt (Fig. 5.1) produced, in general, the oppositely configured epoxides to those obtained with catalysts (**24**) and (**31**).

Several new azepinium salt catalysts, derived from (+)- and (−)-5-amino-2,2-dimethyl-4-phenyl-1,3-dioxane **23** [39] and (−)-isopinocamphenylamine **17** moieties, and fused to *R* or *S* binaphthalene units [40,41], were directly prepared, in good yields, from the bromomethyl carbaldehyde intermediate **32**, which we prepared in turn from commercial *R* or *S* (1,1′)-binaphthalenyl-2,2′-diol (Binol) (Scheme 5.20, Table 5.4).

With the catalysts in hand, we were able to test their effectiveness in several epoxidation reactions. Initially, we screened the catalysts with our usual test substrates, 1-phenylcyclohexene, α-methylstilbene and triphenylethylene (Table 5.5). Catalyst **33a** showed the best reaction profile, being by far the most reactive. For example, in the presence of **33a**, 1-phenylcyclohexene oxide was produced in 69% yield with 91% ee in under 20 min, while the other catalysts (apart from *ent*-**33a**) were less selective, and the reactions were slower. The isopinocamphenyl moiety offers little enantiocontrol, leading to epoxides with only moderate ees. The poor reactivity of catalysts **33b–d** is highlighted by the attempted epoxidations of α-methyl stilbene and triphenyl ethylene, where no epoxides were formed after 4 h.

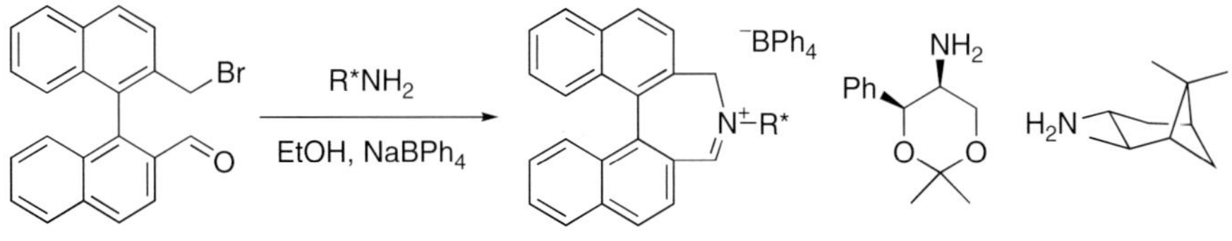

SCHEME 5.20 Formation of azepinium salt catalysts.

TABLE 5.4 Cyclocondensation of primary amines **23** and **17** with the bromoaldehyde **32**

Entry	Bromoaldehyde	Amine	Product	Yield/%
1	(R)	23	33a	66
2	(S)	ent-23	ent-33a	64
3	(S)	23	33b	63
4	(R)	17	33c	71
5	(S)	17	33d	73

TABLE 5.5 Asymmetric epoxidation of unfunctionalized alkenes mediated by catalysts **33a–d**[a]

Alkene	Catalyst	Time (h)	Yield (%)[b]	ee (%)[c]	Configuration[d]
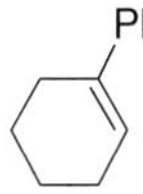 Ph-cyclohexene	33a	0.20	69	91	(−)-1S,2S
	ent-33a	0.20	66	88	(+)-1R,2R
	33b	2.0	54	78	(+)-1R,2R
	33c	2.0	40	53	(−)-1S,2S
	33d	2.0	44	58	(+)-1R,2R
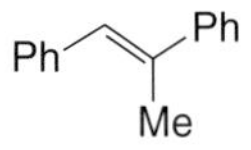 Ph,Ph / Me	33a	0.40	58	49	(−)-1S,2S
	33b	4.0	0	–	–
	33c	4.0	0	–	–
	33d	4.0	0	–	–
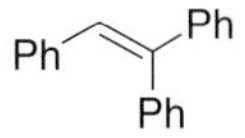 Ph,Ph / Ph	33a	0.50	60	12	(+)-S
	33b	4.0	0	–	–
	33c	4.0	0	–	–
	33d	4.0	0	–	–

[a] Conditions: Iminium salt (5 mol%), Oxone (2 equiv), Na_2CO_3 (4 equiv), $MeCN/H_2O$ (1:1), 0 °C.
[b] Isolated yields.
[c] Enantiomeric excesses were determined by ^1H NMR spectroscopy in the presence of (+)-Eu(hfc)$_3$ (0.1 mol equiv).
[d] The absolute configurations of the major enantiomers were determined by comparison with those reported in the literature.

Catalyst **33a**, however, afforded complete conversion to the corresponding epoxides in a much shorter time, and an isolated yield of ~60%, although the ees were moderate.

Catalyst **33a** was subsequently used to epoxidize several other olefins. Again, the reactivity of the catalyst at (5 mol%) was good, but a wide range of ees was observed (Table 5.6). 1-Phenyl-3,4-dihydronaphthalene was epoxidized with high enantioselectivity (95% ee and 66% yield after 35 min). 4-Phenylstyrene oxide was produced with 29% ee, one of the highest reported ees for the epoxidation of terminal alkenes using iminium salt catalysis.

Having identified a number of successful epoxidation substrates for catalyst **33a** we tested several cycloalkenes of varying ring sizes in the asymmetric epoxidation reaction (Table 5.7). Reactions were carried out using just 1 mol% of catalyst **33a**. Again, good conversions to epoxides were achieved, and, interestingly, the five- and seven-membered ring cycloalkenes were less reactive than was 1-phenylcyclohexene. The reactions took almost five times as long to approach completion, and enantioselectivities were poorer than those observed for 1-phenylcyclohexene: 1-phenylcyclopentene oxide was formed in 55% ee and 1-phenylcycloheptene in 76% ee.

Using catalyst **33a**, we also conducted a catalyst loading study using as test substrate 1-phenylcyclohexene, with catalyst loadings ranging from 0.1 to 5 mol%

TABLE 5.6 Asymmetric epoxidation of various alkenes mediated by catalyst **33a**[a]

Alkene	Time (h)	Yield (%)[b]	ee (%)[c]	Configuration[d]
Ph/Ph stilbene	0.45	58	20	(−)-S,S
Ph-cyclohexene	0.25	63	25	(−)-1S,2S
dihydronaphthalene	0.30	60	17	(+)-1R,2S
1-Ph-dihydronaphthalene	0.35	66	95	(+)-1R,2S
cinnamyl alcohol	2.0	67	38	(−)-2S,3S
4-Ph-styrene	1.0	70	29	(+)-S

[a] Conditions: Iminium salt (5 mol%), Oxone (2 equiv), Na$_2$CO$_3$ (4 equiv), MeCN/H$_2$O (1:1), 0 °C.
[b] Isolated yields.
[c] Enantiomeric excesses were determined by ^{1}H NMR spectroscopy in the presence of (+)-Eu(hfc)$_3$ (0.1 mol equiv) or by chiral HPLC using a Chiracel OD column.
[d] The absolute configurations of the major enantiomers were determined by comparison with those reported in the literature.

TABLE 5.7 Effect of ring size on the epoxidation of several cycloalkenes with catalyst **33a**[a]

Alkene	Time (h)	Yield (%)[b]	ee (%)[c]	Configuration[d]
Ph-cyclopentene	5.0	52	55	(−)-1S,2S
Ph-cyclohexene	1.1	64	91	(−)-1S,2S
Ph-cycloheptene	5.0	57	76	(−)-1S,2S

[a] Conditions: Iminium salt (1 mol%), Oxone (2 equiv), Na$_2$CO$_3$ (4 equiv), MeCN/H$_2$O (1:1), 0 °C.
[b] Isolated yields.
[c] Enantiomeric excesses were determined by ^{1}H NMR spectroscopy in the presence of (+)-Eu(hfc)$_3$ (0.1 mol equiv).
[d] The absolute configurations of the major enantiomers were determined by comparison with those reported in the literature.

TABLE 5.8 Catalyst loading study on the epoxidation of 1-Phenylcyclohexene with catalyst **33a**[a]

Entry	Catalyst (mol%)	Time (h)	Yield (%)[b]	ee (%)[c]	Configuration[d]
1	5.0	0.2	69	91	(−)-1S,2S
2	1.0	1.1	64	91	(−)-1S,2S
3	0.5	2.0	65	91	(−)-1S,2S
4	0.1	6.0	68	88	(−)-1S,2S

[a] Conditions: Oxone (2 equiv), Na$_2$CO$_3$ (4 equiv), MeCN/H$_2$O (1:1), 0 °C.
[b] Isolated yields.
[c] Enantiomeric excesses were determined by ^{1}H NMR spectroscopy in the presence of (+)-Eu(hfc)$_3$ (0.1 mol equiv).
[d] The absolute configurations of the major enantiomers were determined by comparison with those reported in the literature.

(Table 5.8). We were delighted and extremely surprised to observe a high level of asymmetric induction with low catalyst loadings. We have previously reported that it is possible to use just 0.5 mol% of catalyst, but a loss in enantioselectivity is commonly observed. In this case, however, catalyst loadings can be so low that effective epoxidation of 1.0 g of 1-phenylcyclohexene, with 68% yield and 88% ee, can be achieved using just 5 mg of catalyst (0.1 mol%). This loading is extremely low for an organocatalytic system.

2.6. Catalytic asymmetric anhydrous epoxidation mediated by tetraphenylphosphonium monoperoxysulphate and iminium salts

2.6.1. Introduction: Reasons for employing a new stoichiometric oxidant

The standard conditions employed in epoxidation reactions catalysed by iminium salts involve the use of Oxone as stoichiometric oxidant, a base (2 mol equiv of Na$_2$CO$_3$ per equivalent of Oxone) and water/acetonitrile as solvent mixture (Scheme 5.21): the presence of water is essential for Oxone solubility. Under the reaction conditions, there are separate aqueous and organic phases; it is possible that the catalyst acts as a phase transfer agent in these reactions.

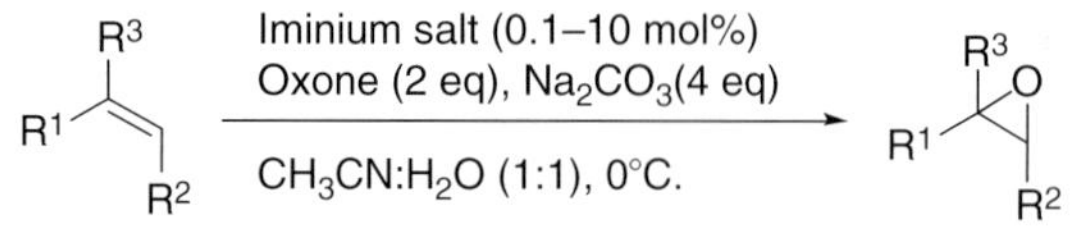

SCHEME 5.21 The standard conditions applied for catalytic asymmetric epoxidation mediated by Oxone and iminium salts.

The principal limitation to this system is the restricted range of temperatures in which the epoxidation can be performed (0 °C to room temperature). The upper limit is determined by the Oxone, which decomposes relatively quickly in the basic medium at room temperature [23]. The lower limit is determined by the use of the aqueous medium; the normal ratio of the water and acetonitrile solvents used is 1:1, and this mixture freezes at around −8 °C.

One potential opportunity to enhance the enantioselectivity of the process would be provided if the reaction could be carried out at lower temperatures. This would require the development of non-aqueous reaction conditions, and because of the solubility profile of Oxone, which has no significant solubility in any organic solvent, this in turn dictates a need for a new stoichiometric oxidant, soluble in organic solvents at low temperatures. Crucially, this oxidant must not oxidize alkenes under the reaction conditions in the absence of the catalyst (background oxidation).

2.6.2. Selection of tetraphenyl phosphonium monoperoxysulphate

Several oxidants were tested in an epoxidation reaction in the presence of iminium salt catalysts to determine which offers the best profile in the absence of water [42]. These reactions were carried out at 0 °C with 1-phenylcyclohexene as substrate and (17) and/or (24) as catalysts (5–20 mol%), in dichloromethane as solvent. Most of the systems examined showed either high levels of background epoxidation (alkaline hydrogen peroxide, peracids, persulphates) or very low rates of reaction, even in the presence of 20 mol% of the catalysts (perselenates, percarbonates, perborates and iodosobenzene diacetate). Tetra-N-butylammonium Oxone, reported by Trost [43], was also unsuccessful as oxidant.

From all of those tested, tetraphenylphosphonium monoperoxysulphate (TPPP), reported by Di Furia in 1994 for oxygen transfer to manganese porphyrins [44], showed the best profile. A modification of Oxone, TPPP is prepared by cation exchange (K^+ to Ph_4P^+) between Oxone and tetraphenyl phosphonium chloride (Scheme 5.22). Further crystallization of the triple salt from CH_2Cl_2 and hexane afforded the desired compound as a colourless solid in 75% yield, which iodometric titration revealed to have ∼85% of the theoretically available oxygen. This composition was also confirmed by ^1H NMR spectroscopy (CD_2Cl_2), by integration of the aromatic hydrogen of the tetraphenyl moiety and the peroxyacidic proton (8.5 ppm).

$$\text{Ph}_4\text{P}^+ \text{ Cl}^- + \text{Oxone (2 KHSO}_5 : \text{KHSO}_4 : \text{K}_2\text{SO}_4\text{)} \rightarrow \text{Ph}_4\text{P}^+ \text{(HSO}_5\text{)}^-$$

SCHEME 5.22 Formation of tetraphenylphosphonium monoperoxysulphate from Oxone.

For epoxidation to proceed, the presence of base is essential under the aqueous conditions when using Oxone as oxidant. We were pleased to discover that, in contrast, the addition of 1 equiv of any of a range of bases (KF, TBAF, CsF, pyridine, 2,6-lutidine, DBN, DBU, DABCO, LiH, NaH) to the test reaction in dichloromethane at 0 °C, with TPPP as oxidant, did not improve the reaction; indeed, the amine bases suppressed epoxidation altogether.

Formation of epoxide was not observed upon treatment of 1-phenylcyclohexene with TPPP in dichloromethane solution at 0 °C in the absence of catalyst. Indeed, such background epoxidation only becomes competitive with the catalysed reaction when the temperature reaches ~40 °C.

The optimum conditions for the asymmetric epoxidation reaction were developed. Because of the exothermic nature of the reaction, a solution of TPPP in the reaction medium is cooled to the desired temperature; the catalysts and substrate are also separately dissolved in the reaction solvent and cooled to the desired temperature. The solution of catalyst is added dropwise to the solution of oxidant, to minimize the increase in reaction temperature, which is allowed to stabilize before dropwise addition of the substrate. The alkene is added last to help maintain the epoxidation process at a constant temperature. The reaction is stopped by high dilution with diethyl ether, in which both the catalyst and oxidant display a low solubility profile.

We have also found that when the catalyst reacts with TPPP, an anionic interchange between the two salts occurs. The tetraphenyl borate $(\text{BPh}_4)^-$ is displaced from the iminium salt by the monoperoxysulphate anion $(\text{HSO}_5)^-$, and the corresponding tetraphenylphosphonium tetraphenylborate is formed. This is rather insoluble in dichloromethane and can be isolated and characterized following filtration from the reaction mixture.

2.6.3. Asymmetric epoxidations using TPPP

2.6.3.1. Temperature studies Several reactions with our test substrate 1-phenylcyclohexene were performed using catalyst (**31**) and our dihydroisoquinolinium catalyst (**24**) under these new conditions. Runs were conducted using 10 mol% of catalyst over a range of temperatures; the results obtained are displayed in Table 5.9. It is important to note here that the corresponding catalysts derived from the (−)-IPC amine (**17** and **31**) gave poor conversions and ees under these conditions. For example, catalyst (**31**) (10 mol%) afforded only 60% conversion to 1-phenylcyclohexene oxide and 13% ee after 2 h at −40 °C. As a comparison, the reported aqueous/acetonitrile results are included [24,34].

TABLE 5.9 Catalytic asymmetric epoxidation of 1-Phenylcyclohexene using TPPP and catalysts (24) and (31)[a]

Entry	Solvent	Catalyst	T (°C)	Time (min)	Conversion (%)[b]	ee (%)[c]
1	DCM	31	−78	120	52	50
2	DCM	31	−60	120	50	36
3	DCM	31	−40	90	100	28
4	DCM	31	0	10	100	26
5	DCM	31	40	<5	100	23
6	DCM	24	−78	480	15	23
7	DCM	24	−40	120	70	37
8	DCM	24	0	10	100	16
9	DCM	24	40	25	75	0
10	DCM/MeCN	31	−78	240	80	70
11	DCM/MeCN	31	−78	480	25	70[d]
12	DCM/MeCN	31	−40	30	100	53
13	DCM/MeCN	31	0	<5	100	51
14	DCM/MeCN	31	40	<5	100	43
15	DCM/MeCN	24	−78	480	35	45
16	DCM/MeCN	24	−40	120	83	46
17	DCM/MeCN	24	0	10	100	34
18	DCM/MeCN	24	40	10	100	25
19	MeCN/H_2O[e]	31	0	5	100	60
20	MeCN/H_2O[e]	24	0	10	55[f]	41

[a] Conditions: Iminium salt (10 mol%), TPPP (2 equiv), solvent, T (°C).
[b] Conversion evaluated from the ^{1}H NMR by integration alkene versus epoxide.
[c] Enantiomeric excess determined by ^{1}H NMR with (+)-Eu(hfc)$_3$ (0.1 mol equiv) as chiral shift reagent.
[d] 2 mol% catalyst.
[e] Conditions: Oxone (2 equiv), sodium carbonate (4 equiv), water/acetonitrile (1:1), 0 °C.
[f] Isolated yield.

We observed that in each case the highest enantiomeric excess is obtained at the lowest temperature [entries 1–5 and 10–14 for catalyst (31), and 6–9 and 15–18 for catalyst (24)]. The increase in ee is coupled with a decrease in epoxide conversion at lower temperatures. A comparison of the two catalysts shows that the biphenyl catalyst (31) is much more reactive (as it is under the Oxone conditions) at lower temperatures and far more selective, regardless of solvent.

The best selectivities are obtained when a mixture of solvents is employed (entries 10–18). A 1:1 ratio of dichloromethane/acetonitrile allows the reaction mixture to be homogeneous at −78 °C (acetonitrile alone freezes at −45 °C), and catalyst (31) has produced one of the best ee values obtained in any epoxidation reaction mediated by iminium salts previously described by us or others (70% ee, entry 10). Even when the catalyst loading is reduced to 2 mol% (entry 11), we are able to gain 25% conversion [a better result than catalyst (24) with 10 mol% at −78 °C], and the ee remains consistent (70%).

With the 1:1 acetonitrile/water solvent system, the biphenyl catalyst (31) outperforms the dihydroisoquinolinium catalyst (24), which is still rather unreactive at −78 °C (entry 15). The levels of enantioselectivity observed for catalyst (24) under the new anhydrous conditions do not generally exceed those obtained using the Oxone/acetonitrile/water conditions, but in contrast it appears that the new experimental conditions enhance selectivity in the case of the biphenyl catalyst (31). This catalyst was therefore selected to screen a range of unfunctionalized alkenes (Table 5.10).

These olefins were not as reactive towards catalyst (31) as the previously tested substrate 1-phenylcyclohexene, therefore the majority of reactions were carried out at −40 °C. Again, the dichloromethane/acetonitrile conditions produce the better results in terms of both enantiomeric excess and epoxide conversion, over the dichloromethane conditions. Triphenylethylene was extremely unreactive when compared with all the other alkenes tested (entries 3, 4, 10 and 11). The best ee obtained was for 1-phenyl-3,4-dihydronaphthalene at −40 °C in dichloromethane/acetonitrile, which in 3 h gave 100% conversion and 65% ee. This is even more remarkable when one considers that 1-phenyl-3,4-dihydronaphthalene also gave the poorest result in dichloromethane (7% ee).

Overall, the differences in enantiomeric excess between the two solvent systems were generally not as vast as those observed in reactions with 1-phenylcyclohexene. However, the conversions to epoxide differed significantly; all the experiments recorded in dichloromethane required longer reaction times, and in some cases still did not reach the level of conversion observed using the dichloromethane/acetonitrile system.

A temperature study was also carried out using dichloromethane as solvent and 2 mol equiv of TPPP over a range of temperatures with the substituted dihydroisoquinolinium catalysts (34), (35), (36) and (37) [45]. The results obtained are collected in Table 5.11. As it can be seen from the table, again we observe that, as the temperature of reaction decreases, the enantioselectivity increases significantly. At the same time, the conversion decreases, probably due to a decrease in reaction rates with the temperature.

TABLE 5.10 Catalytic asymmetric epoxidation of several alkenes mediated by catalyst (31)[a]

Entry	Alkene	Solvent	T (°C)	Time (h)	Conversion (%)[b]	ee (%)[c]	Epoxide Configuration[d]
1	Ph-C(CH₃)=CH-Ph	DCM	−78	8	5	20	(−)-1S,2S
2		DCM/MeCN	−78	2	15	44	(−)-1S,2S
3		DCM	−40	6	100	50	(−)-1S,2S
4		DCM/MeCN	−40	2	86	50	(−)-1S,2S
5	Ph-C(Ph)=CH-Ph	DCM	−78	8	0	–	–
6		DCM/MeCN	−78	8	0	–	–
7		DCM	−40	2	7	47	(+)-S
8		DCM/MeCN	−40	1	30	29	(+)-S
9	Ph-CH=CH-Ph	DCM	−40	4	100	39	(−)-1S,2S
10		DCM/MeCN	−40	2	100	33	(−)-1S,2S
11	1-phenyl-3,4-dihydronaphthalene	DCM	−40	4	100	7	(+)-1R,2S
12		DCM/MeCN	−40	3	100	65	(+)-1R,2S

[a] Conditions: Iminium salt (10 mol%), TPPP (2 equiv), solvent, T (°C).
[b] Conversion evaluated from the ¹H NMR by integration alkene versus epoxide.
[c] Enantiomeric excess determined by ¹H NMR with (+)-Eu(hfc)₃ (0.1 mol equiv) as chiral shift reagent or by Chiral HPLC, on a Chiracel OD column.
[d] The configuration of the major enantiomer was determined by correlation to the known epoxides.

The highest ee (50%) was obtained when the reaction was carried out at −78 °C (entry 18). Under these conditions (using 10 mol% of catalyst), the conversion to epoxide is rather low (15%). Surprisingly, the *p*-methoxyaryl catalyst (**37**) is not as selective as it is under the aqueous conditions, the best ee being only 29% (45% ee in the original aqueous conditions).

It is interesting that, while catalyst **24** (R = H) gives the (−)-(1S,2S) epoxide, when the catalyst contains an electron-withdrawing nitro or sulphone substituent (catalysts **34** and **36**), a change is observed in the absolute configuration of the epoxide enantiomer formed preferentially, to the (+)-(1R,2R) epoxide, in every case.

The thiomethyl catalyst (**35**) gives the (+)-1R,2R enantiomer of 1-phenylcyclohexene between temperatures of 0 °C and temperatures of −50 °C (Table 5.11, entries 3, 8, 12), but at −78 °C the (−)-1S,2S enantiomer is formed (Table 5.15, entry 18). This we believe is due to the sulphide being oxidized to the sulphone at temperatures above −78 °C, therefore altering the electronic nature of the catalyst. Catalyst (**37**), which also has an electron-donating group at C4 (OMe), cannot be altered by the

TABLE 5.11 Catalytic asymmetric epoxidation of 1-Phenylcyclohexene using TPPP in $CH_2Cl_2{}^a$

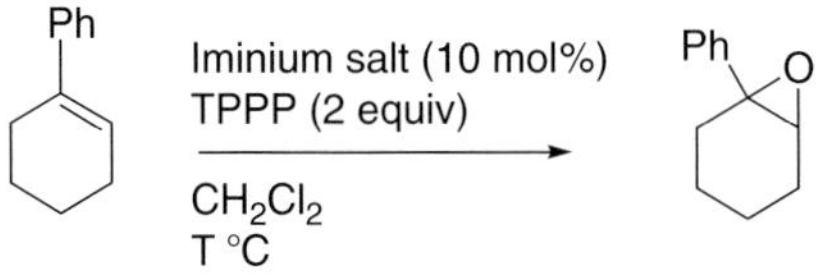

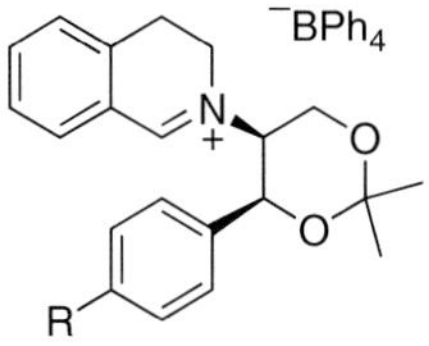

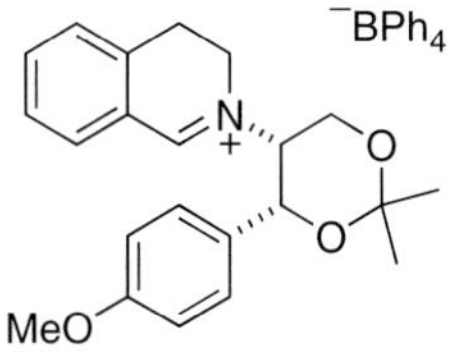

Entry	T (°C)	Catalyst	Conversion (%)[b]	ee (%)[c]	Configuration[d]
1	0	(24) R = H	100	16	(−)-1S,2S
2		(34) R = NO_2	100	25	(+)-1R,2R
3		(35) R = SMe	72	16	(+)-1R,2R
4		(36) R = SO_2Me	100	33	(+)-1R,2R
5		(37)[e] R = OMe	–	–	–
6	−30	(24)	–	–	–
7		(34)	100	28	(+)-1R,2R
8		(35)	92	33	(+)-1R,2R
9		(36)	100	36	(+)-1R,2R
10		(37)[e]	76	25	(+)-1R,2R
11	−45	(24)	70	37[f]	(−)-1S,2S
12		(34)	54	30	(+)-1R,2R
13		(35)	67	33	(+)-1R,2R
14		(36)	96	42	(+)-1R,2R
15		(37)[e]	–	–	–
16	−78	(24)	15	23[g]	(−)-1S,2S
17		(34)	<5	–	–
18		(35)	15	50[g]	(−)-1S,2S
19		(36)	<5	–	–
20		(37)[e]	13	29[g]	(+)-1R,2R

[a] Conditions: Iminium salt (10 mol%), TPPP (2 equiv), DCM, T (°C), 4.0 h.
[b] Conversion evaluated from the ^{1}H NMR spectrum by integration of alkene versus epoxide.
[c] Enantiomeric excess determined by ^{1}H NMR with (+)-Eu(hfc)$_3$ (0.1 mol equiv) as chiral shift reagent.
[d] The absolute configuration of the major enantiomer was determined by comparison to those reported in the literature.
[e] Catalyst configuration (3R,4R).
[f] Reaction carried out at 40 °C.
[g] Reaction time 8 h.

34 R = NO_2

35 R = SMe

36 R = SO_2Me

epoxidation conditions, and regardless of temperature produces the same enantiomer of epoxide. This perhaps indicates that for catalyst 35 at −78 °C, it is the unoxidized sulphide group which is present as the active catalyst.

To prove that the active catalyst at temperatures above −78 °C was the oxidized sulphone, a standard epoxidation reaction was performed in deuteriated dichloromethane and subjected to ^{1}H NMR spectroscopic analysis. We observed that the chemical shift for the thiomethyl group moved from δ 2.42 to 3.03 ppm, indicating that the sulphide had been oxidized to the sulphone, by comparison to authentic catalyst 36. This was an extremely rapid and exothermic process, and the sulphoxide intermediate was not observed. Further to this, the pattern of the AA'BB' aromatic system was consistent with that of a sulphone, the chemical shifts being around δ 8.00 ppm.

X-ray structure analysis may offer one explanation as to why we observe the change in the major enantiomer of epoxide formed (Fig. 5.9). A comparison of catalyst **(24)** and the sulphone catalyst **(36)** shows that in the case of **(24)**, the iminium carbon atom sits above O(1), and the *re* face is blocked by the phenyl ring. In the case of the sulphone catalyst **(36)**, the iminium carbon faces away from the dioxane ring, and so the *re* face is now open to attack from the oxidant, therefore producing an oxaziridinium unit of the opposite configuration. These solid-state conformations, however, may well not be representative of the behaviour of the compounds in solution.

2.6.3.2. The effect of reaction solvent To determine if the electronic effect could enhance enantiomeric excess in the epoxidation reaction several other solvents were screened, using our three most effective dihydroisoquinolinium salt catalysts **(24)**, **(36)** and **(37)** (Table 5.12) [46]. TPPP was found to be insoluble in carbon tetrachloride, ethyl acetate and dimethoxyethane. In dimethylformamide, the TPPP dissolved, but no reaction occurred when employing catalyst **(24)**. However, TPPP was soluble in 1,2-dichloroethane, and epoxidation reactions performed in this solvent gave almost identical results to those obtained with

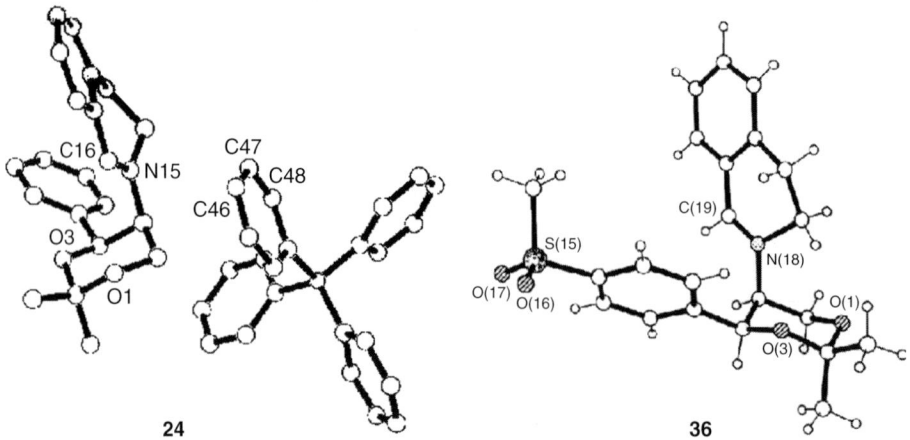

FIGURE 5.9 X-ray crystal structures of catalysts **24** and **36**.

TABLE 5.12 Asymmetric epoxidation of 1-Phenylcyclohexene using various solvents[a]

		Catalyst		
		(24) R = H	(36) R = SO$_2$Me	(37) R = OMe[b]
Entry	Solvent	In all cases: ee (%)[c], conversion (%)[d], time (h), configuration[e]		
1	CH$_3$CN	43[f], 42, 1.0, (−)-1S,2S	45, 89, 2.5, (−)-1S,2S	44, 30, 0.2, (+)-1R,2R
2	CH$_2$Cl$_2$	37[f], 70, 4, (−)-1S,2S	36, 100, 4, (+)-1R,2R	25, 76, 4, (+)-1R,2R
3	C$_2$H$_4$Cl$_2$	24, 29, 24, (−)-1S,2S	32, 97, 4, (+)-1R,2R	17, 87, 24, (+)-1R,2R
4	CHCl$_3$	33, 52, 24, (−)-1S,2S	48, 100, 12, (+)-1R,2R	11, 73, 24, (+)-1R,2R

[a] Epoxidation conditions: Iminium salt (10 mol%), TPPP (2 equiv), solvent, −30 °C, T (°C).
[b] Catalyst configuration (3R,4R).
[c] Enantiomeric excess determined by ^{1}H NMR with (+)-Eu(hfc)$_3$ (0.1 mol equiv) as chiral shift reagent.
[d] Conversion evaluated from the ^{1}H NMR by integration alkene versus epoxide.
[e] The absolute configuration of the major enantiomer was determined by comparison to those reported in the literature.
[f] Reaction carried out at −40 °C.

dichloromethane for catalysts (36) and (37) (Table 5.12, entry 3). Catalyst (24) was far less reactive in this medium (compared with the corresponding reaction in dichloromethane, 70% conversion to 29% after 4 h) and gave a lower ee (24% compared with 37%). When the reactions were repeated in chloroform (Table 5.12, entry 4), we observed a dramatic decrease in ee for catalyst (37), catalyst (24) giving similar results to those obtained in dichloromethane (33% ee). Catalyst (36) in chloroform, however, gave the best ee for 1-phenylcyclohexene oxide that we have observed with this set of iminium salt catalysts. Interestingly, in acetonitrile the change in major epoxide enantiomer formed did not occur with catalyst (36), and at −30 °C we observe a similar degree of selectivity to that of the reaction carried out in chloroform (48% ee, (+)-(1R2R) enantiomer), producing in acetonitrile (−)-(1S2S)-phenylcyclohexene oxide with 45% ee.

The substituted dibenzo[c,e]azepinium salts (31), (38), (39) and (40) were also tested, using the optimum conditions established from Table 5.12, at a temperature of −40 °C (Table 5.13). Under these conditions, acetonitrile appears to be the solvent of choice, with the p-methoxyaryl catalyst (40) producing 1-phenylcyclohexene oxide in 73% ee. However, catalyst (40) was not as effective, in terms of ee, as the corresponding dihydroisoquinolinium salt (36), in either of the solvents. Interestingly, no asymmetric induction was observed in chloroform, previously our most effective reaction solvent. We have previously reported epoxidation of other substrates using catalyst (31) in acetonitrile, but generally the ees are disappointing, with triphenylethylene and 1-phenylcyclohexene being the best substrates (60% ee and 67% ee, respectively) [47].

As previously observed under the aqueous Oxone conditions, the enantioselectivity of the dibenzo[c,e]azepinium salt catalysts is markedly substrate dependent.

TABLE 5.13 Catalytic asymmetric epoxidation of 1-Phenylcyclohexene using the dibenzo[c,e] azepinium salts and TPPP[a]

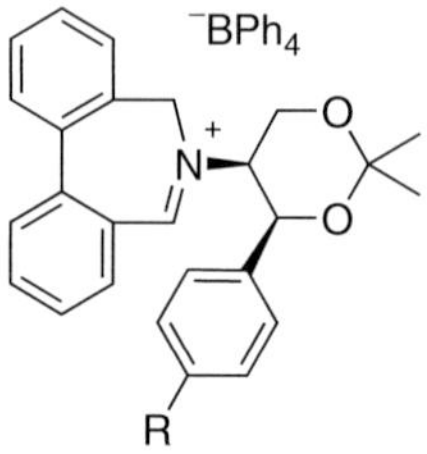

Catalyst	Solvent	Conversion (%)[b]	ee (%)[c]	Configuration[d]
(31) R = H	CH_3CN	100	67	(−)-(1S,2S)
	$CHCl_3$	84	19	(−)-(1S,2S)
(38) R = NO_2	CH_3CN	100	38	(−)-(1S,2S)
	$CHCl_3$	No reaction	–	–
(39) R = SO_2Me	CH_3CN	100	40	(−)-(1S,2S)
	$CHCl_3$	75	0	–
(40) R = OMe[e]	CH_3CN	100	73	(+)-(1R,2R)
	$CHCl_3$	57	5	(+)-(1R,2R)

[a] Epoxidation conditions: Iminium salt (10 mol%), TPPP (2 equiv), solvent, −40 °C T (°C).
[b] Conversion evaluated from the ^1H NMR by integration alkene versus epoxide, numbers in brackets represent isolated yield.
[c] Enantiomeric excess determined by ^1H NMR with (+)-Eu(hfc)$_3$ (0.1 mol equiv) as chiral shift reagent.
[d] The absolute configuration of the major enantiomer was determined by comparison to those reported in the literature.
[e] Catalyst configuration (3R,4R).

31 R = H
38 R = NO_2
39 R = SO_2Me
40 R = OMe (4R,5R)

The corresponding dihydroisoquinolinium salts, however, show less substrate dependency, and, to further explore these catalysts, we were interested to discover if the change in major enantiomer formed is also observed in the epoxidation of alkenes other than 1-phenylcyclohexene. Using the most selective dihydroisoquinolinium catalyst **(36)** from the above study, we were able to epoxidize several unfunctionalized alkenes in acetonitrile and chloroform solution (Table 5.14). Reactions were carried out at a temperature of −40 °C, with 10 mol% of the catalyst.

Again, we observed the switch in major enantiomer formed upon changing the reaction solvent, and again catalyst **36** appeared to be less substrate dependent than did the corresponding dibenzo[c,e]azepinium salts.

However, enantioselectivies were dramatically increased in chloroform (apart from the anomalous reaction with α-methylstilbene), when compared to those achieved in acetonitrile. We found that *trans*-stilbene, usually a poor substrate with our catalysts, is epoxidized with 67% ee in chloroform, whereas the corresponding reaction, performed in acetonitrile, only affords *trans*-stilbene oxide with 30% ee.

TABLE 5.14 Catalytic asymmetric epoxidation of various alkenes using TPPP and catalyst 36[a]

	Solvent					
	CH₃CN			CHCl₃		
Alkene	Yield (%)[b]	ee (%)[c]	Configuration[d]	Yield (%)[b]	ee (%)[c]	Configuration[d]
1-Phenylcyclohexene	73	45	(−)-1S,2S	77	48	(+)-1R,2R
trans-Stilbene	13[e]	30	(−)-S,S	31	67	(+)-R,R
α-Methylstilbene	42	48	(−)-1S,2R	35	2	(+)-1R,2S
1,1-Diphenylethylene (Ph/Ph)	25	27	(+)-S	11[e]	63	(−)-R
1-Phenyl-3,4-dihydronaphthalene	56	38	(+)-1R,2S	98	59	(−)-1S,2R

[a] Epoxidation conditions: Iminium salt (10 mol%), TPPP (2 equiv), solvent, −40 °C, 24 h.
[b] Isolated yield.
[c] Enantiomeric excess determined by ¹H NMR with (+)-Eu(hfc)₃ (0.1 mol equiv) as chiral shift reagent or by Chiral HPLC on a Chiracel OD column.
[d] The absolute configuration of the major enantiomer was determined by comparison to those reported in the literature.
[e] Conversion evaluated from the ¹H NMR by integration alkene versus epoxide.

Further to this work, a range of unfunctionalized *cis*-alkenes was subjected to the reaction conditions (Table 5.15) [47]. Interestingly, we found that no change in configuration of epoxide occurred with this type of substrate. The general trend of increased enantiomeric excess in chloroform over acetonitrile was still present. 1,2-Dihydronaphthylene was epoxidized with 82% ee in 89% yield when chloroform was employed as the reaction solvent. Good ees for the epoxidation of *cis*-β-methylstyrene and indene were also observed (70% ee and 61% ee, respectively). Asymmetric epoxidation of the non-aryl substrate did not occur with any selectivity, giving rise to racemic product.

TABLE 5.15 Catalytic asymmetric epoxidation of various cis-alkenes using TPPP and catalyst **36**[a]

	Solvent							
	MeCN				CHCl$_3$			
Alkene	Time (h)	Yield (%)[b]	ee (%)[c]	Configuration[d]	Time (h)	Yield (%)[b]	ee (%)[c]	Configuration[d]
styrene (Ph-CH=CH-)	24	71	53	(+)-1S,2R	24	85	70	(+)-1S,2R
dihydronaphthalene	16	80	56	(−)-1S,2R	17	89	82	(−)-1S,2R
indene	–	–	–	–	24	83	61	(+)-1S,2R
alkene	–	–	–	–	24	100[e]	0	–

[a] Epoxidation conditions: Iminium salt (10 mol%), TPPP (2 equiv), solvent, −40 °C.
[b] Isolated yield.
[c] Enantiomeric excess determined by ^{1}H NMR with (+)-Eu(hfc)$_3$ (0.1 mol equiv) as chiral shift reagent.
[d] The absolute configuration of the major enantiomer was determined by comparison to those reported in the literature.
[e] Conversion evaluated from the ^{1}H NMR by integration alkene versus epoxide.

The asymmetric epoxidation of cis-alkenes is particularly important when one considers the importance of some of the biologically active compounds that ring opening of the oxirane moiety can give. Levcromakalim (**41**) [48], an antihypertensive agent, is just one of the compounds available from ring opening of the epoxide functionality, and was traditionally synthesized in enantiomerically pure form by resolution [49], until Jacobsen's chiral salen complexes became available [50]. It has been found that the main biological activity resides in this laevorotatory enantiomer, while the dextrorotatory compound exhibits no significant activity [51]. The powerful cyano electron-withdrawing group at C6 of the benzopyran system was found to be essential for good blood pressure-lowering activity [49]. Several other research groups have reported the enantiopure synthesis of the precursor to levcromakalim (**42**) [51]. Shi's fructose-derived ketone catalyst, however, produces the inactive dextrorotatory enantiomer [52,53].

Epoxidation reactions of (43–45) using catalyst (36) in both acetonitrile and chloroform produced the desired (−)-3S,4S enantiomer of (42) in good to excellent enantiomeric excesses (Table 5.16). In acetonitrile the epoxide was formed with 80% ee and 63% yield, but in chloroform the epoxide was formed with 97% ee in 59% yield.

Treatment of compound (42) with pyrrolidone and sodium hydride afforded levcromakalim (41) in 52% yield (Scheme 5.23).

TABLE 5.16 Highly Enantioselective Epoxidation of Benzopyrans Using Catalyst **36** and TPPP[a]

	Solvent							
	CH$_3$CN				CHCl$_3$			
Alkene	Time (h)	Yield (%)[b]	ee (%)[c]	Configuration[d]	Time (h)	Yield (%)[b]	ee (%)[c]	Configuration[d]
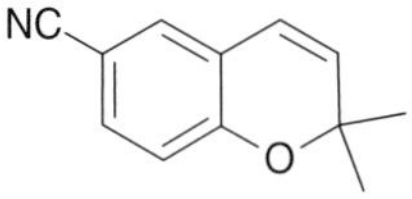 **43** (NC-benzopyran)	24	63	80	(−)-3S,4S	24	59	97	(−)-3S,4S
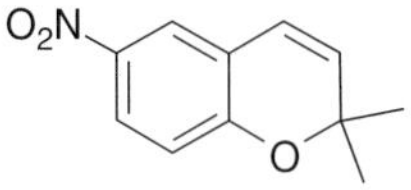 **44** (O$_2$N-benzopyran)	–	–	–	–	24	52	88	(−)-3S,4S
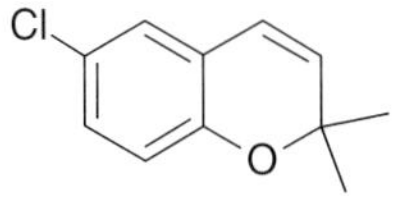 **45** (Cl-benzopyran)	–	–	–	–	24	76	93	(−)-3S,4S

[a] Epoxidation conditions: Iminium salt (10 mol%), TPPP (2 equiv), solvent, −40 °C.
[b] Isolated yield.
[c] Enantiomeric excess determined by Chiral HPLC on a Chiracel OD column.
[d] The absolute configuration of the major enantiomer was determined by comparison to those reported in the literature.

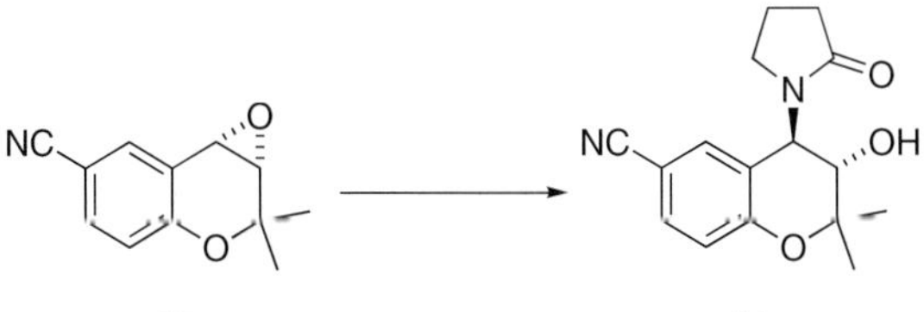

Reagents and Conditions: i: Pyrrolidin-2-one, NaH, DMSO, r.t. 6 h, 52%.

SCHEME 5.23 Ring opening of the epoxide to form the anti-hypertensive agent Levcromakalim.

2.7. Comments on the mechanism

The ability to carry out epoxidation reactions in the absence of water and base has enabled us to investigate the intermediates present during reactions through the medium of NMR spectroscopy [54]. These studies were carried out at various temperatures, employing the iminium tetraphenylborate salt **24** as catalyst, 1-phenylcyclohexene as substrate and TPPP as the stoichiometric oxidant, in deuteriated dichloromethane or chloroform.

Addition of TPPP (1.0 mol equiv) to a solution of iminium salt **24** in CD_2Cl_2 results in an immediate change in the ^1H NMR spectrum: In the new species, all signals are still present, but are shifted by 0.6–1.2 ppm downfield (Fig. 5.10). The proton at the iminium carbon atom is still present, at 9.3 ppm in the ^1H NMR spectrum and at 168.48 ppm in the ^{13}C spectrum. The signal at 1,641 cm^{-1}, which we believe corresponds to the iminium double bond, is still visible in the infrared spectrum.

At the same time, a colourless material is precipitated from the solution. We have characterized this material as tetraphenylphosphonium tetraphenylborate by ^1H NMR spectroscopy and mass spectrometry. The new species formed in solution under these conditions does not induce epoxidation of 1-phenylcyclohexene over 3 h at room temperature. Accordingly, we believe this first intermediate to be iminium salt **46**, the product of ion exchange between the two salts (Scheme 5.24).

	2	**24**	**46**	**48**
	^1H NMR, $CDCl_3$	^1H NMR, $CDCl_3$	^1H NMR, $CDCl_3$	^1H NMR, $CDCl_3$
H1	6.5 ppm, s	8.1 ppm, s	9.3 ppm, s	6.3 ppm, s
H3	4.6 ppm, dq3.5 ppm, m	4.1 ppm, m	4.33 ppm, m	
H1′	3.9 ppm, s	2.9 ppm, m	5.1 ppm, m	3.98 ppm, m

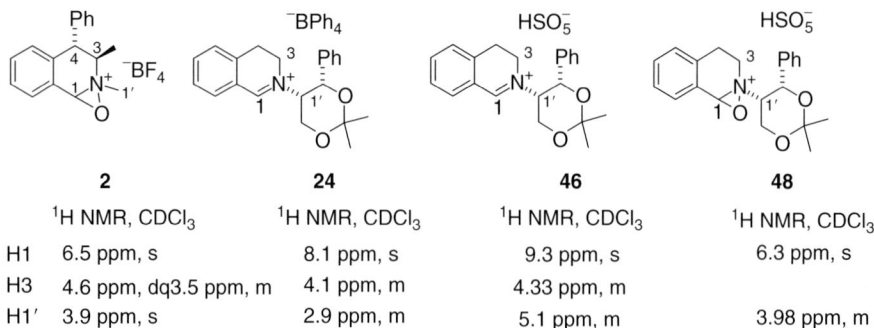

FIGURE 5.10 NMR data for epoxidation reaction intermediates.

24 **46**

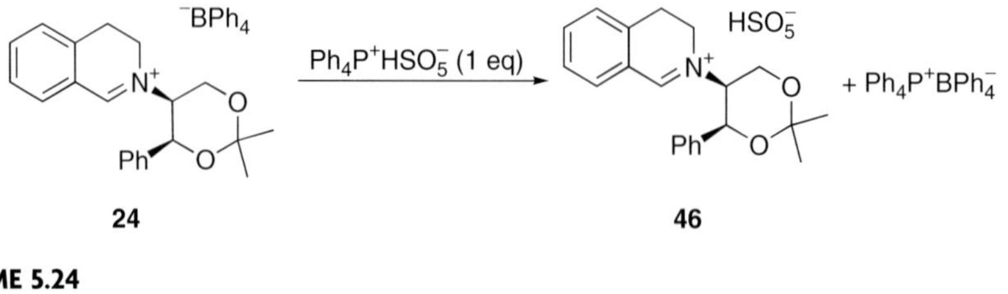

SCHEME 5.24

Treatment of salt **24** with additional TPPP, however, generates a new species in addition to **24**. The new species lacks the iminium proton signal at 9.3 ppm in the ^{1}H NMR spectrum; instead, a new signal at 6.35 ppm is visible. This new species is rather unstable and decomposes relatively quickly (~30 min at −30 °C), but as soon as it is detectable, the addition of the alkene leads to immediate and very rapid epoxide formation. We have previously shown that TPPP does not itself promote epoxidation. Two possible structures for this intermediate may immediately be postulated, both arising from nucleophilic attack of peroxybisulphate at the iminium carbon atom: a neutral intermediate **47**, the product of simple addition, and the corresponding oxaziridium salt **48**, produced by addition and cyclization with expulsion of bisulphate ion (Scheme 5.25).

Comparisons of the ^{1}H chemical shifts of this new intermediate with those of the only oxaziridium salt so far described **2** (Scheme 5.3), coupled with the fact that addition of *m*-chloroperbenzoic acid (*m*-CPBA) to intermediate **46** results in formation of an intermediate with an identical ^{1}H NMR spectrum to that produced by addition of TPPP, suggest that the new species is indeed the oxaziridium salt **48**.

As the relative quantity of TPPP added is increased, so the proportion of **48** present in the mixture increases (Table 5.17).

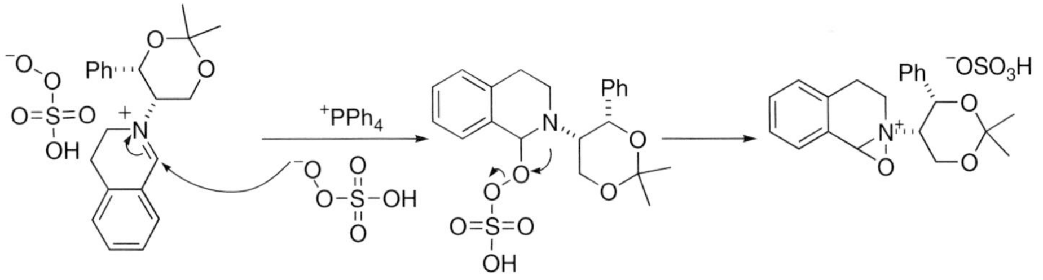

SCHEME 5.25

TABLE 5.17 Comparison of the ratio of intermediates **46** and **48** with increasing equivalents of TPPP

Entry	Equivalents of TPPP	Ratio **46:48**
1	1.0	33.3:1
2	1.5	5.6:1
3	2.0	5.6:1
4	2.5	3.2:1
5	10.0	2.4:1
6	20.0	1.5:1

ACKNOWLEDGEMENTS

We gratefully acknowledge the contributions of Gerry A. Rassias, Laure Finat-Duclos, Adel Ardakani, Patricia Ho-Hune, Suzanne Dilly, Lavril Doswell, Corinne Limousin, Mike McKenzie, David Barros, Mohamed M. Farah, Genna A. Parkes, Phillip Parker and Louise F. Appleby.

REFERENCES

[1] P. Milliet, A. Picot, X. Lusinchi, Formation d'un sel d'oxaziridinium quaternaire par methylation d'un oxazirannemise en evidence de ses proprieties oxydantes, Tetrahedron Lett. 17 (1976) 1573.
[2] P. Milliet, A. Picot, X. Lusinchi, Action de l'acide p-nitroperbenzoïque et de l'eau oxygénée sur un sel d'immonium hétérocyclique stéroïdique et sur l'énamine correspondante, Tetrahedron Lett. 17 (1976) 1577.
[3] P. Milliet, A. Picot, X. Lusinchi, Action comparée de l'eau oxygénée et d'un peracide sur un sel d'immonium pyrrolinique stéroidique et sur l'énamine correspondante. Formation et propriétés d'un sel d'oxaziridinium, Tetrahedron 24 (1981) 4201.
[4] G. Hanquet, X. Lusinchi, P. Milliet, Peracid oxidation of an immonium fluoroborate a new example of oxaziridinium salt, Tetrahedron Lett. 28 (1987) 6061.
[5] G. Hanquet, X. Lusinchi, P. Milliet, Transfert d'oxygene sur la double liaison ethylenique a partir d'un sel d'oxaziridinium, Tetrahedron Lett. 29 (1988) 3941.
[6] X. Lusinchi, G. Hanquet, Oxygen transfer reactions from an Oxaziridinium tetrafluoroborate salt to Olefins, Tetrahedron 53 (1997) 13727.
[7] G. Hanquet, X. Lusinchi, P. Milliet, The stereospecific synthesis of a new chiral oxaziridinium salt, Tetrahedron Lett. 34 (1993) 7271.
[8] L. Bohé, M. Lusinchi, X. Lusinchi, Oxygen atom transfer from a chiral N-alkyl oxaziridine promoted by acid. The asymmetric oxidation of sulfides to sulfoxides, Tetrahedron 55 (1999) 155.
[9] G. Hanquet, X. Lusinchi, Action d'un tétrafluoroborate d'oxaziridinium sur les amines et les imines, Tetrahedron 50 (1994) 12185.
[10] A. Chiaroni, G. Hanquet, M. Lusinchi, C. Riche, First X-ray Determination of an Oxaziridinium Salt: (1S,2R,3R,4S)-2,3-Dimethyl-4-phenyl-1,2,3,4-tetrahydro-1,2-epoxyisoquinolinium Tetrafluoroborate and (1S,2R,3R,4S)-3-Methyl-4-phenyl-1,2,3,4-tetrahydro-2,3-epoxyisoquinoline, Acta Crystallogr., Sect. C, 51 (1995) 2047.
[11] V.K. Aggarwal, M.F. Wang, Catalytic asymmetric synthesis of epoxides mediated by chiral iminium salts, Chem. Commun. (1996) 191.
[12] A. Armstrong, G. Ahmed, I. Garnett, K. Goacolou, Pyrrolidine-Derived iminium salts as catalysts for alkene epoxidation by Oxone®, Synlett (1997) 1075.
[13] A. Armstrong, G. Ahmed, I. Garnett, K. Goacolou, J.S. Wailes, Exocyclic iminium salts as catalysis for alkene epoxidation by Oxone®, Tetrahedron 55 (1999) 2341.
[14] S. Minakata, A. Takemiya, K. Nakamura, I. Ryu, M. Komatsu, Epoxidation of olefins mediated by aliphatic ketiminium salts, Synlett 12 (2000) 1810.
[15] M-K. Wong, L-M. Ho, Y-S. Zheng, C-Y. Ho, D. Yang, Asymmetric epoxidation of olefins catalyzed by chiral iminium salts generated in situ from amines and aldehydes, Org. Lett. 3 (2001) 2587.
[16] A. Armstrong, A.G. Draffan, Intramolecular epoxidation of unsaturated oxaziridines, Synlett (1998) 646.
[17] A. Armstrong, A.G. Draffan, Highly stereoselective intramolecular epoxidation in unsaturated oxaziridines, Tetrahedron Lett. 40 (1999) 4453.
[18] L. Bohé, M. Kammoun, Catalytic oxaziridinium-mediated epoxidation of olefins by Oxone®. A convenient catalyst excluding common side reactions, Tetrahedron Lett. 43 (2002) 803.
[19] G.A. Rassias, PhD Thesis, Loughborough University (1999).
[20] P.C.B. Page, G.A. Rassias, D. Bethell, M.B. Schilling, A new system for catalytic asymmetric epoxidation using iminium salt catalysts, J. Org. Chem. 63 (1998) 2774.
[21] P.C.B. Page, G.A. Rassias, D. Barros, D. Bethell, M.B. Schilling, Dihydroisoquinolinium salts: Catalysts for asymmetric epoxidation, J. Chem. Soc. Perkin Trans. 1 (2000) 3325.

[22] A. Rieche, E. Schmitz, Ringöffnung und Ringschluß bei Isochroman-Abkömmlingen. Zwei neue Umlagerungen (I. Mitteil. über Isochroman), *Chem. Ber.* 89 (1956) 1254.
[23] Dupont product technical information found at: http://www.dupont.com/oxone/techinfo/index.html. Accessed date: 28th February 2008.
[24] P.C.B. Page, G.A. Rassias, D. Barros, A. Ardakani, B. Buckley, D. Bethell, T.A.D. Smith, A.M.Z. Slawin, Functionalized iminium salt systems for catalytic asymmetric epoxidation, *J. Org. Chem.* 66 (2001) 6926.
[25] M. Yamato, K. Hashigaki, S. Ishikawa, N. Qais, Syntheses of chiral oxazolo[2,3-a]tetrahydroisoquinoline and its asymmetric alkylation. Synthesis of (S)-(−)- and (R)-(+)-sa1solidines, *Tetrahedron Lett.* 29 (1988) 6949.
[26] M. Yamato, K. Hashigaki, N. Qais, S. Ishikawa, Asymmetric synthesis of 1-alkytetrahydroisoquinolines using chiral oxazolo[2,3-a]tetrahydroisoquinolines, *Tetrahedron* 46 (1990) 5909.
[27] W. Schneider, B. Müller, Beiträge zur Chemie der Carbinolamine IV. Synthesen des 1-Oxa-7,8-benzo-7,8-didehydro-indolizidins und des 1-Oxa-8,9-benzo-8,9-didehydro-chinolizidins, *Arch. Pharm. Chem.* 294 (1961) 645.
[28] A. Dondoni, D. Perrone, P. Merino, Homologation of L-threonine to α-epimer β-amino-α,γ-dihydroxy aldehydes and acids via stereoselective reduction of 2-thiazolyl amino ketones, *J. Chem. Soc., Chem. Commun.* (1991) 1313.
[29] D.A. Evans, D.L. Rieger, J.R. Gage, ^{13}C NMR chemical shift correlations in 1,3-diol acetonides. Implications for the stereochemical assignment of propionate-derived polyols, *Tetrahedron Lett.* 31 (1990) 7099.
[30] Y. Ohfune, H. Nishio, Acyclic stereocontrolled synthesis of (−)-detoxinine, *Tetrahedron Lett.* 25 (1984) 4133.
[31] R.D. Bach, J.L. Andres, M.D. Su, J.J.W. McDouall, Theoretical model for electrophilic oxygen-atom insertion into hydrocarbons, *J. Am. Chem. Soc.* 115 (1993) 5768.
[32] K.N. Houk, J. Liu, N.C. DeMello, K.R. Condroski, Transition States of Epoxidations: Diradical Character, Spiro Geometries, Transition State Flexibility, and the Origins of Stereoselectivity, *J. Am. Chem. Soc.* 119 (1997) 10147.
[33] I. Washington, K.N. Houk, Transition states and origins of stereoselectivity of epoxidations by oxaziridinium Salts, *J. Am. Chem. Soc.* 122 (2000) 2948.
[34] P.C.B. Page, G.A. Rassias, D. Barros, A. Ardakani, D. Bethell, E. Merifield, New organocatalysts for the asymmetric catalytic epoxidation of alkenes mediated by chiral iminium salts, *Synlett* 4 (2002) 580.
[35] J. Lacour, D. Monchaud, C. Marsol, Effect of the medium on the oxaziridinium-catalyzed enantioselective epoxidation, *Tetrahedron Lett.* 43 (2002) 8257.
[36] J. Vachon, C. Pérollier, D. Monchaud, C. Marsol, K. Ditrich, J. Lacour, Biphasic Enantioselective oletin epoxidation using *tropos* dibenzoazepinium catalysts, *J. Org. Chem.* 70 (2005) 5903.
[37] M.-H. Gonçalves, A. Martinez, S. Grass, P.C.B. Page, J. Lacour, Enantioselective olefin epoxidation using homologous amine and iminium catalysts—a direct comparison, *Tetrahedron Lett.* 47 (2006) 5297.
[38] J. Vachon, S. Rentsch, A. Martinez, C. Marsol, J. Lacour, On the enantioselective olefin epoxidation by doubly bridged biphenyl azepine derivatives – mixed tropos/atropos chiral biaryls, *Org. Biomol. Chem.* 5 (2007) 501.
[39] I.C. Nordin, J.A. Thomas, An improved synthesis of (4S,5S)-2,2-dimethyl-4-phenyl-1,3-dioxan-5-amine, *Tetrahedron Lett.* 25 (1984) 5723.
[40] P.C.B. Page, B.R. Buckley, A.J. Blacker, Iminium salt catalysts for asymmetric epoxidation: The first high enantioselectivities, *Org. Lett.* 6 (2004) 1543.
[41] P.C.B. Page, B.R. Buckley, A.J. Blacker, Corrigenda, *Org. Lett.* 8 (2006) 4669.
[42] P.C.B. Page, D. Barros, B.R. Buckley, A. Ardakani, B.A. Marples, Organocatalysis of asymmetric epoxidation mediated by iminium salts under nonaqueous conditions, *J. Org. Chem.* 69 (2004) 3595.
[43] B.M. Trost, R. Braslau, Tetra-n-butylammonium oxone. Oxidations under anhydrous conditions, *J. Org. Chem.* 53 (1988) 532.
[44] S. Campestrini, F. Di Furia, G. Labat, F. Novello, Oxygen transfer from Ph_4PHSO_5 to manganese porphyrins: Kinetics and mechanism of the formation of the oxo species in homogeneous solution, *J. Chem. Soc. Perkin Trans.* 2 (1994) 2175.

[45] P.C.B. Page, B.R. Buckley, G.A. Rassians, A.J. Blacker, New chiral iminium salt catalysis for asymmetric epoxidation, *Eur. J. Org. Chem.* (2006) 803.
[46] P.C.B. Page, B.R. Buckley, D. Barros, H. Heaney, A.J. Blacker, B.A. Marples, Non-aqueous iminium salt mediated catalytic asymmetric epoxidation, *Tetrahedran* 62 (2006) 6607.
[47] P.C.B. Page, B.R. Buckley, H. Heancy, A.J. Blacker, Asymmetric Epoxidation of *cis*-alkenes mediated by iminium salts: Highly enantioselective synthesis of levcromakalim, *Org. Lett.* 7 (2005) 375.
[48] D. Bell, M.R. Davies, G.R. Geen, I.S. Mann, Copper(I) iodide: A catalyst for the improved synthesis of aryl propargyl ethers, *Synthesis.* (1995) 707.
[49] R. Bergmann, V. Eiermann, R. Gericke, 4-Heterocyclyloxy-2H-1-benzopyran potassium channel activators, *J. Med. Chem.* 33 (1990) 2759.
[50] N.H. Lee, A.R. Muci, E.N. Jacobsen, Enantiomerically pure epoxychromans via asymmetric catalysis, *Tetrahedron Lett.* 32 (1991) 5055.
[51] T. Hashihayata, Y. Ito, T. Katsuki, The first asymmetric epoxidation using a combination of achiral (salen)manganese(III) complex and chiral amine, *Tetrahedron* 53 (1997) 9541.
[52] H. Tian, X. She, L. Shu, H. Yu, Y. Shi, Highly enantioselective epoxidation of *cis*-olefins by chiral dioxirane, *J. Am. Chem. Soc.* 122 (2000) 11551.
[53] H. Tian, X. She, H. Yu, L. Shu, Y. Shi, Designing new chiral ketone catalysts. Asymmetric epoxidation of *cis*-olefins and terminal olefins, *J. Org. Chem.* 67 (2002) 2435.
[54] P.C.B. Page, D. Barros, B.R. Buckley, B.A. Marples, Organocatalysis of asymmetric epoxidation mediated by iminium salts: Comments on the mechanism, *Tetrahedron: Asymmetry* 16 (2005) 3488.

CHAPTER **6**

Selective Aerobic Radical Epoxidation of α-Olefins Catalyzed by N-Hydroxyphthalimide

Carlo Punta,* Davide Moscatelli,* Ombretta Porta,* Francesco Minisci,* Cristian Gambarotti,* and Marco Lucarini[†]

Contents		
	1. Introduction	218
	2. Experimental	219
	2.1. General procedure for the epoxidation of primary olefins	219
	2.2. General procedure for the epoxidation of propylene	220
	3. Results and Discussion	220
	4. Conclusions	227
	Acknowledgments	227
	References	227

Abstract	The combination of N-hydroxyphthalimide (NHPI) and peracids is employed for the free-radical epoxidation of alkenes with high yields. A quite different selectivity is obtained from that of the well-known epoxidation by peracids, which is attributed to a different reaction mechanism and which evidences the molecule-induced homolysis of NHPI by peracids and dioxiranes. Electron paramagnetic resonance (EPR) and thermodynamic studies, as well as the marked solvent effect and the high selectivity in the α-position, strongly support the mechanistic interpretation. The *in situ* generation of acyl peroxyl radicals from acetaldehyde and molecular oxygen and the optimization of the reaction conditions allow the development of an innovative process for the convenient metal-free synthesis of several epoxides, suggesting an intriguing alternative route to the production of propylene oxide.

* Dipartimento di Chimica, Materiali e Ingegneria Chimica "Giulio Natta," Politecnico di Milano, Via Mancinelli 7–I-20131 Milano, Italy
[†] Dipartimento di Chimica Organica "A. Mangini," Università di Bologna, Via San Giacomo 11–40126 Bologna, Italy

Mechanisms in Homogeneous and Heterogeneous Epoxidation Catalysis
DOI: 10.1016/B978-0-444-53188-9.00006-7

© 2008 Elsevier B.V.
All rights reserved.

217

Key Words: α-Olefins, N-hydroxyphthalimide, Peracids, Dioxiranes, Peroxyl radicals, Molecular-induced homolysis, Propylene oxide, Metal-free, Solvent effect, Bond dissociation enthalpy. © 2008 Elsevier B.V.

1. INTRODUCTION

In the last decade, a variety of interesting aerobic oxidations of organic compounds, including alkanes, under mild conditions was achieved by using N-hydroxyphthalimide (NHPI) as free-radical promoter in the presence or absence of metal salts [1]. This organocatalyst was first and widely employed in oxidative processes by Ishii and coworkers [2,3]. Our group has contributed to the understanding of the reaction mechanism [4,5] and developed innovative synthetic routes for the oxidation of a wide range of substrates, including alcohols [4,6], amines [7], amides [8], silanes [9], and alkyl aromatics [10], for the selective functionalization of hydrocarbons [11] and for the acylation and carbamoylation of protonated N-heteroaromatic bases [12,13].

In all these oxidations, the phthalimido-N-oxyl (PINO) radical, generated *in situ* from NHPI in different ways, plays a key role in the catalytic process, acting as a more selective and efficient hydrogen-abstracting species with respect to the peroxyl radicals normally involved in autoxidation processes. According to the general mechanism of radical oxidation, the overall reaction is represented by several key steps that will proceed at an acceptable rate only if the value of the bond dissociation enthalpy (BDE) of the bonds formed is higher or similar to the strength of the bonds cleaved. Thus, the reactivity of NHPI and PINO had to be related to the BDE of the O–H group, that we evaluated to be 88.1 kcal/mol [4,14]. This value is similar to the BDE of O–H in hydroperoxides, suggesting that the different reactivity of PINO compared to peroxyl radicals should be attributed to an enhanced polarity effect involved in the hydrogen abstractions by PINO [5].

The epoxidation of olefins is an important tool for the introduction of an oxygen atom into an organic molecule bearing a double bond. Many methods can be employed for this purpose by using a wide range of reagents, both in the absence and more commonly in the presence of catalysts, especially when dioxygen is the oxidizing agent.

The group of Masui first attempted the direct epoxidation of olefins by using oxygen and NHPI with metalloporphyrins, but they obtained poor results [15]. Ishii and coworkers proposed two different methods. In the first protocol [16,17], the epoxidizing agent is obtained *in situ* by the aerobic oxidation of a suitable alcoholic (benzhydrol) compound in the presence of catalytic amounts of NHPI. The resulting oxidant, which is not able to promote the epoxidation by itself, is then activated in the presence of an olefin by catalytic amounts of hexafluoroacetone (HFA) (Scheme 6.1).

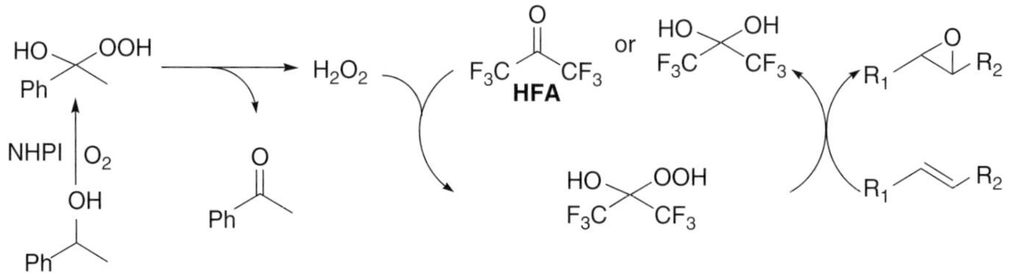

SCHEME 6.1 Mechanism for the epoxidation of alkenes catalyzed by N-hydroxyphthalimide (NHPI) in the presence of alcohols and hexafluoroacetone (HFA).

The second method proposed by Ishii uses hydroperoxides from the *in situ* NHPI-catalyzed oxidation of tetralin or ethylbenzene to epoxidize olefins directly by means of Mo(CO)$_6$ [18] (Scheme 6.2).

Very recently we have developed a new, easier, and selective metal-free NHPI-catalyzed aerobic epoxidation of primary olefins [19] based on the *in situ* generation of peracetic acid from acetaldehyde. In this chapter, we will discuss the reaction mechanism in order to explain the significant differences in selectivity with respect to the epoxidation by peracids and we will show preliminary successful results in the synthesis of propylene oxide.

2. EXPERIMENTAL

The olefins, acetaldehyde, NHPI, and solvents were commercial products and were used without any further purification.

2.1. General procedure for the epoxidation of primary olefins

A solution of 5 mmol of olefins, 15 mmol of acetaldehyde, and 0.5 mmol of NHPI in 10 ml of acetonitrile was stirred at room temperature (RT) in atmospheric pressure of O$_2$ for 14–24 h; the known epoxides were isolated by flash chromatography (hexane:ethyl acetate, 9:1) and characterized by nuclear magnetic resonance (NMR) and mass spectrometry (MS) (by comparison with authentic samples).

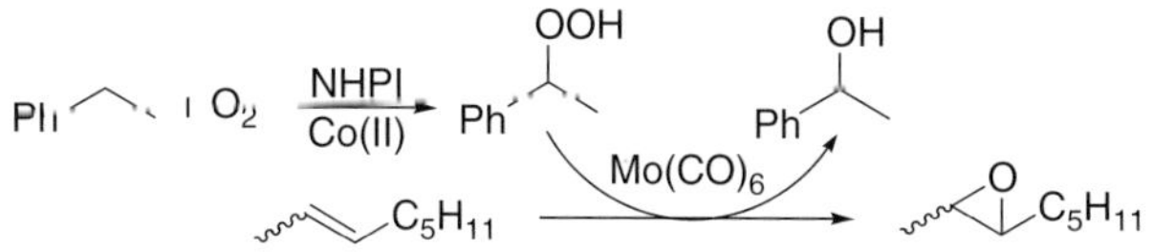

SCHEME 6.2 Mechanism for the epoxidation of alkenes catalyzed by N-hydroxyphthalimide (NHPI) in the presence of ethyl benzene and Mo(CO)$_6$.

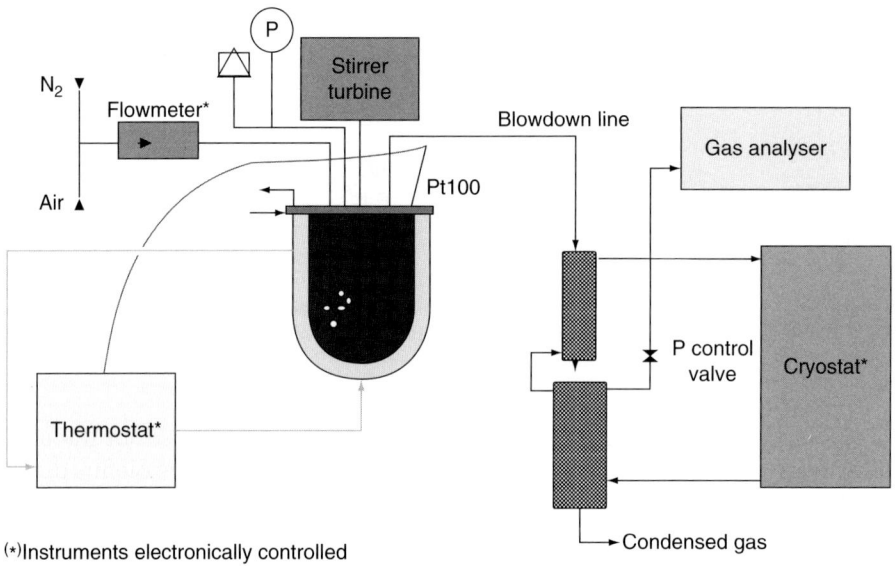

FIGURE 6.1 Schematic layout of the reactor system employed for the epoxidation of propylene.

2.2. General procedure for the epoxidation of propylene

The propylene oxidation was conducted in a Buchi glass vessel (volume 1 liter) provided with a jacket for heat transfer, using diathermic oil. The vessel was equipped with a turbine stirring system that allowed a continuous gas flow, captured at the reactor top and blown into the reacting mixture through the hollow shaft of the stirrer, which had openings at the top and below the impeller. An external circulation thermostat (Haake F6) regulated the temperature of the silicon oil in the jacket, while a Pt100 sensor immersed in the reacting mixture read the internal temperature. A condenser and a gas/liquid separator on the blowdown line were employed to separate and condense the vapors carried out by the outlet gas. In order to minimize this phenomenon, the steel top of the reactor was cooled by water flow. Figure 6.1 shows a schematic of the reaction system layout.

A solution of 500 ml of acetonitrile, 33 g of acetaldehyde, and 3.5 g of NHPI was added to the reactor, which was then purged with nitrogen. After 1 h, the blowdown line was closed and 10.5 g of propylene were added to the system in 0.5 h (Table 6.1).

After the complete addition of propylene, a nonstoichiometric amount of oxygen (air) was added to the mixture following the profile reported in Table 6.2.

3. RESULTS AND DISCUSSION

Several years ago we presented evidence that the oxidation of a variety of organic compounds (hydrocarbons, alcohols, ethers, aldehydes, etc.) by peracids [20] and dioxiranes [21] could be explained by radical mechanism, in contrast with the

TABLE 6.1 Pressure, temperature, and added propylene amount profiles

Time (min)	Charged amount (g)	Relative pressure (atm)	Temperature (°C)
5	1.5	0.33	21.2
10	3.5	0.75	21.7
15	5.5	1.17	21.8
20	7.5	1.61	22.2
25	9.0	1.92	22.4
30	10.5	2.24	22.4

TABLE 6.2 Pressure and temperature profile in the reactor as a consequence of the air addition

Time (h)	Relative pressure (atm)	Temperature (°C)
5	3.0	23.2
10	4.0	23.9
15	5.0	24.5
20	6.0	24.9
25	7.0	25.6
30	8.0	25.9
35	9.0	26.1

generally accepted mechanism of "concerted oxygen insertion" [22,23], which postulates butterfly-type transition states [Eqs. (6.1) and (6.2)], similar to the one originally suggested by Bartlett [24] for alkene epoxidation by peroxides.

$$R-H + \underset{O}{\overset{HOO}{\underset{\|}{C}}}R' \longrightarrow \left[R\overset{H--O}{\underset{H}{\overset{O}{\cdots}}\underset{O}{\overset{\cdots}{C}}}R' \right]^{\neq} \longrightarrow ROH + \underset{O}{\overset{HO}{\underset{\|}{C}}}R' \qquad (6.1)$$

$$R-H + \underset{O}{\overset{O}{\underset{|}{\underset{\|}{C}}}}\underset{Me}{\overset{Me}{\rangle}} \longrightarrow \left[R\overset{O---}{\underset{H}{\overset{\cdots}{\underset{O}{\overset{\cdots}{C}}}}}\underset{Me}{\overset{Me}{\rangle}} \right]^{\neq} \longrightarrow ROH + \underset{O}{\overset{Me}{\underset{\|}{\underset{C}{\overset{Me}{\rangle}}}}} \qquad (6.2)$$

We suggested the occurrence of a "molecule-induced homolysis," process for the formation of radicals pairs in which transition states involving hydrogen abstractions were formed [Eqs. (6.3) and (6.4)].

$$R-H + \overset{O-O}{\underset{H}{|}}\hspace{-2pt}\overset{R'}{\underset{O}{\overset{|}{C}}} \longrightarrow \left[\overset{\delta^-}{R}\text{---}H\text{---}\overset{\delta^-}{O}\text{---}\underset{H}{\overset{\overset{O}{\|}}{O-C-R'}}\right]^{\ne} \longrightarrow R^{\bullet} + H-O-H + \overset{\bullet O}{\underset{O}{\overset{}{C}}}\hspace{-2pt}R' \quad (6.3)$$

$$R-H + \overset{O}{\underset{O}{\overset{}{\bowtie}}} \longrightarrow \left[\overset{\delta^+}{R}\text{---}H\text{---}\overset{O}{\underset{\delta^-}{\overset{}{\bowtie}}}\right]^{\ne} \longrightarrow R^{\bullet} + \overset{\bullet O}{\underset{H-O}{\overset{}{\bowtie}}} \quad (6.4)$$

We ascribed the driving force for Eqs. (6.3) and (6.4) to be the high BDE values of the O–H bonds formed in hydrogen abstractions, which are particularly favorable over weaker C–H bonds (tertiary alkyl, benzyl, RCO–H, etc.).

The relatively low BDE value (88.1 kcal/mol) of the O–H bond for NHPI suggested that peracids and dioxiranes could undergo induced homolysis of NHPI under mild conditions, generating the PINO radical [Eqs. (6.5) and (6.6)], which plays a key role in the aerobic oxidations catalyzed by NHPI.

$$\underset{/}{\overset{\backslash}{N}}-OH + \overset{HOO}{\underset{O}{\overset{}{\bigvee}}}\hspace{-2pt}R \longrightarrow \left[\underset{/}{\overset{\backslash}{N}}-O\text{--}H\text{--}\underset{H}{\overset{\overset{O}{\|}}{O\text{--}O-C-R}}\right]^{\ne} \longrightarrow \underset{/}{\overset{\backslash}{N}}-O^{\bullet} + H-O-H + \overset{\bullet O}{\underset{O}{\overset{}{C}}}\hspace{-2pt}R \quad (6.5)$$

$$\underset{/}{\overset{\backslash}{N}}-OH + \overset{O}{\underset{O}{\overset{}{\bowtie}}} \longrightarrow \left[\underset{/}{\overset{\backslash}{N}}-O\text{--}H\text{--}\overset{O}{\overset{}{\bowtie}}\right]^{\ne} \longrightarrow \underset{/}{\overset{\backslash}{N}}-O^{\bullet} + \overset{\bullet O}{\underset{H-O}{\overset{}{\bowtie}}} \quad (6.6)$$

The hypothetical coupling of the radical pairs generated in Eqs. (6.5) and (6.6) are very likely reversible, so that the PINO radical and the acyloxyl [Eq. (6.7)] or the alkoxyl [Eq. (6.8)] radicals can escape from the solvent cage giving typical free-radical reactions.

$$\left[\underset{/}{\overset{\backslash}{N}}-O^{\bullet} \quad {}^{\bullet}O-\overset{\overset{O}{\|}}{C}-R\right]_{Cage} \underset{Escape}{\rightleftarrows} \begin{array}{l} \underset{/}{\overset{\backslash}{N}}-O-O-\overset{\overset{O}{\|}}{C}-R \\ \\ \underset{/}{\overset{\backslash}{N}}-O^{\bullet} + {}^{\bullet}O-\overset{\overset{O}{\|}}{C}-R \end{array} \quad (6.7)$$

$$\left[\begin{array}{c} \diagdown \\ \diagup \end{array} N-O\cdot \quad \cdot O-\underset{CH_3}{\underset{|}{C}}-CH_3 \\ CH_3 \end{array} \right]_{Cage} \xrightleftharpoons{} \begin{array}{c} \diagdown \\ \diagup \end{array} N-O-O-\underset{CH_3}{\underset{|}{\overset{OH}{\underset{|}{C}}}}-CH_3 \qquad (6.8)$$

$$\xrightarrow{Escape} \begin{array}{c} \diagdown \\ \diagup \end{array} N-O\cdot \; + \; \cdot O-\underset{CH_3}{\underset{|}{\overset{OH}{\underset{|}{C}}}}-CH_3$$

Spectroscopic and chemical investigations support this assumption. The electron paramagnetic resonance (EPR) spectrum of the PINO radical was readily observed by simply adding at RT NHPI to solutions of *m*-chloroperbenzoic acid (*m*-CPBA) in acetonitrile [a (2 H) = 0.46 G; a (N) = 4.77 G] or dimethyldioxirane in acetone [a (2 H) = 0.44 G; a (N) = 4.70 G], as it is shown in Fig. 6.2.

High performance liquid chromatography (HPLC) and gas chromatography (GC) analysis of the products arising from *m*-CPBA and NHPI in acetonitrile at RT revealed the presence of *m*-chlorobenzoic acid as the main reaction product (~90%) and chlorobenzene (~10%) as by-product. Moreover, the same reaction in benzene solution at RT always led to *m*-chlorobenzoic acid as the main reaction product, whereas phenyl *m*-chlorobenzoate and *m*-chlorobiphenyl as by-products.

The only possible explanation for these last by-products and for chlorobenzene is the formation of the acyloxyl radical [Eq. (6.5)], which reacts according to Eqs. (6.9a)–(6.9d).

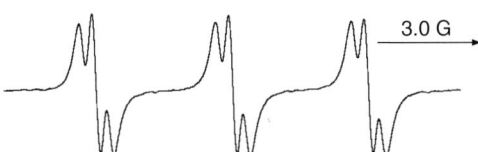

FIGURE 6.2 Electron paramagnetic resonance (EPR) spectrum of phthalimido-*N*-oxyl (PINO) obtained by mixing *N*-hydroxyphthalimide (NHPI) with *m*-chloroperbenzoic acid (*m*-CPBA) in CH$_3$CN at room temperature (RT).

$$\begin{array}{c} & m\text{-Cl}-C_6H_4-COO\cdot & \\ \swarrow^{(b)} & & \searrow^{(a)}\,^{C_6H_6} \\ CO_2 + m\text{-Cl}-C_6H_4\cdot & & m\text{-Cl}-C_6H_4-COOC_6H_5 \\ \swarrow^{NHPI\,(d)} \quad \searrow^{C_6H_6\,(c)} & & \\ C_6H_5-Cl \quad\quad m\text{-Cl}-C_6H_4-C_6H_5 & & \end{array} \qquad (6.9)$$

m-Chlorobenzoic acid was the main reaction product because it was formed by hydrogen abstraction from NHPI by the acyloxyl radical [Eq. (6.10)].

$$\text{(m-ClC}_6\text{H}_4\text{COO}^\bullet) + \text{H-O-N}\genfrac{}{}{0pt}{}{}{} \xrightarrow{k_{10}} \text{(m-ClC}_6\text{H}_4\text{COOH}) + {}^\bullet\text{O-N}\genfrac{}{}{0pt}{}{}{} \qquad k_{10} > 10^9 \ M^{-1}\,s^{-1} \qquad (6.10)$$

These results suggest the possibility of using the aerobic oxidation of aldehydes, catalyzed by NHPI, for the epoxidations of alkenes by peracids generated "*in situ*" under mild conditions [Eq. (6.11)].

Actually the aerobic oxidation of acetaldehyde in acetonitrile solution at RT and atmospheric pressure in oxygen in the presence of alkenes and catalytic amounts of NHPI led to the corresponding epoxides (Table 6.3). No oxidation occurred under the same conditions in the absence of NHPI, clearly indicating that Eq. (6.12) plays a key role in the aerobic epoxidation.

$$\text{>C=C<} + CH_3CHO + O_2 \xrightarrow{NHPI} \text{>C-C<(O)} + CH_3COOH \qquad (6.11)$$

$$CH_3\text{-C(=O)-H} + {}^\bullet\text{O-N}\genfrac{}{}{0pt}{}{}{} \longrightarrow [CH_3\text{-C(=O)-H}^{\delta-}\text{--O-N}\genfrac{}{}{0pt}{}{}{}]^{\ddagger}_{\delta+} \longrightarrow CH_3\text{-C(=O)}^\bullet + HO\text{-N}\genfrac{}{}{0pt}{}{}{} \qquad (6.12)$$

α-Olefins and cyclic olefins give good yields of epoxides, whereas internal acyclic olefins were unreactive under the same conditions; the internal acyclic olefins are, on the contrary, more reactive than α-olefins [25] with peracetic acid and also the stereochemistry for the epoxidation of limonene is different: in the aerobic epoxidation, 67% of the *cis* isomer and 33% of the *trans* isomer are formed, whereas 41% of the *cis* isomer and 59% of the *trans* isomer are formed in the reaction with peracid; moreover, 2-methyl-2-butene is 300 times more reactive than propene with peracetic acid, whereas it is unreactive toward aerobic epoxidation.

Thus, the selectivity observed with several alkenes was quite different from that obtained [25] with peracetic acid (Table 6.3), suggesting that a different reaction mechanism occurred.

Our interpretation of the results is that the acyl radical reacts with O_2 [Eq. (6.13)] to form the acylperoxyl radical which then adds to the olefin and forms a complex which decomposes to produce the epoxide [Eq. (6.14)].

TABLE 6.3 Epoxidation of olefins by aerobic oxidation of acetaldehyde, catalyzed by NHPI

Olefin	Catalyst (%)	Reaction time (h)	Yield of epoxide (%)
1-Hexene	NHPI (10)	24	61
1-Hexene	NHPI (10)	48	70
1-Octene	NHPI (10)	24	80
1-Decene	NHPI (10)	24	81
1-Decene	NHPI (10)	48	94
1-Decene	–	24	–
1-Dodecene	NHPI (10)	23	81
Methyl oleate	NHPI (10)	24	–
cis-2-Hexene	NHPI (10)	24	–
Cyclooctene	NHPI (10)	27	96
Cyclooctene	–	24	–
R(+)-limonene	NHPI (10)	14	cis (67), trans (33) (epoxide at ring double bond)
2-Methyl-2-butene	NHPI (10)	24	–

NHPI, N-hydroxyphthalimide.

$$CH_3-\dot{C}=O + O_2 \longrightarrow CH_3-C(=O)-O-O\cdot \qquad (6.13)$$

$$CH_3-C(=O)-O-O\cdot + H_2C=CHR \longrightarrow CH_3-C(=O)-O-O-CH_2-\dot{C}HR \longrightarrow H_2C\overset{O}{\underset{}{\triangle}}CH-R + \cdot CH_3 + CO_2 \qquad (6.14)$$

The fact that internal acyclic olefins are unreactive toward aerobic epoxidation would suggest that peracetic acid is not formed according to Eq. (6.15).

$$R-C(=O)-OO\cdot + HO-N\big< \xrightarrow{k_{15}} R-C(=O)-OO-H + \cdot O-N\big< \quad k_{15} > 10^4 \text{ M}^{-1}\text{s}^{-1} \qquad (6.15)$$

TABLE 6.4 Solvent effect for the aerobic epoxidation of olefins catalyzed by NHPI

Olefin	Solvent	Reaction time (h)	Yield of epoxide (%)
1-Decene	CH_3CN	24	80
1-Decene	Ph–CN	24	75
1-Decene	CH_3COOH	24	2
1-Decene	CH_3COCH_3	24	24[a]
1-Decene	$CH_2Cl–CH_2Cl$	24	56
1-Decene	Ph–CF_3	24	–

NHPI, N-hydroxyphthalimide. [a] 4% of 2-dodecanone is recovered as a by-product.

We have not evaluated the rate constant k_{15}, but we expect a value higher than 10^4 M^{-1} s^{-1} in comparison to a value of 7.2×10^3 M^{-1} s^{-1} determined for the hydrogen abstraction from NHPI by t-BuOO radical since Eq. (6.15) is about 5 kcal/mol more exothermic (the BDE values of the O–H bonds are respectively 88 and 93 kcal/mol for t-BuOO–H and RCOOO–H). The high rate constant of k_{15} suggests that peracetic acid is actually formed according to Eq. (6.15) in competition with the addition of the olefin [Eq. (6.14)], but its reaction with NHPI [Eq. (6.5)] is faster than the epoxidation step.

We have studied the epoxidation of 1-decene in different solvents, in order to optimize reaction conditions. A marked solvent effect was found, as reported in Table 6.4. Acetonitrile seemed to be the best solvent.

Ingold et al. [26,27] also found a similar significant solvent effect in the reaction between PINO and phenol. They showed that PINO reacts much faster with O–H bonds than with C–H bonds having the same BDE. This conforms to the general observation that H transfer between –OH and –O• is favored kinetically over transfers between –CH and –O•, even where the BDEs for the –OH and –CH are similar. Shaik et al. [28] explained this phenomenon by a valence bond (VB) model in which the H transfer between oxygen atoms involves the formation of a hydrogen bond in the transition state, where each terminal H is oriented toward the lone pair of the oxygen in the other terminus. The hydrogen bond present in the OLHLO system lowers the activation energy relative to that for the OLHLC where such an interaction is not present.

The low conversion obtained in acetic acid is probably due to the strong hydrogen bond, which occurs between the hydroxyl group of the protic solvent and the O–H group present in NHPI, thus inhibiting the molecule-induced homolysis effect. However, a polar solvent is required to make NHPI soluble in the reaction medium: that is why the conversion decreases in dichloroethane, whereas no product is observed in Ph–CF_3, where the organocatalyst remains in suspension.

The by-product observed by carrying out the reaction in acetone was probably because of the lower solubility of O_2 in the reaction medium. Thus, the acyl radical adds to the double bond leading to the formation of 2-dodecanone [Eqs. (6.16) and (6.17)].

$$\text{CH}_3\text{C(O)}^\bullet + \text{CH}_2=\text{CH-(CH}_2)_7\text{CH}_3 \longrightarrow \text{CH}_3\text{C(O)-CH}_2\text{-CH}^\bullet\text{-(CH}_2)_7\text{CH}_3 \quad (6.16)$$

$$\text{R-C(O)-CH}_2\text{-CH}^\bullet\text{-R'} + \text{HO-N}\big\langle \longrightarrow \text{R-C(O)-CH}_2\text{-CH}_2\text{-R'} + \big\rangle\text{N-O}^\bullet \quad (6.17)$$

This catalytic system has been employed in the aerobic epoxidation of propylene, according to the procedure reported in the experimental section.

Preliminary results show complete selectivity to propylene oxide, suggesting a new route for the synthesis of this important epoxide in good yields under mild conditions.

4. CONCLUSIONS

We have reported a new methodology for the selective synthesis of epoxides under mild, eco-friendly aerobic conditions. The catalytic system reported here suggests an innovative procedure to generate PINO radical in the absence of metal salts. This procedure not only shows a new interesting route for the production of simple epoxides (including propylene oxide), but also contributes to the understanding of the reaction mechanism in the oxidation by peracids and dioxiranes, supporting molecule-induced homolysis process for generating radicals.

Future efforts will be focused on the design of lipophilic NHPI derivatives in order to replace the non-green acetonitrile with more benign solvents, or even to carry out the reactions directly in olefins as solvents, including propylene at higher pressures. This will allow improvement of the environmental impact of the process.

ACKNOWLEDGMENTS

Politecnico di Milano and MURST (Research project "Radical Processes in Chemistry and Biology: Synthesis, Mechanism, Applications") are gratefully acknowledged for financial support.

REFERENCES

[1] F. Recupero, C. Punta, Free radical functionalization of organic compounds catalyzed by N-hydroxyphthalimide, *Chem. Rev.* 107 (2007) 3800.

[2] Y. Ishii, S. Sakaguchi, A new strategy for alkane oxidation with O_2 using N-hydroxyphthalimde (NHPI) as a radical catalyst, *Catal. Surv. Jpn.* 3 (1999) 27.

[3] Y. Ishii, S. Sakaguchi, T. Iwahama, Innovation of hydrocarbon oxidation with molecular oxygen and related reactions, *Adv. Synth. Catal.* 343 (2001) 393.

[4] F. Minisci, C. Punta, F. Recupero, F. Fontana, G. F. Pedulli, A new, highly selective synthesis of aromatic aldehydes by aerobic free-radical oxidation of benzylpic alcohols, catalysed by N-hydroxyphtalimide under mild conditions. Polar and enthalpic effects, *Chem. Commum.* (2002) 68.

[5] F. Minisci, F. Recupero, G. F. Pedulli, M. Lucarini, Transition metal salts catalysis in the aerobic oxidation of organic compounds: Thermochemical and kinetic aspects and new synthetic development in the presence of N-hydroxy-derivative catalysts, *J. Mol. Catal. A.* 63 (2003) 204–205.

[6] F. Minisci, F. Recupero, A. Cecchetto, C. Gambarotti, C. Punta, R. Faletti, R. Paganelli, G. F. Pedulli, Mechanism of the aerobic oxidation of alcohols to aldehydes and ketones, catalysed under mild conditions by persistent and non-persistent nitroxyl radicals and transition metal salts-polar, enthalpic and captodative effects, *Eur. J. Org. Chem.* (2004) 109.

[7] A. Cecchetto, F. Minisci, F. Recupero, F. Fontana, G. F. Pedulli, A new selective free radical synthesis of aromatic aldehydes by aerobic oxidation of tertiary benzylamines catalysed by N-hydroxyminides and Co(II) under mild conditions. Polar enthalpic effects, *Tetrahedron Lett.* 43 (2002) 3605.

[8] F. Minisci, C. Punta, F. Recupero, F. Fontana, G. F. Pedulli, Aerobic oxidation of N-alkylamides catalysed by N-hydroxyphatalimide under mild condiditons. Polar and enthaplic effects, *J. Org. Chem.* 67 (2002) 2671.

[9] F. Minisci, F. Recupero, C. Punta, C. Guidarini, F. Fontana, G. F. Pedulli, A new, highly selective, free-radical aerobic oxidation of silanes to silanols catalysed by N-hydroxyphthalimide under mild conditions, *Synlett.* (2002) 1173.

[10] F. Minisci, F. Recupero, A. Cecchetto, C. Gambarotti, C. Punta, R. Paganelli, G. F. Pedulli, F. Fontana, Solvent and temperature effects in the free radical aerobic oxidation of alkyl and acyl aromatics catalysed by transition metal slats and N-hydroxyphatlimide: New processes for the sysnthesis of p-hydroxybenzoic acid, diphenols, and dienes liquid for liquid crystals and cross-linked polymers, *Org. Proc. Res. Devel.* 8 (2004) 163.

[11] F. Minisci, O. Porta, F. Recupero, C. Gambarotti, R. Paganelli, G. F. Pedulli, F. Fontana, New free-radical halogeneations of alkanes, catalysed by N-hydroxyphthalimide. Polar and enthlpic effects on the chemo- and regioselectivity, *Tertrahedron Lett.* 45 (2004) 1607.

[12] F. Minisci, F. Recupero, A. Cecchetto, C. Punta, C. Gambarotti, F. Fontana, G. F. Pedulli, Polar effects in free-radical reactions. A novel homolytic acylation of heteroaromatic bases by aerobic oxidation of aldehydes, catalysed by N-hydroxyphthalimide and Co salts, *J. Heter. Chem.* 40 (2003) 325.

[13] F. Minisci, F. Recupero, C. Putna, C. Gambarotti, F. Antonietti, F. Fontana, G. F. Pedulli, A novel, selective free-radical carbamoylation of heteroaromatic bases by Ce(IV) oxidation of formamide, catalysed by N-hydroxyphthalimide, *Chem. Comm.* (2002) 2496.

[14] R. Amorati, M. Lucarini, V. Mugnaini, G. F. Pedulli, F. Minisci, F. Fontana, F. Recupero, P. Astolfi, L. Greci, Hydroxylamines as oxidation catalysts: Thermochemical and kinetic studies, *J. Org. Chem.* 68 (2003) 1747.

[15] S. Ozaki, T. Hamaguci, K. Tsuchida, Y. Kimata, M. Masui, Epoxidation catalysed by Mn[III]TPPCI using dioxygen activated by a system containing N-hydroxyphthalimide and styrene, 2-norbornene or indene, *J. Chem. Soc. Perkin Trans. II.* (1989) 951.

[16] T. Iwahama, S. Sakaguchi, Y. Ishii, Epoxidation of alkenes using dioxygen in the presence of an alcohol catalyzed by N-hydroxyphthalimide and hexafluoroacetone without any metal catalyst, *Chem. Commun.* (1999) 727.

[17] T. Iwahama, S. Sakaguchi, Y. Ishii, Epoxidation of alkenes with H_2O_2 generated *in situ* from alcohols and molecular oxygen using N-hydroxyphthalimide and hexafluoroacetone as catalysts, *Heterocycles.* 52 (2000) 693.

[18] T. Iwahama, G. Hatta, S. Sakaguchi, Y. Ishii, Epoxidation of alkenes using alkyl hydroperoxides generated *in situ* by catalytic autoxidation of hydrocarbons with dioxygen, *Chem. Commun.* (2000) 163.

[19] F. Minisci, C. Gambarotti, M. Pierini, O. Porta, C. Punta, F. Recupero, M. Lucarini, V. Mugnaini, Molecule-induced homolysis of N-hydroxyphthalimide (NHPI) by peracids and dioxirane. A new, simple, selective aerobic radical epoxidation of alkanes, *Tetrahedron Lett.* 47 (2006) 1421.

[20] A. Bravo, H. R. Bjørsvik, F. Fontana, F. Minisci, A. Serri, Radical versus "oxenoid" oxygen insertion mechanism in the oxidation of alkanes and alcohols by aromatic peracids. New synthetic developments, *J. Org. Chem.* 61 (1996) 9409.
[21] A. Bravo, F. Fontana, G. Fronza, F. Minisci, L. Zhao, Molecule-induced homolysis versus "concerted oxenoid oxygen insertion" in the oxidation of organic compounds by dimethyldioxirane, *J. Org. Chem.* 63 (1998) 254.
[22] W. Adam, R. Curci, L. D'Accolti, A. Dinoi, C. Fusco, F. Gasparrini, R. Kluge, R. Paredes, M. Schulz, A. K. Smerz, L. A. Veloza, S. Weinkötz, R. Winde, Epoxidation and oxygen insertion into alkane CH bonds by dioxirane do not involve detectable radical pathways, *Chem. Eur. J.* 3 (1997) 105.
[23] A. A. Fokin, B. A. Tkachenko, O. I. Korshunov, P. A. Gunchenko, P. R. Schreiner, Molecule-induced alkane homolysis with dioxiranes, *J. Am. Chem. Soc.* 123 (2001) 11248.
[24] P. D. Bartlett, *Rec. Chem. Prog.* 11 (1950) 47.
[25] G. Banillon, C. Lick, K. Schank, The Chemistry of Peroxides, in: S. Patai (Ed.), *The Chemistry of Functional Groups*, John Wiley, New York, (1983); p. 289.
[26] D. V. Avila, K. U. Ingold, J. Lusztyk, W. H. Green, D. R. Procopio, Dramatic solvent effects on the absolute rate constants for abstraction of the hydroxylic hydrogen atom form *tert*-butyl hydroperoxide and phenol by the cumyloxyl radical. The role of hydrogen bonding, *J. Am. Chem. Soc.* 117 (1995) 2929.
[27] D. W. Snelgrove, J. Lusztyk, J. T. Banks, P. Mulder, K. U. Ingold, Kinetic solvent effects on hydrogen-atom abstractions: Reliable, quantitative predictions via a single empirical equation, *J. Am. Chem. Soc.* 123 (2001) 469.
[28] S. Shaik, P. S. de visser, W. Wu, L. Song, P. C. Hiberty, Reply to comment on "*identity hydrogen abstraction reactions, $X \cdot + H\text{-}X' \cdot X\text{-}H + X'$ ($X = X' + CH_3, SiH_3, GeH_3, SnH_3, PbH_3$): A Valence bond modeling*", *J. Phys. Cem. A* 106 (2002) 5043.

SECTION 3
Heterogeneous Catalysis

CHAPTER 7

Investigation of the Origins of Selectivity in Ethylene Epoxidation on Promoted and Unpromoted Ag/α-Al$_2$O$_3$ Catalysts: A Detailed Kinetic, Mechanistic and Adsorptive Study

Kenneth. C. Waugh* and M. Hague[†]

Contents			
	1.	Introduction	234
	2.	The Detailed Kinetics of the Adsorption and Desorption of Oxygen on Silver and the Structure of the Oxide Overlayer and Reaction Conditions	235
	3.	The Adsorption of Ethylene and Ethylene Oxide on Clean and Oxidised Ag(111) and Ag(110)	240
	4.	The Reaction Mechanism	242
	5.	Subsurface Oxygen	247
	6.	Promotion	249
		6.1. Promotion by Cs	249
		6.2. Promotion by Cl	251
	7.	The Detailed Kinetics of the Adsorption and Desorption of Ethylene on an Unoxidised and Oxidised Ag/α-Al$_2$O$_3$ Catalyst	254
	8.	The Detailed Kinetics of the Adsorption and Desorption of Ethylene on an Oxidised Ag/α-Al$_2$O$_3$ Catalyst Subtending the α_1-O State Only	255
	9.	The Detailed Kinetics of the Adsorption and Desorption of Ethylene on the α_2-O State Principally	257
	10.	Ethylene Desorption from an Unoxidised and Oxidised Cs-Promoted Ag/α-Al$_2$O$_3$ Catalyst	257
	11.	The Desorption of Ethylene from an Oxidised Cl-Promoted Cs/Ag/α-Al$_2$O$_3$ Catalyst	258

* School of Chemistry, University of Manchester, Manchester M13 9PL, United Kingdom
[†] BP Oil International, 20 Canada Square, Canary Wharf, London E14 5NJ, United Kingdom

12. Overall Conclusions		259
Acknowledgement		261
References		261

Abstract   The reasons for silver being unique in its ability to oxidise ethylene to ethylene oxide (EO) in high selectivity are examined by critically reviewing the literature accumulated over 40 years and by reporting some new, as yet unpublished, data. It is concluded that silver's uniqueness resides in the strength of its bond to adsorbed O atoms. It is shown that the weaker the Ag–O bond is, the more selective it becomes. The role of Cl promoters is shown to weaken the Ag–O bond. In an industrial Ag/α-Al$_2$O$_3$ catalyst containing Cs, the Cs is held in submonolayer quantities on steps on the Ag surface where it prevents the formation of a strong, unselective Ag–O bond. The selective and unselective reactions derive from two nearly equivalently energetic reaction pathways of a common intermediate. It is concluded here that this intermediate is a CH$_2$–CH$_2$–O–Ag species. This undergoes cyclisation to form EO or another form of cyclisation to form an oxametallacycle which produces acetaldehyde and ultimately CO$_2$. Both reactions are concluded to occur on a nearly completely oxidised Ag surface under industrial reaction conditions. Subsurface oxygen is concluded to play no role in promoting the selectivity of the Ag–O surface state and is unselective of itself. It is considered doubtful that subsurface oxygen is present under industrial reaction conditions.

Key Words: Ethylene oxide, Ethylene, Epoxidation, Silver, Cl promotion, Cs promotion, Promotion, Selectivity, Oxametallacycle, Adsorption, Desorption, Chemisorption, Activation energy, Ag–O bond, Reaction mechanism, Oxidation, Cyclisation, Heterogeneous catalysis, Selective oxidation, Eletrophilic oxygen, Nucleophilic oxygen, Subsurface O atoms, Ag/α-Al$_2$O$_3$ catalyst.   © 2008 Elsevier B.V.

1. INTRODUCTION

It is a truism, bordering on a cliché, in heterogeneous catalysis to state that Ag is unique in its ability to epoxidise ethylene to ethylene oxide (EO). While there is no disagreement about the veracity of this truism, there is considerable disagreement and debate as to the means by which it effects this reaction. The overall reaction is superficially simple, comprising only three reactions: (1) the selective oxidation of ethylene to EO (reaction 7.1), (2) the unselective oxidation of ethylene to CO$_2$ and H$_2$O (reaction 7.2) and (7.3) the over-oxidation of EO to CO$_2$ and H$_2$O (reaction 7.3).

$$C_2H_4 + \tfrac{1}{2}O_2 \leftrightarrow C_2H_4O \qquad (7.1)$$

$$C_2H_4 + 3O_2 \rightarrow 2CO_2 + 2H_2O \qquad (7.2)$$

$$C_2H_4O + \tfrac{5}{2}O_2 \rightarrow 2CO_2 + 2H_2O \qquad (7.3)$$

In spite of their apparent simplicity, every nuance of reactions (7.1)–(7.3) has been the subject of heated debate for almost 70 years.

The main areas of contention relate to the following:

1. the nature of the oxidant, molecular or atomic?
2. if atomic, are there electrophilic and nucleophilic oxygen atoms on the surface?
3. the structure of the oxidising Ag surface under reaction conditions, surface oxide on metallic Ag or surface oxide on a subsurface layer of AgO or Ag_2O?
4. the role of subsurface O atoms
5. the interaction of ethylene with any or all of the oxidants listed above
6. the mechanism of the selective and unselective reactions
7. the effect of promoters Cl and Cs on the oxidant and on the mechanism of the selective and unselective reactions.

In this chapter, we shall review the literature on the detailed kinetics of adsorption and desorption of oxygen on Ag with a view to establishing the nature of the selective oxygen and the surface structure of the oxygen overlayer under reaction conditions. We shall review the literature on the bonding of ethylene to O atoms on Ag and will include new data on the detailed kinetics of the bonding of ethylene to unoxidised and oxidised versions of a commercial Ag/α-Al_2O_3 catalyst and to unoxidised and oxidised versions of these catalysts promoted by Cs and Cl separately, and in combination. We shall review the various theories of the mechanism and of the role of the promoters in relation to these postulated mechanisms.

2. THE DETAILED KINETICS OF THE ADSORPTION AND DESORPTION OF OXYGEN ON SILVER AND THE STRUCTURE OF THE OXIDE OVERLAYER AND REACTION CONDITIONS

In an attempt to encapsulate the essence of the uniqueness of silver's ability to catalyse the epoxidation of ethylene, Linic and co-workers [1] state, 'Silver provides a binding environment that is strong enough to dissociate oxygen but is weak enough to allow for facile surface reaction and desorption of EO.' The implicit concept in this statement, that it is the kinetics of the adsorption and desorption of oxygen on silver, which are particular to silver, and which provide it with its unique ability to epoxidise ethylene, will be the dominant theme of this chapter. The importance of the weakness of the Ag–O bond which allows for the facile surface reaction and desorption of EO is undisputed. However, the suggestion that it is the strength of the Ag–O bond which provides the driving force for the dissociation of oxygen on silver, does not accord with the evidence. The heat of the reaction

$$O_2 + (2Ag)_s \leftrightarrow (2Ag-O)_s \qquad (7.4)$$

where s denotes a surface species has been determined to be 105 kJ/mol for a Cl-promoted Ag surface and 123 kJ/mol for an unpromoted Ag surface [2,3].

The O=O bond strength is 494 kJ/mol [4] and so heats of adsorption of between 105 and 123 kJ/mol are insufficient to drive the thermal dissociation of an oxygen molecule whose bond strength is 494 kJ/mol. Therefore, factors other than the thermodynamics of reaction (4) are responsible for the dissociation of oxygen on silver.

Greenler and co-workers studied the adsorption of atomic and molecular oxygen on a polycrystalline silver ribbon using reflection-absorption infrared spectroscopy (RAIRS) and temperature-programmed desorption (TPD) [2,5]. After dosing O_2 on polycrystalline Ag at 90 K (0.2 Torr, 60 s), they found three peaks in the desorption spectrum at 183, 263 and 604 K. These peaks were in reasonably good agreement with those reported by Barteau and Madix on Ag(110) [6], Grant and Lambert [7] and Campbell [8] on Ag(111). The 604 K peak derived from desorption of atomically held oxygen. The 263 K peak derived from the desorption of a molecularly held oxygen species (a peroxo species, O_2^{2-}); it was identified from its vibrational frequency of 983 cm^{-1} and was held perpendicular to the surface. The 183 K peak derived from the desorption of a molecularly held oxygen species; it was held parallel to the surface and had a vibrational frequency of 622 cm^{-1} from which the force constant of 152 N/m for the O–O bond was determined. This value of the force constant corresponds to the transfer of 1.7 electrons to the π^* orbital, giving an O–O bond order significantly less than unity and a substantial weakening of the O–O bond strength. Greenler and co-workers calculated an O–O bond strength of 67 kJ/mol on the assumption that the O–O bond strength correlated with the bond order. This evidence caused Greenler and co-workers to conclude that this molecular oxygen species was held parallel to the surface and was the precursor to dissociative adsorption of oxygen [5]. Heats of adsorption of oxygen on Ag of 105–123 kJ/mol could easily drive the thermal dissociation of the weakened O_2 molecule having a bond strength of 67 kJ/mol.

The uniqueness of silver therefore resides not only in the weakness of the Ag–O bond, but also in its facility for lowering the energy barrier for breaking the O=O bond by donation of electrons to the π^* orbital of the O_2 molecule. Silver has the lowest work function of the group Cu, Ag, Au, having a value of 4.26 eV for polycrystalline silver, Ag(P), and 4.74 eV for Ag(111) compared with values of 4.65 and 4.94 eV for Cu(P) and Cu(111), respectively, and 5.1 and 5.31 eV for Au(P) and Au(111) [9]. It is the combination of the lowest work function combined with the weakness of the Ag–O bond which confers the uniqueness on Ag.

Greenler and co-workers calculated a value of 155 ± 4 kJ/mol for the activation energy for the desorption of the atomically held oxygen by line-shape analysis of the leading edge of the 604 K desorption peak. The atomically held oxygen had a vibrational frequency of 351 cm^{-1} and Greenler and co-workers were able to use this to construct O atom adsorption isotherms by plotting the intensity of the 351 cm^{-1} adsorption peak as a function of O_2 pressure at different temperatures. The isosteric heat of adsorption calculated from these isotherms was 105 ± 15 kJ/mol (per O_2 molecule), a value which was constant over the coverage range 0.25–0.5 of saturation. From this, the activation energy for the dissociative adsorption of oxygen, which is the difference in energies of the desorption activation energy and the heat of adsorption, is 50 kJ/mol. This value is slightly lower than the value of the O=O bond strength of 67 kJ/mol calculated from the oxygen bond order.

Dean and Bowker studied the kinetics of the dissociative adsorption of oxygen on an Ag/α-Al$_2$O$_3$ catalyst prepared by impregnation of α-Al$_2$O$_3$ with AgNO$_3$ [10]. Different oxygen atom coverages of the Ag were produced by varying the adsorption temperature for a constant O$_2$ dose, and by varying the O$_2$ dose for a constant adsorption temperature. They calculated the sticking probabilities from the isotherms produced and activation energies to dissociative adsorption of oxygen of 16 kJ/mol at 0.05 of saturation coverage rising to 50 kJ/mol at 0.5 of saturation from the temperature dependence of the sticking probabilities.

The activation energy of desorption determined by line-shape analysis of the leading edge of the desorption peak varied from 117 kJ/mol ($\theta = 0.05$) to 150 kJ/mol ($\theta = 0.5$). A desorption activation energy of 130 ± 25 kJ/mol was obtained by varying the heating rate and plotting ($2 \ln T_m - \ln \beta$) against $1/T$. The value of the oxygen-sticking probability was 10^{-6} at 300 K from which the authors concluded that the surface of their Ag/α-Al$_2$O$_3$ catalyst was almost exclusively Ag(111) [10].

Atkins and co-workers studied the detailed kinetics of the adsorption of oxygen on a Ag/α-Al$_2$O$_3$ catalyst prepared from a silver oxalate precursor [3,11]. They found two states in the oxygen desorption spectrum at 523 K (250 °C) and 573 K (300 °C) (designated hereafter as the α_1 and α_2 states, respectively) after having dosed pure O$_2$ (1 bar) for 1 h on to the catalyst which was held at 513 K (240 °C). The high-temperature state was lost on heating to 773 K (500 °C) for the first desorption, but the low-temperature state was unaffected by this heating (Fig. 7.1).

The authors determined the activation energy to adsorption to the α_1 state by measuring the coverage of that state after having dosed pure oxygen on to a catalyst subtending only that state for 5 min at 353 K (80 °C), 383 K (110 °C), 423 K (150 °C) and 453 K (180 °C) [3] (Fig. 7.2).

An activation energy to adsorption of 17 ± 1 kJ/mol was obtained from a plot of the logarithm of the coverage versus the reciprocal of the adsorption temperature (Fig. 7.2). This value is the same low-coverage activation energy to adsorption obtained by Dean and Bowker [10].

Line-shape analysis of the desorption peaks shown in Fig. 7.2 was carried out by Atkins and co-workers from a plot of ln (desorption rate/coverage^2) versus $1/T$ from which a value of 140 ± 1 kJ/mol was obtained for the desorption activation energy [3]. The α_1 peak, therefore, derived from desorption of oxygen from the Ag (111) face and the heat of adsorption of oxygen on that face is 123 kJ/mol.

Atkins and co-workers simulated the oxygen desorption spectrum containing the 523 and 573 K peaks using the desorption activation energies 140, 145 and 155 kJ/mol by fitting to the peak maximum temperatures and the full-width at half-maximum. The 140 and 145 kJ/mol desorption activation energies are the literature values for oxygen desorption from Ag(111) and Ag(110) [12–14]. The 155 kJ/mol is the value used to fit the 573 K desorbing peak and is thought to derive from oxygen desorbing from a stepped surface. The simulation required areas of the silver surface corresponding to these faces to be included. These are

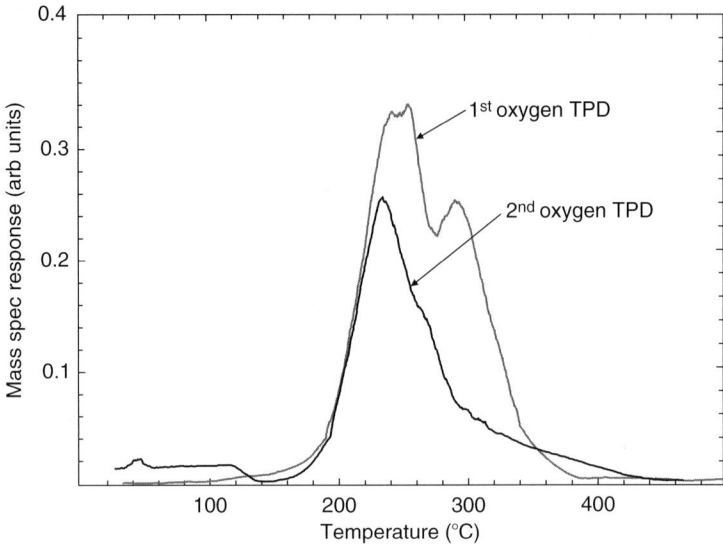

FIGURE 7.1 The temperature-programmed desorption (TPD) spectra of O_2 from an Ag/α-Al_2O_3 catalyst: O_2 dosing, O_2 101 kPa, 25 cm^3/min, 1 h, 513 K.

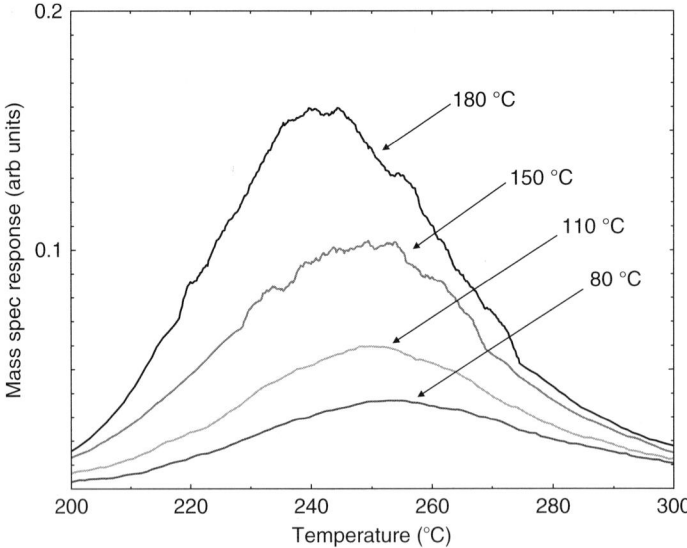

FIGURE 7.2 The temperature-programmed desorption (TPD) spectra of O_2 from an Ag/α-Al_2O_3 catalyst, having dosed O_2 (101 kPa, 25 cm^3/min) for 5 min at the temperatures shown on the figure.

listed in Table 7.1 from which it can be seen that the original morphology of the Ag/α-Al_2O_3 catalyst is Ag(111) (1.1 m^2/g), Ag(110) (0.2 m^2/g) and Ag (stepped) (0.7 m^2/g). Simulation of the morphology of the catalyst that had lost the 573 K

TABLE 7.1 Desorption activation energies, pre-exponential terms and surface areas used as inputs for oxygen TPD simulations

Peak	Ag(111)	Ag(110)	High index face/defect sites
Activation energy to desorption (kJ/mol)	140	145	155
A-factor (cm^2/molecule·s)	2×10^{-3}	2×10^{-3}	2×10^{-3}
First desorption (m^2/g)	1.1	0.2	0.7
Second desorption (m^2/g)	0.96	0.18	0.06

peak was achieved with the same areas of the Ag(111) and Ag(110) faces, but with virtually no area corresponding to the stepped face [3].

Norskov and co-workers have calculated the adsorption activation energy and the heat of adsorption of oxygen on Ag(111) using density functional theory (DFT) [15]. They obtained a value of 145 kJ/mol for the adsorption activation energy and a value of 28 kJ/mol for the heat of adsorption, corresponding to a desorption activation energy of 173 kJ/mol. These values do not accord with those found experimentally, suggesting that some refinement of the method is required.

The EO is produced commercially using ethylene/oxygen mixtures in a ratio of 3:1 or 4:1 at 513–543 K (240–270 °C) and between 15 and 25 atm [16]. The fractional coverage (θ_O) of the silver by oxygen atoms under operating conditions can be calculated from these data and the heat of adsorption of 123 kJ/mol using the Langmuir isotherm

$$\theta_O = \frac{K[O_2]^{1/2}}{1 + K[O_2]^{1/2}} \tag{7.5}$$

where $[O_2]$ (mol/cm^3) is the concentration of oxygen in the gas phase and $K = A\,e^{\Delta H/RT}$ is the adsorption equilibrium constant (ΔH is the heat of adsorption and A is the A-factor ratio). The value of θ_O so obtained is unity. Under steady-state industrial conditions, the silver surface will be virtually saturated with adsorbed oxygen atoms. There will be some removal of the adsorbed oxygen atoms by reaction with ethylene. However, using an activation energy of 63 kJ/mol for the formation of EO [1], it is possible to calculate a reactive sticking probability of ethylene with the oxidised surface of $\sim 10^{-7}$ at 500 K. This should be compared with a dissociative sticking probability of O_2 of $\sim 10^{-6}$ [10] and so the adsorbed oxygen atoms that are removed by reaction will be rapidly replenished by adsorption and so will maintain the silver surface in a near saturated oxidised state. The reaction conforms essentially to a Mars and van Krevelen mechanism [17].

Campbell and Paffett noted a new state desorbing at 565 K when they dosed O_2 (50 Torr) on to Ag(110) at 485 K [18]. This state was in addition to the one at 605 K observed when O_2 (1,900 L) was dosed on to Ag(110) at 135 K. A c(6 × 2)–O overlayer accompanied the formation of the 565 K desorption peak. When the 565 K peak was removed by flashing to 575 K, a p(2 × 1)–O low energy electron diffraction (LEED) pattern, observed when the 605 K peak alone existed, was restored.

Couves and co-workers dosed O_2 (1 bar, 25 cm^3/min) at 513 K for 1 h on to the Ag(111) face of a sintered Ag/α-Al$_2$O$_3$ catalyst. They observed one dominant peak at 523 K, corresponding to oxygen desorption from Ag(111) with a high temperature trailing edge resulting from the 5% coverage of the Ag surface by Ag(110) [11]. [The difference in the peak maximum temperatures reported by Campbell and Paffett and those of Couves and co-workers is a result of the different heating rates employed—11 K/s (Campbell and Paffett) and 0.17 K/s (Couves).] There were no low temperature desorption states resulting from a $c(6 \times 2)$–O overlayer. Couves and co-workers dosed the O_2 on to the Ag/α-Ag$_2$O$_3$ catalyst at conditions virtually identical to those operating in the industrial reactor. Couves and co-workers dosed the oxygen on at 1-bar pressure and at 513 K, while industrial operating conditions dose the oxygen on at the same temperature but at 3 bar partial pressure and so the oxidised state of the Ag observed by Couves and co-workers will probably be the same as that existing on the Ag surface under industrial reaction conditions. It is probable that under industrial reaction conditions, the Ag will be [Ag(111)] covered with a $p(2 \times 1)$–O overlayer. Additionally, Couves and co-workers saw no evidence of subsurface oxygen from their dosing method. Indeed, Atkins and co-workers found that it required dosing O_2 (1 bar, 25 cm^3/min) for 1 h at temperatures in excess of 653 K to populate the subsurface states of adsorbed oxygen [3]. These temperatures are far in excess of those used industrially and so the catalyst operating industrially is unlikely to have any subsurface oxygen, in contradiction to the generally held view. The slightly higher oxygen partial pressure (3 bar) used industrially at 513 K is unlikely to force dissolution of O atoms in the Ag when 1-bar pressure of O_2 at 513 K used by Couves and co-workers does not.

3. THE ADSORPTION OF ETHYLENE AND ETHYLENE OXIDE ON CLEAN AND OXIDISED Ag(111) AND Ag(110)

The adsorption of ethylene on clean and oxidised Ag(111) and Ag(110) has been studied by a variety of surface science techniques, TPD, RAIRS, electron energy loss spectrometry (EELS), LEED, ultraviolet photoelectron spectrometry (UPS) and work function measurements [19–21]. Tysoe and co-workers found that ethylene adsorbed weakly and molecularly when dosed on to Ag(111) at 80 K [19]. The peak maximum temperature for its desorption was 138 K, corresponding to a desorption activation energy of 39 kJ/mol. The desorption activation energy increased to 46 kJ/mol on a Ag(111) surface predosed with O atoms ($\theta_0 = 0.1$). Kruger and Benndorf found a desorption activation energy of 42 kJ/mol for ethylene adsorption on Ag(110) [20]. Backx and co-workers, using EELS, found that ethylene adsorbed parallel to the surface of clean Ag(110) and that the amount adsorbed increased by a factor of 13 when ethylene was adsorbed on oxidised Ag(110) [21].

EO desorbs intact at a peak maximum temperature of 158 K (45 kJ/mol) when dosed on to clean Ag (111) at 80 K. It is held with the molecular plane held perpendicular to the surface [19]. The desorption activation energy of EO increased to 51 kJ/mol for EO dosed at 80 K on to oxidised ($\theta_0 = 0.1$) Ag(111). When EO is dosed on to oxidised Ag(111) at 300 K, an acetaldehyde species is formed on the surface, the infrared features of which decrease coincidentally with acetaldehyde/EO desorption [19].

In the opinion of the authors, one of the most important and illuminating papers on the nature of the bonding of ethylene to an oxidised silver surface is that of Cant and Hall [22]. Although the paper was a microreactor study of the kinetics and mechanism of ethylene epoxidation over a silver sponge, by using *cis*- and *trans*-1,2-d$_2$-ethylene as a reactant and by their finding of 92% equilibration of the isomers in the EO product, the authors showed, quite unequivocally, that at some stage in the reaction coordinate in the formation of EO, the C=C double bond is broken and there is near free rotation about the C–C single bond formed. (Cant and Hall considered that the lack of complete *cis–trans* equilibration was probably due to interference with the catalyst surface. However, the adsorbed ethylene has probably formed H-bonds to the oxygen overlayer of the catalysts formed under reaction conditions and this would inhibit, or cause an energy barrier to, free rotation.) Since the prevailing view at the time was that it was adsorbed molecular oxygen (O_2^-) which was the oxidant [23], Cant and Hall proposed that it was the ethylene peroxy species (CH_2–CH_2–O–O–Ag) which was responsible for the *cis–trans* isomerisation. This intermediate was also considered to be common to the formation of both EO and CO_2 [22].

In an equally unambiguous paper, Grant and Lambert showed that it was atomically held oxygen which was responsible for the epoxidation of ethylene and that the molecularly held species, though present, was merely a spectator [24]. The intermediate responsible for isomerisation was, therefore, CH_2–CH_2–O–Ag, and Grant and Lambert stated as much in their paper [24]. They produced the reaction mechanism in their paper, in which they proposed that 'selective oxidation results from an electrophilic attack by an $O_{(a)}$ on the olefinic π-bond'.

The sequence above, however, was ambiguous and could be taken to suggest that the reaction $O_{(a)}$ with ethylene was concerted even though in the paper they stated that it was 'consistent with the observation of no retention of configuration upon epoxidation of either *cis*- or *trans*-1,2-d$_2$-ethylene' [24].

The reaction of an adsorbed O atom with the C atom at one end of an ethylene molecule has a precedent in homogeneous gas-phase chlorination reactions of olefins. It was originally thought that in the photo-induced chlorination of olefins, the first Cl atom added symmetrically to the double bond to form the Cl equivalent of EO [25].

$$\underset{Cl}{\overset{H}{>}}C=C\underset{H}{\overset{Cl}{<}} + Cl \longrightarrow \underset{Cl}{\overset{H}{>}}C\underset{Cl}{\overset{}{-}}C\underset{H}{\overset{Cl}{<}}$$

However, in the chlorination of *cis*- and *trans*-1,2-dichloroethylene, Knox and co-workers showed that decomposition of the hot radical produced by the addition of the first Cl atom to the olefin produced an equilibrium mixture of *cis*- and *trans*-1,2-dichloroethylene [26,27]. The hot radical had a lifetime of $\sim 10^{-8}$ s while the rotational lifetime was $\sim 10^{-12}$ s. The first Cl atom therefore added on to the end carbon atom of the olefin, allowing free rotation of the single bond produced. The only hindrance to free rotation in the hot radical was in the chlorination of tetrachloroethylene when steric hindrance of the five chlorine atoms prevented it [26,27]. A similar form of restricted rotation obviously applies in the epoxidation reaction where, in the epoxidation of *cis*- and *trans*-1,2-d$_2$-ethylene, the product EO is only 92% equilibrated. Cant and Hall suggested that the interference may have been due to interaction with the surface [22], but as we have argued above, this could also be due to H-bonding of the adsorbed intermediate to the surface Ag–O.

4. THE REACTION MECHANISM

Barteau with numerous co-workers has made a significant contribution to our understanding of the mechanism of the epoxidation of ethylene over silver in a body of work which spans over a quarter of a century since his original publications with Madix in 1980. His most recent papers relate to the nature of the intermediates formed on the surface of the silver and their role in the selective and unselective oxidation of ethylene [28–34]. Linic and Barteau investigated the adsorption and decomposition of EO on Ag(111) using TPD, high-resolution electron energy loss spectroscopy (HREELS) and DFT [28]. Dosing EO on to Ag (111) at 130 K resulted in its being adsorbed molecularly; it desorbed intact at 200 K, on temperature programming, corresponding to a desorption activation energy of 59 kJ/mol. However, dosing EO on to Ag(111) at 250 K produced a surface intermediate from which, on temperature programming, EO (principally), ethanol and water evolved coincidently at 300 K with ethylene (larger than the H_2O and C_2H_5OH peaks) evolving at 325 K [28]. Two intermediates were considered to have been possible sources for the evolution of the EO, C_2H_5OH and H_2O. They were the species singly bonded to the Ag, intermediate 1, or the species held on the Ag with two bonds, intermediate 2.

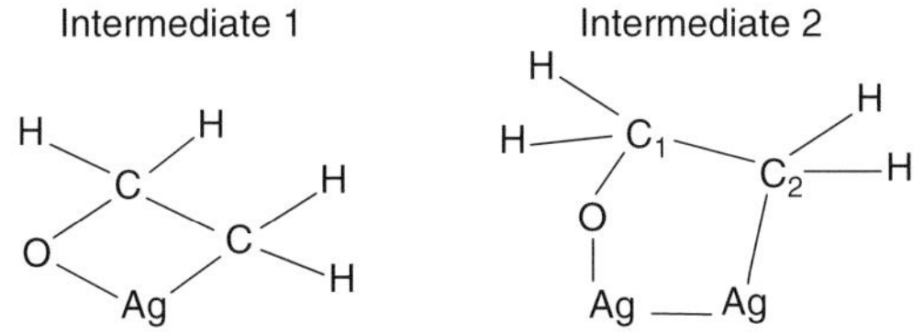

Intermediate 1

H₂C-CH₂ bridging O-Ag (structure diagram)

Intermediate 2

H₂C₁-C₂H₂ with O bonded to C₁ and Ag-Ag (structure diagram)

DFT calculations of the vibrational frequencies of these adsorbed species produced spectra of which those deriving from intermediate 1 more closely resembled the HREEL spectrum of the surface. The authors calculate an activation energy of 55 kJ/mol for cyclisation of intermediate 1 from the peak maximum temperature of 300 K for EO desorption using the Redhead equation [35] and an assumed value of 10^{13} s^{-1} for the desorption pre-exponential term and a heating rate of 1.8 K/s. (In their later paper in reference to this work, they state that the value is 77 kJ/mol [1].)

Experimentally, Couves *et al.* determined an activation energy of 60 kJ/mol for EO production by line-shape analysis of the EO peak produced by temperature-programmed reaction of ethylene with O atoms adsorbed on the Ag(111) face of a Ag/α-Al₂O₃ catalyst [11]. Grant and Lambert obtained an activation energy of 45 kJ/mol for the selective oxidation of ethylene to EO on Ag(111) by steady-state rate measurements using a high pressure reaction cell coupled to a ultra-high vacuum chamber [24]. In the same experiment, they found an activation energy of 50 kJ/mol for the unselective reaction [24].

In a follow-up paper, Linic and Barteau constructed a reaction coordinate and a microkinetic model for ethylene epoxidation on silver from DFT calculations [1]. The calculations were based on an Ag₁₅ cluster to represent the silver surface. The reaction coordinate produced had TS1 as the species formed by the initial interaction of ethylene with an adsorbed oxygen atom. This is in agreement with species proposed by Grant and Lambert [24], by Force and Bell [36] and by the analysis presented earlier in this chapter for the initial interaction of ethylene with an oxidised silver surface.

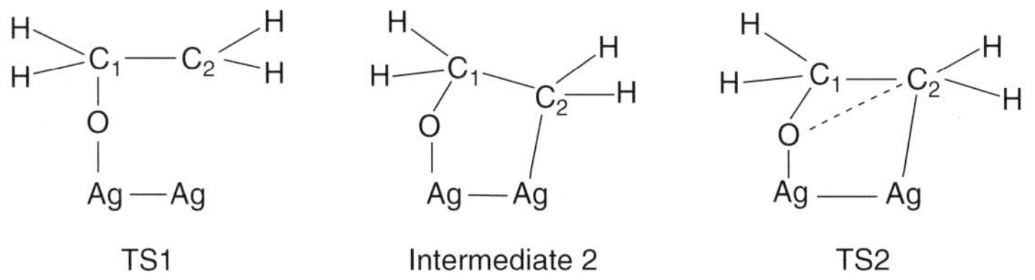

TS1 **Intermediate 2** **TS2**

(three structural diagrams of reaction intermediates)

The calculations then predicted that the next step along the reaction coordinate was the formation of the stable oxametallacycle, intermediate 2. The final species formed before EO is TS2 in which the Ag–O bond is stretched and the O–C₂ bond is beginning to be formed. The authors calculate an activation energy of 67 kJ/mol for ring closure of intermediate 1 in reasonably good agreement with the experimental values of Couves and co-workers [11] and Grant and Lambert [24]. The DFT calculations predicted that the unselective route is 1,2-H migration in intermediate 2 to form acetaldehyde whose oxidation to methyl acetate and ultimately to CO_2 is considered to be facile [29,37]. The difference in the values of the activation energy barriers for TS2 and the intermediate forming acetaldehyde is calculated to be 1.3 kJ/mol that would predict selectivities to EO of 41–43% at 400–500 K [29]. These values are in good agreement with those found industrially.

Linic and Barteau calculated that the heat of adsorption of intermediate 2 was greater than that of intermediate 1 and justified its inclusion in the reaction coordinate even though intermediate 1 was the preferred structure for EO adsorption and decomposition by stating that the energy difference between the adsorbed species is so small as to make the two interchangeable [1]. Even so, when the oxametallacycle (intermediate 2) is synthesised on the surface of Ag(110) or Ag(111) by adsorption of 2-iodoethanol, acetaldehyde and no EO is produced on temperature programming. The authors explain that this is because of the I atoms which are co-adsorbed with the oxametallacycle. However, Lambert and co-workers found that the adsorbed I atoms had a mild promoting effect in ethylene epoxidation [38] and so its co-adsorption with the oxametallacycle (intermediate 2) is unlikely to have a detrimental effect on the unimolecular cyclisation of that molecule to form EO. It is entirely possible that intermediate 2 is not involved in the selective reactive pathway and that intermediate 2 is the intermediate which results in total oxidation. Another point relating to these DFT calculations [1] is that they are carried out on an Ag_{15} slab. The kinetic and thermodynamic calculation carried out here predict that the surface of the $Ag/\alpha\text{-}Al_2O_3$ catalyst at steady state is fully oxidised, calling the validity of DFT calculations on Ag_{15}, as an indicator of the structure of the intermediates under realistic operating conditions, into question.

Couves and co-workers studied the mechanism of ethylene epoxidation on an oxidised $Ag/\alpha\text{-}Al_2O_3$ catalyst by temperature-programmed reaction of ethylene with the adsorbed O atoms [11]. The reaction of ethylene with an oxidised catalyst subtending saturation levels of the 523 K (250 °C) (the α_1 state) and 573 K (300 °C) (the α_2 state) desorbing O species (Fig. 7.1), produced two peaks at 373 (100 °C) and 473 K (200 °C) for the co-evolution of EO and CO_2 (Fig. 7.3). The selectivity of the 373 K peak to EO was 57% while that of the 473 K peak was 33%. (Both selectivities are corrected for the contribution of the EO cracking fraction to the $m/z = 44$ peak.)

The authors demonstrated that it was the more weakly held α_1 state which was the more selective in the epoxidation of ethylene in two experiments. They first desorbed most of the α_1 state by heating a catalyst dosed with both the α_1 and α_2 states to 503 K for 10 min in He, leaving the α_2 state unaffected and the α_1 state reduced to 30% of its original value (Fig. 7.4).

They then reacted the catalyst with ethylene and found a significant reduction in the 373 K (100 °C) reacting peak while the 473 K (200 °C) peak was unaffected (Fig. 7.5).

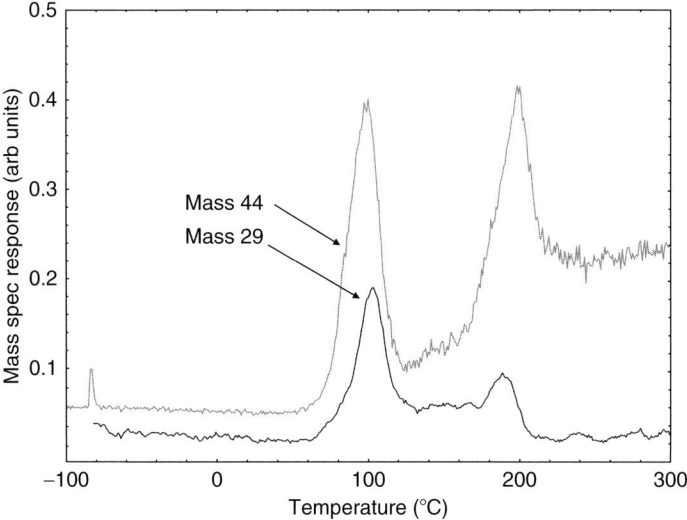

FIGURE 7.3 The temperature-programmed reaction of *ethylene* with an oxidised Ag/α-Al$_2$O$_3$ catalyst subtending the α_1-O and the α_2-O states on the surface: $m/z = 29$ is EO; $m/z = 44$ is CO$_2$ and 0.5 EO.

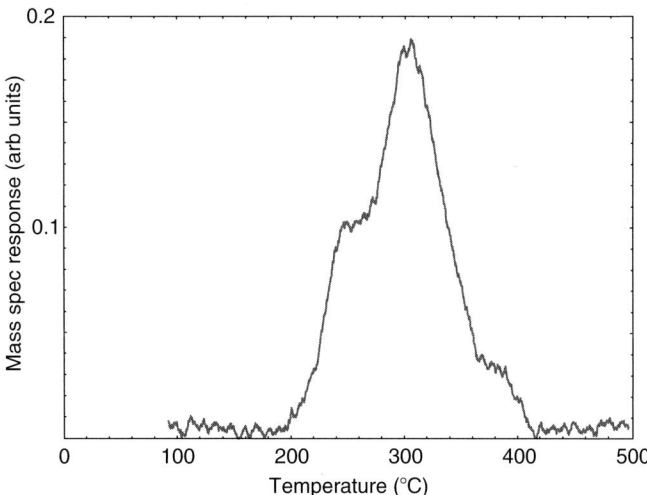

FIGURE 7.4 The temperature-programmed desorption (TPD) spectrum of O$_2$ from the Ag/α-Al$_2$O$_3$ catalyst subtending the α_2-O state principally on the surface.

Close examination of Fig. 7.5 reveals that the temperatures of the two EO/CO$_2$ co-existing peaks are now 363 K (90 °C) and 463 K (190 °C) (10 K lower than that in Fig. 7.3). The selectivity of the higher temperature peak remains the same at 34% while that of the 363 K peak is now 75%. The low coverage of the adsorbed O atoms on the Ag(111) face (~0.15 ML) has increased the selectivity to EO

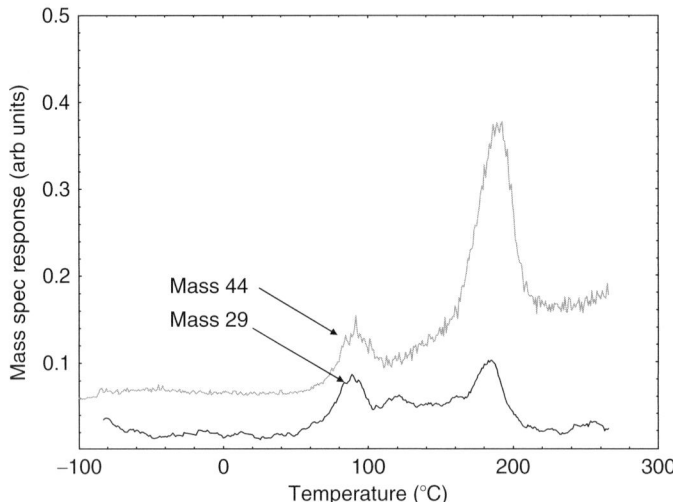

FIGURE 7.5 The temperature-programmed reaction of *ethylene* with an oxidised Ag/α-Al$_2$O$_3$ catalyst subtending the α$_2$-O state principally on the surface: $m/z = 29$ is EO; $m/z = 44$ is CO$_2$ and 0.5 EO.

apparently by lowering the activation energy to cyclisation evidenced by the lowering of the peak maximum temperature for EO/CO$_2$ co-evolution. This contradicts other studies which claim that higher coverages of the silver by O atoms result in higher EO selectivities [37].

The second experimental method used to demonstrate that the 373 K EO/CO$_2$ co-evolving peak derived from reaction of ethylene with the α$_1$-O state was to react ethylene with an Ag/α-Al$_2$O$_3$ catalyst containing only the α$_1$ state by dosing O$_2$ on to a sintered catalyst produced by heating the catalyst to 773 K (Fig. 7.1). The temperature-programmed reaction spectrum obtained is shown in Fig. 7.6. EO and CO$_2$ co-evolve at 363 K (90 °C) with a selectivity to EO of 50%, a value which is essentially unchanged from that obtained by reaction of ethylene with a catalyst containing both the α$_1$ and α$_2$-O states [11].

Couves and co-workers therefore reached the following important conclusions: (1) the O atoms adsorbed on Ag(111) and on a stepped surface, Ag surface which co-existed on their catalyst react independently of each other with ethylene, (2) the α$_1$ state (the weaker Ag–O bond) is more selective in the reaction with ethylene to form ethylene epoxide; generally then since the weaker Ag–O bond is more selective, it would appear possible that higher selectivities could be achieved by including additives in/on the Ag which would weaken the Ag–O bond strength and (3) the co-evolution of EO and CO$_2$ at nearly identical onset and peak maximum temperatures (EO may have a lower onset temperature of 5–10 K and a lower peak maximum of 5 K than CO$_2$) indicates that the formation of EO and CO$_2$ derives from two reaction pathways available to a common intermediate; the activation energies for these reaction pathways are therefore nearly the same with that for CO$_2$ formation being slightly higher (by between 5 and 10 kJ/mol)

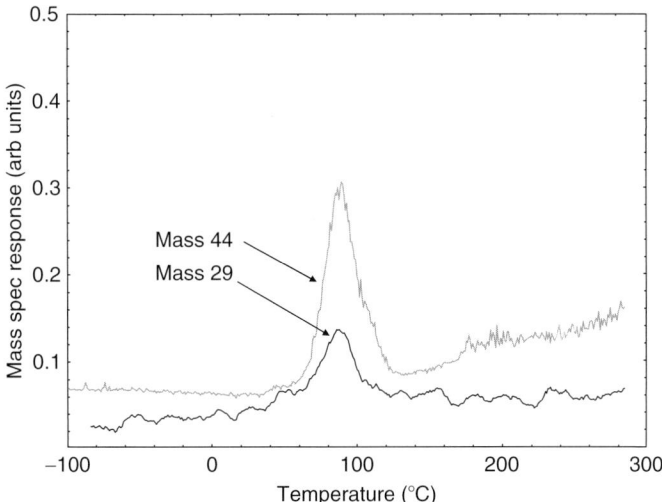

FIGURE 7.6 The temperature-programmed reaction of *ethylene* with an oxidised Ag/α-Al$_2$O$_3$ catalyst subtending the α$_1$-O state only: $m/z = 29$ is EO; $m/z = 44$ is CO$_2$ and 0.5 EO.

than the activation energy for EO formation, these values being estimated from 5 to 10 K difference in the onset and peak maximum temperatures. (Grant and Lambert found activation energies of 45 and 50 kJ/mol for the selective and unselective reactions, respectively [24], in agreement with the estimate of Couves and co-workers.)

Another point to note is that whereas the formation of EO from TS1 or one of the oxametallacycles intermediate 1 or 2 involves a simple unimolecular rearrangement, the total oxidation of the species formed by a slight change in the structure of one of these intermediates, requires six O atoms. It must involve reactions which have a low activation energy and which occur simultaneously rather than sequentially, otherwise EO and CO$_2$ would not have been found to co-evolve in the temperature-programmed reaction experiments [3, 11]. Therefore, the slightly changed version of TS1 or intermediates 1 or 2 which gives rise to CO$_2$ must react rapidly with the sea of surface oxygen atoms which exist on the surface of the Ag under reaction conditions.

5. SUBSURFACE OXYGEN

It has been claimed that subsurface oxygen is essential for selective oxidation of ethylene on silver but that it is not required for total oxidation [24,39]. Atkins and co-workers tested this claim by dosing O$_2$ (101 kPa) on to their Ag/α-Al$_2$O$_3$ catalyst for 1 h (25 cm^3/min) at 633 K (360 °C), 693 K (420 °C), 753 K (480 °C) and 813 K (540 °C), producing 1.6, 2.0, 2.3 and 2.8 monolayers of subsurface O atoms [3]. (These figures were obtained using a value of 1.19×10^{15} atom/cm^2 for the surface Ag atom density and an Ag area of 1 m²/g.) This amount of subsurface oxygen was measured by TPD and is shown in Fig. 7.7.

The figure shows large zero-order peaks originating at ~590 K (317 °C) for all adsorption temperatures and maximising at ~760 K (487 °C) for the lowest coverage and at 873 K (600 °C) for the highest coverage. This was the oxygen desorbing from the subsurface O atoms [3]. However, Atkins and co-workers found that the surface oxygen peak maximum temperature was unaffected by the existence of the subsurface oxygen atoms. Its value remained fixed at 520 K (247 °C), regardless of the amount of subsurface oxygen in the range of 1.6–2.8 monolayers.

Since the surface Ag–O bond strength was unaffected by the presence of subsurface oxygen, it was predicted that the selectivity of the surface oxygen in its reaction with ethylene would also be unaffected. Atkins and co-workers confirmed this by reacting ethylene with an Ag/α-Al$_2$O$_3$ catalyst containing surface O atoms and 2.3 ML of subsurface O atoms [3]. The temperature-programmed reaction spectrum they obtained is shown in Fig. 7.8, showing one peak for the co-evolution of EO and CO$_2$ at 407 K (134 °C). This temperature is ~30 K higher than that observed with no subsurface oxygen, indicating that subsurface O atoms may be deleterious for catalyst activity. The reaction of ethylene with the subsurface oxygen, however, was found to be completely unselective, producing only CO$_2$ and H$_2$O, the temperature dependence of which followed that of O$_2$ evolution from the subsurface regions.

The authors therefore concluded that subsurface oxygen was not essential for the selective oxidation of ethylene on an Ag/α-Al$_2$O$_3$ catalyst. Its presence had no effect on the activity or selectivity of the surface O atoms [3].

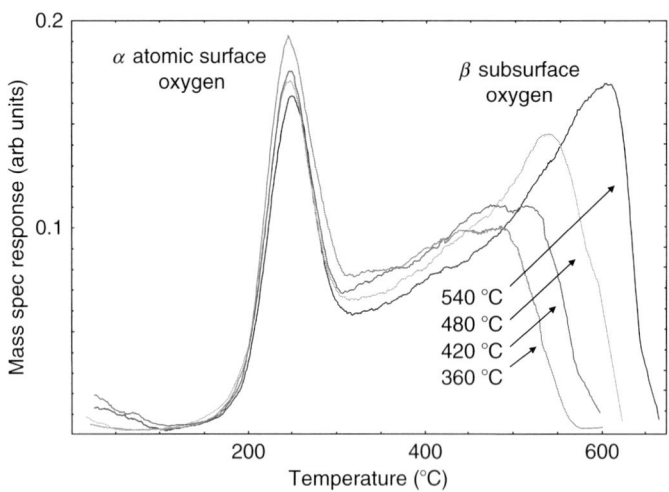

FIGURE 7.7 The temperature-programmed desorption (TPD) spectra of an oxidised Ag/α-Al$_2$O$_3$ catalyst containing 1.6, 2.0, 2.3 and 2.8 monolayers of subsurface O atoms produced by dosing O$_2$ (101 kPa, 25 cm³/min) for 1 h at the temperatures indicated.

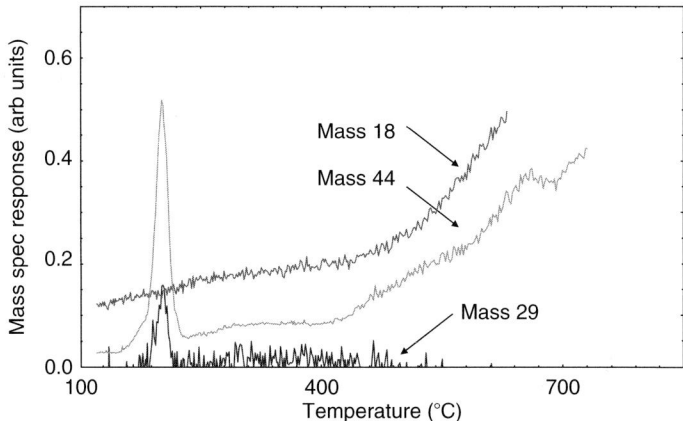

FIGURE 7.8 The temperature-programmed reaction of *ethylene* with an oxidised Ag/α-Al$_2$O$_3$ catalyst containing both surfaces and 2.3 ML of subsurface O atoms: $m/z = 29$ is EO; $m/z = 44$ is CO$_2$ and 0.5 EO; $m/z = 18$ is H$_2$O.

6. PROMOTION

In the absence of additives, Ag/α-Al$_2$O$_3$ catalysts typically produce selectivities in the conversion of ethylene to EO of ~50%. The addition of the promoters, Cs and Cl in combination, increases the selectivity to ~80–85%. The use of Cl as a promoter alone increases the selectivity from 50 to ~75% while their combined use raises the overall selectivity to ~85%. The Cl promoter is added continuously from the gas phase. The Cs promoter is added during the catalyst preparation. We shall deal first with Cs promotion because of its simplicity and because it allows for an unambiguous definition of the role of the Cl promoter.

6.1. Promotion by Cs

The Cs-promoted catalyst studied by Atkins and co-workers was prepared by adding CsOH solution to the Ag/ethylene diamine liquid. The α-Al$_2$O$_3$ support was added to the Cs/Ag/ethylene diamine liquid and the material was dried overnight at 373 K. The catalyst so produced was 10 wt% Ag on Al$_2$O$_3$ with a Cs loading of 300 ppm w/w [3].

The effect of that Cs loading was studied by O$_2$ TPD and by ethylene temperature-programmed reaction [3]. The O$_2$ desorption spectrum obtained by Atkins and co-workers for the Cs/Ag/α-Al$_2$O$_3$ catalyst is shown in Fig. 7.9, lower curve. The upper curve with two peak maxima at 523 K (250 °C) and 573 K (300 °C) is that shown previously for O$_2$ desorption from a fresh, unpromoted Ag/α-Al$_2$O$_3$ catalyst (Fig. 7.1).

Adding Cs to the Ag/α-Al$_2$O$_3$ catalyst was found to have lowered the amount of O$_2$ desorbing at 573 K by a factor of 3. The α$_1$ O state was unaffected. Since the 573 K O$_2$ peak was considered to derive from oxygen desorption from a stepped

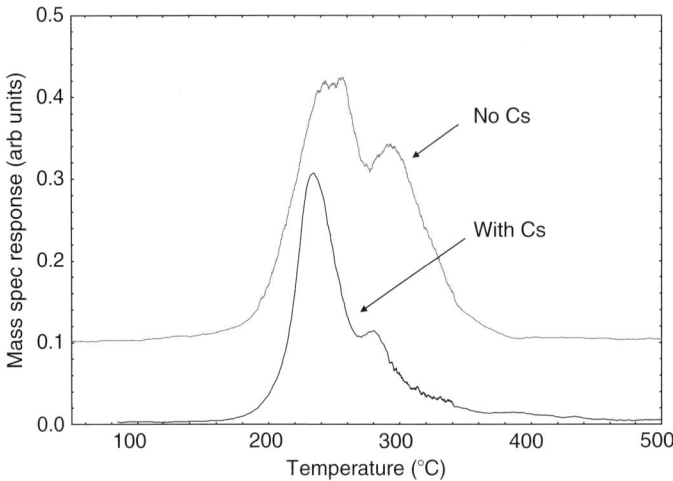

FIGURE 7.9 The O_2 temperature-programmed desorption (TPD) spectrum from a Cs-promoted Ag/α-Al_2O_3 catalyst and from a Cs-free Ag/α-Al_2O_3 catalyst.

Ag surface, it was concluded that the Cs was bound to the stepped surface. The Cs loading was 300 ppm w/w corresponding to 1.4×10^{18} atom/g. The area of the stepped surface had been estimated to be 0.7 m^2/g [11] and so the coverage of the stepped surface by Cs was 2×10^{14} atom/cm. This coverage is slightly less than saturation and so accounts for the small residual 570 K peak observed [11]. An interesting side effect of the Cs doping was that it stabilised the surface morphology of the Ag, evidenced by the second O_2 desorption being identical to the first, whereas in the absence of Cs, the α_2-O desorption peak was virtually lost after the first desorption [11].

The ethylene temperature-programmed reaction spectrum of an oxygen-covered Cs/Ag/α-Al_2O_3 catalyst produced a peak for the co-evolution of EO and CO_2 at 373 K (100 °C) with a selectivity to EO of 44%. The Cs had had no effect on the kinetics of desorption of oxygen from Ag(111) nor on the amount of oxygen desorbing from that surface, nor on its selectivity in its oxidising ethylene to EO (Fig. 7.10).

There is a small CO_2 peak at 473 K (200 °C) in Fig. 7.10 with no discernable coincident EO evolution. The CO_2 peak had been lowered to 20% of that on the unpromoted Ag/α-Al_2O_3 catalyst, commensurate with the Cs blocking roughly 70% of the stepped surface.

Barteau and co-workers used the concept that it was the difference in energy for the formation of the transition state leading to the cyclisation of intermediate 2 relative to that for 1,2-H shift to calculate the effect of additives which would increase the energy difference in favour of the cyclising transition state [30–34]. The DFT calculations predicted that the addition of Cs would increase selectivity by a through-space interaction. These calculations also predicted that the addition of Cu would increase selectivity by a through-surface interaction [33] and this was

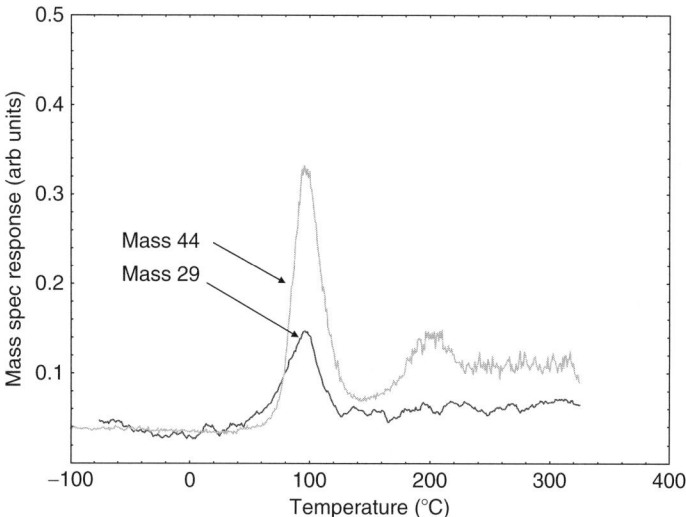

FIGURE 7.10 The temperature-programmed reaction spectrum of *ethylene* with a surface-oxidised Cs/Ag/α-Al$_2$O$_3$ catalyst ($m/z = 29$) is EO; $m/z = 44$ is CO$_2$ and 0.5 EO.

indeed found to be the case [32]. Indeed, a Cs- and Cl-promoted Cu/Ag bimetallic catalyst was found to be the most selective [33]. Therefore, even though the calculations are based on the premise of an intermediate adsorbed on Ag metal and are therefore inapplicable under operating conditions where the surface of the catalyst will be virtually totally oxidised; nevertheless, they have produced a useful outcome.

6.2. Promotion by Cl

The principal promoter used in EO production is the Cl atom. It is introduced to the Ag by adding vinyl chloride or dichloroethane in ppm quantities to the feed. Its use raises the selectivity from ~50 to ~75%. There is a story which has circulated for many years about how it was discovered. It may be apocryphal, but it does have a ring of truth. In the 1940s, Union Carbide, which was the original company to manufacture EO by air oxidation over an Ag catalyst, noticed that, on occasion, the selectivity improved from ~50 to ~70%. The plant manager correlated this improved selectivity to the wind direction, noting that the improved selectivity occurred when the wind came from the direction of a vinyl chloride plant. Obviously, vinyl chloride in ppm quantities was being introduced to the reactor in the air used for the oxidation.

Until Atkins and co-workers published that Cl atoms lowered the O$_2$ desorption peak maximum temperature from 513 K (240 °C) to 481 K (208 °C) [3], there had been nothing published linking the effect of Cl atoms on any measurable physical property of the catalyst. Generally, the comments were nuanced along the lines that, because Cl atoms (which were considered to be on the surface of the Ag) were more electronegative than O atoms, they withdrew electrons from the surface O atoms in their vicinity, causing these O atoms to be more electrophilic [1,24]. O atoms outside the sphere of influence of these surface Cl atoms were regarded as nucleophilic and so were involved in H abstraction and in the ultimate formation of H_2O and CO_2 [24]. Indeed, this formed a highly plausible description of the role of the Cl promoter and of the mechanism of the reaction presented by Grant and Lambert [24]. Lambert and co-workers extended the idea of Cl causing O to become more electrophilic by determining the promoting effect of the other halogens F, Br and I. They found that the promoting effect was directly related to the electron affinity (EA) of the halogen with Cl (EA = 350 kJ/mol) being the most promoting, followed by F (EA = 330 kJ/mol) and Br (EA = 320 kJ/mol) and finally I (EA = 290 kJ/mol) [38]. They concluded that valence charge withdrawal from $O_{(a)}$ by the co-adsorbed halogen enhanced the electrophilicity of the $O_{(a)}$ and therefore promoted its effectiveness as an epoxidising agent [38]. They did not measure the effect of the halogen on the Ag–O bond strength.

Campbell and Paffett reported that adsorbed Cl atoms had no effect on the kinetics of adsorption and desorption of oxygen on Ag(110) [40]. To an extent, this result was not surprising since Campbell and Paffett dosed the Cl atoms on to the Ag(110) from Cl_2 gas at 300 K, producing ordered overlayers of Cl, discernable by LEED, and areas that were Cl free [40].

Atkins and co-workers chlorided their Cs-promoted Ag/α-Al_2O_3 catalyst in a pilot plant operated by BP under industrial conditions [500 K, 1.5 MPa, 4,750 h (GHSV), C_2H_4/air (2% C_2H_4) and dichloroethane (ppm)] [3]. The catalyst was discharged from the reactor, following stopping of the reaction by switching flows to N_2 at 500 K and cooling to ambient in the N_2 stream. A sample of the catalyst was crushed and loaded into their microreactor where it was dosed with O_2 (101 kPa, 513 K, 1 h, 25 cm^3/min). The O_2 desorption spectrum they obtained from it had one peak at 481 K compared with the Cs alone promoted catalyst which had an O_2 peak at 513 K (Fig. 7.11). The O atom desorption activation energy had been lowered from 140 kJ/mol for the Cs-promoted Ag/α-Al_2O_3 catalyst to 129 kJ/mol for the Cl-promoted Cs/Ag/α-Al_2O_3 catalyst. It should be recalled that the Cs is held on the stepped Ag surface and has no effect on the kinetics of O desorption from the Ag(111) face of the catalyst. An important point to note here is that the O atom coverage of the Cl-promoted Cs/Ag/α-Al_2O_3 catalyst was the same as that of the Cl-free catalyst for the same dosage. Therefore, the Cl atoms did not block the O atom adsorption and could not have been on the external surface of the Ag(111). Bowker and Waugh [41–43] and Piao and co-workers [44] have shown that Cl atoms migrate into the bulk of the Ag and that this migration is activated.

Atkins and co-workers determined the activation energy for adsorption of oxygen on a Cl-promoted Cs/Ag/α-Al_2O_3 catalyst by measuring the dependence of the O atom coverage of that catalyst as a function of the adsorption temperature for the same O_2 dose (O_2, 101 kPa, 5 min, 25 cm^3/min) at 382 K (109 °C), 403 K (130 °C), 413 K (140 °C) and 423 K (150 °C) (Fig. 7.12) [3]. The activation energy of adsorption was increased from 17 kJ/mol (Cl free) to 24 kJ/mol for Cl promoted. The increased adsorption activation energy is consistent with the subsurface Cl

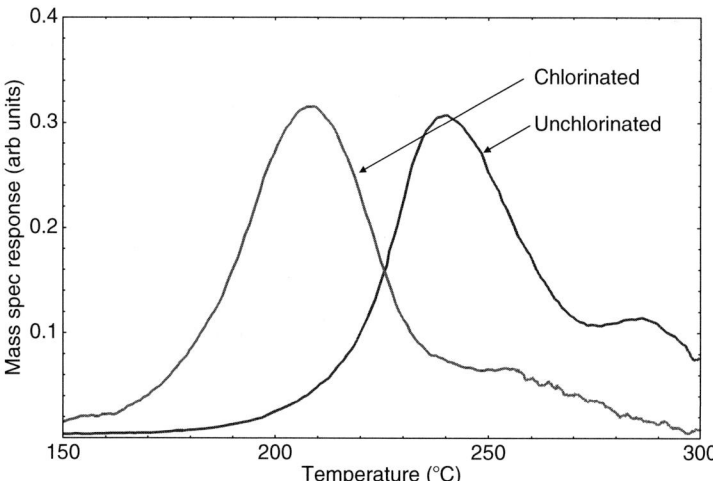

FIGURE 7.11 The O_2 temperature-programmed desorption (TPD) spectrum from a Cl-promoted Cs/Ag/α-Al$_2$O$_3$ catalyst (Peak max temperature = 481 K, 207 °C) and from a Cl-free Cs/Ag/α-Al$_2$O$_3$ catalyst (Peak max temperature = 520 K, 247 °C).

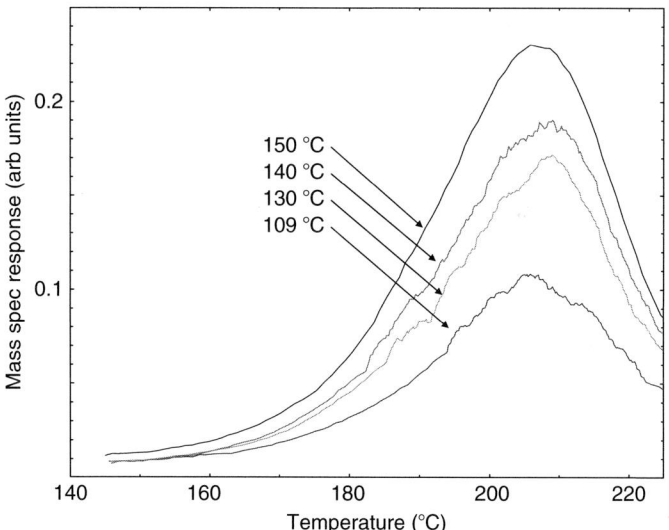

FIGURE 7.12 The dependence of the O atom coverage of a Cl-promoted Cs/Ag/α-Al$_2$O$_3$ catalyst for the adsorption temperatures 382 K (109 °C), 403 K (130 °C), 413 K (140 °C) and 423 K (150 °C).

atoms withdrawing electrons from the surface Ag atoms and so increasing the Ag work function. Line-shape analysis of the Cl-promoted O_2 desorption peaks gave a desorption activation energy of 129 kJ/mol. The overall effect of Cl promotion is to lower the heat of adsorption of O atoms from 123 kJ/mol (Cl free) to 105 kJ/mol. Therefore, at any given temperature, this will lower the O atom surface population and so lower the activity of the catalyst. Crucially, however, it will also lower the activation necessary for the cyclisation of the surface intermediate TS1 or intermediate 1 responsible for EO production while having no effect on the unselective reaction.

7. THE DETAILED KINETICS OF THE ADSORPTION AND DESORPTION OF ETHYLENE ON AN UNOXIDISED AND OXIDISED Ag/α-Al$_2$O$_3$ CATALYST

Figures 7.13 and 7.14 are the TPD spectra of ethylene from an unoxidised (Fig. 7.13) and oxidised (Fig. 7.14) Ag/α-Al$_2$O$_3$ catalyst (0.25 g). The Ag is oxidised to saturation coverage by dosing O$_2$ (101 kPa) at 513 K for 1 h at a flow rate of 25 cm^3/min, producing a surface having the α$_1$-O and α$_2$-O states. The ethylene (2.02 kPa in He, total pressure 101 kPa) is dosed at 183 K for 5 min at a flow rate of 25 cm^3/min after which the temperature is lowered to 158 K, the flow switched to He for 2 min and the temperature is raised at 10 K min^{-1} under He.

One peak with a maximum at 203 K (−70 °C) and a long trailing edge is observed for ethylene desorption on the unoxidised Ag/α-Al$_2$O$_3$ catalyst. This peak, together with three other peaks at 263 K (−10 °C), 273 K (0 °C) and 283 K (10 °C), are observed for ethylene desorption from the oxidised Ag/α-Al$_2$O$_3$ catalyst. [The α$_1$-O and α$_2$-O desorption states are observed at 523 K (250 °C) and 573 K (300 °C) and are unaffected by the adsorption of ethylene—no reaction has occurred.] The additional peaks at 263, 273 and 283 K, corresponding to desorption activation energies of 73, 75 and 78 kJ/mol, must result from the

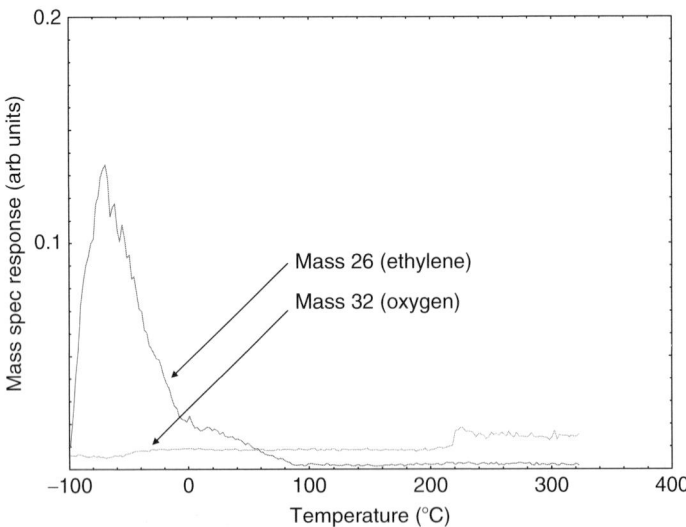

FIGURE 7.13 The *ethylene* desorption from an unoxidised, unpromoted Ag/α-Al$_2$O$_3$ catalyst: m/z = 26, ethylene; m/z = 32, O$_2$.

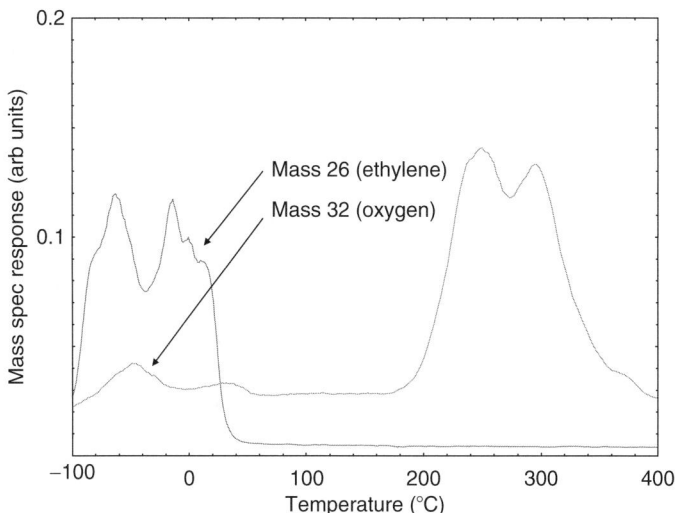

FIGURE 7.14 The *ethylene* desorption from an oxidised, unpromoted surface containing the α_1-O and the α_2-O states: $m/z = 26$, ethylene; $m/z = 32$, O_2.

bonding of the ethylene to the α_1-O and α_2-O states, while the common peak at 203 K results from ethylene desorption from the α-Al$_2$O$_3$ support.

The amount of ethylene adsorbed on the α_1-O and α_2-O sites is 1.05×10^{17} molecule (0.25 g catalyst) or a coverage of 2.2×10^{13} molecule/cm^2. The amount of ethylene dosed was 6.3×10^{19} molecule. The desorption half-life of the 75 kJ/mol state at 158 K is 4.3×10^{11} s and so there is negligible desorption of the ethylene during the 120 s He purge at 158 K. The low coverage must therefore be due to the adsorption being activated, on the basis of which, since only 1.67×10^{-3} of the ethylene molecules dosed at 183 K is adsorbed, the adsorption activation energy for ethylene on to the α_1-O and α_2-O oxygen states is 9.7 kJ/mol.

8. THE DETAILED KINETICS OF THE ADSORPTION AND DESORPTION OF ETHYLENE ON AN OXIDISED Ag/α-Al$_2$O$_3$ CATALYST SUBTENDING THE α_1-O STATE ONLY

Figure 7.15 is the TPD spectrum of ethylene from an oxidised Ag/α-Al$_2$O$_3$ catalyst subtending the α_1-O state only. [This oxidised catalyst was produced, as explained earlier, by dosing O$_2$ (101 kPa, 1 h, 513 K 25 cm^3/min) on to the Ag/α-Al$_2$O$_3$ catalyst which had lost its stepped surface by heating to 773 K.] The α_1-O state is observed to desorb at 523 K as seen before [11], while the ethylene desorbs at 203 K (α-Al$_2$O$_3$) with a second peak at ~253 K. This peak can be seen more clearly in Fig. 7.16 which is the difference spectrum between the ethylene desorption peaks in Fig. 7.15 and the ethylene desorption peak from the unoxidised Ag/α-Al$_2$O$_3$ catalyst shown in Fig. 7.13.

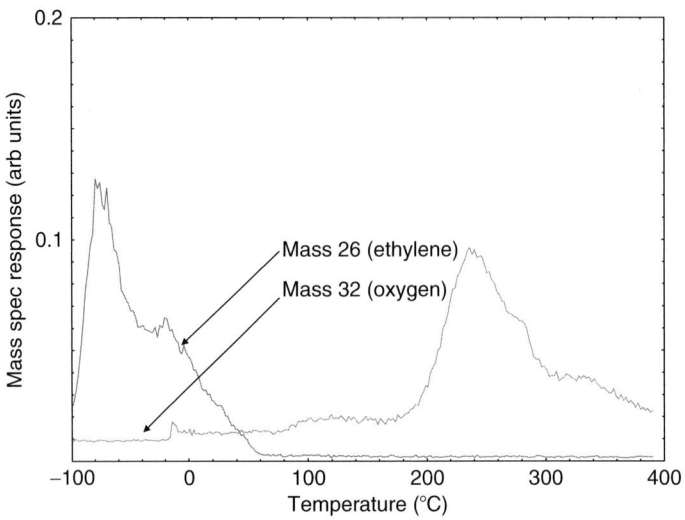

FIGURE 7.15 The *ethylene* desorption from an unpromoted Ag/α-Al$_2$O$_3$ catalyst containing the α$_1$-O state only: $m/z = 26$, ethylene; $m/z = 32$, O$_2$.

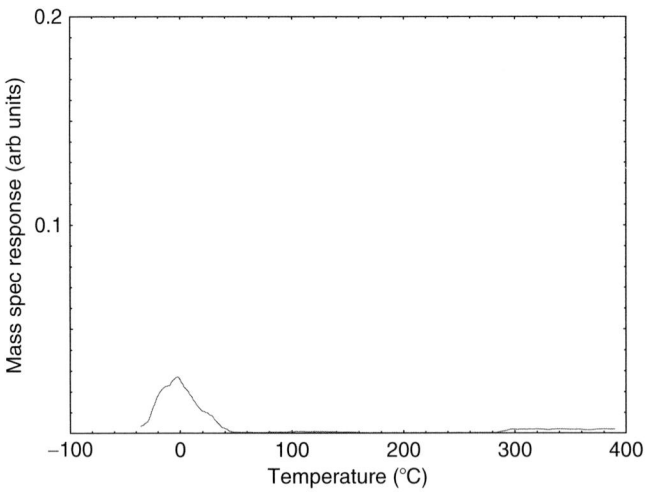

FIGURE 7.16 The *ethylene* desorption difference spectrum between an unoxidised and α$_1$-O oxidised Ag/α-Al$_2$O$_3$ catalyst.

The difference spectrum shows two desorption states at 263 K (−10 °C) (73 kJ/mol) and 273 K (0 °C) (75 kJ/mol). The amount of ethylene adsorbed is 3.1×10^{16} molecule (0.25 g) corresponding to a coverage of 1.2×10^{13} molecule/cm^2. The fraction of the 6.3×10^{19} molecules of ethylene dosed, which is adsorbed, is 5×10^{-4} from which an activation energy to adsorption of 11.6 kJ/mol is determined.

9. THE DETAILED KINETICS OF THE ADSORPTION AND DESORPTION OF ETHYLENE ON THE α_2-O STATE PRINCIPALLY

Figure 7.17 is the TPD spectrum of ethylene from an oxidised Ag/α-Al$_2$O$_3$ catalyst subtending the α_2-O state principally. The method for producing this state was to desorb the α_1-O state by heating the catalyst containing both states to 503 K for 10 min in He. Figure 7.4 shows that this has removed roughly 80% of the α_1-O state. Here, the α_2-O state is seen to desorb at 573 K (300 °C) with a shoulder at 523 K (250 °C) corresponding to the remaining α_1-O state. In addition to the ethylene peak desorbing from the α-Al$_2$O$_3$ at 203 K (−70 °C), peaks at ~253 K (−20 °C) and ~303 K (30 °C) are observed. The difference spectrum shown in Fig. 7.18 shows ethylene peaks at 263 K (−10 °C) and 273 K (0 °C) from the residual α_1-O state and an additional peak at 303 K from the α_2-O state. The coverage of the α_2-O state by ethylene is 5×10^{13} molecule/cm^2. The reaction probability of ethylene on to this state is 5.2×10^{-4}, corresponding to an adsorption activation energy of 11.5 kJ/mol.

10. ETHYLENE DESORPTION FROM AN UNOXIDISED AND OXIDISED Cs-PROMOTED Ag/α-Al$_2$O$_3$ CATALYST

The desorption spectrum of ethylene from the unoxidised Cs-promoted Ag/α-Al$_2$O$_3$ catalyst is identical to that shown in Fig. 7.13 for ethylene desorption from an unoxidised, unpromoted Ag/α-Al$_2$O$_3$ catalyst. The ethylene desorption spectrum from an oxidised Cs/Ag/α-Al$_2$O$_3$ catalyst is identical to that from an

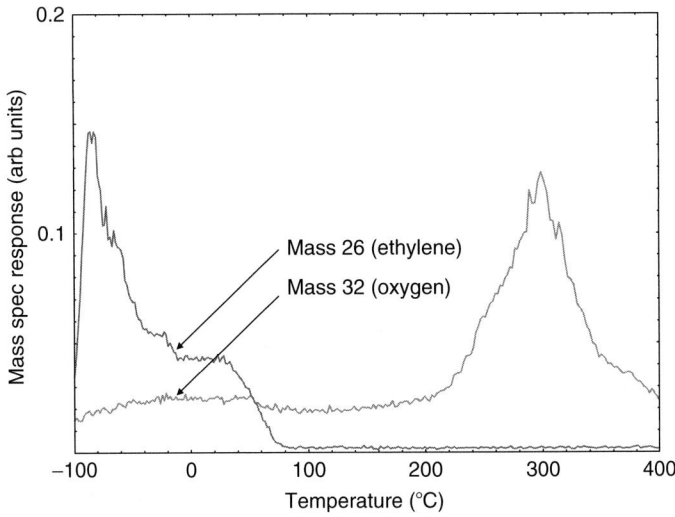

FIGURE 7.17 The *ethylene* desorption spectrum from an oxidised Ag/α-Al$_2$O$_3$ catalyst containing the α_2-O state only: $m/z = 26$, ethylene; $m/z = 32$, O$_2$.

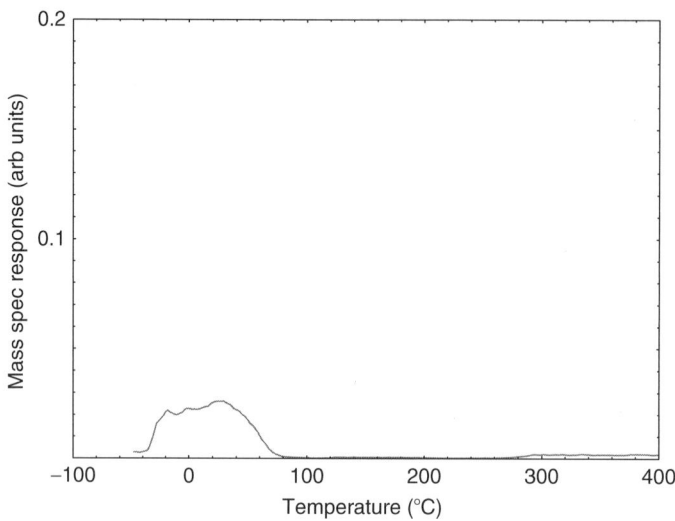

FIGURE 7.18 The difference spectrum between *ethylene* desorption profiles from an unoxidised and an oxidised Ag/α-Al$_2$O$_3$ catalyst containing the α$_2$-O state only.

oxidised Ag/α-Al$_2$O$_3$ catalyst subtending the α$_1$-O state only (Fig. 7.15). These results are consistent with the Cs being held on the stepped Ag surface only.

11. THE DESORPTION OF ETHYLENE FROM AN OXIDISED Cl-PROMOTED Cs/Ag/α-Al$_2$O$_3$ CATALYST

Figure 7.19 is the ethylene desorption spectrum from an oxidised Cl-promoted Cs/Ag/α-Al$_2$O$_3$ catalyst. It shows one peak at 193 K (−80 °C). No peak is seen at this temperature from an unoxidised Cl-promoted Cs/Ag/α-Al$_2$O$_3$ catalyst, indicating that the process of chlorinating that catalyst in the pilot plant has contaminated the α-Al$_2$O$_3$.

The 193 K ethylene peak derives from ethylene desorption from the Cl-weakened Ag–O bond. Its desorption activation energy is 53 kJ/mol; its coverage is 3×10^{13} molecule/cm^2 and its activation energy to adsorption is 12 kJ/mol.

It is possible to make the following conclusions from these data:

1. The stronger the Ag–O bond, the higher is the desorption activation energy of ethylene from the C$_2$H$_4$/O–Ag complex.
2. The Cl promoter lowers the Ag–O bond strength to its lowest measured value of 105 kJ/mol. The corollary of this is that the heat of adsorption of ethylene on to the Cl-promoted, oxidised Ag/α-Al$_2$O$_3$ catalyst is also at its lowest value so that the coverage of Cl-promoted the oxidised catalyst by ethylene at operating temperatures will be reduced, thus lowering the activity of the catalyst. The effect of Cl is kinetic and thermodynamic on the chemisorption of ethylene and not site-blocking in oxygen chemisorption.

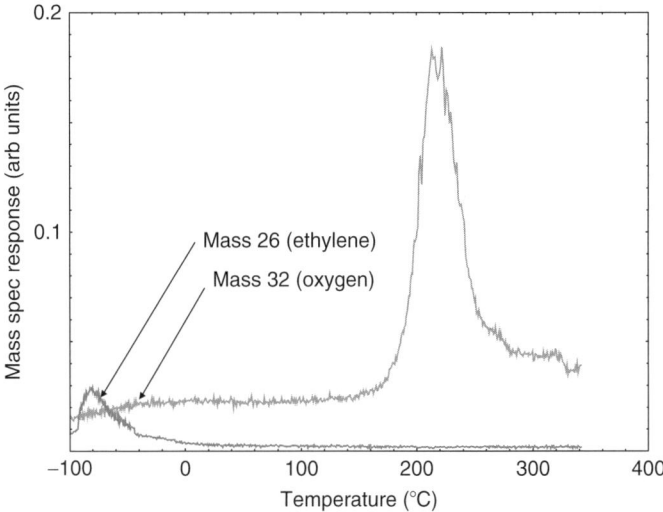

FIGURE 7.19 The *ethylene* desorption spectrum from an oxidised Cl- and Cs-promoted Ag/α-Al$_2$O$_3$ catalyst: $m/z = 26$, ethylene; $m/z = 32$, O$_2$.

TABLE 7.2 The relationship between the strength of the Ag–O bond and the adsorption and desorption activation energies of *ethylene* on to and from that bond

O–Ag state	Temperature (K)	E_d(O$_2$) (kJ/mol)	E_{ads}(C$_2$H$_4$/O–Ag) (kJ/mol)	Temperature (C$_2$H$_4$) (K)	E_d(C$_2$H$_4$) (kJ/mol)
α$_2$-O	573	155	11.5	303	84
α$_1$-O	523	140	11.6	263, 273	73, 75
O/Cl	481	129	12	193	53

3. The Cs promoter has no effect on the bonding of ethylene to the α-Al$_2$O$_3$ support. It is located, at submonolayer amounts, on the stepped Ag surface where it blocks the chemisorption of the less selective, more strongly held, O atom.

These results are summarised in Table 7.2.

12. OVERALL CONCLUSIONS

1. The defining factor for the determination of selectivity on α-Al$_2$O$_3$-supported Ag catalysts is the strength of the Ag–O bond. The lower the Ag–O bond strength, the more selective is the adsorbed O atom.
2. The Cl promoter acts to weaken the Ag–O bond strength and so increases the selectivity. The Cl effect is electronic and not site-blocking. The Cl is not located on the surface of the Ag but in the bulk. There it removes electrons from the

surface Ag atoms and so increases the activation energy for the dissociative chemisorption of O_2 from 17 to 24 kJ/mol. It also decreases the desorption activation energy from 140 to 129 kJ/mol.

3. Subsurface oxygen plays no role in promoting the selectivity of the Ag–O surface state. Its reaction with ethylene is completely unselective. It is also unlikely to be present under industrial operating conditions.
4. In the industrial Ag/α-Al_2O_3 catalyst, the Cs promoter is located on the stepped Ag surface at submonolayer (probably sub-optimal) levels. It is not held on the α-Al_2O_3 support.
5. Ethylene is adsorbed both on the α-Al_2O_3 support and on the O atoms of the surface oxidised Ag. It is most strongly held on the strongest Ag–O bond.
6. Temperature-programmed reaction experiments of ethylene with a surface oxidised Ag containing the $α_1$-O and the $α_2$-O species have shown that, regardless of the strength of the Ag–O bond, EO and CO_2 are formed simultaneously. This confirms the often postulated thesis that EO and CO_2 derive from two near equivalently energetic reaction pathways of a common intermediate. It is generally agreed that the initial interaction of ethylene with an oxidised Ag surface is to form the species which we have termed TS1 here. The next step in the reaction mechanism is probably cyclisation of this species by a simple intramolecular rearrangement involving the species below which we have termed TS3. This is the selective reaction pathway proposed by Grant and Lambert [24].

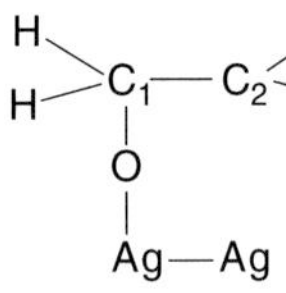

TS1

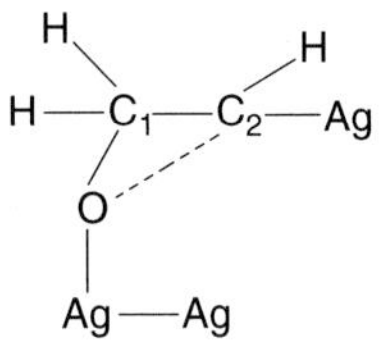

TS3

The species TS1 is the heterogeneous analogy of the 'hot' trichloroethyl radical (CHClCHCl–Cl˙) which is formed by the addition of a Cl atom to *cis*- and *trans*-dichloroethylene [26,27]. The radical is hot because it contains within it the C–Cl bond energy which must be distributed among the available oscillators for it to exist. It has a lifetime of $\sim 10^{-8}$ s at 300 K which is far in excess of the rotational lifetime of $\sim 10^{-11}$ s. The CH_2CH_2–O–Ag species (TS1) will not be 'hot' because of all of the oscillators available to it through bonding to the solid and so the C_1–O bond energy released on its formation will be easily accommodated. It is therefore not a transition state but a stable entity.

The unselective pathway must also involve an equally simple unimolecular rearrangement of TS1. It is entirely possible that this rearrangement forms intermediate 2 which has been shown not to form EO but to form acetaldehyde, whose oxidation to CO_2 and H_2O on the surface oxide overlayer would be facile.

Intermediate 2

7. DFT has been extremely useful in predicting that a Cu/Ag bimetallic catalyst would be more selective than Ag alone. However, the validity of the method of using DFT calculations of a reaction pathway on Ag_{15} clusters as a model for the real catalyst which would be surface oxidised under reaction conditions is questionable.

ACKNOWLEDGEMENT

The authors would like to acknowledge financial support from BP Chemicals Plc for M. Hague.

REFERENCES

[1] S. Linic, M. A. Barteau, Construction of a reaction coordinate and a microkinetic model for ethylene epoxidation on silver from DFT calculations and surface science experiments, *J. Catal.* 214 (2003) 200.
[2] X -D. Wang, W. T. Tysoe, R. G. Greenler, K. Truszkowska, A reflection-adsorption infrared spectroscopy study of the adsorption of atomic oxygen on silver, *Surf. Sci.* 257 (1991) 335.
[3] M. Atkins, J. Couves, M. Hague, B. H. Sakakini, K. C. Waugh, On the role of Cs, Cl and subsurface O in promoting selectivity in Ag/α-Al_2O_3 catalysed oxidation of ethene to ethene epoxide, *J. Catal.* 235 (2005) 103.
[4] T. L. Cottrell, *The Strengths of Chemical Bonds*. Butterworths, London, 1958.
[5] X- D. Wang, W. T. Tysoe, R. G. Greenler, K. Truszkowska, A reflection-adsorption infrared spectroscopy study of the adsorption of dioxygen species on a silver surface, *Surf. Sci.* 258 (1991) 335.
[6] M. A. Barteau, R. J. Madix, The adsorption of molecular oxygen species on Ag(110), *Surf. Sci.* 97 (1980) 101.
[7] R. B. Grant, R. M. Lambert, Basic studies of the oxygen surface chemistry of silver : Chemisorbed and molecular species on pure Ag(111), *Surf. Sci.* 146 (1984) 256.
[8] C. T. Campbell, Atomic and molecular oxygen adsorption on Ag(111), *Surf. Sci.* 157 (1985) 43.
[9] CRC Handbook of Chemistry and Physics, 36th ed., CRC Press, Boca Raton, FL, 1995/1996.

[10] M. Dean, M. Bowker, Adsorption studies on catalysts under UHV/HV conditions, *App. Surf. Sci.* 35 (1988) 27.
[11] J. Couves, M. Atkins, M. Hague, B. H. Sakakini, K. C. Waugh, The activity and selectivity of oxygen atoms adsorbed on a Ag/α–Al_2O_3 catalyst in ethene epoxidation, *Catal. Lett.* 99 (2005) 45.
[12] S. Bare, K. Griffiths, W. N. Lennard, H. T. Tang, Generation of atomic oxygen on Ag(111) and Ag (110) using NO_2: A TPD, LEED, HREELS, XPS and NRA study, *Surf. Sci.* 342 (1995) 185.
[13] A. Raukema, D. A. Butler, F. M. A. Box, A. W. Kleyn, Dissociative and non-dissociative sticking of O_2 at the Ag(111) surface, *Surf. Sci.* 347 (1996) 151.
[14] A. Raukema, D. A. Butler, A. W. Kleyn, The interaction of oxygen with the Ag(110) surface, *J. Phys.: Condens. Matter* 8 (1996) 2247.
[15] J. K. Norskov, T. Bligaard, A. Logadottir, S. Bahn, L. B. Hansen, M. Bollinger, H. Bengaard, B. Hammer, Z. Sljiuancanin, M. Maurickakis, Y. Xu, S. Dahl, C. J. H. Jacobsen, Universality in heterogeneous catalysis, *J.Catal.* 209 (2002) 275.
[16] J. Lacson, *Chemical Economics Handbook.* SRI International, 2003.
[17] P. Mars, D. W. van Krevelen, Oxidations carried out by means of vanadium oxide catalysts, *Chem. Eng. Sci.* 3(Special Suppl.) (1954) 41.
[18] C. T. Campbell, M. T. Paffett, The interactions of O_2, CO and CO_2 with Ag(110), *Surf. Sci.* 143 (1984) 517.
[19] D. Stacchiola, G. Wu, M. Kalchev, W. T. Tysoe, A reflection-adsorption infrared spectroscopic study of the adsorption of ethylene and ethylene oxide on oxygen-covered Ag(111), *Surf. Sci.* 486 (2001) 9.
[20] B. Kruger, C. Benndorf, Ethylene and ethylene-oxide adsorption on Ag(110), *Surf. Sci.* 178 (1986) 704.
[21] C. Backx, C. P.M deGroot, P. Biloen, Adsorption of Oxygen on Ag(110) studied by high resolution ELS and TPD, *Surf. Sci.* 104 (1981) 300.
[22] N. W. Cant, W. K. Hall, Catalytic oxidation: VI. Oxidation of labelled olefins over silver, *J. Catal.* 52 (1978) 81.
[23] P. A. Kilty, N. C. Rol, W. M. H. Sachtler, Identification of oxygen complexes adsorbed on silver and their function in the catalytic oxidation of ethylene, *Proceedings of the 5th Int. Congress Catalysis*, Paper 64, p. 929, North-Holland, Amsterdam, 1973.
[24] R. B. Grant, R. M. Lambert, A single crystal study of the silver-catalysed selective oxidation and total oxidation of ethylene, *J. Catal.* 92 (1985) 364.
[25] P. Sykes, *A Guidebook to Mechanisms in Organic Chemistry*, 1963, Longmans, London.
[26] J. H. Knox, K. C. Waugh, Activated chloroalkyl radicals in the chlorination of trichloroethylene and other olefins, *Trans. Faraday Soc.* 65 (1969) 1585.
[27] P. C. Beadle, J. H. Knox, F. Placido, K. C. Waugh, Decomposition of Activated chloroalkyl radicals. Calculations by the Marcus-Rice theory, *Trans. Faraday Soc.* 65 (1969) 1571.
[28] S. Linic, M. A. Barteau, Formation of a stable surface oxametallacycle that produces ethylene oxide, *J. Am. Chem. Soc.* 124 (2002) 310.
[29] S. Linic, M. A. Barteau, Control of ethylene epoxidation selectivity by surface oxametallacycles, *J. Am. Chem. Soc.* 125 (2003) 4034.
[30] S. Linic, J. Jankowiak, M. A. Barteau, Selectivity driven design of bimetallic ethylene epoxidation catalysts from first principles, *J. Catal.* 224 (2004) 489.
[31] S. Linic, M. A. Barteau, On the mechanism of Cs promotion in ethylene epoxidation on Ag, *J. Am. Chem. Soc.* 126 (2004) 8086.
[32] J. Jankowiak, M. A. Barteau, Ethylene epoxidation over silver and copper-silver bimetallic catalysts. I. Kinetics and selectivity, *J. Catal.* 236 (2005) 366.
[33] J. Jankowiak, M. A. Barteau, Ethylene epoxidation over silver and copper-silver bimetallic catalysts. II. Cs and Cl promotion, *J. Catal.* 236 (2005) 379.
[34] J. C. Dellamorte, J. Lauterbach, M. A. Barteau, Rhenium promotion of Ag and Cu-Ag bimetallic catalysts for ethylene epoxidation, *Catal. Today* 120 (2007) 182.
[35] P. A. Redhead, Thermal desorption of gases, *Vacuum* 12 (1962) 203.
[36] E. L. Force, A. T. Bell, The relationship of adsorbed species observed by infrared spectroscopy to the mechanism of ethylene oxidation over silver, *J. Catal.* 40 (1975) 356.

[37] R. A. van Santen, H. P. C. Kuipers, The mechanism of ethylene epoxidation, *Adv. Catal.* 35 (1987) 265.
[38] R. M. Lambert, R. L. Cropley, A. Husain, M. S. Tikhov, Halogen-induced selectivity in heterogeneous epoxidation is an electronic effect—fluorine, chlorine, bromine and iodine in the Ag-catalysed selective oxidation of ethene, *Chem. Comm.* (2003) 1184.
[39] C. Backx, J. Moolhuysen, P. Greener, R. A. van Santen, Reactivity of oxygen adsorbed on silver power in the epoxidation of ethyl, *J. Catal.* 72 (1981) 364.
[40] C. T. Campbell, M. T. Paffett, The role of chlorine promoters in catalytic ethylene epoxidation over the Ag(110) surface, *Appl. Surf. Sci.* 19 (1984) 28.
[41] M. Bowker, K. C. Waugh, The adsorption of chlorine and chloridation of silver(111), *Surf. Sci.* 134 (1983) 639.
[42] M. Bowker, K. C. Waugh, Chlorine adsorption and chloridation of silver(110), *Surf. Sci.* 155 (1985) 1.
[43] M. Bowker, K. C. Waugh, B. Wolfindale, G. Lamble, D. A. King, The adsorption of chlorine and chloridation of silver(100), *Surf. Sci.* 179 (1987) 254.
[44] H. Piao, K. Adib, M. A. Barteau, A temperature-programmed X-ray photoelectron spectroscopy (TPXPS) study of chlorine adsorption and diffusion on Ag(111), *Surf. Sci.* 557 (2004) 13.

CHAPTER 8

Computational Strategies for Identification of Bimetallic Ethylene Epoxidation Catalysts

A. B. Mhadeshwar and M. A. Barteau

Contents		
	1. Introduction	266
	2. Computational Details	267
	2.1. Periodic slab calculations using DACAPO	267
	2.2. Cluster calculations using ADF	267
	2.3. Application of both methods for ethylene epoxidation on Ag and bimetallic catalysts	268
	3. Activation Energies on Ag	270
	4. Extension to Ag-Based Bimetallic Catalysts	272
	4.1. Assumptions	272
	4.2. Selectivity trends for bimetallic catalysts	272
	4.3. Linear correlations	275
	5. Predictions Using Microkinetic Modeling	276
	5.1. Model development	276
	5.2. Selectivity trends for different bimetallic catalysts	277
	6. Conclusions	279
	Acknowledgments	279
	References	279

Abstract A variety of catalyst combinations involving silver, including AgCu, AgPd, AgPt, AgCd, AgAu, AgRh, AgIr, AgZn, AgNi, and AgOs, were explored computationally for their ability to improve the selectivity of ethylene epoxidation. Using quantum mechanical density functional theory (DFT), activation energies for the selective and nonselective reactions of the key oxametallacycle (OME) intermediate were computed. While consideration of only the branching reactions of the OME suggests several candidate bimetallics, results of a

Center for Catalytic Science and Technology, Department of Chemical Engineering, University of Delaware, Newark, Delaware 19716

Mechanisms in Homogeneous and Heterogeneous Epoxidation Catalysis
DOI: 10.1016/B978-0-444-53188-9.00008-0

© 2008 Elsevier B.V.
All rights reserved.

more complete microkinetic model based on the DFT parameters point to AgCu as the sole bimetallic among those considered that can achieve higher selectivities than pure silver.

Key Words: Ethylene epoxidation, Silver, Bimetallic catalyst design, Copper, Density functional theory, Microkinetic modeling.   © 2008 Elsevier B.V.

1. INTRODUCTION

The process of ethylene epoxidation has been well established for the last 50 years or more [1]. Ethylene oxide (EO) is produced in large quantities and is primarily utilized for the production of ethylene glycol. Silver has been employed as a unique catalyst in EO production so far, since it (unpromoted) provides ~50% selectivity to the desired product, EO, and promoted catalysts have been reported to achieve selectivities approaching 90% [2]. Understanding the mechanism of this process on Ag and finding improved catalysts are two major objectives of continuing research.

Significant advances in understanding this chemistry have been made over the last two decades in understanding the mechanistic aspects on Ag. The nature of adsorbed oxygen responsible for epoxidation has been confirmed as atomic rather than molecular [3]. By combining surface science experiments and quantum mechanical density functional theory (DFT) calculations, it has been shown that an oxametallacycle (OME) intermediate is formed on Ag from adsorbed ethylene and oxygen [4–7]. Recent work in our group has indicated that reactions of the OME control selectivity [8,9]. The predicted kinetic isotope effect in EO selectivity [9] is also consistent with the corresponding experimental observations [10].

As far as finding improved catalysts or operating conditions is concerned, a variety of experimental studies have been conducted. Promotion by alkali metal salts [11], especially cesium [12,13], has been shown to increase the EO selectivity. Similarly, it is observed that addition of chlorine to the catalyst results in higher EO selectivity [14–18]. Atmospheric pressure experiments in our laboratory have indicated that Cu–Ag bimetallic catalysts are more effective than pure Ag [19]. Furthermore, it is observed that after addition of Cs and Cl to the Cu–Ag bimetallic catalyst, it outperforms the corresponding promoted silver catalyst [20]. Published experimental studies have typically focused on a limited number of catalyst components, whereas one could potentially explore a wide variety of catalyst combinations using theoretical tools.

Motivated by the idea of rational catalyst design, we take advantage of the computational efficiency of DFT methods to examine a variety of Ag-containing bimetallic catalysts. We focus initially on the branching reactions of the OME since these have been shown to control selectivity [9]. Initial screening based solely on OME reactions suggests several promising bimetallic combinations. However, if one considers a more complete microkinetic model, copper stands out for its ability to enhance the selectivity of silver-based bimetallic catalysts.

2. COMPUTATIONAL DETAILS

Both periodic slabs and finite clusters were employed in our DFT calculations. DACAPO (with the new ASE2 python interface) [21] and Amsterdam Density Functional (ADF) [22–24] packages were used for the slab and cluster calculations, respectively. Details of the calculations are presented below.

2.1. Periodic slab calculations using DACAPO

A periodic two-layer 2×2 Ag(111) slab was used with eight vacuum layers. The adsorbates as well as gas-phase molecules were relaxed, whereas the slab layers were fixed, consistent with our previous work [8]. A plane wave basis set with an energy cutoff of 350 eV was employed with an 18-k Chadi–Cohen point mesh. The PW91 exchange–correlation functional was applied and the Vanderbilt ultrasoft pseudopotential [25] was used to describe the core electrons. The experimentally measured lattice constant of 4.09 Å [26] was used for constructing the Ag slab. While calculating the energies of the gas-phase molecules, a Fermi temperature ($k_B T$) of 0.0001 eV was used. The binding energy (Q) of a molecule was calculated as

$$Q = E_{\text{Adsorbate molecule+Slab}} - E_{\text{Slab}} - E_{\text{Gas molecule}} \tag{8.1}$$

where E stands for the energy calculated in DACAPO. Zero-point energies were not included. A sample picture of the slab with an OME is shown in Fig. 8.1a.

Given the optimized configurations of adsorbed reactants and products, the Nudged Elastic Band (NEB) method in DACAPO was employed to find the transition state and the corresponding activation energy barrier. Ten images were considered between reactant and products. The maximum force criterion was set to 0.05 eV per image, but this criterion was too strict for the current system and all images were finally converged within 0.1 eV per image. NEB iterations of reaction coordinates were continued until either the force criterion was satisfied or the activation energies did not change significantly (<1 kcal/mol) over the last ~5–6 iterations of the reaction coordinate. These calculations are significantly CPU intensive (CPU ~15,000 h, e.g., see Fig. 8.2), and all such calculations have been performed on the Pacific Northwest National Laboratory (PNNL) supercomputer cluster.

2.2. Cluster calculations using ADF

The one-electron Kohn–Sham equations were solved using the Vosko–Wilk–Nusair (VWN) functional [27] to obtain the local potential. Gradient correlations for the exchange (Becke functional) [28] and correlation (Perdew functional) [29] energy terms were included self-consistently. ADF represents molecular orbitals as linear combinations of Slater-type atomic orbitals. The double-ζ basis set was employed and all calculations were spin unrestricted. Integration accuracies of 10^{-3}–10^{-4} and 10^{-6} were used during the single-point and vibrational frequency calculations, respectively. The cluster size chosen for Ag or any bimetallic was

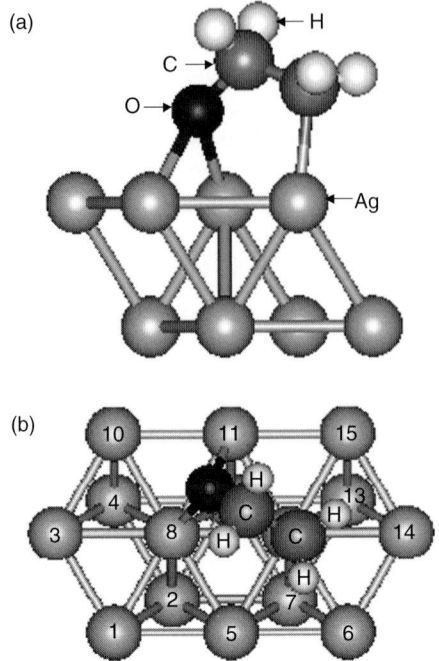

FIGURE 8.1 Schematics of (a) Ag slab and (b) Ag_{15} cluster used in DACAPO and Amsterdam Density Functional (ADF) calculations, respectively. Numbers are shown on the Ag atoms in (b) to indicate precisely which Ag atoms are substituted with other metals in remaining figures and tables.

15 atoms, with 10 atoms in the top layer and 5 atoms in the bottom layer, to give adequate representation of the surface and to minimize edge effects. A sample picture of the cluster is shown in Fig. 8.1b. The same lattice constant (4.09 Å) was used while constructing these clusters. For CPU comparison, each single-point energy calculation in ADF takes of the order of 30 min, whereas the optimization and vibrational frequency calculations require ~30 and ~100 h, respectively.

2.3. Application of both methods for ethylene epoxidation on Ag and bimetallic catalysts

Since the periodic DFT calculations, especially the NEB calculations, are extremely time consuming, we use a hybrid approach for rational catalyst design. The OME was taken as the reactant, whereas adsorbed EO and acetaldehyde were considered as the products of different competitive reactions. ADF was used to provide initial guesses for species geometries to enhance efficiency. As a first step, structures of the reactants and products were optimized on the periodic Ag slab. Two NEB calculations were performed for the formation of adsorbed EO and acetaldehyde starting from the common OME intermediate. The difference between these two activation energies is the main factor responsible for the selectivity to the

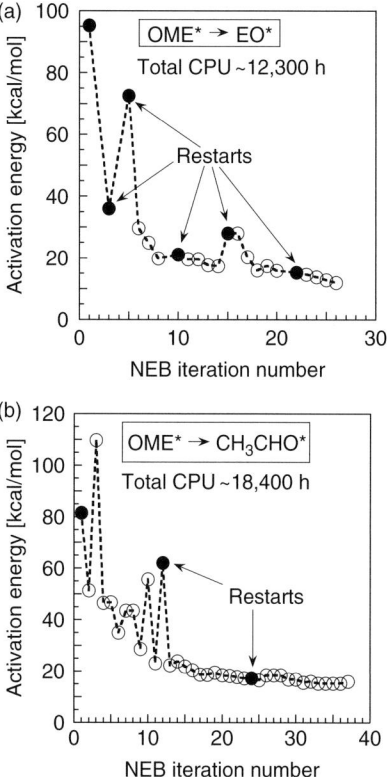

FIGURE 8.2 Progression of Nudged Elastic Band (NEB) calculations over multiple iterations for (a) ethylene oxide (EO) formation and (b) acetaldehyde formation from oxametallacycle (OME). Multiple restarts are necessary due to limited CPU availability.

desired product (EO), provided that the corresponding pre-exponential factors are similar and the activation energies are not very high (low conversions or yields) [9].

In order to extend this idea to different bimetallic catalysts involving Ag, one needs to repeat this exercise for each catalyst. Because of the limited CPU availability, we make the important assumption that the structure of the transition state on any bimetallic catalyst is the same as that on pure Ag. This assumption should be safest for bimetallic catalysts in which the concentration of the second metal is low compared to Ag. However, in order to explore the effect of such low loadings, one needs a larger unit cell in periodic calculations, which again increases the computational load significantly. To solve this problem, we transferred the transition states from the NEB calculations to the 15-atom clusters of Ag or any bimetallic and carried out single-point calculations in ADF to get the reaction coordinate as well as the activation energies. The transition states were verified using vibrational frequency calculations on the Ag cluster. Last, by changing the type of metal in the cluster, we have explored the effect on the DFT-predicted selectivity.

3. ACTIVATION ENERGIES ON Ag

The activation energies from NEB calculations for EO and acetaldehyde formation are shown at each NEB iteration in Fig. 8.2a and b, respectively. Because of the limited CPU availability, restarts of NEB calculations were required at multiple occasions utilizing the results of the previous calculations. On an Ag slab, the activation energy for EO formation from the OME is estimated to be 12.0 kcal/mol, whereas that for acetaldehyde formation is 16.1 kcal/mol. As described in Section 2, next we take the converged images from the NEB calculations and transfer them onto 15-atom clusters in order to perform single-point energy calculations in ADF. Figures 8.3 and 8.4 show the top and side views of the images in both reactions. Both O–Ag and C–Ag bonds in the OME weaken and ring formation occurs to produce weakly adsorbed EO, as shown in Fig. 8.3. Image 6 is the transition state here and it is closer to EO; that is, it is a late transition state. In acetaldehyde formation, transfer of an H atom from one carbon to another occurs followed by slight rotation of H atoms on the terminal carbon to form the weakly adsorbed acetaldehyde, as shown in Fig. 8.4. The transition state is shown in Image 3 and it is an early transition state, as it more closely resembles the OME. Comparison of DACAPO- and ADF-based reaction coordinates is shown in Fig. 8.5. Using ADF single-point calculations (with no further geometry optimization), we find that the corresponding activation energies for EO formation and acetaldehyde formation are 13.5 and 15.9 kcal/mol, respectively. This level of agreement between DACAPO- and ADF-based energetics also provides support for using two different methods together in our approach.

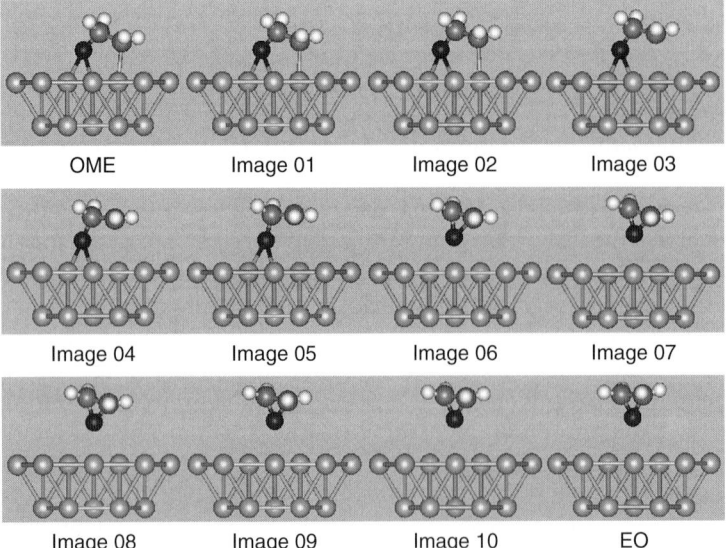

FIGURE 8.3 Side views of images along the final reaction coordinate for OME→EO*. Image 6 is the transition state.

Identification of Bimetallic Ethylene Epoxidation Catalysts 271

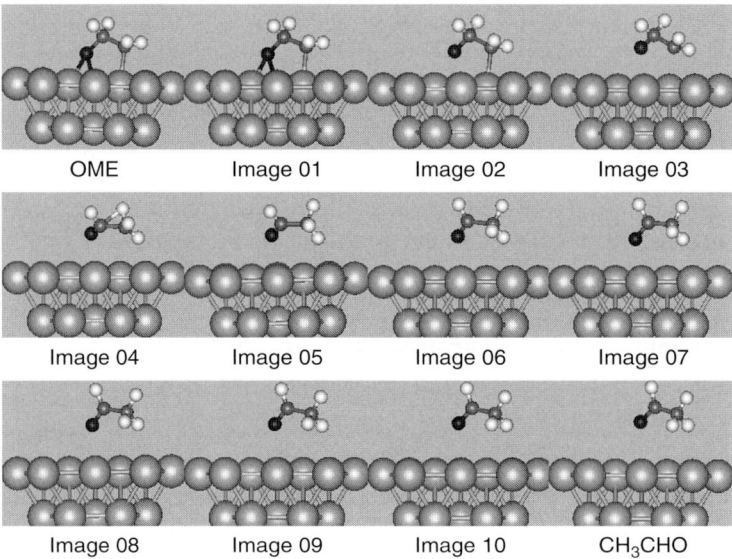

FIGURE 8.4 Side views of images along the final reaction coordinate for OME→CH$_3$CHO*. Image 3 is the transition state.

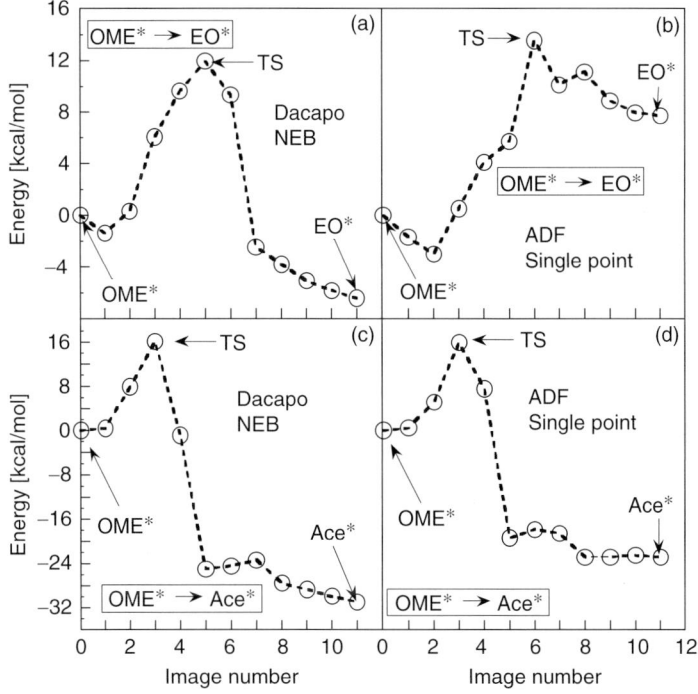

FIGURE 8.5 Reaction coordinates for formation of adsorbed ethylene oxide (EO*) (a and b) and adsorbed acetaldehyde (ACE*) (c and d) formation using both DACAPO (a and c) and Amsterdam Density Functional (ADF) (b and d). The activation energies are quite similar in both methods.

In our previous paper [8], the activation energy for ring closure of the OME was calculated to be 16 kcal/mol. The difference between the previous calculation and those reported here is due to different convergence criteria and DACAPO versions. In the previous work, the energy cutoff was 340 eV, the vacuum spacing between slabs was 15 Å, and the NEB convergence criterion was 0.05 eV/Å in the direction perpendicular to the reaction coordinate. In the present work, we have used an energy cutoff of 350 eV, a vacuum spacing of 20 Å, and an NEB convergence criterion of 0.1 eV image in the new ASE/DACAPO version.

4. EXTENSION TO Ag-BASED BIMETALLIC CATALYSTS

4.1. Assumptions

Traditionally, Ag has been used for EO manufacture. As mentioned previously, the selectivity to EO is primarily driven by the difference in activation energies between OME reaction channels. Therefore, in order to find a better catalyst, we explore the differences in activation energies on bimetallic catalysts. Metals around Ag in the periodic table (e.g., Cu, Pd, Pt, Cd, Au, Rh, Ir, Ni, Zn, and Os) are considered in this study. The approach is fairly simple. As mentioned previously, we assume that the structure of the transition state remains the same on a bimetallic AgM catalyst if the loading of the metal M is much less than that of Ag. In this study, we have replaced one Ag atom (specifically, atom number 11 in Fig. 8.1b) with an M atom to generate a bimetallic $Ag_{14}M_1$ cluster, and single-point energy calculations are carried out using ADF for the reactant, 10 images, and the product for both parallel reactions.

4.2. Selectivity trends for bimetallic catalysts

Table 8.1 and Figs. 8.6 and 8.7 show the results of single-point energy calculations. The curves are similar in shape for each reaction on different bimetallic catalysts, but differ in activation energies. Only six cases are shown in Figs. 8.6 and 8.7 for clarity. From Table 8.1, it is observed that $Ag_{14}Zn_1$ and $Ag_{14}Pd_1$ represent the bimetallic catalysts with the highest and lowest activation energies for EO formation, respectively. Similarly, $Ag_{14}Os_1$ and $Ag_{14}Au_1$ have the highest and lowest activation energies for acetaldehyde formation, respectively.

In order to determine the trends for selectivity toward EO formation, we define the variable $\Delta\Delta E$ as follows.

$$\Delta\Delta E_{Ag_{14}M_1} = \Delta E_{Ag_{14}M_1} - \Delta E_{Ag_{15}} \tag{8.2}$$

with

$$\Delta E_{\text{any catalyst}} = E_{\text{acetaldehyde formation}} - E_{\text{ethylene oxide formation}} \tag{8.3}$$

TABLE 8.1 Activation energies for ring closure (E_1) and acetaldehyde formation (E_2) from OME for different bimetallic $Ag_{14}M_1$ catalysts

Catalyst	Ag	AgCu	AgPd	AgPt	AgCd	AgAu	AgRh	AgIr	AgZn	AgNi	AgOs
E_1	13.54	16.56	11.28	10.70	23.79	11.37	18.22	19.11	25.47	22.73[a]	25.83[a]
ΔH_1	7.73	11.97	5.99	4.03	17.47	4.33	15.31	13.78	20.06	17.87	21.30[b]
E_2	15.91	20.53	16.17	14.42	18.80	13.13	19.47	18.59	22.44	23.33	26.59
ΔH_2	−22.85	−18.55	−24.82	−27.44	−12.81	−26.36	−16.74	−17.78	−10.24	−12.85	−9.43
ΔE	2.38	3.97	4.88	3.72	−4.99	1.76	1.25	−0.52	−3.03	0.60	0.75
$\Delta\Delta E$	0.00	1.60	2.50	1.34	−7.37	−0.62	−1.12	−2.90	−5.41	−1.77	−1.62

[a] Based on the linear correlation for ring closure in Fig. 8.9.
[b] Based on a linear correlation $\Delta H_2 = 1.017\Delta H_1 - 31.099$, $R^2 = 0.99$. This correlation simply means that heats of adsorption for EO and acetaldehyde on all catalysts are linearly proportional and do not vary significantly.

These values are from single-point energy calculations using ADF, with converged geometries taken from NEB calculations on pure Ag. Corresponding heats of reactions (ΔH_1 and ΔH_2), ΔE, and $\Delta\Delta E$ are also shown (see text for more details). All energies are in kcal/mol.

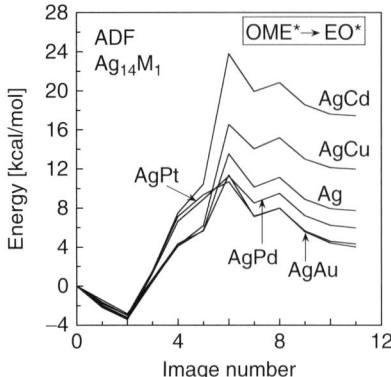

FIGURE 8.6 Reaction coordinates for ethylene oxide (EO) formation on different bimetallic $Ag_{14}M_1$ catalysts using Amsterdam Density Functional (ADF). Only six cases are shown for clarity.

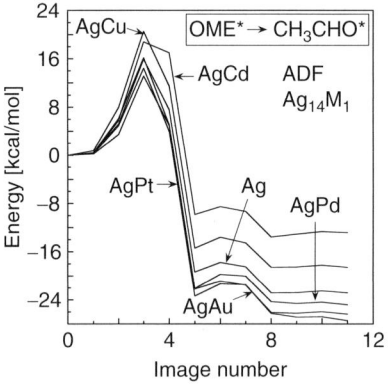

FIGURE 8.7 Reaction coordinates for acetaldehyde formation on different bimetallic $Ag_{14}M_1$ catalysts using Amsterdam Density Functional (ADF). Only six cases are shown for clarity.

Here, E represents the activation energy, ΔE is the difference in activation energies, and $\Delta\Delta E$ is the difference of a difference. If $\Delta\Delta E$ is positive for a bimetallic catalyst, it is more selective for EO than pure Ag and vice versa. Note that $\Delta\Delta E$ is zero for pure Ag. Figure 8.8 shows the comparison of various bimetallic catalysts in this study. The $\Delta\Delta E$ values are also given in Table 8.1. It is found that $Ag_{14}Cu_1$, $Ag_{14}Pd_1$, and $Ag_{14}Pt_1$ are more selective than pure Ag, whereas all other bimetallic catalysts are more prone to form acetaldehyde. Au addition has not been found to be beneficial in experimental studies, and our DFT results predict that it should decrease EO selectivity. From the previous experimental studies in our group, Cu–Ag bimetallic catalysts were found to be more selective than pure Ag [19,30] and this computational study also confirms that. Clearly, this simple idea of assuming the same transition state structures on different

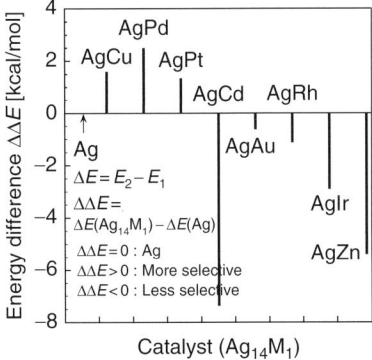

FIGURE 8.8 Comparison of difference in relative activation energies for a variety of bimetallic $Ag_{14}M_1$ catalysts. It is proposed that the ethylene oxide (EO) selectivity is associated with this difference.

low-loading bimetallic catalysts provides trends consistent with available experimental observations and simultaneously results in tremendous CPU savings.

4.3. Linear correlations

Even though we have explored bimetallic catalyst combinations around Ag in the periodic table, one strategy is to extend this approach to other possible bimetallic catalysts or for bimetallic catalysts of different composition ratios by extrapolation. Rather than performing the same exercise of computing energies for reactant and intermediate structures along the 2 reaction coordinates (a total of 20 images), and 2 products for each catalyst, we explore the possibility of correlation to obtain an approximate idea about the selectivity. Different research groups have proposed linear relationships between activation energies and heats of adsorption or heats of surface reactions [31–34]. In Fig. 8.9, we have plotted the activation energies (E_1 and E_2) in Table 8.1 versus the corresponding heats of surface reactions (ΔH_1 and ΔH_2), where subscripts 1 and 2 correspond to EO formation and acetaldehyde formation, respectively. For the ring closure (EO formation), a total of 9 points (out of 11 combinations, since we had convergence problems in ADF for $Ag_{14}Os_1$ and $Ag_{14}Ni_1$) lie on a straight line with a regression of 0.96, indicating that there is a strong correlation between the activation energy and the heat of surface reaction (energy of adsorbed EO – energy of OME). For acetaldehyde formation, the 11 points are fairly well represented by a straight line with a regression of 0.84. Both activation energies increase with the corresponding heats of reactions, indicating that the more the endothermicity (or the less the exothermicity) on a given catalyst, the higher is the activation barrier. In fact, since we had convergence problems to get the ring closure activation energies on $Ag_{14}Os_1$ and $Ag_{14}Ni_1$, we used these linear correlations to obtain those activation energies and to determine the effect of Os or Ni addition on EO selectivity. On the basis of our predictions, both should decrease the EO selectivity compared to pure Ag.

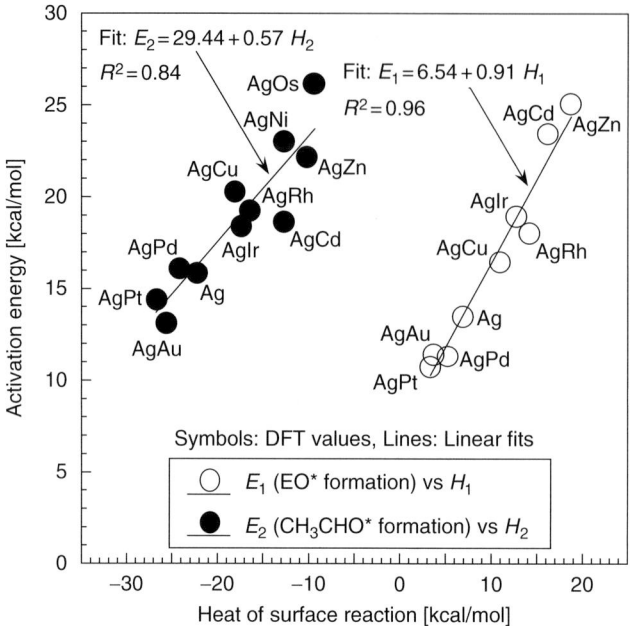

FIGURE 8.9 Linear correlations for activation energies as a function of heats of surface reactions on a variety of bimetallic $Ag_{14}M_1$ catalysts. It is proposed that the ethylene oxide (EO) selectivity is associated with this difference.

5. PREDICTIONS USING MICROKINETIC MODELING

Even though DFT results are useful in obtaining trends in EO selectivity by calculating activation energies, there are additional factors controlling the selectivity. For example, the pre-exponential factors and the absolute magnitude of activation energies can also be important. Our DFT calculations have suggested several bimetallic catalyst combinations to be better than pure Ag, as shown in Fig. 8.8. Here, we also explore the trends suggested by microkinetic modeling in order to test those results.

5.1. Model development

The microkinetic model for ethylene epoxidation is shown schematically in Fig. 8.10. Ethylene and oxygen adsorb and react to produce the OME, which further isomerizes to EO and acetaldehyde. The microkinetic model has seven

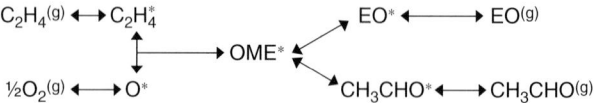

FIGURE 8.10 Schematic of the microkinetic model for ethylene epoxidation.

reversible reactions. For each bimetallic catalyst, we have calculated heats of adsorption of oxygen, ethylene, EO, and acetaldehyde as well as stability of the OME with respect to EO (gas) using ADF. The heats of atomic oxygen adsorption from ADF are normalized with the recent DACAPO results on pure Ag [35]. Stabilities of ethylene, EO, and acetaldehyde do not vary by more than 1.5 kcal/mol on all catalysts, but those of O and OME show significant variation. Table 8.2 shows the stabilities of O and OME. The semiempirical unity bond index–quadratic exponential potential (UBI–QEP) method [36,37] is used to calculate the activation energies for oxygen adsorption–desorption and OME formation from ethylene and oxygen. Activation energies for EO and acetaldehyde formation from OME are taken from our DFT calculations. The same nominal values of pre-exponential factors (10^{13} s^{-1} for desorption and 10^{11} s^{-1} for Langmuir–Hinshelwood-type surface reactions) are assumed for all catalysts. The pre-exponential factors are shown in Table 8.3. A detailed microkinetic model that captures a variety of experimental data sets on Ag was recently proposed by the Stoltze group [38–40]. However, the goal of our microkinetic model is mainly to capture the trends in EO selectivity on different catalysts.

5.2. Selectivity trends for different bimetallic catalysts

Plug flow reactor (PFR) simulations were carried out for the following operating conditions based on experiments of Jankowiak and Barteau [19] on Ag/α-Al_2O_3 monoliths: pressure = 1.34 atm, temperature = 490 K, reactor length = 1 cm, tube diameter = 1.8 cm, flow rate = 100 cm^3/min at STP, and inlet C_2H_4:O_2:N_2 = 1:1:8. Pre-exponential factors for the parallel ring closure and isomerization reactions were adjusted to match the experimental EO selectivity of 32%. By appropriately changing the heats of adsorption, stability of the OME, and activation energies for isomerization, we then calculate EO selectivity on different bimetallic $Ag_{14}M_1$

TABLE 8.2 Oxygen heat of adsorption QO and stability of OME with respect to $EO^{(gas)}$ on different bimetallic $Ag_{14}M_1$ catalysts

Catalyst	QO(kcal/mol)	OME stability (kcal/mol)
	Effect of metal substitution	
Ag	90.800^a	4.860
$Ag_{14}Cu_1$	106.031	12.608
$Ag_{14}Pd_1$	90.182	19.467
$Ag_{14}Pt_1$	87.731	18.702
$Ag_{14}Cd_1$	98.580	10.768
$Ag_{14}Au_1$	86.884	10.946

EO, ethylene oxide; OME, oxametallacycle.
[a] ADF-based heat of oxygen adsorption is 102.222 kcal/mol on pure Ag, but we have chosen the DACAPO-based value of 90.8 kcal/mol from the periodic DFT calculations of Li et al. [35]. Other values for oxygen adsorption are correspondingly normalized by subtracting a factor of 11.422 (=102.222 − 90.8) kcal/mol.
These values are from single-point energy calculations using ADF, with converged geometries taken from NEB calculations on pure Ag.

TABLE 8.3 Pre-exponential factors (A) and sticking coefficients (S) in the microkinetic model

Reaction	S (unitless) or A (s^{-1})
$O_2 + 2* \rightarrow 2O*$	1.0
$2O* \rightarrow O_2 + 2*$	1.0×10^{13}
$C_2H_4 + * \rightarrow C_2H_4*$	1.0
$C_2H_4* \rightarrow C_2H_4 + *$	1.0×10^{13}
$C_2H_4O + * \rightarrow C_2H_4O*$	1.0
$C_2H_4O* \rightarrow C_2H_4O + *$	1.0×10^{13}
$CH_3CHO + * \rightarrow CH_3CHO*$	1.0
$CH_3CHO* \rightarrow CH_3CHO + *$	1.0×10^{13}
$OME* + * \rightarrow C_2H_4* + O*$	1.0×10^{11}
$C_2H_4* + O* \rightarrow OME* + *$	1.0×10^{11}
$C_2H_4O* \rightarrow OME*$	1.0×10^{11}
$OME* \rightarrow C_2H_4O*$	1.0×10^{11}
$CH_3CHO* \rightarrow OME*$	2.5×10^{12}
$OME* \rightarrow CH_3CHO*$	2.5×10^{12}

OME, oxametallacycle.

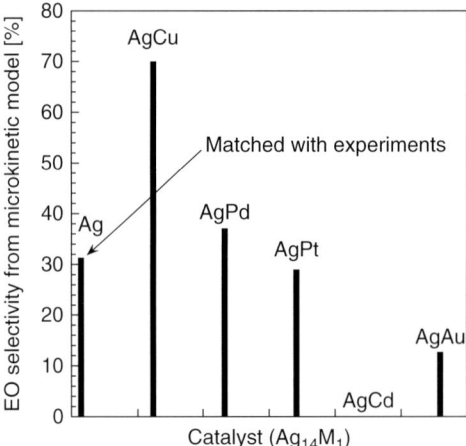

FIGURE 8.11 Predictions of ethylene oxide (EO) selectivity on different bimetallic $Ag_{14}M_1$ catalysts from the microkinetic model. Two pre-exponential factors were adjusted to correctly capture the experimental EO selectivity on pure Ag.

catalysts. The results are shown in Fig. 8.11. AgCu shows higher EO selectivity than that on pure Ag, consistent with the DFT calculations. Similarly, AgCd and AgAu show lower EO selectivity than does pure Ag, also consistent with the DFT calculations. However, AgPd and AgPt have moderate selectivities, similar to pure Ag. It is observed that apart from the differences in activation energies, pre-exponential factors, and magnitudes of activation energies, the stability of the OME is another factor controlling the selectivity. As shown in Table 8.2, OME stability on $Ag_{14}Pd_1$ and $Ag_{14}Pt_1$ is much higher than that on other catalysts, which results in almost zero activation energy for OME formation from adsorbed EO. In comparison, the activation energy for OME formation from adsorbed acetaldehyde is more than that for acetaldehyde desorption, so the EO selectivity decreases significantly on these two catalysts. Overall, AgCu is the only bimetallic catalyst showing significantly higher EO selectivity than pure Ag, based on microkinetic analysis.

6. CONCLUSIONS

A combination of quantum mechanical techniques and microkinetic modeling has been employed to explore improved catalysts than Ag for the ethylene epoxidation process. A combination of 10 bimetallic catalysts and pure Ag was investigated. Rather than carrying out a large number of computationally expensive NEB calculations on periodic slabs, a cluster approach was employed by assuming the same transition state on different bimetallic catalysts as that on pure Ag. This approach results in tremendous computational savings, while maintaining reasonable accuracy. On the basis of the DFT calculations for the OME branching reactions, only AgCu, AgPd, and AgPt are predicted to be better catalysts than pure Ag for EO selectivity. Microkinetic modeling was used to incorporate the DFT results into a more comprehensive model and AgCu was found to be the only bimetallic catalyst that results in significantly higher EO selectivity than pure Ag. According to the microkinetic analysis, AgPd and AgPt do not lead to improved selectivity because of the high stability of the OME on these catalysts. Overall, this study confirms that AgCu is the only promising bimetallic catalyst out of the 10 combinations considered.

ACKNOWLEDGMENTS

We gratefully acknowledge the support of the US Department of Energy, Office of Basic Energy Sciences, Division of Chemical Sciences, Geosciences, and Biosciences (Grant FG02–03ER15468) for this research. We also acknowledge The William R. Wiley Environmental Molecular Sciences Institute at Pacific Northwest National Laboratories for access to the Molecular Science Computing Facility through Grand Challenge Project gc3568.

REFERENCES

[1] R. A. van Santen, H. P. C. E. Kuipers, The mechanism of ethylene epoxidation, *Adv. Catal.* 35 (1987) 265.
[2] J. R. Monnier, The selective epoxidation of non-allylic olefins over supported silver catalysts in *3rd World Congress on Oxidation Catalysis.* (R. K. Grasselli, S. T. Oyama, A. M. Gaffney, and J. E. Lyons, Eds.), *Studies in Surface Science and Catalysis*, Vol. 110, Elsevier, Amsterdam, 1997, p. 135.
[3] R. A. van Santen, C. P. M. de Groot, The mechanism of ethylene epoxidation, *J. Catal.* 98 (1986) 530.
[4] G. S. Jones, M. Mavrikakis, M. A. Barteau, J. M. Vohs, First synthesis, experimental and theoretical vibrational spectra of an oxametallacycle on a metal surface, *J. Am. Chem. Soc.* 120 (1998) 3196.

[5] S. Linic, M. A. Barteau, Formation of a stable surface oxametallacycle that produces ethylene oxide, *J. Am. Chem. Soc.* 124 (2002) 310.
[6] M.-L. Bocquet, A. Michaelides, D. Loffreda, P. Sautet, A. Alavi, D. A. King, New insights into ethene epoxidation on two oxidized Ag (111) surfaces, *J. Am. Chem. Soc.* 125 (2003) 5620.
[7] S. Linic, H. Piao, K. Adib, M. A. Barteau, Ethylene epoxidation on Ag: Identification of the crucial surface intermediate by experimental and theoretical investigation of its electronic structure, *Angew. Chem. Int. Ed.* 43 (2004) 2918.
[8] S. Linic, M. A. Barteau, Construction of a reaction coordinate and a microkinetic model for ethylene epoxidation on silver from DFT calculations and surface science experiments, *J. Catal.* 214 (2003) 202.
[9] S. Linic, M. A. Barteau, Control of ethylene epoxidation selectivity by surface oxametallacycles, *J. Am. Chem. Soc.* 125 (2003) 4034.
[10] N. W. Cant, W. K. Hall, Catalytic-oxidation 6. oxidation of labeled olefins over silver, *J. Catal.* 52 (1978) 81.
[11] J. C. Zomerdijk, M. W. Hall, Technology for the manufacture of ethylene-oxide, *Catal. Rev. - Sci. Eng.* 23 (1981) 163.
[12] R. P. Nielsen, J. H. La Rochelle, Ethylene oxide process, United States Patent 4,012,425, (1977) to Shell Oil.
[13] W. D. Mross, Alkali doping in heterogeneous catalysis, *Catal. Rev. - Sci. Eng.* 25 (1983) 591.
[14] C. T. Campbell, M. T. Paffett, The role of chlorine promoters in catalytic ethylene epoxidation over the Ag(110) surface, *Appl. Surf. Sci.* 19 (1984) 28.
[15] C. T. Campbell, B. E. Koel, Chlorine promotion of selective ethylene oxidation over Ag(110)- kinetics and mechanism, *J. Catal.* 92 (1985) 272.
[16] C. T. Campbell, Chlorine promoters in selective ethylene epoxidation over Ag(111): A comparison with Ag(110), *J. Catal.* 99 (1985) 28.
[17] S. A. Tan, R. B. Grant, R. M. Lambert, Chlorine oxygen interactions and the role of chlorine in ethylene oxidation over Ag(111), *J. Catal.* 100 (1986) 383.
[18] K. L. Yeung, A. Gavriilidis, A. Varma, M. M. Bhasin, Effects of 1,2 dichloroethane addition on the optimal silver catalyst distribution in pellets for epoxidation of ethylene, *J. Catal.* 174 (1998) 1.
[19] J. T. Jankowiak, M. A. Barteau, Ethylene epoxidation over silver and copper-silver bimetallic catalysts: I. Kinetics and selectivity, *J. Catal.* 236 (2005) 366.
[20] J. T. Jankowiak, M. A. Barteau, Ethylene epoxidation over silver and copper-silver bimetallic catalysts: II. Cs and Cl promotion, *J. Catal.* 236 (2005) 384.
[21] B. Hammer, O. H. Nielsen, J. J. Mortensen, L. Bengtsson, L. B. Hansen, A. C. E. Madsen, Y. Morikawa, T. Bligaard, A. Christensen, *DACAPO version 2.7 (CAMP, Technical University, Denmark)*
[22] G. te Velde, F. M. Bickelhaupt, S. J. A. van Gisbergen, C. Fonseca Guerra, E. J. Baerends, J. G. Snijders, T. Ziegler, Chemistry with ADF, *J. Comput. Chem.* 22 (2001) 931.
[23] C. Fonseca Guerra, J. G. Snijders, G. te Velde, E. J. Baerends, Towards an order-N DFT method, *Theor. Chem. Acc.* 99 (1998) 391.
[24] *ADF2004.01, SCM, Theoretical Chemistry, Vrije Universiteit, Amsterdam, The Netherlands, http://www.scm.com.*
[25] D. Vanderbilt, Soft Self-Consistent pseudopotentials in a generalized eigenvalue formalism, *Phys. Rev. B* 41 (1990) 7892.
[26] N. W. Ashcroft, N. D. Mermin, *Solid State Physics*, Holt, Rinehart and Winston: New York (1976).
[27] S. H. Vosko, L. Wilk, M. Nusair, Accurate spin-dependent electron liquid correlation energies for local spin-density calculations - A critical analysis, *Can. J. Phys.* 58 (1980) 1200.
[28] A. D. Becke, Density-functional exchange-energy approximation with correct asymptotic-behavior, *Phys. Rev. A* 38 (1988) 3098.
[29] J. P. Perdew, Density-functional approximation for the correlation-energy of the inhomogeneous Electron-Gas, *Phys. Rev. B* 33 (1986) 8822.
[30] S. Linic, J. T. Jankowiak, M. A. Barteau, Selectivity driven design of bimetallic ethylene epoxidation catalysts from first principles, *J. Catal.* 224 (2004) 489.

[31] A. Michaelides, Z.-P. Liu, C. J. Zhang, A. Alavi, D. A. King, P. Hu, Identification of general linear relationships between activation energies and enthalpy changes for dissociation reactions at surfaces, *J. Am. Chem. Soc.* 125 (2003) 3704.
[32] Z.-P. Liu, P. Hu, General trends in the barriers of catalytic reactions on transition metal surfaces, *J. Chem. Phys.* 115 (2001) 4977.
[33] J. K. Norskov, T. Bligaard, A. Logadottir, S. Bahn, E. W. Hansen, M. Bollinger, H. Bengaard, B. Hammer, Z. Sljivancanin, M. Mavrikakis, Y. Xu, S. Dahl, C. J. H. Jacobsen, Universality in heterogeneous catalysis, *J. Catal.* 209 (2002) 275.
[34] S. Dahl, A. Logadottir, C. J. H. Jacobsen, J. K. Norskov, Electronic factors in catalysis: The volcano curve and the effect of promotion in catalytic ammonia synthesis, *Appl. Catal. A: Gen.* 222 (2001) 19.
[35] W.-X. Li, C. Stampfl, M. Scheffler, Oxygen adsorption on Ag(111): A density-functional theory investigation, *Phys. Rev. B* 65 (2002) 075407.
[36] E. Shustorovich, The bond-order conservation approach to chemisorption and heterogeneous catalysis: Applications and implications, *Adv. Catal.* 37 (1990) 101.
[37] E. Shustorovich, H. Sellers, The, UBI-QEP method: A practical theoretical approach to understanding chemistry on transition metal surfaces, *Surf. Sci. Rept.* 31 (1998) 1.
[38] C. Stegelmann, N. C. Schiodt, C. T. Campbell, P. Stoltze, Microkinetic modeling of ethylene oxidation over silver, *J. Catal.* 221 (2004) 630.
[39] C. Stegelmann, P. Stoltze, Microkinetic analysis of transient ethylene oxidation experiments on silver, *J. Catal.* 226 (2004) 129.
[40] C. Stegelmann, P. Stoltze, Isotope effect and selectivity promotion in ethylene oxidation on silver, *J. Catal.* 232 (2005) 444.

CHAPTER 9

Effect of Support on Ethylene Epoxidation on Ag, Au, and Au–Ag Catalysts

Sumaeth Chavadej,[*] **Siriphong Rojluechai,**[*] **Johannes W. Schwank,**[†] **and Vissanu Meeyoo**[‡]

Contents

1. Introduction — 284
2. Experimental — 285
   - 2.1. Catalyst preparation — 285
   - 2.2. Catalyst characterization — 286
   - 2.3. Ethylene epoxidation reaction experiments — 286
3. Results and Discussion — 287
   - 3.1. Catalyst characterization — 287
   - 3.2. Ethylene conversion — 289
   - 3.3. Ethylene oxide selectivity and CO_2 selectivity — 290
4. Conclusions — 295
   - Acknowledgments — 295
   - References — 295

Abstract

Au, Ag, and Au–Ag catalysts on different alumina, titania, and ceria supports were studied for their catalytic activity in the ethylene oxidation reaction. The addition of an appropriate amount of Au to an Ag/γ-Al_2O_3 catalyst was found to enhance the catalytic activity because Au acts as a diluting agent on the Ag surface creating new single silver sites which favor molecular oxygen adsorption. The Ag catalysts on both titania and ceria supports exhibited very poor catalytic activity toward the epoxidation reaction; so, pure Au catalysts on these two supports were investigated. The Au/TiO_2 catalysts provided the highest selectivity to ethylene oxide at a relatively low ethylene conversion, whereas the Au/CeO_2 catalysts were found to favor total oxidation

---

[*] The Petroleum and Petrochemical College, Chulalongkorn University, Bangkok 10330, Thailand
[†] Department of Chemical Engineering, The University of Michigan, Ann Arbor, Michigan 48109
[‡] Department of Chemical Engineering, Mahanakorn University, Bangkok 10530, Thailand

even at low temperatures. Among the studied catalysts, the bimetallic Au–Ag/γ-Al$_2$O$_3$ catalyst showed the best activity for the ethylene epoxidation. The catalytic activity of the gold catalysts was found to depend on the support material and the catalyst preparation method which governs the Au particle size and the interaction between the Au particles and the support.

Key Words: Ethylene epoxidation, Silver catalyst, Gold catalyst, Au–Ag catalyst. © 2008 Elsevier B.V.

1. INTRODUCTION

The most widely used process for ethylene oxide manufacture is the direct catalytic oxidation of ethylene with air or oxygen over supported silver catalysts [1]. A unique support material for silver catalysts is commercial α-alumina because it provides highly selective ethylene epoxidation. This is due to its inertness for the isomerization of ethylene oxide to acetaldehyde [2]. Low–surface-area α-alumina supports with high silver loadings are typically used for the commercial production of ethylene oxide. In these commercial silver catalysts, silver is poorly dispersed over the support [3]. As a result, these catalysts give relatively low yields of ethylene oxide. Seyedmonir et al. [4] conducted a systematic study of the catalytic activity of Ag on different supports such as η-Al$_2$O$_3$, TiO$_2$, α-Al$_2$O$_3$, and SiO$_2$. They reported that both ethylene oxide selectivities over Ag/η-Al$_2$O$_3$ and Ag/TiO$_2$ were very low, about 10%, as compared to a high value of about 60% over the Ag/α-Al$_2$O$_3$ catalysts in the presence of 0.5 ppm EDC (ethylene dichloroethane). In contrast, in the absence of EDC and CO$_2$ at 523 K, ethylene oxide selectivities of 17% and 55% were obtained over 4.4 and 7.6 nm Ag crystallites on SiO$_2$, respectively, compared to 23% over 1-μm Ag crystallites on α-Al$_2$O$_3$. Mao and Vannice [5] reported that a high–surface-area (HSA) α-alumina (about 100 m^2/g) was a poor support for the ethylene epoxidation reaction due to the formation of small nanoparticles of Ag. The reaction rate was found to increase with increasing surface area of catalyst, but the ethylene oxide selectivity decreased.

TiO$_2$ and CeO$_2$ have some special properties which could enhance the catalytic activity of ethylene oxidation reactions. Both TiO$_2$ and CeO$_2$ have gained a great deal of attention because they are nonstoichiometric oxide materials which allow oxygen migration onto the surfaces of supported metallic particles, which, in turn, promote oxidative reactions [6]. It has been reported that group VIII noble metals supported on TiO$_2$ exhibit a strong metal–support interaction (SMSI). Seyedmonir et al. [4] studied the ethylene oxide reaction over Ag/TiO$_2$, and Yong et al. [2] reported that silver supported on TiO$_2$ showed zero ethylene oxide selectivity. It was also concluded that the zero selectivity of the silver catalyst supported on TiO$_2$ for the ethylene epoxidation is due to the isomerization of ethylene oxide to acetaldehyde on the support followed by the complete oxidation reaction. Shastri et al. [7] studied the catalytic behavior of gold supported on TiO$_2$. They pointed out that high gold dispersion on TiO$_2$ was possible up to 973 K and that the agglomeration of gold into large particles coincided with the phase transformation of the titania into rutile at 1,073 K. The stability of the gold dispersion was explained to be not due to the SMSI effect. A temperature of 973 K appeared sufficient to accomplish the complete phase transformation of anatase to rutile in blank TiO_2. Mallick and Scurrell [8] reported that introducing ZnO onto TiO_2 caused a surface modification of the titania and was associated with a negative effect on the catalytic activity. Compared with Au/TiO_2, Au/TiO_2–ZnO was found to behave as a moderately good catalyst up to a certain temperature, but appeared to suffer from severe deactivation with an increase in the time on stream at higher temperatures.

Ceria is known to have good thermal stability, resulting in the maintenance of its high surface area at high temperatures. Thus, the use of ceria as a support can minimize the sintering of loaded catalyst particles. Interestingly, ceria also behaves as an oxygen reservoir so that it has been widely employed for CO oxidation and water-gas shift reactions [9,10]. Bera et al. [11] reported that the Au dispersed on a CeO_2 surface was found to be metallic in structure (Au^0) as well as ionic in form (Au^{3+}) and both Au^0 and Au^{3+} species were catalytically active. All the oxidation reactions of NO, CO, and hydrocarbons over 1 wt.% Au/CeO_2 with heat treatment (1,073 K for 100 h) were found to occur at significantly lower temperatures, as compared to those of 1 wt.% Au/TiO_2 and 1 wt.% Au/Al_2O_3.

In this study, the effect of support material on the ethylene epoxidation reaction over Au and Au–Ag/HSAγ-Al_2O_3 catalysts was determined. TiO_2 and CeO_2, as reducible oxide supports, were also investigated, in comparison with the HSA alumina support.

2. EXPERIMENTAL

2.1. Catalyst preparation

In this work, silver catalysts were prepared by the incipient wetness method using aluminum oxide (fumed gamma alumina, Degussa C, 85–115 m^2/g, Degussa AG) and silver nitrate precursor solutions to achieve various nominal silver loadings. From our previous results [12], an optimum Ag loading of 13.2 wt.% was found to provide maximum selectivity to ethylene oxide at a relatively high ethylene conversion. Hence, this Ag loading was used to prepare Au–Ag catalysts of different Au contents using the impregnation method with a chloroauric acid precursor (Sigma-Aldrich Co.) [18].

Apart from alumina, titania and ceria were also used to prepare catalysts in order to determine the effect of support material on the ethylene epoxidation reaction. Initially, silver loaded on either titania or ceria was prepared at different silver loadings using a silver nitrate precursor. All of these silver catalysts on both supports were found to have no activity toward ethylene epoxidation. Hence, bimetallic Au–Ag catalysts on these two supports were not studied, but instead only pure gold catalysts on these reducible oxide supports were investigated. Au catalysts on TiO_2 (Degussa P25, Degussa AG) and CeO_2 (sol–gel urea hydrolysis) were prepared from aqueous gold precursor solutions to obtain different gold loadings by using the impregnation method. Then, the catalyst precursors were dried at 383 K overnight followed by calcination in air at 773 K for 5 h. Other Au/TiO$_2$ catalysts were prepared by the deposition–precipitation and single-step, sol–gel methods [13–15].

The CeO$_2$ support was synthesized by the hydrolysis of a cerous nitrate solution with a urea solution. A first solution was prepared by dissolving 6.51 g of Ce(NO$_3$)$_3$6H$_2$O (purity 99%, Sigma-Aldrich Co.) in 150 ml of distilled water. Next, a second solution was prepared by dissolving 1.20 g of urea, (NH$_2$)$_2$CO (purity 99%, Sigma-Aldrich Co.) in 50 ml of distilled water. Then, these two solutions were mixed in a 250-ml Pyrex bottle with a screw cap. The resultant mixture was placed in an oven at 373 K for 50 h to achieve the hydrolysis step. The mixture solution was then allowed to cool to room temperature and was centrifuged to separate the precipitate. The separated precipitate was washed with hot distilled water (353 K) four to five times and finally was washed with ethanol. The resulting precipitate was dried overnight at 383 K and calcined at 773 K for 4 h to obtain CeO$_2$. Two methods were used to deposit the gold component. For the single-step sol–gel method with urea, 0.0447 g of HAuCl$_4$ was added to the mixture of the two prepared solutions of cerous nitrate and urea. For the second preparation method, the CeO$_2$ synthesized by the sol–gel method was impregnated with different Au loadings. The same procedure was used for the preparation of the Au/TiO$_2$ catalysts.

2.2. Catalyst characterization

The specific surface areas of all catalyst samples prepared were determined by N$_2$ adsorption at 77 K [Brunauer–Emmette–Teller (BET) method] using a surface-area analyzer (Autosorb 1, Quantachrome Instruments). The metal contents in all of the catalyst samples were analyzed by an atomic absorption spectrophotometer (Spectr AA-300, Varian Inc.). The crystalline structures of all catalyst samples were examined by X-ray diffraction (XRD) on a RINT 2000 diffractometer (Rigaku Corp.). Mean crystallite sizes of the studied catalysts were calculated by the Scherrer equation from X-ray line broadening using the full-line width at half-maximum of the highest intensity X-ray peaks. The morphology of the catalyst samples was investigated by transmission electron microscopy (TEM) (2010, JEOL Ltd.). The existence of Au and Ag particles on these three supports was verified by using energy dispersive spectroscopy (EDS) in conjunction with the TEM. The particle sizes of Au and Ag were determined by the statistical data analysis of the TEM images.

2.3. Ethylene epoxidation reaction experiments

Ethylene epoxidation reaction experiments over all studied catalysts were conducted in a differential flow reactor, which was operated at a constant pressure of 3.6 MPa and different reaction temperatures. The tubular reactor having 10-mm internal diameter was placed in a furnace equipped with a temperature controller. Typically, 30 mg of a catalyst sample was placed inside the Pyrex tube reactor and secured with Pyrex glass wool plugs. The packed catalyst was initially pretreated with oxygen at 473 K for 2 h in order to remove carbonaceous impurities and residual moisture from the catalyst. The feed gas was a mixture of oxygen and ethylene in helium. All of the gases were high-purity grade and their flow rates were regulated by mass flow controllers. From preliminary results, the levels of 6% of both oxygen and ethylene were found to yield the maximum selectivity to ethylene oxide over the 13.2 wt.% Ag/γ-Al$_2$O$_3$ catalyst [12]. Therefore, this composition of feed gas mixture was selected for further experiments in order to determine the effect of support material on the ethylene epoxidation reaction. The feed gas was passed through the reactor at a constant space velocity of 6,000 h^{-1} and the reaction temperature was varied from 493 to 543 K for all catalysts, except for the Au/CeO$_2$ catalysts for which a temperature range of 413–473 K was used. At temperatures lower than these studied ranges, the ethylene conversion was extremely low and all reaction products were lower than detectable concentrations. Inlet and exit gases were analyzed by using an online gas chromatograph (Hewlett-Packard 5890 Series II) equipped with a HaYeseb D 100/120-packed column (Valco Instruments Co. Inc.), capable of separating CO$_2$, CO, C$_2$H$_4$, and O$_2$. Under the studied conditions for all prepared catalysts, the concentrations of CO were below the detectable limit, indicating that the formation of CO can be neglected. The ethylene oxide selectivity was calculated from the carbon material balance with 0.25% carbon atom error [16,17]. Moreover, the calculated values of ethylene oxide produced were confirmed by performing O$_2$ mass balance with 0.3% oxygen atom error.

3. RESULTS AND DISCUSSION

3.1. Catalyst characterization

The measured values of the BET surface areas of all the studied catalysts prepared with the different support materials at the optimum catalyst loadings are shown in Table 9.1. For either commercial γ-alumina or titania, the addition of either silver or gold did not affect its surface area significantly. Interestingly, the BET surface area of the γ-alumina support was not altered significantly even though a large amount of silver was loaded up to 13.2 wt.%, implying that the silver particles are well dispersed without sintering. The good dispersion of the silver and gold particles on γ-Al$_2$O$_3$ is clearly verified by the existence of nanosized particles in all studied catalysts measured by both XRD and TEM, as shown in Table 9.1. For the HSA γ-alumina support with pure Ag and Au–Ag, the mean crystallite sizes of Ag and Au–Ag obtained by XRD were found to be much lower than those obtained by TEM. The difference probably arises because the XRD technique measures the mean effective size of each crystallite, while the TEM method measures the mean diameter of particles in which each particle may consist of several crystallites. As expected, no Au peak was detectable by XRD because the crystallite sizes of the Au particles present on all of the studied supports were too small. Therefore, the mean Au crystallite sizes were obtained only by using TEM enhanced with EDS. From Table 9.1, the Au particles on the TiO$_2$

TABLE 9.1 Structural characteristics of prepared catalysts on different support materials at optimum catalyst loadings

Support	Ag content (wt.%)	Au content (wt.%)	BET surface area (m^2/g)	Metal crystallite size (nm) XRD	Metal crystallite size (nm) TEM
Al$_2$O$_3$^{a} (imp)	13.2	0	90	19 (Ag)	30 (Ag)
Al$_2$O$_3$^{a} (imp)	13.2	0.54	89	18 (Au–Ag)	30 (Ag) 4 (Au)
TiO$_2$^{b} (imp)	0	0.96	60	–	3.2 (Au)
TiO$_2$ (dp)	0	1.28	58	–	2.5 (Au)
TiO$_2$ (sol–gel)	0	0.96	60	–	1.2 (Au)
CeO$_2$ (sol–gel/imp)	0	0.69	105	–	6.2 (Au)
CeO$_2$ (single-step sol–gel)	0	1.03	102	–	5.6 (Au)

Note: imp, impregnation method; dp, deposition–precipitation method; XRD, X-ray diffraction; TEM, transmission electron microscopy.
(Ag) indicates Ag particles.
(Au) indicates Au particles.
(Au–Ag) indicates Au–Ag crystallites.
^{a} BET surface area of Al$_2$O$_3$-Degussa is 98 m^2/g.
^{b} BET surface area of TiO$_2$-Degussa P25 is 61 m^2/g.

prepared by the impregnation method have the largest crystallite size, as compared to those prepared on the TiO$_2$ prepared by the deposition–precipitation and single-step sol–gel methods. For the Ag/CeO$_2$ catalysts, the sol–gel/impregnation method gave a higher mean crystallite size of Au particles than the single-step sol–gel technique did. This is because the single-step sol–gel technique gives a better dispersion than the sol–gel/impregnation method. Interestingly, all Au catalysts on three studied supports provided Au particles in the nanometer size range.

The TEM/EDS technique was used to image both Ag and Au particles on the studied supports. On the basis of both XRD and TEM results, the Ag particles prepared on the commercial γ-alumina were in the nanometer size range (Table 9.1 and Fig. 9.1a), indicating that silver is highly dispersed on the alumina support. The addition of Au on the 13.2 wt.% Ag/Al$_2$O$_3$ catalyst resulted in the formation of small particles of Au on the Ag particles with the Au particle size being about 4 nm (Table 9.1 and Fig. 9.1b). As mentioned in our previous publication, at a low Au loading not higher than 0.54 wt.%, separate Au and Ag particles were formed, while at a high Au loading greater than 0.54 wt.%, an alloy was formed [18]. Figure 9.2a and c reveals the presence of the gold particles as dark spots with highly uniform dispersion in the nanometer size range on the three TiO$_2$ supports. In contrast, Fig. 9.3a shows an inhomogeneous dispersion of gold particles on the CeO$_2$ support prepared by the impregnation method as seen by the random collection of dark spots. Interestingly, Fig. 9.3b shows that the single-step sol–gel technique provides good dispersion of Au on CeO$_2$, indicating a strong interaction between CeO$_2$ and Au. From these results, it can be concluded that a good dispersion of Au particles with nanometer size range can be obtained on the HSA

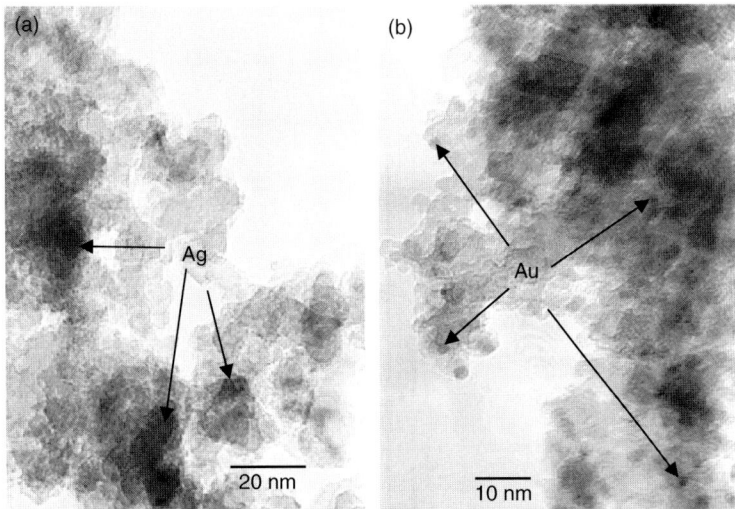

FIGURE 9.1 Transmission electron microscopy (TEM) micrographs of (a) 13.2 wt.% Ag/γ-Al$_2$O$_3$ and (b) 0.63 wt.% Au-13.2 wt.% Ag/γ-Al$_2$O$_3$.

γ-alumina, titania, and ceria supports. In comparing the three supports, ceria gave the largest particle size of Au. From the results, it can be concluded that Au particle size depends on the preparation method and the nature of the support.

3.2. Ethylene conversion

The effect of reaction temperature on the ethylene conversion over the various catalysts having optimum loadings of Ag and Au on the different supports is shown in Table 9.2. The optimum catalyst loading on each support is reported elsewhere [12,18]. Both Ag/γ-Al$_2$O$_3$ and Au–Ag/γ-Al$_2$O$_3$ catalysts were operated in the temperature range of 493–573 K. The ethylene conversion was low from 493 to 528 K but increased significantly with increasing reaction temperatures above 528 K. However, for any given temperature, the Au–Ag/γ-Al$_2$O$_3$ catalyst provided nearly the same ethylene conversion as the Ag/γ-Al$_2$O$_3$ catalyst.

For the Au/TiO$_2$ catalysts, the ethylene conversion gradually increased with increasing temperature in the range of 493–543 K, with no activity below 493 K. For any given reaction temperature, the 0.96 wt.% Au on TiO$_2$ catalyst prepared by the impregnation method was found to give a higher ethylene conversion than those prepared by the deposition–precipitation or single-step sol–gel methods.

Interestingly, the ethylene conversion over the Au/CeO$_2$ catalysts was found in the temperature range of 413–433 K, which is considerably lower than that over the Au–Ag/γ-Al$_2$O$_3$ and Au/TiO$_2$ catalysts. The ethylene conversion over each Au/CeO$_2$ catalyst substantially increased with increasing reaction temperature, and the 0.69 wt.% Au/CeO$_2$ catalyst prepared by the sol–gel/impregnation method had a higher ethylene conversion than the Au/CeO$_2$ catalyst prepared by the single-step sol–gel method.

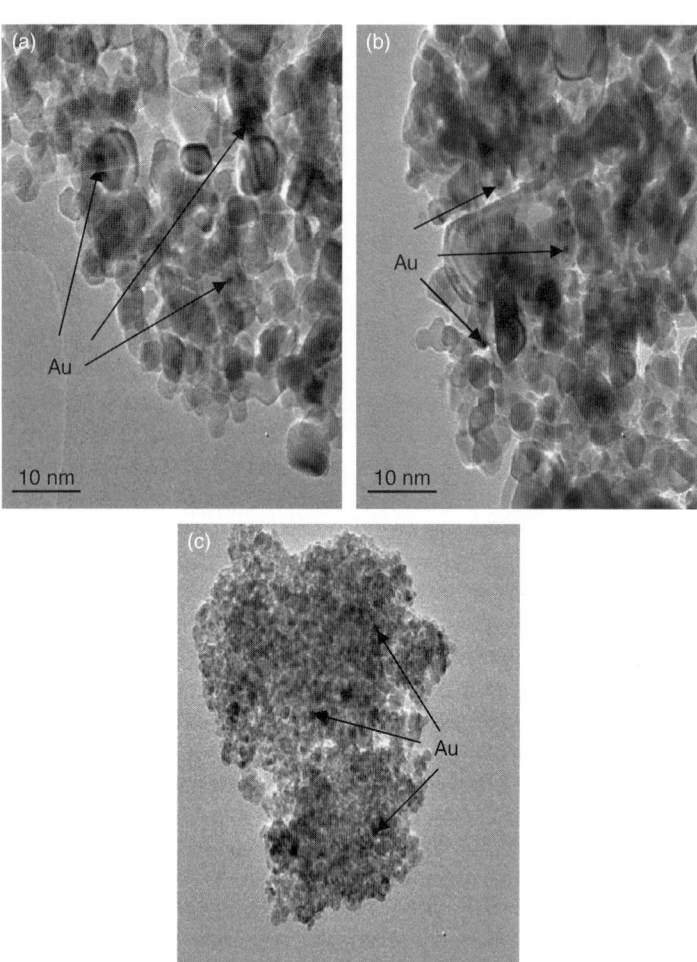

FIGURE 9.2 Transmission electron microscopy (TEM) micrographs of (a) 0.96 wt.% Au/TiO$_2$ (impregnation), (b) 1.28 wt.% Au/TiO$_2$ (deposition–precipitation), and (c) 0.96 wt.% Au/TiO$_2$ (sol–gel).

Among these three supports, the Au/TiO$_2$ catalysts provided a slightly lower ethylene conversion than the catalysts on γ-Al$_2$O$_3$ and CeO$_2$. In addition, for any given ethylene conversion, the CeO$_2$ required a much lower reaction temperature since CeO$_2$ has an oxygen storage property or is reducible and can supply atomic oxygen species to react with ethylene.

3.3. Ethylene oxide selectivity and CO$_2$ selectivity

Figure 9.4 illustrates the effect of Au loading of the 13.2 wt.% Ag/γ-Al$_2$O$_3$ catalyst on the ethylene epoxidation reaction by plotting Au loading versus the normalized turnover number (the turnover number ratio of Au–Ag catalyst to Ag catalyst). The normalized turnover number of ethylene oxide increased slightly with

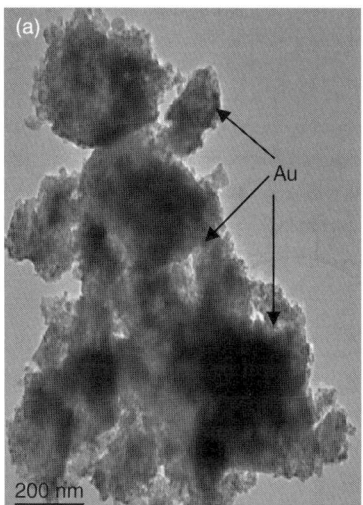 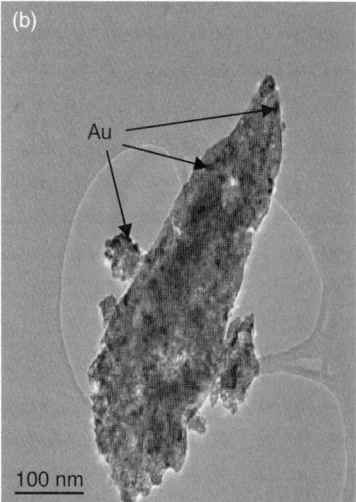

FIGURE 9.3 Transmission electron microscopy (TEM) micrographs of (a) 0.69 wt.% Au/CeO$_2$ (sol–gel/impregnation) and (b) 1.03 wt.% Au/CeO$_2$ (single-step sol–gel).

increasing Au loading and reached the maximum value at an Au loading between 0.54 and 0.63 wt.%. It decreased with increasing Au loading above 0.63 wt.%. The results indicate that addition of gold with an appropriate amount on Ag catalyst can promote the ethylene epoxidation reaction by weakening the Ag–O bond.

Table 9.2 compares the selectivity over all studied catalysts. The ethylene oxide selectivity of either the Ag/γ-Al$_2$O$_3$ or Au–Ag/γ-Al$_2$O$_3$ catalysts drastically decreased with increasing reaction temperature. This is because the total oxidation of ethylene is favorable at high temperatures. The addition of an appropriate amount (<0.63 wt.%) of Au to the 13.2 wt.% Ag catalyst (which exists in separate Au particles on Ag particles) was found to promote the ethylene epoxidation reaction by weakening the Ag–O bond in the reaction temperature range of 493–528 K [18,19]. When the Au loading was higher than 0.63 wt.%, the activity of the ethylene epoxidation decreased because of the formation of an Au–Ag alloy which resulted in a decrease in the adsorption capacity of molecular oxygen [18].

According to our temperature-programmed desorption (TPD) results [18], addition of a small amount of Au to the Ag catalyst does not alter the peak maximum temperature of oxygen, as shown in Fig. 9.5. However, a significant shift to a lower temperature of the peak maximum temperature was found at the highest Au loading of 0.93 wt.%. As shown in Table 9.3, the calculated amount of oxygen adsorbed on the catalysts decreases with increasing Au content. Interestingly, the oxygen adsorption on the Ag catalyst decreased when the Au loading increased. This indicates that the interaction between silver and oxygen is weakened remarkably in the presence of gold. The presence of Au atoms has been found to affect the electronic properties of Ag [19,20]. In general, the dissociative adsorption of oxygen on Ag requires a charge transfer from Ag to oxygen. Hence, it is not unexpected that the electron deficiency induced on Ag atoms by the

TABLE 9.2 The activity of each prepared catalyst at space velocity of 6,000 h^{-1}, $P = 3.6$ MPa and 6% O_2 and 6% C_2H_4 balance with He

Support	Ag content (wt.%)	Au content (wt.%)	Temperature (K)	C_2H_4 conversion (%)	EO selectivity (%)	CO_2 selectivity (%)
Al_2O_3	13.2	–	493	0.9	84	16
			513	1.6	83	17
			528	1.8	62	38
			543	3.1	38	62
Al_2O_3	13.2	0.63	493	1.2	93	7
			513	1.6	89	11
			528	1.8	79	21
			543	3.9	32	68
TiO_2 (imp)	–	0.96	493	1.0	99	1
			513	1.1	96	4
			528	1.3	88	12
			543	1.3	76	24
TiO_2 (dp)	–	1.28	493	0.9	88	12
			513	0.9	82	14
			528	1.1	77	23
			543	1.2	56	44
TiO_2 (sol–gel)	–	0.96	493	0.6	86	14
			513	0.8	78	22
			528	0.9	66	34
			543	1.1	42	58
CeO_2 (sol–gel/imp)	–	0.69	413	0.7	55	45
			433	1.2	0	100
			453	2.1	0	100
			473	4.3	0	100
CeO_2 (single-step sol–gel)	–	1.03	413	0.5	77	23
			433	0.9	51	49
			453	1.1	18	82
			473	2.6	0	100

imp, impregnation method; dp, deposition–precipitation method.

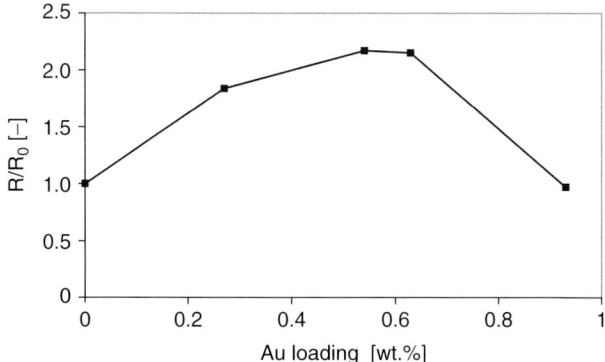

FIGURE 9.4 Normalized turnover number of ethylene epoxidation on 13.2 wt.% Ag/Al$_2$O$_3$ at various gold loadings and reaction temperature of 513 K [turnover number obtained on 13.2 wt.% Ag/Al$_2$O$_3$ with various gold loadings (R) over that obtained on 13.18 wt.% Ag/Al$_2$O$_3$ with no Au loading (R_0)].

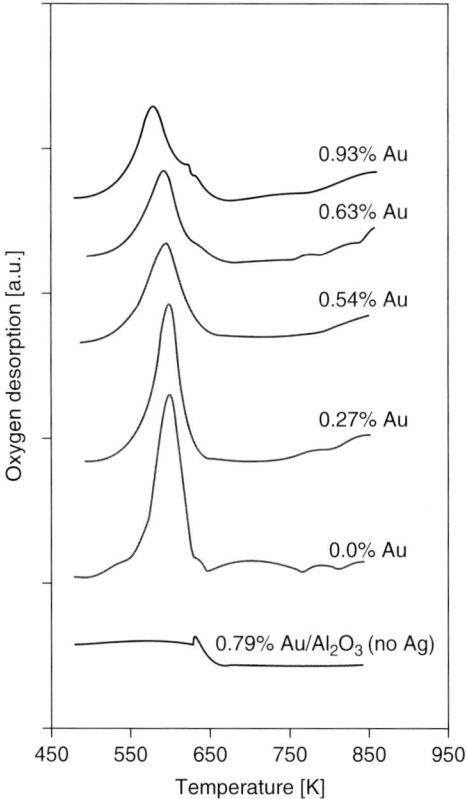

FIGURE 9.5 TPD profiles of O$_2$ of 13.2 wt.% Ag/Al$_2$O$_3$ at various gold loadings and 0.79 wt.% Au/Al$_2$O$_3$.

TABLE 9.3 Total oxygen desorption of 13.2 wt.% Ag/γ-Al$_2$O$_3$ at various Au loadings

Catalyst	Total oxygen desorption ($\times 10^{-6}$ mol/gcat)
13.2 wt.% Ag/γ–Al$_2$O$_3$	220
0.27 wt.% Au–13.2 wt.% Ag/γ-Al$_2$O$_3$	202
0.54 wt.% Au–13.2 wt.% Ag/γ-Al$_2$O$_3$	194
0.63 wt.% Au–13.2 wt.% Ag/γ-Al$_2$O$_3$	187
0.93 wt.% Au–13.2 wt.% Ag/γ-Al$_2$O$_3$	139

presence of neighboring Au atoms would result in weakening the Ag–O bond. The result confirms that under the presence of small amounts of Au on Ag catalyst, the interaction between gold and silver affects significantly the oxygen adsorption. Kondarides and Verykios [19] also reported that added Au weakened the bond strength between silver and oxygen.

As mentioned earlier, there was zero selectivity to ethylene oxide over Ag catalysts on both TiO$_2$ and CeO$_2$. Therefore, only gold loaded on both supports was studied in this work. Table 9.2 shows that a catalyst composed of 0.96 wt.% Au on commercial TiO$_2$ prepared by the impregnation method gives the highest ethylene oxide selectivity among Au/TiO$_2$ catalysts prepared by the impregnation, deposition–precipitation, and single-step sol–gel methods. In contrast, the 0.96 wt.% Au on the TiO$_2$ prepared by the single-step sol–gel method provided the highest CO$_2$ selectivity. From the results, it is clearly seen that there is a good correlation between the particle size of gold, which is determined by TEM (Table 9.1) and the ethylene epoxidation reaction. The ethylene oxide selectivity increased when the gold particle size became larger. It has been reported that oxygen species are formed at the perimeter interface between the gold particles and the TiO$_2$ support when the particle size is greater than 2 nm [21]. The oxygen species located at the perimeter interface is mostly molecular oxygen [22,23] which is believed to react directly with ethylene in the gas phase to produce ethylene oxide. Among the three preparation methods of Au catalysts on TiO$_2$, the impregnation method was found to give the largest gold particle size (3.2 nm) compared with the other two methods (2.5 and 1.2 nm). Our present work has confirmed that ethylene oxide selectivity depends on the particle size of gold and the interaction between gold and the support. Again, the impregnation method also provides both the highest ethylene conversion and ethylene oxide selectivity as compared to the other two preparation techniques. The impregnation method likely provides more active reaction sites to generate more active oxygen species than the other two preparation methods.

For the Au/CeO$_2$ catalysts, the CO$_2$ selectivity increased substantially with increasing reaction temperature, whereas the ethylene oxide selectivity decreased (see Table 9.2). The Au on CeO$_2$ prepared by the sol–gel/impregnation method was found to favor the total oxidation reaction over the epoxidation reaction as compared to that prepared by the single-step sol–gel method. This is because the impregnation method provides more active Au reaction sites than single-step sol–gel method. Table 9.1 shows that the Au particle sizes of both catalyst preparation methods are less than 10 nm which are believed to favor the total oxidation reaction according to the literature [21]. It is also believed that the high oxygen mobility of reducible CeO_2 is responsible for the enhancement of the total oxidation reaction at much lower temperatures as compared to the other supports, γ-Al_2O_3 and TiO_2. The same phenomenon was also reported by Wootsch et al. [24] whereby Pt/CeO_2 catalysts were more active at lower temperatures for both CO and H_2 oxidation reactions than Pt/Al_2O_3. Pozdnyakova et al. [25] studied selective CO oxidation in hydrogen-rich environments over Pt/CeO_2 catalysts. Complete CO oxidation was observed at over 1% Pt/CeO_2 at a very low temperature of 370 K. Another possibility is the further oxidation of ethylene oxide by the bulk atomic oxygen of the CeO_2 support.

4. CONCLUSIONS

Among the three support materials, γ-Al_2O_3, TiO_2, and CeO_2, the ethylene epoxidation was found to occur over the Ag catalysts on γ-Al_2O_3, while Ag catalysts on both TiO_2 and CeO_2 gave only the total oxidation reaction. For the Au catalysts on γ-Al_2O_3, no activity was found toward the ethylene epoxidation reaction. The addition of an appropriate quantity of Au on 13.2% Ag/γ-Al_2O_3 catalyst was found to enhance both the ethylene conversion and the selectivity to ethylene oxide. Under an Au loading lower than the optimum gold loading of 0.63 wt.%, the existence of the separate Au–Ag structure creates new single silver sites which favor molecular oxygen adsorption, leading to the enhancement of the ethylene oxide selectivity. Interestingly, the Au catalysts on TiO_2 also were shown to be a good candidate for the ethylene epoxidation reaction, provided that the Au particle size is greater than 2 nm. CeO_2 is a poor support for Au catalysts because it is easily reduced and promotes total oxidation instead of epoxidation. The results also show that the catalytic activity of Au catalysts depends not only on the size of the Au particles and the support material, but also on the catalyst preparation method.

ACKNOWLEDGMENTS

The authors would like to gratefully acknowledge The Thailand Research Fund for providing a Royal Golden Jubilee scholarship for Mr. Siriphong Rojluechai and a Basic Research Grant for the corresponding author. The Research Unit of Petrochemical and Environmental Catalysis under The Ratchadapisakesompok Fund, Chulalongkorn University, and The National Excellence Center for Petroleum, Petrochemicals and Advanced Materials under The Ministry of Education are also acknowledged for their partial financial support and for providing all analytical instruments.

REFERENCES

[1] G. Ertl, H. Knözinger, J. Weitkamp, In: Handbook of Heterogeneous Catalysis, Vch Weinheim (1997) p. 2244.
[2] Y. S. Yong, E. M. Kennedy, N. W. Cant, Oxide catalysed reactions of ethylene oxide under conditions relevant to ethylene epoxidation over supported silver, Appl. Catal. A: Gen. 76 (1991) 31.

[3] S. Matar, M. J. Mirbach, H. A. Tayim, *In*: Catalysis in Petrochemical Processes, Kluwer Academic Publishers Dordrecht; The Netherlands (1989) p. 85.
[4] S. R. Seyedmonir, J. K. Plischke, M. A. Vannice, H. W. Young, Ethylene oxidation over small silver crystallites, *J. Catal.* 123 (1990) 534.
[5] C.-F. Mao, M. A. Vannice, High surface area α-alumina. III. Oxidation of ethylene, ethylene oxide, and acetaldehyde over silver dispersed on high surface area α-alumina, *Appl. Catal. A: Gen.* 122 (1995) 61.
[6] M. J. Holgado, A. C. Inigo, V. Rives, Effect of preparation conditions on the properties of highly reduced Rh/TiO_2, *Appl. Catal. A: Gen.* 175(1–2) (1998) 33.
[7] A. G. Shastri, A. K. Datye, J. Schwank, Gold-titania interactions: Temperature-dependence of surface-area and crystallinity of TiO_2 and gold dispersion, *J. Catal.* 87(1) (1984) 265.
[8] K. Mallick, M. S. Scurrell, CO oxidation over gold nanoparticles supported on TiO_2 and TiO_2-ZnO: Catalytic activity effects due to surface modification of TiO_2 with ZnO, *Appl. Catal. A* 253 (2003) 527.
[9] S. Imamura, H. Yamada, K. Utani, Combustion activity of Ag/CeO_2, *Appl. Catal. A: Gen.* 192 (2000) 221.
[10] J. Kaspar, P. Fornasiero, M. Graziani, Use of CeO_2-based oxides in the three-way catalysis, *Catal. Today* 50 (1999) 285.
[11] P. Bera, M. S. Hegde, Characterization and catalytic properties of combustion synthesized Au/CeO_2 catalyst, *Catal. Lett.* 79(1–4) (2002) 75.
[12] S. Rojluechai, *Selective oxidation of ethylene over supported ag and bimetallic Au-Ag catalysis*, Ph.D. Dissertation, Petroleum and Petrochemical College, Chulalongkorn University, Bangkok, Thailand, 2006.
[13] S. Tsubota, D. A. H. Cunningham, Y. Bando, M. Haruta, Preparation of nanometer gold strongly interacted with TiO_2 and the structure sensitivity in low-temperature oxidation of CO, *Prep. Catal.* 91 (1995) 227.
[14] B. S. Uphade, M. Okumura, S. Tsubota, M. Haruta, Effect of physical mixing of CsCl with Au/Ti-MCM-41 on the gas-phase epoxidation of propane using H_2 and O_2, *Appl. Catal. A: Gen.* 190(1–2) (2000) 43.
[15] M. Schneider, D. G. Duff, T. Mallat, M. Wildberger, A. Baiker, High-surface-area platinum-titania aerogels-preparation, structural-properties, and hydrogenation activity, *J. Catal.* 147(2) (1994) 500.
[16] D. Lafarga, M. A. Al-Juaied, C. A. Bondy, A. Varma, Ethylene epoxidation on Ag-Cs/α-Al_2O_3 catalyst: Experimental results and strategy for kinetic parameter determination, *Ind. Eng. Chem. Res.* 39 (2000) 2148.
[17] K. L. Yeung, A. Gavriilidis, A. Varma, M. M. Bhasin, Effects of 1, 2 dichloroethane addition on the optimal silver catalyst distribution in pellets for epoxidation of ethylene, *J. Catal.* 174 (1998) 1.
[18] S. Rojluechai, S. Chavadej, J. Schwank, V. Meeyoo, Activity of ethylene epoxidation over high surface area alumina support Au-Ag catalysts, *J. Chem. Eng. Japan* 39 (2006) 321.
[19] D. I. Kondarides, X. E. Verykios, Interaction of oxygen with supported Ag-Au alloy catalysts, *J. Catal.* 158 (1996) 363.
[20] N. Tories, X. E. Verikios, The oxidation of ethylene over silver-based alloy catalysts: Silver-gold alloys, *J. Catal.* 123 (1987) 161.
[21] T. Hayashi, K. Tanaka, M. Haruta, Selective vapor-phase epoxidation of propylene over Au/TiO_2 catalysts in the presence of oxygen and hydrogen, *J. Catal.* 178(2) (1998) 566.
[22] J. Schwank, Catalytic gold: Application of element gold in heterogeneous catalysis, *Gold Bull.* 16(4) (1983) 103.
[23] M. Haruta, M. Date, Advances in the catalysis of Au nanoparticles, *Appl. Catal. A: Gen.* 222 (2001) 427.
[24] A. Wootsch, C. Descorme, D. Duprez, Preferential oxidation of carbon monoxide in the presence of hydrogen (PROX) over ceria–zirconia and alumina-supported Pt catalysts, *J. Catal.* 225 (2004) 259.
[25] O. Pozdnyakova, D. Teschner, A. Wootsch, J. Kröhnert, B. Steinhauer, H. Sauer, L. Toth, F. C. Jentoft, A. Knop-Gericke, Z. Paál, R. Schlögl, Preferential CO oxidation in hydrogen (PROX) on ceria-supported catalysts, part I: Oxidation state and surface species on Pt/CeO_2 under reaction conditions, *J. Catal.* 237 (2006) 1.

CHAPTER **10**

Epoxidation of Propylene with Oxygen–Hydrogen Mixtures

Masatake Haruta and **Jun Kawahara**

Contents		
	1. Introduction	298
	2. Liquid-Phase Epoxidation of Propylene	300
	3. Gas-Phase Epoxidation of Propylene	301
	3.1. Mechanistic study of Au/TiO_2 catalysts	303
	3.2. Improvements in the catalytic performance of Au/TS-1	307
	3.3. Improvement of the catalyst life of Au/3D mesoporous Ti-SiO_2	308
	3.4. Replacement of Au with Ag	310
	4. Conclusions	310
	References	311

Abstract

Propylene oxide (PO) can be formed by a single step either in the liquid phase or in the gas phase through the reaction of propylene with hydrogen–oxygen mixtures. In a semibatch reactor using methanol or butanol as a solvent, Pd/TS-1 catalysts promoted by Pt selectively produce PO with a space time yield (STY) above 120 g_{PO} kg_{cat}^{-1} h^{-1}, which is comparable to that of industrial ethylene oxide production. In the gas phase, gold nanoparticles deposited on anatase TiO_2 exhibit selectivities to PO higher than 90%. The high STY is obtained when the support for gold nanoparticles is changed from TiO_2 to microporous crystalline TS-1 or mesoporous Ti-SiO_2. In both the liquid- and gas-phase epoxidation, H_2O_2 is formed by the catalysis of noble metals and transforms at isolated tetrahedrally coordinated Ti cation sites into Ti-OOH species, which produces PO by the reaction with propylene.

Key Words: Propylene, Propylene oxide, Liquid-phase epoxidation, Semibatch reactor, Supercritical CO_2, Gas-phase epoxidation, Pd/TS-1 catalyst, Au/TiO_2 catalyst, Au/Ti-SiO_2 catalyst, Ag catalyst, Mesoporous

Department of Applied Chemistry, Graduate School of Urban Environmental Sciences, Tokyo Metropolitan University, Hachioji 192–0397, Tokyo, Japan, and Japan Science and Technology Agency, CREST, Kawaguchi 332-0012, Saitama, Japan

Mechanisms in Homogeneous and Heterogeneous Epoxidation Catalysis
DOI: 10.1016/B978-0-444-53188-9.00010-9

© 2008 Elsevier B.V.
All rights reserved.

titanium silicates, Space time yield, Deposition–precipitation method, Gold nanoparticles, Gold clusters, Density functional theory, Titanium hydroperoxide, H_2O_2 formation, Promoter, Trimethylsilylation, Trimethylamine. © 2008 Elsevier B.V.

1. INTRODUCTION

The two major methods of propylene oxide (PO) synthesis are the chlorohydrin process and the hydroperoxidation process. In the chlorohydrin process, propylene is reacted with chlorine and water to generate propylene chlorohydrin, which is reacted with a base such as calcium hydroxide to produce PO [1]. In the hydroperoxidation processes, propylene is reacted with an organic hydroperoxide derived from isobutane, ethylbenzene, or cumene to produce PO and an alcohol [2]. These processes require multiple steps and suffer the additional drawback of not producing the desired PO alone. The chlorohydrin process produces by-product salts and chlorinated side-products, and the hydroperoxidation processes produce coproducts that require additional processing or recycle. The deficiencies of both processes have spurred research into the production of PO as a single product derived from the oxidation of propylene [3]. The direct oxidation with oxygen does not produce sufficient selectivity [4–6], and considerable interest has arisen in the oxidation of propylene with H_2/O_2 mixtures. While the energy required to activate O_2 for reaction by directly splitting it into its constituent atoms is 498 kJ mol^{-1}, which is larger than C–H bond energies, the copresence of H_2 can activate O_2 at relatively mild conditions with energy input of less than 10 kJ mol^{-1}. This leads to a feasible alternative method of controlling the reactivity of oxygen species so as to produce valuable oxygenated organic compounds [7].

Table 10.1 compares the reaction conditions and the process performances for propylene epoxidation with O_2 and H_2. In the liquid phase with MeOH or BuOH as a solvent, Pd-based catalysts supported on TS(titanosilicalite)-1 have been used. In batch reactors, as shown in Fig. 10.1 [8], the selectivity to PO is lower than 50% and hydrogenation to produce propane prevails [9], whereas PO selectivity is improved to 88–99% when semibatch reactors [8,10,11] or flow reactors are used [12]. Accordingly, the space time yield (STY) for PO exceeds 100 g_{PO} kg_{cat}^{-1} h^{-1}, which is comparable to the STY for the current industrial production of ethylene oxide with molecular oxygen and an Ag/α-Al$_2$O$_3$ catalyst. It should be noted that the STY for PO has recently been reaching a level of 100 g_{PO} kg_{cat}^{-1} h^{-1} for the gas-phase epoxidation of C_3H_6 with H_2 and O_2. In the gas-phase epoxidation, only the coinage metals, Ag and Au, are selective to PO, and only when they are deposited on anatase TiO$_2$ or titanium silicates by the deposition–precipitation (DP) method. In contrast, Pd and Pt are selective not to oxidation but to hydrogenation to form propane in the gas phase.

TABLE 10.1 Direct propylene epoxidation with O_2 and H_2: representative catalytic performance data

Phase	Solvent, carrier	Catalyst[a]	Temperature, pressure[b]	$C_3H_6/O_2/H_2/$ carrier, SV ml $g_{cat}^{-1}\,h^{-1}$	C_3H_6 conversion %	PO selectivity %	PO STY $g_{PO}\,kg_{cat}^{-1}\,h^{-1}$	Reactor	Notes	References
Liquid	t-BuOH, N_2	0.5%Pd/TS-1	318 K, 0.1	17/11/11/29, 7,970	3.7	99.8	129	Semi-batch	Si/Ti = 15	[10]
Liquid	MeOH+ H_2O, N_2	1%Pd–0.01%Pt/ TS-1	316 K, 0.7	2.4/31.7/31.7/34.7, MeOH 15 g + H_2O 5 g, 20,500	16	88	182	Semi-batch	NaBr as promoter	[8,9]
Liquid	MeOH+ H_2O	1%Pd–0.02%Pt/ TS-1	316 K, 5	18.7/11.2/4.6/ CO_2 33+ MeOH 23+H_2O 13.2, 7,250	3.5 (0.5 h), 0.8 (30 h)	99 (0.5 h), 28 (30 h)	122 (0.5 h), 7.1 (30 h)	Flow, high pressure	Catalytic deactivation	[12]
Liquid	SC[c] CO_2	0.47%Pd/TS-1	318 K, 13.1	29.5/61.2/9.3/ excess[d]	7.5 (4.5 h)	94 (4.5 h)	12	Batch	MeOH + H_2O+ [CO_2 or N_2] is not good	[14,15]
Gas	Ar	1%Au/TiO_2 (P-25)	323 K, 0.1	10/10/10/70, 4,000	1.1	99<	12	Flow	DP method	[16]
Gas	He	1% Au/TiO_2 (P-25)	323 K, 0.1	10/10/10/70, 6,000	0.9	99	14	Flow	DP method, FT-IR study	[29]
Gas	Ar	5% Au/TiO_2 (P-25)	343 K, 0.1	33/33/33/0, 1,800	0.6	83	7.6	Flow	Au colloid, 4.6 nm calcined at 573 K	[22]
Gas	N_2	0.05%Pt–0.95% Au/TiO_2/ SiO_2	373 K, 0.1	10/10/10/70, 3,787	1.0<	<90	7.1[e]	Flow	Increased H_2 efficiency	[20]
Gas	He	0.081%Au/TS-1	473 K, 0.1	10/10/10/70, 7,000	10	76	134 (steady state)	Flow	No catalytic deactivation	[41]
Gas	Ar	0.3%Au/Ti-SiO_2	423 K, 0.1	10/10/10/70, 4,000	8.5	91	80 (steady state)	Flow	Silylation, mesopore, $(CH_3)_3N$	[45]
Gas	N_2	2% Ag/TS-1	423 K, 0.1	5/11/17/67, 4,000	1.4	94	6.8	Flow	DP method. Oxidized Ag of 8 nm is optimum.	[51]
Gas	N_2	2% Ag/TiO_2	323 K, 0.1	10/10/10/70, 4,000	0.4	92	3.8	Flow	Only DP method is effective. Ag: 2–4 nm	[52]

[a] Metals in wt. %.
[b] Pressure in MPa.
[c] SC: supercritical.
[d] Excess: more than 10 times as the total mols of reactants.
[e] Estimated by the present authors.

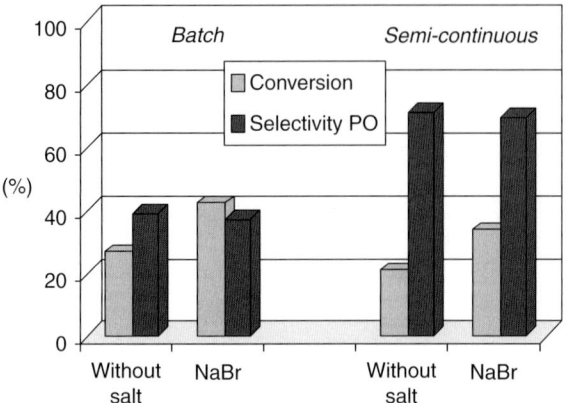

FIGURE 10.1 Influence of reactor systems and effect of addition of NaBr to the reaction of MeOH, H_2O, H_2, O_2, and propylene [8]. Reprinted with permission from [8]. Copyright (2001) Elsevier.

2. LIQUID-PHASE EPOXIDATION OF PROPYLENE

The liquid-phase epoxidation of propylene with O_2-H_2 mixtures was first reported by Sato and Miyake of Tosoh Corp., who used BuOH as a solvent and Pd as the catalytic metal [10]. In liquid-phase epoxidation, TS-1 is usually used as a catalyst support for Pd because it exhibits the best performance for the epoxidation of propylene with H_2O_2 [13]. When the reaction is conducted at temperatures below 323 K, the selectivity to PO is above 90%, whereas the conversion of propylene is not sufficiently high and below 4%.

Later, Hoelderich used an aqueous solution of MeOH as a solvent because H_2O is useful for the efficient formation of H_2O_2, and MeOH is effective for the reaction of H_2O_2 with propylene to form PO [8,9,11]. The addition of a small amount of Pt to Pd (Pt/Pd = 1–2 mol%) was especially effective for improving selectivity to PO. The results of X-ray photoelectron spectroscopy (XPS) analyses indicated that the addition of Pt to Pd can maintain the presence of oxidic Pd, most likely Pd^{2+}, which might be responsible for the epoxidation of propylene [9]. The impregnation of NaBr on TS-1 also improved selectivity to PO by suppressing the ring-opening reaction of PO (Fig. 10.1). The final catalytic performance under atmospheric pressure was high giving a propylene conversion of 16% and a PO selectivity of 88%, which corresponded to a STY of PO of 182 $g_{PO}\,kg_{cat}^{-1}\,h^{-1}$ [8]. However, under a high pressure of 5 MPa, catalyst deactivation became severe and the selectivity to PO dropped from 99% at 30 min after onset of the reaction to 28% after 30 h [12]. This was explained as occurring from the formation of formic acid from the methanol co-solvent under high pressure. Formic acid is a strongly acidic species which catalyzes the formation of several by-products, such as methyl formate, acetone, and acrolein.

In supercritical solvents, as can be presumed from the results obtained in a high-pressure flow reactor, methanol is not a good solvent because it is readily transformed into methyl formate, suppressing the selectivity to PO [14,15].

TABLE 10.2 Results for batch epoxidation of propylene in three different supercritical solvents [14]. Reprinted with permission from [14]. Copyright (2003) Wiley.

				Propylene selectivity	
Solvent system[a]	Catalyst mass (g)	Propylene conversion (%)	PO (%)	C_3H_8 (%)	Ring-opened by-products (%)
CO_2	0.1502	9.5	77.1	22.9	_[b]
CO_2	0.1952	6.5	91.2	8.8	_[b]
CO_2	0.2998	7.5	94.3	5.7	_[b]
MeOH + H_2O + CO_2	0.1565	3.5	17.4	75.0	7.6
MeOH + H_2O + CO_2	0.2063	4.7	41.1	46.5	12.4
MeOH + H_2O + N_2	0.1993	16	3.5	95.2	1.4

[a] Total pressure = 13.1 MPa, T = 318 K, reaction time 4.5 h. Concentrations of propylene 4 mM, H_2 1.26 mM, O_2 8.3 mM, in 3.84×10^{-5} m^3 CO_2.
[b] No PO ring-opened by-products were detected by gas chromatography (GC).

As seen from Table 10.2 [14], when a mixture of methanol, H_2O, and N_2 is used at supercritical conditions, PO selectivity is only 3.5%. When N_2 is replaced by CO_2, PO selectivity is increased to 41%. The high PO selectivity of around 94% can be obtained in neat CO_2 at supercritical conditions, however, the STY is about one order of magnitude smaller than that for semibatch reactors under normal pressure in MeOH solvent.

3. GAS-PHASE EPOXIDATION OF PROPYLENE

Hayashi and Haruta found that finely dispersed gold deposited on TiO_2 could produce PO with selectivities above 90% in a gas stream containing C_3H_6, O_2, and H_2 at temperatures below 373 K [16]. Among a variety of metal oxide supports, TiO_2 with the anatase structure was effective, but not rutile or amorphous titania. The requirement for the size of Au particles was also very strict. Figure 10.2 shows that a diameter of 2–5 nm is optimum for the production of PO, whereas smaller Au clusters below 2 nm produce almost exclusively propane [16–18]. This phenomenon suggests that small Au clusters behave like Pd and Pt in the presence of O_2; Au clusters can dissociate H_2 molecules at low temperatures.

The preparation method and synthesis conditions are crucial to the catalytic performance of Au/TiO_2 [16,19–22]. As seen from Fig. 10.3, the impregnation method produced large spherical Au particles simply loaded on TiO_2, and resulted in the combustion of C_3H_6 and H_2 yielding a large amount of H_2O and a small amount of CO_2 at a relatively high temperature. Ion exchange by using [Au(ethylenediamine)$_2$]Cl$_3$ produced smaller sizes of Au particles on Ti-SiO$_2$ at a

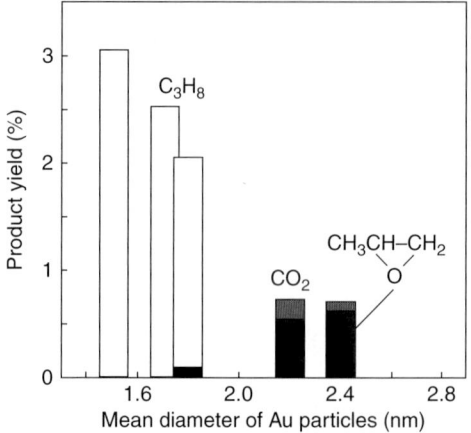

FIGURE 10.2 Product yield as a function of the mean diameter of Au particles deposited on anatase TiO_2 in the reaction of propylene with O_2 and H_2.

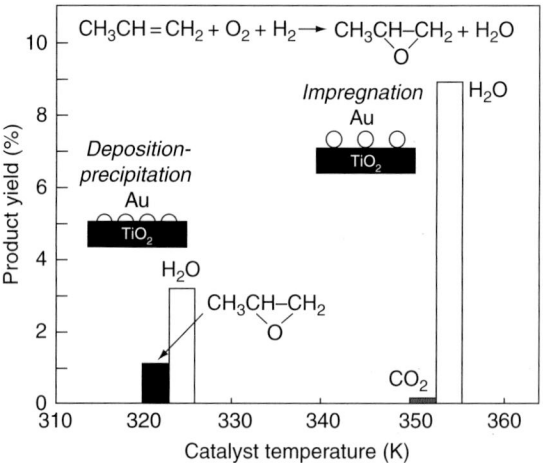

FIGURE 10.3 Product yield of the reaction of propylene with O_2 and H_2 over Au/TiO_2 catalysts prepared by the deposition–precipitation (DP) and impregnation methods.

limited loading of Au, and resulted in the production of propane [20]. Mixing of size-controlled Au colloids with TiO_2 powder did not produce PO, but produces propane when tetrakis(hydroxymethyl)phosphonium chloride (THPC) was used as a stabilizer for the Au colloids [20,21]. In contrast, Au colloids obtained in the presence of poly-vinylpyrrolidone (PVP) or dodecylthiol led to Au/TiO_2 catalysts selective to PO [21–23], although their catalytic performances were inferior to that of catalysts prepared by DP. The DP method produced hemispherical Au particles attached to the TiO_2 surfaces at their basal plane and formed the most efficient catalysts for PO production.

Over Au/TiO$_2$ (anatase), the reaction should be carried out at a temperature below 373 K, because otherwise PO (boiling point 307 K) is further oxidized to acetone and CO$_2$ and H$_2$O. Accordingly, the STY of PO is below 10 g$_{PO}$ kg$_{cat}^{-1}$ h^{-1}, which is one order of magnitude lower than that for industrial ethylene oxide production. Much larger STY of PO can be obtained by replacing TiO$_2$ with TiO$_2$ highly dispersed on SiO$_2$ surfaces, and with titanium silicalites and by operating the reaction at a temperature from 423 to 473 K.

Currently, there are four major lines of research in the gas-phase epoxidation of propylene: (1) mechanistic studies of Au/TiO$_2$ catalysts through kinetics, spectroscopic identification of adsorbed species, and surface science, (2) experimental and theoretical investigation of Au/TS-1 catalysts, (3) improvement of catalyst life of Au/3D mesoporous Ti-SiO$_2$ catalysts, and (4) replacement of Au with Ag on TS-1 supports.

3.1. Mechanistic study of Au/TiO$_2$ catalysts

Relatively detailed studies have been done for the reaction pathways over Au/TiO$_2$ catalysts mainly because of the simplicity of the catalytic material components. The rate of PO formation at temperatures around 323 K does not depend on the partial pressure of C$_3$H$_6$ up to 20 vol%, and then decreases with an increase in C$_3$H$_6$ pressure, while it increases monotonously with the partial pressures of O$_2$ and H$_2$ [16]. A kinetic isotope effect of H$_2$ and D$_2$ was also observed [23]. These rate dependencies indicate that active oxygen species are formed by the reaction of O$_2$ and H$_2$ and that this reaction is rate-determining [17,24].

Propylene is adsorbed on the surfaces of both Au nanoparticles and the TiO$_2$ support, which was indicated by temperature-programmed desorption (TPD) experiments for an Au/TiO$_2$ catalyst and the TiO$_2$ support treated similarly as in the catalyst preparation [16]. The adsorption of propylene occurs nearly to saturation and tends to inhibit the epoxidation at higher partial pressures. Propylene adsorbs on Au(111) and Au(100) surfaces with its molecular plane tilted slightly with respect to the surface plane [25]. The desorption activation energy is 39.3 kJ mol^{-1} on Au(111) and Au(100), which is slightly lower than the 45.1–52.7 kJ mol^{-1} found on Ag(110) and is appreciably lower than the 79.4 kJ mol^{-1} observed on Pd (111). On oxygen-covered surfaces of Au, propylene adsorbs more tightly than on the bare metal surfaces. Density functional theory (DFT) calculations suggest that propylene binds to a corner atom on Au clusters [26]. The binding involves an electron transfer from the highest occupied molecular orbital (HOMO) of propylene to one of the lowest unoccupied molecular orbitals (LUMOs) of Au, thus leading to a stronger bond to Au$_n^+$ than to Au$_n$. It is surprising that the binding energy of propylene to positively charged monoatomic Ag and Au clusters is calculated to be 159.7 and 274.6 kJ mol^{-1}, respectively, showing stronger bonding for Au(C$_3$H$_6$)$^+$ [27]. Campbell reported that propylene adsorbs weakly on Au surfaces and adsorbs moderately on TiO$_2$(110) with a desorption activation energy of 47.2 kJ mol^{-1} [28] and that propylene adsorbs most strongly at the perimeter of Au islands on TiO$_2$(110).

Nijhuis and coworkers assumed, based on a Fourier transform infrared (FTIR) investigation, that propylene is adsorbed on the surfaces of Au particles but not on TiO_2 surfaces [29]. On the other hand, they suggested that bidentate propoxy species which were identified by FTIR (Fig. 10.4) are adsorbed on the TiO_2 surfaces. They finally proposed that propylene is most likely adsorbed at the perimeter interfaces between Au nanoparticles and the TiO_2 support to explain the formation of bidentate propoxy species on TiO_2 surfaces aided by Au nanoparticles [30]. They explained the role of H_2O in the reactant gas stream as enhancing the desorption of PO from the catalyst surfaces, but H_2O in turn suppressed the adsorption of C_3H_6 resulting in minor changes in steady-state catalytic activity after 5 h [31].

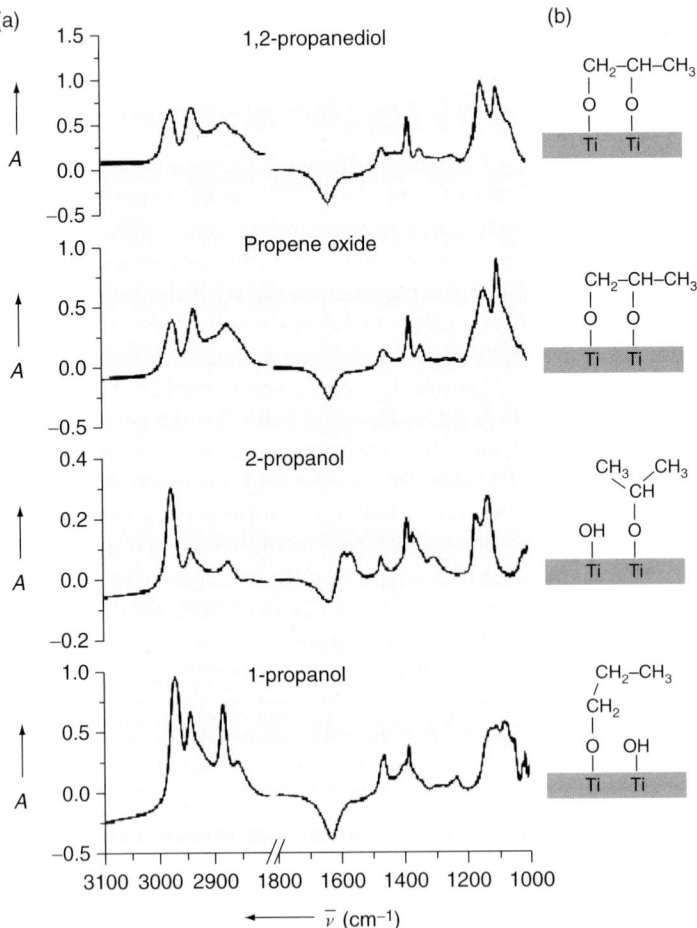

FIGURE 10.4 (a) IR spectra after adsorption and subsequent desorption of 1,2-propanediol, propylene oxide (PO), 2-propanol, and 1-propanol on titania (323 K). (b) Most likely structures of the adsorbate species that correspond to the IR spectra in (a) [29]. Reprinted with permission from [29]. Copyright (2005) Wiley VCH.

The identification of surface-adsorbed species has been carried out with FTIR [29] and Raman spectroscopy [30] during reaction and with gas chromatography–mass spectrometry (GC–MS) analysis after the epoxidation reaction [32]. The aggregation of Au nanoparticles is not appreciable during reaction at temperatures below 473 K [31,32]. Catalyst deactivation, which happens within a few hours with a decrease in C_3H_6 conversion down to 50%, can be accounted for by the accumulation of successively oxidized compounds after isomerization and cracking of propylene oxygenates and their oligomerized compounds [32]. Nijhuis et al. have also identified intermediate species during reaction: the major adsorbates are bidentate carbonate/carboxylate/formate species [29,30].

Due to the accumulation of the oligomerized products of PO and carbonate-derived intermediates, the propylene conversion tended to decrease sharply soon after the reaction and then after 10–20 min, it gradually decreased with time onstream. Quasi–steady-state conversions were usually obtained after 30–60 min. The selectivity to PO and hydrogen conversion reached steady state much earlier. The used catalysts could be almost completely recovered by treating at 523 K in a stream of air, which allowed repeated experiments.

As for oxygen species, significantly important knowledge has been accumulated both by experimental and theoretical investigations. Goodman and his co-workers reported inelastic neutron scattering (INS) evidence (Fig. 10.5) for the formation of OOH and H_2O_2 species from O_2 and H_2 on the Au/TiO_2 catalyst [33]. Barton and Podkolzin proposed based on experimental and theoretical investigation that water formation from O_2 and H_2 over gold deposited on SiO_2, MFI (Zeolite Socony-<u>M</u>obile <u>f</u>ive) zeolite, and TS-1 proceeds through the formation of OOH and H_2O_2 intermediates [34]. The catalytic activity of Au on MFI or TS-1 is higher than that of Au/SiO_2 by 60–70 times, which can be ascribed to the higher

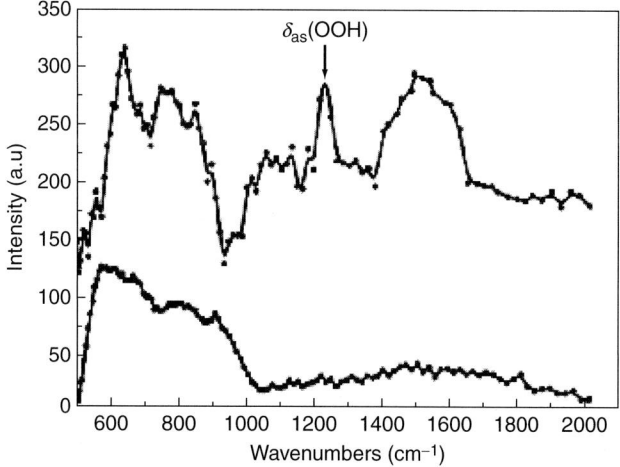

FIGURE 10.5 (Top) Inelastic neutron scattering (INS) spectrum of Au/TiO_2 reacted with O_2 and H_2 at 523 K for 4 h in flowing O_2:H_2:He (1:1:7). (Bottom) INS spectrum of water at 523 K adsorbed on Au/TiO_2 for comparison [33]. Reprinted with permission from [33]. Copyright (2004) American Chemical Society.

concentration of 13-atom clusters of Au on MFI and TS-1 supports than on SiO_2 supports. As shown in Fig. 10.6 [34], 13-atom Au clusters can be incorporated in the pore intersections in the MFI zeolite structure. Gold clusters smaller than that are less reactive due to the instability of the OOH intermediate whereas larger Au particles are less reactive due to the instability of adsorbed oxygen. In the epoxidation as well as H_2O_2 formation, the rate-determining step is assumed to be the addition of H_2 in gas phase to the surface-adsorbed OOH to form H_2O_2 [24,30,35]. Nijhuis and coworkers emphasize the importance of two other reaction steps, a reactive adsorption of propylene on TiO_2 to produce bidentate propoxy species, and a reactive desorption of this adsorbed species to form PO [24,30,35].

Intensive work on DFT calculations has been done to seek the probable pathways for the formation of H_2O_2 [36,37] and PO [38,39] through a collaboration of theoreticians and experimentalists. The Au clusters composed of 3–55 atoms having positive or negative electric charges were studied for H_2O_2 synthesis as the key step in propylene epoxidation [34,36,37]. Thomson and Delgass carried out calculations for Au_3, A_4^+, Au_5, and $Au_5^=$ clusters in the gas phase and noted that both neutral and charged Au clusters are active for the formation of H_2O_2 [37]. They suggested that in the gas phase, A_4^+ is the most active, while Barton and Podkolzin reported for MFI-supported Au clusters that intermediate-size Au_{13} clusters are the most active for water synthesis [34].

The DFT calculations indicated that the rate-determining step is the attack of Au-OOH on the C=C double bond to form PO with an activation barrier of 81.9 kJ mol^{-1} [39]. Since a Ti-based support is indispensable for obtaining high selectivity to PO, it is reasonably assumed that isolated tetrahedrally coordinated Ti species participate in the epoxidation. Ti sites located adjacent to Si vacancies in

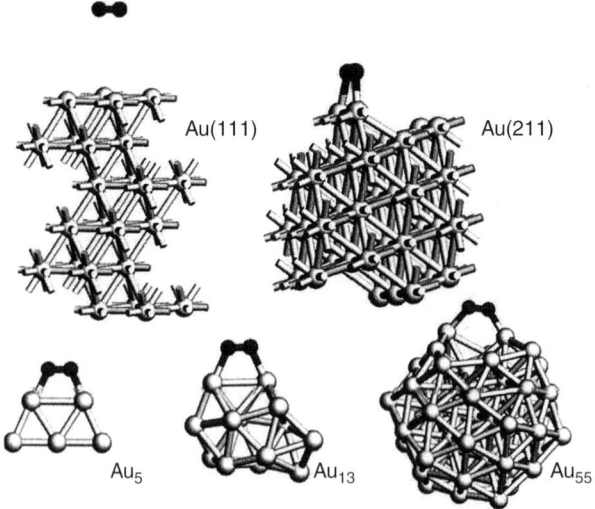

FIGURE 10.6 Computational models of Au surfaces. Models are shown with optimized O_2 geometries. For Au(111) and Au(211), a single unit cell of the periodic slab is shown [34]. Reprinted with permission from [34]. Copyright (2005) American Chemical Society.

the TS-1 lattice are more reactive than fully coordinated Ti sites and may be the site for epoxidation. Another pathway is also probable, namely, epoxidation on the surfaces of Au [39].

3.2. Improvements in the catalytic performance of Au/TS-1

Although Au/TS-1 catalysts possess good catalyst stability owing to the well-crystallized structure and hydrophobicity of TS-1, the catalytic performance has been inferior to Au supported on 3D mesoporous titanosilicates in terms of propylene conversion, PO selectivity, and H_2 utilization efficiency [40]. However, recently Delgass has obtained a high PO STY of 134 g_{PO} kg_{cat}^{-1} h^{-1} (propylene conversion 10%, PO selectivity 76%) using a 0.081 wt.% Au/TS-1 catalyst [41], which is comparable to that of ethylene oxide in commercial plants. The key to the success appears to be a pretreatment of the TS-1 support in a 1 M aqueous solution of NH_4NO_3 at 353 K for 15 h. The remarkable enhancement seems to be due to the selective deposition of Au near the Ti sites as well as to an increase in actual Au loadings, presumably because of the preferential formation of an Au–amine complex.

Although Au particles with mean diameters around 5–6 nm were observed by transmission electron microscopy (TEM), Delgass suggests based on DFT calculations that such large particles are not active and selective, but that small Au clusters such as Au_3 are responsible for the epoxidation of propylene [36–38]. According to this hypothesis, TS-1 is useful as a support to confine small Au clusters into the spaces of microcages. Figure 10.7 shows a sharp contrast in research strategy between Delgass's group and Haruta's group in terms of the

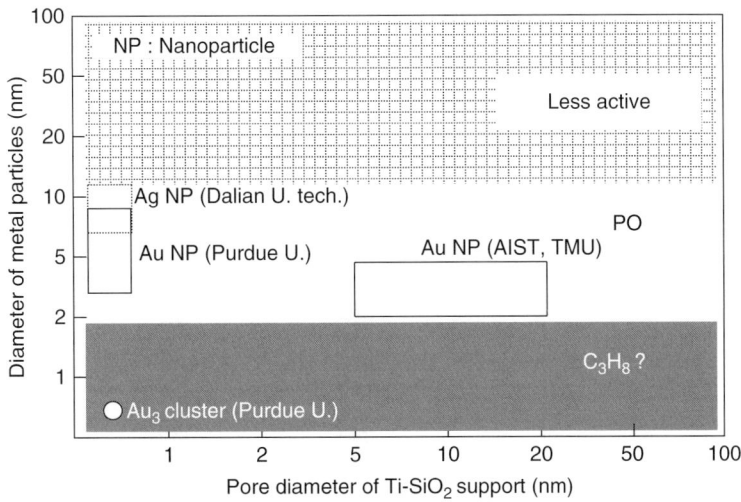

FIGURE 10.7 Silver and Au supported on titanium silicates for propylene epoxidation with O_2 and H_2: three lines of approaches.

size of Au particles and pores of the Ti-SiO$_2$ supports. It also shows the position of Ag/TS-1 catalysts studied by Guo's group.

3.3. Improvement of the catalyst life of Au/3D mesoporous Ti-SiO$_2$

Figure 10.8 shows that the yield of PO increases with an increase in the diameter of pores of titanium silicate supports including TS-1, Ti-MCM-41, and Ti-MCM-48 [32,42,43]. This suggests that larger pores are advantageous, especially for the smooth diffusion of reactants and rapid escape of the product, PO [42]. Although nonporous supports were expected to facilitate the desorption of PO, Ti deposited on superfine nonporous SiO$_2$ (diameter 10–140 nm, specific surface area 77 m^2 g^{-1}) did not result in higher yields of PO [44]. TEM observations showed that the population density of Au particles over the surfaces of nonporous supports was small in comparison with those for micro- and mesoporous supports, probably because of the lack of defect sites.

A catalyst consisting of Au deposited on a 3D mesoporous Ti-SiO$_2$ with a pore diameter of 9 nm has given a high STY. At a space velocity (SV) of 4,000 h^{-1} ml gcat^{-1}, the catalyst gave a propylene conversion above 8%, a PO selectivity of 91%, and a steady STY of 80 g$_{PO}$ kg$_{cat}^{-1}$ h^{-1} [45]. The surface of the 3D mesoporous Ti-SiO$_2$ were trimethylsilylated to give a hydrophobic character which allowed the use of a higher temperature of reaction [46]. As solid-phase promoters, alkaline or alkaline earth metal chlorides have been found to be efficient, however, chloride anions markedly enhance the coagulation of Au particles in a short period [47]. The best promoter was found to be Ba(NO$_3$)$_2$, which might be transformed into BaO after calcination at 573 K and might function by neutralizing acid sites on the catalyst surfaces [45].

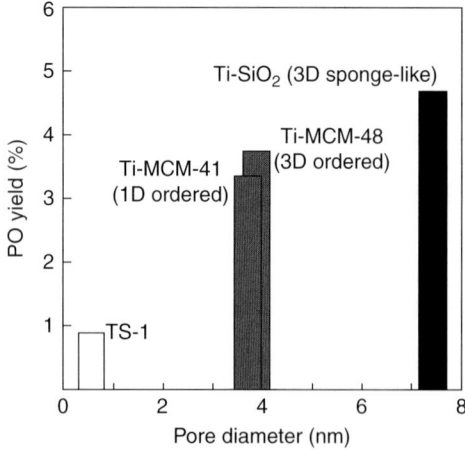

FIGURE 10.8 PO yield of the reaction of propylene with O$_2$ and H$_2$ as a function of pore diameter of porous titanium silicate supports for Au catalysts.

An interesting feature in promoting reagents is that trimethylamine, a strong Lewis base with a pK_a value of 9.9, introduced to the reactant gas stream at an concentration of 10–20 ppm appreciably improves the catalytic performances [48]. In every performance item, improvement was observed to a certain degree in propylene conversion, PO selectivity, H_2 utilization efficiency, and catalyst life. It is worth noting that trimethylamine makes used catalysts better than fresh catalysts in catalytic performance. Trimethylamine might kill the mobile acid sites which appear and disappear intermittently, suppressing by-product formation from PO. Trimethylamine can also adsorb on the surfaces of Au and depress the combustion of H_2 to form H_2O, thus leading to improved H_2 utilization efficiency.

In this catalytic system, as schematically shown in Fig. 10.9 [48], a most probable pathway is that over the Au surfaces O_2 and H_2 react with each other to form H_2O_2, which then move to isolated sites of Ti cations to form Ti-OOH species [49]. This oxidic species react with propylene adsorbed on the support surfaces to form PO. It has recently been verified that Ti-OOH species is a true reaction intermediate and that a bidentate propoxy species is probably a spectator on the surface [50]. The coverage, θ, of the Ti-hydroperoxo species was determined from the area of the pre-edge peak in the Ti K-edge XANES spectra at reaction conditions. Measurement of the changes in Ti-hydroperoxo coverage, $d\theta/dt$, under transient experiments at reaction conditions with $H_2/O_2/Ar$ and $C_3H_6/H_2/O_2$ gas mixtures, allowed the estimation of initial net rate of propylene epoxidation (3.4×10^{-4} s^{-1}), which closely matched the TOF (2.5×10^{-4} s^{-1}) obtained for the same catalyst at steady-state conditions.

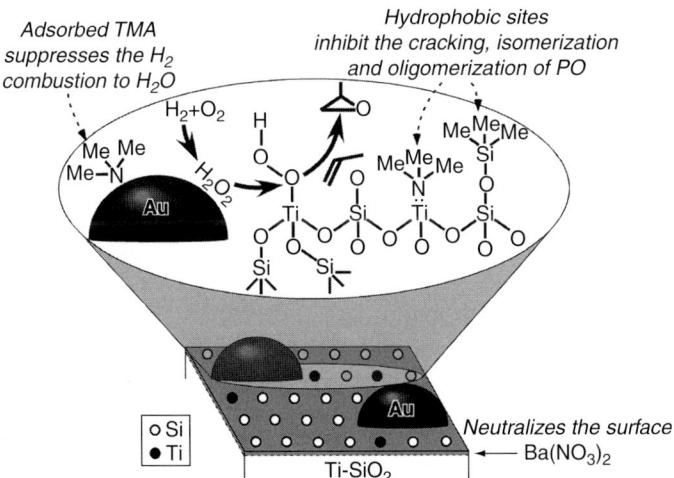

FIGURE 10.9 Probable reaction pathways for propylene epoxidation with O_2 and H_2 in the presence of trimethylamine on trimethylsilylated Au–Ba(NO$_3$)$_2$/titanosilicate catalyst [48]. The accumulation of PO and its derivatives on the catalyst surfaces can be minimized by changing the surface nature into a hydrophobic one to prevent catalyst deactivation. Hydrogen combustion over Au surfaces can also be suppressed by the adsorption of trimethylamine to improve the atom efficiency of the overall reaction. Reprinted with permission from [48]. Copyright (2005) Wiley VCH.

3.4. Replacement of Au with Ag

It is interesting to note that Ag can also do a similar job to that of Au under the same reaction conditions. In 2001, Shueth showed that Ag deposited on TiO_2 (Degussa P25) could also produce PO in the gas phase with a reactant mixture of propylene, O_2, H_2, and N_2 with selectivity above 90% at a propylene conversion of 0.36% [51]. Soon afterwards, Guo reported that Ag deposited on TS-1 exhibited better performances (propylene conversion 1.4%, PO selectivity 93.5%), while he claimed that Ag on TiO_2 was not active for PO synthesis [52–54].

A common feature for both Ag/TiO_2 and Ag/TS-1 catalysts is that catalyst preparation is crucial and among impregnation, ion exchange, microemulsion, sol–gel, and DP techniques, only the DP techinque led to activity at a temperature of 323 K or above and PO selectivity higher than 60%. Since the DP technique forms hemispherical metal particles strongly attached to the metal oxide supports, the significantly large influence of preparation methods implies that the contact structure of Ag particles with the TiO_2 support and selective deposition of Ag on Ti sites but not on SiO_2 surfaces are important for PO synthesis, as in the case of supported Au catalysts. The requirements of Ag are that Ag should not be fully reduced to metallic species and that Ag particles should be in the size range of 2–4 nm in diameter (on TiO_2 supports) or at around 8 nm (on a TS-1 support). Accordingly, there is an optimum Ag loading of 2 wt.%, above which the major product switches to propanal. The above requirements for Ag catalysts are almost identical to those of Au catalysts except for the necessity of oxidic species, indicating that catalytic mechanism may be similar between Ag and Au. It has not yet been reported that propane is produced over Ag catalysts instead of PO.

4. CONCLUSIONS

(1) Propylene epoxidation with O_2/H_2 mixture in MeOH or BuOH solvent can proceed with a selectivity above 88% over Pd/TS-1 catalyst in semibatch reactors. The addition of a small amount of Pt is effective to maintain cationic Pd and enhances the catalytic performance.
(2) In the gas-phase epoxidation of propylene with O_2/H_2 mixture, silver or gold supported on TiO_2 or titanosilicates exhibits high selectivities to PO, which change from 94% to 76% depending on the reaction temperature and propylene conversion.
(3) In both the liquid-phase and the gas-phase reactions, the STYs of PO have reached a level of $100\ g_{PO}\ kg_{cat}^{-1},\ h^{-1}$ which is comparable to that of ethylene oxide synthesis in industrial processes. However, there still remain some barriers against commercialization, namely catalyst life and hydrogen utilization efficiency.
(4) In supported gold catalysts for the gas-phase propylene epoxidation, two hypotheses are competing. One assumes that gold clusters inside the cages of TS-1 (crystalline microporous titanium silicates) may act as the working

sites for epoxidation, whereas another describes that gold nanoparticles larger than 2 nm in diameter dispersed on mesoporous titanium silicates is responsible for the formation of hydrogen peroxide.

REFERENCES

[1] D. L. Trent, Propylene oxide, In: *Kirk Othmer Encyclopedia of Chemical Technology*, on-line edition, John Wiley & Sons, New York, 2001.
[2] H. Mimoun, M. Mignard, P. Brechot, L. Saussine, Selective epoxidation of olefins by oxo[N-(2-oxidophenyl)salicylidenaminato]vanadium(V) alkylperoxides. On the mechanism of the Halcon epoxidation process, *J. Am. Chem. Soc.* 108 (1986) 3711.
[3] J. R. Monnier, The direct epoxidation of higher olefins using molecular oxygen, *Appl. Catal. A. Gen.* 221 (2001) 73.
[4] S. T. Oyama, K. Murata, M. Haruta, Direct oxidation of propylene to propylene oxide with molecular oxygen, *Shokubai*. 46 (2004) 13.
[5] J. Lu, J. J. Bravo-Suárez, A. Takahashi, M. Haruta, S. T. Oyama, In situ UV-vis studies of the effect of particle size on the epoxidation of ethylene and propylene on supported silver catalysts with molecular oxygen, *J. Catal.* 232 (2005) 85.
[6] J. Lu, J. J. Bravo-Suárez, M. Haruta, S. T. Oyama, Direct propylene epoxidation over modified Ag/$CaCO_3$ catalysts, *Appl. Catal. A: Gen.* 302 (2006) 283.
[7] M. Haruta, Gold rush, *Nature* 437 (2005) 1098.
[8] W. Laufer, W. F. Hoelderich, Direct oxidation of propylene and other olefins on precious metal containing Ti-catalysts, *Appl. Catal. A: Gen.* 213 (2001) 163.
[9] W. F. Hoelderich, 'One-pot' reactions: A contribution to environmental protection, *Appl. Catal. A. Gen.* 194–195 (2000) 487.
[10] A. Sato, T. Miyake, *Jpn. Pat. Tokkaihei* 4–352771, and *Shokubai* 34 (1992) 132.
[11] R. Meiers, U. Dinagerdissen, W. F. Hoelderich, Synthesis of propylene oxide from propylene, oxygen, and hydrogen catalyzed by palladium–platinum–containing titanium silicalite, *J. Catal.* 176 (1998) 376.
[12] G. Jenzer, T. Mallat, M. Maciejewski, F. Eigenmann, A. Baiker, Continuous epoxidation of propylene with oxygen and hydrogen on a Pd–Pt/TS-1 catalyst, *Appl. Catal. A: Gen.* 208 (2001) 125.
[13] M. G. Clerici, G. Belussi, U. Romano, Synthesis of propylene oxide from propylene and hydrogen peroxide catalyzed by titanium silicalite, *J. Catal.* 129 (1991) 159.
[14] T. Danciu, E. J. Beckman, D. Hancu, R. N. Cochran, R. Grey, D. M. Hajnik, J. Jewson, Direct synthesis of propylene oxide with CO_2 as the solvent, *Angew. Chem. Int. Ed.* 42 (2003) 1140.
[15] E. J. Beckman, Production of H_2O_2 in CO_2 and its use in the direct synthesis of propylene oxide, *Green. Chem.* 5 (2003) 332.
[16] T. Hayashi, K. Tanaka, M. Haruta, Selective vapor-phase epoxidation of propylene over Au/TiO_2 catalysts in the presence of oxygen and hydrogen, *J. Catal.* 178 (1998) 566.
[17] M. Haruta, Catalysis of gold nanoparticles deposited on metal oxides, *CATTECH* 6 (2002) 102.
[18] M. Haruta, When gold is not noble: Catalysis by nanoparticles, *Chem Record*. 3 (2003) 75.
[19] E. E. Stangland, B. Taylor, R. P. Andres, W. N. Delgass, Direct vapor phase propylene epoxidation over deposition-precipitation gold-titania catalysts in the presence of H_2/O_2: Effects of support, neutralizing agent, and pretreatment, *J. Phys. Chem. B* 109 (2005) 2321.
[20] A. Zwijnenburg, M. Saleh, M. Makkee, J. A. Moulijn, Direct gas-phase epoxidation of propene over bimetallic Au catalysts, *Catal. Today* 72 (2002) 59.
[21] A. Zwijnenburg, A. Goossens, W. G. Sloof, M. W. J. Craje, A. M. van der Kraan, L. Jos de Jongh, M. Makkee, J. A. Moulijn, XPS and Mössbauer characterization of Au/TiO_2 propene epoxidation catalysts, *J. Phys. Chem. B* 106 (2002) 9853.
[22] J. Chou, E. W. McFarland, Direct propylene epoxidation on chemically reduced Au nanoparticles supported on titania, *Chem. Commun*. (2004) 1618.

[23] E. E. Stangland, K. B. Stavens, R. P. Andres, W. N. Delgass, Characterization of gold–titania catalysts via oxidation of propylene to propylene oxide, *J. Catal.* 191 (2000) 332.
[24] T. A. Nijhuis, M. Makkee, J. A. Moulijn, B. M. Weckhuysen, The production of propene oxide: catalytic processes and recent developments, *Ind. Eng. Chem. Research.* 45 (2006) 3447.
[25] K. A. Davis, D. W. Goodman, Propene adsorption on clean and oxygen-covered Au(111) and Au(100) surfaces, *J. Phys. Chem. B* 104 (2000) 8557.
[26] S. Chretien, H. Metiu, Binding of propene on small gold clusters and on Au(111): Simple rules for binding sites and relative binding energies, *J. Chem. Phys.* 121 (2004) 3756.
[27] R. Olson, S. Varganov, M. S. Gordon, H. Metiu, The binding of the noble metal cations Au^+ and Ag^+ to propene, *Chem. Phys. Lett.* 412 (2005) 416.
[28] H. M. Ajo, V. A. Bondzie, C. T. Campbell, Propene adsorption on gold particles on TiO_2(110), *Catal. Lett.* 78 (2002) 359.
[29] T. A. Nijhuis, T. Visser, B. M. Weckhuysen, The role of gold in gold-titania epoxidation catalysts, *Angew. Chem. Int. Ed.* 44 (2005) 1115.
[30] T. A. Nijhuis, T. Visser, B. M. Weckhuysen, Mechanistic study into the direct epoxidation of propene over gold/titania catalysts, *J. Phys. Chem. B* 109 (2005) 19309.
[31] T. A. Nijhuis, B. M. Weckhuysen, The role of water in the epoxidation over gold–titania catalysts, *Chem. Commun.* (2005) 6002.
[32] B. S. Uphade, T. Akita, T. Nakamura, M. Haruta, Vapor-phase epoxidation of propene using H_2 and O_2 over Au/Ti-MCM-48, *J. Catal.* 209 (2002) 331.
[33] C. Sivadinarayana, T. V. Choudhary, L. L. Daemen, J. Eckert, D. W. Goodman, The Nature of the surface species formed on Au/TiO_2 during the reaction of H_2 and O_2: An inelastic neutron scattering study, *J. Am. Chem. Soc.* 126 (2004) 38.
[34] D. G. Barton, S. G. Podkolzin, Kinetic study of a direct water synthesis over silica-supported gold nanoparticles, *J. Phys. Chem. B* 109 (2005) 2262.
[35] T. A. Nijhuis, T. Q. Gardner, B. M. Weckhuysen, Modeling of kinetics and deactivation in the direct epoxidation of propene over gold-titania catalysts, *J. Catal.* 236 (2005) 153.
[36] D. H. Wells, Jr., W. N. Delgass, K. T. Thomson, Formation of hydrogen peroxide from H_2 and O_2 over a neutral gold trimer: A DFT study, *J. Catal.* 225 (2004) 69.
[37] A. M. Joshi, W. N. Delgass, K. T. Thomson, Comparison of the catalytic activity of Au_3, Au_4^+, Au_5, and Au_5 in the gas-phase reaction of H_2 and O_2 to form hydrogen peroxide: A Density Functional Theory investigation, *J. Phys. Chem. B* 109 (2005) 22392.
[38] D. H. Wells, Jr., W. N. Delgass, K. T. Thomson, Evidence of defect-promoted reactivity for epoxidation of propylene in titanosilicate (TS-1) catalysts: A DFT study, *J. Am. Chem. Soc.* 126 (2004) 2956.
[39] A. M. Joshi, W. N. Delgass, K. T. Thomson, Partial oxidation of propylene to propylene oxide over a neutral gold trimer in the gas phase: A Density Functional Theory study, *J. Phys. Chem. B* 110 (2006) 2572.
[40] N. Yap, T. P. Andress, W. N. Delgass, Reactivity and stability of Au in and on TS-1 for epoxidation of propylene with H_2 and O_2, *J. Catal.* 226 (2004) 156.
[41] L. Gumaranatunge, W. N. Delgass, Enhancement of Au capture efficiency and activity of Au/TS-1 catalysts for propylene epoxidation, *J. Catal.* 232 (2005) 38.
[42] A. K. Sinha, S. Seelan, S. Tsubota, M. Haruta, Catalysis by gold nanoparticles: epoxidation of propene, *Top. Catal.* 29 (2004) 95.
[43] Y. A. Kalvachev, T. Hayashi, S. Tsubota, M. Haruta, Vapor-phase selective oxidation of aliphatic hydrocarbons over gold deposited on mesoporous titanium silicates in the co-presence of oxygen and hydrogen, *J. Catal.* 186 (1999) 228.
[44] C. Qi, T. Akita, M. Okumura, M. Haruta, Epoxidation of propylene over gold catalysts supported on non-porous silica, *Appl. Catal. Gen.* 218 (2001) 81.
[45] A. K. Sinha, S. Seelan, S. Tsubota, M. Haruta, A three-dimensional mesoporous titanosilicate support for gold nanoparticles: Vapor-phase epoxidation of propene with high conversion, *Angew. Chem. Int. Ed.* 43 (2004) 1546.
[46] C. Qi, T. Akita, M. Okumura, M. Haruta, Effect of surface chemical properties and texture of mesoporous titanosilicates on direct vapor-phase epoxidation of propylene over Au catalysts at high reaction temperature, *Appl. Catal. A: Gen.* 253 (2003) 75.

[47] B. S. Uphade, M. Okumura, S. Tsubota, M. Haruta, Effect of physical mixing of CsCl with Au/Ti-MCM-41 on the gas-phase epoxidation of propene using H_2 and O_2: Drastic depression of H_2 consumption, *Appl. Catal. A: Gen.* 190 (2000) 43.
[48] B. Chowdhury, J. J. Bravo-Suarez, M. Date, S. Tsubota, M. Haruta, Trimethylamine as a gas-phase promoter: Highly efficient epoxidation of propylene over supported gold catalysts, *Angew. Chem. Int. Ed.* 45 (2006) 412.
[49] B. Chowdhury, J. J. Bravo-Suarez, N. Mimura, J. Lu, K. K. Bando, S. Tsubota, M. Haruta, *In situ* UV-vis and EPR study on the formation of hydroperoxide species during direct gas phase propylene epoxidation over Au/Ti-SiO2 catalyst, *J. Phys. Chem. B* 110 (2006) 22995.
[50] J. J. Bravo-Suarez, K. K. Bando, J. Lu, M. Haruta, T. Fujitani, S. T. Oyama, Transient technique for identification of true reaction intermediates: Hydroperoxide species in propylene epoxidation on gold/titanosilicate catalysts by X-ray absorption fine structure spectroscopy, *J. Phys. Chem. C.* 112 (2008) 1115.
[51] A. Lange de Oliveira, A. Wolf, F. Schueth, Highly selective propene epoxidation with hydrogen/oxygen mixtures over titania-supported silver catalysts, *Catal. Lett.* 73 (2001) 157.
[52] R. Wang, X. Guo, X. Wang, J. Hao, G. Li, J. Xiu, Effects of preparation conditions and reaction conditions on the epoxidation of propylene with molecular oxygen over Ag/TS-1 in the presence of hydrogen, *Appl. Catal. A: Gen.* 261 (2004) 7.
[53] X. Guo, R. Wang, X. Wang, J. Hao, Effects of preparation method and precipitator on the propylene epoxidation over Ag/TS-1 in the gas phase, *Catal. Today.* 93–95 (2004) 217.
[54] C. Wang, X. Guo, X. Wang, R. Wang, J. Hao, Gas-phase propylene epoxidation over Ag/TS-1 prepared in W/O microemulsion: Effects of the molar ratio of water to surfactant and the reaction temperature, *Catal. Lett.* 96 (2004) 79.

CHAPTER 11

Propylene Epoxidation by $O_2 + H_2$ over Au Nanoparticles on Ti-Nanoporous Supports

Ajay M. Joshi, Bradley Taylor, Lasitha Cumaranatunge, Kendall T. Thomson, and W. Nicholas Delgass

Contents

1. Introduction — 316
2. Propylene Epoxidation over Au/TS-1 Catalysts — 317
   2.1. Activity, selectivity, and stability of Au/TS-1 catalysts — 318
   2.2. Nature of active Au sites in Au/TS-1 catalysts — 319
   2.3. Control of Au deposition — 320
   2.4. Importance of defects in TS-1 — 321
3. Au on Ti-Containing Mesoporous Supports — 322
   3.1. Ti–MCM-41 materials — 322
   3.2. Ti–MCM-48 materials — 323
   3.3. Ti–TUD materials — 323
   3.4. Ti–SBA-15 materials — 323
   3.5. Amorphous mesoporous titanosilicates and mixed oxides — 324
4. Promoters and Postsynthesis Support Treatments — 324
   4.1. Alkali earth metals — 324
   4.2. Titanium grafting — 325
   4.3. Silylation — 326
   4.4. Trimethylamine cofeeding — 327
5. Reaction Kinetics — 327
   5.1. "Sequential" versus "simultaneous" epoxidation mechanism — 330
6. Conclusions and Future Outlook — 331
Acknowledgments — 333
References — 333

School of Chemical Engineering, Purdue University, West Lafayette, Indiana 47907

Mechanisms in Homogeneous and Heterogeneous Epoxidation Catalysis
DOI: 10.1016/B978-0-444-53188-9.00011-0

© 2008 Elsevier B.V.
All rights reserved.

Abstract

Production of propylene oxide in a single step with no side products has been a long-sought industrial target. While a liquid-phase H_2O_2/TS-1-based route appears to be imminent, due to handling problems and cost associated with H_2O_2, researchers have also focused on propylene epoxidation using H_2 and O_2 over Au/Ti catalysts. Au nanoparticles on mesoporous and nanoporous Ti supports are promising due to their remarkable stability and commercially interesting activity, but significant improvements in H_2 efficiency are desired for commercialization. This review summarizes the advances in propylene epoxidation using H_2 and O_2 over Au/TS-1 and Au/Ti-mesoporous supports. Implications of several interesting findings, such as Au particle size and support effects, effect of catalyst pretreatments, effect of gas-phase additives and catalyst promoters, reaction kinetics, and mechanistic insights from quantum chemical calculations, are discussed.

Key Words: Direct propylene epoxidation, Propylene oxide, Gold, Titanium, Propene, Au/Ti catalysts, Catalysis by gold, Titanium silicalite, TS-1, Gold/TS-1, Hydrogen peroxide, Kinetics, Design of experiments, Deposition–precipitation, Ammonium nitrate, Selective oxidation, Alkene epoxidation, Density functional theory, DFT calculations, QM/MM calculations. © 2008 Elsevier B.V.

1. INTRODUCTION

Propylene oxide (PO) is a valuable chemical intermediate in the commercial manufacture of propylene glycol and polyurethanes which are used to produce important products such as adhesives, paints, and cosmetics [1]. Commercial production of PO has been dominated by two technologies: the chlorohydrin process [2–5] and the hydroperoxide process [6–9]. However, the chlorohydrin process is not ecofriendly and the hydroperoxide process leads to the formation of large volumes of a coproduct, either *t*-butanol or styrene depending on which hydroperoxide is used, thus reducing the flexibility of commercial operation. Therefore, the evolving choice of the industry is to selectively oxidize propylene to PO using liquid-phase H_2O_2 [10,11], through integration of H_2O_2 production with propylene epoxidation for economic viability [12]. Sequential reduction and oxidation of anthraquinones produces dilute H_2O_2 solutions in a CH_3OH–water mixture [13], which is subsequently used (without separating H_2O_2) for propylene epoxidation over very active and selective titanium silicalite-1 (TS-1) catalysts under mild conditions (1 atm, $\sim$40 °C) [14–17]. On the basis of this H_2O_2-based integrated propylene epoxidation technology, commercial-scale plants are currently being built by DOW–BASF [10,11]. Additional details of different PO technologies are available in a recent review by Nijhuis *et al.* [18].

While industry has selected liquid-phase H_2O_2-based technology for imminent PO plants [11], due to handling issues and cost associated with making H_2O_2, a direct gas-phase propylene epoxidation process, analogous to the commercial ethylene oxide (EO) process [19,20], has long been desired [21]. While ethylene epoxidation using O_2 is catalyzed by Ag/α-Al_2O_3 catalysts, the analogous process for propylene suffers from poor selectivity due to combustion pathways which involve abstraction of allylic H atoms in propylene by atomic O on Ag [22,23], and both academia [24–29] and industry [30] are still seeking improvements. However, using quite different catalysts, Haruta and coworkers [31] achieved a breakthrough by showing direct propylene epoxidation using H_2 and O_2 over catalysts prepared by deposition–precipitation (DP) of Au [32] on TiO_2 (>90% selectivity toward PO, 1 atm, 30–120 °C). These findings were later confirmed by several laboratories [33–36]. It was soon realized that Au/TiO_2 catalysts suffer from rapid deactivation due to oligomerization of PO species on adjacent Ti sites [37–39] and isolation of active Ti centers can improve the catalyst stability [33,35,40]. Following this lead, both academic [33,35,37,40–44] and industrial [45–52] research has been focused on Ti-silica and titanosilicate materials.

Haruta and coworkers have estimated that the minimum requirements for a potential commercial process are a propylene conversion of 10%, a PO selectivity of 90%, and a H_2 efficiency (100 × moles of PO formed/moles of H_2 consumed) of 50% [53]. This condition translates to a PO production rate of $\sim$100 g_{PO} kg_{cat}^{-1} h^{-1} at a propylene feed concentration of 10%. While >90% PO selectivity can be routinely achieved, most of the research is targeted to improve the catalytic activity, stability, and H_2 efficiency. In this regard, Au nanoparticles on mesoporous and nanoporous Ti-supports are especially promising due to excellent catalyst stability and good activity [33,40–42,54–57], but further improvements in the catalytic activity and H_2 efficiency are desired for commercialization. Therefore, this review is focused on direct propylene epoxidation using H_2 and O_2 over Au/TS-1 and Au/Ti-mesoporous supports. In what follows, we discuss several interesting findings on these catalysts such as Au particle-size dependence, support effects, effect of catalyst pretreatments, effect of gas-phase additives and catalyst promoters, reaction kinetics, and mechanistic insights from quantum chemical calculations. Implications of these findings for further improvements of Au/Ti-based direct propylene epoxidation process are also discussed.

2. PROPYLENE EPOXIDATION OVER Au/TS-1 CATALYSTS

The first synthesis of TS-1 was reported over two decades ago [58], but considerable interest in using this material as a catalyst began only after Clerici et al. [14] showed that TS-1 can catalyze the liquid-phase epoxidation of propylene with very high selectivities toward PO (90–97%) using a dilute solution of H_2O_2. This inherent epoxidation activity, in combination with the highly dispersed tetrahedrally coordinated Ti contained within a silica matrix, has made TS-1 promoted with nanoscale Au a promising candidate for the direct gas-phase epoxidation of propylene using H_2 and O_2 [33,51,52,59]. In addition, more economical synthesis of TS-1 in micellar media has also been reported [60]. The best Au/TS-1 catalysts have been made using the DP [32] of Au on TS-1. Therefore, in most of the studies discussed below, DP of Au was used, unless otherwise stated.

2.1. Activity, selectivity, and stability of Au/TS-1 catalysts

Early efforts using Au/TS-1 by Haruta and coworkers [61] for the gas-phase epoxidation of propylene over a 0.5 g of catalyst bed at a space velocity of 4,000 ml g_{cat}^{-1} h^{-1} produced propanal (a structural isomer of PO) with 70% selectivity, especially when the catalyst had been thoroughly washed in distilled water after Au deposition by the DP method. They attributed this observation to the acidity associated with TS-1 that catalyzed the ring opening of PO and resulted in an aldehyde. On the contrary, Nijhuis et al. [33] were able to produce PO (>95% selectivity) over a 1 wt.% Au/TS-1 catalyst at a space velocity of 6,600 ml g_{cat}^{-1} h^{-1} (70–150 °C). The propylene conversion, however, remained at <2% with PO formation rates ranging from 4 to 18 g_{PO} kg_{cat}^{-1} h^{-1}. They found that the Au/TS-1 catalysts are more resistant to deactivation and produced slightly higher propylene conversions compared to the Au/TiO$_2$ catalysts. However, the H$_2$ efficiency of these Au/TS-1 catalysts was only 5–6% compared to 22–26% for the Au/TiO$_2$ catalysts [33]. Nijhuis et al. [33] also performed a post-reaction thermogravimetric analysis (TGA) on both Au/TiO$_2$ and Au/TS-1 catalysts in order to investigate the deactivation behavior. Since no change in Au particle size between fresh and spent catalyst was observed from TEM micrographs, deactivation by sintering of Au was excluded. They found two forms of carbonaceous deposits, one combusting at 230 °C and the other at 330 °C for the Au/TiO$_2$ catalyst which showed significant deactivation with time-on-stream [33]. On the other hand, the Au/TS-1 sample which did not exhibit any deactivation showed only one carbonaceous deposit combusting at 330 °C. Hence, they concluded that the carbonaceous deposit that combusted at 230 °C, perhaps a polymer of either propylene or PO, caused the deactivation of Au/TiO$_2$ catalysts.

Stangland et al. [62] compared a series of Ti-supports with varying degrees of Ti-connectivity including TiO$_2$, monolayer TiO$_2$ on SiO$_2$, submonolayer TiO$_2$ on SiO$_2$, and TS-1. They concluded that TS-1 was both the most active and the most robust of the examined supports yielding the highest conversion and selectivity. Yap et al. [54] achieved propylene conversions of 2.5–6.5% and PO selectivities of 60–85% (50–80 g_{PO} kg_{cat}^{-1} h^{-1}) at 170 °C and a space velocity of 7,000 ml g_{cat}^{-1} h^{-1}. Their time-dependent kinetic measurements showed that the stability of Au/TS-1 catalysts is dependent on the Au and Ti loading, and that dilute Au and Ti systems produced more stable and active Au/TS-1 catalysts. This important finding has been recently exploited by Taylor et al. [59]. An Au/TS-1 catalyst prepared with Si/Ti = 36 and an Au loading of 0.05 wt.% produced 116 g_{PO} kg_{cat}^{-1} h^{-1} at 200 °C, which is one of the highest rates reported for a TS-1-based catalyst, with no evidence of deactivation during the 40-h temperature program [59]. In addition, use of lower Ti and Au contents resulted in very active catalysts when rates were normalized to the total Au content, 350 g_{PO} g_{Au}^{-1} h^{-1} at 200 °C for 0.01 wt.% Au/TS-1 (Si/Ti = 500), indicative of the more efficient use of Au and Ti for epoxidation [59]. The low Au loadings coupled with the absence of Au particles in TEM micrographs make it likely that, in these materials, significant activity is due to <1-nm Au entities [59].

2.2. Nature of active Au sites in Au/TS-1 catalysts

One of the interesting issues in direct propylene epoxidation using H_2 and O_2 over Au/Ti catalysts is the Au particle-size dependence of the catalytic activity and selectivity. Haruta and coworkers have suggested that ~2–5-nm Au particles prevalent in Au/TiO_2 catalysts containing >1.0 wt.% Au are active in formation of PO, while smaller particles in catalysts containing <0.1 wt.% Au form propane [31]. However, it was not possible for them to characterize particles of <1 nm due to limitations of their TEM instrument. Interestingly, density functional theory (DFT) calculations suggest that small Au clusters, such as Au_3 [63,64], Au_5 [64], and Au_{13} [65], can activate H_2 and O_2 to make H_2O_2 which then decomposes to form water [65]. Because of the instability of adsorbed O_2, larger clusters, such as Au_{55} and extended Au surfaces, were predicted to be inactive [65]. Therefore, whether few-atom Au clusters are active in propylene epoxidation or not has been an important question open for further investigation.

In this direction, Yap et al. [54] varied the external surface area of TS-1 by synthesizing ~170- and ~519-nm TS-1 particles. Using these supports, they prepared a variety of Au/TS-1 catalysts with Au loadings of 0.06–0.74 wt.% and observed average Au particle diameters of 2–7 nm resulting from Au deposition by the DP method and investigated the catalytic activity over a reaction time of 24–36 h at a space velocity of 7,000 ml g_{cat}^{-1} h^{-1} (10/10/10/70 vol% of propylene/O_2/H_2/He) and temperatures of 140, 170, and 200 °C. Interestingly, comparable PO formation rates per Au atom were observed over Au/TS-1 (170 nm) and Au/TS-1 (519 nm) catalysts suggesting that the rate does not scale with external surface area of TS-1 and that the active Au–Ti sites are not exclusively the 2–5-nm Au particle/Ti sites on the external surface of TS-1 [54]. Moreover, TEM studies suggested that the visible (>1 nm) Au particles on TS-1 cannot account for the total Au content of these catalysts, indicating that ~70% of the Au is in form of invisible (<1 nm) Au sites, perhaps few-atom Au clusters, located either on the external surface of TS-1 or inside the TS-1 pores (diameter ~ 5.5 Å) containing a majority of Ti sites (high internal surface area). Also, increasing the Au loading up to 0.74% did not increase the PO rates proportionally, suggesting that the active Au–Ti PO-forming centers are limited [54].

As stated earlier, Taylor et al. [59] have reported commercially interesting PO formation rates for Au/TS-1 catalysts containing very low Au loading (0.01–0.05 wt.%). Careful examination of TEM micrographs was performed, but no nanometer-sized particles were observed. Again, this means that most of the catalytic activity is due to invisible Au clusters (<1 nm). Thus, *invisible* few-atom Au clusters, perhaps located inside the TS-1 pores, on Au/TS-1 catalysts with 0.01–0.05 wt.% Au as well as *visible* nanometer-sized Au particles prevalent on the external surface of Au/TS-1 catalysts containing >1 wt.% Au are catalytically active in direct propylene epoxidation.

In the same study [59], the authors varied the Si:Ti ratio from 36 to 1,143 and found that deposition of Au via DP onto these TS-1 supports using similar deposition solution concentrations resulted in Au loadings that directly track the drop in Ti loading (i.e., these catalyst have an inherent Au uptake that depends on Ti content). Increasing the concentration of the deposition solution resulted in catalysts with higher Au loadings but these catalysts showed inferior activity, selectivity, and stability. These findings imply that small Au particles may preferentially deposit around Ti sites, consistent with the predictions of quantum mechanics/molecular mechanics (QM/MM) calculations [66], and the number of such Au–Ti active sites is likely to have an optimal value for a given Ti loading.

2.3. Control of Au deposition

The effect of neutralizing agent (LiOH, NaOH, KOH, RbOH, and CsOH) and the pH (6–8) during DP on the catalytic activity in direct propylene epoxidation was investigated previously for Au/Ti–MCM-48 catalysts [42]. The best results were obtained with NaOH as the neutralizing agent and with a pH of the Au solution of 7.0 ± 0.1. In fact, NaOH was also found to be better than urea, $NaHCO_3$, Na_2CO_3, and NH_3 [67]. On the contrary, DP with Na_2CO_3 in the pH range of 7–9 was found to be the best for Au/TiO_2 catalysts [62], suggesting that the optimal neutralizing agent depends on the characteristics of the support material. It is essential here to point out that the nature of the Au species in the DP solution depends on the pH; hydrolyzed (chlorine free) anionic Au species were found to be dominant around a pH of 9 [68,69]. At lower pH values, Au particles deposited on TiO_2 were larger and exhibited relatively poor activity for CO oxidation. However, Au uptake was found to decrease progressively as the final pH was raised above 8 [68]. This means that the lower Au particle size and better activity in CO oxidation over Au/TiO_2 catalysts are observed in the pH range where Au uptake from the solution is far from complete, and that higher uptake does not necessarily imply better activity [68].

The ability to enhance the catalyst site density by increasing the deposition of useful Au onto TS-1 during DP has been a critical problem. Attempts at getting higher Au loadings by increasing the concentration of the deposition solution resulted in inferior activity for direct propylene epoxidation. In fact, Yap et al. [54] reported only a mere 1–3% of the available Au in the DP solution being deposited on TS-1 support materials. Cumaranatunge et al. [56] reported that TS-1 supports pretreated (before Au deposition) in 1 M NH_4NO_3 solution at 80 °C resulted in a significant enhancement in the Au capture efficiency and activity of Au/TS-1 catalysts for propylene epoxidation. The Au capture efficiency was defined as the fraction of Au from the $HAuCl_4 \cdot xH_2O$ precursor solution that was deposited on the support during DP. The NH_4NO_3-treated TS-1 materials, on average, captured four times more Au than that captured by the untreated TS-1 materials (same batch) from the same deposition solution. This increase in capture efficiency suggests that it is possible to use lower concentrations of the rather expensive Au precursor to attain the same target Au loadings.

Cumaranatunge et al. [56] also compared the Au/TS-1 catalysts, with and without NH_4NO_3 treatment (same batch of TS-1), containing similar Au loading, and found a significant enhancement in activity for the NH_4NO_3-treated catalysts. One such NH_4NO_3-treated 0.058 wt.% Au/TS-1 catalyst showed 5% propylene conversion with 83% selectivity at 200 °C and maintained its activity over the 40-h test period. At the space velocity of 7,000 ml g_{cat}^{-1} h^{-1}, this corresponds to a rate of

76 g_{PO} kg_{cat}^{-1} h^{-1}, which implies a remarkably high turnover rate (TOR) of over 0.1 molecules per Au atom per s [56]. In addition, utilizing the NH_4NO_3 pretreatment, the authors produced a catalyst with an Au loading of 0.081 wt.% which resulted in a rate of 134 g_{PO} kg_{cat}^{-1} h^{-1} at 200 °C which is the highest PO formation rate reported in the open literature. However, the H_2 efficiency of this catalyst was <20% which is well below the target of >50%. The authors also noted an optimal Au loading that was support dependent for both NH_4NO_3-treated and untreated catalysts; but Au loadings of >0.1 wt.% were all detrimental to catalyst stability.

While the use of NH_4OH instead of NH_4NO_3 enhanced the Au capture efficiency, it did not enhance the PO formation rate [56]. Hence, NH_4^+ ions increase the Au capture efficiency, but the mechanism of this enhancement is still unclear. Regarding improvement in the activity, Solsona *et al.* [70] found that small quantities of nitrates enhanced the CO oxidation rate over Au catalysts while excess nitrates resulted in deactivation. Similarly, small quantities of nitrates remaining on the TS-1 supports after the NH_4NO_3 treatment could account for the enhancement in PO formation rates for the Au/TS-1 catalysts. Along these lines, Haruta and coworkers have previously reported comparable activity on catalysts containing Au nanoparticles on three-dimensional (3D), silylated mesoporous titanosilicate supports with >7-nm pores in the presence of impregnated $Ba(NO_3)_2$ promoter [71]. It is important to note that the Au concentrations utilized with the NH_4NO_3 treated TS-1 supports at the lab-scale were small enough to prevent the formation of "fulminating Au"—a brown-colored compound that is explosive on contact [72]. However, this safety issue must be addressed carefully in the potential commercial-scale operation.

2.4. Importance of defects in TS-1

Even well-made TS-1 contains a small fraction of Si-vacancy defects [73,74]. Consistent with FTIR results on H_2O_2/TS-1 [75], previous DFT calculations on nondefect (tetrapodal) and metal-vacancy defect (tripodal) Ti sites in TS-1 suggested that H_2O_2 attack on Ti-defect sites leads to Ti-OOH species (and water), while H_2O_2 attack on Ti-nondefect sites is kinetically and thermodynamically less favorable [76]. Moreover, Ti-OOH species can catalyze propylene epoxidation to PO [76–78]. Recent QM/MM calculations on adsorption of Au_{1-5} clusters inside the TS-1 pores suggest that the Ti-defect site is also the most favorable binding site for small Au clusters [66]. Therefore, defects in TS-1 are likely to stabilize adsorbed Au clusters and prevent sintering.

Encouraged by these computational predictions, Taylor *et al.* [79] followed a previously published technique [80] and added 12–20-nm carbon pearls in the TS-1 synthesis gel. These pearls were combusted along with the templating agent during calcination of the support material at 535 °C. The directed growth of the zeolite around the carbon pearls potentially enhances the number of Ti-defect sites which have been predicted to be active in epoxidation using H_2O_2/TS-1 [76,77]. While direct measurement of the defect density was not possible, the materials produced appeared significantly less regular in TEM images, compared to their normally prepared counterparts [79]. Catalysts prepared using these modified materials were consistently active and stable, with a 0.33 wt.% Au/TS-1 catalyst producing 132 g_{PO} kg_{cat}^{-1} h^{-1} at 200 °C, one of the highest PO formation rates reported, despite relatively high Au loadings and contamination with octahedral Ti species as seen in DRUV-Vis analysis [79]. This implied that modified synthesis may have resulted in encapsulation of the octahedral Ti species, muting their normal detrimental effect.

3. Au ON Ti-CONTAINING MESOPOROUS SUPPORTS

Shortly after the initial foray into the use of microporous titanosilicates as the highly dispersed Ti-supports for propylene epoxidation, interest shifted to mesoporous titanosilicates. Mesoporous Ti-containing materials are similar to microporous materials in that they offer highly dispersed Ti centers and reasonably well-defined tetrahedral Ti sites incorporated in a silicious framework. Moreover, the existence of a mesoporous pore system of sufficient dimensions to incorporate Au species in the range of 2 nm allows for Au entities to access essentially the entirety of the support surface area and enhances transport of reactants and products to and from the sites.

3.1. Ti–MCM-41 materials

The initial investigation of Ti-containing mesoporous materials for the gas-phase epoxidation of propylene began in the late 1990s using the one-dimensional hexagonal MCM-41 materials (Space Group p6m) [81]. With dodecyltrimethylammonium chloride (C12TMACl) as the organic templating agent, typical pore sizes range from 2 to 2.5 nm. Deposition of Au via DP onto these materials (Ti/Si = 0.020, 0.028, and 0.048) resulted in a particle-size distribution in which 70% of the Au particles were capable of accessing the pore system [81]. As compared to similarly prepared Au on dispersed TiO_2 on SiO_2 catalysts, the Au/Ti–MCM-41 catalysts were more active (17 g_{PO} kg_{cat}^{-1} h^{-1}), more selective to PO (96%), more stable, and more efficient in utilizing H_2 (4%) at 100 °C [81]. The enhanced stability was ascribed, at least partially, to the larger pore system, which was less easily occluded by PO degradation products. Similarly to the Au/TS-1 catalysts, the Ti loading resulting in most active catalyst was Ti/Si = 0.028 [81].

Ti–MCM-41 prepared using C12TMACl and cetyltrimethylammonium chloride (C16TMACl) as organic structure directing agents generated materials with an average pore size of 4.55 and 6.12 nm, respectively [40]. Upon DP of Au, the larger pore catalysts were slightly more active (28 g_{PO} kg_{cat}^{-1} h^{-1} vs 27 g_{PO} kg_{cat}^{-1} h^{-1}), more selective to PO (96% vs 90%), and more H_2 efficient (11% vs 6%). These beneficial traits were ascribed to the increased accessibility of a larger fraction of deposited Au species to the pore system and the ability of the larger pore system to better accommodate deposits of PO degradation products without completely blocking internal active sites [40]. In comparison to mixed TiO_2–SiO_2 supports, which act as a reasonable approximation of the structure of the wall material of the

MCM-type materials, the introduction of an ordered pore system greatly enhanced catalytic activity, PO selectivity, and H_2 efficiency.

In general, the reactivity of Au/Ti–MCM-41 materials can be summarized in terms of Ti content. With increasing Ti content up to the framework incorporation limit of Ti/Si $\approx$ 0.02, Au deposited during DP increases, PO production rate increases, PO selectivity decreases, and H_2 efficiency decreases [40,44,81–83].

3.2. Ti–MCM-48 materials

Whereas MCM-41 materials have a one-dimensional pore system, MCM-48 (Space Group Ia3d) materials possess a three-dimensional system. While MCM-41 materials were an improvement over microporous materials due to the incorporation of larger pores, MCM-48 materials allow for even greater access to the internal crystallite surface area through the interconnectivity of the three-dimensional channel system. In comparison to Au/Ti–MCM-41 materials, Au/Ti–MCM-48 materials exhibit slightly higher activity (53 g_{PO} kg_{cat}^{-1} h^{-1} vs 47 g_{PO} kg_{cat}^{-1} h^{-1}), PO selectivity (92% vs 88%), and H_2 efficiency (10% vs 7%), which are attributed primarily to the increased access to internal active sites [41,42,44]. Similar to other dispersed TiO_2 systems, the optimum Ti loading was found to be Ti/Si $\approx$ 0.02, beyond which extra-framework octahedral Ti species are formed. Increases in Ti loading up to this limit have been shown to result in higher PO production rates, lower PO selectivity, and decreased H_2 efficiency.

3.3. Ti-TUD materials

While MCM-41- and 48-based materials dominate as the primary mesoporous materials explored for gas-phase propylene epoxidation, a recent article examines the reactivity of Au deposited on Ti-TUD containing 3 mol% Ti [57]. Ti-TUD consists of a sponge-like structure with an average pore size of about 13 nm. Although the specific surface area of this material is less than that of MCM-41 or MCM-48, the larger pore system allowed for essentially all of the deposited Au to have access to the pore system. A maximum rate of 53.7 g_{PO} kg_{cat}^{-1} h^{-1} at 170 °C (using a 20/20/20/40 vol% mixture of H_2/O_2/propylene/He) is similar to that of other catalysts prepared with Ti–MCM-41 and Ti–MCM-48, however, Ti-TUD catalysts were one of the rare Au–Ti-based catalysts to show stable PO production over the course of 10 h on stream [57]. Moreover, these materials have PO selectivities above 85% and H_2 efficiency ranging from 5% to 14% [57].

3.4. Ti–SBA-15 materials

The recent examination of titanium containing Au/SBA-15 catalysts was carried out comparing titanium incorporated hydrothermally and by postsynthesis grafting using titanium (IV) oxyacetylacetonate monohydrate [84]. Titanium addition by grafting resulted in more active and selective catalysts, though overall the catalysts had lower activity relative to others prepared using mesoporous titanosilicates. This is in direct contrast to the results found for Au/Ti–MCM-41 catalysts [82], though this is likely due to titanium incorporation at inaccessible sites within the thick silicate walls that make up SBA-15. X-ray absorption spectroscopy showed that the average bulk gold particle size was 1.4–3.0 nm, roughly 3 nm smaller than the average size as determined by TEM, implying that again, significant portions of gold deposited on these supports consists of entities of <1 nm [84]. Further refinement of the X-ray fine structure implied that these small gold entities exist as relatively flat sheets with two dimensions being much larger than the third. Ultimately, lower titanium contents were shown to stabilize smaller gold particle sizes, an observation consistent with the implied activity of small gold entities on Au/TS-1 [59].

3.5. Amorphous mesoporous titanosilicates and mixed oxides

Likely a result of higher activity of amorphous Au/Ti–MCM-41 materials relative to crystalline materials [44], one of the exciting materials to be developed as Ti-supports are the three-dimensionally mesoporous wormhole titanosilicates [71,85] consisting of a disordered system of 5–10-nm pores which allow significant access of Au particles to the pore system. Catalysts prepared from these materials have exhibited some of the highest PO production rates published thus far. Even more remarkable than a rate of 51.7–91.6 g_{PO} kg_{cat}^{-1} h^{-1} at 150 °C is the observation that despite significant deactivation, these materials have been shown to recover >80% of their activity following treatments in H_2 and O_2 at 250 °C [71,85]. These materials have been shown to allow isomorphous incorporation of Ti to much higher levels than other microporous materials (6 mol%). Unlike TiO_2–SiO_2 prepared by sol–gel techniques, this material has an enhanced Ti content near the external surface (as measured by XPS) which likely contributes to the substantial improvement in activity over Ti–MCM materials. Similar to other mesoporous materials, with increasing Ti content come increased PO production rates, decreased PO selectivity, and decreased H_2 efficiency. Overall, H_2 efficiency exceeds that seen for other mesoporous materials, ranging from mid-20s to low 40s depending on catalyst preparation.

Recently, Dai et al. [86] investigated 4 wt.% Au catalysts supported on TiO_2–SiO_2 mixed oxides prepared by nonhydrolytic sol–gel synthesis route for direct propylene epoxidation. High-resolution TEM studies showed that the Au particle size was around 2–4 nm. The Ti loading was varied systematically, and the best performance was observed over a catalyst containing 10 mol% Ti: 61.3 g_{PO} kg_{cat}^{-1} h^{-1} (after 1 h on stream) and 36.0 g_{PO} kg_{cat}^{-1} h^{-1} (after 4 h on stream) at quite low temperatures (120 °C).

4. PROMOTERS AND POSTSYNTHESIS SUPPORT TREATMENTS

4.1. Alkali earth metals

A concern regarding the industrial viability of Au/Ti-based catalysts for the epoxidation of propylene has always been the inefficient use of H_2. The first example of improved H_2 efficiency as a result of the addition of a promoter was the physical mixing of CsCl with Au/Ti–MCM-41 [41,87]. A 1 wt.% physical mixture of CsCl with Au/Ti–MCM-41 resulted in an increase in H_2 efficiency from about 6.5% to 36% while only dropping the PO production rate from 29.5 to 17.1 g_{PO} kg_{cat}^{-1} h^{-1} at 100 °C [87]. Direct combustion of H_2 was not observed over these physical mixtures until the temperature exceeded 350 °C, a temperature well above the temperatures necessary for propylene epoxidation. A portion of the loss in catalytic activity is attributed to Au particle agglomeration in the presence of Cl ions. Preparation of catalysts using CsOH as the neutralizing agent during DP did not result in catalysts with improved H_2 efficiency [40,87], perhaps because the formation of the active Au species is hampered by the relatively weak basic nature of CsOH over the preferred neutralization agent, NaOH.

Addition of $Ba(NO_3)_2$ to Au/Ti-based catalysts has also been shown to increase propylene conversion, presumably by mitigating any catalyst surface acidity and aiding the production of hydroperoxy-like oxidizing agents [71]. The addition of 1 wt.% $Ba(NO_3)_2$ resulted in an increase in PO production from 64.9 to 91.6 g_{PO} kg_{cat}^{-1} h^{-1} at 150 °C with only a slight loss in PO selectivity and H_2 efficiency. The addition of 2.4 wt.% $Ba(NO_3)_2$ to Au/Ti-TUD catalysts did not appear to generate as substantial an increase in activity [57].

4.2. Titanium grafting

Postsynthesis addition of Ti to mesoporous support materials using titanocene dichloride and Ti(IV) isopropoxide has been seen as a method of increasing Ti content of support materials beyond the tetrahedral substitutional limit of Si/Ti ≈ 33 without the introduction of octahedral Ti species [82,83]. Catalysts prepared from MCM-41-type materials containing Ti incorporated both hydrothermally and by postsynthesis grafting were more active for propylene epoxidation than support materials prepared solely by hydrothermal incorporation or grafting [82]. The higher activity, which is attributed to the additional tetrahedral Ti content of the catalyst, increased from 47.3 g_{PO} $kg_{cat}^{-1}h^{-1}$ for a catalyst prepared hydrothermally with 3.0 mol% Ti to 53.4 g_{PO} kg_{cat}^{-1} h^{-1} for a catalyst prepared hydrothermally with 1.5 mol% Ti followed by 3 mol% Ti by grafting using Ti(IV) isopropoxide. Despite the increase in PO production rate, Ti deposited by grafting is less active than that incorporated by hydrothermal methods. A catalyst prepared by hydrothermal method with 1.5 mol% Ti had a PO production rate of 45.6 g_{PO} kg_{cat}^{-1} h^{-1} compared to 33.8 g_{PO} kg_{cat}^{-1} h^{-1} for a catalyst prepared by Ti grafting onto MCM-41. Catalysts prepared using Ti(IV) isopropoxide were more active than catalysts prepared using titanocene dichloride. Ultimately, catalysts prepared by hydrothermal Ti incorporation followed by grafting were found to have slower deactivation, higher H_2 efficiency, and higher PO production rates than catalysts prepared by only a single-step Ti incorporation. The H_2 efficiency is likely improved by the addition of extra Ti centers which can more quickly utilize oxidant generated on Au sites. Improved stability is attributed to the elimination/replacement of hydrophilic and slightly acidic terminal silanol groups with Ti species that are active for the epoxidation reaction.

4.3. Silylation

The deactivation of catalysts in the Au–Ti family of materials is generally ascribed to the degradation of PO via oligomerization and polymerization through the formation of an adsorbed propoxy intermediate [33,37,88]. This intermediate forms readily over titanol groups available on extended octahedral TiO_2 phases and its formation has been hypothesized for external silanol groups present on mixed oxide supports [42]. An attempt at mitigating any strong interactions between surface silanols and PO has been the replacement of terminal silanols with hydrophobic methylsilyl groups. First attempted with Au/Ti–MCM-48 materials [42], the capping of external silanol groups with methoxytrimethylsiloxane following Au deposition resulted in a catalyst with improved PO selectivity (95% vs 90%) and H_2 efficiency (12% vs 7.5%) but lower PO rate (27 $g_{PO}\ kg_{cat}^{-1}\ h^{-1}$ vs 43 $g_{PO}\ kg_{cat}^{-1}\ h^{-1}$). The decreased activity is likely a result of site blocking by the capping agent.

A subsequent study introduced the trimethylsilylating agent prior to Au deposition [44]. Ti–MCM-48 material was capped using N-methyl-N-(trimethylsilyl)trifluoroacetamide (MSTFA), though the resulting hydrophobic material did not allow for Au deposition via deposition impregnation. Instead, a liquid grafting method was applied using $(CH_3)_2Au(O_2C_5H_7)$ as the Au source. Silylated Ti–MCM-41 was prepared in a single-step process, where the silylating agent was incorporated into the hydrothermal sol–gel synthesis as methyltriethoxysilane (MTEOS). Variations in the ratio of MTEOS with tetraethylorthosilicate (TEOS) allowed for control over the hydrophobicity of the resulting Ti–MCM-41 [44]. As MTEOS content in the growth liquid increased, normalized surface area and crystallinity of the product decreased, culminating in an amorphous material prepared using mixtures exceeding MTEOS/(MTEOS + TEOS) = 0.5 in the growth liquor. The most crystalline materials were obtained using mixtures below 0.35. As shown by FTIR, these materials were not ultimately hydrophobic, as evidenced by adsorbed water even at high MTEOS contents [44]. Au was deposited on the Au/Ti–MCM-41 material by DP.

The catalytic activity of silylated Au/Ti–MCM-48 catalysts was low, ranging from 2.1 to 15 $g_{PO}\ kg_{cat}^{-1}\ h^{-1}$ (rates after 1 h on stream at 250 °C) depending on the amount of N-methyl-N-(trimethylsilyl)trifluoroacetamide used in their preparation [44]. This lower activity was partially a result of the low Au loadings (~0.5 wt.%) of these catalysts, though more likely a consequence of the use of a method other than DP, which changed the characteristics of the deposited gold ultimately resulting in a limited number of active sites. Catalysts prepared by Au grafting were more than an order of magnitude less active than the methylsilylated Au/Ti–MCM-41 catalysts discussed later when rates are normalized to Au loading. Silylation did eliminate the rapid initial deactivation of Au/Ti–MCM-48, though at longer times, the deactivation rate appeared comparable. In this case, silylation did not improve the H_2 efficiency (<4%), perhaps a result of the indiscriminate deposition of Au on silica surfaces which can catalyze water formation in the absence of epoxidation activity [65]. Examination of these catalysts via FTIR following reaction showed that it was possible to rehydroxylate the surface, which likely ultimately aided catalyst deactivation. Of significant interest is the finding that the PO selectivity of these catalysts remained above 80% despite the high reaction temperature [44].

The one-step methylsilylation of Ti–MCM-41 resulted in catalysts that were reasonably stable, though the PO production rates were low even at 200 °C (2.5–10 g_{PO} kg_{cat}^{-1} h^{-1} after 1 h on stream). The H_2 efficiency did increase with MTEOS content in the growth mixture from 6.3% to 11%. PO selectivity over these catalysts was comparable to the silylated Au/Ti–MCM-48 catalysts, though at a reaction temperature 100 °C lower. Catalyst stability improved as crystallinity of the support decreased [44]. Although these supports are not completely hydrophobic, DP resulted in only ~0.02 wt.% gold deposited onto these supports.

Silylated catalysts prepared from methoxytrimethylsilane [71,85] contacted with amorphous, disordered mesoporous titanosilicate (2 mol%) followed by deposition of Au via DP resulted in an increase in PO production rate from 52 to 67 g_{PO} kg_{cat}^{-1} h^{-1} over the unsilylated catalyst at 150 °C. There was no significant effect on H_2 efficiency (an increase from 33.3% to 35.3%). Catalyst deactivation rate appears to have been decreased slightly, 58% retention of activity over the first 4 h relative to 44% for the unsilylated catalyst. Silylation in combination with Ba promotion and the high Ti content of the disordered mesoporous materials has resulted in some of the most active catalysts thus reported with a PO rate of 92 g_{PO} kg_{cat}^{-1} h^{-1} at 150 °C.

4.4. Trimethylamine cofeeding

Despite the improvements in catalyst stability through the use of highly dispersed Ti-mesoporous supports and the replacement of terminal silanol groups via silylation, catalyst lifetime for PO production remains a concern. The most recent abatement of rapid catalyst deactivation has been the incorporation of small quantities of trimethylamine (TMA) in the feed [53]. The introduction of TMA, in concentrations as low as 10–20 ppm, has been shown to reduce water production, likely through interaction with Au entities via the amine group, as well as to increase the hydrophobicity of the catalyst surface by bonding to Ti Lewis acid sites, aiding in PO desorption. The combination of TMA cofeeding with Ba(NO$_3$)$_2$ doping and silylation has allowed Au/disordered-mesoporous-titanosilicate catalysts to produce PO at 64–81 g_{PO} kg_{cat}^{-1} h^{-1}, a H_2 efficiency of 35% and only little deactivation over the course of 4 h on stream.

5. REACTION KINETICS

The introduction of microporous and mesoporous supports with well-dispersed Ti has allowed for some detailed kinetic studies into the PO and water generation mechanisms primarily as a result of the outstanding stability of catalysts prepared from these materials. Prior to the synthesis of stable catalysts, a number of reaction mechanisms were proposed based on observed trends in reactivity, DFT calculations [63,64,76,78], and analogs to liquid-phase epoxidation reactions over Ti-based catalysts [14,15,89] rather than kinetic analysis. The first proposed mechanisms were constructed for the Au/TiO$_2$ and Au/TiO$_2$/SiO$_2$ system.

Haruta and coworkers first proposed the formation of a hydroperoxy or peroxy intermediate near Au/Ti interfaces [31]. Nijhuis *et al.* expanded upon this basic mechanism, proposing three mechanisms dependent on the spillover of either H or H_2O_2 from an Au site to a Ti center [33]. The third of these mechanisms, the formation of a hydroperoxy intermediate at an Au site followed by spillover and epoxidation at a Ti site, has become the focus of most mechanistic discussions especially after the experimental identification of hydroperoxy and H_2O_2 species on Au/TiO$_2$ catalysts in the presence of H_2 and O_2 [90]. Haruta and coworkers had suggested that for Au/Ti–MCM-48 catalysts, H_2O_2 is formed on Au entities followed by migration to an adjacent Ti site. This peroxy intermediate then reacted with propylene adsorbed on an adjacent Ti site to form adsorbed PO and water [42]. The final reaction mechanism proposed by Haruta and coworkers backed away from the reactivity on adjacent Ti sites in favor of propylene adsorbing and reacting on a hydroperoxy site already adsorbed to a tetrahedral Ti site [53].

One of the first comprehensive kinetic analyses over Au/TS-1 concerned the oxidation of H_2 to form water [65]. DFT calculations [63,65] showed the RDS to be the formation of adsorbed H_2O_2 which ultimately decomposes to form water [65]. In the presence of propylene, the generated H_2O_2 is expected to perform the epoxidation, therefore similarities between the production of PO and water can likely be drawn. Traditional kinetic analysis [65] produced a power rate law for water production of $r_{H_2O} = k_0 \exp[-(37.1 \pm 1.1 \text{ kJ mol}^{-1})/RT][H_2]^{0.76 \pm 0.02}[O_2]^{0.17 \pm 0.02}$. Development of a series of elementary steps [Eqs. (11.1–11.5)] capable of reproducing the observed experimental orders and consistent with DFT calculations proved to require two active sites: one capable of nondissociative adsorption of O_2 and dissociative adsorption of H_2 and a second available for only dissociative adsorption of H_2 [65]. The resulting rate expression [65] is presented as Eq. (11.6).

$$O_2 + ^* \leftrightarrow O_2^* \tag{11.1}$$

$$H_2 + 2^* \leftrightarrow 2H^* \tag{11.2}$$

$$H_2 + 2g \leftrightarrow 2Hg \tag{11.3}$$

$$O_2^* + Hg \leftrightarrow OOH^* + g \tag{11.4}$$

$$OOH^* + H_2 + g \rightarrow H_2O_2^* + Hg \tag{11.5}$$

$$r_{H_2O} = \frac{k K_{OOH} K_{O_2} P_{O_2} \sqrt{K_{H_2} P_{H_2}} P_{H_2}}{(1 + \sqrt{K_{H_2} P_{H_2}} + K_{O_2} P_{O_2} + K_{OOH} K_{O_2} P_{O_2} \sqrt{K_{H_2} P_{H_2}})(1 + \sqrt{K_{H_2} P_{H_2}})} \tag{11.6}$$

Although the microporous nature of TS-1 would not be expected to play a large role in the generation of H_2O_2, Au on MFI-type supports was shown to be more than an order of magnitude more active than Au deposited on SiO_2 for the oxidation of H_2 [65]. The apparent benefit of the three-dimensional pore system, in conjunction with DFT results that Au_{13} is very active for water formation [65], suggests that entities much smaller than 2 nm may contribute significantly to the activity of H_2O_2 generating catalysts [63,64]. This also implies that the considerable improvements in selectivity observed over mesoporous Ti-materials may be more a result of the material's tolerance toward PO degradation products than making available a larger fraction of the support surface area to 2-nm Au particles.

A factorial design of kinetic experiments utilizing Au/TS-1 was the first study of propylene epoxidation kinetics in the absence of catalyst deactivation [55]. The adoption of a factorial design allowed for the examination of the largest statistically significant number of reaction conditions centered around the standard reaction mixture of 10/10/10/70 vol% H_2/O_2/propylene/diluent while avoiding flammable conditions. The experimental results [55] from the evaluation of three Au/TS-1 catalysts showed that PO production could be approximated using the power rate law expression $r_{PO} = k\,[H_2]^{0.60 \pm 0.03}[O_2]^{0.31 \pm 0.04}[C_3H_6]^{0.18 \pm 0.04}$. The activation energy was found to be catalyst specific and ranged from 35 to 54 kJ mol^{-1}. A mechanistic model to reproduce experimentally observed orders was constructed primarily from mechanistic steps proposed via DFT calculations [63–65]. In Eqs. (11.7–11.14), S_1 denotes an Au site and S_2 denotes a Ti-containing site.

$$O_2 + S_1 \leftrightarrow O_2 - S_1 \tag{11.7}$$

$$H_2 + O_2 - S_1 + S_1 \leftrightarrow HOO - S_1 + H - S_1 \tag{11.8}$$

$$O_2 + 2H - S_1 \leftrightarrow HOO - S_1 + H - S_1 \tag{11.9}$$

$$H_2 + HOO - S_1 \leftrightarrow HOOH - S_1 + H - S_1 \tag{11.10}$$

$$C_3H_6 + S_2 \leftrightarrow C_3H_6 - S_2 \tag{11.11}$$

$$C_3H_6 - S_2 + HOOH - S_1 \rightarrow H_2O - S_1 + PO - S_2 \tag{11.12}$$

$$PO - S_2 \leftrightarrow PO + S_2 \tag{11.13}$$

$$H_2O - S_1 \leftrightarrow H_2O + S_1 \tag{11.14}$$

$$r_{PO} = \frac{k_6 z}{L_1} \left[\frac{L_1 K_5 [C_3H_6]}{1 + K_5 [C_3H_6]} \right] \left[\frac{L_2 K_3^{1/2} K_4 [H_2]^{1/2}}{K_1^{1/2} K_2^{1/2}} \right] \left[\frac{K_1^{1/2} K_2^{1/2} K_3^{1/2} [H_2]^{1/2} [O_2]}{1 + K_1^{1/2} K_2^{1/2} K_3^{1/2} [H_2]^{1/2} [O_2]} \right]$$

$$= k[C_3H_6]^n [H_2]^{m+1/2} [O_2]^{2m} \quad (11.15)$$

Although the mechanism is similar to the mechanistic steps presented in reference [65], Taylor et al. [55] used enhanced H_2 dissociation in the presence of O_2, consistent with both previous experimental observation [91] and DFT calculations [63,64]. Moreover, the rate determining step (Eq. 11.12) is unique because the fractional propylene order required the inclusion of adsorbed propylene in the RDS [55]. This mechanism and its associated rate expression (Eq. 11.15) were the simplest means of reproducing the observed reaction orders ($n = 0.18$, $m = 0.14$). It does imply a relation between H_2 and O_2 orders ($m + 1/2$, $2m$), however. The addition of a third active site for dissociative H_2 adsorption [65] would provide independent control over all three reaction orders.

Recently, another kinetic study using a stable Ba-doped, mesoporous Au/Ti-TUD catalyst determined the power rate law expression for PO, CO_2, and water production [57]. CO_2 formation was found to result from the further reaction of PO rather than directly from propylene. The experimental reaction orders for water production were very similar to those obtained in reference [65], $r_{H_2O} = k_1 [H_2]^{0.67 \pm 0.07} [O_2]^{0.16 \pm 0.07}$. The PO production rate took the form $r_{PO} = k_2 [H_2]^{0.54 \pm 0.06} [O_2]^{0.24 \pm 0.06} [C_3H_6]^{0.36 \pm 0.06}$, which is similar to the results from reference [55]. The power rate law expression retains the roughly 2:1 relationship between H_2 and O_2 orders, though the orders cannot be well fitted by the rate expression derived in reference [55]. Propylene order on Au/Ti-TUD (0.36 ± 0.06) catalysts [57] is larger than that on Au/TS-1 (0.18 ± 0.04) catalysts [55]. This difference can perhaps be rationalized by noting that adsorption of linear hydrocarbons is strongest around a pore size of 4–5 Å and larger pore sizes result in weaker adsorption of hydrocarbons due to weaker dispersion interactions with the pore walls [92]. Therefore, relatively stronger propylene adsorption in TS-1 pores (5.5 Å) than that in Ti-TUD pores (13 nm) is consistent with a larger apparent propylene reaction order for the later. To explain the kinetic data on Au/Ti-TUD catalysts, Lu et al. proposed a detailed reaction mechanism with three active sites, two Au and one Ti [57].

5.1. "Sequential" versus "simultaneous" epoxidation mechanism

It has been popularly speculated that the mechanism of direct propylene epoxidation using H_2 and O_2 is "sequential" involving (1) *in situ* H_2O_2 formation on Au [63–65,90,93–96], (2) spillover of H_2O_2 to Ti sites, and (3) propylene epoxidation using H_2O_2 on these Ti sites [76–78,97–102]. However, Joshi et al. [103] have reported a quite different DFT-based Au-only epoxidation pathway where propylene attack on H-Au-OOH species is the RDS. Interestingly, Taylor et al. have explained their kinetic data over 0.02–0.06 wt.% Au/TS-1 catalysts using a unique mechanism that involves attack of propylene (adsorbed on Ti sites) on H_2O_2–Au species to form PO in the RDS [55]. The Au and Ti sites are involved in the single RDS and hence this mechanism is termed as "simultaneous" [55]; the possibility of S_2 being an Au–Ti interface site capable of adsorbing propylene cannot be ruled out [104]. On the other hand, in their *in situ* UV–vis study, Haruta and coworkers [105] observed the formation of Ti-OOH species on 4.9 wt.% Au/TS-1 catalysts. However, due to the absence of transient kinetic data, they could not confirm whether Ti-OOH is just a spectator or a true reactive intermediate [105]. Nevertheless, these results strongly support the DFT predictions that H_2 and O_2 react on Au to form H_2O_2 [63–65,93], but it is still unclear whether the direct propylene epoxidation occurs primarily through "sequential" or "simultaneous" or both the mechanisms. In any case, the fractional orders for O_2, H_2, and propylene suggest that all three species must be adsorbed, thereby ruling out an Eley-Rideal step for propylene incorporation.

Very recently, Joshi *et al.* [106] have performed DFT-based mechanistic studies on Au_3 supported on TS-1 to examine whether the "sequential" mechanism is viable if Au and Ti sites are in proximity inside the TS-1 pores. On a bare Ti-defect site (no Au), the calculated ΔE_{act} for Ti-OOH formation (70.3 kJ mol^{-1}) and for subsequent propylene epoxidation (84.9 kJ mol^{-1}) suggests that epoxidation is viable. However, the Ti-defect site is also the most favorable binding site for small Au clusters [66]. Interestingly, the ΔE_{act} for Ti-OOH formation on an Au_3/Ti-defect site is 134.3 kJ mol^{-1}, suggesting that the "sequential" mechanism is kinetically inhibited due to the proximity between Au clusters and the Ti-defect sites [106]. Perhaps even higher activation energies for Ti-OOH formation are likely on Au_{4-5}/Ti-defect sites where steric effects will be more severe. Joshi *et al.* [106] also suggest that in the "simultaneous" mechanism, propylene is likely to be adsorbed on Au–Ti interface sites ($\Delta E_{ads} \sim -83.7$ kJ mol^{-1}) rather than on Ti sites ($\Delta E_{ads} \sim -41.8$ kJ mol^{-1}). The predicted adsorption energies are consistent with the reaction order of propylene (0.18 ± 0.04) in the power law model. These results indicate that the "simultaneous" mechanism dominates if Ti-defect sites are covered by Au clusters [106]. The "sequential" mechanism can operate through a combination of Au sites and distal Ti defect sites.

6. CONCLUSIONS AND FUTURE OUTLOOK

A decade of stimulating research has resulted in significant advances in direct propylene epoxidation using H_2 and O_2 over Au on mesoporous and nanoporous Ti-supports. While the selectivity and activity of these catalysts are now in a commercially interesting range, poor H_2 efficiency is still a major concern. Moreover, although many of these catalysts have been shown to be stable over a period of 40–100 h, substantially longer stability testing is needed. Despite high-quality data and well-designed factorial experiments, there is still insufficient information to formulate a reasonable reaction mechanism capable of explaining all of the published data. Isotopic transient kinetic studies, perhaps with some guidance from DFT calculations, should be performed to identify the mechanism(s) of direct propylene epoxidation over different Au/Ti catalysts.

In our view, it is necessary to expand the research on direct propylene epoxidation in following different directions. (1) It is critical to improve the H_2 efficiency of current Au/TS-1 catalysts (to at least 50%) without compromising the catalyst activity, selectivity, and stability. Liquid-phase studies on formation of H_2O_2 from H_2 and O_2 suggest that supported Au–Pd [107,108] and to some extent Au–Pt [109] catalysts are better than Au catalysts. Recent DFT calculations also suggest that few-atom Au-alloy clusters can activate H_2 and O_2 [110] to make H_2O_2 [111]. Such alloy catalysts may be useful in improving the $TOR_{H_2O_2}/TOR_{H_2O}$ in direct propylene epoxidation as well and may improve the H_2 efficiency. In fact, improved H_2 efficiencies were reported upon addition of Pt to Au/TiO$_2$/SiO$_2$ catalysts for gas-phase propylene epoxidation using H_2 and O_2 [112], and such efforts merit further exploration. (2) In light of the promotional effect of nitrates [53,56] and gas-phase additives, such as TMA [53], systematic optimization of activity, selectivity, stability, and H_2 efficiency of various Au/Ti catalysts for direct propylene epoxidation is possible, and such studies should be performed. Moreover, a combination of NH_4NO_3 pretreatment of TS-1 and introduction of mesoporous-scale defects in TS-1 should be examined. (3) While DP [32] has been the most popular method for making good Au/Ti catalysts, from a commercial standpoint, simpler methods such as impregnation are more viable and should be explored further [113,114] particularly in conjunction with the use of other promoters. From a fundamental standpoint, alternative methods for making supported Au catalysts, such as pulsed laser ablation [115], chemical vapor deposition (CVD) [116], solution deposition of phosphine-stabilized Au clusters [117], etc., need to be examined further. Recently, Zheng *et al.* [118] have achieved one-phase synthesis of monodisperse noble-metal nanoparticles in a single step. They have also developed a general method to prepare metal nanoparticles supported on acidic and basic oxides, which allows precise control of the metal-particle size, size-distribution, and loading [119]. Supported Au catalysts synthesized using this route should be tested for direct propylene epoxidation. (4) The search for better direct propylene epoxidation catalysts should not be restricted to Ti-supports. Non-Ti-supports, such as vanadium-containing zeolites [52] and ZrO_2 [120], have been examined in some patents, and may merit wider study. (5) While it seems to be established that rates follow Ti loading as long as the Ti stays isolated and tetrahedral, less is known about optimal Au loading. For TS-1 supports, the highest Au efficiency is at the lowest loadings and it is clear that excess Au is detrimental to stability. More study of effects of Au loading on mesoporous materials is needed. (6) All catalysts for PO production show an optimum in yield with temperature because eventually PO decomposition accelerates. A benefit of the catalysts discussed here is that many of them delay the PO degradation until higher temperatures are reached, thus allowing higher temperatures to enhance PO desorption and avoiding fouling by PO oligomerization. The temperature at which the maximum PO rate occurs is, therefore, a characterization parameter that can be important in the search for higher performance.

ACKNOWLEDGMENTS

This work was funded by the United States Department of Energy, Office of Basic Energy Sciences, through the Grant DE-FG02–01ER-15107 (W.N.D.), and by the National Science Foundation through the Grant CTS-0238989-CAREER (K.T.T.). Computational resources were obtained through a Grant from the National Computational Science Alliance (MCA04N010) and through the supercomputing resources at Purdue University. The authors thank Prof. S. Ted Oyama and an anonymous reviewer for carefully examining this manuscript and for their valuable suggestions.

REFERENCES

[1] D. L. Trent, Propylene Oxide, in: *Kirk-Othmer: Encyclopedia of Chemical Technology*, Wiley, New York, 2001.
[2] E. Bartolome, W. Koehler, G. Stoeckelman, A. May, Continuous manufacture of propylene oxide from propylene chlorohydrine, *U.S. Patent No.* 3,886,187, 1975, Assigned to BASF Corporation.
[3] W. F. Richey, *Chlorohydrins*, in: *Kirk-Othmer: Encyclopedia of Chemical Technology*, Wiley, New York, 1994.
[4] M. D. Cisneros, M. T. Holbrook, L. N. Ito, Hydrodechlorination process and catalyst for use ttherein, *U.S. Patent No.* 5,476,984, 1995, Assigned to DOW Chemical Company.
[5] E. M. Jorge, Chlorohydrin process, *U.S. Patent No.* 6,043,400, 2000, Assigned to DOW Chemical Company.
[6] J. Kollar, Epoxidation Process, *U.S. Patent No.* 3,351,635, 1967, Assigned to Halcon International, Inc.
[7] M. Pell, E. I. Korchak, Epoxidation using ethylbenzene hydroperoxide with alkali or adsorbent treatment of recycle ehtylbenzene, *U.S. Patent No.* 3,439,001, 1969, Assigned to Halcon International, Inc.
[8] W. S. Dubner, R. N. Cochran, Propylene oxide-styrene monomer process, *U.S. Patent No.* 5,210,354, 1993, Assigned to ARCO Chemical Technology.
[9] J. J. van der Sluis, Process for the preparation of styrene and propylene oxide, *U.S. Patent No.* 6,504,038, 2003, Assigned to Shell Corporation.
[10] Peroxide-based propylene oxide production makes headway, *Chem. Eng. Prog.*, (October, 2003) 14.
[11] A. Tullo, Dow and BASF to build propylene oxide, *Chem. Eng. News*, 82(36) (2004) 15.
[12] M. G. Clerici, P. Ingallina, Process for producing olefin oxides, *U.S. Patent No.* 5,221,795, 1993, Assigned to Enichem.
[13] W. T. Hess, Hydrogen Peroxide, in: *Kirk-Othmer: Encyclopedia of Chemical Technology*, Wiley, New York, 1995.
[14] M. G. Clerici, G. Bellussi, U. Romano, Synthesis of propylene oxide from propylene and hydrogen peroxide catalyzed by titanium silicalite, *J. Catal.* 129 (1991) 159.
[15] M. G. Clerici, P. Ingallina, Epoxidation of lower olefins with hydrogen peroxide and titanium silicalite, *J. Catal.* 140 (1993) 71.
[16] M. G. Clerici, P. Ingallina, *Oxidation reactions with in situ generated oxidants*, Catal. Today 41 (1998) 351.
[17] B. S. Lane, K. Burgess, Metal-catalyzed epoxidations of alkenes with hydrogen peroxide, *Chem. Rev.* 103 (2003) 2457.
[18] T. A. Nijhuis, M. Makkee, J. A. Moulijn, B. M. Weckhuysem, The production of propene oxide: Catalytic processes and recent developments, *Ind. Eng. Chem. Res.* 45 (2006) 3447.
[19] J. P. Dever, K. F. George, W. C. Hoffman, H. Soo, Ethylene Oxide, in: *Kirk-Othmer: Encyclopedia of Chemical Technology*, Wiley, New York, 1995.
[20] J. G. Serafin, A. C. Liu, S. R. Seyedmonir, *Surface science and the silver-catalyzed epoxidation of ethylene:* An industrial perspective, *J. Mol. Catal. A: Chem.* 131 (1998) 157.
[21] J. R. Monnier, *The direct epoxidation of higher olefins using molecular oxygen*, Appl. Catal. A: Gen. 221 (2001) 73.

[22] E. A. Carter, W. A. Goddard, The surface atomic oxyradical mechanism for Ag-catalyzed olefin epoxidation, *J. Catal.* 112 (1988) 80.
[23] R. M. Lambert, F. J. Williams, R. L. Cropley, A. Palermo, Heterogeneous alkene epoxidation: Past, present, and future, *J. Mol. Catal. A: Chem.* 228 (2005) 27.
[24] R. Wang, X. Guo, X. Wang, J. Hao, Propylene epoxidation over silver supported on titanium silicalite zeolite, *Catal. Lett.* 90 (2003) 57.
[25] D. Sullivan, P. Hooks, M. Mier, J. W. van Hal, X. K. Zhang, Effect of support and preparation on silver-based direct olefin epoxidation catalyst, *Top. Catal.* 38 (2006) 303.
[26] J. Lu, J. J. Bravo-Suárez, A. Takahashi, M. Haruta, S. T. Oyama, In situ uv-vis studies of the effect of particle size on the epoxidation of ethylene and propylene on supported silver catalysts with molecular oxygen, *J. Catal.* 232 (2005) 85.
[27] J. Q. Lu, J. J. Bravo-Suárez, M. Haruta, S. T. Oyama, Direct propylene epoxidation over modified Ag/CaCo$_3$ catalysts, *Appl. Catal. A: Gen.* 302 (2006) 283.
[28] R. L. Cropley, F. J. Williams, O. P. H. Vaughan, A. J. Urquhart, M. S. Tikhov, R. M. Lambert, Copper is highly effective for epoxidation of a "difficult" alkene, whereas silver is no, *Surf. Sci.* 578 (2005) L85.
[29] J. R. Monnier, K. T. Peters, G. W. Hartley, The selective epoxidation of conjugated olefins containing allylic substituents and epoxidation of propylene in the presence of butadiene, *J. Catal.* 225 (2004) 374.
[30] B. Cooker, A. M. Gaffney, J. D. Jewson, W. H. Onimus, *Propylene*, Epoxidation using chloride-containing Silver Catalysts, U.S. Patent No. 5,780,657, 1998, Assigned to ARCO Chemical Technology.
[31] T. Hayashi, K. Tanaka, M. Haruta, Selective vapor-phase epoxidation of propylene over Au/TiO$_2$ catalysts in the presence of oxygen and hydrogen, *J. Catal.* 178 (1998) 566.
[32] S. Tsubota, D. A. H. Cunningham, Y. Bando, M. Haruta, Preparation of highly dispersed gold on titanium and magnesium oxide, *Stud. Surf. Sci Catal.* 91 (1995) 227.
[33] T. A. Nijhuis, B. J. Huizinga, M. Makkee, J. A. Moulijin, Direct epoxidation of propylene using gold dispersed on TS-1 and other titanium-containing supports, *Ind. Eng. Chem. Res.* 38 (1999) 884.
[34] E. E. Stangland, K. B. Stavens, R. P. Andres, W. N. Delgass, Propylene epoxidation over gold-titania catalysts, *Stud. Surf. Sci. Catal.* 130 (2000) 827.
[35] E. E. Stangland, K. B. Stavens, R. P. Andres, W. N. Delgass, Characterization of gold-titania catalysts via oxidation of propylene to propylene oxide, *J. Catal* 191 (2000) 332.
[36] J. Chou, E. W. McFarland, Direct propylene epoxidation on chemically reduced au nanoparticles supported on titania, *Chem. Commun.* (2004) 1648.
[37] G. Mul, A. Zwijnenburg, B. van der Linden, M. Makkee, J. A. Moulijn, Stability and selectivity of Au/TiO$_2$ and Au/TiO$_2$/SiO$_2$ catalysts in propene epoxidation: An *in-situ* ft-ir study, *J. Catal.* 201 (2001) 128.
[38] A. Zwijnenburg, A. Goossens, W. G. Sloof, M. W. J. Crajé, A. M. van der Kraan, L. J. de Jongh, M. Makkee, J. A. Moulijn, XPS mossbauer characterization of Au/TiO$_2$ propene epoxidation catalysts, *J. Phys. Chem. B* 106 (2002) 9853.
[39] A. Zwijnenburg, M. Makkee, J. A. Moulijn, Increasing the low propene epoxidation product yield of gold/titania-based catalysts, *Apply. Catal. A: Gen.* 270 (2004) 49.
[40] B. S. Uphade, Y. Yamada, T. Akita, T. Nakamua, M. Haruta, Synthesis and characterization of Ti-MCM-41 and vapor-phase epoxidation of propylene using H$_2$ and O$_2$ over Au/Ti-MCM-41, *Appl. Catal. A: Gen.* 215 (2001) 137.
[41] B. S. Uphade, M. Okumura, N. Yamada, S. Tsubota, M. Haruta, Vapor-phase epoxidation of propene using H$_2$ and O$_2$ over Au/Ti-MCM-41 and Au/Ti-MCM-48, *Stud. Surf. Sci. Catal.* 130 (2000) 833.
[42] B. S. Uphade, T. Akita, T. Nakamura, M. Haruta, Vapor-phase epoxidation of propene using H$_2$ and O$_2$ over Au/Ti-MCM-48, *J. Catal.* 209 (2002) 331.
[43] M. P. Kapoor, A. K. Sinha, S. Seelan, S. Inagaki, S. Tsubota, H. Yoshida, M. Haruta, Hydrophobicity induced vapor-phase oxidation of propene over gold supported on titanium incorporated hybrid mesoporous silsequioxane, *Chem. Commum.* (2002) 2902.

[44] C. Qi, T. Akita, M. Okumura, K. Kuraoka, M. Haruta, Effect of surface chemical properties and texture of mesoporous titanosilicates on direct vapor-phase epoxidation of propylene over Au catalysts at high reaction temperature, *Appl. Catal. A: Gen.* 253 (2003) 75.
[45] H. W. Clark, R. G. Bowman, J. J. Maj, S. R. Bare, G. E. Hartwell, Process for the Direct Oxidation of Olefins to Olefin Oxides, U.S. Patent No. 5,965,754, 1999, Assigned to DOW Chemical Company.
[46] G. H. Grosch, U. Muller, M. Schulz, N. Rieber, H. Wurz, *Oxidation* catalyst and process for the production of epoxides from olefines, hydrogen and oxygen using said oxidation catalyst, U.S. Patent No. 6,008,389, 1999, Assigned to BASF Corporation.
[47] C. A. Jones, Direct epoxidation process using a mixed catalyst system, U.S. Patent No. 6,307,073, 2001, Assigned to ARCO Chemical Technology.
[48] R. G. Bowman, J. L. Womack, H. W. Clark, J. J. Maj, G. E. Hartwell, Process for the direct oxidation of olefins to olefin oxides, U.S. Patent No. 6,031,116, 2000, Assigned to DOW Chemical Company.
[49] A. Kuperman, R. G. Bowman, H. W. Clark, G. E. Hartwell, B. J. Schoeman, H. E. Tuinstra, G. R. Meima, Process for the hydro-oxidation of olefins to olefin oxides using oxidized gold catalyst, U.S. Patent No. 6,255,499, 2001, Assigned to DOW Chemical Company.
[50] R. G. Bowman, A. Kuperman, H. W. Clark, G. E. Hartwell, G. R. Meima, Process for the direct oxidation of olefins to olefin oxides, U.S. Patent No. 6,323,351, 2001, Assigned to DOW Chemical Company.
[51] G. R. Meima, H. W. Clark, R. G. Bowman, A. Kuperman, G. E. Hartwell, Process for the direct oxidation of olefins to olefin oxides, U.S. Patent No. 6,646,142, 2003, Assigned to DOW Chemical Company.
[52] P. J. Whitman, J. F. Miller, J. John, H. Speidel, R. N. Cochran, Direct epoxidation process, U.S. Patent No. 7,138,535, 2006, Assigned to Lyondell Chemical Technology.
[53] B. Chowdhury, J. J. Brávo-Suarez, M. Daté, S. Tsubota, M. Haruta, Trimethylamine as a gas-phase promoter: Highly efficient epoxidation of propylene over supported gold catalysts, *Angew. Chem. Int. Ed.* 45 (2006) 412.
[54] N. Yap, R. P. Andres, W. N. Delgass, Reactivity and stability of Au in and on TS-1 for epoxidation of propylene with H_2 and O_2, *J. Catal.* 226 (2004) 156.
[55] B. Taylor, J. Lauterbach, G. E. Blau, W. N. Delgass, Reaction kinetic analysis of the gas-phase epoxidation of propylene over Au/TS-1, *J. Catal.* 242 (2006) 142.
[56] L. Cumaranatunge, W. N. Delgass, Enhancement of Au Capture efficiency and activity of Au/TS-1 catalysts for propylene epoxidation, *J. Catal.* 232 (2005) 38.
[57] J. Lu, X. Zhang, J. J. Bravo-Suarez, S. Tsubota, J. Gaudet, S. T. Oyama, Kinetics of propylene epoxidation using H_2 and O_2 over a Gold/mesoporous titanosilicate catalyst, *Catal. Today* 123 (2007) 189.
[58] M. Taramasso, G. Perego, B. Notari, Preparation of porous crystalline synthetic material comprised of silicon and titanium oxides, U.S. Patent No. 4,410,501, 1983, Assigned to Snamprogetti S.p.A.
[59] B. Taylor, J. Lauterbach, W. N. Delgass, Gas phase epoxidation of propylene over small gold ensembles on TS-1, *Appl. Catal. A Gen.* 291 (2005) 188.
[60] R. B. Khomane, B. D. Kulkarni, A. Paraskar, S. R. Sainkar, Synthesis, characterization and catalytic performance of titanium silicalite-1 prepared in micellar media, *Mater. Chem. Phys.* 76 (2002) 99.
[61] B. S. Uphade, S. Tsubota, T. Hayashi, M. Haruta, Selective oxidation of propylene to propylene oxide or propionaldehyde over Au supported on titanosilicates in the presence of H_2 and O_2, *Chem. Lett.* (1998) 1277.
[62] E. E. Stangland, B. Taylor, R. P. Andres, W. N. Delgass, Direct vapor phase propylene epoxidation over deposition-precipitation gold-titania catalyst in the presence of H_2/O_2: Effects of support, neutralizing agent, and pretreatment, *J. Phys. Chem. B* 109 (2005) 2321.
[63] D. H. Wells, W. N. Delgass, K. T. Thomson, Formation of hydrogen peroxide from H_2 and O_2 over a neutral gold trimer, A DFT study, J Catal. 225 (2004) 69.
[64] A. M. Joshi, W. N. Delgass, K. T. Thomson, Comparison of the catalytic activity of Au_3, Au_4^+, Au_5, and Au_5^- in the gas phase reaction of H_2 and O_2 to form hydrogen peroxide: A density functional theory investigation, *J. Phys. Chem. B* 109 (2005) 22392.

[65] D. G. Barton, S. G. Podkolzin, Kinetic study of a direct water synthesis over silica-supported gold nanoparticles, *J. Phys. Chem. B* 109 (2005) 2262.
[66] A. M. Joshi, W. N. Delgass, K. T. Thomson, Adsorption of small Au_n (n=1–5) and Au-Pd clusters inside the Ts-1 and S-1 Pores, *J. Phys. Chem. B* 110 (2006) 16439.
[67] A. K. Sinha, S. Seelan, S. Tsubota, M. Haruta, Catalysis by Gold nanoparticles: Epoxidation of propene, *Top. Catal* 29 (2004) 95.
[68] F. Moreau, G. C. Bond, A. O. Taylor, Gold on titania catalysts for the oxidation of carbon monoxide: Control of pH during preparation with various gold contents, *J. Catal.* 231 (2005) 105.
[69] F. Moreau, G. C. Bond, Preparation and reactivation of Au/TiO_2 catalysts, *Catal. Today* 122 (2007) 260.
[70] B. Solsona, M. Conte, Y. Cong, A. Carley, G. Hutchings, Unexpected Promotion of Au/TiO_2 by nitrate for CO oxidation, *Chem. Commun.* (2005) 2351.
[71] A. K. Sinha, S. Seelan, S. Tsubota, M. Haruta, A three-dimensional mesoporous titanosilicate support for gold nanoparticles: Vapor-phase epoxidation of propene with high conversion, *Angew. Chem. Int. Ed.* 43 (2004) 1546.
[72] J. M. Fisher, Beware! fulminating gold, *Chem. Brit.* 39 (2003) 12.
[73] P. F. Henry, M. T. Weller, C. C. Wilson, Structural investigation of TS-1. Determination of the true nonrandom titanium framework substitution and silicon vacancy distribution from powder neutron diffraction studies using isotopes, *J. Phys. Chem. B* 105 (2001) 7452.
[74] C. Lamberti, S. Bordiga, A. Zecchina, G. Artioli, G. Marra, G. Spano, Ti location in the MFI framework of Ti-silicalite-1: A neutron powder diffraction study, *J. Am. Chem. Soc.* 123 (2001) 2204.
[75] W. Lin, H. Frei, Photochemical and FT-IR probing of the active site of hydrogen peroxide in Ti silicalite sieve, *J. Am. Chem. Soc.* 124 (2002) 9292.
[76] D. H. Wells, W. N. Delgass, K. T. Thomson, Evidence of defect-promoted reactivity for epoxidation of propylene in titanosilicate (TS-1) eatalysts: A DFT study, *J. Chem. Soc* 126 (2004) 2956.
[77] P. E. Sinclair, C. R. A. Catlow, Quantum chemical study of the mechanism of partial oxidation reactivity in titanosilicate catalysts: Active site formation, oxygen transfer, and catalyst deactivation, *J. Phys. Chem. B* 130 (1999) 1084.
[78] D. H. Wells, A. M. Joshi, W. N. Delgass, K. T. Thomson, A quantum chemical study of comparison of various propylene epoxidation mechanisms using H_2O_2 and TS-1 catalyst, *J. Phys. Chem. B* 110 (2006) 14627.
[79] B. Taylor, J. Lauterbach, W. N. Delgass, The effect of mesoporous scale defects on the activity of Au/Ts-1 for the epoxidation of propylene, *Catal. Today* 123 (2007) 50.
[80] C. J. H. Jacobsen, C. Madsen, J. Houzvicka, I. Schmidt, A. Carlsson, Mesoporous zeolite single crystals, *J. Am. Chem. Soc.* 122 (2000) 7116.
[81] Y. A. Kalvachev, T. Hayashi, S. Tsubota, M. Haruta, Vapor-phase selective oxidation of aliphatic hybrocarbons over gold deposited on mesoporous titanium silicates in the co-presence of oxygen and hydrogen, *J. Catal.* 186 (1999) 228.
[82] A. K. Sinha, S. Seelan, T. Akita, S. Tsubotta, M. Haruta, Vapor phase propylene epoxidation over Au/Ti-MCM-41 catalysts prepared by different Ti incorporation modes, *Appl. Catal. A Gen.* 240 (2003) 243.
[83] A. K. Sinha, S. Seelan, T. Akita, S. Tsubota, M. Haruta, Vapor-phase epoxidation of propene over Au/Ti-MCM-41 catalysts. Influence of Ti content and Au content, *Catal. Lett.* 85 (2003) 223.
[84] E. Sacaliuc, A. M. Beale, B. M. Weckhuysen, T. A. Nijhuis, Propene epoxidation over Ti-SBA-15 catalysts, *J. Catal.* 248 (2007) 235.
[85] A. K. Sinha, S. Seelan, M. Okumura, T. Akita, S. Tsubota, M. Haruta, Three-dimensional mesoporous titanosilicates prepared by modified sol-gel method: Ideal gold catalyst supports for enhanced propene epoxidation, *J. Phys. Chem. B* 109 (2005) 3956.
[86] M. Dai, D. Tang, Z. Lin, H. Yang, Y. Yuan, Ti-Si mixed oxides by non-hydrolytic Sol-Gel synthesis as potential gold catalyst supports for Gas-phase epoxidation of propylene in H_2 and O_2, *Chem. Lett.* 35 (2006) 878.
[87] B. S. Uphade, M. Okumura, S. Tsubota, M. Haruta, Effect of physical mixing of CsCl with Au/Ti-MCM-41 on the Gas-phase epoxidation of propene using H_2 and O_2: Drastic depression of H_2 Consumption, *Appl. Catal. A. Gen.* 190 (2000) 43.

[88] T. A. Nijhuis, T. Visser, B. M. Weckhuysen, The role of gold in gold-titania epoxidation catalysts, *Angew. Chem. Int. Ed.* 44 (2005) 1115.
[89] V. N. Shetti, P. Manikandan, D. Srinivas, P. Ratnasamy, Reactive oxygen species in epoxidation reactions over titanosilicate molecular sieves, *J. Catal.* 216 (2003) 461.
[90] C. Sivadinarayana, T. V. Choudhary, L. L. Daemen, J. Eckert, D. W. Goodman, The nature of the surface species formed on Au/TiO_2 during the reaction of H_2 and O_2: An inelastic neutron scattering study, *J. Am. Chem. Soc.* 126 (2004) 38.
[91] S. Naito, M. Tanimoto, Oxygen enhanced hydrogen exchange and hydrogenation over supported gold catalysts, *Chem. Commun.* (1988) 832.
[92] S. P. Bates, W. J. M. van Well, R. A. van Santen, B. Smit, Energetics of n-alkanes in zeolites: A configurational-bias monte carlo investigation into pore size dependence, *J. Am. Chem. Soc.* 118 (1996) 6753.
[93] P. P. Olivera, E. M. Patrito, H. Sellers, Hydrogen peroxide synthesis over emtallic catalysts, *Surf. Sci.* 313 (1994) 25.
[94] P. Landon, P. J. Collier, A. J. Papworth, C. J. Kiely, G. J. Hutchings, Direct formation of hydrogen peroxide from H_2/O_2 using a gold catalyst, *Chem. Commun.* (2002) 2058.
[95] M. Okumura, Y. Kitagawa, K. Yamagcuhi, T. Akita, S. Tsubota, M. Haruta, Direct production of hydrogen peroxide using H_2 and O_2 over highly dispersed Au catalysts, *Chem. Lett.* 32 (2003) 822.
[96] S. Ma, G. Li, X. Wang, The Direct synthesis of hydrogen peroxide from H_2 and O_2 over Au/TS-1 and application in oxidation of thiophene *in situ*, *Chem. Lett.* 35 (2006) 428.
[97] H. Munakata, Y. Oumi, A. Miyamoto, A DFT study on peroxo-complex in titanosilicate catalyst: Hydrogen peroxide activation on titanosilicalite-1 catalyst and reaction mechanisms for catalytic olefin epoxidation and for hydroxylamine formation from ammonia, *J. Phys. Chem. B* 105 (2001) 3493.
[98] M. Neurock, L. E. Manzer, Theoretical insights on the mechanism of alkene epoxidation by H_2O_2 with titanium silicalite, *Chem. Commun.* (1996) 1133.
[99] G. Tozzola, M. A. Mantegazza, G. Ranghino, G. Petrini, S. Bordiga, G. Ricchiardi, C. Lamberti, R. Zullian, A. Zecchina, On the structure of the active site of Ti-silicalite in reactions with hydrogen peroxide: A vibrational and computational study, *J. Catal.* 179 (1998) 64.
[100] G. N. Vayssilov, R. A. van Santen, Catalytic activity of titanium silicalites-a DFT study, *J. Catal.* 175 (1998) 170.
[101] I. V. Yudanov, P. Gisdakis, C. D. Valentin, N. Rösch, Activity of peroxo and hydroperoxo complexes of Ti(IV) in olefin expoxidation: A density functional model study of energetics and mechanism, *Eur. J. Inorg. Chem.* (1999) 2135.
[102] W. Panyaburapa, T. Nanok, J. Limtrakul, Epoxidation reaction of unsaturated hydrocarbons with H_2O_2 over defect TS-1 investigated by ONIOM method: Formation of active sites and reaction mechanisms, *J. Phys. Chem. C* 111 (2007) 3433.
[103] A. M. Joshi, W. N. Delgass, K. T. Thomson, Partial oxidation of propylene to propylene oxide over a neutral gold trimer in the gas phase: A density functional theory study, *J. Phys. Chem. B* 110 (2006) 2572.
[104] H. M. Ajo, V. A. Bondzie, C. T. Campbell, Propene adsorption on gold particles on $TiO_2(110)$, *Catal. Lett.* 78 (2002) 359.
[105] B. Chowdhury, J. J. Bravo-Suárez, N. Mimura, J. Lu, K. K. Bando, S. Tsubota, M. Haruta, In situ UV-vis and EPR study on the formation of hydroperoxide species during direct gas phase propylene epoxidation over $Au/Ti-SiO2$ Catalyst, *J. Phys. Chem. B* 110 (2006) 22995.
[106] A. M. Joshi, W. N. Delgass, K. T. Thomson, , Mechanistic implications of Au_n/Ti-lattice proximity for propylene epoxidation, *J. Phys. Chem. C* 111 (2007) 7841.
[107] P. Landon, P. J. Collier, A. F. Carley, D. Chadwick, A. J. Papworth, A. Burrows, C. J. Kiely, G. J. Hutchings, Direct synthesis of hydrogen peroxide from H_2 and O_2 using Pd and Au catalysts, *Phys. Chem. Chem. Phys.* 5 (2003) 1917.
[108] J. K. Edwards, B. E. Solsona, P. Landon, A. F. Carley, A. Herzing, C. J. Kiely, G. J. Hutchings, Direct synthesis of hydrogen peroxide from H_2 and O_2 using TiO_2-supported Au-Pd catalysts, *J. Catal.* 236 (2005) 69.
[109] G. Li, J. Edwards, A. F. Carley, G. J. Hutchings, Direct synthesis of hydrogen peroxide from H_2 and O_2 using zeolite-supported Au-Pd catalysts, *Catal. Today* 122 (2007) 361.

[110] A. M. Joshi, W. N. Delgass, K. T. Thomson, Analysis of O_2 adsorption on binary-alloy clusters of gold: Energetics and correlations, *J. Phys. Chem. B* 110 (2006) 23373.
[111] A. M. Joshi, W. N. Delgass, K. T. Thomson, Investigation of gold-silver, gold-copper, and gold-palladium dimers and trimers for hydrogen peroxide formation from H_2 and O_2, *J. Phys. Chem. C* 111 (2007) 7384.
[112] A. Zwijnenburg, M. Saleh, M. Makkee, J. A. Moulijn, Direct gas-phase epoxidation of propylene over bimetallic Au catalysts, *Catal. Today* 72 (2002) 59.
[113] A. Kuperman, R. G. Bowman, H. W. Clark, G. E. Hartwell, G. R. Meima, Method of preparing a catalyst containing Gold and Titanium, *U.S. Patent No.* 6,821,923, 2004, Assigned to DOW Chemical Company.
[114] A. Kuperman, R. G. Bowman, H. W. Clark, G. E. Hartwell, G. R. Meima, Method of preparing a catalyst containing Gold and Titanium, *U.S. Patent No.* 6,984,607, 2006, Assigned to DOW Chemical Company.
[115] S. Senkan, M. Kahn, S. Duan, A. Ly, C. Leidhom, High-throughput metal nanoparticle catalysis by pulsed laser ablation, *Catal. Today* 117 (2006) 291.
[116] M. Okumura, S. Tsubota, M. Iwamoto, M. Haruta, Chemical vapor deposition of gold nanoparticles on MCM-41 and their catalytic activities for the low-temperature oxidation of CO and of H_2, *Chem. Lett.* (1998) 315.
[117] C. C. Chusuei, X. Lai, K. A. Davis, E. K. Bowers, J. P. Fackler, D. W. Goodman, A nanoscale model catalyst preparation: Solution deposition of phosphine-stabilized gold clusters onto a planar TiO_2(110) Support, *Langmuir* 17 (2001) 4113.
[118] N. Zheng, J. Fan, G. D. Stucky, One-step one-phase synthesis of monodisperse noble-metallic nanoparticles and their colloidal crystals, *J. Am. Chem. Soc.* 128 (2006) 6550.
[119] N. F. Zheng, G. D. Stucky, A general strategy for oxide-supported metal nanoparticle catalysts, *J. Am. Chem Soc.* 128 (2006) 14278.
[120] C. A. Jones, R. A. Grey, Epoxidation Process, *U.S. Patent No.* 5,939,569, 1999, Assigned to ARCO Chemical Technology.

CHAPTER 12

The Epoxidation of Propene over Gold Nanoparticle Catalysts

T. Alexander Nijhuis,* Elena Sacaliuc,† and **Bert M. Weckhuysen†**

Contents		
	1. Introduction	340
	2. Experimental	341
	2.1. Catalyst preparation	341
	2.2. Catalyst characterization	342
	2.3. Catalyst activity testing	342
	3. Results and Discussion	343
	3.1. Characterization	343
	3.2. Catalytic performance	343
	3.3. Catalyst deactivation	348
	4. Conclusions	352
	Acknowledgments	353
	References	353

Abstract	Different gold nanoparticle catalysts on titania, silica, and titanosilicate supports are compared in the hydro-epoxidation of propene. All catalysts tested were active in the propene epoxidation, with Au/TiO$_2$ showing the highest activity at low temperature, but also a high rate of deactivation. It is shown that the deactivation of the catalysts is directly related to a side reaction of a bidentate propoxy reaction intermediate. This species can react to produce propene oxide, but as a side reaction it can also produce very strongly adsorbed species, most likely carbonates, which cause a reversible deactivation. There are no indications that the catalyst deactivation is caused by changes in the size or the state of the gold nanoparticles. Catalysts containing a lower amount of titania dispersed in or on a silica support are more stable,

* Laboratory for Chemical Reactor Engineering, Department of Chemical Engineering and Chemistry,
  Eindhoven University of Technology, P.O. Box 513, 5600 MB Eindhoven, The Netherlands
† Laboratory for Inorganic Chemistry and Catalysis, Department of Chemistry, Utrecht University, Sorbonnelaan 16,
  3584 CA Utrecht, The Netherlands

Mechanisms in Homogeneous and Heterogeneous Epoxidation Catalysis　　　© 2008 Elsevier B.V.
DOI: 10.1016/B978-0-444-53188-9.00012-2　　　All rights reserved.

but require a higher reaction temperature for a similar activity. Ti-SBA-15 is the most promising support material, but the gold deposition inside the structure requires further optimization. A key factor, that needs to be solved for all gold catalysts, is the efficiency in which hydrogen is used as a co-reactant. The currently obtained hydrogen efficiencies of up to 10% are insufficient to run a process profitably.

Key Words: Gold, Propene oxide, Titania, SBA-15, Silica, Titanosilicate, Hydro-oxidation, Deactivation, Hydrogen efficiency, Ti-SBA-15, Gold hydrazide, Fulminating gold. © 2008 Elsevier B.V.

1. INTRODUCTION

Propene oxide is a very important chemical intermediate, produced at about six million tons per year (2006) with demand still growing by ~5% annually [1,2]. One of the important new developments for the production of propene oxide are epoxidation catalysts based on gold nanoparticles [1,3–7]. Gold–titania catalysts are capable of very selectively epoxidizing propene in one step at mild conditions using molecular oxygen in the presence of hydrogen as a sacrificial reductant. Because of the use of hydrogen as a co-reactant, this epoxidation is better called a hydro-oxidation rather than a simple oxidation reaction. The two main processes in operation for the production of propene oxide, the chlorohydrin process and the hydroperoxide processes, have disadvantages which make an alternative desirable. A recent review on all processes in use and under development for the propene epoxidation is available [1]. The chlorohydrin process produces a large salty wastewater stream as well as chlorinated side products. The hydroperoxide processes, SM-PO (styrene monomer-propene oxide) and PO-TBA (propene oxide–*tert*-butyl alcohol), produce a co-product in a fixed quantity (typically—two to four times by weight the amount of propylene oxide produced), making the process economics heavily dependent on the value of the co-product. The alternative processes that have been developed in recent years[1,8], the new Dow-BASF hydrogen peroxide combination process (together with the similar Degussa-Headwaters process) and the Sumitomo hydroperoxide process, are complex (using three different reactors) and use, similarly to gold–titania catalysts, sacrificial hydrogen. The hydrogen peroxide processes consume hydrogen in the production of the hydrogen peroxide, which is the oxidant in the epoxidation. The Sumitomo process is a hydroperoxide process, which produces cumyl alcohol out of cumene and uses the hydrogen to convert the alcohol co-product back to cumene. Therefore, the simple (single reactor process) and selective hydro-epoxidation of propene over gold–titania catalysts has good potential for being applied in a highly competitive process.

Despite this attractiveness, gold–titania-based catalyst systems have a number of weaknesses which need to be improved: the propene conversion levels remain low, often the catalyst stability is insufficient, and the hydrogen efficiency is low [1,9,10]. The hydrogen efficiency (defined as the molar amount of propene oxide produced divided by the amount of hydrogen consumed) is dominated by the water produced by the direct hydrogen oxidation over the catalyst. The problem of low catalyst stability is mostly understood and can be solved largely by making sure of a very low level of residual chlorine [10,11] (the catalysts are commonly produced using a gold chloride precursor and residual chlorine makes the catalyst very sensitive to sintering of the gold particles) and by dispersing the titania in a silica matrix [9,10,12,13]. The low conversion levels can also be improved by using a Ti–Si support [9,10], which can produce propene oxide at a higher temperature than gold on titania, or improved even further by using a surface modified Ti–Si support [7,14]. For the activity, it should be noted that even though conversion levels remain low, the catalyst activity for a 1 wt.% gold catalyst is not exceptionally low (the more active gold/titania catalysts produce in the order of 0.1 $mol_{PO}/m^3_{reactor}/s$) compared to other chemical processes applied in industry [15]. To improve the low hydrogen efficiency (typically less than 10%), however, no solutions have been published in the literature yet. It has been claimed that mixing the catalyst with CsCl results in a significant improvement in the hydrogen efficiency [16]; however, there have been no reports confirming these results. Furthermore, the large amount of chloride contributed by the CsCl is detrimental to the catalyst stability toward sintering. Therefore, solving the problem of the low hydrogen efficiency can be seen as the main last hurdle before this type of catalyst can be applied. A hydrogen efficiency in the order of 50% is desirable for a profitable process.

In this chapter, a comparison is made between different gold-based catalysts designed for the propene epoxidation: gold supported on titania, gold supported on dispersed titania supports [Ti on amorphous silica and Ti-SBA-15 (mesoporous silica)], and gold on silica for comparatory purposes. Attention will be given to the activity, selectivity, stability, and hydrogen efficiency.

2. EXPERIMENTAL

2.1. Catalyst preparation

The supports used for the preparation of the catalysts were titania (Degussa-P25, 45 m^2/g, primarily anatase), silica [two types: Davisil 645 (Grace Davison, 295 m^2/g) and OX50 (Degussa, 50 m^2/g)], 0.4 wt.% Ti on OX50 silica [11], and Ti-SBA-15 (Si/Ti = 40, Ti deposited by grafting, preparation according to Sacaliuc et al. [17]).

Catalysts were prepared by means of a deposition precipitation method using ammonia [9].[1] A quantity of 10 g of support was dispersed in 100 ml of

---

[1] When preparing gold catalysts using deposition precipitation using ammonia, care should be taken as the possibility exists for the formation of gold hydrazide (fulminating gold). In the preparations in this chapter the risks are very minor considering the small quantities of gold and the low loadings of the catalysts prepared. Care is advisable, however, considering a reported incident [Fisher, *Gold Bull.* 36 (2003) 155]. It is recommended that readers take the advantages (ease of making stable catalysts without chloride or sodium present) and disadvantages of this preparation method in consideration.

demineralized water with a magnetic stirrer. The pH of the slurry was 3.8 for the catalysts prepared on titania and 7 for the silica support. The pH was raised to 9.5 using 2.5% ammonia. The target loading in gold was 1 wt.%, for which 172 mg of hydrogen tetrachloroaurate(III) solution ($HAuCl_4$, Aldrich—30 wt.% solution in dilute HCl) was diluted in 40 ml of demineralized water and added gradually over a 0.25-h period to the support, while keeping the pH between 9.4 and 9.6 by periodically adding ammonia. After addition of all the gold, the dispersion was stirred for 1 h after which it was filtered and washed three times with 200 ml of demineralized water. It was found that aging the dispersion had a beneficial effect on the broadness of the particle size distribution. The yellow catalyst was dried overnight in air at 333 K and then calcined. Calcination was carried out by heating to 393 K (5 K/min heating) for 2 h followed by 4 h at 673 K (5 K/min heating and cooling). The thus-obtained catalysts had an intense dark color originating from the plasmon bands of the gold nanoparticles: purple for the titania-supported catalysts, red-brown for the silica-supported catalysts, salmon pink for the Ti-OX50-supported catalysts, and red-orange for the Ti-SBA-15-supported catalyst.

2.2. Catalyst characterization

Scanning electron microscopy (SEM) and transmission electron microscopy (TEM) micrographs were taken of the catalysts to determine the gold particle size and distribution on the catalysts, both before and after use in catalytic experiments. X-ray fluorescence (XRF) analysis was used to determine the gold loading on the catalysts and the presence of contaminants affecting the activity (e.g., chloride). X-ray photoelectron spectroscopy (XPS) analysis was performed with a Thermo VG Scientific XPS system with a reaction chamber for sample pretreatment allowing the performance of quasi *in situ* measurements, which were used to determine changes in the oxidation state of the gold.

2.3. Catalyst activity testing

A flow reactor was used to determine the catalytic performance of the different catalysts. The experiments were carried out with typically 0.30 g of catalyst and a gas flow of 50 Nml/min (GHSV 10,000 h^{-1}). In the epoxidation experiments, a gas mixture was used similar to that in most research in the literature: 10% of oxygen, 10% of hydrogen, and 10% of propene in helium (all gas compositions given in vol.%). The pressure was 1.1 bar. In this study, the activity was determined at 323, 373, 423, and 473 K; the lowest temperature being most appropriate for the titania-supported catalyst, the higher temperatures being more optimal for the Ti–silica-supported catalysts.

The analysis of the gas leaving the reactor was carried out using an Interscience Compact-GC (gas chromatography) system, equipped with a Molsieve 5A and a Porabond Q column, each with a thermal conductivity detector (TCD). Gas samples were analyzed every 3 min. The catalysts were regenerated prior to measuring at the next reaction condition: the catalysts were heated for 1 h at 573 K (10 K/min) in a 50-ml/min gas stream consisting of 10% of oxygen in helium. Experiments performed with the bare supports did not show any catalytic activity. The catalytic tests were performed in a fully automated system over a period of typically 5–10 days during which multiple reaction conditions were applied, including repeat conditions to verify catalyst stability.

3. RESULTS AND DISCUSSION

3.1. Characterization

The XRF analysis of the catalysts showed that for all samples the catalyst loading was close to the desired loading of 1 wt.%. No residual chloride could be detected on the catalysts. For both the Ti-OX50 and the Ti-SBA-15, SEM and TEM analysis showed a homogeneous distribution of the Ti over the supports with no amorphous titania present. The particle sizes were determined by measuring all particles on a number of TEM pictures until over 200 particles were measured. A selection of the TEM pictures is shown in Fig. 12.1. The TEM analysis showed that the gold particle size was support independent and similar (∼4 nm average particle size), with the exception of the gold particles supported on the mesoporous Ti-SBA-15. For the Ti-SBA-15-supported catalysts, the gold particle size was larger. Also, the gold particles were clearly elongated, as a result of the gold particle growth being restricted by the SBA-15 mesopores. This larger gold particle size was most likely the result of mass transfer limitations during the deposition, drying, or calcination, and currently is being investigated further.

The difference in color of the catalysts, even though the gold particle size and distribution was similar, can be explained by a different interaction between the gold particles and the supports. The most relevant results of the catalyst characterization are provided in Table 12.1. The results of the catalytic tests are summarized in Table 12.2.

3.2. Catalytic performance

Figure 12.2 shows the catalytic performance for the various catalysts at two different temperatures. It can be seen that all tested gold catalysts have propene epoxidation activity, but that the support very strongly influences this activity. The titania-supported catalyst had the highest activity at low temperatures (323 K), the titania–silica-supported catalysts were most active at 423 K, while the silica-supported catalysts were most active at 473 K (not shown in the figure, see Table 12.2). The titania-supported catalyst, even though active at low temperature, lost activity very rapidly (loss of activity is 75% in 5 h). This rapid deactivation is also the reason why the conversion of the catalyst dropped as the temperature was increased from 323 to 373 K. As the temperature increased, both the rates of epoxidation and deactivation increased. At longer reaction times, this caused the propene oxide yield to be lower with increasing temperature. This is made clear in Fig. 12.3, which shows that for an increasing reaction temperature, the time at which the maximum propene oxide yield is reached

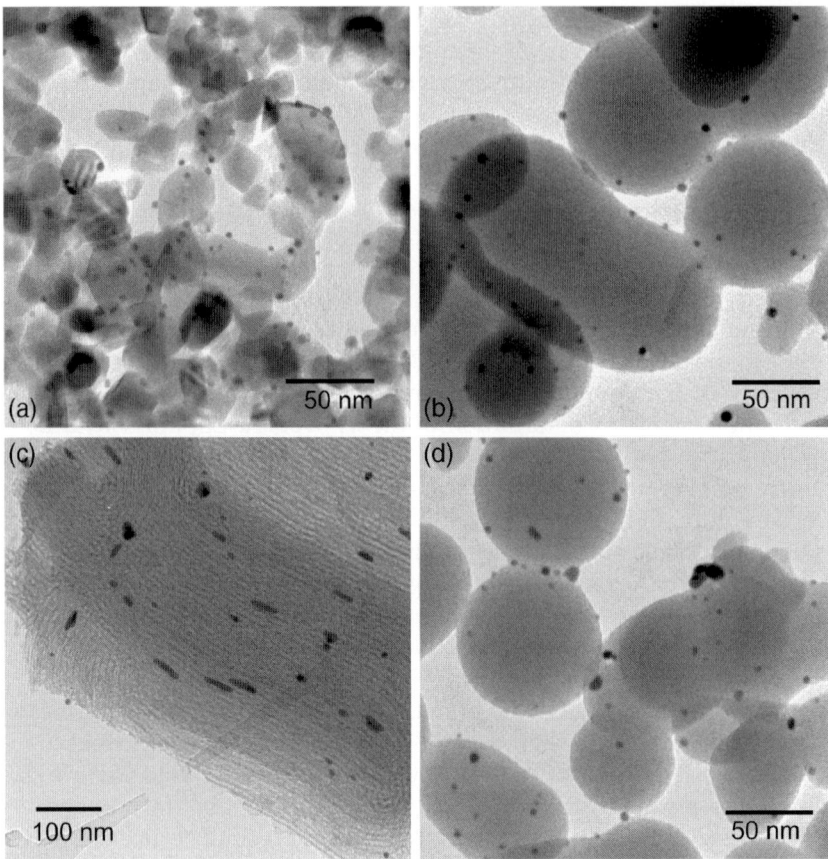

FIGURE 12.1 Selected transmission electron microscopy (TEM) pictures. (a) Au/TiO$_2$, (b) Au/Ti-SiO$_2$, (c) Au/Ti-SBA-15, and (d) Au/SiO$_2$ (OX50).

TABLE 12.1 Summary of catalyst characterization results

	Au/TiO$_2$	Au/Ti-SiO$_2$	Au/ Ti-SBA-15	Au/SiO$_2$ (Davisil)	Au/SiO$_2$ (OX50)
BET area (m^2/g)	45	49	555	295	50
Au loading (wt.%)	0.92	0.94	1.01	1.04	1.02
Metal particle size (nm) (S.D.)	4.0 (1.1)	4.8 (1.8)	7.7 (1.8)[a]	3.5 (1.5)	4.0 (1.8)
Support Ti content (wt.% TiO$_2$)	99.8	0.76	2.91	0.04	0.05

[a] The gold particles size for the Au/Ti-SBA-15 sample is determined by the SBA-15 pores. The particle diameter for most particles is equal to the SBA-15 channel diameter. The channel restriction makes the particles elongated; the length of the particles is up to five times the diameter (see Fig. 12.1C). For these elongated particles, the diameter was taken.

TABLE 12.2 Summary of catalytic performance of the catalysts. [0.3 g of catalyst, 50 Nml/min gas feed rate (10% H_2, O_2, and propene), total pressure 1.1 bar(a).]

Catalyst	Propene conversion (%)			PO selectivity (%)				Hydrogen efficiency[c] (%)			
	323 K	373 K	423 K	473 K	323 K	373 K	423 K	473 K	323 K	373 K	423 K	473 K
(A) Average during 1–5 h of catalytic cycle												
Au/TiO_2	0.18	0.10	1.36	2.78[b]	99.9	77.6	2.2	0.0	11.3 (1.5)	1.7 (4.3)	0.1 (26.9)	0.0 (77.4)
Au/Ti-SiO_2	0.03	0.14	0.45	0.80	100	99.9	87.6	61.2	8.6 (0.4)	5.9 (2.4)	4.2 (9.6)	1.8 (26.6)
Au/Ti-SBA-15	0.01	0.26	0.61	1.38	100	93.9	83.8	42.2	1.6 (0.8)	8.4 (2.9)	5.3 (9.6)	2.3 (24.8)
Au/SiO_2 (Davisil)	0	0.06	0.48[b]	1.78[b]	—	40.8	61.5	59.0	0 (6.7)	0.1 (31.4)	0.2[a] (100)	0.7[a] (100)
Au/SiO_2 (OX50)	0	0.04	0.18[b]	1.31[b]	—	7.2	51.3	62.9	0 (2.2)	0 (25.2)	0.1 (73.5)	0.7[a] (100)
(B) Average during 6–20 h of catalytic cycle												
Au/TiO_2	0.05	0.08	1.45	2.53[b]	100	74.5	1.7	0.0	7.1 (0.8)	1.5 (4.1)	0.1 (27.9)	0.0 (75.0)
Au/Ti-SiO_2	0.02	0.10	0.42	0.73	100	100.0	85.8	62.8	7.9 (0.3)	5.2 (1.9)	4.3 (8.3)	1.8 (24.9)
Au/Ti-SBA-15	0.04	0.17	0.46	1.28	100	94.9	84.2	43.9	7.8 (0.5)	7.6 (2.1)	4.8 (8.1)	2.4 (23.8)
Au/SiO_2 (Davisil)	0	0.06	0.49[b]	1.79[b]	—	23.2	60.4	59.2	0 (4.3)	0 (29.1)	0.2[a] (100)	0.8[a] (100)
Au/SiO_2 (OX50)	0	0.03	0.29[b]	1.31[b]	—	0.0	47.2	62.0	0 (2.5)	0 (21.2)	0.3 (54.7)	0.7[a] (100)

[a] Hydrogen efficiency restricted by hydrogen supply.
[b] For these experiments the actual catalyst temperature was significantly higher (15–20 K) than the oven temperature indicated as a result of the high conversion of hydrogen.
[c] Values between brackets for hydrogen efficiency denote hydrogen conversion.

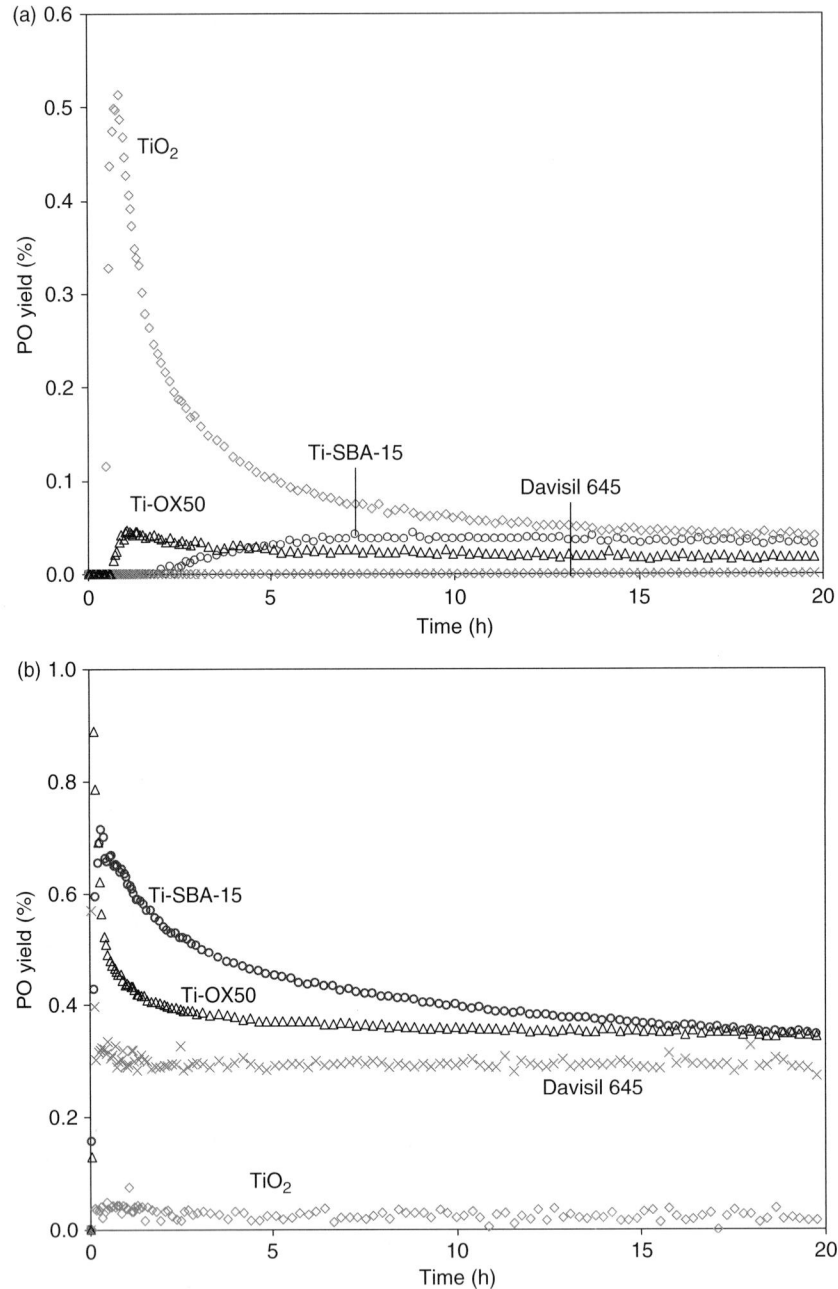

FIGURE 12.2 Catalytic performance at two selected reaction temperatures (A: 323 K, B: 423 K) for Au/TiO$_2$ (◇), Au/Ti-SBA-15 (○), Au/Ti-OX50 (△), and Au/Davisil 645 (×). [0.3 g of catalyst, 50 Nml/min gas feedrate (10% H$_2$, O$_2$, and propene), total pressure 1.1 bar(a)].

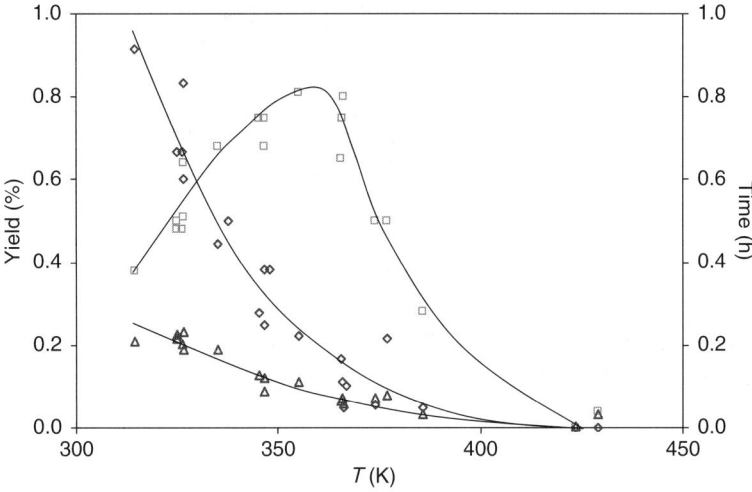

FIGURE 12.3 Average yield (Δ), maximum yield (□), and time at which the maximum yield is obtained (◊) for propene epoxidation over 1 wt.% Au/TiO$_2$ at different reaction temperatures. [Lines are drawn as a guide to the eye; average conversion from 30 to 270 min reaction time, 0.3 g of catalyst, 50 Nml/min gas feedrate (10% H$_2$, O$_2$, and propene), total pressure 1.1 bar(a)].

decreases, and up to 373 K, the maximum obtained amount of propene oxide yield increases. The average propene oxide yield (from 0.5- to 4.5-h reaction time), however, decreased monotonously with increasing temperature. In Fig. 12.3 it can also be seen that for the gold on titania catalysts, there is a large difference between the average yield and the maximum yield. In most papers, the maximum propene oxide yield is reported, whereas in this work the emphasis is on the average yield over a few hours time, which explains why the catalyst activity is lower than other reported values.

The highest propene oxide yields were obtained with both the Ti-SBA-15- and the Ti–silica-supported catalysts, although a higher reaction temperature was needed in comparison to the titania-supported catalyst. The deactivation for these catalysts was also considerably less. At lower temperatures (up to 423 K), all catalysts had an inhibition period for both propene oxide and water formation, which is explained by product adsorption on the support. The side products produced by all catalysts were similar. Primarily, carbon dioxide and acetaldehyde were produced as side products and, in smaller quantities, also propanal, acrolein, acetic acid, and formaldehyde. Propanol (both 1- and 2- as well as propanediol), acetone, carbon monoxide, and methanol were only observed in trace amounts.

Given the fact that the gold particles were the largest for the Ti-SBA-15-supported catalysts and also for a large part inaccessible, as they are tightly restrained in the SBA-15 channels, it can be stated that this support is more promising for active propene epoxidation catalysts, once a different gold deposition method has been developed in which smaller uniform gold nanoparticles are formed in the channels. Optimally, uniform gold particles slightly larger than 2 nm should be prepared in the SBA-15 channels, allowing for a high surface area and a good accessibility in the channels (no channel blockage). Smaller particles might be undesirable, considering the observation that on titania they started to catalyze the unwanted hydrogenation of propene to propane [18,19]. In another report for gold supported on titanosilicate, this shift in activity toward propene hydrogenation for smaller gold particles was not observed [12]. However, in our first attempts to produce smaller gold particles using an alternative preparation method, in which we produced a catalyst containing a significant amount of small (<2 nm) gold particles inside Ti-SBA-15, we also observed a strong shift toward a hydrogenation activity.

3.3. Catalyst deactivation

The deactivation for all catalysts could be reversed by a high temperature cycle at 573 K in 10% oxygen in helium, which indicates that the catalyst deactivation has to do with adsorbed species and not with sintering of the gold nanoparticles, since this would have been irreversible. This is confirmed by TEM analysis, which did not show a significant increase in the gold particle size of the "spent" catalysts. In Fig. 12.4, the particle size distributions for fresh and used (>200 h) samples of the titania- and silica (OX50)-supported catalysts are shown (>200 particles measured per sample). For the titania-supported catalyst a small increase is observed in the average particle size, for the silica-supported catalyst a small decrease is observed, both changes are considered to be within the accuracy of the method used.

Attempts to regenerate the catalyst by a high-temperature (573 K) treatment in helium, only partially restored the catalytic activity, indicating that the catalyst is cleaned primarily by combusting the adsorbed species rather than just desorbing them. Quasi *in situ* XPS measurements showed no changes in the oxidation state of gold of a calcined catalyst after it was exposed to hydrogen or oxygen at a temperature of up to 573 K. Testing the catalysts over a longer period of time (up to 10 days in multiple cycles) showed that the catalyst activity after an intermediate regeneration was stable [no loss in activity compared to the first cycle, or at worst only a minor loss in activity (the Au/Ti-SBA-15 was the least stable, this catalyst lost less than 5% of its activity)].

The link between the deactivation and the epoxidation is made clear in Fig. 12.5, in which the hydrogen oxidation reaction (no propene) is observed to be stable for 25 h. Once propene is added to the feed thereafter, the catalyst starts deactivating rapidly. Removal of the propene in the feed stops the deactivation process and the catalytic activity gradually increases. The fact that even after 25 h, the activity is not back to its original level, indicates that the deactivating species are bonded strongly to the catalyst.

Figure 12.6 shows a plot for the titania-supported catalyst of the catalytic activity versus the amount of propene oxide produced. It can be seen that initially (from the maximum activity to half of the maximum activity) the activity decreases linearly with the amount of propene oxide produced and only later on, that the rate of deactivation decreases. This behavior can be explained on the basis of the deactivation model based on spectroscopic data published

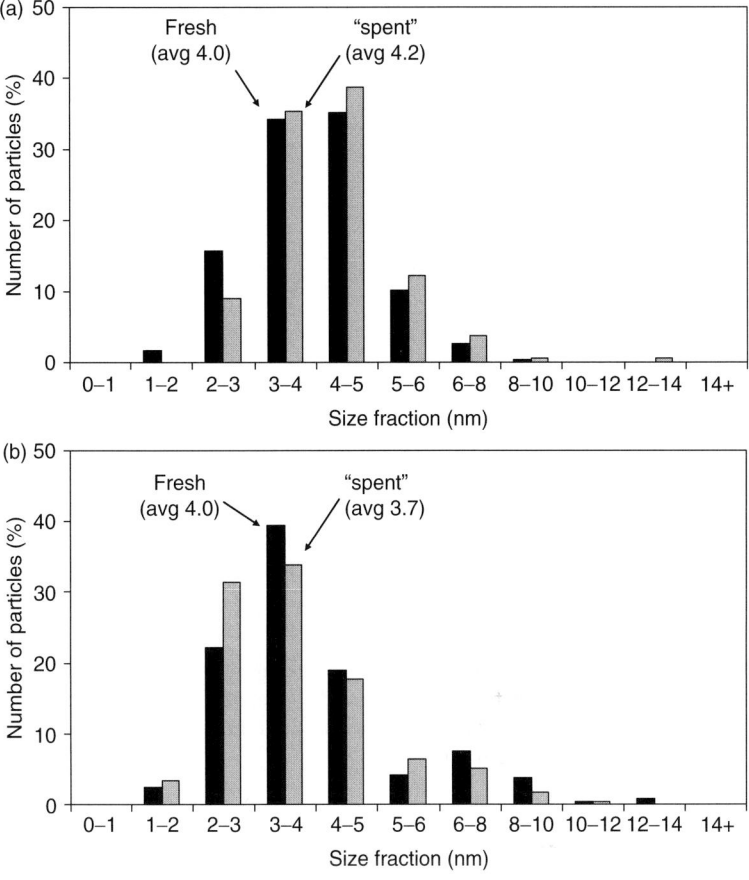

FIGURE 12.4 Particle size distributions of freshly prepared and "spent" (>200 h in operation) gold on titania (a) and gold on OX50 silica (b) catalysts as determined by transmission electron microscopy (TEM).

previously [20]. In this model, as a first step in the epoxidation, a bidentate propoxy species is formed on the titania (or titania/silica) support. The fact that the bidentate propoxy species is removed by a hydrogen/oxygen mixture from the catalyst surface indicates that it is indeed a reaction intermediate [6,11]. Both propene oxide and deactivated sites are formed in competing parallel reaction pathways from this bidentate propoxy species on the catalyst surface. Since the amount of propene oxide produced is related directly to the amount of bidentate propoxy species that were present on the surface and this same species produces the deactivating carbonate species, deactivation would be expected to be linked directly to the amount of propene oxide produced. At longer times, this direct relationship no longer applies because of a slow desorption of the deactivating carbonates. This is in agreement with the gradual increase in the activity observed for the hydrogen oxidation in Fig. 12.5, once the propene feed has been removed.

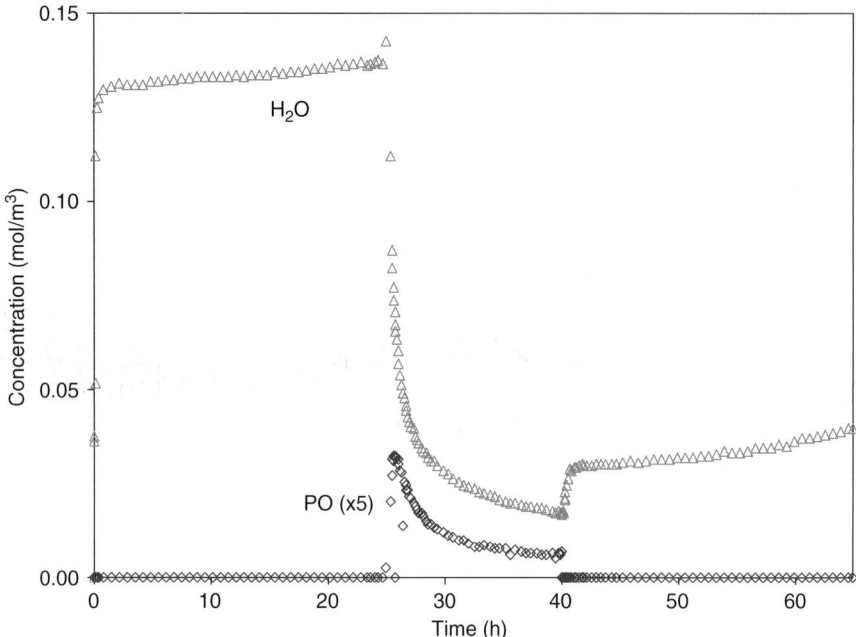

FIGURE 12.5 Deactivation during epoxidation and hydrogen oxidation over 1 wt.% Au/TiO$_2$. Twenty-five hours hydrogen oxidation only, followed by epoxidation for 15 h, followed by hydrogen oxidation only for 25 h. [0.3 g of catalyst, 50 Nml/min gas feedrate (6% H$_2$, O$_2$, and from $t = 25$ to 40 h propene), total pressure 1.1 bar(a).]

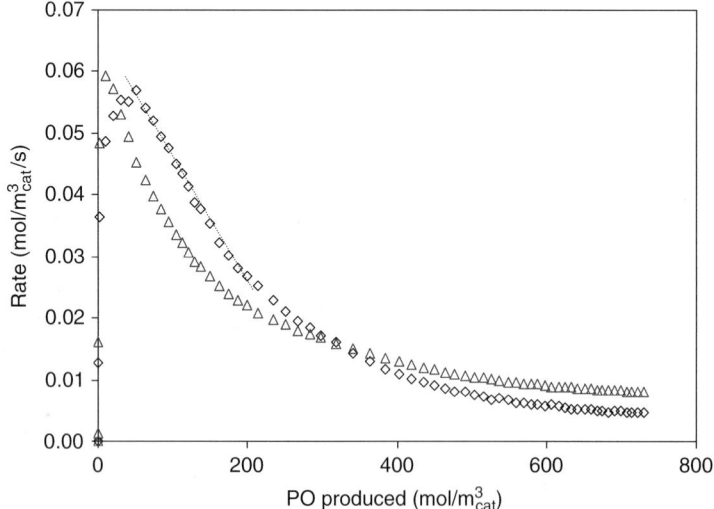

FIGURE 12.6 Deactivation versus the amount of propene oxide produced during the propene epoxidation over 1 wt.% Au/TiO$_2$. ◇: propene oxide formation rate; △: water formation rate/10. [0.3 g of catalyst, 50 Nml/min gas feedrate (10% H$_2$, O$_2$, and propene), total pressure 1.1 bar(a).]

In contrast to the bidentate propoxy species, *in situ* infrared experiments showed that the carbonate species could not be removed from the catalyst by exposing it to hydrogen/oxygen at reaction temperature [6,11].

To further verify the proposed deactivation model, Fig. 12.7 shows a very simple numerical model fitted to the propene epoxidation data shown in Fig. 12.6. The rate of deactivation in this model is assumed to be first order in the epoxidation rate, while the reactivation is first order in the deactivated sites. Numerically this simplified deactivation/reactivation model can be expressed by Eq. (12.1):

$$\frac{\partial r_{PO}}{\partial t} = -k_{deact} \times r_{PO} + k_{react} \times (r_{PO,0} - r_{PO}) \quad (12.1)$$

with t, time (s); r_{PO}, rate of propene epoxidation (mol/g$_{cat}$/s)—subscript 0 denotes the rate at $t = 0$; k_{deact}, rate constant for catalyst deactivation (formation of deactivating species) (s^{-1}); and k_{react}, rate constant for catalyst reactivation (desorption of deactivating species) (s^{-1}).

In Fig. 12.7, it can be seen that this simple model describes the observed activity pattern well. Only in the first half hour of the experiment, a discrepancy exists between the rate predicted by the model and the observed rate. This can be explained by product adsorption on the support as discussed previously [20], which was not included in this model. A more extensive version of this simplified deactivation model, which includes the occupancies of all surface species, is published elsewhere [21].

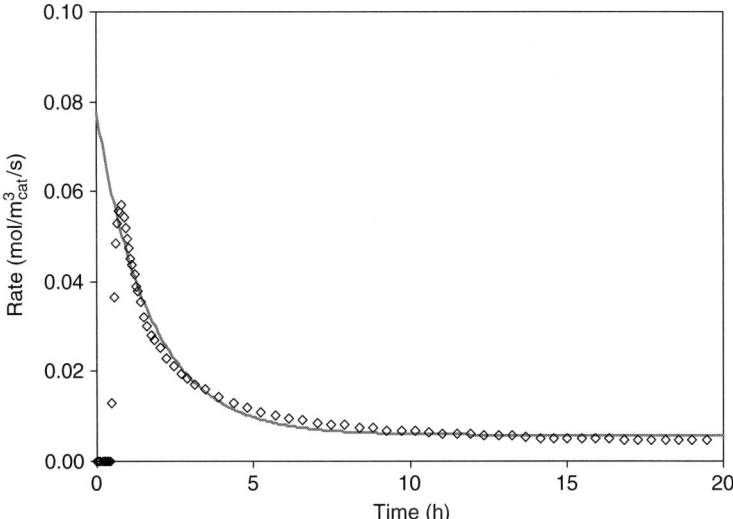

FIGURE 12.7 Deactivation modeled based on a first-order deactivation in the amount of propene oxide and a first-order reactivation of the deactivated sites [according to Eq. (1)]. Model fitted to data from $t = 0.75$ to 20 h. Propene epoxidation over 1 wt.% Au/TiO$_2$. ◊: measured propene oxide formation rate; line: modeled rate. [0.3 g of catalyst, 50 Nml/min gas feedrate (10% H$_2$, O$_2$, and propene), total pressure 1.1 bar(a).]

If the rate of hydrogen oxidation (water formation) is examined versus the amount of propene oxide produced, it can be seen that the hydrogen oxidation also decreases, but not linearly with the propene oxide produced, and also more slowly than the epoxidation. This indicates that the direct hydrogen oxidation also proceeds over sites other than those involved in the epoxidation. The fact that the hydrogen oxidation rate as such does decrease, points to the support playing a role in this reaction as well, although according to other studies this reaction proceeds exclusively over the gold nanoparticles [22]. Several spectroscopic studies (using both Raman and infrared spectroscopy) carried out by us showed that no (strongly) adsorbed species on the gold particles could be detected under reaction conditions. This role of the support in the hydrogen oxidation should be seen as, for example, the formation of peroxide species on the gold, which then subsequently decompose over the support, since experiments performed over the bare support materials showed no activity.

For the catalysts supported on Ti–Si carriers or on pure silica, the lower deactivation rate can be explained in two ways. First, a bidentate propoxy intermediate on the surface is no longer bonded to two Ti sites on the catalyst (due to the low Ti loading), but to both a Ti and an Si site, or only to silica sites. This affects the relative ratio between the deactivation and propene oxide formation reactions. This reduced deactivation indicates that the energetics are relatively more favorable toward the epoxidation. The bidentate propoxy species most likely is more susceptible to C–C bond breakage, which would trigger the deactivating consecutive oxidation of this species, because of the higher acidity of the titania support compared to the Ti–silica and silica supports. Second, the silica and Ti–silica catalysts require a higher reaction temperature to produce propene oxide. At this higher temperature, the desorption of deactivating carbonate species is faster. The need for the higher reaction temperature can be explained by the weaker reactivity of the support to produce the bidentate propoxy intermediate species compared to the titania-supported catalysts. *In situ* spectroscopic infrared experiments [11] show that the largest amount of bidentate propoxy species could be observed on titania-supported catalysts, while considerably smaller amounts could be observed on Ti–silica-supported catalysts, and no measurable amount could be detected on silica-supported catalysts.

Both catalysts supported on silica produce a small amount of propene oxide at the highest temperatures; however, the selectivities remain low, proving the necessity of titania on the support to have an effective epoxidation catalyst. Furthermore, these catalysts by far have the lowest hydrogen efficiency because of the direct oxidation of hydrogen into water, making the economics of these catalysts very unattractive.

4. CONCLUSIONS

Gold nanoparticle catalysts have a good potential for a future process for the direct epoxidation of propene. In a single reactor, propene can be selectively epoxidized using a hydrogen–oxygen mixture. Catalysts containing titanium dispersed on or in a silica support yield the highest amount of propene oxide at a high selectivity and reasonably good stability. If the maximum propene oxide yield is determined per gold surface site, the Ti-SBA-15-supported catalysts are the most active; however, the deposition of small gold particles in the mesopore structure requires further optimization. Titania-supported catalysts are intrinsically the most active; however, they severely suffer from deactivation by a sequential oxidation of the bidentate propoxy reaction intermediate. Catalyst deactivation is caused by carbonate formation on the catalyst, which can be removed easily by means of a simple regeneration in air at 573 K.

A major challenge that remains is the hydrogen efficiency of the catalyst. Since the propene epoxidation is performed in the presence of hydrogen, it is desirable that the hydrogen is used only in the epoxidation reaction and is not converted directly into water. At this time none of the catalysts have a sufficiently high hydrogen efficiency to be able to run a process profitable and, therefore, this remains one of the key challenges to be solved for this catalyst system for propene epoxidation.

ACKNOWLEDGMENTS

STW/NWO is kindly acknowledged for the VIDI grant supporting the research of T.A.N. and E.S. NWO/CW is kindly acknowledged for the VICI grant supporting the research of B.M.W.

REFERENCES

[1] T. A. Nijhuis, M. Makkee, J. A. Moulijn, B. M. Weckhuysen, The production of propene oxide: Catalytic processes and recent developments, *Ind. Eng. Chem. Res.* 45 (2006) 3447.
[2] D. L. Trent, Propylene Oxide, Kirk-Othmer: Encyclopedia of Chemical Technology, Wiley, New York, NY 2001.
[3] M. Haruta, Size- and support-dependency in the catalysis of gold, *Catal. Today* 36 (1997) 153.
[4] B. Taylor, J. Lauterbach, W. N. Delgass, Gas-phase epoxidation of propylene over small gold ensembles on TS-1, *Appl. Catal. A* 291 (2005) 188.
[5] A. K. Sinha, S. Seelan, M. Okumura, T. Akita, S. Tsubota, M. Haruta, Three-dimensional mesoporous titanosilicates prepared by modified sol-gel method: Ideal gold catalyst supports for enhanced propene epoxidation, *J. Phys. Chem. B* 109 (2005) 3956.
[6] T. A. Nijhuis, T. Visser, B. M. Weckhuysen, The role of gold in gold-titania epoxidation catalysts, *Angew. Chem. Int. Ed.* 44 (2005) 1115.
[7] A. K. Sinha, S. Seelan, S. Tsubota, M. Haruta, A three-dimensional mesoporous titanosilicate support for gold nanoparticles: Vapor-phase epoxidation of propene with high conversion, *Angew. Chem. Int. Ed.* 43 (2004) 1546.
[8] A. H. Tullo, P. L. Short, Propylene oxide routes take off, *Chem. Eng. News* 84 (2006) 22.
[9] T. A. Nijhuis, B. J. Huizinga, M. Makkee, J. A. Moulijn, Direct epoxidation of propene using gold dispersed on TS-1 and other titanium-containing supports, *Ind. Eng. Chem. Res.* 38 (1999) 884.
[10] A. Zwijnenburg, M. Makkee, J. A. Moulijn, Increasing the low propene epoxidation product yield of gold/titania-based catalysts, *Appl. Catal. A* 270 (2004) 49.
[11] T. A. Nijhuis, T. Visser, B. M. Weckhuysen, A mechanistic study into the direct epoxidation of propene over gold-titania catalysts, *J. Phys. Chem. B* 109 (2005) 19309.
[12] N. Yap, R. P. Andres, W. N. Delgass, Reactivity and stability of Au in and on TS-1 for epoxidation of propylene with H_2 and O_2, *J. Catal.* 226 (2004) 156.

[13] A. K. Sinha, S. Seelan, T. Akita, S. Tsubota, M. Haruta, Vapor phase propylene epoxidation over Au/Ti-MCM-41 catalysts prepared by different Ti incorporation modes, *Appl. Catal. A* 240 (2003) 243.
[14] M. Weisbeck, C. Schild, G. Wegener, G. Wiessmeier, Surface-modified mixed oxides containing precious metal and titanium for the selective oxidation of hydrocarbons, *US Patent* 6734133, 2004, Assigned to BASF.
[15] G. Mul, A. Zwijnenburg, B. van der Linden, M. Makkee, J. A. Moulijn, Stability and selectivity of Au/TiO$_2$ and Au/TiO$_2$/SiO$_2$ catalysts in propene epoxidation: An *in situ* FT-IR study, *J. Catal.* 201 (2001) 128.
[16] B. S. Uphade, M. Okumura, S. Tsubota, M. Haruta, Effect of physical mixing of CsCl with Au/Ti-MCM-41 on the gas-phase epoxidation of propene using H$_2$ and O$_2$: Drastic depression of H$_2$ consumption, *Appl. Catal. A* 190 (2000) 43.
[17] E. Sacaliuc, A. Beale, B. M. Weckhuysen, T. A. Nijhuis, Propene epoxidation over Au-Ti-SBA-15 catalysts, *J. Catal.* 248 (2007) 234.
[18] M. Haruta, M. Date, Advances in the catalysis of Au nanoparticles, *Appl. Catal. A* 222 (2001) 427.
[19] T. Hayashi, K. Tanaka, M. Haruta, Selective vapor-phase epoxidation of propylene over Au/TiO$_2$ catalysts in the presence of oxygen and hydrogen, *J. Catal.* 178 (1998) 566.
[20] T. A. Nijhuis, B. M. Weckhuysen, The direct epoxidation of propene over gold-titania catalysts—A study into the kinetic mechanism and deactivation, *Catal. Today* 117 (2006) 84.
[21] T. A. Nijhuis, T. Q. Gardner, B. M. Weckhuysen, Modeling of kinetics and deactivation in the direct epoxidation of propene over gold-titania catalysts, *J. Catal.* 236 (2005) 153.
[22] D. G. Barton, S. G. Podkolzin, Kinetic study of a direct water synthesis over silica-supported gold nanoparticles, *J. Phys. Chem. B* 109 (2005) 2262.

CHAPTER 13

Propylene Epoxidation via Shell's SMPO Process: 30 Years of Research and Operation

J. K. F. Buijink, Jean-Paul Lange, A. N. R. Bos, A. D. Horton, and F. G. M. Niele

Contents

1.	Introduction	356
	1.1. General process description	356
2.	Catalytic Epoxidation	358
	2.1. The nature of the active site	358
	2.2. Mechanism of epoxidation	359
	2.3. Catalyst deactivation mechanisms	362
3.	SMPO Process Improvement	363
	3.1. Trace component chemistry	363
	3.2. Air-oxidation improvements	366
	3.3. Catalytic dehydration	367
4.	Conclusions	369
	Acknowledgment	369
	References	369

Abstract

Shell has coproduced propylene oxide (PO) and styrene using its proprietary styrene monomer propylene oxide (SMPO) process for three decades. Research, development, and plant trials have been performed on a continuous basis in order to improve its efficiency and cost competitiveness. We report here some of the key fundamental and technological learnings gathered over various parts of the process.

The heart of the process is formed by the catalytic epoxidation of propylene with ethylbenzene hydroperoxide using a silica-supported titanium catalyst. Fundamental studies have provided models for the active sites, the reaction mechanism, and the cause of catalyst deactivation. Other process elements have been investigated and improved as well. Minor by-products

Shell Global Solutions International B.V., P.O. Box 38000, 1030 BN Amsterdam, The Netherlands

Mechanisms in Homogeneous and Heterogeneous Epoxidation Catalysis
DOI: 10.1016/B978-0-444-53188-9.00013-4

© 2008 Elsevier B.V.
All rights reserved.

have been tracked across the process to unravel their impact on the various process steps. Computational flow dynamic (CFD) studies of the air-oxidation step, as well as broad catalytic studies of the dehydration step, have led to significant performance improvements.

Key Words: Styrene, Active site, Ultrafine titania particles, Process improvement, SMPO, Trace components, Catalytic dehydration, Computational flow dynamics, Quantum mechanical calculations, Oxygen starvation, Sharpless mechanism, Titanium peroxide, Oxygen transfer, Proton parking place, Leaching, Peroxolysis, Reactive distillation. © 2008 Elsevier B.V.

1. INTRODUCTION

Propylene oxide (PO) is a versatile chemical intermediate used in a wide range of industrial and commercial products. Current world production is over 6 million metric tons a year. While several processes exist, the Shell Chemicals companies have derived a strong competitive advantage by using and continually developing their proprietary styrene monomer propylene oxide (SMPO) technology, a process in which propylene and ethylbenzene (EB) are converted into PO and styrene monomer (SM), respectively. Worldwide, there are now five world-scale SMPO plants based on Shell technology, the most recent one started up in 2006 in China.

We will describe here the SMPO process, with particular attention to the propylene epoxidation step, and exemplify numerous process improvements that stem from a deep understanding of all aspects of the process. We will start with a general process description to provide the necessary context.

1.1. General process description

The SMPO process comprises four main reaction steps that are schematically drawn in Fig. 13.1. A simplified process flow sheet is found in Fig. 13.2:

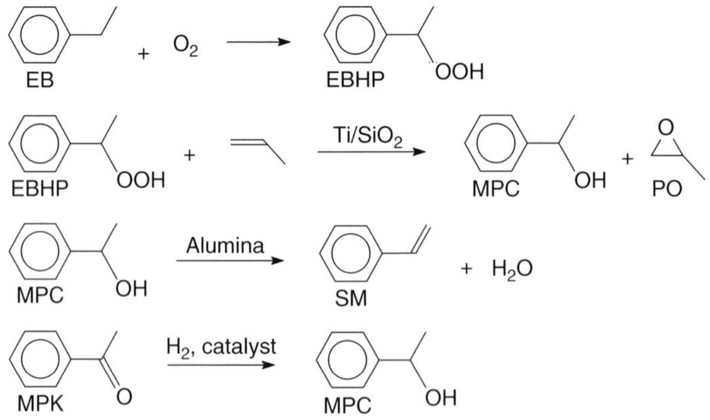

FIGURE 13.1 Main reaction steps of the SMPO process.

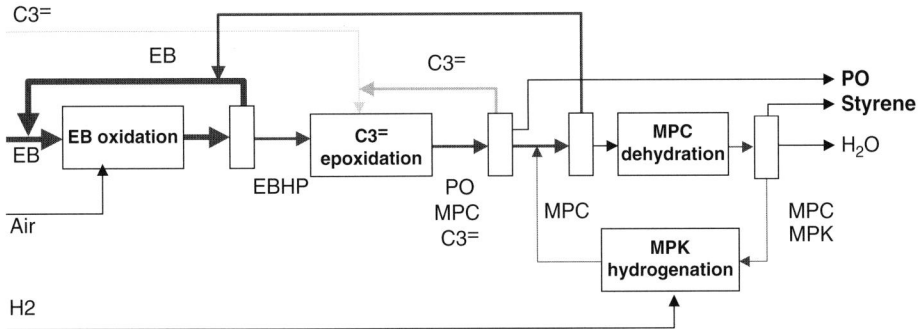

FIGURE 13.2 Simplified process flow scheme for the SMPO process.

1. Oxidation of EB with air to ethylbenzene hydroperoxide (EBHP)
2. Epoxidation of propylene with EBHP to produce PO and methyl phenyl carbinol (MPC)
3. Dehydration of MPC to SM
4. Hydrogenation of methyl phenyl ketone (MPK), which is a side-product of the first two reactions, to MPC, for subsequent conversion to SM.

The first step is the air-oxidation of EB to EBHP. This is performed by Shell in cross-flow operation in a series of large *horizontal* bubble column reactors with a very low aspect ratio, which are equipped with baffles and heating/cooling coils. Air is introduced via separate middle and side sparger systems. The gas outlet stream contains besides unconverted oxygen a very significant amount of EB from evaporation/stripping. This EB is recovered in a condensing column and recycled to the reactor train.

The oxidation reactor effluent is subjected to a separate stage where part of the EB is removed for recycling, the reactor effluent cooled, and the EBHP concentration of the reactor effluent is raised. This product is then neutralized to remove acidic by-products and washed with water.

The subsequent step, the epoxidation of propylene by EBHP, is carried out in the liquid phase over a proprietary heterogeneous catalyst, to produce crude PO and MPC. The feed to the reactors consists of make-up and recycle propylene and EBHP in EB. The reaction train consists of a number of adiabatic fixed bed reactors with interstage cooling. Deactivated catalyst is replaced, incinerated to remove residual hydrocarbons, and dispersed in a landfill.

The overall conversion of EBHP is virtually 100% and the PO selectivity is typically around 95% based on EBHP and even higher on propylene. Small amounts of MPK, phenol, and aldehydes are also produced.

The product from the epoxidation reactor is sent to the crude PO recovery unit. The unit contains a number of distillation columns in which this product is separated into unreacted propylene for recycle to the epoxidation section, crude PO, EB, and styrene precursors (mainly MPC and MPK). Light hydrocarbons going overhead in the various distillation columns can be sent to a hot oil furnace.

The crude PO obtained from the crude PO recovery unit is purified in a finishing unit, which consists of a number of distillation columns in which water is removed by azeotropic distillation with normal butane and aldehydes, and light- and heavy ends are also removed from the crude PO.

Prior to entering the MPC dehydration step, the EB, MPC, and MPK stream from the PO recovery unit is washed and EB is removed by distillation. The MPC and MPK are sent to the dehydration reactors, where MPC is dehydrated to styrene using one of the commercially available catalysts.

The reactor product is separated into crude styrene, which is sent to the SM finishing unit, and MPK with traces of unconverted MPC, which is sent to a catalytic hydrogenation unit—the fourth and last process step of SMPO. The formation of by-products other than EB is insignificant. The product containing MPC and some EB and MPK is recycled to the SM reaction unit.

2. CATALYTIC EPOXIDATION

The catalytic epoxidation step may be considered to be the heart of the SMPO process. The catalyst is prepared in a multistep gas-phase process by treatment of a silica carrier with titanium tetrachloride, heating the obtained material, followed by steaming and silylation. Possible improvements to this process and research on the support material have already been reviewed [1]. Here, we discuss the nature of the active site, the mechanism of epoxidation, and the various catalyst deactivation mechanisms that exist for this unique catalyst.

2.1. The nature of the active site

Heterogeneous olefin epoxidation over solid titania-silica catalysts has been the subject of numerous publications in the open literature. The general picture that emerges is that isolated titanium (IV) species on a silica surface or in a zeolite matrix are responsible for the high epoxidation activity [2]. This picture is supported by model catalyst work on titanium silasesquioxane complexes [3,4] that form active homogeneous epoxidation catalysts [5] and by various successful attempts to prepare well-defined, site-isolated titanium complexes by grafting molecular precursors on mesoporous silica [6–9]. These site-isolated titanium complexes have been shown to possess catalytic activity in olefin epoxidation.

This simple picture can only be partly correct, as it has been shown that both unsupported ultrafine titania nanoparticles [10] and silica-supported monodispersed subnanometric titania particles [11] are active catalysts for olefin epoxidation with hydroperoxides. In these small titania particles, the titanium atoms possess coordination states between four and six. Quantum chemical calculations derived similar activation energies for oxygen transfer on mononuclear four- and five-coordinated titanium sites [12,13].

We therefore propose that "the active site" of the SMPO titania-on-silica epoxidation catalyst is actually a continuum of unsaturated mono- and polynuclear titanium species supported by silica. The most active of these titanium

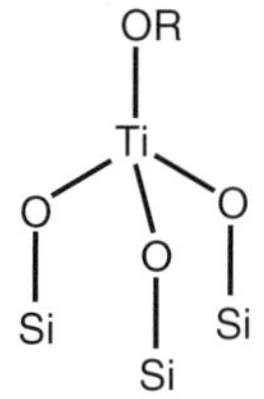

$$\text{Si-O}\diagdown\underset{\underset{\underset{\text{Si}}{|}}{\text{O}}}{\overset{\overset{\text{OR}}{|}}{\text{Ti}}}\diagup\text{O-Si}$$

FIGURE 13.3 Proposed most active site in SMPO titanium-on-silica epoxidation catalyst (R = Me$_3$Si) Si denotes Silica. Reprinted with permission from [1]. Copyright (2004) Elsevier.

species is most likely the four-coordinated titanium species possessing tripodal geometry as drawn schematically in Fig. 13.3.

2.2. Mechanism of epoxidation

A principle problem in mechanistic heterogeneous catalysis research is that a catalyst surface generally comprises a large range of structurally different catalytic sites. These sites are inherently difficult to characterize with surface analysis techniques. Homogeneous catalysts, on the other hand, usually contain one type of site bonded to one or more ligands. Here, we make explicit use of knowledge on homogeneous catalysts, and also the chemistry of organometallic compounds, in proposing a consistent catalytic cycle for the heterogeneously catalyzed propylene epoxidation in the SMPO process.

A close and well-researched homogeneous analogue of the epoxidation catalyst is the so-called Sharpless catalyst, a homogeneous titanium (IV) catalyst for the asymmetric epoxidation of allylic alcohols with t-butyl hydroperoxide (TBHP) [14]. A mechanism has been proposed for this type of epoxidation [15], which can be expressed in a generic form as shown in Fig. 13.4.

A titanium peroxide, Ti(OOR), is proposed as active intermediate. The molecular structure of this titanium peroxide moiety is rather exotic, as the anionic peroxide is η^2 coordinated to titanium with both its oxygen atoms, the distal oxygen formally forming a covalent bond and the proximal oxygen a dative bond. The olefinic bond directly attacks the activated, distal oxygen atom of the Ti(η^2-OOR) moiety, and, via a "bi-triangular" transition state and a labile titanium (epoxo)(alkoxide) complex, it removes the oxygen atom as an epoxide and leaves a titanium alkoxide behind. The resulting titanium alkoxide is protolyzed by another hydroperoxide molecule, ROOH, releasing the free alcohol and regenerating the titanium peroxide species. This reaction is driven by the great pK_a difference between ROH (around 18) and ROOH (around 11) [16]. Clearly, with the formation of a new titanium peroxide complex, the cycle is closed. The oxidation state of titanium remains IV for all species in the catalytic cycle.

Although exotic, the structure of the proposed titanium peroxide has been confirmed independently. In 1996, the first X-ray crystal structure of a titanium alkyl peroxide was reported [17]. This structure clearly shows an η^2 bonding mode for the peroxo ligand. The authors demonstrated the capacity of their titanium alkyl peroxo complex to oxidize nucleophiles, which indicated the electrophilic

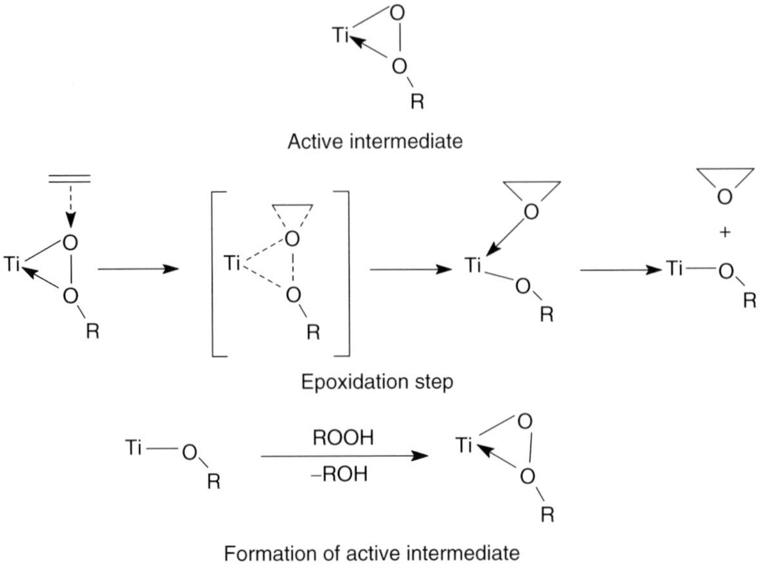

FIGURE 13.4 Proposed intermediates and reaction steps in Sharpless mechanism.

character of the activated oxygen atom. The electrophilic character is further supported by various studies involving quantum mechanical calculations of metalated hydroperoxides, M(η^2-OOR), concluding that the M(η^2-OOR) species is in principle an oxenoid [18].

Wu and Lai executed a quantum chemical study on the Sharpless epoxidation catalyst [12,13]. They showed the chemical pathway of the Sharpless mechanism to be sound according to the quantum chemical method applied, that is, a nonlocal density functional methodology. The authors concluded that the η^2 bonding mode for the Ti–O–OR moiety in its equilibrium state does exist, and provided support for the bi-triangular transition state. Furthermore, they calculated an η^2 structure for the Ti–O–OR unit in the transition structure. The electrophilic nature of the reaction is indicated by a small energy gap between the highest occupied molecular orbital (HOMO) of the olefin and the lowest unoccupied molecular orbital (LUMO) of the Ti(η^2-OOR) unit as well as by a negative charge transfer of about 0.2 from the C=C moiety to TiOOR in the transition structure.

Our own quantum mechanical calculations point to a transition state for oxygen transfer to the olefin being partly stabilized by electron back-donation from an oxygen lone pair to a π* orbital on the C=C double bond. Of the two oxygen lone pairs, the one in the TiOO plane is already involved in O to Ti σ donation and is therefore less available for C–O bond formation than the one perpendicular to the TiOO plane (see Fig. 13.5).

The calculated activation energy for olefin epoxidation occurring via the above transition state structure is quite low, which is in agreement with the observation that the titanium-catalyzed epoxidation is a highly efficient reaction. The barrier calculated at the MP2 level is 15.1 kcal/mol. This value compares well with the 12–15 kcal/mol reported by Wu and Lai (see above).

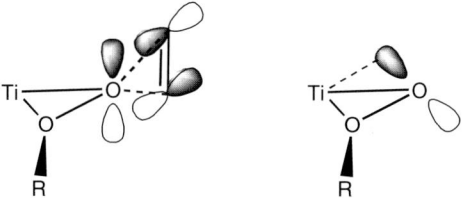

FIGURE 13.5 Simplified molecular orbital analysis of transition state in titanium-catalyzed olefin epoxidation.

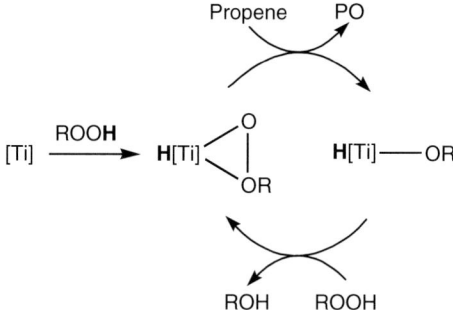

FIGURE 13.6 Simplified catalytic cycle for propylene epoxidation with hydroperoxides. [Ti] denotes coordinatively unsaturated silica-supported titanium species.

Titanium (IV)-catalyzed, Sharpless-type olefin epoxidation is best considered as a nucleophilic substitution at the distal oxygen of a titanium peroxide unit. The quantum mechanical calculations show that both oxygen atoms of the peroxotitanium moiety are electrophilic: as compared to a free hydroperoxide, both oxygen atoms display electron deficiency. This electron deficiency obviously is induced by the Lewis acidic titanium (IV) center. Furthermore, and of great significance, the distinct η^2 bonding mode is already close to the transition state structure for oxygen transfer—in η^2 coordination, the peroxo ligand is as it were preorganized for oxygen transfer.

For our heterogeneous system, the active intermediate Ti(η^2-OOR) is proposed to be generated from the active sites (or rather, precursor sites). Reaction with an organic hydroperoxide ROOH involves proton transfer to one of the siloxy ligands and forms a neutral coordinated silanol group together with the desired titanium peroxide. The Ti–O–Si moieties acting as a "proton parking place" are seriously weakened by this so-called peroxolysis.

We believe that the mechanism for the Sharpless epoxidation can be generalized, and that the heterogeneously catalyzed propylene epoxidation step in the SMPO process proceeds in a similar fashion, as depicted in the simplified mechanism shown in Fig. 13.6.

Further supporting evidence for this proposed mechanism has been obtained in collaboration with Prof. J. M. Basset (Centre National de la recherché Scientifique, Lyon). The aim was to synthesize the various intermediates proposed in the mechanism and to look at their reactivity in propylene epoxidation with organic hydroperoxide. The results of this study will be published separately.

2.3. Catalyst deactivation mechanisms

The heterogeneous titanium-on-silica catalyst used in the SMPO epoxidation step deactivates over the course of its operational life: faster at the start of the operation and more slowly towards the end. Various analyses provided evidence for leaching of Ti-species as well as agglomeration of highly dispersed Ti-species. These and other observations led us to develop the following model for the deactivation of the catalyst. The most active sites at a fresh titanium-on-silica catalyst surface are likely to be tripodally coordinated tetrahedral titanium centers. These sites are unstable due to a combination of a low coordination number and high Lewis acidity. Immediately after exposure to the reaction feed, expansion of the coordination shell would cause loss of activity. Moreover, a local excess of EBHP would lead to peroxolysis of the Si–O–Ti linkage creating soluble titanium complexes [5]. These leached species could be redeposited on the silica or titania surface further downstream. This process would lead to a redistribution of activity [1] and the growth of titania particles. Alternatively, water can assume a similar role as hydroperoxide, leading to hydrolysis of the active sites.

After a certain period of operation, depending on the operating conditions, tripodal sites would no longer be present at the titanium-on-silica catalyst surface. At this stage, ultrafine titania clusters would do the catalytic work. The larger the titania cluster, the lower its catalytic activity, but the higher its stability (Fig. 13.7). For this reason, catalyst deactivation would slow down with time. When peroxolysis or hydrolysis becomes less important, poisoning, probably by by-products in the epoxidation process, starts to have a significant catalyst deactivation effect.

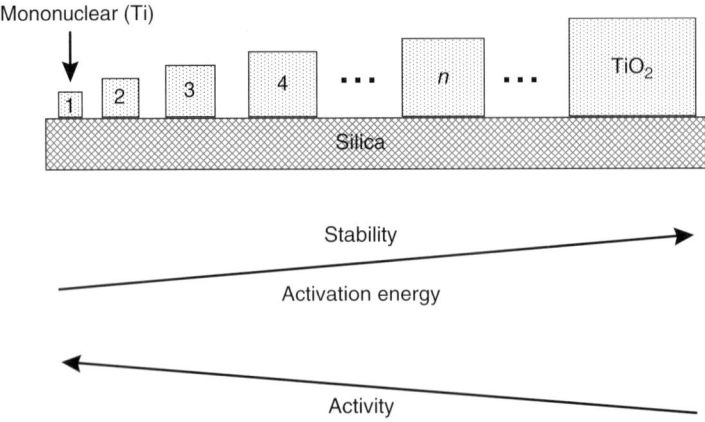

FIGURE 13.7 Representation of titanium-on-silica catalyst in SMPO as a range of silica-supported—and therefore stabilized—ultrafine titania particles containing catalytically active, coordinatively unsaturated titanate sites.

3. SMPO PROCESS IMPROVEMENT

Similar to the heterogeneously catalyzed propylene epoxidation step discussed above, a detailed understanding of all conversion steps has been obtained in 30 years of operation of the SMPO process. This has led to numerous process improvements that helped optimize the overall conversion of propylene and EB to PO and SM. In turn, this has led to much reduced specific capital employed for the production of these bulk chemicals, as is shown for PO in Fig. 13.8.

3.1. Trace component chemistry

Despite the complexity of the process, SMPO is very selective, with an EB consumption only slightly higher than the competing (coproduct-free) EB dehydrogenation route to styrene. In order to optimize the process, a thorough understanding of the trace by-product chemistry is essential. The formation of by-products does, of course, represent a direct yield loss, but further reaction with one of the desired products, MPC or PO (or precursors), later in the process, can magnify this loss. Moreover, traces of by-products may result in contamination of the styrene or PO products (necessitating expensive purification procedures), or negatively influence specific process steps (catalyst degradation). These effects are accentuated when the components are not effectively bled and therefore build up in recycle streams.

Individual reaction steps in SMPO have been (re)examined in laboratory and bench scale tests, which included spiking experiments, often with "labeled" reagents. These experiments interrogated the chemistry of all process steps,

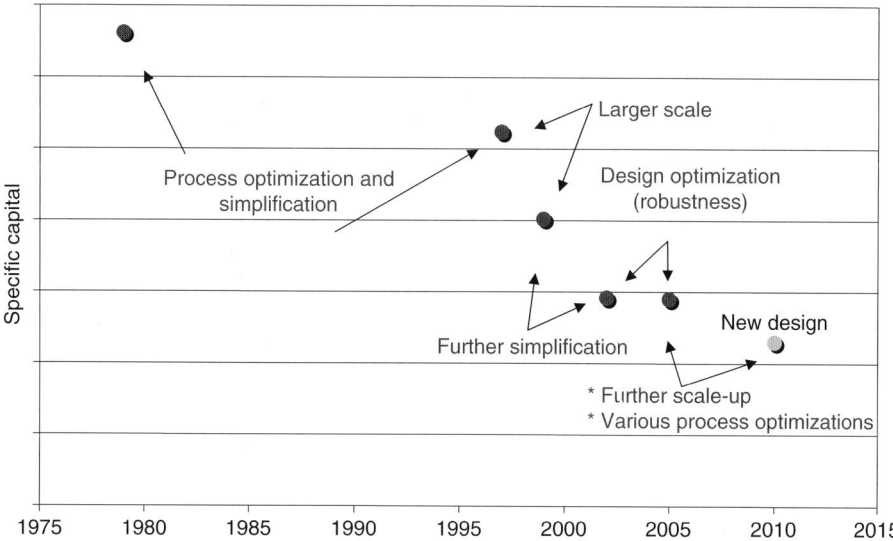

FIGURE 13.8 Historic overview of effect of SMPO process development on specific capital employed for SMPO plants.

namely, the EB oxidation, propylene epoxidation, MPC dehydration, and MPK hydrogenation sections. Extensive sampling exercises in SMPO plants, combined with advanced analytical methods, allowed all significant by-products to be tracked. The improved understanding of by-product chemistry developed by this "dual-track" approach will be illustrated by following particular aromatic components across the SMPO process.

Of course, PO is a highly reactive molecule and the by-product chemistry, often involving the titanium-on-silica catalyst, including isomerization, dimerization, hydrolysis, and alcoholysis, has also been (re)investigated. Not surprisingly, PO yield loss (and MPC, via adduct formation) and the separation of PO from trace components with similar boiling points, such as lower aldehydes, are the main by-product issues here. Due to space limitations, this chemistry will not be discussed further here.

The air-oxidation of EB is characterized by lower selectivity to the alkyl hydroperoxide, than, for example, cumene (in the phenol/acetone process). However, this is acceptable in SMPO because the major by-products, MPC and MPK, although not involved in propylene epoxidation, are styrene precursors (MPK, via hydrogenation to MPC). Loss of additional (potential) PO yield occurs to a lesser degree in the epoxidation section, via consumption of EBHP in reactions other than propylene epoxidation, most significantly dehydration to MPK. Formation of MPC and MPK in "nonproductive" (in terms of PO) side reactions is the most important reason why the product SM/PO molar ratio is higher than the theoretical value of 1.0, although many other side reactions, involving both aromatic components and PO, also play a role.

The following discussion will cover specific aromatic by-products, which, unlike MPC and MPK, represent a true styrene yield loss, as they are not precursors of this product. These include 2-phenylethanol (*beta*-phenylethanol, BPEA), ethylphenols and phenol, benzaldehyde, benzyl alcohol, and benzoic acid.

Although the benzylic hydrogen of EB is most susceptible to autoxidation (giving EBHP), activation of both the methyl and aromatic hydrogens also proceeds, giving the *primary* isomer of EBHP ($PhCH_2CH_2OOH$) and ethylphenols, respectively. We have shown that the selectivity for *primary*-EBHP relative to *secondary*-EBHP increases with conversion in the range of 0.5–1.3%. Similar to the major isomer, *primary*-EBHP epoxidizes propylene to PO in the presence of the titanium-on-silica catalyst, forming the *primary* alcohol, BPEA. Although BPEA, like MPC, might be considered a styrene precursor, dehydration of the primary alcohol is much slower. In fact, careful spiking studies with ^{13}C-labeled material under realistic process conditions, allowed the balance between extra styrene formation from BPEA, and loss of styrene via reaction of BPEA with MPC (precursors) to give additional heavy ends, to be established. The alcohol, which builds up in the recycle stream of the dehydration reactor, forms the cross-ether with MPC, and gives (via reversible dehydrogenation) aldol condensation products with MPK, over the alumina catalyst. The BPEA may be recovered for use as a fragrance in detergents.

Phenol and the ethylphenols are formed in both the oxidation and epoxidation sections. The Lewis acidic titanium-on-silica catalyst is implicated in the heterolytic cleavage of EBHP to phenol and acetaldehyde, and the hydroxylation of EB to ethylphenols, both minor reactions competing with the desired propylene epoxidation with EBHP. Although total phenols formation is low relative to MPC and MPK, the phenols build up in recycle streams due to limited bleeding. Phenol is known to be a potent inhibitor of the autoxidation of EB and levels in the recycle EB should be as low as possible. Phenols consume up to two to three molecules of MPC (styrene), in the dehydration section, giving alkylated phenols (similar to cumylphenols formation in the phenol/acetone process). Loss of these alkylated phenols via the heavy ends stream effectively removes the phenol and ethylphenols from the process. It has been shown that recycling benzyl alcohol (see below) to the dehydration section reduces the amount of MPC consumed on reaction with phenols [19].

C_7 by-products derive from reactions in oxidation, and, to a lesser extent, epoxidation, with the radical decomposition of EBHP to benzaldehyde being the major route. Benzaldehyde, like phenol, is an undesirable component in the recycle EB to oxidation. Most of the benzaldehyde enters the dehydration reactors with the MPC and is removed from styrene by distillation.

Part of the benzaldehyde stream undergoes hydrogenation to benzyl alcohol (parallel to MPK → MPC), which is returned to the dehydration reactor. Benzyl alcohol is partially consumed by phenol alkylation and other reactions over the alumina catalyst. As is also the case for BPEA, the rather low conversion leads to a significant benzyl alcohol recycle in dehydration and hydrogenation.

Benzoic acid is formed in oxidation at a roughly similar level to *primary*-EBHP (up to 1% of EB consumed), via the facile oxidation of benzaldehyde. To mitigate possible corrosion effects or catalyst deterioration downstream, the acid is partially removed from the EBHP before the epoxidation step, using an alkaline washing step. The remaining benzoic acid is removed via the heavy end stream, partly in ester form.

Products containing two or more aromatic rings are removed, together with high-boiling monoaromatics, from the process via the heavy ends bleed. As already mentioned, some by-products in the MPC feed react with one or more equivalents of MPC (or precursor MPK) in the dehydration section, resulting in magnification of the yield loss. In fact, heavy ends formation is several times *lower* when pure MPC is used as feed compared to the normal process feed. The major culprits in this yield loss "magnification" are the phenols, primary alcohols, and benzaldehyde, which are implicated in the formation of (poly)alkylated phenols, ethers, aldol condensation products, and esters, as already discussed. The ether, 2,3-DPEE, which is formed in both the oxidation (radical route) and dehydration sections (from MPC or styrene), and a complex mixture of styrene "oligomers" formed over the alumina catalyst are other important components of the heavy ends stream.

Increased understanding of the trace chemistry of SMPO allows process modifications to be introduced to reduce by-product formation, or, at least, to allow the effect of such changes to be predicted. For example, "oxygen starvation" has been shown in bench scale EB oxidation experiments to lead to higher selectivities for MPC, MPK, benzaldehyde, benzoic acid, and phenol by-products. Higher by-product formation reflects the anaerobic decomposition of EBHP, which reacts with EB solvent to give up to two equivalents of MPC. Higher levels of benzaldehyde (rapidly oxidized to benzoic acid) reflect increased formation of the alkoxyl radical via this route. The knowledge of the effect of "oxygen starvation" was a key input in the fluid dynamics work discussed in the next section, which led to modification of the oxidation reactors and ultimately higher EBHP selectivity.

3.2. Air-oxidation improvements

The selectivity to EBHP observed in commercial plants falls up to a few percent short of what is achieved at ideal laboratory conditions. It was therefore hypothesized that this may be due to "O_2 starvation" in poorly aerated regions of the large air-oxidation reactors. With computational fluid dynamics (CFD), we modeled the complex hydrodynamics, including mass transfer and chemical reaction. Key CFD predictions were the presence of very strong liquid circulation patterns in the transverse direction and a fast upward flow of gas bubbles along quite narrow paths with very high gas hold up. For certain configurations, the gas was predicted to mainly rise through two paths near the walls, that is, the liquid in the center of the reactor would not be effectively aerated and is predicted to suffer from "starvation" [20].

To validate the model predictions, we performed radioactive tracer experiments in one of our commercial plants during normal production. The ^{41}Ar tracer was injected in the air feed at different positions. Fourteen detectors on the reactor wall yielded a tracking of the gas tracer, from which among other things the gas rise velocities and the liquid circulation patterns could be determined.

Figure 13.9 illustrates some of the important observations from the gas tracer experiments in the commercial plant. First of all, these experiments confirmed the key CFD prediction that the major part of the gas injected at the middle does indeed largely rise sideways, that is, mostly near the wall rather than through the middle. Second, the rise velocity of the bubbles injected at the side was observed

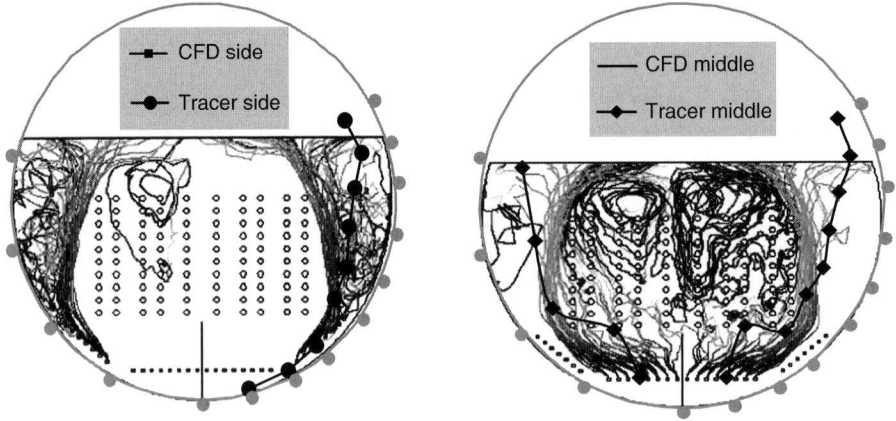

FIGURE 13.9 Bubble trajectories of gas from the side spargers (left picture, $v = 0.8$ m/s) and middle spargers (right picture, $v = 2.5$ m/s): comparison computational fluid dynamic (CFD) prediction (lines) and tracer experiments (dots or diamonds) [20]. Reprinted with permission from from [32]. Copyright (2006) Elsevier.

to be about twice as high as the bubbles injected near the center, which also implies that, even though these two "plumes" are actually quite close to each other, they are largely segregated. In retrospect, CFD had already predicted this, as shown by the respective trajectories in Fig. 13.9.

After the successful commercial plant-scale validation, CFD was used to evaluate a number of potential measures to reduce the "starvation" predicted to occur for certain configurations and operating conditions. Options included repositioning of the gas inlets and installing additional baffles at specific locations [21]. After implementation, the reaction temperature could be lowered significantly resulting in improved selectivity and thus plant yields.

3.3. Catalytic dehydration

The dehydration of MPC to SM was in commercial practice long before the SMPO process. In the 40s, Union Carbide commercialized a styrene process based on the oxidation of EB to MPK, followed by selective hydrogenation to MPC and finally dehydration to styrene [22,23]. The dehydration step proceeded in the vapor phase by passing the alcohol over a titania or alumina catalyst at 200–300 °C. Poor overall yields and corrosion problems made this oxidation route to styrene obsolete. Shell adopted this dehydration technology for its SMPO process and has improved it over the years.

3.3.1. Gas-phase processes

The peculiarity of the TiO_2 and Al_2O_3 catalysts does not rely on their activity but rather on their styrene selectivity and their moderate tendency to produce oligomeric materials that deactivate the catalyst. Indeed, many catalysts are much more active, some being so active that they could be operated in the liquid phase below 150 °C [24,25]. However, these active catalysts usually exhibited lower styrene yields and/or faster deactivation than do TiO_2 and Al_2O_3 [26,27].

Over the years, Shell has explored process and catalyst improvements to increase yields and catalyst life. For instance, the activity of an Al_2O_3 catalyst could be sustained for longer runs upon cofeeding steam with MPC [28]. The initial activity could also be recovered through a simple wash with hot EB [28]. As for the catalyst improvements, surface area, pore volume, and catalyst shape appeared to be critical parameters for optimization. An Al_2O_3 catalyst having a similar surface area could see its activity and selectivity improved upon increasing its pore volume, for example, from 0.3 to 0.6 mL/g and changing its shape from cylindrical tablets to star-shaped extrudates [29]. Additional improvements were achieved upon developing smaller catalyst particles and increasing their pore volume further [30]. This novel formulation provided significant improvement in stability as well. The importance of these factors suggests that the reaction is affected by diffusion limitations of MPC and SM into and out of the catalyst pores.

Intrinsically more active catalysts were also investigated in an attempt to perform the reaction at significantly lower temperature and, thereby, simplify the engineering of the heat supply needed to drive the reaction. Various zeolite powders showed a high activity and selectivity when operated at milder temperature, for example, 220 °C. Pore-diffusion appeared to be critical here again: extruded H-ZSM-5 catalysts showed satisfactory selectivity (>90%) and stability (>150 h) only upon operating with small extrudates and at moderate rate, for example, low temperature and/or low zeolite content [31,32]. These factors were handily combined in the parameter $\Phi^*(D_{eff})^{1/2}$ defined as $d_p/4^*(5 \times \text{zeolite content})^{1/2}$, with d_p representing the extrudate diameter (see Fig. 13.10).

3.3.2. Liquid-phase processes

As the zeolite catalysts can operate at low temperature, we also investigated the possibility of carrying out the reaction under reactive distillation conditions to withdraw styrene and water as they are produced [33,34]. Medium-pore zeolites such as H-ZSM-5 and H-ZSM-11 were very active and selective at 170 °C [33]. High activity and selectivity required small zeolite crystals (<50 nm), however (Fig. 13.11).

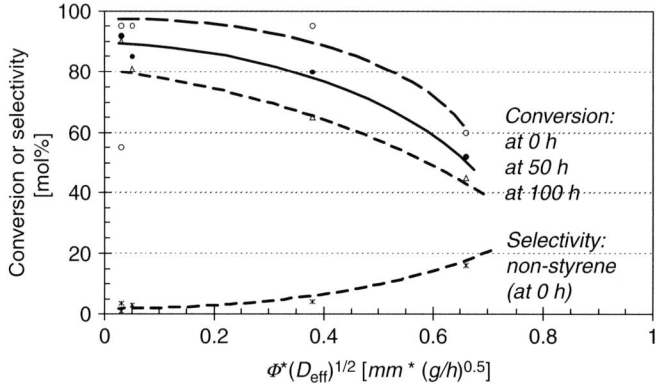

FIGURE 13.10 Influence of mass transport on the activity, selectivity, and stability of H-ZSM-5/SiO$_2$ catalysts operating in gas phase adapted from reference [32].

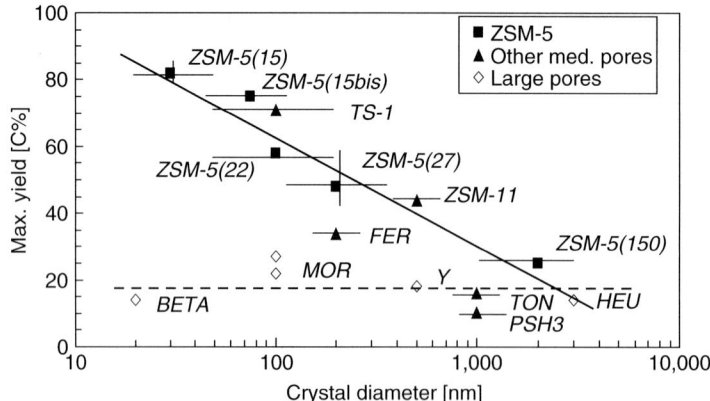

FIGURE 13.11 Maximum styrene monomer yield during the dehydration of methyl phenyl carbinol over zeolites catalysts operating under reactive distillation conditions [33].

In contrast, large-pore zeolites and mesoporous solid acids did not produce much SM under these conditions. This suggests that the reaction is sensitive to shape selectivity. However, the SM selectivity rapidly degraded with time on stream, leading to the formation of ethers and SM oligomers. The formation of heavy products could not be depressed by process conditions such as adding bulky nitrogen-bases or radical scavengers, diluting the feed with an organic solvent, or stripping styrene with steam [34].

The reactive distillation operation is obviously not limited to zeolite catalysts. It can also be carried out with homogeneous acids such as sulfuric acid or *p*-toluene-sulfonic acid. Since they lack shape selectivity, these catalysts first convert phenyl-ethanol to the corresponding ether and only then to styrene. Hence, the reaction proceeds in a solution of heavy products that have accumulated over time. Additives have been developed to control the oligomerization reactions and keep the liquid viscosity at a workable level [35]. The heavy liquid medium needs to be bled. Its contamination with strong acid makes its disposal costly, however.

4. CONCLUSIONS

The SMPO process is an efficient way of making two chemical commodities, PO and SM. Continuous and diverse efforts in research, development, and plant trials have brought a deep understanding of numerous process aspects, which have resulted in significant improvements in efficiency and cost competitiveness of the process. These investigations varied from catalytic studies (e.g., with real and model systems) to computational studies (fluid dynamics or quantum mechanics), to spiking experiments in laboratory and commercial reactors (e.g., with labeled components) as well as measurements of basic data, detailed process flow sheeting, and many more. Selected examples reported here illustrate the breath that is needed to keep a technology at the top of its league. More information on Shell technologies can be found in the recent Shell Technology report [36].

ACKNOWLEDGMENT

The authors would like to acknowledge all Shell employees who have contributed over the past decades to the build-up of knowledge on the SMPO process.

REFERENCES

[1] J. K. F. Buijink, J. J. M. VanVlaanderen, M. Crocker, F. G. M. Niele, Propylene epoxidation over titanium-on-silica catalyst-the heart of the SMPO process, *Catal. Today* 93–95 (2004) 199.
[2] R. Hutter, T. Mallet, D. Dutoit, A. Baiker, Titania-silica aerogels with superior catalytic performance in olefin epoxidation compared to large pore Ti-molecular sieves, *Top. in Catal.* 3 (1996) 421.
[3] M. Crocker, R. H. M. Herold, A. G. Orpen, M. T. A. Overgaag, Synthesis and structural characterisation of tripodal titanium silsesquioxane complexes: A new class of highly active catalysts for liquid phase alkene epoxidation, *Chem. Commun.* (1997) 2411.

[4] M. Crocker, R. H. M. Herold, A. G. Orpen, M. T. A. Overgaag, Synthesis and characterisation of titanium silasesquioxane complexes: Soluble models for the active site in titanium silicate epoxidation catalysts, *J. Chem. Soc., Dalton Trans.* (1999) 3791.

[5] H. C. L. Abbenhuis, S. Krijnen, R. van Santen, Modelling the active sites of heterogeneous titanium epoxidation catalysts using titanium silasequioxanes: Insight into specific factors that determine leaching in liquid-phase processes, *Chem. Commun.* (1997) 331.

[6] T. Maschmeyer, F. G. Rey Sankar, J. M. Thomas, Heterogeneous catalysts obtained by grafting metallocene complexes onto mesoporous silica, *Nature* 378 (1995) 159.

[7] J. Jarupatrakorn, T. D. Tilley, Silica-supported, single-site titanium catalysts for olefin epoxidation. A molecular precursor strategy for control of catalyst structure, *J. Am. Chem. Soc.* 124 (2002) 8380.

[8] C. Rosier, G. Niccolai, J. M. Basset, Catalytic hydrogenolysis and isomerization of light alkanes over the silica-supported titanium hydride complex ($\equiv$SiO)$_3$TiH, *J.Am. Chem. Soc.* 119 (1997) 12408.

[9] F. Bini, C. Rosier, R. Petroff Saint-Arroman, E. Neumann, C. Dablemont, A. deMallman, F. Lefebvre, G. P. Niccolai, J.-M. Basset, M. Crocker, J.-K Buijink, Surface organometallic chemistry of titanium: Synthesis, characterization, and reactivity of (=Si-O)$_n$Ti(CH$_2$C(CH$_3$)$_3$)$_{4-n}$(n=1,2) grafted on aerosil silica and MCM-41, *Organometallics* 25 (2006) 3743.

[10] S. Inamura, T. Nakai, K. Utani, H. J. Kania, Epoxidation activity of coordinatively unsaturated titanium oxide, *Catalysis* 161 (1996) 495.

[11] A. Tuel, L. G. Hubert-Pfalzgraf, Nanometric monodispersed titanium oxide particles on mesoporous silica: synthesis, characterization, and catalytic activity in oxidation reactions in the liquid phase, *J. Catal.* 217 (2003) 343.

[12] Y-D. Wu, D. K. W. Lai, A density functional study on the stereocontrol of the sharpless epoxidation, *J. Am. Chem. Soc.* 117 (1995) 11327.

[13] Y-D. Wu, D. K. W. Lai, Transition structure for the epoxidation mediated by titanium(IV) peroxide. A density functional study, *J. Org. Chem.* 60 (1995) 673.

[14] T. Katsuki, K. B. Sharpless, The first practical method for asymmetric epoxidation, *J. Am. Chem Soc.* 102 (1980) 5974.

[15] K. B. Sharpless, S. S. Woodward, M. G. Finn, On the mechanism of titanium-tartrate catalyzed asymmetric epoxidation, *Pure & Appl. Chem.* 55 (1983) 1823.

[16] M. Johnson, Thermochemical properties of peroxides and Peroxly Radicals, *J. Phys. Chem.* 100 (1996) 6814.

[17] G. Boche, K. Möbus, K. Harms, M. March, [((η^2-*tert*-Butylperoxo)titanatrane)$_2$. 3 Dichloromethane]: X-ray crystal structure and oxidation reactions, *J. Am. Chem. Soc.* 118 (1996) 2770.

[18] G. Boche, J. C. W. Lohrenz, The electrophilic nature of carbenoids, nitrenoids, and oxenoids, *Chem. Rev.* 101 (2001) 697.

[19] M. Boelens, A. D. Horton, T. M. Nisbet, A. B. van Oort, Process for preparing styrene WO2005054157 (2005) assigned to Shell Int. Res..

[20] P. A. A. Klusener, G. Jonkers, F. During, E. D. Hollander, C. J. Schellekens, I. H. J. Ploemen, A. Othman, A. N. R. Bos, Horizontal cross-flow bubble column ractors: CFD and validation by plant scale tracer experiments, *Chem. Eng. Sci.* 62 (2007) 5495.

[21] E. D. Hollander, P. A. A. Klusener, I. H. J. Ploemen, C. J. Schellekens, Horizontal reactor vessel WO2006024655 (2006) assigned to, *Shell Int. Res.*

[22] W. L. Faith, D. B. Keyes, R. L. Clark, *Industrial Chemicals*. 2nd Ed., Wiley (1957).

[23] H. J. Sanders, H. F. Keag, H.S McCullough, H.S., Acetophenone, *Ind. Eng. Chem.* 45 (1953) 2.

[24] G. G. Overberger, J. H. Saunders, m-Chlorostyrene, *Organic Syntheses*, Collect , Vol. III, p. 204–206. Wiley, 1953.

[25] R. V. Hoffman, R. D. Bishop, P. M. Fitch, R. Hardenstein, Anhydrous copper(II) sulfate: An efficient catalyst for the liquid-phase dehydration of alcohols, *J. Org. Chem.* 45 (1980) 917.

[26] G. Csomontayi, A. Panovici, Dehydration and dehydrogenation of methylphenylcarbinol on various catalysts, *Rev. Rowmaine Chim.* 17 (1972) 525.

[27] T. Takahashi, T. Kai, M. Tashiro, Dehydration of l-Phenylethanol over solid acidic cataysts *Can, J. Chem. Eng.* 66 (1988), 433.

[28] J. R. Skinner, C. E. Sanborn, Verfahren zur herstellung styrol durch wasserabspaltung aus α-methylbenzylalkohol, DE2146919 (1972) assigned to Shell, *Int. Res.*
[29] H. Dirkzwager, M. Van Zwienen, *Process for the preparation of styrenes*, WO1999058480 (1999) assigned to Shell, *Int. Res.*
[30] J. A. M. van Brokekhoven, C. M. A. M. Mesters, *Process for the preparation of styrene*, WO2004076389 (2004) to Shell, *Int. Res.*
[31] G. C. van Giezen, J.-P. Lange, C. M. A. M. Mesters, *Process for the preparation of styrenes*, WO19999042425 (1999) to Shell, *Int. Res.*
[32] J,-P. Lange, C. M. A. M. Mesters, Mass transport limitations in zeolite catalysts: The dehydration of 1-phenyl-ethanol to styrene, *Appl. Catal. A: Gen.* 210 (2001) 247.
[33] J.-P. Lange, V. Otten, Dehydration of pheny-ethanol to styrene: Zeolite catalysis under reactive distillation, *J. Catal.* 238 (2006) 6.
[34] J-P. Lange, V. Otten, Dehydration of phenyl ethanol to styrence under reactive distillation conditions: Understanding the catalyst deactivation, *Ind. Eng. Chem.* 46 (2007) 6899.
[35] A. D. Horton, T. M. Nisbet, M. van Zwienen, *Process for preparing styrene*, WO2004000766 (2004) to Shell, *Int. Res.*
[36] *Shell Technology Report.* (2007) www.shell.com/technology.

CHAPTER **14**

Propylene Epoxidation with Ethylbenzene Hydroperoxide over Ti-Containing Catalysts Prepared by Chemical Vapor Deposition

Kuo-Tseng Li, Chia-Chieh Lin, and **Ping-Hung Lin**

Contents
1. Introduction 374
2. Experimental 375
   2.1. Catalyst preparation and characterization 375
   2.2. Catalytic property measurements 375
3. Results and Discussion 376
   3.1. Ti/SiO$_2$ catalysts 376
   3.2. Packed-bed reactor data 378
   3.3. Ti/MCM-41 and Ti/MCM-48 catalysts 380
   3.4. Kinetics of propylene epoxidation with EBHP 384
4. Conclusions 385
   Acknowledgment 385
   References 385

Abstract

Chemical vapor deposition (CVD) using TiCl$_4$ was used to prepare Ti/SiO$_2$, Ti/MCM-41, and Ti/MCM-48 catalysts. These catalysts were characterized by inductively coupled plasma-atomic emission spectroscopy (ICP-AES), X-ray photoelectron spectroscopy (XPS), Fourier transform infrared (FTIR) spectroscopy, nitrogen adsorption, and were used to catalyze the epoxidation of propylene to propylene oxide (PO) with *in situ* prepared ethylbenzene hydroperoxide (EBHP). CVD time and CVD temperature affected the catalyst performance significantly. The optimum temperature range was 800–900 °C, and the optimum deposition time was 2.5–3 h. The maximum PO yields obtained in a batch reactor were 87.2, 94.3, and 88.8% for Ti/SiO$_2$, Ti/MCM-41, and Ti/MCM-48, respectively. Ti/MCM-41 had higher titanium

Department of Chemical Engineering, Tunghai University, Taichung, Taiwan, ROC

Mechanisms in Homogeneous and Heterogeneous Epoxidation Catalysis
DOI: 10.1016/B978-0-444-53188-9.00014-6

© 2008 Elsevier B.V.
All rights reserved.

concentration and better epoxide selectivity than the corresponding Ti/MCM-48, which suggested that pore structure had a strong effect on the diffusion of TiCl$_4$ molecules inside the supports. The rate of EBHP disappearance was found to exhibit first-order dependence on the EBHP concentration with activation energy of 42.8 kJ/mol. The CVD-prepared Ti/SiO$_2$ catalyst was stable in continuous packed-bed reactor operation.

Key Words: Propylene oxide synthesis, Propylene epoxidation with ethylbenzene hydroperoxide, *tert*-butyl hydroperoxide, cumene hydroperoxide, Preparation of Ti/SiO$_2$, Ti/MCM-41, Ti/MCM-48 catalysts, Chemical vapor deposition, 1,2-Epoxyoctane synthesis. © 2008 Elsevier B.V.

1. INTRODUCTION

Propylene oxide (PO) is an important chemical intermediate, which is mainly used in the manufacture of polyols, propylene glycols, and propylene glycol ethers [1]. The world annual production capacity of PO is about 7 million metric tons [2]. PO is mainly produced commercially by either the chlorohydrin (about 43%) or organic hydroperoxide processes. The chlorohydrin route produces large amounts of salt by-product, and new plants have used the hydroperoxide processes [3].

In the organic hydroperoxide processes, PO (C$_3$H$_6$O) is synthesized by the catalytic reaction of propylene (C$_3$H$_6$) and an alkyl hydroperoxide (ROOH), according to the following reaction scheme:

$$C_3H_6 + ROOH \xrightarrow{\text{Catalyst}} \underset{CH_3CH-CH_2}{\overset{O}{\triangle}} + ROH \qquad (14.1)$$

Commercial organic hydroperoxides used include *tert*-butyl hydroperoxide (TBHP), ethylbenzene hydroperoxide (EBHP), and cumene hydroperoxide (CHP). About 60% of the commercial PO is produced by using EBHP to oxidize propylene, catalyzed mainly by a Ti/SiO$_2$ catalyst [2]. Commercial Ti/SiO$_2$ catalysts for propylene epoxidation are prepared by impregnating silica with TiCl$_4$ or an organic titanium compound, followed by filtration, drying, calcination, and silylation [4]. Recently, we used a chemical vapor deposition (CVD) method to prepare Ti/SiO$_2$ and Ti/MCM-41 catalysts in the temperature range 400–1,100 °C, and found that the Ti/SiO$_2$ catalyst prepared at the deposition temperature of 900 °C exhibited the best PO yield for propylene epoxidation with TBHP and CHP [5–7]. The CVD process for the preparation of epoxidation catalyst is a single-step procedure, which is much simpler than the conventional multistep epoxidation catalyst preparation method. Since EBHP is the most frequently used oxidant for propylene epoxidation, the major purpose of this chapter is to use CVD-prepared Ti-containing catalysts (including Ti/SiO$_2$, Ti/MCM-41, and Ti/MCM-48) for catalyzing the reaction between propylene and EBHP to produce PO.

2. EXPERIMENTAL

2.1. Catalyst preparation and characterization

Three supports were used for chemical deposition of $TiCl_4$ vapor, including an amorphous silica gel (H_2SiO_3, Strem Chemicals, Newburyport, MA, USA large pore, microspheroidal white powder), purely siliceous MCM-41, and purely siliceous MCM-48. Si-MCM-41 was synthesized using sodium silicate solution (Sigma Aldrich, St. Louis, MO, USA 27% as SiO_2), tetrapropylammonium bromide (Lancaster, Morecambe, England >98%) and cetyltrimethylammonium bromide (Lancaster, 98%) according to a literature method [8]. Si-MCM-48 was synthesized using tetraethyl orthosilicate (Lancaster, 98%), cetyltrimethylammonium bromide, and sodium hydroxide following procedures described earlier [9]. The fresh prepared MCM-41 and MCM-48 had BET surface areas of 1,071 and 1,387 m^2/g, respectively.

Chemical vapor deposition of $TiCl_4$ was carried out in a 0.007-m I.D. quartz tube packed with 1.5 g of the supports. An electrically heated tube furnace in which the temperature could be controlled to within 1 °C was used to house the CVD reactor. Before the deposition of $TiCl_4$, the support powder was dried under 100 ml/min N_2 at 400 °C for 1 h. After the drying stage, gaseous feed of 1.0 vol.% $TiCl_4$ (ACROS, Geel, Belgium 99.9%) vapor in nitrogen was introduced into the CVD reactor at a flow rate of 50 ml/min for 0.5–3 h.

The titanium contents of the resulting catalyst samples were determined with an inductively coupled plasma-atomic emission spectrometer (ICP-AES) (Kontron, Germany Model S-35) after HF acid digestion of the solid. N_2 adsorption/desorption isotherms at 77 K were obtained using a Micromeritics ASAP 2020 apparatus. Catalyst crystalline structure was examined by X-ray diffraction (XRD) on a Shimadzu XRD-6000 diffractometer with Cu Kα radiation. X-ray photoelectron spectroscopy (XPS) data were acquired on a VG Microtech MT-500 spectrometer using Al Kα X-ray radiation (1,486.6 eV). Fourier transform infrared (FTIR) data were obtained on a Shimadzu IR Prestige FTIR spectrophotometer.

2.2. Catalytic property measurements

Propylene epoxidation experiments were performed in a stirred high-pressure batch reactor (Parr). In a typical experiment, 0.005 g mol of EBHP (in 50 ml ethylbenzene) and 0.3 g of Ti-containing catalyst (prepared by the CVD method mentioned above) were charged to the batch reactor. EBHP was prepared from the reaction of ethylbenzene and oxygen in a stainless steel reactor at 140 °C for 2 h. Propylene at 8 kg/cm^2 was introduced into the epoxidation reactor at 5 °C and saturated with the reaction mixture. Unless specified otherwise, the epoxidation reaction temperature was 100 °C, the agitator speed was 150 rpm, and the reaction time was 1 h. The PO concentration in the product solution was analyzed with a Shimadzu GC-17A gas chromatograph using a 60-m DB-Waxeter column, and the concentration of EBHP was determined by iodometric analysis.

EBHP conversion was defined as the percentage of EBHP in the feed that had reacted. Propylene oxide yield was defined as the moles of PO formed per mole of

EBHP in the feed. Propylene oxide selectivity was defined as the percentage of EBHP reacted to PO (i.e., propylene oxide selectivity = propylene oxide yield/EBHP conversion).

3. RESULTS AND DISCUSSION

3.1. Ti/SiO$_2$ catalysts

Six Ti/SiO$_2$ epoxidation catalysts were prepared by chemical deposition of TiCl$_4$ vapor on silica gel at 900 °C using different reaction times ranging from 0.5 to 3 h. Figure 14.1 shows EBHP conversion and PO yield as a function of TiCl$_4$ deposition time for the CVD-prepared Ti/SiO$_2$ samples. CVD time had little effect on EBHP conversion, but had a strong influence on PO yield and PO selectivity (shown in row 2 of Table 14.1). PO selectivity and yield increased rapidly with increasing deposition time, and reached a maximum yield of 87.2% and a maximum selectivity of 90% for the catalyst prepared with a deposition time of 2.5 h. The other product was acetophenone, which was generated from the decomposition of EBHP [10]. When CHP and TBHP were used as the oxidants for propylene epoxidation, the Ti/SiO$_2$ catalyst prepared with the 2.5 h CVD time also had the maximum PO yield and PO selectivity, as shown in rows 3 and 4 of Table 14.1.

Figure 14.2 shows surface titanium concentrations (in μmol/m^2) as a function of deposition time for Ti/SiO$_2$ catalysts, which were calculated from the titanium content (obtained from ICP-AES measurements) and surface area data (obtained from BET measurements). Titanium concentration increased with increase in deposition time, and reached a maximum value of 1.12 μmol/m^2 at a deposition time of 2.5 h. Further increase in deposition time resulted in a decrease in titanium concentration. It has been reported that the concentration of surface silanol groups is around 1 μmol/m^2 at 900 °C [11]; therefore, all of the silanol groups on the silica should have been converted to SiOTiCl$_3$ (via the reaction SiOH + TiCl$_4$ →

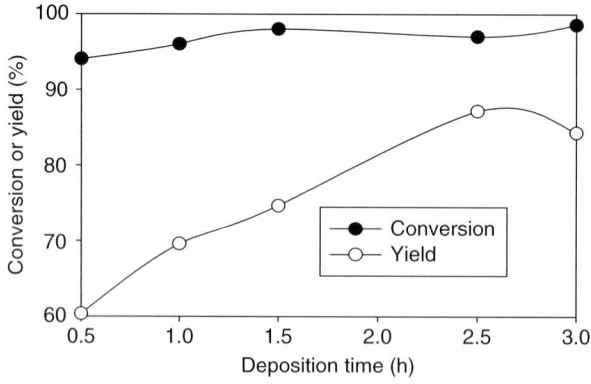

FIGURE 14.1 Influence of TiCl$_4$ deposition time on ethylbenzene hydroperoxide (EBHP) conversion and propylene oxide yield for Ti/SiO$_2$ catalysts.

TABLE 14.1 Effects of oxidant and chemical vapor deposition (CVD) time on epoxide yield (Y), epoxide selectivity (S), and turnover frequencies (TOF; units: moles of epoxide product per mole of Ti per hour)

Oxidant	Time (h)						
	0.5	1.0	1.5	2.0	2.5	3.0
EBHP (in ethylbenzene)[a]	Y = 60.4% S = 64.2% TOF = 175.3	Y = 69.6% S = 72.5% TOF = 98.5	Y = 74.7% S = 76.2% TOF = 70.2	—	Y = 87.2% S = 90% TOF = 80.6	Y = 84.3% S = 85.5% TOF = 100.9
CHP (in cumene)[a]	Y = 64.6% S = 86.6% TOF = 187.5	Y = 67.8% S = 81.7% TOF = 95.9	Y = 83.6% S = 91.3% TOF = 78.6	Y = 89.1% S = 97.3% TOF = 83.6	Y = 93.3% S = 96.6% TOF = 86.3	Y = 88.8% S = 92.2% TOF = 106.2
TBHP (in benzene)[a]	Y = 61.7% S = 72.2% TOF = 179	—	—	Y = 85.7% S = 88.3% TOF = 80.4	Y = 92.3% S = 97.2% TOF = 85.4	—
TBHP reaction with 1-octene in packed-bed reactor	S = 50.9%	S = 69.3%	S = 66.6%	S = 95.1%	S = 98.5%	S = 90.4%
TBHP reaction with 1-octene in batch reactor[b]	Y = 58.8% S = 70.6% TOF = 170.6	Y = 60.8% S = 73.2% TOF = 86	Y = 77.8% S = 80.8% TOF = 73	Y = 83.9% S = 85.4% TOF = 78.7	Y = 91.3% S = 94.5% TOF = 84.4	Y = 91.4% S = 92.3% TOF = 109.4

EBHP, ethylbenzene hydroperoxide; CHP, cumene hydroperoxide; TBHP, tert-butyl hydroperoxide.
[a] Propylene epoxidation conditions: catalyst weight: 0.3 g; hydroperoxide: 5 mmol; solvent: 100 ml; epoxidation time: 1 h; epoxidation temperature: 100 °C; propylene pressure: 8 bar.
[b] 1-Octene epoxidation conditions: catalyst weight: 0.3 g; hydroperoxide: 5 mmol; 1-octene: 100 ml; epoxidation time: 1 h; epoxidation temperature: 110 °C.

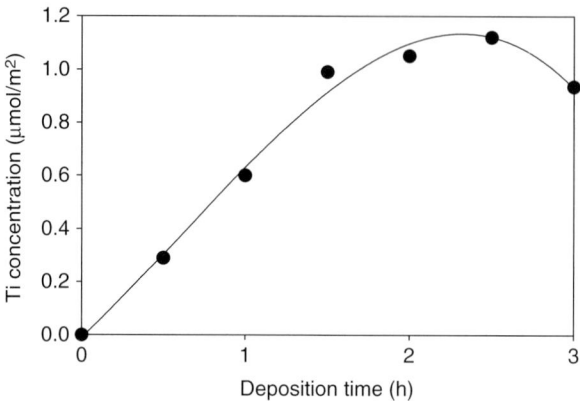

FIGURE 14.2 Titanium concentration as a function of deposition time for Ti/SiO$_2$ catalysts.

SiOTiCl$_3$ + HCl [5]) at the deposition time of 2.5 h because this Ti/SiO$_2$ sample contained 1.12 μmol Ti/m^2. That is, the results in Fig. 14.2 suggest that the silica surface was saturated with titanium for the Ti/SiO$_2$ sample prepared at 2.5 h.

For the Ti/SiO$_2$ catalyst prepared at 2.5 h (containing a Ti concentration of 1.12 μmol/m^2), Table 14.1 indicates that EBHP gave lower PO yield and PO selectivity than CHP and TBHP. PO selectivity was 90% for EBHP, and was around 97% for TBHP and CHP. This might be due to the fact that the EBHP used here was laboratory prepared from the autoxidation of ethylbenzene, while the TBHP and CHP used were high purity commercial grades. Impurities (including water, acetophenone, and 1-methylbenzyl alcohol) in the laboratory-prepared EBHP might slightly affect the PO selectivity.

3.2. Packed-bed reactor data

Packed-bed reactor operation was used to test the stability of the CVD-prepared Ti/SiO$_2$ catalysts. For simplification, the reaction used was the epoxidation of 1-octene with TBHP. It is much easier and simpler to carry out 1-octene epoxidation (than propylene epoxidation) in a packed-bed reactor because 1-octene (boiling point, b.p. = 122 °C) is a liquid under the reaction temperature (110 °C). No decay of 1,2-epoxyoctane yield was observed during 12 h of continuous operation with the packed-bed reactor, therefore, the CVD-prepared Ti/SiO$_2$ catalysts were stable under continuous operation conditions.

For the packed-bed reactor operation, Fig. 14.3 and row 5 of Table 14.1 show TBHP conversion, 1,2-epoxyoctane yield, and selectivity versus TiCl$_4$ deposition time. Batch reactor data obtained for the reaction between 1-octene and TBHP are shown in row 6 of Table 14.1. Both the packed-bed reactor data and the batch reactor data show that epoxide yield and epoxide selectivity increased with the increase in deposition time. The maximum 1,2-epoxyoctane selectivity obtained with the packed-bed reactor was 98.5% (for the catalyst prepared with 2.5 h deposition time), which was slightly higher than that (selectivity = 94.5%)

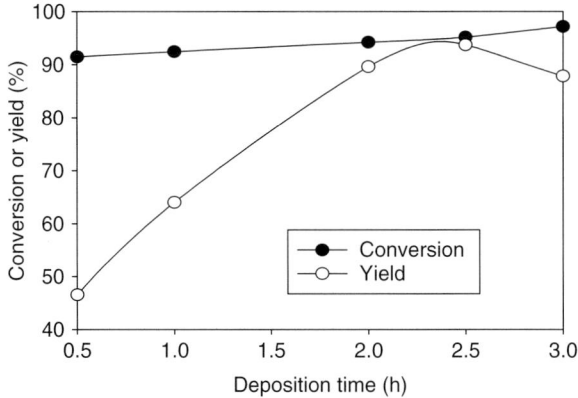

FIGURE 14.3 Influence of TiCl$_4$ deposition time on *tert*-butyl hydroperoxide (TBHP) conversion and 1,2-epoxyoctane yield for Ti/SiO$_2$ catalysts in a packed-bed reactor (catalyst: 0.1 g, 1-octene flow rate = 1 ml/h, reaction temperature = 110 °C, TBHP concentration: 0.05 mol/liter).

obtained with the batch reactor. Similar to the EBHP data presented in Fig. 14.1, Fig. 14.3 also shows that deposition time had little effect on TBHP conversion.

It is interesting to note that the shape of epoxide yield curves in Figs. 14.1 and 14.3 is similar to the shape of the Ti concentration curve in Fig. 14.2. The epoxide yield, epoxide selectivity (presented in Table 14.1), and titanium concentration all increased with increasing deposition time, and all reached maximum values at the deposition time of 2.5 h. The maximum epoxide selectivity obtained for the catalyst prepared at 2.5 h should be mainly due to the fact that it had the maximum Ti concentration and therefore the minimum residual silanol concentration. It is known that the hydroxyl group in SiOH may behave as a Bronsted acid site (SiOH + B = SiO$^-$ + HB$^+$) [12]. These Bronsted acid sites can catalyze the decomposition of hydroperoxide (an undesired reaction) and decrease the selectivity to epoxide [13]. The Bronsted acid catalyzed hydroperoxide decomposition may involve electrophilic proton attack on either hydroperoxide oxygen, which results in formation of H$_2$O$_2$ or H$_2$O and the corresponding carbenium cation [14,15]. The decrease in Ti concentration (shown in Fig. 14.2) and the decrease in epoxide selectivity (shown in Table 14.1) at 3 h deposition time might be due to the diffusion of internal OH groups to the surface which resulted in the breakage of the surface SiO–Ti bonds.

It was proposed that there are two types of titanium species in Ti/SiO$_2$ catalysts [10]: tetrahedrally coordinated titanium and octahedrally coordinated titanium. The former had higher epoxide selectivity than the latter, and the percentage of the former increased with increasing Ti content. Therefore, the highest epoxide selectivity observed for the Ti/SiO$_2$ catalyst prepared with 2.5 h should be partly because it had the largest percentage of tetrahedrally coordinated titanium.

Table 14.1 also presents the turnover frequencies (TOF) of the CVD-prepared Ti/SiO$_2$ catalysts. For 1-octene epoxidation with TBHP in a batch reactor at 110 °C and 1 h, the calculated TOF ranging from 73 to 170 moles of epoxide product per mole of Ti per hour for the CVD-prepared Ti/SiO$_2$ catalysts with Ti contents ranging from 0.275 to 0.863 wt.%. For Ti/SiO$_2$ catalysts prepared with impregnation method and contained 0.4 and 4 wt.% Ti, the reported TOF under similar reaction conditions (1-octene epoxidation with TBHP at 107 °C and 1 h) were 47 and 120 moles of epoxide product per mole of Ti per hour, respectively [16]. Therefore, the activities of our CVD-prepared Ti/SiO$_2$ catalysts were similar to those of the Ti/SiO$_2$ catalysts prepared with impregnation method.

3.3. Ti/MCM-41 and Ti/MCM-48 catalysts

Five Ti/MCM-41 and four Ti/MCM-48 catalysts were prepared by chemical deposition of TiCl$_4$ on MCM-41 and MCM-48 of different temperatures in the range of 700–900 °C with 3 h deposition time. These CVD-prepared samples were used to catalyze the reaction between propylene and EBHP. Figures 14.4 and 14.5 show EBHP conversion and PO yield as a function of the TiCl$_4$ deposition temperature for Ti/MCM-41 and Ti/MCM-48, respectively. Table 14.2 shows PO selectivity as a function of CVD temperature for Ti/MCM-41 and Ti/MCM-48 with three different oxidants (EBHP, TBHP, and CHP). All of the EBHP conversions shown in Figs. 14.4 and 14.5 are between 90 and 99%, which indicate that the catalyst activity was not sensitive to the change of the deposition temperature, and was also not sensitive to the change of the supports. In Fig. 14.4, the significant decrease in the conversion for the catalyst prepared at the deposition temperature of 900 °C (about 4% less than that at 850 °C) should be due to the rapid decrease in titanium content and the decrease in surface area at this high temperature. For both Ti/MCM-41 and Ti/MCM-48 catalysts, PO yield (shown in Figs. 14.4 and 14.5) and PO selectivity (shown in Table 14.2) increased rapidly with the increase in the deposition temperature, and both catalysts reached a maximum PO yield (94.3% for Ti/MCM-41 and 88.8% for Ti/MCM-48) and a maximum PO selectivity (97% for Ti/MCM-41 and 90–91% for Ti/MCM-48) at the deposition temperature of 800 °C. Ti/MCM-41 catalysts showed slightly higher PO selectivity than the

TABLE 14.2 Effects of deposition temperature (T_d), oxidants, and supports on propylene oxide selectivity (%)

Catalyst	Oxidant	T_d (°C)		
		750	800	850
Ti/MCM-41	EBHP	96	97	95
Ti/MCM-41	TBHP	84	92	89
Ti/MCM-41	CHP	84	92	81
Ti/MCM-48	EBHP	86	90	82
Ti/MCM-48	TBHP	83	91	89
Ti/MCM-48	CHP	85	91	75

EBHP, ethylbenzene hydroperoxide; TBHP, tert-butyl hydroperoxide; CHP, cumene hydroperoxide.

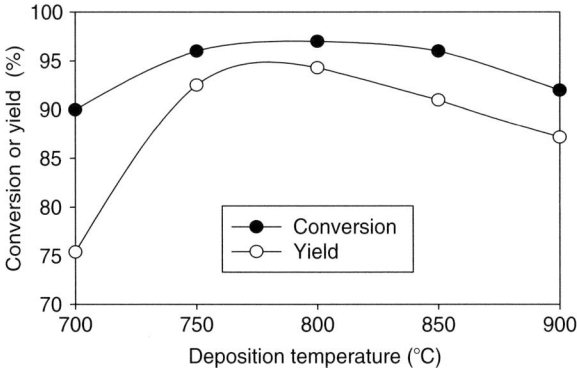

FIGURE 14.4 Influence of TiCl$_4$ deposition temperature on ethylbenzene hydroperoxide (EBHP) conversion and propylene oxide yield for Ti/MCM-41 catalysts.

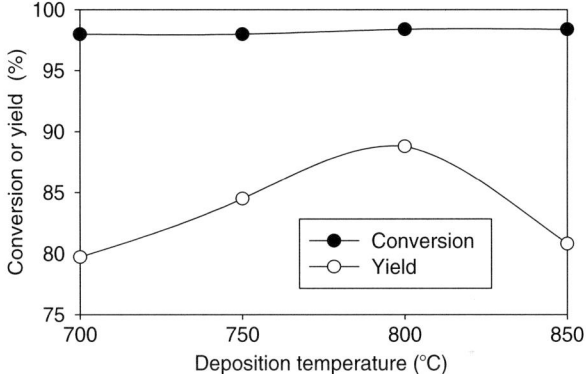

FIGURE 14.5 Influence of TiCl$_4$ deposition temperature on ethylbenzene hydroperoxide (EBHP) conversion and propylene oxide (PO) yield for Ti/MCM-48 catalysts.

corresponding Ti/MCM-48, which should be due to the higher Ti concentration and therefore the lower residual silanol groups for Ti/MCM-41. It might be also due to the fact that Ti/MCM-41 had higher concentration of tetrahedrally coordinated titanium.

Table 14.3 shows titanium concentration as function of deposition temperature for Ti/MCM-41 and Ti/MCM-48 catalysts. At the same deposition temperature, Ti/MCM-41 had higher Ti concentration than Ti/MCM-48. Figure 14.6 shows Ti 2p XPS spectra and Fig. 14.7 shows infrared spectra of Ti/MCM-41 and Ti/MCM-48 catalysts prepared at the deposition temperature of 800 °C. In Fig. 14.7, the band due to the stretching vibration of SiO units bonded to Ti atoms is observed at the wavenumber of around 960 cm^{-1} [17] in the IR spectra. Figures 14.6 and 14.7 also indicate that Ti/MCM-41 had higher Ti concentration than Ti/MCM-48.

The Ti concentration difference observed in Table 14.3 and in Figs. 14.6 and 14.7 might be caused by the difference in structure between MCM-41 and

TABLE 14.3 Titanium concentration (C_{Ti}) and surface area (S_a) as a function of deposition temperature

	Temperature (°C)		
	750	800	850
MCM-41	$C_{Ti} = 0.53\ \mu mol/m^2$ $S_a = 768\ m^2/g$	$C_{Ti} = 0.68\ \mu mol/m^2$ $S_a = 558\ m^2/g$	$C_{Ti} = 0.84\ \mu mol/m^2$ $S_a = 356\ m^2/g$
MCM-48	$C_{Ti} = 0.51\ \mu mol/m^2$ $S_a = 731\ m^2/g$	$C_{Ti} = 0.54\ \mu mol/m^2$ $S_a = 673\ m^2/g$	$C_{Ti} = 0.59\ \mu mol/m^2$ $S_a = 548\ m^2/g$

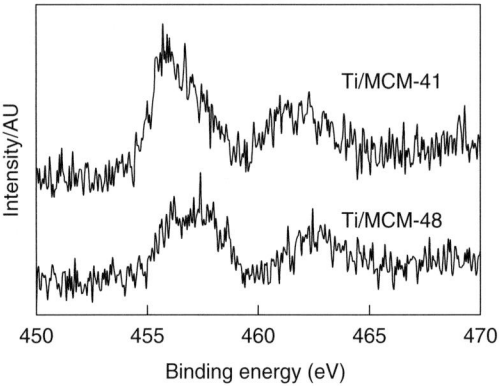

FIGURE 14.6 Ti 2p X-ray photoelectron spectroscopy (XPS) spectra obtained for Ti/MCM-41 and Ti/MCM-48 samples prepared at 800 °C.

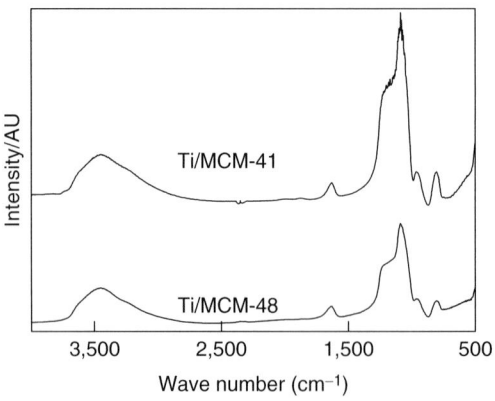

FIGURE 14.7 Infrared spectra of Ti/MCM-41 and Ti/MCM-48 samples prepared at 800 °C.

MCM-48. It is well known that the effective diffusion coefficient (D_e) of a component A in a porous pellet is proportional to the pellet porosity (ϵ), and is inversely proportional to pore tortuosity (τ) [18]. That is,

$$D_e = D_{KA}\varepsilon/\tau \tag{14.2}$$

D_{KA} in Eq. (14.2) is the Knudsen diffusivity, which is proportional to pore radius (R_{pore}) [19]:

$$D_{KA} = 9{,}700 R_{pore}(T/M_A)^{1/2} \text{cm}^2/\text{s} \tag{14.3}$$

where R_{pore} is the pore radius in centimeter, T is the absolute temperature in Kelvins, and M_A is the molecular weight of component A (i.e., $TiCl_4$). Pore size distribution measurements (results shown in Fig. 14.8) indicated that pore radius and pore volume (i.e., porosity) of MCM-48 were smaller than those of MCM-41. In addition, the tortuosity of MCM-48 is larger than that of MCM-41 because MCM-48 has a significantly more complex structure than the straightforward case of hexagonal MCM-41. It is known that MCM-41 has regular one-dimensional, hexagonal array of uniform channels with each pore surrounded by six neighbors; MCM-48 has a cubic pore system, which is indexed in the space group Ia3d [20]. Therefore, the effective diffusion coefficient (D_e) of $TiCl_4$ in MCM-48 should be lower than that in MCM-41 because of the smaller pore volume, smaller pore radius, and larger tortuosity of MCM-48 (compared to MCM-41). It is more difficult for $TiCl_4$ molecules to arrive at the interior silanol (SiOH) sites of MCM-48 for the occurrence of the deposition reaction, which resulted in the lower Ti concentration in the Ti/MCM-48 sample (compared to Ti/MCM-41), as observed in Table 14.3. In Table 14.3, the difference of Ti concentration between Ti/MCM-41 and Ti/MCM-48 increased with the increase in CVD temperature, which indicates that diffusion resistance became more important at the higher CVD temperature. This should be caused by the fact that the intrinsic rate constant

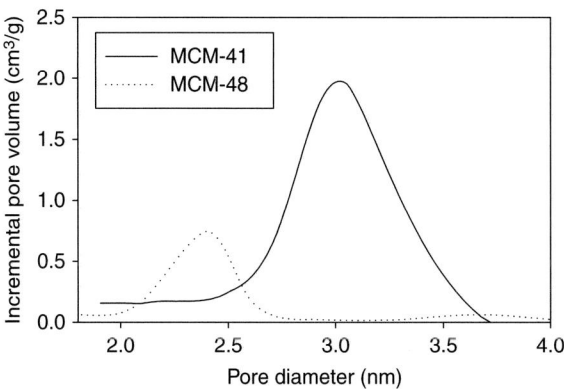

FIGURE 14.8 Pore size distributions of MCM-41 and MCM-48 materials.

for the reaction between TiCl$_4$ and SiOH is more temperature sensitive than the effective diffusion coefficient.

3.4. Kinetics of propylene epoxidation with EBHP

The Ti/MCM-41 catalyst prepared at the deposition temperature of 800 °C was used for a kinetic study. The amount of EBHP used was 5 mmol and the amount of propylene used was in large excess. Kinetic measurements were carried out over a temperature of 70–100 °C and a reaction time of 1–3 h with 0.03 g catalyst. The EBHP conversion obtained was in the range of 10–70%.

To determine reaction rate parameters from the experimental data, the following differential equation was used to describe the reaction system in a constant-volume batch reactor assuming a pseudo-first-order equation for propylene epoxidation:

$$-\ln(1 - X) = kt \tag{14.4}$$

where X is the conversion of EBHP.

Experimental results were plotted according to Eq. (14.4), and straight lines passing through zero fit the experimental points quite well, as illustrated in Fig. 14.9. Therefore, propylene epoxidation with EBHP can be treated as pseudo-first order with respect to EBHP concentration (C_A). The rate equation can therefore be written as

$$-r_A = kC_A \tag{14.5}$$

An Arrhenius plot k indicated that the frequency factor, A, and the activation energy, E, were 4.5×10^5 h^{-1} and 42.9 kJ/mol, respectively. The calculated activation energy for the reaction between propylene and EBHP is slightly higher than

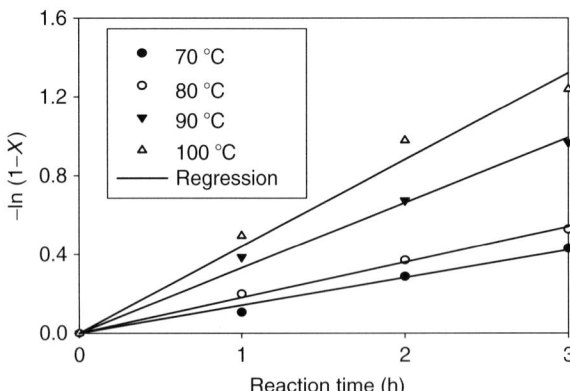

FIGURE 14.9 Test of pseudo-first-order kinetic model for propylene epoxidation with ethylbenzene hydroperoxide (EBHP) over Ti/MCM-41.

that (36.9 kJ/mol) obtained for the reaction of propylene with CHP, which suggests that the propylene epoxidation rate with EBHP was more temperature sensitive than that with CHP.

4. CONCLUSIONS

A series of titanium-based epoxidation catalysts were prepared by the CVD of $TiCl_4$ vapor on silica gel (SiO_2) and on silicious mesoporous molecular sieves (MCM-41 and MCM-48). The catalysts were used to catalyze the epoxidation of propylene and 1-octene with organic hydroperoxides. The deposition time and temperature affected the catalyst performance significantly. The optimum deposition temperature was 800 °C for Ti/MCM-41 and Ti/MCM-48 (with 3 h deposition time). The optimum deposition time was 2.5 h for Ti/SiO_2 at 900 °C. The best yield of PO obtained with the optimum Ti/MCM-41 catalyst was over 94%. The Ti/MCM-41 catalysts had higher Ti concentration than the corresponding Ti/MCM-48 catalysts, which was ascribed to the smaller pore size/volume and tortuous pore structure of MCM-48 (compared to the easier diffusion of $TiCl_4$ molecules in the one-dimensional, uniform pores of MCM-41). Kinetic studies indicated that the propylene epoxidation on the Ti/MCM-41 catalyst was first order in EBHP with an activation energy of 42.9 kJ/mol. Packed-bed reactor data suggest that the CVD-prepared catalysts were stable under continuous operation conditions. The CVD-prepared Ti/SiO_2 catalysts had TOF similar to those of impregnation prepared Ti/SiO_2 catalysts.

ACKNOWLEDGMENT

We gratefully acknowledge the National Science Council of the Republic of China for financial support (Grant No. NSC-90–2214-E-029–002).

REFERENCES

[1] D. L. Trent, Propylene oxide, in: J. I. Kroschwitz, M. Howe-Grant (Eds.), Encyclopedia of Chemical Technology, Vol. 20, Wiley, New York, 1996, pp. 271–302.
[2] A. H. Tullo, P. L. Short, Propylene oxide routes take off, *Chem. Eng. News* 86(41) (2006) 22–23.
[3] M. Mccoy, New routes to propylene oxide, *Chem. Eng. News* 79(43) (2001) 19.
[4] Y. Z. Han, E. Marales, R. G. Gastinger, K. M. Carroll, Heterogeneous epoxidation catalyst, U.S. Patent 6,114,552 *Assigned to Arco Chemical Technology* 2000.
[5] K. T. Li, I. C. Chen, Epoxidation of propylene on Ti/SiO_2 catalysts prepared by chemical vapor deposition, *Ind. Eng. Chem. Res.* 41 (2002) 4028.
[6] K. T. Li, C. C. Lin, Propylene epoxidation over Ti/MCM-41 catalysts prepared by chemical vapor deposition, *Catal. Today* 97 (2004) 257.
[7] K. T. Li, P. H. Lin, S. W. Lin, Preparation of Ti/SiO_2 catalyst by chemical vapor deposition method for olefin epoxidation with cumene hydroperoxide, *Appl. Catal. A: Gen.* 301 (2006) 59–65.
[8] D. Das, C. M. Tsai, S. Cheng, Improvement of hydrothermal stability of MCM-41 mesoporous molecular sieve, *Chem. Commum.* (1999) 473.
[9] S. Wang, D. Wu, Y. Sun, B. Zhong, The synthesis of MCM-48 with high yields, *Mater. Res. Bull.* 36 (2001) 1717.

[10] G. Blanco-Brieva, M. C. Capel-Sanchez, J. M. Campos-Martin, J. L. G. Fierro, Effect of precursor nature on the behavior of titanium-polysiloxane homogeneous catalysts in primary alkene epoxidation, *J. Mol. Catal. A* 269 (2007) 133.
[11] K. Tanabe, M. Misono, Y. Ono, H. Hattori, *New Solid Acids and Bases*, Elsevier, Amsterdam, 1989.
[12] H. H. Kung, *Transition Metal Oxides: Surface Chemistry and Catalysis*, Elsevier, Amsterdam, 1989.
[13] R. A. Sheldon, Synthetic and mechanistic aspects of metal-catalyzed epoxidations with hydroperoxides, *J. Mol. Catal.* 7 (1980) 107.
[14] M. S. Kharasch, J. G. Burt, The chemistry of hydroperoxides. VIII. The acid-catalyzed decomposition of certain hydroperoxides, *J. Org. Chem.* 16 (1951) 150.
[15] M. Stojanova, C. Karshalykov, G. L. Price, V. Kanazirev, On the reactivity of H-, Ga-, and Cu-MFI zeolites towards t-butylhydroperoxide (TBHP), *Appl. Catal. A.* 143 (1996) 175.
[16] D. E. De Vos, B. F. Sels, P. A. Jacobs, Practical heterogeneous catalysis for epoxide production, *Adv. Synth. Catal.* 345 (2003) 457.
[17] A. Thangaraj, R. Kumar, S. P. Mirajkar, P. Ratnasamy, Catalytic properties of crystalline titanium silicates. 1. Synthesis and characterization of titanium-rich zeolites with MFI structure, *J. Catal.* 130 (1991) 1.
[18] H. S. Fogler, *Elements of Chemical Reaction Engineering* 4th ed, Prentice Hall, Upper Saddle River, NJ, 2005, p. 815.
[19] M. E. Davis, R. J. Davis, *Fundamentals of Chemical Reaction Engineering,* McGraw-Hill, NY, 2003, p. 191.
[20] J. C. Vartuli, C. T. Kresge, W. J. Roth, S. B. McCullen, J. S. Beck, K. D. Schmitt, M. E. Leonowicz, J. D. Lutner, E. W. Sheppard, Designed synthesis of mesoporous molecular sieve systems using surfactant directing agents, in: W. R. Moser(Ed.), Advanced Catalysts and Nanostructured Materials: Modern Synthetic Methods, Academic Press, San Diego, 1996, pp. 1–19.

CHAPTER 15

Metal Species Supported on Organic Polymers as Catalysts for the Epoxidation of Alkenes

Ulrich Arnold

Contents		
	1. Introduction	388
	2. Supported Manganese Catalysts	389
	2.1. Supported manganese-salen complexes	389
	2.2. Supported manganese-porphyrin complexes	395
	3. Supported Molybdenum Catalysts	396
	4. Supported Ruthenium and Iron Catalysts	398
	5. Supported Titanium Catalysts	399
	6. Supported Tungsten Catalysts	400
	7. Supported Rhenium Catalysts	401
	8. Supported Cobalt, Nickel, and Platinum Catalysts	402
	9. Supported BINOL-Complexes of Lanthanoids and Calcium	402
	10. Conclusion	403
	Acknowledgment	407
	References	407

Abstract   Recent developments in the field of immobilized metal catalysts for liquid-phase alkene epoxidation are reviewed. Progress since 2000 is summarized considering organic polymers as supports with a focus on polymer types, catalyst preparation, and performance rather than physicochemical parameters of the modified polymers. A broad variety of metals such as manganese, molybdenum, titanium, ruthenium, iron, tungsten, rhenium, cobalt, nickel, platinum, lanthanum, ytterbium, and calcium are considered and several immobilization strategies are described. Recent advances comprise new catalyst systems for asymmetric heterogeneous or homogeneous

Department of Chemical Engineering (ITC-CPV), Forschungszentrum Karlsruhe GmbH, Hermann-von-Helmholtz-Platz 1, D-76344 Eggenstein-Leopoldshafen, Germany

epoxidation, new polymeric supports, and new polymer-based epoxidation catalysts with high long-term activities in the range of months.

Key Words: Polymer-supported catalysts, Manganese(II) complexes, Manganese-salen complexes, Manganese-porphyrin complexes, Molybdenum catalysts, Ruthenium porphyrins, Ruthenium(III) complexes, Iron catalysts, Titanium catalysts, Sharpless epoxidation, Tungsten catalysts, Methyltrioxorhenium, Cobalt, Nickel, Platinum, Aerobic epoxidation, Lanthanum, Ytterbium, Calcium, BINOL-complexes.   © 2008 Elsevier B.V.

1. INTRODUCTION

Immobilization of catalytically active species on suitable supports is one crucial approach to highly active, selective, and recyclable catalyst systems. Organic polymers have been widely explored as supports during the last decades and various strategies for the attachment of metal species have been developed [1–6]. A broad variety of low-cost monomers together with a series of well-established polymerization techniques as well as polymers that can easily be modified and adjusted to the needed requirements makes such catalyst systems easily accessible and renders this concept highly attractive. Many polymer-supported catalyst systems exhibit performances similar or even superior to those of comparable nonsupported catalysts and various immobilized catalysts have been described that have proven to be easily separable from the reaction mixtures and recyclable in principle. However, one should take into account that a polymer-supported metal catalyst that shows an excellent initial performance is, at least from a technological standpoint, worthless if it deactivates after some cycles. Despite intensive research in this area, there is still a lack of highly stable catalyst systems for long-term applications, for example, in a continuously operating process and detailed data on the long-term performance of polymer-supported catalysts exceeding periods of a few hours or days are mostly unavailable. High catalyst durability is as desirable as easy accessibility, high activity and selectivity, and implies thermal, mechanical and chemical resistance as well as resistance to metal leaching from the polymer matrix, catalyst poisoning, and, particularly in the case of oxidation catalysis, resistance to oxidative attack. The development of adequate catalyst systems is certainly an intriguing challenge and a variety of techniques is available to meet the above-mentioned criteria.

Numerous attempts have been made to develop polymer-supported epoxidation catalysts and major reviews on heterogeneous liquid-phase epoxidation catalysts in general have been published [7,8]. Concerning polymer-supported metal complex epoxidation catalysts in particular, a comprehensive review appeared in 2000 [9]. In the meantime, several reviews have appeared that cover some aspects such as advances in heterogeneous asymmetric epoxidation [10–14], new methods for the recycling of chiral catalysts [15], hybrid organic-inorganic catalysts [16], applications of catalysts on soluble supports [17], polymerizable transition metal complexes [18], immobilized catalysts for industrial application [19], or supported metalloporphyrin complexes [20].

Here a compact survey on recent developments in the field of polymer-supported metal species for catalytic epoxidation is given. Progress since 2000 is considered focusing on catalyst preparation, catalytic performance, and catalyst recyclability rather than physicochemical properties of the polymers.

2. SUPPORTED MANGANESE CATALYSTS

The field of polymer-supported manganese complexes is dominated by supported manganese(III)-salen complexes for asymmetric epoxidation and supported manganese(III) porphyrin catalysts. Besides these immobilized manganese(III) species, some attempts have been made to attach manganese(II) complexes on organic polymers. Polymers containing 1,4,7-triazacyclononane moieties have been prepared by ring-opening metathesis polymerization (ROMP) of norbornene attached to the azacycle [21]. The manganese-loaded polymers were tested as catalysts in the epoxidation of numerous alkenes using H_2O_2 as oxidant and high activity under mild reaction conditions comparable to or even superior than that of the monomeric complex was observed.

In another approach, β-diketonate complexes of various metals were attached to a divinyl benzene (DVB) cross-linked polystyrene [22] and a DVB-methyl methacrylate copolymer [23]. Their catalytic performances were evaluated in the epoxidation of several alkenes with H_2O_2. Best activities were observed using manganese(II)-containing systems and the catalysts were shown to be recyclable. Manganese(II) Schiff base complexes supported on a styrene-DVB copolymer were also investigated but showed low and decreasing activity upon recycling in the epoxidation of cyclooctene and norbornene with *tert*-butyl hydroperoxide (TBHP) [24]. Similar activities were reported on a styrene-DVB copolymer modified with L-valine and loaded with manganese(II) acetate [25].

The main strategies of supporting manganese epoxidation catalysts on organic polymers are outlined in Table 15.1. These techniques are given as examples and have also been employed for several other metal catalysts aside from manganese.

2.1. Supported manganese-salen complexes

Between 2000 and 2005, several major reviews on asymmetric organic synthesis were published [10–14] which also covered some advances in the dynamic field of polymer-immobilized manganese-salen complexes. In 2000, immobilization of Jacobsen's epoxidation catalyst [26] on polystyrene and polymethacrylate resins was reported [27]. Catalytic performances were evaluated using 1,2-dihydronaphthalene, indene, 1-phenyl-3,4-dihydronaphthalene and 1-phenylcyclohexene as substrates, and *m*-chloroperbenzoic acid (*m*-CPBA) and *N*-methylmorpholine-*N*-oxide (NMO) as oxidant/co-oxidant. Epoxide yields up to 61% and ee values up to 91%

TABLE 15.1 Polymer-supported manganese species for catalytic epoxidation

Immobilization strategy[a]	References	Immobilization strategy[a]	References
	[21]		[29]
	[22,23]		[33]

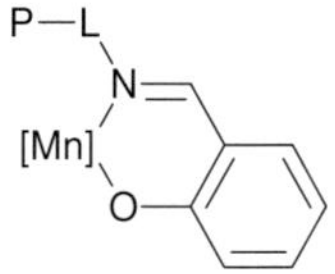

[24]

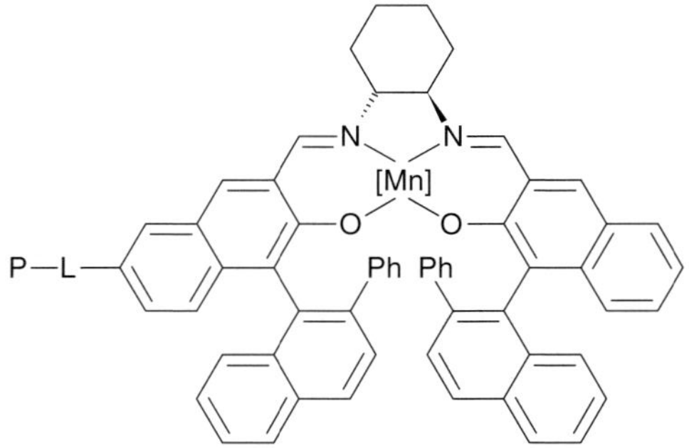

[25]

[27,28,32,36]

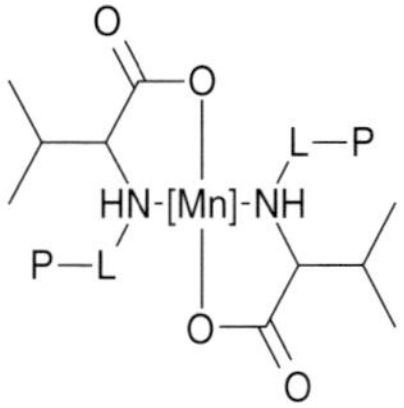

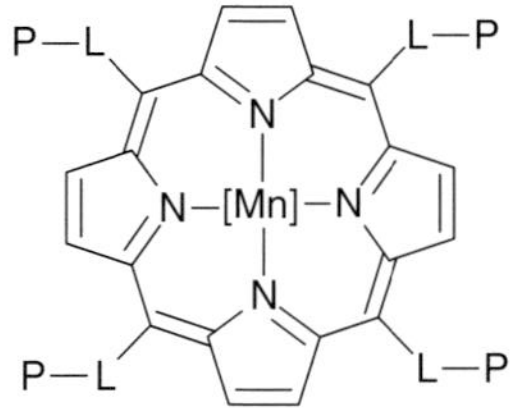

[40,41]

[43,47–49,51,53]

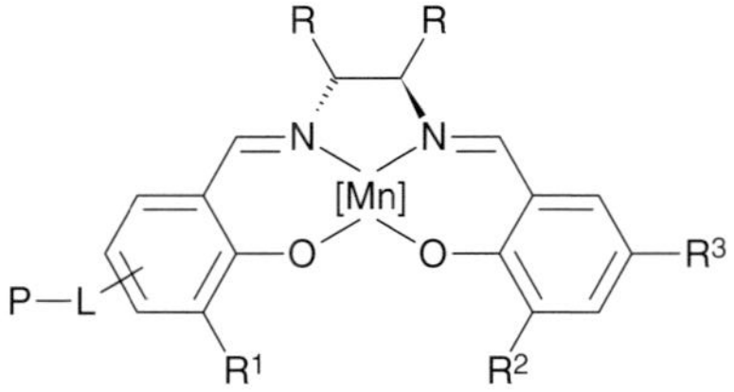

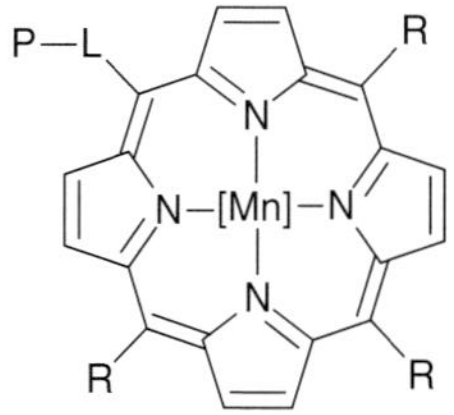

[44–46,51,52]

(continued)

TABLE 15.1 (continued)

Immobilization strategy[a]	References	Immobilization strategy[a]	References
(Mn-salen complex with P–L linker at para positions)	[30,31,34,37,42]	(Mn-bromoporphyrin with P–L–imidazole axial ligand)	[50]
(Mn-salen complex with axial L–P ligand and R1–R4 substituents)	[35,38,39]		

[a] L, linker; P, polymer.

were reached in the epoxidation of 1-phenylcyclohexene. The catalysts showed significant loss of activity and enantioselectivity upon recycling and reuse.

Also in 2000, attachment of the Jacobsen catalyst to polymeric supports such as poly(ethylene glycol) and different polystyrene-based resins through a glutarate spacer was described [28]. Soluble as well as insoluble polymer-bound complexes were employed as catalysts in the epoxidation of styrene, cis-2-methylstyrene, and dihydronaphthalene with m-CPBA/NMO. Results were similar to those achieved with the nonsupported catalyst. Catalyst recycling was shown to be possible either by filtration or by precipitation and one catalyst system could be used for three cycles without significant loss of activity and enantioselectivity.

High enantioselectivity in the epoxidation of chromenes by NaClO/4-phenyl-pyridine-N-oxide (PPNO) or m-CPBA/NMO has been achieved by use of a salen-type catalyst with a chiral pyrrolidine backbone [29]. The manganese complex was attached via a glutaric linker to a hybrid resin of low cross-linked polystyrene and poly(ethylene glycol) in which the polymer chains were terminally functionalized with an amino group (NovaSyn® TG amino resin). The supported catalyst showed a comparable performance to that of its homogeneous analogue. However, partial decomposition of the catalyst under epoxidation conditions was reported.

Dendritic and nondendritic polystyrene-bound manganese-salen complexes were described by Seebach and coworkers [30]. The supported catalysts were prepared by suspension copolymerization of styrene with the vinyl-substituted complexes and employed in the epoxidation of phenyl-substituted alkenes by m-CPBA/NMO. Activities and selectivities were similar to those obtained with the monomeric complexes. High catalyst stabilities were observed and it was demonstrated that the immobilized catalysts can be recycled up to 10 times without loss of performance. Laser ablation inductively coupled plasma mass spectrometry was used to monitor the manganese content in repeatedly used polystyrene beads and a correlation between metal leaching from the support and catalytic activity was disclosed [31].

Copolymerization of an acryloyl-substituted manganese-salen complex with ethylene glycol dimethacrylate and styrene [32] yielded catalysts that showed low enantioselectivity in the epoxidation of styrene by m-CPBA/NMO or iodosylbenzene (PhIO). The influence of porosity and cross-linking degree on activity was explored and the catalysts could be used repeatedly with no loss of activity for at least three cycles.

Since 2005 several new catalysts based on manganese(III)-salen complexes have been described. A supported salen-type catalyst derived from 2,3-diamino-D-glucose and anchored to a CHO-functionalized Wang resin by acetalization was reported [33]. Four different oxidants, namely, m-CPBA/NMO, H_2O_2, NaClO, and (nBu$_4$N)HSO$_5$, were used for the epoxidation of cis-2-methylstyrene and a conversion of 99% with 80% ee was achieved using m-CPBA/NMO. However, significant metal leaching from the polymer matrix was observed, which prevented catalyst recycling.

Gothelf and coworkers described chiral manganese-salen-bridged polymers obtained by condensation of a trialdehyde with chiral diamines in the presence of Mn(OAc)$_2$ [34]. The polymers were tested as catalysts in the epoxidation of cis-2-methylstyrene with m-CPBA/NMO and conversions of up to 84% with high diastereoselectivity and enantioselectivities of up to 67% ee were observed. The catalysts were reused up to six times without decrease in activity or selectivity.

Axial bonding of manganese-salen complexes to polystyrene containing phenoxide or sulfonate groups yielded catalysts for the asymmetric epoxidation of styrene and styrene derivatives with NaClO/PPNO [35]. Catalyst performances were similar to the corresponding nonsupported systems and epoxide yields reached 93% with 70% ee using 1-phenylcyclohexene as substrate. Catalyst recyclability was investigated and the catalysts could be used up to three times in the epoxidation of α-methylstyrene.

A modular approach for the development of supported salen catalysts, similar to the strategy employed for supported triazacyclononane catalysts [21], was reported [36]. For this purpose, a manganese-salen complex attached to a norbornene monomer was synthesized and polymerized by ring-opening metathesis polymerization. The resulting polymers and copolymers showed high catalytic activity and enantioselectivity using m-CPBA/NMO as oxidant but a significant decline of activity and selectivity was observed upon catalyst recycling in the epoxidation of 1,2-dihydronaphthalene.

An achiral manganese-salen complex modified with phosphonium groups at the 5 and 5' positions of the salen ligand was supported on a commercially available ion-exchange resin (Dowex MSC-1) via ionic bonding and was used for the epoxidation of various alkenes with $NaIO_4$ as oxidant [37]. The effect of different oxidants such as $KHSO_5$, H_2O_2, H_2O_2/urea, NaClO, TBHP, and Bu_4NIO_4 was studied and $NaIO_4$ was shown to be the most efficient. The same group also used imidazole- [38] and 1,4-phenylenediamine-modified polystyrene [39] as supports for achiral Mn(salophen)Cl. Improved selectivities, epoxide yields, and stabilities compared with the Mn(salen)-Dowex catalyst system were reported.

In 2006, Smith et al. gave a full account of a formerly described unsymmetrical Katsuki-type manganese-salen complex bearing two binaphthyl units in the ligand and attached to polystyrene via one binaphthyl group by an ester link [40,41]. The polymers were shown to be highly enantioselective and recoverable catalysts for the epoxidation of 1,2-dihydronaphthalene by NaClO/PPNO. Enantioselectivity remained high in up to six consecutive runs using the recycled catalyst.

Immobilization of a sulfonated chiral manganese-salen catalyst on a functionalized Merrifield resin yielded a remarkably active epoxidation catalyst [42]. Its activity and enantioselectivity was examined by epoxidation of 6-cyanochromene, indene, styrene, 4-methylstyrene, and trans-stilbene using m-CPBA/NMO and quantitative yields were obtained in less than 5 min. Enantioselectivities were between 33% (4-methylstyrene) and 96% ee (6-cyanochromene). The same complex was also supported on silica and a layered double hydroxide (LDH) and the catalytic performances of the systems were compared. Recycling experiments were carried out and the silica-based system showed metal leaching combined with a significant decrease in yield and ee. The layered double hydroxide- and resin-catalysts exhibited a slight decrease in activity and constant ee values in five consecutive reactions.

2.2. Supported manganese-porphyrin complexes

Alkene epoxidation catalyzed by manganese porphyrins is an intensively investigated research area and several supported catalysts have been described since 2000. Manganese(III) complexes of tetracationic and tetraanionic porphyrins have been supported on countercharged ion-exchange resins (Dowex MSC-1 and Amberlyst A-27) and surface-modified silica supports [43]. Performances of the supported catalysts and analogous uncharged homogeneous systems have been investigated in the epoxidation of cyclooctene and (E)- and (Z)-4-methylpent-2-ene with PhIO. The catalysts were shown to be recyclable and cyclooctene epoxide yields from 85 up to 100% were reached in 10 consecutive reactions using the Amberlyst-supported catalyst. The supported catalysts can be superior to their homogeneous analogues and very high stereoretention in the epoxidation of 4-methylpent-2-ene was observed.

Covalently anchored manganese porphyrins were obtained via reaction of a hydroxyphenyl-modified porphyrin with an Argogel chloride [polystyrene-poly(ethylene glycol) copolymer] and a chloromethylated Merrifield resin, respectively, followed by treatment of the polymers with $MnCl_2$ [44]. The systems were tested in the epoxidation of a series of alkenes with imidazole as axial ligand for the metalloporphyrin and $NaIO_4$ as oxidant. Yields varied between 22% (1-dodecene) and 98% (cyclooctene). Reuse of the catalysts showed a superior performance of the Merrifield resin and decomposition of the Argogel system. Two years later, an extension of this study was reported [45]. Polystyrene resins (Merrifield as well as Wang resins) and the same immobilization strategy were employed. Dienes were chosen as substrates and the highest activities and stabilities were obtained using a carboxy-functionalized Wang resin. Recently, the same authors reported manganese-porphyrin catalysts tethered to a Wang resin via different peptide linkers [46]. Limonene was used as substrate and it was found that a peptide linker incorporating histidine could act as axial ligand via the nitrogen donor atom of the imidazole group leading to good chemoselectivity and improved stability. The catalyst could be used in a second run without loss of activity.

Immobilization of a sulfonated manganese octabromoporphyrin derivative on an ion-exchange resin (Amberlite IRA-400) yielded an active catalyst for the epoxidation of various alkenes and hydroxylation of alkanes with $NaIO_4$ and imidazole as additional ligand [47]. Shortly thereafter, the same authors described a covalently immobilized manganese porphyrin obtained via reaction of manganese tetra(4-aminophenyl) porphyrin with poly(4-styrylmethyl-acylchloride) [48]. Its catalytic performance was very similar to the Amberlite-based system. Catalyst recycling experiments using styrene as substrate showed only a slight decrease in activity over four consecutive reactions. A similar concept was employed by reacting manganese tetra(4-pyridyl) porphyrin with chloromethylated styrene-DVB copolymer [49]. Compared with the previously reported systems, significantly higher conversions of a series of alkenes and higher selectivities were observed. Further improvement in terms of catalyst activity and reusability was recently reported [50]. A manganese octabromoporphyrin complex was attached to imidazole-modified polystyrene via coordinative bonding, as described earlier for the immobilization of a manganese salophen complex [38]. Ultrasonic irradiation was shown to enhance the activity of the catalyst and led to shorter reaction times and higher product yields. Catalyst recyclability was tested in the epoxidation of cyclooctene and only a slight decrease in conversion was observed over four runs.

Poly(ethylene glycol)-supported manganese porphyrins were tested in the epoxidation of cyclooctene, 1-dodecene, cyclohexene, styrene, and indene with PhIO or H_2O_2 in the presence of N-alkylimidazoles as axial ligands [51]. The polymers were soluble in the reaction mixtures and could be precipitated and reused. Epoxide yields from 80 to 100%, except for 1-dodecene (38% yield), were obtained using PhIO as oxidant.

Catalysts similar to those described by de Miguel and Brulé [44] were employed for the epoxidation of cholest-5-ene derivatives with PhIO and imidazoles as additional ligands [52]. Yields of up to 93% and high diastereoselectivity (β/α isomer ratios of >99%) have been reported.

A remarkable approach was reported in 2004 by Simonneaux and coworkers [53]. Manganese complexes of spirobifluorenyl-substituted porphyrins were electropolymerized by anodic oxidation and the resulting poly(9,9'-spirobifluorene manganese porphyrin) films were shown to be efficient epoxidation catalysts in the presence of imidazole. The polymers were tested in the epoxidation of cyclooctene and styrene using PhIO or $PhI(OAc)_2$ as oxidants. Epoxide yield reached 95% in the case of cyclooctene and 77% in the case of styrene. The electrosynthesized polymers could be recovered by filtration and reused up to eight times without loss of activity and selectivity.

Encapsulation of manganese-porphyrin complexes in a polystyrene matrix by physical interaction was described [54] rendering the catalysts highly dispersible in organic solvents. Different oxidants, such as $NaIO_4$, $KHSO_5$, and NaClO, were compared using imidazole or pyridine as axial ligands. Conversions of up to 99% were obtained in the epoxidation of styrene and α-methylstyrene using $NaIO_4$. The catalysts were found to be stable and could be recycled at least two times without loss of activity.

3. SUPPORTED MOLYBDENUM CATALYSTS

Among some early attempts to develop epoxidation catalysts based on polymer-supported molybdenum species, the work of Sherrington *et al.* is outstanding. Highly active and recyclable catalysts based on polybenzimidazole and polyimides have been reported [9]. In the meantime, supported catalysts based on commercially available ion-exchange resins [55,56] and a Merrifield resin [57] have been described. The latter has been functionalized by reacting the chloromethylated polystyrene with deprotonated 2-(3-pyrazolyl)pyridine. The resulting polymer was loaded with oxodiperoxo molybdenum(VI) species via coordinative bonding to the chelating ligand and the system was shown to be a recyclable catalyst in the epoxidation of cyclooctene with TBHP as oxidant.

Cross-linked poly(4-vinylpyridine-co-styrene) was synthesized by radical polymerization and varying amounts of DVB were added as cross-linking agent [58]. The polymers were loaded with molybdenyl acetylacetonate [$MoO_2(acac)_2$] and tested as cyclohexene epoxidation catalysts with TBHP. The effect of the degree of cross-linking was investigated and the highest activity was observed for a system with a medium cross-linking degree of 4%.

In another approach, benzimidazole-functionalized dendrons were used as supports for molybdenum species [59]. Metal loading was carried out by treatment with $Mo(CO)_6$ or $MoO_2(acac)_2$ and the dendritic complexes were used as catalysts for the epoxidation of cyclohexene with TBHP. Reactions were shown to be heterogeneously catalyzed and recyclability of the catalysts was demonstrated.

Suspension polycondensation of pyromellitic dianhydride and 3,5-diamino-1,2,4-triazole yielded triazole-containing polyimide beads that were used as a support for $MoO_2(acac)_2$ [60]. The resulting catalyst showed high activity and selectivity in the epoxidation of cyclohexene and cycloctene as well as in the epoxidation of noncyclic alkenes such as styrene, 1-octene, and 1-decene with TBHP. The catalyst could be recycled 10 times and activity decreased significantly in the case of 1-octene epoxidation whereas activity remained high in the epoxidation of cyclic alkenes.

A series of polymer-anchored epoxidation catalysts was obtained by modifying Merrifield resin with imidazole [61], diphosphines [62], or piperazine [63] followed by treatment with UV-activated $Mo(CO)_6$. High activities in the epoxidation of cyclic (cyclooctene, cyclohexene, indene, and α-pinene) as well as linear alkenes (styrene, α-methylstyrene, 1-heptene, 1-dodecene, cis- and trans-stilbene) were observed using TBHP as oxidant. The catalysts were recovered and reused up to 10 times in the epoxidation of cyclooctene without loss of activity.

Recently, thermosetting epoxy resins such as the tetraglycidyl derivative of 4,4'-methylenedianiline or the triglycidyl derivative of 4-aminophenol were introduced as matrices for catalytically active metal species [64,65]. Molybdenyl acetylacetonate, molybdenum ethoxide, or 2-ethylhexanoate were used as initiators for anionic polymerization of resin monomers and these initiators acted simultaneously as precursors for catalytically active species in the polymerized materials. Thus, a series of epoxidation catalysts could be obtained in a convenient time- and cost-saving one-step procedure by simple heating of resin/initiator mixtures. The catalytic performance of these metal-doped thermosets was evaluated in the epoxidation of cyclohexene, 1- and 2-octene, styrene, (R)-(+)-limonene, and 1,2-dihydronaphthalene using TBHP as oxidant. Catalyst recycling tests with repeated use in up to 120 reactions revealed unprecedented long-term activities over periods of months and catalyst lifetimes of years can be expected. Metal leaching was investigated by metal enrichment techniques combined with sensitive atomic spectroscopy. Metal losses of the catalysts were extremely low but mechanistic investigations suggest a superposition of heterogeneous and homogeneous catalysis. Organic–inorganic hybrid catalysts can be prepared by adding inorganic components to the liquid resins and subsequent polymerization. Their properties can be easily controlled by various parameters, for example, choice of resin, initiator and filling material, ratio of the components, and polymerization conditions.

Up to now, only a few catalyst systems based on organic polymers such as molybdenum compounds supported on benzimidazole, polystyrene, or poly(glycidyl methacrylate) [9] as well as micelle-incorporated manganese-porphyrin catalysts [66] have been tested in the epoxidation of propene. Molybdenum-doped epoxy resins were also employed in the epoxidation of propene with TBHP and propene oxide yields of up to 88% were obtained [65]. The catalysts were employed repeatedly in up to 10 reactions without significant loss of activity and metal leaching proved to be very low.

4. SUPPORTED RUTHENIUM AND IRON CATALYSTS

Ruthenium porphyrins are predominantly employed in the field of ruthenium-catalyzed epoxidation and several attempts have been made to gain efficient catalysts by immobilization of ruthenium porphyrins onto organic polymers. In 2000, Che and coworkers described a carbonyl ruthenium(II) porphyrin covalently attached to a Merrifield resin that efficiently catalyzes epoxidation of a wide variety of alkenes with 2,6-dichloropyridine-N-oxide (Cl$_2$pyNO) [67]. Complete diastereoselectivity in the epoxidation of a glycal and a protected α-amino alkene was reported. Catalyst reusability was investigated in up to nine reactions using styrene as substrate, and epoxide yield remained stable after a drop in the second run. The same porphyrin system was attached to poly(ethylene glycol), and soluble polymer-supported ruthenium catalysts for epoxidation, cyclopropanation, and aziridination of alkenes were obtained [68]. To evaluate activity and selectivity, numerous alkenes including electron-deficient chalcone were epoxidized and the catalytic performance was comparable to the ruthenium porphyrin supported on Merrifield resin. The catalyst was reused five times and its activity decreased only slightly upon recycling.

Copolymerization of a vinyl-substituted carbonyl ruthenium(II) porphyrin with ethylene glycol dimethacrylate yielded an efficient catalyst for the epoxidation of several alkenes with Cl$_2$pyNO [69]. Styrene conversions of >99% were reached and a decrease in activity of around 15% was found upon recycling of the catalyst by filtration. The same strategy was employed to synthesize chiral polymer-supported porphyrin complexes of ruthenium and iron for asymmetric epoxidation [70]. Metalloporphyrins bearing chiral vinyl-substituted octahydrodimethanoanthracene moieties were copolymerized with styrene using DVB or ethylene glycol dimethacrylate as cross-linking agents. The resulting polymers were tested as catalysts in the asymmetric epoxidation of styrenes by Cl$_2$pyNO and yields of up to 72% with up to 74% ee were observed in the epoxidation of nonsubstituted styrene. A recycling test comprising three reactions showed decreasing yields from 70 to 19% and a slight decrease in enantioselectivity from 71 to 64%. Using an analogous polymeric iron porphyrin as catalyst, significantly lower yields and ee values compared with ruthenium porphyrins were obtained. Furthermore, the monomeric iron complex showed better performance than the polymeric counterpart.

Apart from polymer-supported ruthenium(II) porphyrins, immobilized ruthenium(III) complexes were investigated as epoxidation catalysts. Chloromethylated poly(styrene-co-DVB) was reacted with 2-aminopyridine [71] or L-valine [72], thus anchoring chelating moieties, and metal loading of the polymers was carried out by treatment with $RuCl_3$. Epoxide yield reached 51% in cyclooctene epoxidation with TBHP and low epoxide selectivity was observed using styrene and cyclohexene as substrates. Polymer-anchored L-valine was also used to immobilize copper(II). The polymers were tested as catalysts for asymmetric epoxidation of terminal olefins with m-CPBA and moderate ee values up to 32% were reached in the epoxidation of 1-octene [73]. The same group functionalized poly(styrene-co-DVB) with Schiff base ligands and tested iron(III)-loaded polymers as catalysts for cyclooctene and styrene epoxidation by TBHP [74]. Low epoxide yields and decreasing activity upon recycling were observed.

5. SUPPORTED TITANIUM CATALYSTS

Since 2000 several attempts have been made to develop efficient polymer-supported titanium catalysts comprising immobilized titanates, soluble polymer-bound Sharpless catalysts for asymmetric epoxidation and inorganic–organic hybrid systems. Macroporous supports containing N-(p-hydroxyphenyl) and N-(3,4-dihydroxybenzyl)maleimide have been prepared by suspension copolymerization of N-(p-acetoxyphenyl) or N-(piperonyl)maleimides with styrene and DVB [75]. Subsequent deprotection of hydroxyl groups and treatment with Ti(OiPr)$_4$ or TiCl$_4$ yielded titanium-loaded polymers that were tested as cyclohexene epoxidation catalysts with TBHP as oxidant. The system based on N-(p-hydroxyphenyl)maleimide and Ti(OiPr)$_4$ was shown to be more active than nonsupported Ti(OiPr)$_4$ and epoxide yields up to 90% were observed. Recycling experiments revealed a drop of activity after the first run but stable performance in the following three runs. Significant leaching of about 20% of the initially loaded titanate was observed.

A very similar methodology has been employed for the preparation of a titanium-loaded poly(p-hydroxystyrene) [76]. Copolymerization of p-acetoxystyrene with styrene and varying amounts of DVB yielded resins with different cross-link levels. Hydroxyl deprotection and reaction with Ti(OiPr)$_4$ gave recyclable catalysts that were tested in the epoxidation of cyclohexene, styrene, and 1-dodecene with TBHP. Conversions of linear alkenes were significantly lower compared with cyclohexene epoxidation. Activities were higher than those observed with nonsupported Ti(OiPr)$_4$ and decreased with increasing cross-linking degree of the support.

Soluble polymer-supported Sharpless epoxidation catalysts were obtained using substituted tartrate ligands [77,78]. Esterification of L-(+)-tartaric acid with poly(ethylene glycol) monomethylether (MPEG) and various alcohols yielded a series of ligands for the asymmetric epoxidation of several allylic alcohols using Ti(OiPr)$_4$/TBHP. Moderate epoxide yields and good ee values were observed, significantly depending on substituents and the Ti/ligand molar ratio.

Enantioselectivity decreased upon ligand recycling, but ligand recovery by simple precipitation and filtration facilitated product isolation. Surprisingly, a tartrate ester ligand prepared from L-(+)-tartaric acid and MPEG with a molecular weight of 2,000 gave an enantioselectivity contrary to that of L-(+)-diethyltartrate or a L-(+)-tartrate ligand prepared from MPEG with a lower molecular weight. However, these results could not be reproduced and were thoroughly reinvestigated by Janda and coworkers [79]. It was shown that enantioselectivity of the reaction can be reversed as a function of the molecular weight of the attached achiral MPEG. Enantioreversal was explained by the occurrence of two different titanium–ligand complexes, $Ti_2(Tartrate)(OiPr)_6$ and $Ti_2(Tartrate)_2(OiPr)_4$. Formation of the latter could be suppressed by long MPEG chains and formation of the former was favored by MPEG tartrates with lower molecular weights. These findings could offer a new approach for the control of asymmetric reactions by means of achiral appended polymers.

Very recently, inorganic–organic hybrid catalyst systems were developed by copolymerization of styrene, DVB, and a vinyl-substituted titanium silsesquioxane on mesoporous silica SBA-15 [80]. The materials exhibited advantageous properties, for example, large specific surface areas and pore volumes, high accessibility of active sites, and a hydrophobic environment around the active centers. They behaved as interfacial catalysts due to the hydrophobicity of the organic component and the hydrophilicity of the inorganic component. In two-phase epoxidation of cyclooctene with aqueous H_2O_2, the heterogeneous hybrid catalysts showed much higher activity than the homogeneous counterpart and epoxide yields up to 70% were obtained. Catalyst recycling tests showed a significant decrease in activity after the first run using H_2O_2 but activity was maintained over eight consecutive reactions by using TBHP as oxidant.

6. SUPPORTED TUNGSTEN CATALYSTS

Several polymer-supported tungsten catalysts have been reported that showed good performances in epoxidation reactions with aqueous H_2O_2 as oxidant. Polyglycidylmethacrylate resins have been modified by amino and ammonium groups using two different strategies [81]: the epoxide ring was opened directly by reaction with trimethylamine or by reaction with methyliodide followed by amination. Subsequently, the aminoalkyl derivatives were quaternized or phosphorylated to afford grafted phosphotriamides. The polymers were loaded with peroxotungstic species by treatment with peroxotungstic acid $H_2W_2O_{11}$, peroxotungstate $HW_2O_{11}^-$, phosphoperoxo tungstate $PW_4O_{24}^{3-}$, or phosphonoperoxo tungstate $C_6H_5-PO_3(WO_5)_2^{2-}$. The catalyst systems were tested in the epoxidation of cyclohexene and yields up to 85% were obtained. Polystyrene-supported phosphine oxide, phosphonamide and phosphoramide ligands as well as polybenzimidazole and polymethacrylate-based phosphoryl ligands were also compared [82]. Loading with peroxotungstic species yielded a series of catalysts that were evaluated in the epoxidation of cyclohexene and activities higher than those of analogous soluble catalysts were reached. Recyclability of some systems was demonstrated in up to five consecutive reactions and no significant loss of activity was observed.

Electrostatic immobilization of the Venturello anion $PW_4O_{24}^{3-}$ [83] on anion exchange resin Amberlite IRA-900 was described [84] and the tungsten-loaded polymer showed good activity in the epoxidation of several cyclic alkenes including the terpenic substrates γ-terpinene and terpinolene [85]. Catalyst recycling tests with cyclooctene as substrate revealed that the catalyst completely maintains activity over three consecutive reactions. Reaction mixtures exhibited no catalytic activity after catalyst removal by filtration, suggesting that the catalytically active species do not leach from the resin.

Recently, a series of polymer-anchored tungsten carbonyl catalysts based on modified polystyrenes was prepared [86]. Polymer modification was carried out by reaction of chloromethylated polystyrene (2% cross-linked with DVB) with diphosphines, di- and triamines, pyrazine, 4,4'-bipyridine, and imidazole. The polymers were treated with $W(CO)_6(THF)$ and their catalytic performance was evaluated in the epoxidation of cyclooctene. Different solvents and oxidants were tested and epoxide yields up to 98% were obtained using the system CH_3CN/H_2O_2. A detailed catalyst recycling study was carried out and the catalyst containing 4,4'-bipyridine units kept constant activity over 10 reactions whereas other catalysts revealed deactivation.

7. SUPPORTED RHENIUM CATALYSTS

Recent work in the field of polymer-based rhenium catalysts concentrated on the immobilization of methyltrioxorhenium (MTO) to gain highly efficient and recoverable catalysts for the epoxidation of alkenes with H_2O_2. Poly(4-vinylpyridine), poly(4-vinylpyridine-N-oxide), and unmodified polystyrene, all cross-linked with DVB, were used as supports and the MTO-loaded polymers showed high catalytic activity in the epoxidation of cyclohexene, cyclooctene, styrene, α-methylstyrene, and trans-stilbene under mild reaction conditions [87]. Catalyst recycling tests in the epoxidation of cyclohexene showed constant activity in five successive runs except for microencapsulated MTO with polystyrene. The same catalysts were used for the epoxidation of monoterpenes such as geraniol, nerol, S-(+)-carene, (+)-α-pinene, and R-(+)-limonene [88]. Very recently, these MTO-loaded polymers were shown to be efficient catalysts for the domino epoxidation-methanolysis of various glycals with high facial diastereoselectivity [89].

Various nitrogen-functionalized polymers, for example, aminated polystyrene and polyacrylate resins as well as copolymers of 4-vinylpyridine, butyl- or methylmethacrylate, and ethylene glycol dimethacrylate have been used as supports for MTO [90]. N-oxidation of tertiary amine and pyridine groups was carried out by treatment with H_2O_2 and the oxidized supports were compared with their nonoxidized counterparts. Epoxidation of α-pinene was investigated and metal leaching from oxidized supports was found to be higher than from nonoxidized supports. Low conversions were reported using polystyrene, polyacrylate, or polyvinylpyridine supports. Improved conversions but significant formation of campholenic aldehyde were observed using copolymer supports.

8. SUPPORTED COBALT, NICKEL, AND PLATINUM CATALYSTS

Several polymer-bound salts and complexes of cobalt, nickel, and platinum have been shown to be efficient catalysts for the aerobic epoxidation of alkenes with an aldehyde as coreactant. Polyaniline, poly-o-toluidine, and poly-o-anisidine have been used as supports for cobalt(II) acetate and cobalt(II)-salen and the metal-loaded polymers catalyzed the epoxidation of *trans*-stilbene by isobutyraldehyde/O_2 under mild reaction conditions affording the epoxide up to a yield of 93% [91]. The same supports have been used for bis(8-hydroxyquinoline)cobalt(II) and the immobilized catalyst was used in the epoxidation of α-pinene, limonene, and 1-decene. High conversions were obtained and the maximum epoxide selectivity in the case of α-pinene was 59% [92]. Metal leaching from the polymers was observed but no catalyst recycling tests were reported. A derivative of 5-(2-pyridylmethylidene)-hydantoin was anchored on polystyrene and used as bidentate ligand for cobalt(II) [93]. The catalyst was tested in the epoxidation of cyclohexene, norbornene, and 1-heptene with PhIO or H_2O_2 as oxidants. Higher epoxide yields compared with reactions catalyzed by the nonsupported complex were observed but no data concerning catalyst recycling were reported.

Besides pyridine-containing polystyrene and polypropylene resins, polybenzimidazole has been employed as support for nickel(II) acetylacetonate [94]. The nickel-loaded polymer was shown to be an efficient catalyst for the epoxidation of (S)-(−)-limonene, α-pinene, and 1-octene using isobutyraldehyde/O_2 as coreactant/oxidant. However, significant metal leaching from the support associated with a loss of activity upon recycling was reported. It was shown that the reaction is heterogeneously catalyzed, and leached metal species did not contribute to the catalytic activity.

With respect to polymer-bound platinum catalysts, a dinuclear amidate-bridged platinum(III) complex was supported on polyvinylpyridine [95] and tested as catalyst in the epoxidation of cyclohexene and styrene by isobutyraldehyde/O_2. Its catalytic activity was found to be significantly lower compared to the same complex immobilized on inorganic supports.

9. SUPPORTED BINOL-COMPLEXES OF LANTHANOIDS AND CALCIUM

Since 2000 a few catalysts for asymmetric epoxidation based on polymer-anchored chiral 1,1'-bi-2-naphthol (BINOL) have been developed. Polystyrene-supported BINOL was prepared by radical copolymerization of styrene with BINOL, bearing 4-vinylbenzyloxy groups in the 3- or 6-position [96]. Immobilization of lanthanum or ytterbium was accomplished by treatment of the polymers with La(OiPr)$_3$ or Yb(OiPr)$_3$. The lanthanoid-loaded polymers were employed as catalysts for the enantioselective epoxidation of the α,β-unsaturated ketones chalcone and benzalacetone using cumene hydroperoxide (CMHP) or TBHP as oxidants. Epoxidation of chalcone catalyzed by a polymer-supported lanthanum complex afforded the epoxide up to a yield of 96% with 98% ee. Epoxide yields up to 90% with 88% ee were obtained in the epoxidation of benzalacetone using a supported ytterbium–BINOL complex. A significant decrease in activity but constant ee values were observed upon catalyst recovery and reuse in three consecutive reactions.

In another approach, epoxidation of chalcone and substituted chalcones by TBHP was catalyzed by a calcium–BINOL complex bearing a poly(ethylene glycol) chain in 6-position [97]. Chalcone oxide yields between 92 and 95% were obtained with ee values ranging from 40 to 47% using a catalyst loading of 10 mol%. The polymeric catalyst is soluble in the reaction mixtures and can be precipitated and recovered. Reuse in three consecutive reactions showed a continuous decrease in activity and enantioselectivity.

10. CONCLUSION

Progress in the field of polymer-supported epoxidation catalysts since the year 2000 was surveyed. Significant advances in the development of catalyst systems for asymmetric epoxidation, new catalyst preparation techniques, as well as catalysts with high long-term activity are apparent. Some representative catalyst systems and their most important features are summarized in Table 15.2.

Considering asymmetric epoxidation, catalytic performances of supported manganese complexes with salen-type ligands could be improved and concepts such as immobilized Katsuki-type complexes or complexes with new chiral backbones have emerged. In this context, conjugated salen-cross-linked polymers, obtained by condensation of a trialdehyde with chiral diamines in the presence of a manganese salt, are a promising approach toward active, selective, and stable catalyst systems for multiple use. In the field of supported Sharpless-epoxidation catalysts, discovery of enantioreversal as a function of the molecular weight of polymer-modified tartrate ligands could offer a new approach for control of asymmetric reactions.

With respect to the widely investigated metalloporphyrins for catalytic epoxidation, progress was made in the area of polymer-supported ruthenium porphyrins for asymmetric epoxidation. Manganese-porphyrin complexes attached via peptide linkers to organic polymers showed enhanced selectivity and catalyst stability due to donor atoms in the linker that could coordinate to the metal center. This shows that improvement can be achieved not only by optimization of the polymer or metal complex but also by appropriate choice of the linker. Furthermore, electropolymerization by anodic oxidation of suitable manganese-porphyrin complexes proved to be a promising technique for the preparation of efficient immobilized epoxidation catalysts.

TABLE 15.2 Catalytic performances of polymer-supported epoxidation catalysts

Catalyst	Alkene	Oxidant	Time (h)	T (°C)	Yield (%)	TON[a]	ee (%)	Runs[b]	Reference
Mn (triazacyclononane)-polynorbornene	Styrene	H_2O_2	3	0	80	80	—	1	[21]
	Cyclohexene	H_2O_2	3	0	90	90	—		
Mn(salen)-MPEG	Styrene	m-CPBA/NMO	<0.25	−78	82	20	52	3	[28]
Mn(salen)-polystyrene	Styrene	m-CPBA/NMO	0.5	−20	100	5	62	10	[30]
Mn(salen)-cross-linked triethynylbenzene	cis-2-Methylstyrene	m-CPBA/NMO	2	0	84	6	64	6	[34]
Mn(salen)-phenoxy-polystyrene	Styrene	NaClO/PPNO	24	0	76	153	60	3	[35]
Mn(salen)-polynorbornene	Styrene	m-CPBA/NMO	0.083	−20	100	25	33	3	[36]
Mn(sulfonato-salen)-polystyrene	Styrene	m-CPBA/NMO	<0.083	−20	100	100	48	5	[42]
Mn(porphyrin)-polystyrene	Styrene	$NaIO_4$	12	25	86	20	—	3	[44]
Mn(porphyrin)-peptide-polystyrene	Limonene	$NaIO_4$	36	25	85	20	—	2	[46]
Mn(bromo-porphyrin)-imidazole-polystyrene	Styrene	$NaIO_4$	2	25	96	80	—	4	[50]

Catalyst	Substrate	Oxidant							Ref.
Mn(porphyrin)-poly(ethylene glycol)	Styrene	PhIO	1.5	25	80	400	—	7	[51]
Electropolymerized Mn(9,9′-spirobifluor- enylporphyrin)	Cyclohexene Styrene	PhIO PhI(OAc)$_2$	18 1.5	25 25	82 77	410 77	— —	8	[53]
Mn(porphyrin)- polystyrene- microcapsules	Styrene	NaIO$_4$	3.5	88	95	190	—	3	[54]
Mo(CO)$_5$-piperazine- polystyrene	Styrene	TBHP	10	76	92	84	—	8	[63]
Mo-doped epoxy resins	Cyclohexene Styrene	TBHP TBHP	4.5 24	76 90	100 99	91 127	— —	120	[65]
Ru(porphyrin)- polystyrene	Cyclohexene Propene Styrene	TBHP TBHP Cl$_2$pyNO	24 24 24	90 90 25	95 75 96	121 240 1344	— — —	9	[67]
Ru(porphyrin)-poly (ethylene glycol dimethacrylate)	Cyclohexene Styrene	Cl$_2$pyNO Cl$_2$pyNO	24 24	25 25	66 >99	924 500	— —	2	[69]
Chiral Ru(porphyrin)- polystyrene	Styrene	Cl$_2$pyNO	24	25	72	238	72	3	[70]
Ti(OiPr)$_4$-MPEG- tartrate	trans-Hex-2-en- 1-ol	TBHP	8	−20	84	17	92	4	[78]

(continued)

TABLE 15.2 (continued)

Catalyst	Alkene	Oxidant	Time (h)	T (°C)	Yield (%)	TON[a]	ee (%)	Runs[b]	Reference
Ti(silsesquioxane)-polystyrene-SBA-15	Cyclooctene	H_2O_2	17	60	70	175	—	8	[80]
(WO$_5$)-polymethacrylate	Cyclohexene	TBHP	17	60	88	227	—	1	[81]
		H_2O_2	0.75	70	85	172	—		
(PW$_4$O$_{24}^{3-}$)-Amberlite	Cyclohexene	H_2O_2	24	32	51	167	—	3	[85]
W(CO)$_m$-polystyrene	Cyclooctene	H_2O_2	1.5	82	96	27	—	10	[86]
MeReO$_3$-poly(4-vinylpyridine)	Styrene	H_2O_2	6.5	25	>96	7	—	5	[87]
Co(8-hydroxyquinoline)-polyaniline	Cyclohexene	H_2O_2	1.5	25	89	6	—	1	[92]
	Limonene	O_2	1	65	86	126	—		
Ni(acac)$_2$-polybenzimidazole	Styrene	O_2	4	25	23	23	—	2	[94]
Pt(amidate)-Poly(4-vinylpyridine)	Limonene	O_2	4	25	74	74	—	1	[95]
	Styrene	O_2	14	25	27	111	—		
La(BINOL)-polystyrene	Cyclohexene	O_2	8	25	12	51	—	3	[96]
	Chalcone	CMHP	20	25	96	19	98		

CPBA, chloroperbenzoic acid; NMO, N-methylmorpholine-N-oxide; PPNO, 4-phenylpyridine-N-oxide; TBHP, tert-butyl hydroperoxide; BINOL, 1,1′-bi-2-naphthol; CMHP, cumene hydroperoxide.
[a] TON, turnover number (moles of epoxide per moles of catalyst).
[b] Maximum number of reactions (initial run + recycling experiments) carried out with the catalyst systems.

Catalysts with unprecedented long-term activities in the range of at least some months were obtained by use of thermosetting epoxy resins as supports. The catalysts were prepared in a convenient one-step procedure employing metal complexes that act simultaneously as polymerization initiators as well as precursors for catalytically active species in the resulting polymers. With respect to polymeric supports, it should be pointed out that about 65% of the reviewed catalyst systems are based on polystyrene and ~15% are derived from polymethacrylate and poly(ethylene glycol). The latter is predominantly used for the preparation of soluble polymer-supported catalysts, a research area that has gained increasing attention in recent years. Taking into account that about 80% of the reported epoxidation catalysts are based on only three polymer types it can be assumed that there is still a great potential for catalyst systems based on other polymers. Promising results were already obtained using polynorbornene, polybenzimidazole, polyimide, or thermosetting resins as supports. Furthermore, hybrid catalyst systems comprising organic polymers together with inorganic components, such as recently reported titanium catalysts based on polystyrene-silica or molybdenum catalysts supported on epoxy resin-silica composites, offer new possibilities for the development of efficient polymer-supported epoxidation catalysts.

ACKNOWLEDGMENT

Financial support from the "Bundesministerium für Bildung und Forschung" (BMBF) is gratefully acknowledged.

REFERENCES

[1] B. M. L. Dioos, I. F. J. Vankelecom, P. A. Jacobs, Aspects of immobilisation of catalysts on polymeric supports, *Adv. Synth. Catal.* 348 (2006) 1413.
[2] S. Kobayashi, R. Akiyama, Renaissance of immobilized catalysts. New types of polymer-supported catalysts, 'microencapsulated catalysts', which enable environmentally benign and powerful high-throughput organic synthesis, *Chem. Commun.* (2003) 449.
[3] N. E. Leadbeater, M. Marco, Preparation of polymer-supported ligands and metal complexes for use in catalysis, *Chem. Rev.* 102 (2002) 3217.
[4] B. Clapham, T. S. Reger, K. D. Janda, Polymer-supported catalysis in synthetic organic chemistry, *Tetrahedron* 57 (2001) 4637.
[5] D. C. Sherrington, Polymer-supported reagents, catalysts, and sorbents: Evolution and exploitation—A personalized view, *J. Polym. Sci. Part A: Polym. Chem.* 39 (2001) 2364.
[6] Y. R. de Miguel, E. Brulé, R. G. Margue, Supported catalysts and their applications in synthetic organic chemistry, *J. Chem. Soc., Perkin Trans.* 1 (2001) 3085.
[7] M. Dusi, T. Mallat, A. Baiker, Epoxidation of functionalized olefins over solid catalysts, *Catal. Rev. Sci. Eng.* 42 (2000) 213.
[8] D. E. De Vos, B. F. Sels, P. A. Jacobs, Practical heterogeneous catalysts for epoxide production, *Adv. Synth. Catal.* 345 (2003) 457.
[9] D. C. Sherrington, Polymer-supported metal complex alkene epoxidation catalysts, *Catal. Today* 57 (2000) 87.
[10] P. K. Dhal, B. B. De, S. Sivaram, Polymeric metal complex catalyzed enantioselective epoxidation of olefins, *J. Mol. Catal. A* 177 (2001) 71.

[11] W. Sun, C. G. Xia, Application of chiral metal-salen complexes in asymmetric catalysis, *Prog. Chem.* 14 (2002) 8.
[12] Q. H. Fan, Y. M. Li, A. S. C. Chan, Recoverable catalysts for asymmetric organic synthesis, *Chem. Rev.* 102 (2002) 3385.
[13] S. Bräse, F. Lauterwasser, R. E. Ziegert, Recent advances in asymmetric C–C and C-heteroatom bond forming reactions using polymer-bound catalysts, *Adv. Synth. Catal.* 345 (2003) 869.
[14] Q. H. Xia, H. Q. Ge, C. P. Ye, Z. M. Liu, K. X. Su, Advances in homogeneous and heterogeneous catalytic asymmetric epoxidation, *Chem. Rev.* 105 (2005) 1603.
[15] U. Kragl, T. Dwars, The development of new methods for the recycling of chiral catalysts, *Trends Biotechnol.* 19 (2001) 442.
[16] M. H. Valkenberg, W. F. Hölderich, Preparation and use of hybrid organic-inorganic catalysts, *Catal. Rev. Sci. Eng.* 44 (2002) 321.
[17] D. E. Bergbreiter, Applications of catalysts on soluble supports, *Top. Curr. Chem.* 242 (2004) 113.
[18] P. Mastrorilli, C. F. Nobile, Supported catalysts from polymerizable transition metal complexes, *Coord. Chem. Rev.* 248 (2004) 377.
[19] N. End, K. U. Schöning, Immobilized catalysts in industrial research and application, *Top. Curr. Chem.* 242 (2004) 241.
[20] E. Brulé, Y. R. de Miguel, Supported metalloporphyrin catalysts for alkene epoxidation, *Org. Biomol. Chem.* 4 (2006) 599.
[21] A. Grenz, S. Ceccarelli, C. Bolm, Synthesis and application of novel catalytically active polymers containing 1,4,7-triazacyclononanes, *Chem. Commun.* (2001) 1726.
[22] V. A. Nair, K. Sreekumar, Polymer supported catalysts for epoxidation reactions, *Curr. Sci.* 81 (2001) 194.
[23] V. A. Nair, K. Sreekumar, Poly(methyl methacrylate) supported β-diketone linked metal complexes: Heterogeneous epoxidation catalysts, *J. Polym. Mater.* 19 (2002) 155.
[24] S. A. Patel, S. Sinha, A. N. Mishra, B. V. Kamath, R. N. Ram, Olefin epoxidation catalysed by Mn(II) Schiff base complex in heterogenised-homogeneous systems, *J. Mol. Catal. A* 192 (2003) 53.
[25] V. B. Valodkar, G. L. Tembe, M. Ravindranathan, R. N. Ram, H. S. Rama, Synthesis, characterization, and catalytic activity of polymer anchored amino acid Mn(II) complexes, *J. Macromol. Sci.-Pure Appl. Chem. A* 41 (2004) 839.
[26] W. Zhang, J. L. Loebach, S. R. Wilson, E. N. Jacobsen, Enantioselective epoxidation of unfunctionalized olefins catalyzed by (salen)manganese complexes, *J. Am. Chem. Soc.* 112 (1990) 2801.
[27] L. Canali, E. Cowan, H. Deleuze, C. L. Gibson, D. C. Sherrington, Polystyrene and polymethacrylate resin-supported Jacobsen's alkene epoxidation catalyst, *J. Chem. Soc., Perkin Trans.* 1 (2000) 2055.
[28] T. S. Reger, K. D. Janda, Polymer-supported (salen)Mn catalysts for asymmetric epoxidation: A comparison between soluble and insoluble matrices, *J. Am. Chem. Soc.* 122 (2000) 6929.
[29] C. E. Song, E. J. Roh, B. M. Yu, D. Y. Chi, S. C. Kim, K. J. Lee, Heterogeneous asymmetric epoxidation of alkenes catalysed by a polymer-bound (pyrrolidine salen)manganese(III) complex, *Chem. Commun.* (2000) 615.
[30] H. Sellner, J. K. Karjalainen, D. Seebach, Preparation of dendritic and non-dendritic styryl-substituted salens for cross-linking suspension copolymerization with styrene and multiple use of the corresponding Mn and Cr complexes in enantioselective epoxidations and hetero-Diels-Alder reactions, *Chem. Eur. J.* 7 (2001) 2873.
[31] H. Sellner, K. Hametner, D. Günther, D. Seebach, Manganese distribution in polystyrene beads prepared by copolymerization with cross-linking dendritic salens using laser ablation inductively coupled plasma mass spectrometry, *J. Catal.* 215 (2003) 87.
[32] D. Disalvo, D. B. Dellinger, J. W. Gohdes, Catalytic epoxidations of styrene using a manganese functionalized polymer, *React. Funct. Polym.* 53 (2002) 103.
[33] C. Borriello, R. Del Litto, A. Panunzi, F. Ruffo, A supported Mn(III) catalyst based on D-glucose in the asymmetric epoxidation of styrenes, *Inorg. Chem. Commun.* 8 (2005) 717.
[34] M. Nielsen, A. H. Thomsen, T. R. Jensen, H. J. Jakobsen, J. Skibsted, K. V. Gothelf, Formation and structure of conjugated salen-cross-linked polymers and their application in asymmetric heterogeneous catalysis, *Eur. J. Org. Chem.* (2005) 342.

[35] H. Zhang, Y. Zhang, C. Li, Asymmetric epoxidation of unfunctionalized olefins catalyzed by Mn(salen) axially immobilized onto insoluble polymers, *Tetrahedron: Asymmetry* 16 (2005) 2417.
[36] M. Holbach, M. Weck, Modular approach for the development of supported, monofunctionalized, salen catalysts, *J. Org. Chem.* 71 (2006) 1825.
[37] B. Bahramian, V. Mirkhani, M. Moghadam, S. Tangestaninejad, Selective alkene epoxidation and alkane hydroxylation with sodium periodate catalyzed by cationic Mn(III)-salen supported on Dowex MSC1, *Appl. Catal. A* 301 (2006) 169.
[38] V. Mirkhani, M. Moghadam, S. Tangestaninejad, B. Bahramian, Polystyrene-bound imidazole as a heterogeneous axial ligand for Mn(salophen)Cl and its use as biomimetic alkene epoxidation and alkane hydroxylation catalyst with sodium periodate, *Appl. Catal. A* 311 (2006) 43.
[39] V. Mirkhani, M. Moghadam, S. Tangestaninejad, B. Bahramian, Polystyrene-bound 1,4-phenylenediamine as a heterogeneous axial ligand for Mn(salophen)Cl and its use as biomimetic alkene epoxidation and alkane hydroxylation catalyst with sodium periodate, *Polyhedron* 25 (2006) 2904.
[40] K. Smith, C. H. Liu, Asymmetric epoxidation using a singly-bound supported Katsuki-type (salen)Mn complex, *Chem. Commun.* (2002) 886.
[41] K. Smith, C. H. Liu, G. A. El-Hiti, A novel supported Katsuki-type (salen)Mn complex for asymmetric epoxidation, *Org. Biomol. Chem.* 4 (2006) 917.
[42] B. M. Choudary, T. Ramani, H. Maheswaran, L. Prashant, K. V. S. Ranganath, K. V. Kumar, Catalytic asymmetric epoxidation of unfunctionalised olefins using silica, LDH and resin-supported sulfonato-Mn(salen) complex, *Adv. Synth. Catal.* 348 (2006) 493.
[43] H. C. Sacco, Y. Iamamoto, J. R. L. Smith, Alkene epoxidation with iodosylbenzene catalysed by polyionic manganese porphyrins electrostatically bound to counter-charged supports, *J. Chem. Soc., Perkin Trans.* 2 (2001) 181.
[44] E. Brulé, Y. R. de Miguel, Supported manganese porphyrin catalysts as P450 enzyme mimics for alkene epoxidation, *Tetrahedron Lett.* 43 (2002) 8555.
[45] E. Brulé, Y. R. de Miguel, K. K. M. Hii, Chemoselective epoxidation of dienes using polymer-supported manganese porphyrin catalysts, *Tetrahedron* 60 (2004) 5913.
[46] E. Brulé, K. K. M. Hii, Y. R. de Miguel, Polymer-supported manganese porphyrin catalysts—Peptide-linker promoted chemoselectivity, *Org. Biomol. Chem.* 3 (2005) 1971.
[47] S. Tangestaninejad, M. H. Habibi, V. Mirkhani, M. Moghadam, Mn (Br$_8$TPPS) supported on Amberlite IRA-400 as a robust and efficient catalyst for alkene epoxidation and alkane hydroxylation, *Molecules* 7 (2002) 264.
[48] S. Tangestaninejad, M. H. Habibi, V. Mirkhani, M. Moghadam, Manganese(III) porphyrin supported on polystyrene as a heterogeneous alkene epoxidation and alkane hydroxylation catalyst, *Synth. Commun.* 32 (2002) 3331.
[49] M. Moghadam, S. Tangestaninejad, M. H. Habibi, V. Mirkhani, A convenient preparation of polymer-supported manganese porphyrin and its use as hydrocarbon monooxygenation catalyst, *J. Mol. Catal. A* 217 (2004) 9.
[50] V. Mirkhani, M. Moghadam, S. Tangestaninejad, H. Kargar, Mn(Br$_8$TPP)Cl supported on polystyrene-bound imidazole: An efficient and reusable catalyst for biomimetic alkene epoxidation and alkane hydroxylation with sodium periodate under various reaction conditions, *Appl. Catal. A* 303 (2006) 221.
[51] M. Benaglia, T. Danelli, G. Pozzi, Synthesis of poly(ethylene glycol)-supported manganese porphyrins: Efficient, recoverable and recyclable catalysts for epoxidation of alkenes, *Org. Biomol. Chem.* 1 (2003) 454.
[52] C. P. Du, Z. K. Li, X. M. Wen, J. Wu, X. Q. Yu, M. Yang, R. G. Xie, Highly diastereoselective epoxidation of cholest-5-ene derivatives catalyzed by polymer-supported manganese(III) porphyrins, *J. Mol. Catal. A* 216 (2004) 7.
[53] C. Poriel, Y. Ferrand, P. Le Maux, J. Rault-Berthelot, G. Simonneaux, Organic cross-linked electropolymers as supported oxidation catalysts: Poly((tetrakis(9,9'-spirobifluorenyl)porphyrin) manganese) films, *Inorg. Chem.* 43 (2004) 5086.
[54] R. Naik, P. Joshi, S. Umbarkar, R. K. Deshpande, Polystyrene encapsulation of manganese porphyrins: Highly efficient catalysts for oxidation of olefins, *Catal. Commun.* 6 (2005) 125.

[55] S. V. Kotov, S. Boneva, T. Kolev, Some molybdenum-containing chelating ion-exchange resins (polyampholites) as catalysts for the epoxidation of alkenes by organic hydroperoxides, *J. Mol. Catal. A* 154 (2000) 121.
[56] S. V. Kotov, E. Balbolov, Comparative evaluation of the activity of some homogeneous and polymeric catalysts for the epoxidation of alkenes by organic hydroperoxides, *J. Mol. Catal. A* 176 (2001) 41.
[57] M. J. Hinner, M. Grosche, E. Herdtweck, W. R. Thiel, A Merrifield resin functionalized with molybdenum peroxo complexes: Synthesis and catalytic properties, *Z. Anorg. Allg. Chem.* 629 (2003) 2251.
[58] P. Reyes, G. Borda, J. Gnecco, B. L. Rivas, $MoO_2(acac)_2$ immobilized on polymers as catalysts for cyclohexene epoxidation: Effect of the degree of crosslinking, *J. Appl. Polym. Sci.* 93 (2004) 1602.
[59] S. Chavan, W. Maes, J. Wahlen, P. Jacobs, D. De Vos, W. Dehaen, Benzimidazole-functionalized dendrons as molybdenum supports for selective epoxidation catalysis, *Catal. Commun.* 6 (2005) 241.
[60] J. H. Ahn, J. C. Kim, S. K. Ihm, C. G. Oh, D. C. Sherrington, Epoxidation of olefins by molybdenum (VI) catalysts supported on functional polyimide particulates, *Ind. Eng. Chem. Res.* 44 (2005) 8560.
[61] G. Grivani, S. Tangestaninejad, M. H. Habibi, V. Mirkhani, Epoxidation of alkenes by a highly reusable and efficient polymer-supported molybdenum carbonyl catalyst, *Catal. Commun.* 6 (2005) 375.
[62] S. Tangestaninejad, M. H. Habibi, V. Mirkhani, M. Moghadam, G. Grivani, Readily prepared polymer-supported molybdenum carbonyls as novel reusable and highly active epoxidation catalysts, *Inorg. Chem. Commun.* 9 (2006) 575.
[63] G. Grivani, S. Tangestaninejad, M. H. Habibi, V. Mirkhani, M. Moghadam, Epoxidation of alkenes by readily prepared and highly active and reusable heterogeneous molybdenum-based catalysts, *Appl. Catal. A* 299 (2006) 131.
[64] U. Arnold, W. Habicht, M. Döring, Metal-doped epoxy resins—New catalysts for the epoxidation of alkenes with high long-term activities, *Adv. Synth. Catal.* 348 (2006) 142.
[65] U. Arnold, F. Fan, W. Habicht, M. Döring, Molybdenum-doped epoxy resins as catalysts for the epoxidation of alkenes, *J. Catal.* 245 (2007) 55.
[66] J. H. M. Heijnen, V. G. de Bruijn, L. J. P. van den Broeke, J. T. F. Keurentjes, Micellar catalysis for selective epoxidations of linear alkenes, *Chem. Eng. Process.* 42 (2003) 223.
[67] X. Q. Yu, J. S. Huang, W. Y. Yu, C. M. Che, Polymer-supported ruthenium porphyrins: Versatile and robust epoxidation catalysts with unusual selectivity, *J. Am. Chem. Soc.* 122 (2000) 5337.
[68] J. L. Zhang, C. M. Che, Soluble polymer-supported ruthenium porphyrin catalysts for epoxidation, cyclopropanation, and aziridination of alkenes, *Org. Lett.* 4 (2002) 1911.
[69] O. Nestler, K. Severin, A ruthenium porphyrin catalyst immobilized in a highly cross-linked polymer, *Org. Lett.* 3 (2001) 3907.
[70] Y. Ferrand, R. Daviaud, P. Le Maux, G. Simonneaux, Catalytic asymmetric oxidation of sulfide and styrene derivatives using macroporous resins containing chiral metalloporphyrins (Fe, Ru), *Tetrahedron: Asymmetry* 17 (2006) 952.
[71] R. Antony, G. L. Tembe, M. Ravindranathan, R. N. Ram, Synthesis and catalytic property of poly (styrene-*co*-divinylbenzene) supported ruthenium(III)-2-aminopyridyl complexes, *Eur. Polym. J.* 36 (2000) 1579.
[72] V. B. Valodkar, G. L. Tembe, M. Ravindranathan, H. S. Rama, Catalytic epoxidation of olefins by polymer-anchored amino acid ruthenium complexes, *React. Funct. Polym.* 56 (2003) 1.
[73] V. B. Valodkar, G. L. Tembe, R. N. Ram, H. S. Rama, Catalytic asymmetric epoxidation of unfunctionalized olefins by supported Cu(II)-amino acid complexes, *Catal. Lett.* 90 (2003) 91.
[74] R. Antony, G. L. Tembe, M. Ravindranathan, R. N. Ram, Synthesis and catalytic activity of Fe(III) anchored to a polystyrene-Schiff base support, *J. Mol. Catal. A* 171 (2001) 159.
[75] H. Deleuze, X. Schultze, D. C. Sherrington, Synthesis of porous supports containing *N*-(*p*-hydroxyphenyl)- or *N*-(3-4-dihydroxybenzyl) maleimide-anchored titanates and application as catalysts for transesterification and epoxidation reactions, *J. Polym. Sci. Pol. Chem.* 38 (2000) 2879.
[76] H. Deleuze, X. Schultze, D. C. Sherrington, Reactivity of some polymer-supported titanium catalysts in transesterification and epoxidation reactions, *J. Mol. Catal. A* 159 (2000) 257.

[77] H. C. Guo, X. Y. Shi, Z. Qiao, S. Hou, M. Wang, Efficient soluble polymer-supported Sharpless alkene epoxidation catalysts, *Chem. Commun.* (2002) 118.
[78] H. C. Guo, X. Y. Shi, X. Wang, S. Z. Liu, M. Wang, Liquid-phase synthesis of chiral tartrate ligand library for enantioselective Sharpless epoxidation of allylic alcohols, *J. Org. Chem.* 69 (2004) 2042.
[79] N. N. Reed, T. J. Dickerson, G. E. Boldt, K. D. Janda, Enantioreversal in the Sharpless asymmetric epoxidation reaction controlled by the molecular weight of a covalently appended achiral polymer, *J. Org. Chem.* 70 (2005) 1728.
[80] L. Zhang, H. C. L. Abbenhuis, G. Gerritsen, N. N. Bhriain, P. C. M. M. Magusin, B. Mezari, W. Han, R. A. van Santen, Q. Yang, C. Li, An efficient hybrid, nanostructured, epoxidation catalyst: Titanium silsesquioxane-polystyrene copolymer supported on SBA-15, *Chem. Eur. J.* 13 (2007) 1210.
[81] G. Gelbard, F. Breton, M. Quenard, D. C. Sherrington, Epoxidation of cyclohexene with polymethacrylate-based peroxotungstic catalysts, *J. Mol. Catal. A* 153 (2000) 7.
[82] G. Gelbard, Epoxidation of alkenes with tungsten catalysts immobilised on organophosphoryl macroligands, *C. R. Acad. Sci. Paris, Série IIc, Chimie/Chemistry* 3 (2000) 757.
[83] C. Venturello, R. D'Aloisio, Quaternary ammonium tetrakis(diperoxotungsto)phosphates(3-) as a new class of catalysts for efficient alkene epoxidation with hydrogen peroxide, *J. Org. Chem.* 53 (1988) 1553.
[84] D. Hoegaerts, B. F. Sels, D. E. de Vos, F. Verpoort, P. A. Jacobs, Heterogeneous tungsten-based catalysts for the epoxidation of bulky olefins, *Catal. Today* 60 (2000) 209.
[85] B. F. Sels, A. L. Villa, D. Hoegaerts, D. E. De Vos, P. A. Jacobs, Application of heterogenized oxidation catalysts to reactions of terpenic and other olefins with H_2O_2, *Top. Catal.* 13 (2000) 223.
[86] S. Tangestaninejad, M. H. Habibi, V. Mirkhani, M. Moghadam, G. Grivani, Simple preparation of some reusable and efficient polymer-supported tungsten carbonyl catalysts and clean epoxidation of cis-cyclooctene in the presence of H_2O_2, *J. Mol. Catal. A* 255 (2006) 249.
[87] R. Saladino, V. Neri, A. R. Pelliccia, R. Caminiti, C. Sadun, Preparation and structural characterization of polymer-supported methylrhenium trioxide systems as efficient and selective catalysts for the epoxidation of olefins, *J. Org. Chem.* 67 (2002) 1323.
[88] R. Saladino, V. Neri, A. R. Pelliccia, E. Mincione, Selective epoxidation of monoterpenes with H_2O_2 and polymer-supported methylrheniumtrioxide systems, *Tetrahedron* 59 (2003) 7403.
[89] A. Goti, F. Cardona, G. Soldaini, C. Crestini, C. Fiani, R. Saladino, Methyltrioxorhenium-catalyzed epoxidation-methanolysis of glycals under homogeneous and heterogeneous conditions, *Adv. Synth. Catal.* 348 (2006) 476.
[90] L. M. González R., A. L. Villa de P., C. Montes de C., G. Gelbard, Immobilization of methyltrioxorhenium onto tertiary amine and pyridine N-oxide resins, *React. Funct. Polym.* 65 (2005) 169.
[91] G. Kowalski, J. Pielichowski, M. Jasieniak, Polymer supported cobalt(II) catalysts for alkene epoxidation, *Appl. Catal. A* 247 (2003) 295.
[92] E. Blaz, J. Pielichowski, Polymer-supported cobalt (II) catalysts for the oxidation of alkenes, *Molecules* 11 (2006) 115.
[93] E. K. Beloglazkina, A. G. Majouga, R. B. Romashkina, N. V. Zyk, A novel catalyst for alkene epoxidation: A polymer-supported $Co^{II}LCl_2$ {L = 2-(alkylthio)-3-phenyl-5-(pyridine-2-ylmethylene)-3,5-dihydro-4H-imidazole-4-one} complex, *Tetrahedron Lett.* 47 (2006) 2957.
[94] B. B. Wentzel, S. M. Leinonen, S. Thomson, D. C. Sherrington, M. C. Feiters, R. J. M. Nolte, Aerobic epoxidation of alkenes using polymer-bound Mukaiyama catalysts, *J. Chem. Soc., Perkin Trans.* 1 (2000) 3428.
[95] W. Chen, J. Yamada, K. Matsumoto, Catalytic olefin epoxidation with molecular oxygen over supported amidate-bridged platinum blue complexes, *Synth. Commun.* 32 (2002) 17.
[96] D. Jayaprakash, Y. Kobayashi, S. Watanabe, T. Arai, H. Sasai, Enantioselective epoxidation of α,β-unsaturated ketones using polymer-supported lanthanoid-BINOL complexes, *Tetrahedron: Asymmetry* 14 (2003) 1587.
[97] G. Kumaraswamy, N. Jena, M. N. V. Sastry, G. V. Rao, K. Ankamma, Synthesis of 6,6'- and 6-MeO-PEG-BINOL-Ca soluble polymer bound ligands and their application in asymmetric Michael and epoxidation reactions, *J. Mol. Catal. A* 230 (2005) 59.

SECTION 4
Phase-Transfer Catalysis

CHAPTER 16

Fine-Tuning and Recycling of Homogeneous Tungstate and Polytungstate Epoxidation Catalysts

Paul L. Alsters,* Peter T. Witte,* Ronny Neumann,[†]
Dorit Sloboda Rozner,[†] Waldemar Adam,[‡,§] Rui Zhang,[‡]
Jan Reedijk,** Patrick Gamez,** Johan E. ten Elshof,[††]
and Sankhanilay Roy Chowdhury[††]

Contents		
	1. Introduction	416
	2. Characteristics and Preparation of Sandwich POMs	417
	3. Benchmarking Sandwich POM-Catalyzed Epoxidations	418
	4. Effect of Carboxylic Acids as Cocatalysts in Tungstate-Catalyzed Epoxidations	420
	5. Epoxidations that Afford Acid-Sensitive Products	421
	6. Sandwich POM Catalyst Recycling	425
	7. Conclusions	426
	Acknowledgments	427
	References	427

Abstract	This chapter reviews our work on epoxidations with aqueous dihydrogen peroxide catalyzed by tungstate and polytungstate catalysts. The activity of the $[WZn_3(ZnW_9O_{34})_2]^{12-}$ sandwich polyoxometalate (sandwich POM) in cyclooctene epoxidation is compared to other W-based catalyst systems under ceteris paribus conditions. Catalyst systems based on H_2WO_4 display the highest activity of the 11 W-catalyst systems tested. Replacing H_2WO_4

* DSM Pharma Products, Advanced Synthesis, Catalysis, and Development, P.O. Box 18, 6160 MD Geleen, The Netherlands
[†] Department of Organic Chemistry, Weizmann Institute of Science, Rehovot 76100, Israel
[‡] Institute of Organic Chemistry, University of Wurzburg, Am Hubland, 97074 Wurzburg, Germany
[§] Department of Chemistry, Facundo Bueso FB-110, University of Puerto Rico, Rio Piedras, Puerto Rico 00931
** Leiden Institute of Chemistry, Gorlaeus Laboratories, Leiden University, P.O. Box 9502, 2300 RA, Leiden, The Netherlands
[††] University of Twente, MESA + Institute for Nanotechnology, P.O. Box 217, 7500 AE Enschede, The Netherlands

Mechanisms in Homogeneous and Heterogeneous Epoxidation Catalysis
DOI: 10.1016/B978-0-444-53188-9.00016-X

© 2008 Elsevier B.V.
All rights reserved.

with Na_2WO_4 in these systems results in a strong decrease of catalyst activity, and this points to the importance of catalyst acidity. An experimental screening of various carboxylic acids cocatalysts has generated a novel W-based epoxidation system comprising Na_2WO_4/H_2WO_4 + $ClCH_2CO_2H$ + Aliquat 336 as catalyst components. The sandwich POM is among the most efficient of the nonacidic W-based catalyst systems. Examples where the pH neutral nature of the sandwich POM is used advantageously in the preparation of acid-sensitive epoxides are given. Allylic alcohols are epoxidized efficiently with very high reactor yields and very low sandwich POM catalyst loadings. Allylic alcohol epoxidation also proceeds very efficiently with the vanadium containing $[WZn(VO)_2(ZnW_9O_{34})_2]^{12-}$ POM catalyst and an organic hydroperoxide instead of H_2O_2 as the terminal oxidant. The diastereoselectivity of the epoxidation of chiral allylic alcohols is rationalized via 1,2- and/or 1,3-allylic strain effects. Fine-tuning of catalyst activity and selectivity is achieved by varying the relative amount and nature of phase-transfer cocatalysts, for example, $[(n-C_8H_{17})_3NMe]Cl$ or $[n-C_{16}H_{31}(HOCH_2CH_2)NMe_2]H_2PO_4$. Attention is paid to practical aspects that are of relevance for large-scale, industrial use of the sandwich POM catalyst, such as catalyst preparation, handling, and recycling. For the latter, an efficient method based on nanofiltration employing an α-alumina-supported γ-alumina membrane has been developed.

Key Words: Epoxidation, Tungstate, Polyoxometalate, Dihydrogen peroxide, Recycling, Nanofiltration.   © 2008 Elsevier B.V.

1. INTRODUCTION

Among the large variety of catalytic methods available for the epoxidation of alkenes, those based on tungstate catalysts hold a prominent place when it comes to industrial relevance. Several positive features of W-based catalyst systems explain their popularity in industry:

- Oxidations, including epoxidations, with tungstate catalysts can be run with inexpensive and "green" aqueous H_2O_2 as the terminal oxidant.
- Tungstate is readily available, inexpensive, and not very toxic. Active W-based catalysts are usually devoid of any expensive organic ligands, and merely assembled from simple inorganic components plus a phase-transfer catalyst when required.
- W-based catalysts have a very low activity for unproductive H_2O_2 decomposition into water and dioxygen, even at elevated temperatures. The evolution of large amounts of O_2 poses implementation problems with regard to process safety. In addition, the need for a large excess of diluted aqueous H_2O_2 to reach full conversion also negatively affects process economics in terms of high variable costs and decreased space–time yield, thus also increasing the fixed costs.

On the down side, catalyst activity of W-based systems is often only high enough to meet technoeconomic demands in case of reactive alkenes, such as electron-rich alkenes and allylic alcohols.

We review here our results on alkene epoxidation with $[WZnM_2(ZnW_9O_{34})_2]^{12-}$ polytungstozincate anions (M = Zn, Mn, or VO) as the catalyst and

H_2O_2 or organic hydroperoxides as the terminal oxidants. Our interest in these polytungstozincate species stems from an observation of Tourné and Zonnevijlle, who reported in 1991 that concentrated aqueous solutions of the $[WZn_3(ZnW_9O_{34})_2]^{12-}$ polytungstozincate anion could be stored for 20 years at room temperature without degradation [1]. The apparent hydrolytic stability of this polytungstozincate contrasts with the susceptibility of many Keggin and Wells–Dawson-type polyoxometalates toward hydrolytic degradation. Although hydrolytic stability is obviously a favorable property of a catalyst species under aqueous conditions, the catalytic properties of the $[WZn_3(ZnW_9O_{34})_2]^{12-}$ polytungstozincate anion were not known prior to our investigations on the use of this species as an oxidation catalyst for reactions with aqueous dihydrogen peroxide. The lack of interest in this species in literature related to oxidation catalysis is surprising in view of the large number of reports on the use of other polyoxometalates (POMs) as catalysts for oxidation reactions [2–5]. Interest in the use of POMs as oxidation catalysts is a result of their inorganic nature, which typically makes these species very tolerant toward strongly oxidizing conditions and high temperatures. The structure of the $[WZn_3(ZnW_9O_{34})_2]^{12-}$ anion is illustrated in Fig. 16.1, which clarifies the designation of this species as a "sandwich POM."

We have measured the performance of the $[WZn_3(ZnW_9O_{34})_2]^{12-}$ sandwich POM as a catalyst relative to other known tungsten-based catalyst systems for alkene epoxidation. Attention will also be paid to the fulfillment of large-scale processing requirements by sandwich POM catalysts, including ease of catalyst preparation and recycling. Besides the use of sandwich POMs as epoxidation catalysts, we will also summarize our recent results on a new W-based catalyst system based on tungstate, a carboxylic acid, and a phase-transfer catalyst.

2. CHARACTERISTICS AND PREPARATION OF SANDWICH POMs

Other than the hydrolytic stability of the sandwich POM, its ready availability presents another attractive feature. Its preparation in the form of the sodium salt merely involves slow addition of $Zn(NO_3)_2$ to an aqueous solution of Na_2WO_4 and nitric acid. The yield is nearly quantitative when these reagents are used in the amounts dictated by the stoichiometry of Eq. (16.1) [1]. Thus, formation of the

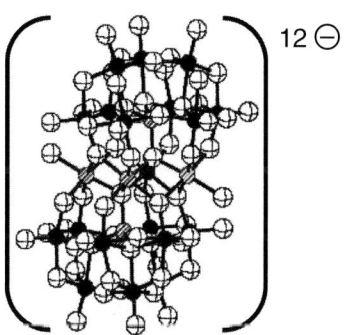

FIGURE 16.1 Structure of the $[WZn_3(ZnW_9O_{34})_2]^{12-}$ "sandwich" polyoxometalate anion.

sandwich POM is carried out by one-step "self-assembly" from readily available, inexpensive chemicals [6]. This obviously adds positively to its relevance for large-scale, industrial applications.

$$19Na_2WO_4 + 5Zn(NO_3)_2 + 16HNO_3 \rightarrow Na_{12}[WZn_3(ZnW_9O_{34})_2] \\ + 26NaNO_3 + 8H_2O \quad (16.1)$$

The Zn atoms in the interlayer of the sandwich POM can be replaced by other metal atoms via simple exchange reactions [Eq. (16.2)]. These newly introduced metal atoms may modify the catalytic activity compared to the parent $[WZn_3(ZnW_9O_{34})_2]^{12-}$ species. Examples of anionic species that are accessible through such exchange reactions [1] are $[WZnMn_2(ZnW_9O_{34})_2]^{12-}$ [7] and $[WZn(VO)_2(ZnW_9O_{34})_2]^{12-}$ [8].

$$Na_{12}[WZn_3(ZnW_9O_{34})_2] + 2M^{2+} \rightarrow Na_{12}[WZnM_2(ZnW_9O_{34})_2] + 2Zn^{2+} \quad (16.2)$$

Besides modifying the chemical properties of the sandwich POM via metal exchange reactions, its physical properties, in particular the solubility in organic solvents, can also be modified readily by simply exchanging the sodium counter cations for more lipophilic quaternary onium ions. In this way, the sandwich POM based on for example, methyltri-n-octylammonium counter cations can be dissolved in highly apolar media such as toluene.

Like several other tungsten species, $[WZn_3(ZnW_9O_{34})_2]^{12-}$ exhibits alkene epoxidation activity in the presence of aqueous dihydrogen peroxide. ^{183}W NMR measurements on an aqueous solution of $Na_{12}[WZn_3(ZnW_9O_{34})_2]$ and H_2O_2 (2,000 mol equiv) are in line with the formation of monoperoxo derivatives of the sandwich POM, with two distinct monoperoxo units being present as indicated by two new peaks around −700 ppm [9]. No evidence for mononuclear tungsten species that result from degradation of the POM structure was obtained. Thus, these measurements demonstrate the robustness of the sandwich POM framework under oxidative, hydrolytic conditions. As will be discussed below, this robustness is further illustrated by multiple catalyst recycle without activity drop after epoxidation runs with aqueous H_2O_2.

3. BENCHMARKING SANDWICH POM-CATALYZED EPOXIDATIONS

We have carried out an experimental comparison of the epoxidation activity of various W-based catalysts in order to assess the performance of the $[WZn_3(ZnW_9O_{34})_2]^{12-}$ sandwich POM relative to other W-based epoxidation catalysts [10]. The 11 catalyst systems used for this experimental study are listed below. The catalysts were either added as an isolated tungsten compound or formed without isolation *in situ*:

- 12[(n-C_8H_{17})$_3$NMe]Cl + $Na_{12}[WZn_3(ZnW_9O_{34})_2]$, that is, *in situ* sandwich POM
- [(n-C_8H_{17})$_3$NMe]$_{12}$[$WZn_3(ZnW_9O_{34})_2$], that is, isolated sandwich POM
- [(n-C_8H_{17})$_3$NMe]$_{12}$[$WZnMn_2(ZnW_9O_{34})_2$], that is, isolated Mn sandwich POM [7]

- 3[N-(n-C$_{16}$H$_{31}$)pyridinium]Cl + H$_3$[PO$_4$(WO$_3$)$_{12}$], that is, in situ Ishii [11]
- [N-(n-C$_{16}$H$_{31}$)pyridinium]$_3$[PO$_4$(WO$_3$)$_{12}$], that is, isolated Ishii
- 3[(n-C$_8$H$_{17}$)$_3$NMe]Cl + H$_3$PO$_4$ + 4H$_2$WO$_4$, that is, in situ Venturello [12]
- [(n-C$_8$H$_{17}$)$_3$NMe]$_3$[PO$_4${WO(O$_2$)$_2$}$_4$], that is, isolated Venturello
- 2[(n-C$_8$H$_{17}$)$_3$NMe]Cl + 2H$_2$WO$_4$, that is, in situ Prandi [13]
- [(n-C$_8$H$_{17}$)$_3$NMe]$_2$[{WO(O$_2$)$_2$}$_2$O], that is, isolated Prandi
- [(n-C$_8$H$_{17}$)$_3$NMe]HSO$_4$ + PO(OH)$_2$CH$_2$NH$_2$ + 2Na$_2$WO$_4$, that is, in situ Noyori [14]
- [(n-C$_8$H$_{17}$)$_3$NMe]HSO$_4$ + PO(OH)$_2$CH$_2$NH$_2$ + 2H$_2$WO$_4$, that is, in situ Noyori-plus

As is evident from this list, these catalysts display a large structural variety, from simple mononuclear H$_2$WO$_4$ to the [WZn$_3$(ZnW$_9$O$_{34}$)$_2$]$^{12-}$ sandwich POM with 19 W atoms. Activities were expressed on a "per W atom" basis rather than on a "per mol catalyst" basis. This choice is justified by the industrial practice to express catalyst loadings in wt.% rather than mol%, and in this respect it should be kept in mind that the molecular weights of these catalysts are largely determined by the number of W atoms in the molecular formula.

For this comparative study, the catalytic epoxidation of cyclooctene was used as a representative transformation, since it allows for accurate determinations of conversion and yield because cyclooctene is not prone to allylic oxidation and its corresponding epoxide is not prone to hydrolysis. Conversion-time profiles for each catalyst system were measured under identical conditions, that is, a catalyst loading corresponding to 0.1 mol% W, 1.5 mol equiv 50% H$_2$O$_2$, and toluene as solvent at 60 °C. The results are shown in Fig. 16.2.

The highest activity is displayed by acidic catalyst systems formed in situ from H$_2$WO$_4$. These systems generate mineral acid on formation of the catalytically active peroxido tungsten species by reaction of H$_2$WO$_4$ with H$_2$O$_2$. This mineral acid liberation is demonstrated below by the stoichiometry of Eqs. (16.3) and (16.4) for the in situ Venturello and Prandi systems (Q = quaternary ammonium cation):

$$3QCl + H_3PO_4 + 4H_2WO_4 + 8H_2O_2 \rightarrow Q_3[PO_4\{WO(O_2)_2\}_4] + 12H_2O + 3HCl \qquad (16.3)$$

$$2QCl + 2H_2WO_4 + 4H_2O_2 \rightarrow Q_2[\{WO(O_2)_2\}_2O] + 5H_2O + 2HCl \qquad (16.4)$$

The higher acidity of the in situ Ishii, Venturello, and Prandi catalysts systems also accounts for their higher activity compared to their isolated analogs. For the in situ Prandi- and Venturello-catalyzed epoxidation of cyclooctene, the importance of the acidity of the W-source is further underlined by the observation that conversion drops to negligible values on replacing H$_2$WO$_4$ by Na$_2$WO$_4$. Similarly, replacing Na$_2$WO$_4$ by H$_2$WO$_4$ in the well-known Noyori system generates the "Noyori-plus" system and results in a strong activity boost. The importance of the nature of the W-source (Na$_2$WO$_4$ or H$_2$WO$_4$) is often overlooked in the literature.

The fact that the Noyori system does show appreciable activity despite being based on Na_2WO_4 instead of H_2WO_4 can probably be ascribed to the acidity of the HSO_4^- anion present in the phase-transfer catalyst.

4. EFFECT OF CARBOXYLIC ACIDS AS COCATALYSTS IN TUNGSTATE-CATALYZED EPOXIDATIONS

Remarkably, activity in the foregoing nonactive Na_2WO_4-based systems cannot be restored by adding 2 mol equiv of HCl relative to the Na_2WO_4 catalyst instead of using H_2WO_4. These observations illustrate that tungsten-based catalyst systems can be fine-tuned by varying the nature of acid cocatalysts. This triggered us to explore the effect of various organic acids as cocatalysts in the presence of tungstate and a phase-transfer catalyst for alkene epoxidation with H_2O_2. These acid screening experiments were carried out with a 1/1 molar mixture of Na_2WO_4 and H_2WO_4. This combination provides the ideal compromise between the positive features of both W-sources, that is, solubility imparted by Na_2WO_4 and activity imparted by H_2WO_4. Soluble catalyst precursors are a desirable property in order to ensure fast formation of catalytically active peroxido species on contact with H_2O_2. The poor solubility of H_2WO_4 causes considerable lag times before epoxidation proceeds efficiently, and a catalyst activation step is required in order to avoid such an induction period,[1] since formation of peroxido tungsten species from insoluble H_2WO_4 proceeds only slowly even in hot concentrated H_2O_2. Large-scale processing of exothermic reactions like epoxidations is considerably facilitated by avoiding a considerable induction period, since this minimizes the risk of nonisothermal processing conditions caused by an inefficient and slow start of the reaction. In contrast to pure H_2WO_4, 1/1 Na_2WO_4/H_2WO_4 is soluble in water, thus enabling the epoxidations to take off rapidly on addition of H_2O_2.

We have screened a number of carboxylic acids as cocatalysts besides 1/1 Na_2WO_4/H_2WO_4 as the catalyst and Aliquat 336 as phase-transfer catalyst [15,16]. Catalyst loadings relative to the alkene in these experiments were typically 0.2 mol% Na_2WO_4, 0.2 mol% H_2WO_4, 1.6 mol% carboxylic acid, and 0.4 mol% Aliquat 336. The carboxylic acids that were screened include various substituted acetic acids, benzoic acids, and salicylic acids, with cyclooctene as the substrate and 1.5 mol equiv of H_2O_2 as the oxidant at 60 °C without organic solvent [16]. Catalyst activity was lowest for metal-chelating acids, that is, glycine and (substituted) salicylic acids. In the benzoic acid series, activity was lowered by the presence of electron-donating OH or NH_2 substituents in the *meta-* or *para*-position. Activity was also negatively influenced by increasing the acidity to too high values in the acetic acid series, with an optimum activity being found for $ClCH_2CO_2H$ and CH_3CO_2H, Cl_2CHCO_2H, and Cl_3CCO_2H, all performing less. Other acetic or benzoic acids performed less well than $ClCH_2CO_2H$, though not dramatically. The importance of an optimum activity rather than a too high acidity is also illustrated by the fact that replacing 0.2 mol% Na_2WO_4 + 0.2 mol% H_2WO_4 by either 0.4 mol% Na_2WO_4 or 0.4 mol% H_2WO_4 in the presence of

---

[1] This catalyst activation step is not included in Fig. 16.2.

ClCH$_2$CO$_2$H and Aliquat 336 negatively affected the conversion-time profile. Accordingly, the system Na$_2$WO$_4$/H$_2$WO$_4$ + ClCH$_2$CO$_2$H + Aliquat 336 was found to provide the optimum performance in cyclooctene epoxidation. For this highly reactive alkene, the TON reached 1,800 after 4 h at total W-catalyst loading of 0.05 mol%. Less reactive alkenes could also be epoxidized fairly efficiently, including terminal alkenes, with TOF of ~40 h^{-1} at 60 °C at 0.4 mol% total W-catalyst loading [15]. All these reactions start without an induction period.

5. EPOXIDATIONS THAT AFFORD ACID-SENSITIVE PRODUCTS

Synthetically, highly acidic epoxidation systems are only of limited relevance because many epoxides, unlike cyclooctene epoxide, do not tolerate a high acidity of the reaction medium. Figure 16.2 illustrates that the sandwich POMs are among the fastest of the neutral catalyst systems devoid of acidity generated by the catalyst on its reaction with H$_2$O$_2$. As a result of their pH neutral nature, the sandwich POMs have a broader synthetic scope compared to the more acidic *in situ* Venturello, Prandi, and Noyori-plus systems. When elevated temperatures are used, epoxide selectivity may still be low even when using the pH neutral POM catalyst. This was found to be the case in the epoxidation of geraniol using a

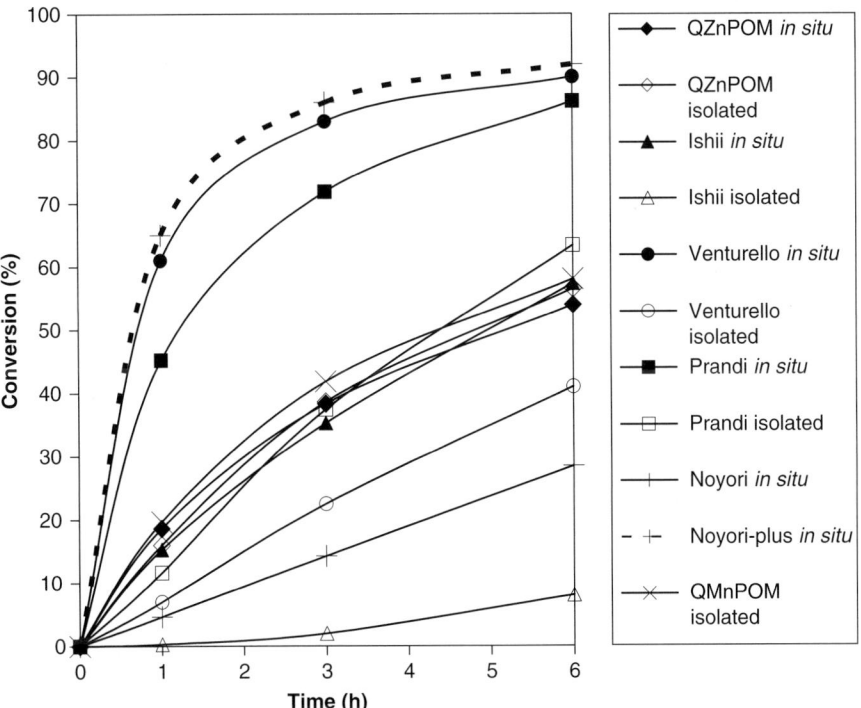

FIGURE 16.2 Conversion versus time for the epoxidation of cyclooctene catalyzed by various W-based catalyst systems (0.1 mol% W) at 60 °C in toluene in the presence of 1.5 equiv 50% H$_2$O$_2$. Reprinted with permission from [10]. Copyright 2004 American Chemical Society.

self-assembled aqueous $Na_{12}[WZn_3(ZnW_9O_{34})_2]$ solution [Eq. (16.1)] without isolating the catalyst. At 60 °C, the rate of the reaction was found to be high even when the catalyst loading was drastically reduced to 0.0005 mol% sandwich POM (i.e., 0.01 mol% W). Addition of some toluene (36 vol.% excluding aqueous H_2O_2) was found to increase the selectivity, since without additional solvent, extensive epoxide hydrolysis occurs because 50% H_2O_2 mixes completely with geraniol under the reaction conditions to afford a monophasic system that promotes hydrolysis. Under the two-phase conditions in the presence of toluene and H_2O, efficient conversion required the addition of $[(n\text{-}C_8H_{17})_3NMe]Cl$ (QCl) as a phase-transfer catalyst. Remarkably, not only the rate of conversion but also the selectivity was found to depend strongly on the amount of QCl, and upon optimizing the QCl/W ratio, an optimum in both the rate and the selectivity was observed at QCl/W = 10 (Fig. 16.3). The higher selectivity obtained in the presence of QCl suggests that chloride lowers the Lewis acidity of the POM catalyst, thus suppressing epoxide hydrolysis.

Note that the epoxidation of geraniol proceeds with a very high process efficiency: 1 kg of $Na_{12}[WZn_3(ZnW_9O_{34})_2]$ is enough to convert 6 m^3 of geraniol into 2,3-geranyl oxide within 4 h. This corresponds to a TOF of 43,385 h^{-1}, a TON of 17,3538, and a space–time yield of 95 kg m^3/h. When expressed per mol W rather than per mol W-catalyst, the corresponding values for $Na_{12}[WZn_3(ZnW_9O_{34})_2]$ are TOF = 2,283 h^{-1} and TON = 9,134.

We have also explored the use of vanadium substituted $Q_{12}[WZn(VO)_2(ZnW_9O_{34})_2]$, prepared via metal exchange [Eq (16.2)], as a catalyst for allylic alcohol epoxidations with nonchiral and chiral organic hydroperoxides instead of H_2O_2 as the terminal oxidant [8,17]. These conversions are assumed to proceed via V-alkylperoxido species. Besides the substantial asymmetric induction for a number of substrates with a chiral TADDOL-derived hydroperoxide, a particularly

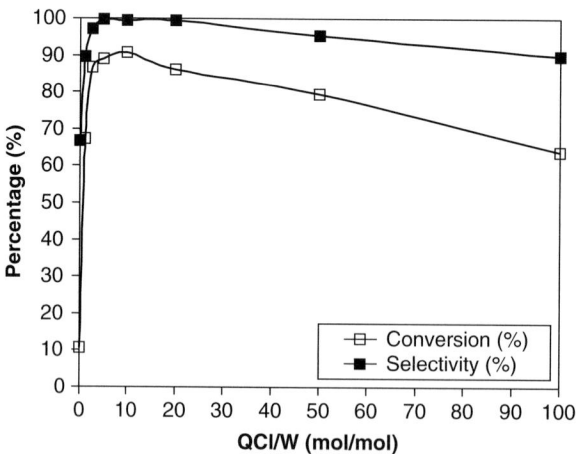

FIGURE 16.3 Conversion and selectivity after 4 h versus QCl/W ratio in the self-assembled sandwich polyoxometalate (0.0005 mol%; 0.01 mol% W)-catalyzed epoxidation of geraniol with 1.06 equiv 50% H_2O_2 at 60 °C in toluene. Reprinted with permission from [10]. Copyright 2004 American Chemical Society.

noteworthy feature is the very high TON that is achievable with this vanadium-based sandwich POM. Expressed per mol V, the TON reached 21,000, and the employed catalyst loading is 2–3 orders of magnitude lower than those commonly used in V-catalyzed allylic alcohol epoxidations with organic hydroperoxides [18–22].

The stereochemical outcome of $Q_n[WZnM_2(ZnW_9O_{34})_2]$ [M = Zn(II), Mn(II), Ru(III), Fe(III)]-catalyzed H_2O_2 epoxidation of various allylic alcohols with the OH group attached to a chiral center is controlled by allylic strain effects [23,24]. Thus, allylic alcohols with only 1,2-allylic strain were found to afford *erythro*-epoxides with excellent diastereoselectivity (Fig. 16.4, alkene A). In contrast, *threo*-epoxides predominate strongly in case of 1,3-allylic strain (alkene B). Diastereoselectivity drops to very low values in the absence of allylic strain (alkene C) or when both 1,2- and 1,3-allylic strain are present in the substrate (alkene D).

For acyclic allylic alcohols, very little α,β-unsaturated enone formation was observed besides epoxidation. Chemoselectivity was much less for cyclic allylic alcohols, for which oxidation of the allylic alcohol group competed significantly with epoxidation. In the case of 2-cyclohexenol as the substrate, the enone was even found to be the main product. A comparative sandwich POM-catalyzed epoxidation study of various (substituted) cycloalkenols revealed that the enone versus epoxide chemoselectivity is controlled by the C=C–C–OH dihedral angle $α_A$ in the allylic alcohol substrate. The more this dihedral angle deviates from the optimum C=C–C–OW dihedral angle $α_W$ for allylic acohol epoxidation, the more enone is formed (Fig. 16.5).

The optimum $α_W$ angle for the tungsten–alcoholate epoxidation template is estimated from a comparison of the sandwich POM diastereoselectivities with those observed for $VO(acac)_2$/TBHP and $Ti(OiPr)_4$/TBHP [23]. These two allylic alcohol epoxidation systems also operate via metal–alcoholate templates, with

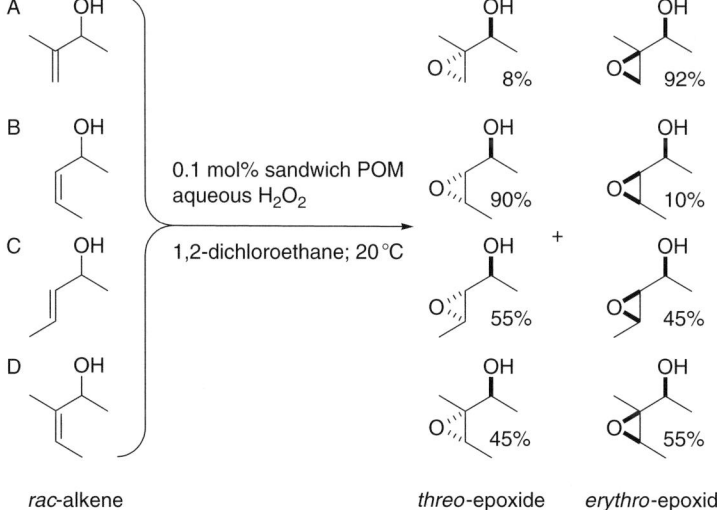

FIGURE 16.4 Effect of 1,2- and/or 1,3-allylic strain on the diastereoselectivity observed with sandwich polyoxometalate (POM)-catalyzed epoxidations of chiral allylic alcohols.

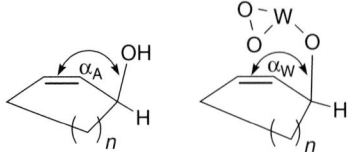

FIGURE 16.5 Dihedral C=C−C−OM (M = H or W) α angles in cyclic allylic alcohols and the corresponding peroxido tungstate–alcoholate template.

dihedral α angles of ~45° and ~80°, respectively. The intermediate diastereoselectivities observed with the sandwich POM suggest that it epoxidizes with a dihedral angle of about 60°, that is, in between those of the V and Ti systems.

The optimum QCl/W = 10 ratio required for the sandwich POM-catalyzed epoxidation of geraniol is much higher than the QCl/W = 12/19 expected from the stoichiometry $Q_{12}[Zn_5W_{19}O_{68}]$. The high amount of QCl may be caused by the use of a self-assembled sandwich POM solution, which contains substantial amounts of $NaNO_3$ [see Eq. (16.1)], and nitrate may compete with the POM anion in transfer to the organic phase by the quaternary ammonium cation. This conclusion is substantiated by other QCl/W optimization experiments on the epoxidation of cyclooctene with the systems $[(n\text{-}C_8H_{17})_3NMe]Cl + Na_{12}[WZn_3(ZnW_9O_{34})_2]$ and $[(n\text{-}C_8H_{17})_3NMe]Cl + H_2WO_4$, where isolated $Na_{12}[WZn_3(ZnW_9O_{34})_2]$ was used in the former case. In toluene at 60 °C, the optimum QCl/W ratios for these catalyst systems were found to be 0.3 and 0.6, respectively. These ratios are less than expected on the ground of the stoichiometry of the sandwich POM and Prandi systems (i.e., QCl/W = 12/19 or 1/1, respectively, with a 1/1 ratio expected for the Prandi system based on $Q_2[W_2O_{11}]$).

Selectivity in reactions that yield acid-sensitive epoxides with the onium salt + sandwich POM catalyst system can be increased further by using $[n\text{-}C_{16}H_{31}(HOCH_2CH_2)NMe_2]H_2PO_4$ instead of $[(n\text{-}C_8H_{17})_3NMe]Cl$ as the phase-transfer catalyst [10]. The former is an inexpensive quaternary ammonium compound sold under the trade name "Luviquat mono CP." The unusual $H_2PO_4^-$ anion buffers the pH of the reaction mixture, whereas the quaternary ammonium cation emulsifies the two-phase system[2] (Fig. 16.6). Luviquat mono CP is widely used in hair and skin care product formulations [25]. Besides being readily available on a large scale, it is also biocompatible. We have used the Luviquat mono CP + self-assembled sandwich POM system for the epoxidation of 3-carene, which is much less reactive than cyclooctene or geraniol and yields a fairly acid-sensitive carene epoxide. In toluene at 60 °C with 0.26 mol% sandwich POM, 4.4 mol% Luviquat mono CP, and 1.5 eq H_2O_2, near quantitative formation of the desired epoxide was obtained in a few hours. Remarkably, a very stable emulsion is present when the reaction is ongoing, but at the end of the reaction, the aqueous and organic phases separate readily within a few minutes, thus facilitating work-up and product isolation (Fig. 16.6).

---

[2] In addition, phosphate might influence catalyst performance through interaction with the polyoxotungstate, analogous to the Venturello system.

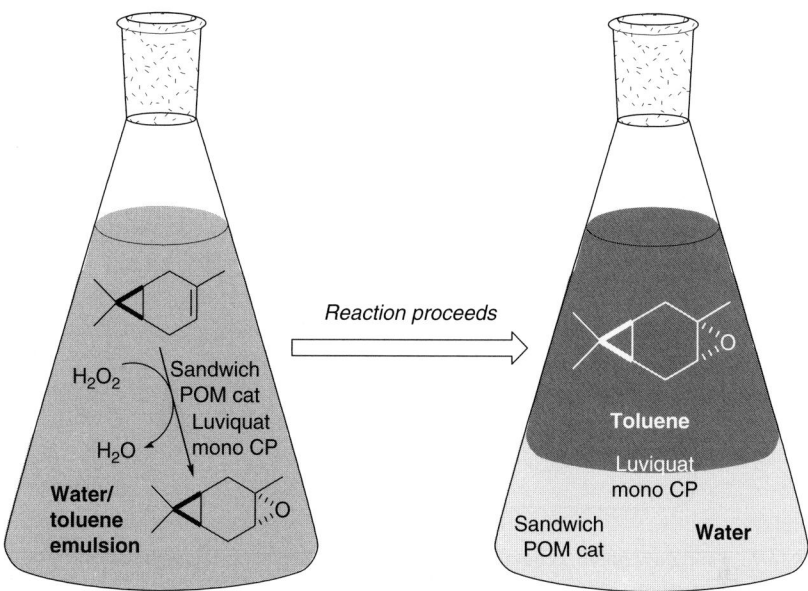

FIGURE 16.6 Epoxidation of 3-carene with H_2O_2 catalyzed by self-assembled sandwich POM plus Luviquat mono CP as cocatalyst, which accelerates the two-phase reaction by forming a stable emulsion that separates into a water and organic layer at full conversion.

6. SANDWICH POM CATALYST RECYCLING

In the foregoing examples based on phase-transfer catalysis, the catalyst and product reside in the same phase. For that reason, catalyst recycling cannot be achieved by simply phase separation. The latter can be applied for catalyst recycling when aqueous $Na_{12}[WZn_3(ZnW_9O_{34})_2]$ is employed as the catalyst without additional phase-transfer catalyst. We have used such aqueous biphasic oxidations for effecting oxidative transformations of various substrates, such as alcohols, diols, pyridines, amines, and anilines [6,9], and catalyst recycling through simple separation of the aqueous phase has been demonstrated. We reasoned that POM anions present in the organic phase due to the use of a lipophilic phase–transfer catalyst might be efficiently retained by nanofiltration. This reasoning was based on the fact that high retentions through nanofiltration require not only a sufficiently large molecular size but also a high shape persistency of the molecular structure [26]. POM anions nicely fulfill both requirements, since they are large and have a highly rigid, shape-persistent structure. In addition, their high negative charge may aid to obtain high retentions through nanofiltration. Near quantitative catalyst retentions are of particular importance for continuous rather than batch process conditions in order to minimize gradual catalyst supply over time to compensate for the loss of catalyst caused by a low retention. We investigated the use of ceramic membranes for sandwich POM recycling through nanofiltration because ceramic membranes display a high robustness toward elevated temperatures, organic solvents, and oxidizing

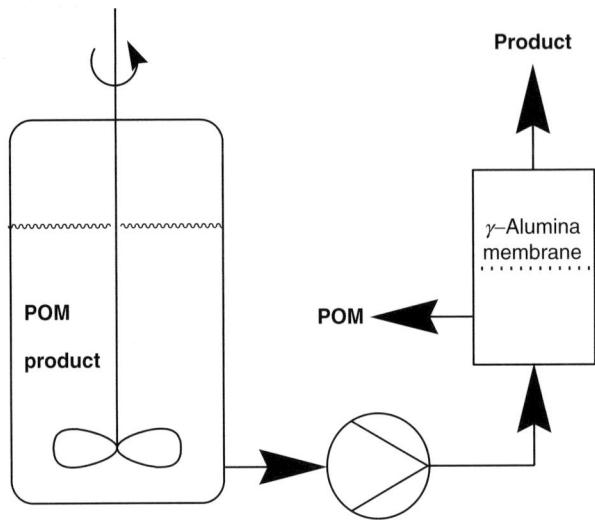

FIGURE 16.7 Sandwich polyoxometalate (POM) catalyst recycling through nanofiltration based on a ceramic alumina membrane.

conditions. We were pleased to obtain indeed near quantitative retentions (>99.9%) in toluene as the solvent with an α-alumina-supported γ-alumina membrane (Fig. 16.7) [27,28].

It was demonstrated that the catalyst could be recycled efficiently over six runs without loss in activity for the epoxidation of cyclooctene under $[(n\text{-}C_8H_{17})_3NMe]_{12}$ $[WZn_3(ZnW_9O_{34})_2]$ catalysis as a model reaction.

Besides the accessibility through self-assembly, the recycling possibility through nanofiltration, and the pH neutral character of the sandwich POM, another feature that adds positively to its industrial relevance is the lack of an induction period in oxidations with aqueous H_2O_2 (see Fig. 16.2). As outlined above, the efficient start of a reaction without substantial lag period is a desirable property for catalysts of highly exothermic reactions. Apparently, and in line with the ^{183}W NMR measurements described earlier, the active epoxidizing peroxido tungsten species is formed rapidly from the sandwich POM precursor on contact with H_2O_2.

7. CONCLUSIONS

Epoxidation of alkenes proceeds efficiently with aqueous H_2O_2 in the presence of catalysts containing the $[WZn_3(ZnW_9O_{34})_2]^{12-}$ sandwich POM. Under identical conditions and at equal level of 0.1 mol% W atoms, the sandwich POM is among the most efficient of the neutral, nonacidic W-catalysts. The pH neutral nature of the sandwich POM catalyst has synthetic relevance when acid-sensitive epoxides are desired. Allylic alcohols are particularly efficiently epoxidized by the $[WZn_3(ZnW_9O_{34})_2]^{12-}$ sandwich POM. Allylic alcohol epoxidation is also highly efficient with $[WZn(VO)_2(ZnW_9O_{34})_2]^{12-}$ as the catalyst and an organic hydroperoxide instead of H_2O_2 as the terminal oxidant. Fine-tuning of catalyst activity and selectivity can be achieved by optimizing the nature and amount of the phase-transfer cocatalyst.

H_2WO_4-based acidic catalyst systems (Venturello, Prandi, and Noyori-plus) display the highest activity of the 11 W-catalyst systems tested for cyclooctene epoxidation with H_2O_2. When H_2WO_4 is replaced with Na_2WO_4 in Venturello, Prandi, and Noyori systems, catalyst activity drops strongly. On screening various carboxylic acids as cocatalysts, a novel W-based epoxidation system comprising $Na_2WO_4/H_2WO_4 + ClCH_2CO_2H$ + Aliquat 336 as catalyst components has been found.

An aqueous catalyst solution of $Na_{12}[WZn_3(ZnW_9O_{34})_2]$ without phase-transfer catalyst can be recycled through simple phase separation. Solutions of lipophilic $[(n-C_8H_{17})_3NMe]_{12}[WZn_3(ZnW_9O_{34})_2]$ in organic solvents can be recycled very efficiently through nanofiltration based on α-alumina-supported γ-alumina membranes.

ACKNOWLEDGMENTS

We thank the European Commission (SUSTOX grant, G1RD-CT-2000–00347) and the Dutch Economy, Ecology, Technology (EET) program for generous financial support. Support from the NRSC Catalysis (a Research School Combination of HRSMC and NIOK) is also kindly acknowledged. The EET program is a joint program of the Ministry of Economic Affairs, the Ministry of Education, Culture and Science, and the Ministry of Housing, Spatial Planning and the Environment.

REFERENCES

[1] C. M. Tourné, G. F. Tourné, F. Zonnevijlle, Chiral polytungstometalates $[WM_3(H_2O)_2(XW_9O_{34})_2]^{12-}$ (X = M = zinc or cobalt) and their M-substituted derivatives. Syntheses, chemical, structural and spectroscopic study of some D,L sodium and potassium salts, *J. Chem. Soc. Dalton Trans.* (1991) 143.
[2] N. Mizuno, M. Misono, Heterogeneous catalysis, *Chem. Rev.* 98 (1998) 199.
[3] C. L. Hill, C. M. Prosser-McCartha, Homogeneous catalysis by transition metal oxygen anion clusters, *Coord. Chem. Rev.* 143 (1995) 407.
[4] I. V. Kozhevinikov, Catalysis by heteropoly acids and multicomponent polyoxometalates in liquid-phase reactions, *Chem. Rev.* 98 (1998) 171.
[5] R. Neumann, Polyoxometalate complexes in organic oxidation chemistry, *Prog. Inorg. Chem.* 47 (1998) 317.
[6] D. Sloboda-Rozner, P. L. Alsters, R. Neumann, A water-soluble and "self-assembled" polyoxometalate as a recyclable catalyst for oxidation of alcohols in water with hydrogen peroxide, *J. Am. Chem. Soc.* 125 (2003) 5280.
[7] R. Neumann, M. Gara, The manganese-containing polyoxometalate, $[WZnMn^{II}_2(ZnW_9O_{34})_2]^{12-}$, as a remarkably effective catalyst for hydrogen peroxide mediated oxidations, *J. Am. Chem. Soc.* 117 (1995) 5066.
[8] W. Adam, P. L. Alsters, R. Neumann, C. R. Saha-Möller, D. Seebach, A. K. Beck, R. Zhang, Chiral hydroperoxides as oxygen source in the catalytic stereoselective epoxidation of allylic alcohols by sandwich-type polyoxometalates: Control of enantioselectivity through a metal-coordinated template, *J. Org. Chem.* 68 (2003) 8222.
[9] D. Sloboda-Rozner, P. Witte, P. L. Alsters, R. Neumann, Aqueous biphasic oxidation: A water-soluble polyoxometalate catalyst for selective oxidation of various functional groups with hydrogen peroxide, *Adv. Synth. Cat.* 346 (2004) 339.
[10] P. T. Witte, P. L. Alsters, W. Jary, R. Müllner, P. Pöchlauer, D. Sloboda-Rozner, R. Neumann, Self-assembled $Na_{12}[WZn_3(Zn W_9O_{34})_2]$ as an industrially attractive multi-purpose catalyst for oxidations with aqueous hydrogen peroxide, *Org. Process Res. Dev.* 8 (2004) 524.

[11] Y. Ishii, K. Yamawaki, T. Ura, H. Yamada, T. Yoshida, M. Ogawa, Hydrogen peroxide oxidation catalyzed by heteropoly acids combined with cetylpyridinium chloride. Epoxidation of olefins and allylic alcohols, ketonization of alcohols and diols, and oxidative cleavage of 1,2-diols and olefins, *J. Org. Chem.* 53 (1988) 3587.

[12] C. Venturello, R. D'Aloisio, Quaternary ammonium tetrakis(diperoxotungsto)phosphates(3-) as a new class of catalysts for efficient alkene epoxidation with hydrogen peroxide, *J. Org. Chem.* 53 (1988) 1553.

[13] J. Prandi, H. B. Kagan, H. Mimoun, Epoxidation of isolated double bonds with 30% hydrogen peroxide catalyzed by pertungstate salts, Originally, [Ph$_3$PBn]Cl instead of [(n-C$_8$H$_{17}$)$_3$NMe]Cl was used as phase-transfer catalyst, *Tetrahedron Lett.* 27 (1986) 2617.

[14] K. Satu, M. Aoki, M. Ogawa, T. Hashimoto, R. Noyori, A practical method for epoxidation of terminal olefins with 30% hydrogen peroxide under halide-free conditions, *J. Org. Chem.* 61 (1996) 8310.

[15] P. U. Maheswari, P. de Hoog, R. Hage, P. Gamez, A Na$_2$WO$_4$/H$_2$WO$_4$-based highly efficient biphasic catalyst towards alkene epoxidation, using dihydrogen peroxide as oxidant, *J. Reedijk Adv. Synth. Cat.* 347 (2005) 1759.

[16] P. U. Maheswari, X. Tang, R. Hage, P. Gamez, The role of carboxylic acids on a Na$_2$WO$_4$/H$_2$WO$_4$-based biphasic homogeneous alkene epoxidation, using H$_2$O$_2$ as oxidant, *J. Reedijk J. Mol. Cat. A* 258 (2006) 295.

[17] W. Adam, P. L. Alsters, R. Neumann, C. R. Saha-Möller, D. Seebach, R. Zhang, Highly efficient catalytic asymmetric epoxidation of allylic alcohols by an oxovanadium-substituted polyoxometalate with a regenerative TADDOL-Derived hydroperoxide, *Org. Lett.* 5 (2003) 725.

[18] A. L. Villa, D. E. De Vos, F. Verpoort, B. F. Sels, P. A. Jacobs, A study of V-pillared layered double hydroxides as catalysts for the epoxidation of terpenic unsaturated alcohols, *J. Catal.* 198 (2001) 223.

[19] M. J. Haanepen, J. H. C. Van Hooff, VAPO as catalyst for liquid phase oxidation reactions. Part I: preparation, characterization and catalytic performance, *Appl. Catal. A* 152 (1997) 183.

[20] M. J. Haanepen, A. M. Elemans-Mehring, J. H. C. Van Hooff, VAPO as catalyst for liquid phase oxidation reactions. Part II: stability of VAPO-5 during catalytic operation, *Appl. Catal. A* 152 (1997) 203.

[21] K. B. Sharpless, R. C. Michaelson, High stereo- and regioselectivities in the transition metal catalyzed epoxidations of olefinic alcohols by tert-butyl hydroperoxide, *J. Am. Chem. Soc.* 95 (1973) 6136.

[22] R. C. Michaelson, R. E. Palermo, K. B. Sharpless, Chiral hydroxamic acids as ligands in the vanadium catalyzed asymmetric epoxidation of allylic alcohols by tert-butyl hydroperoxide, *J. Am. Chem. Soc.* 99 (1977) 1990.

[23] W. Adam, P. L. Alsters, R. Neumann, C. R. Saha-Möller, D. Sloboda-Rozner, R. Zhang, A highly chemoselective, diastereoselective, and regioselective epoxidation of chiral allylic alcohols with hydrogen peroxide, catalyzed by sandwich-type polyoxometalates: Enhancement of reactivity and control of selectivity by the hydroxy group through metal-alcoholate bonding. *J. Org. Chem.* 68 (2003) 1721.

[24] W. Adam, P. L. Alsters, R. Neumann, C. R. Saha-Möller, D. Sloboda-Rozner, R. Zhang, A new highly selective method for the catalytic epoxidation of chiral allylic alcohols by sandwich-type polyoxometalates with hydrogen peroxide, *Synlett* (2002) 2011.

[25] See: www.basf.com/businesses/consumer/cosmeticingredients/html/mquaternary.html.

[26] H. P. Dijkstra, G. P. M. van Klink, G. van Koten, The use of ultra- and nanofiltration techniques in homogeneous catalyst recycling, *Acc. Chem. Res.* 35 (2002) 798.

[27] P. T. Witte, S. Roy Chowdhury, J. E. ten Elshof, D. Sloboda-Rozner, R. Neumann, P. L. Alsters, Highly efficient recycling of a ''sandwich'' type polyoxometalate oxidation catalyst using solvent resistant nanofiltration, *Chem. Commun.* (2005) 1206.

[28] S. Roy Chowdhury, P. T. Witte, D. H. A. Blank, P. L. Alsters, J. E. ten Elshof, Recovery of homogeneous polyoxometalate catalysts from aqueous and organic media by a mesoporous ceramic membrane without loss of catalytic activity, *Chem. Eur. J.* 12 (2006) 3061.

CHAPTER 17

Reaction-Controlled Phase-Transfer Catalysis for Epoxidation of Olefins

Shuang Gao and **Zuwei Xi**

Contents		
	1. Introduction	430
	2. Reaction-Controlled Phase-Transfer Catalyst Based on Quaternary Ammonium Phosphotungstates	431
	3. Influence of the Composition of the Heteropolyphosphotungstate Anion [40]	432
	3.1. Preparation of the catalyst	432
	3.2. Characterization of catalyst	432
	4. Influence of Different Quaternary Ammonium Cations [43]	433
	5. Epoxidation of Propylene with *In Situ* Generated H_2O_2 as the Oxidant	435
	6. Epoxidation of Propylene with Aqueous H_2O_2 [45] as the Oxidant	438
	7. Epoxidation of Cyclohexene and Others Olefins	439
	8. Epoxidation of Allyl Chloride	440
	9. Conclusion	444
	Acknowledgments	444
	References	444

Abstract This chapter reviews recent progress made in reaction-controlled phase-transfer catalysis for the epoxidation of olefins using quaternary ammonium heteropolyphosphotungstate catalysts. The system exhibits high conversion and selectivity as well as excellent catalyst stability in the epoxidation of olefins using H_2O_2 as the oxidant. For example, for cyclohexene epoxidation in an aqueous/oil biphasic system, the conversion based on H_2O_2 was 92% and the selectivity was 94%. This catalytic process has been commercialized in China. Using *in situ* H_2O_2 generated by the oxidation of 2-ethylanthrahydroquinone (EAHQ), the selectivity for propylene oxide (PO) based on propylene was 95%, and the yield based on EAHQ was 85%. The catalytic system is homogeneous

Dalian Institute of Chemical Physics, Chinese Academy of Sciences, Dalian 116023, China

Mechanisms in Homogeneous and Heterogeneous Epoxidation Catalysis
DOI: 10.1016/B978-0-444-53188-9.00017-1

© 2008 Elsevier B.V.
All rights reserved.

during the epoxidation; however, after the H_2O_2 is used up, the catalyst can be recovered as a precipitate and can be reused. Thus, the advantages of both homogeneous and heterogeneous catalysts are combined in one system through reaction-controlled phase transfer.

Key Words: Reaction-controlled phase-transfer catalysis, 2-ethylanthrahydroquinone, Hydrogen peroxide, Propylene, Propylene oxide (PO), Cyclohexene, Cyclohexene oxide, Allyl chloride, Epichlorohydrin, Epoxidation. © 2008 Elsevier B.V.

1. INTRODUCTION

Alkene epoxidation is a very useful reaction in industry and organic synthesis. The resultant epoxides are essential precursors in the synthesis of various important substances like plasticizers, perfumes, and epoxy resins [1]. For example, over 5,000,000 and 70,000 metric tonnes of propylene and butene oxides, respectively, are produced per year [2]. Current commercial production of propylene oxide (PO) usually employs the chlorohydrin process or the Halcon process, which gives rise to disposal problem for the resultant salts or large amounts of coproducts. As a result of increasing stringent environment legislation, there is currently much interest in the research and development of environmentally friendly methods for preparation of PO without any coproduct.

In contrast to such classical processes, the catalytic epoxidation with hydrogen peroxide as an oxidant offers advantages because (1) it generates only water as a by-product and (2) it has a high content of active oxygen species [3–6].

Many transition-metal catalysts such as titanosilicates [7], metalloporphyrins [8], methyltrioxorhenium [9], tungsten compounds [10–12], polyoxometalates [13,14], manganese complexes [15,16], and nonheme iron complexes [17,18] have been used as effective catalysts for heterogeneous and homogeneous epoxidation with hydrogen peroxide. Clerici *et al.* has used the redox reaction of 2-ethylanthraquinone (EAQ)/2-ethylanthrahydroquinone (EAHQ), with O_2 as the oxidant, to produce H_2O_2 *in situ*, which then undergoes propylene epoxidation in the presence of TS-1 zeolite1 [19]. However, because of the restriction imposed by the properties of the TS-1 zeolite, its reaction medium in the EAQ/EAHQ system is more complicated than that of the normal industrial EAQ/EAHQ (polymethylbenzenes plus trialkylphosphate) system for H_2O_2 production, thus resulting in a substantial decrease in reaction efficiency. For the above homogeneous catalytic systems, there is a common problem, the difficulty of catalyst separation and reuse. In 2001, Xi *et al.* [20] reported a novel reaction-controlled phase-transfer catalyst system based on quaternary ammonium heteropolyoxotungstates. This catalytic system can be applied to the homogeneous catalytic epoxidation of most olefins (such as linear terminal olefins, internal olefins, cyclic olefins, styrene, and allyl chloride) and the oxidation of alcohols [21]. After reaction, the catalyst can be filtered and reused just like a heterogeneous catalyst. The concept of a reaction-controlled phase-transfer catalyst can be summarized as follows. The catalyst itself is insoluble in the reaction medium, but under the action of one of the reactants, it can form soluble active species that subsequently react with another reactant to selectively generate the desired product. When the first reactant is used up, the catalyst returns to its original composition and precipitates from the reaction medium, so that the catalyst can be easily separated and reused. This type of catalyst system has also been found to be applicable to other reactions such as (1) $K_3PV_4O_{24}$ and heteropoly blue for the synthesis of phenol from benzene as reported by the group of Jian [22, 23], (2) selenium dioxide for the reductive carbonylation of nitroaromatics as reported by the group of Lu [24,25], and (3) quaternary ammonium decatungstate for alcohol oxidation and cyclohexene oxidation to adipic acid by the group of Guo [26,27].

This chapter will present recent progress made in reaction-controlled phase-transfer catalysis for the epoxidation of olefins, focusing on work with heteropolyphosphotungstates and quaternary ammonium ions from our group. We have systemically investigated the influence of composition of the heteropoly anion and various quaternary ammonium ions on the catalyst activity. The epoxidation of propylene, allyl chloride, and others olefins and the stability of the catalyst in recycle will be summarized and discussed in detail.

2. REACTION-CONTROLLED PHASE-TRANSFER CATALYST BASED ON QUATERNARY AMMONIUM PHOSPHOTUNGSTATES

The diversity of polyoxometalates has led to numerous applications in the fields of structural chemistry, analytical chemistry, surface science, medicine, electrochemistry, and photochemistry [28]. The catalytic properties of polyoxometalates have also attracted much attention [29–34] because their acidic and redox properties can be controlled at the molecular level. Various catalytic systems for H_2O_2-based epoxidation catalyzed by heteropolyoxotungstates have been developed. Venturello and coworkers [35] reported the synthesis of tetraalkylammonium heteropolyperoxotungstates, such as $[(C_6H_{13})_4N]_3[PO_4[W_4O(O_2)_2]_2]_{24}$, for the epoxidation of olefins. Ishii et al. [11] found that the system composed of $H_3PW_{12}O_{40}$ and cetylpyridinium chloride can catalyze epoxidation of alkenes with commercially available H_2O_2 solution as oxidant. Duncan et al. [36] proposed that $PO_4[WO(O_2)_2]_4^{3-}$ is the active intermediate in the Ishii–Venturello system. While reacting with 1-octene, the active intermediate is transformed into a mixture of three phosphotungstates, PW_4, PW_3, and PW_2, which reformed the $\{PO_4[WO(O_2)_2]_4\}^{3-}$ structure upon interaction with H_2O_2. Crystals of $\{HPO_4[WO(O_2)_2]_2\}^{2-}$ [37] and $\{PO_4[WO(O_2)_2]_4\}^{3-}$ [35] were prepared and investigated for the epoxidation of limonene by H_2O_2, and was found to be 30 times more active than the molybdenum analog [37]. Gresley et al. [38] also reported their studies on the catalytic epoxidation activity of $\{PO_4[WO(O_2)_2]_2[WO(O_2)_2(H_2O)]\}^{3-}$. Salles et al. [39] pointed out that there is a dynamic equilibrium in the

$\{PO_4[WO(O_2)_2]_4\}^{3-}$ structure, where the positions of the four $WO(O_2)_2$ structural units in $\{PO_4[WO(O_2)_2]_4\}^{3-}$ is exchangeable. Here, the mechanism of the epoxidation by the catalyst $Q_3[PW_4O_{16}]$ is investigated.

3. INFLUENCE OF THE COMPOSITION OF THE HETEROPOLYPHOSPHOTUNGSTATE ANION [40]

3.1. Preparation of the catalyst

Catalyst A $[\pi\text{-}C_5H_5NC_{16}H_{33}]_3[PW_4O_{16}]$ was prepared according to Sun et al. [41]. Catalyst C was prepared in a similar manner as catalyst A, but in the step of extraction of $\{PO_4[WO(O_2)_2]_4\}^{3-}$, the quaternary ammonium salt was replaced by $(C_{18}H_{37}(30\%) + C_{16}H_{33}(70\%))N(CH_3)_3Cl^-$ with a molar ratio PO_4^{3-}/Q^+ of 1:3 and CH_2Cl_2 was replaced by CH_2ClCH_2Cl. The catalyst precipitated in the course of stirring. The catalyst was washed with water until the pH of the washed liquid was 4, and then was dried under an infrared lamp.

Catalysts B and D were obtained from a commercial plant for catalyst production. The production methods were similar to catalyst C but more simple. Catalyst E was obtained in the same manner as catalyst A, except that cetylpyridinium chloride was replaced by cetyltrimethyl ammonium chloride.

3.2. Characterization of catalyst

All five catalysts were characterized with Fourier transform infrared (FTIR) spectroscopy. As shown in Fig. 17.1, a feature of the catalysts is the lack of the peroxo band at $v(O-O) = 842$ cm^{-1} [42], which is different from the catalysts of Venturello et al. [35]. For catalysts A and E, the O–O group was removed in the distillation of the solvent at 60 °C. For catalyst C, the drying of the catalyst under an infrared lamp had the same effect.

All five catalysts were characterized with inductively coupled plasma (ICP) elemental analyses. As shown in Table 17.1, the tungsten-phosphor (W/P) ratio of catalysts A, E, C, and D is between 3.3 and 4.8, but that of catalyst B is 7.5.

The catalysts were also characterized with ^{31}P magic angle spinning nuclear magnetic resonance (MAS-NMR) spectra (Fig. 17.2) which showed that they were all mixtures of several heteropolyphosphotungstate species.

Salles et al. [39] reported that the ^{31}P NMR bands of PW_4, PW_3, and PW_2 are located at $\delta = -3.5, -1.6$, and -0.5 ppm, respectively. Hill et al. [36] found that the bands of PW_4 and PW_{12} are at $\delta = 0.6$ and -18.5 ppm, respectively. Although Salles and Hill obtained different chemical shifts for PW_4 because of the use of different D-solvents, their results show that species with high W/P ratio have chemical shifts located upfield compared to species with low W/P ratio.

Catalysts A, E, C, and D all have a ^{31}P MAS-NMR band at ca. $\delta = 5$ ppm, which corresponds to a phosphotungstate with low W/P ratio. As discussed earlier, the method of preparation of these catalysts results in destruction of the peroxidic O–O bond, and some decomposition of their structure results in a mixture of

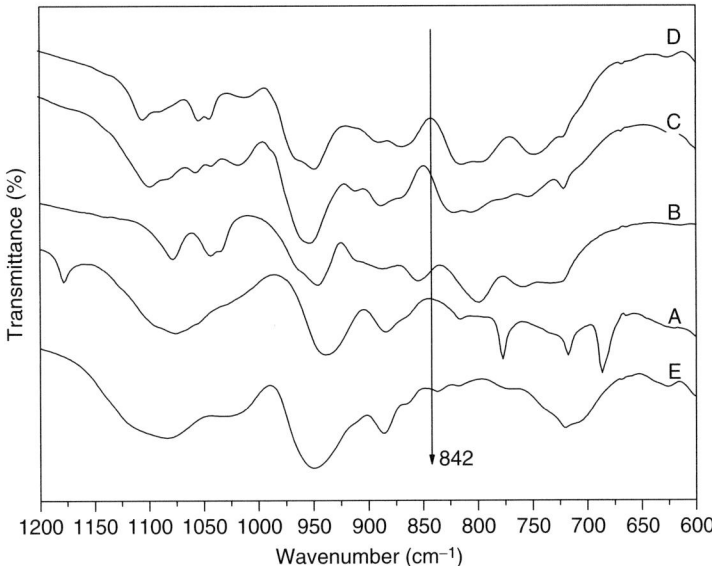

FIGURE 17.1 Fourier transform infrared (FTIR) spectra of catalysts.

TABLE 17.1 Results of ICP characterization

Catalyst	P (wt.%)	W (wt.%)	Molar ratio of W/P
A	1.53	38.1	4.3:1
E	1.91	37.0	3.3:1
B	0.83	37.1	7.5:1
C	1.36	38.9	4.8:1
D	1.46	35.7	4.1:1

phosphotungstate with different W/P ratio. The individual amounts of heteropolyphosphotungstates are all small except for the species with the ^{31}P MAS-NMR band at ca. $\delta = 5$ ppm. In the spectrum of catalyst B, there are two distinct bands, but no band at ca. $\delta = 5$ ppm. The W/P ratio of catalyst B is high.

The propylene epoxidation results on the above catalysts showed that only heteropolyphosphotungstates with low W/P ratio (ca. 4) have high reactivity. The band at ca. $\delta = 5$ ppm in the ^{31}P MAS-NMR spectra corresponds to an entity with low W/P ratio, which is likely a precursor which can be efficiently converted to the active species, $\{PO_4[WO(O_2)_2]_4\}^{3-}$.

4. INFLUENCE OF DIFFERENT QUATERNARY AMMONIUM CATIONS [43]

Table 17.2 shows that the structure of the quaternary ammonium cation (e.g., the number of carbon atoms) also has an important effect on the formation of a reaction-controlled phase-transfer catalyst and on its catalytic performance.

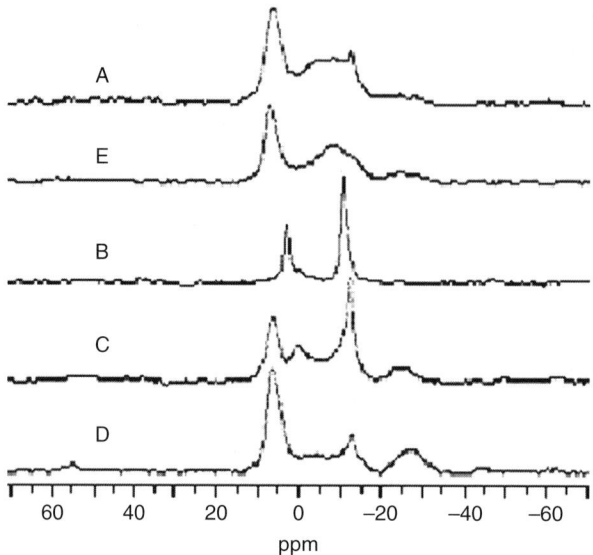

FIGURE 17.2 ^{31}P magic angle spinning nuclear magnetic resonance (MAS-NMR) spectra of catalysts.

TABLE 17.2 Effect of Q^+ in $Q_3[PO_4(WO_3)_4]$ on cyclohexene epoxidation

Catalyst	Q^+	During epoxidation	After epoxidation	Conversion[a] (%)	Selectivity[b] (%)
I	$[(n\text{-}Pr)_4N]^+$	Insoluble	Insoluble	60.6	60.2
A	$[\pi\text{-}C_5H_5NC_{16}H_{33}]^+$	Soluble	Insoluble	90.6	96.7
II	$[(C_{18}H_{37})Me_2NCH_2Ph]^+$	Soluble	Soluble	96.5	85.8

[a] The conversion of cyclohexene was based on H_2O_2.
[b] The selectivity to cyclohexene oxide was based on cyclohexene.
Reaction conditions: cyclohexene:H_2O_2 = 2:1 (molar ratio), 16 ml of CH_2ClCH_2Cl at 35 °C for 1.5 h.

Catalyst I, containing the small tetrapropyl ammonium ion, was insoluble in the reaction medium during the epoxidation, so both the conversion and selectivity were low, only 60.6% and 60.2%, respectively. Catalyst A is a reaction-controlled phase-transfer catalyst with high catalytic activity and selectivity. Although catalyst II also has good catalytic performance, it was totally soluble in the reaction system during and after the epoxidation, because it contains a big octadecyl benzyl methyl ammonium ion. This makes catalyst recovery difficult.

5. EPOXIDATION OF PROPYLENE WITH *IN SITU* GENERATED H_2O_2 AS THE OXIDANT

PO is an important chemical feedstock useful for producing polyurethanes, resins, surfactants, etc. This section presents results on the epoxidation of propylene with a reaction-controlled phase-transfer catalyst with *in situ* H_2O_2 and aqueous H_2O_2 as the oxidants.

When a combination of WO_4^{2-}/PO_4^{3-} was used as the catalyst under phase-transfer conditions [10] with aqueous hydrogen peroxide as the oxidant, the yield of PO based on hydrogen peroxide is low (30–40%) since PO is water soluble and undergoes hydrolysis easily.

To improve the situation, a green route to the PO (Scheme 17.1) was proposed in 2001 [20] using reaction-controlled phase-transfer catalysis

$$EAQ + H_2 \xrightarrow{Pd\ catalyst} EAHQ \qquad (17.1)$$

$$EAHQ + O_2 \rightarrow H_2O_2 + EAQ \qquad (17.2)$$

$$CH_3CH=CH_2 + H_2O_2 \xrightarrow{catalyst\ A} CH_3CH\overset{O}{-}CH_2 + H_2O \qquad (17.3)$$

$$Net\quad CH_3CH=CH_2 + O_2 + H_2 \rightarrow CH_3CH\overset{O}{-}CH_2 + H_2O \qquad (17.4)$$

The industrial process using the EAQ/EAHQ redox system for H_2O_2 production was coupled with the propylene epoxidation reaction, so the net reaction consumed only O_2 (air), H_2, and propylene, and produced PO with high selectivity. In this process, trimethyl benzenes (TMB) and tributylphosphate (TBP) were used as a mixed solvent. The results [44] were shown in Table 17.3–17.5.

TABLE 17.3 Epoxidation of propylene of different temperatures[a]

Temperature (°C)	C_3H_6:EAHQ (molar ratio)	Conversion of H_2O_2 (%)	Conversion[b] (%)	Selectivity[c] (%)	Yield[b] (%)
45	2.5:1	75	75	80	60
55	2.6:1	95	80	96	78
65	2.7:1	99.7	89	95	85
70	2.7:1	100	89	89	79

[a] Reaction condition: 80 ml of 0.38 mol/l 2-ethylanthrahydroquinone (EAHQ) solution was oxidized with O_2, and 30 mmol of H_2O_2 was formed. The reaction was maintained at the temperature for 8 h.
[b] The conversion of propylene and the yield of PO were based on EAHQ.
[c] The selectivity for PO was based on propylene.

Under 45 °C, the conversion of H_2O_2 was only 75%, the yield of PO based on EAHQ was 60%, and the selectivity to PO based on propylene was 80%. As the catalyst was not active enough at 45 °C, only part of the H_2O_2 was consumed after 8 h. By increasing the temperature to 65 °C, the conversion of H_2O_2 increased to 99.7%, the yield of PO to 85%, and the selectivity to PO based on propylene was 95%. At 70 °C, the conversion of H_2O_2 was 100%, but the yield to PO decreased to 79% and the selectivity to PO based on propylene decreased to 89%. At higher temperatures, owing to the decomposition of H_2O_2 and the hydrolysis of PO, the yields should be lower. The main by-product in this reaction was propylene glycol. A temperature of 65 °C was close to optimal for the epoxidation reaction.

Table 17.4 showed that when the molar ratio of EAHQ to the catalyst was increased from 300:1 to 800:1, the reaction time and the turnover number (TON) were prolonged correspondingly, the conversion of propylene was virtually unchanged from 90% to 88%, and the selectivity suffered little loss. Since the conversion of propylene and the selectivity for PO were not sensitive to the amount of the catalyst, the catalyst was shown to be stable under the reaction conditions.

Results of recycling of catalyst A were listed in Table 17.5. For the fresh catalyst, the molar ratio of EAHQ to the catalyst was 200:1, the conversion of

TABLE 17.4 Effect of the amount of $[C_5H_5NC_{16}H_{33}]_3PW_4O_{16}$ on the epoxidation of propylene[a]

EAHQ:catalyst (molar ratio)	C_3H_6:EAHQ (molar ratio)	Time (h)	Conversion[b] (%)	Selectivity[c] (%)	TON[d]
300:1	2.4:1	5	90	94	255
600:1	2.4:1	8	90	91	492
800:1	2.7:1	10	88	90	640

[a] Reaction conditions: 75.0 ml of 0.45 mol/l 2-ethylanthrahydroquinone (EAHQ) solution is oxidized with O_2, and 33.7 mmol of H_2O_2 is formed.
[b] The conversion of propylene was based on EAHQ.
[c] The selectivity for PO was based on propylene.
[d] Turnover number.

TABLE 17.5 Results of recycle of catalyst A[a]

Entry	C_3H_6:EAHQ (molar ratio)	Conversion[b] (%)	Selectivity[c] (%)
Fresh	2.5:1	91	94
Cycle 1	2.7:1	87	96
Cycle 2	2.4:1	90	92
Cycle 3	2.4:1	88	94

[a] Reaction conditions: (1) 75.0 ml of 0.42 mol/l EAHQ solution was oxidized with O_2, and 31.5 mmol H_2O_2 is formed. (2) EAHQ:fresh catalyst = 200:1. At the end of the reactions, the catalyst was separated by centrifugation, washed with toluene, and used in the next reaction without addition of fresh catalyst.
[b] Conversion of propylene was based on EAHQ.
[c] Selectivity for PO was based on propylene.

propylene based on EAHQ was 91%, and the selectivity to PO based on propylene was 94%. At the end of the reaction, the recovery efficiency of the catalyst was about 90% by weight. In subsequent runs using the recycled catalyst only, the conversion of propylene based on EAHQ was 87%, 90%, and 88%, respectively, and the selectivity for PO was about 94% average. The catalyst activity remained unchanged in the recycle runs.

Recently, considerable effort has been expended on commercializing this technology. A new-generation reaction-controlled phase-transfer catalyst (catalyst D) was developed with lower production cost and a higher recovery yield (>95%) of the catalyst which makes it more economical.

The stability of the catalyst to recycle was studied (Table 17.6).

At the end of the reaction, when most of the H_2O_2 was consumed, catalyst D was separated as a precipitate. It was found that the catalyst activity remained unchanged after reuse of the catalyst in seven cycles. The recycled catalysts were characterized by FTIR and ^{31}P MAS-NMR.

Figure 17.3 compares the FTIR spectra of the fresh catalyst and of the recovered sample after the seventh cycle. It can be seen that there are little differences in the IR spectra between the catalysts. No ν(O–O) band at 842 cm^{-1} was observed in either catalyst. The IR band at 887 cm^{-1} can be attributed to the ν(W–Ob–W) (corner-sharing).

The ^{31}P MAS-NMR (Fig. 17.4) results show that the structure of the recovered catalyst undergoes some change after the first run, but that the catalyst maintained a similar structure and tends to form single species with low W/P ratio after several recycle times. The above results show that catalyst D has good stability. The exact composition of the single species is under investigation.

TABLE 17.6 Effect of the recycle of catalyst D on propylene epoxidation[a]

Catalyst	Yield[b] (%)	Recovery yield (wt.%)
Fresh	82.6	95.3
Cycle 1	84.6	96.1
Cycle 2	83.3	95.8
Cycle 3	82.0	97.5
Cycle 4	81.7	96.6
Cycle 5	82.7	96.5
Cycle 6	82.5	94.5
Cycle 7	81.5	98.2

[a] Reaction conditions: (1) 70 ml of 0.32 mol/l EAHQ solution produced in a fixed-bed reactor with a Pd/Al$_2$O$_3$ catalyst was oxidized with O$_2$, and 20 mmol H$_2$O$_2$ was formed. (2) EAHQ: fresh catalyst = 300:1. At the end of each cycle, the catalyst was separated by centrifugation, washed with toluene, and used in the next reaction with addition of fresh catalyst to make up for losses.
[b] Yield of PO was based on EAHQ.

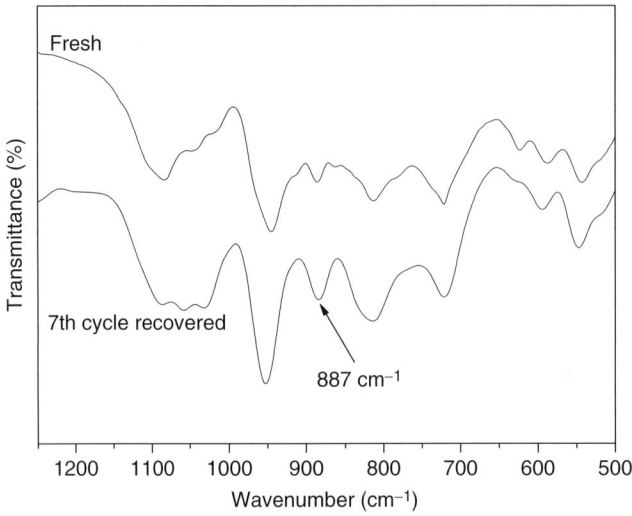

FIGURE 17.3 Fourier transform infrared (FTIR) spectra of fresh and recovered catalyst D in propylene epoxidation with *in situ* H_2O_2 as oxidant.

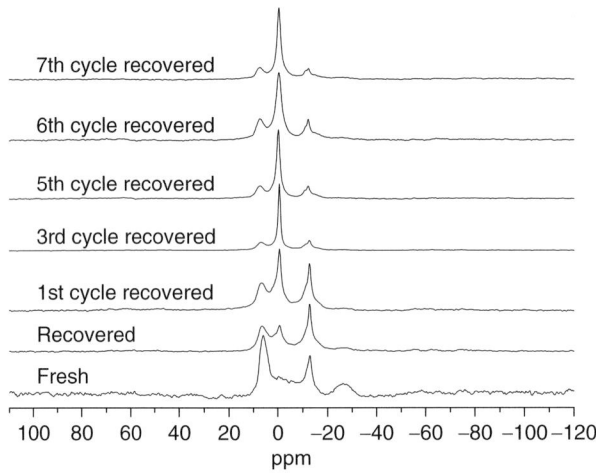

FIGURE 17.4 ^{31}P magic angle spinning nuclear magnetic resonance (MAS-NMR) spectra of fresh and recovered catalyst D in propylene epoxidation with *in situ* H_2O_2 as oxidant.

6. EPOXIDATION OF PROPYLENE WITH AQUEOUS H_2O_2 [45] AS THE OXIDANT

Since PO is water soluble, a few homogeneous catalytic systems employing aqueous H_2O_2 as oxidant have been reported [34]. This section report results on the homogeneous catalytic epoxidation of propylene to PO with 52% H_2O_2 using the reaction-controlled phase-transfer catalyst A [45]. The catalyst is easily recovered with 94% recovery yield (by weight). The selectivity for the PO is 92.0% based on propylene, and the yield is 90.9% based on H_2O_2. The effect of the TBP/toluene volume ratio on epoxidation is reported in Table 17.7.

Table 17.7 shows that the yield of PO, the selectivity to PO, and the recovered yield of the catalyst all increase with increase in the TBP/toluene volume ratio reaching a maximum at a ratio of 3:4 and then decrease. It is considered that the water phase is unfavorable for the epoxidation reaction due to the hydrolysis of PO to propylene glycol. When the volume ratio of TBP/toluene reaches 3:4, the polarity of TBP makes the aqueous/oil biphasic mixture a monophasic system, and no aqueous phase dissociates from the oil phase. Thus, the epoxidation reaction is little affected by the water in the added 52% H_2O_2.

Interestingly, higher amounts of TBP had the opposite effect and decreased the PO yield. This is because the catalyst, $[\pi\text{-}C_5H_5NC_{16}H_{33}]_3[PW_4O_{16}]$, consists of two parts, the heteropoly anion and the quaternary ammonium cation which have nearly opposite solubility behavior. When the amount of TBP was in excess, the catalyst could not dissolve completely in the reaction system even under the action of H_2O_2, and the catalytic activity was lower.

7. EPOXIDATION OF CYCLOHEXENE AND OTHERS OLEFINS

The results of the epoxidation of various nonfunctionalized olefins by catalyst B with aqueous H_2O_2 are shown in Table 17.8.

Cyclohexene oxide is an important intermediate used, for example, in synthesis of pesticides. A process for catalytic epoxidation of cyclohexene to cyclohexene oxide by reaction-controlled phase-transfer catalysis has been commercialized in China since 2003. It is an environment-friendly process compared to the polluting traditional chlorohydrin method.

TABLE 17.7 Effect of different volume ratios of TBP/Toluene[a] on PO yield

TBP/toluene (volume ratio)	Yield[b] (%)	Conversion[c] (%)	Selectivity[d] (%)	Recovery yield (wt.%)
Toluene	18.1	66.9	27.0	70.4
1:3	68.6	84.5	80.9	86.8
1:2	82.0	98.4	83.3	86.2
3:4	90.4	98.2	92.0	94.0
1:1	85.8	97.2	87.8	85.8
2:1	56.3	72.2	78.0	27.7
3:1	33.5	46.4	72.2	59.1
TBP	31.5	39.8	79.6	66.0

[a] Reaction conditions: solvent volume 70 ml, reaction time 4.5 h, reaction temperature 65 °C
[b] Yield of PO based on H_2O_2.
[c] Conversion of propylene based on H_2O_2.
[d] Selectivity to PO based on propylene.

TABLE 17.8 Epoxidation of olefins catalyzed by catalyst B[a]

Olefins	Reaction temperature (°C)	Reaction time (h)	Conversion[b] (%)	Selectivity[c] (%)
Cyclohexene	60	1	99.3	99.5
1-Octene	60	4	90.6	94.4
1-Dodecene	60	5	81.9	96.8
Styrene[d]	60	6	81.3	100
[e]	35	1.5	83.7	98.5

[a] Reaction conditions: olefins 15 mmol, catalyst B 0.05 g, olefins:H_2O_2:catalyst II = 600:200:1 (molar ratio).
[b] Conversion was based on H_2O_2.
[c] Selectivity was based on cyclohexene.
[d] 0.0192 g Na_2HPO_4–NaH_2PO_4 was added.
[e] The product was [structure].

The epoxidation of cyclohexene exhibited the highest conversion (99.5%, based on H_2O_2) and the shortest reaction time (1 h, entry 1). The recycled catalysts kept a similar structure as the fresh catalyst and a good stability during reaction (Fig. 17.5).

But if the epoxidation conditions used is not suitable for this catalytic system, such as no additive added, part of the used catalyst will convert to a heteropolyphosphotungstate with the Keggin structure [42], and then the recycle catalytic activity will decrease.

Comparing the epoxidation results of terminal olefins having different carbon atoms, it can be seen that the conversion based on H_2O_2 went up from 81.9% to 90.6% with decrease in carbon atoms from 1-dodecene to 1-octene (entries 2 and 3) and the reaction time required dropped from 5 to 4 h.

For styrene epoxidation, the conversion based on H_2O_2 was 81.3% with $Na_2HPO_4 \cdot 12H_2O$ added and the selectivity was high (100%).

The selective epoxidation of different double bonds in a single compound was investigated with limonene as a substrate. It was found that the epoxidation of the terminal double bond in the limonene molecule did not occur but epoxidation of the endo double bond occurred when the reaction temperature was kept under 35 °C.

The above results show that this catalyst system has good catalytic activity for various nonfunctionalized olefins. Even for the less-active terminal olefins, the conversion based on H_2O_2 was over 80% and selectivity was over 94%, and the catalyst could be reused easily.

8. EPOXIDATION OF ALLYL CHLORIDE

Epichlorohydrin is also an important chemical. Presently, epichlorohydrin is mainly produced by the chlorohydrin process (90%) which is problematic because it releases large amounts of effluent containing $CaCl_2$.

Allyl chloride is relatively difficult to epoxidize because of electron withdrawal by chlorine atom in adhesion the double bond, and for this reason has

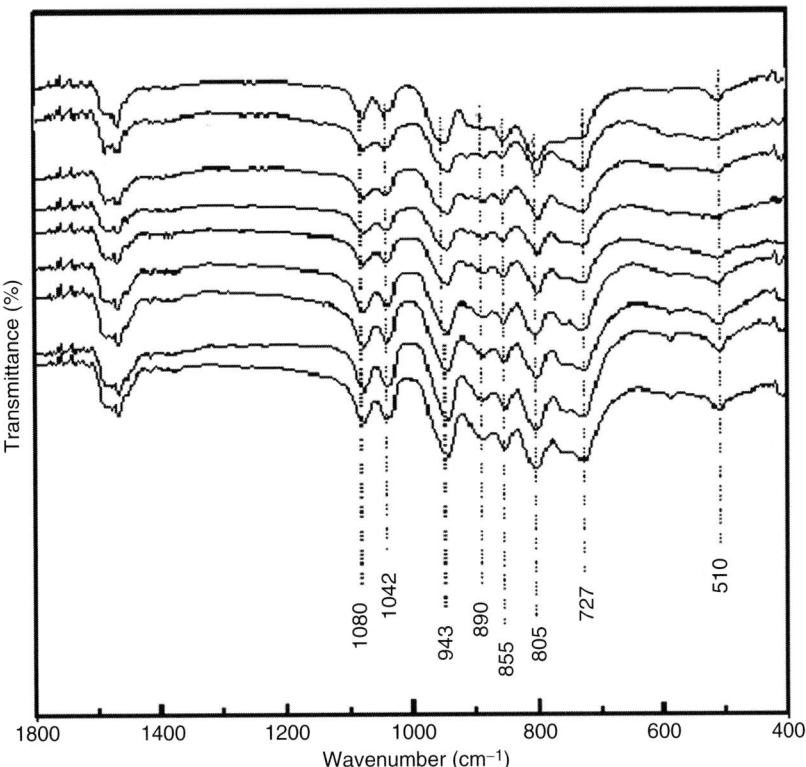

FIGURE 17.5 Fourier transform infrared (FTIR) spectra of fresh and recovered catalyst B in cyclohexene epoxidation.

been less studied. Current interest is on the direct epoxidation of allyl chloride with H_2O_2 as the oxidant [10, 41, 46–48]. First, the epoxidation of allyl chloride with aqueous hydrogen peroxide by catalyst E under biphasic conditions using 1,2-dichloroethane (DCE) as solvent was investigated [49]. The experimental results showed that the yield of epichlorohydrin is influenced by the amount of solvent. The reaction system give good catalytic properties, the highest epichlorohydrin yield reaches 88%. Toluene is not a good solvent for the epoxidation of allyl chloride.

A major shortcoming of the above systems is the use of organic solvents. The solvents are harmful to the environment, and make the separation of the product rather difficult. The best solvent is no solvent for green chemistry [50]. A method to produce epichlorohydrin using no organic solvent was developed using catalyst E. It shows high catalytic activity for the epoxidation of allyl chloride, especially in the presence of additives, such as $Na_2HPO_4 \cdot 12H_2O$ and $NaHCO_3$. The results of the influence of temperature are shown in Fig. 17.6.

The peak value of epichlorohydrin yield was 88.4%, attained at 65 °C. The catalyst E is very active, an epichlorohydrin yield of 76.4% can be reached at 45 °C when reaction time is prolonged to 3 h.

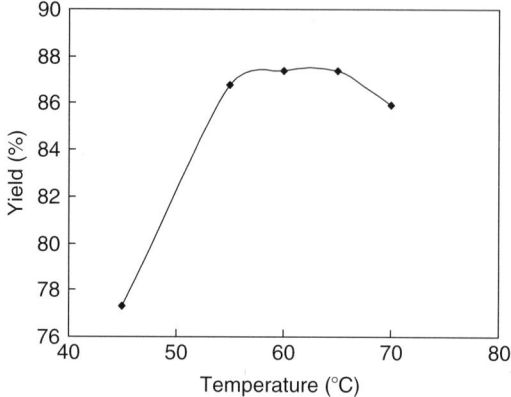

FIGURE 17.6 Influence of temperature on the epoxidation of allyl chloride. Reaction conditions: 1.65 g 33.5 wt.% H_2O_2 aqueous solution, 10 g allyl chloride, 3 g internal standard, 0.31 g catalyst E, 0.0075 g $Na_2HPO_4 \cdot 12H_2O$, 2 h except the reaction time was 3 h at 45 °C.

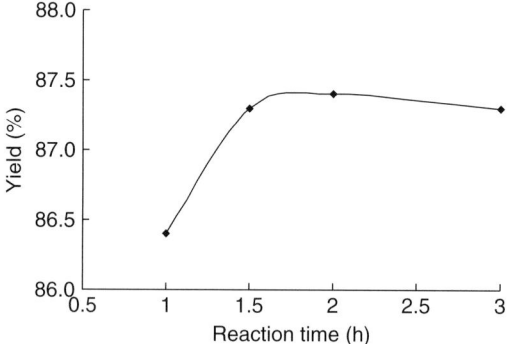

FIGURE 17.7 Influence of reaction time on the epoxidation of allyl chloride. Reaction conditions: 1.65 g 33.5 wt.% H_2O_2 aqueous solution, 10 g allyl chloride, 3 g internal standard, 0.31 g catalyst E, 0.0075 g $Na_2HPO_4 \cdot 12H_2O$, and reaction temperature 65 °C.

The effect of reaction time upon the activity of catalyst E is shown in Fig. 17.7. At 65 °C, after only 1 h, the catalyst achieves an epichlorohydrin yield of 86.4%. When reaction time prolonged to 2 h, the epichlorohydrin yield is raised to around 88.5% and then there is a little decrease. For effective catalyst recovery, the ideal reaction time is 2 h.

The concentration change of aqueous hydrogen peroxide in the biphasic reaction system as a function of reaction time was investigated. Figure 17.8 shows that the rate of change of the concentration of the aqueous hydrogen peroxide is constant. This means that the reaction rate in the epoxidation of allyl chloride is zero-order with respect to hydrogen peroxide

Catalyst D was used to develop this commercialized allyl chloride epoxidation technology. The recycling of the catalyst in the solvent-free reaction system was studied (Table 17.9).

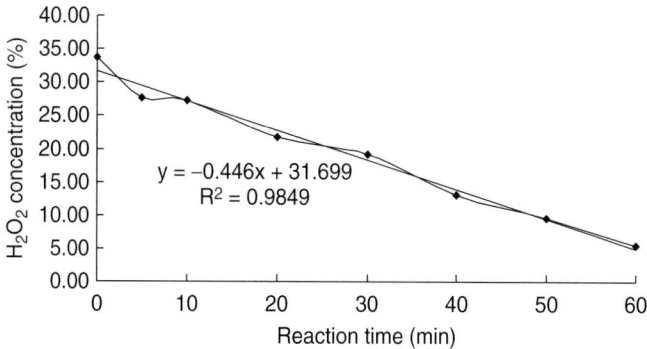

FIGURE 17.8 Dynamics of epoxidation of allyl chloride at 50 °C. Reaction conditions: 0.31 g catalyst E, 1.65 g 33.5% H_2O_2 aqueous solution, 3.0 g internal standard, 10.0 g allyl chloride, and reaction temperature 50 °C.

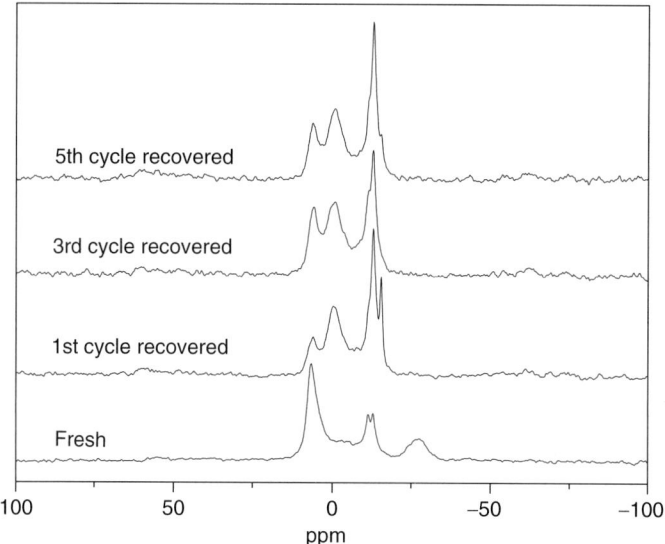

FIGURE 17.9 ^{31}P magic angle spinning nuclear magnetic resonance (MAS-NMR) spectra of fresh and recovered catalyst D in the epoxidation of allyl chloride.

The results are shown in Table 17.9; the epichlorohydrin yield was stable at about 84%, and the recovery efficiency of catalyst remained 100% after five cycles.

The fresh and recovered catalysts were studied by ^{31}P MAS-NMR (Fig. 17.9). There were changes after the first run, but subsequently the spectra did not change. This indicates that the catalyst was still a mixture maintaining a similar composition after recycle, and explains the stable activity.

TABLE 17.9 Effect of Recycle of Catalyst D on Allyl Chloride Epoxidation[a]

Catalyst	Yield[b] (%)	Recovery[c] (wt.%)
Fresh	87.8	100
Cycle 1	86.5	105
Cycle 2	85.8	107
Cycle 3	85	109
Cycle 4	83.5	112
Cycle 5	83.6	111

[a] Reaction conditions: 10 g allyl chloride, 1.34 g 51% H_2O_2 aqueous solution, 3.0 g internal standard, 0.38 g catalyst D and relevant $Na_2HPO_4 \cdot 12H_2O$, 65 °C, 2 h.
[b] Epichlorohdrin yield based on H_2O_2.
[c] The precipitated catalyst was washed twice with petroleum ether, dried under an infrared lamp, and used in the next reaction with addition of 5% of fresh catalyst D.

9. CONCLUSION

This chapter reviewed reaction-controlled phase-transfer catalysis work for the epoxidation of olefins mainly carried out in our research group. The catalyst consists of quaternary ammonium heteropolyphosphotungstates and exhibits high conversion and selectivity as well as excellent catalyst stability in the epoxidation of nonfunctionalized and functionalized olefins using H_2O_2 as the oxidant. A unique feature is that the catalytic system is homogeneous during the epoxidation but the catalyst can be recovered as a precipitate after the H_2O_2 is used up. The factors influencing the catalytic activity are studied in detail. It is found that the catalyst is actually a mixture of several heteropolyphosphotungstate species with different W/P ratio and that only species with low W/P ratio (ca. 4) have high reactivity toward propylene epoxidation. Also, the structure of the quaternary ammonium cation has an important influence on the reaction-controlled phase-transfer catalytic performance. The catalyst could be kept stable during the reaction. But if the reaction condition is not suitable, the catalyst forms less-active heteropolytungstates with the Keggin structure. In summary, a green route for the production of olefins oxides, such as PO, epichlorohydrin, and cyclohexene oxide, has been developed.

ACKNOWLEDGMENTS

This work was supported by the National Nature Science Foundation of China (No. 20143002, No. 20233050), the Innovation Fund of the Dalian Institute of Chemical Physics, Chinese Academy of Sciences (No. K2001E2), the National Basic Research Program of China (Grant No. 2003CB615805), and SINOPEC.

REFERENCES

[1] G. Grivani, S. Tangestaninejad, M. H. Habibi, V. Mirkhani, M. Moghadam, Epoxidation of alkenes by a readily prepared and highly active and reusable heterogeneous molybdenum-based catalyst, *Appl. Catal., A: Gen.* 299 (2006) 131.

[2] A. Tullo, A dow, BASF to build propylene oxide, *Chem. Eng. News* 82 (2004) 15.
[3] B. S. Lane, K. Burgess, Metal-catalyzed epoxidations of alkenes with hydrogen peroxide, *Chem. Rev.* 103 (2003) 2457.
[4] R. Noyori, M. Aoki, K. Sato, Green oxidation with aqueous hydrogen peroxide, *Chem. Commun.* 16 (2003) 1977.
[5] D. E. De Vos, B. F. Sels, P. A. Jacobs, Practical heterogeneous catalysts for epoxide production, *Adv. Synth. Catal.* 345 (2003) 457.
[6] J.-M. Brégeault, Transition-metal complexes for liquid-phase catalytic oxidation: Some aspects of industrial reactions and of emerging technologies, *Dalton Trans.* 17 (2003) 3289.
[7] P. Battioni, J. P. Renaud, J. F. Bartoli, M. Reina-Artiles, M. Fort, D. Mansuy, Monooxygenase-like oxidation of hydrocarbons by hydrogen peroxide catalyzed by manganese porphyrins and imidazole: Selection of the best catalytic system and nature of the active oxygen species, *J. Am. Chem. Soc.* 110 (1988) 8462.
[8] B. Notari, Microporous crystalline titanium silicates, *Adv. Catal.* 41 (1996) 253.
[9] C. C. Romao, F. E. Kuhn, W. A. Herrmann, Rhenium(VII) oxo and imido complexes: Synthesis, structures, and applications, *Chem. Rev.* 97 (1997) 3197.
[10] C. Venturello, E. Alneri, M. Ricci, A new, effective catalytic system for epoxidation of olefins by hydrogen peroxide under phase-transfer conditions, *J. Org. Chem.* 48 (1983) 3831.
[11] Y. Ishii, K. Yamawaki, T. Ura, H. Yamada, T. Yoshida, M. Ogawa, Hydrogen peroxide oxidation catalyzed by heteropoly acids combined with cetylpyridinium chloride. Epoxidation of olefins and allylic alcohols, ketonization of alcohols and diols, and oxidative cleavage of 1,2-diols and olefins, *J. Org. Chem.* 53 (1988) 3587.
[12] K. Sato, M. Aoki, M. Ogawa, T. Hashimoto, D. Panyella, R. Noyori, A halide-free method for olefin epoxidation with 30% hydrogen peroxide, *Bull. Chem. Soc. Jpn.* 70 (1997) 905.
[13] R. Neumann, M. Gara, The manganese-containing polyoxometalate, [WZnMnII2(ZnW9O34)2]12-, as a remarkably effective catalyst for hydrogen peroxide mediated oxidations, *J. Am. Chem. Soc.* 117 (1995) 5066.
[14] N. Mizuno, C. Nozaki, I. Kiyoto, M. Misono, Highly efficient utilization of hydrogen peroxide for selective oxygenation of alkanes catalyzed by diiron-substituted polyoxometalate precursor, *J. Am. Chem. Soc.* 120 (1998) 9267.
[15] D. E. De Vos, J. L. Meinershagen, T. Bein, Highly selective catalysts derived from intrazeolite trimethyltriazacyclononane-manganese complexes, *Angew. Chem. Int. Ed. Engl.* 35 (1996) 2211.
[16] B. S. Lane, M. Vogt, V. J. DeRose, K. Burgess, Manganese-catalyzed epoxidations of alkenes in bicarbonate solutions, *J. Am. Chem. Soc.* 124 (2002) 11946.
[17] M. C. White, A. G. Doyle, E. N. Jacobsen, A synthetically useful, self-assembling MMO mimic system for catalytic alkene epoxidation with aqueous H_2O_2, *J. Am. Chem. Soc.* 123 (2001) 7194.
[18] K. Chen, M. Costas, L. Que Jr, Spin state tuning of non-heme iron-catalyzed hydrocarbon oxidations: Participation of Fe^{III}–OOH and Fe^V=O intermediates, *J. Chem. Soc., Dalton Trans.* 5 (2002) 672.
[19] P. Ingallina, M. G. Clerici, L. Rossi, Catalysis with TS-1: New perspectives for the industrial use of hydrogen peroxide, *Stud. Surf. Sci. Catal.* 92 (1995) 31.
[20] Z. W. Xi, N. Zhou, Y. Sun, K. L. Li, Reaction-controlled phase-transfer catalysis for propylene epoxidation to propylene oxide, *Science* 292 (2001) 1139.
[21] S. J. Zhang, S. Gao, Z. W. Xi, J. Xu, Solvent-free oxidation of alcohols catalyzed by an efficient and reusable heteropolyphosphatotungstate, *Catal. Commun.* 7 (2006) 731.
[22] M. Li, X. Jian, T. M. Han, L. Liu, Y. Shi, Reaction-controlled phase-transfer catalyst K3PV4O24 for synthesis of phenol from benzene, *Chin. J. Catal.* 25 (2004) 681.
[23] M. Li, X. Jian, T. M. Han, Y. An, Heteropoly blue as reaction-controlled phase-transfer catalyst, *Acta Chim. Sinica* 62 (2004) 540.
[24] J. Z. Chen, S. W. Lu, Synthesis of substituted pyridyl ureas by selenium dioxide-catalyzed carbonylation, *Appl. Catal., A: Gen.* 261 (2004) 199.
[25] J. Z. Chen, G. Ling, S. Lu, Synthesis of N-phenyl-N-pyrimidylurea derivatives by selenium- or selenium dioxide-catalyzed reductive carbonylation of nitroaromatics, *Eur. J. Org. Chem.* 17 (2003) 3446.

[26] M. Guo, Catalytic oxidation of cyclohexene to adipic acid with a reaction-controlled phase transfer catalyst, *Chin. J. Catal.* 24 (2003) 483.
[27] M. Guo, Quaternary ammonium decatungstate catalyst for oxidation of alcohol, *Green Chem.* 6 (2004) 271.
[28] C. L. Hill, Introduction: Polyoxometalates-multicomponent molecular vehicles to probe fundamental issues and practical problems, *Chem. Rev.* 98 (1998) 1.
[29] T. Okuhara, N. Mizuno, M. Misono, Catalytic chemistry of heteropoly compounds, *Adv. Catal.* 41 (1996) 113.
[30] N. Mizuno, M. Misono, Heterogeneous catalysis, *Chem. Rev.* 98 (1998) 199.
[31] R. Neumann, Polyoxometallate complexes in organic oxidation chemistry, *Prog. Inorg. Chem.* 47 (1998) 317.
[32] C. L. Hill, C. Chrisina, M. Prosser-McCartha, Homogeneous catalysis by transition metal oxygen anion clusters, *Coord. Chem. Rev* 143 (1995) 407.
[33] I. V. Kozhevnikov, Catalysis by heteropoly acids and multicomponent polyoxometalates in liquid-phase reactions, *Chem. Rev.* 98 (1998) 171.
[34] K. Kamata, K. Yonehara, Y. Sumida, K. Yamaguchi, S. Hikichi, N. Mizuno, Highly selective catalysts derived from intrazeolite trimethyltriazacyclononane-manganese complexes, *Science* 300 (2003) 964.
[35] C. Venturello, R. D'Aloisio, J. C. J. Bart, M. Ricci, Quaternary ammonium tetrakis(diperoxotungsto)phosphates(3-) as a new class of catalysts for efficient alkene epoxidation with hydrogen peroxide, *J. Mol. Catal.* 32 (1985) 107.
[36] D. C. Duncan, R. C. Chambeers, E. Hecht, C. L. Hill, Mechanism and dynamics in the $H_3[PW_{12}O_{40}]$-catalyzed selective epoxidation of terminal olefins by H_2O_2. Formation, reactivity, and stability of $\{PO_4[WO(O_2)_2]_4\}^{3-}$, *J. Am. Chem. Soc.* 117 (1995) 681.
[37] L. Salles, C. Aubry, R. Thouvenot, F. Robert, C. Doremieux-Morin, G. Chottard, H. Ledon, Y. Jeannin, J.-M. Bregeault, ^{31}P and ^{183}W NMR spectroscopic evidence for novel peroxo species in the "$H_3[PW_{12}O_{40}].yH_2O/H_2O_2$" system. Synthesis and x-ray structure of tetrabutylammonium (m-Hydrogen phosphato)bis(m-peroxo)bis(oxoperoxotungstate)(2-): A catalyst of olefin epoxidation in a biphase medium, *Inorg. Chem.* 33 (1994) 871.
[38] N. Melaine Gresley, W. P. Griffth, A. C. Lammel, H. I. S. Noqueria, B. C. Parkin, Studies on polyoxo and polyperoxo-metalates part 5: Peroxide-catalysed oxidations with heteropolyperoxotungstates and -molybdates, *J. Mol. Catal.* 117 (1997) 185.
[39] L. Salles, J.-Y. Piquemal, R. Thouvenot, C. Minot, J.-M. Brégeault, Catalytic epoxidation by heteropolyoxoperoxo complexes: From novel precursors or catalysts to a mechanistic approach, *J. Mol. Catal., A: Chem.* 117 (1997) 375.
[40] J. Li, S. Gao, M. Li, R. Zhang, Z. W. Xi, Influence of composition of heteropolyphosphatotungstate catalyst on epoxidation of propylene, *J. Mol. Catal., A: Chem.* 218 (2004) 247.
[41] Y. Sun, Z. W. Xi, G. Cao, Epoxidation of olefins catalyzed by $[\pi\text{-}C_5H_5NC_{16}H_{33}]_3[PW_4O_{16}]$ with molecular oxygen and a recyclable reductant 2-ethylanthrahydroquinone, *J. Mol. Catal.* 166 (2001) 219.
[42] J. Gao, Y. Chen, C. Li, S. Gao, N. Zhou, Z. W. Xi, A spectroscopic study on the reaction-controlled phase transfer catalyst in the epoxidation of cyclohexene, *J. Mol. Catal.* 210 (2004) 197.
[43] K. L. Li, N. Zhou, Z. W. Xi, Effects od solvents and quaternary ammonium ions in heteropolyoxotungstates on reaction-controlled phase-transfer catalysis for cyclohexene epoxidatio, *Chin. J. Catal.* 23 (2003) 125.
[44] N. Zhou, Z. W. Xi, G. Cao, S. Gao, Epoxidation of propylene by using $[p\text{-}C_5H_5NC_{16}H_{33}]_3[PW_4O_{16}]$ as catalyst and with hydrogen peroxide generated by 2-ethylanthrahydroquinone and molecular oxygen, *Appl. Catal., A: Gen.* 250 (2003) 239.
[45] S. Gao, M. Li, Y. Lv, N. Zhou, Z. W. Xi, Epoxidation of propylene with aqueous hydrogen peroxide on a reaction-controlled phase-transfer catalyst, *Org. Process Res. Dev.* 8 (2004) 131.
[46] M. G. Clerici, P. Ingllina, Epoxidation of lower olefins with hydrogen peroxide and titanium silicalite, *J. Catal.* 140 (1993) 71.
[47] H. Gao, G. Lu, J. Suo, S. Li, Epoxidation of allyl chloride with hydrogen peroxide catalyzed by titanium silicalite 1, *Appl. Catal., A: Gen.* 138 (1996) 27.

[48] C. Venturello, R. D'Aloisio, Quaternary ammonium tetrakis(diperoxotungsto)phosphates(3-) as a new class of catalysts for efficient alkene epoxidation with hydrogen peroxide, *J. Org. Chem.* 53 (1988) 1553.
[49] J. Li, Z. W. Xi, S. Gao, Epoxidation of allyl chloride catalyzed by heteropolyphosphatotungstate under oil/water biphasic conditions, *J. Mol. Catal. (China)* 20 (2006) 395.
[50] R. A. Sheldon, Green solvents for sustainable organic synthesis: State of the art, *Green Chem.* 7 (2005) 267.

SECTION 5
Biomimetic Catalysis

CHAPTER 18

Bio-Inspired Iron-Catalyzed Olefin Oxidations: Epoxidation Versus *cis*-Dihydroxylation

Paul D. Oldenburg, Rubén Mas-Ballesté, and Lawrence Que, Jr

Contents		
	1. Introduction	452
	2. Structure–Reactivity Correlation of Catalysts	453
	3. Toward Synthetically Useful Applications	457
	4. Mechanistic Landscape	459
	4.1. Fe^{III}–OOH versus Fe^V–O oxidants	459
	4.2. Fe^{IV}=O versus Fe^V=O oxidants	464
	Acknowledgment	466
	References	466

Abstract   A number of nonheme iron complexes have recently been identified that catalyze the epoxidation and *cis*-dihydroxylation of olefins with H_2O_2 as oxidant. These catalysts have been inspired by a class of arene *cis*-dihydroxylating enzymes called the Rieske dioxygenases, the active sites of which consist of an iron center ligated by two histidines and a bidentate aspartate residue. The two remaining sites are *cis* to each other and are utilized for oxygen activation. The most effective biomimetic catalysts thus far have polydentate ligands that provide two such *cis*-oriented labile sites to activate the H_2O_2 oxidant. This chapter summarizes recent developments in this area of bio-inspired oxidation catalysis and discusses the evolution of the mechanistic pathways proposed to rationalize the new experimental results and the dichotomy between olefin epoxidation versus *cis*-dihydroxylation.

Key Words: Nonheme, Iron, Biomimetic, Bio-inspired, Cytochrome P450, Rieske dioxygenases, Hydrogen peroxide, Peroxide activation, Homogeneous catalysis, *cis*-Dihydroxylation, Mechanism, Catalytic additives, Asymmetric

Department of Chemistry and Center for Metals in Biocatalysis, University of Minnesota, Minneapolis, Minnesota 55455

catalysis, DFT calculations, High-valent iron, Catalytic intermediates, Nitrogen-containing ligands, Nitrogen/oxygen-containing ligands, Isotopic labeling studies. © 2008 Elsevier B.V.

1. INTRODUCTION

A number of iron-based enzymes catalyze the stereospecific oxidation of C=C bonds [1,2]. The most widely studied of these, the cytochromes P450, have active sites consisting of a heme that is attached to the protein backbone through coordination of a thiolate residue at one of the axial positions, leaving the opposite position of the iron octahedron available for oxygen binding and activation (Fig. 18.1) [3,4]. These, as well as biomimetic iron-porphyrin analogs of the P450 systems, have been extensively studied, and much has been learned about their oxidative mechanisms. In this generally accepted mechanism, molecular oxygen coordinates to the iron, forming a ferric-hydroperoxo intermediate species, which, following heterolysis of the O–O hydroperoxo bond, generates an $Fe^{IV}=O(por\bullet)$ as the oxidant for olefin epoxidation (Fig. 18.1) [5,6].

More recently, nonheme iron enzymes have become better characterized and some have been shown to promote similar oxidative transformations [2,7]. Of particular interest is the Rieske dioxygenase family that catalyzes the *cis*-dihydroxylation of arene double bonds [8,9]. The active sites of these enzymes consist of an iron center facially ligated by two histidines and an aspartate residue [9,10], a common binding motif for oxygen-activating nonheme iron enzymes (Fig. 18.1) [11]. In stark contrast to the heme-based systems, this ligand arrangement results in the availability of two *cis* sites for dioxygen binding and activation. Further differentiating these from the heme systems, these enzymes carry out the *cis*-dihydroxylation of C=C bonds, instead of epoxidation [12,13]. Detailed experimental studies suggest that the mechanism of oxygen activation begins with O_2

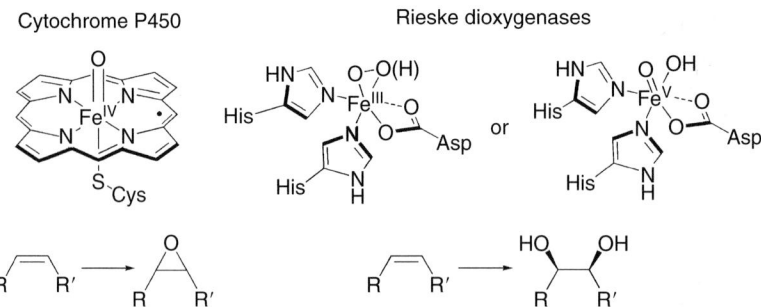

FIGURE 18.1 Structures of the proposed active oxidants from heme (left) and nonheme (right) iron oxygenases for C=C bond oxidation and the product observed in these reactions.

coordination to the iron center, forming an Fe^{III}-OO(H) intermediate [14,15], with the peroxo coordinated in a side-on fashion as observed in the crystal structure of the ESO_2 adduct (Fig. 18.1) [16]. This species can either attack the arene directly or undergo heterolysis of the O–O bond, forming an HO–Fe^V=O species as the active oxidant [2,14,17,18].

These developments have spurred efforts to design bio-inspired nonheme iron catalysts for olefin oxidation reactions [19–21]. The epoxidation and cis-dihydroxylation of olefins are important chemical transformations in both natural product [22,23] and drug synthesis [24,25]. While epoxidation typically involves either peracids or metal-based catalysts with H_2O_2 [26], olefin cis-dihydroxylation involves the use of OsO_4. These nonheme iron catalysts therefore potentially offer an environmentally more benign alternative to peracid-mediated epoxidations and osmium-catalyzed cis-dihydroxylation reactions. The most extensively studied ones thus far are complexes composed of tetradentate N4 ligands that have cis-oriented available coordination sites, analogous to those observed for the Rieske dioxygenase enzymes. These complexes have been shown to catalyze both the epoxidation and cis-dihydroxylation of olefins, but many factors affect which of these products is favored, revealing a surprisingly complex reaction landscape for catalysis. This chapter will summarize the strategies followed in the design of such catalysts, their activities in olefin oxidation catalysis, and their oxidative mechanisms.

2. STRUCTURE–REACTIVITY CORRELATION OF CATALYSTS

Iron complexes of many nitrogen-rich polydentate ligands have been investigated as potential catalysts for olefin oxidation (Fig. 18.2). Those of pentadentate ligands, analogous to cytochrome P450, have only one labile site. Those of tetradentate ligands have two labile sites. These sites can be either cis or trans relative to each other depending on the topology that the tetradentate ligand adopts (Fig. 18.3). For those with cis sites, the two may be equivalent or inequivalent depending on the nature of the ligating groups trans to these sites. The only complex of a tridentate ligand that has been isolated has three facially oriented labile sites.

Table 18.1 compares the reactivities of all the nonheme iron olefin oxidation catalysts reported in the literature to date with limiting H_2O_2 oxidant. In these studies, the H_2O_2 was introduced by syringe pump in order to minimize potential iron-catalyzed peroxide disproportionation. For example, in the case of **9**, yields decreased by 50% when H_2O_2 was added all at once [27]. Among all of the catalysts examined, complexes **6** and **9** give rise to the highest oxidant conversions, 84% for **6** [27] and 90% for **9** [20]. With cis-dialkyl olefins as substrates, the stereochemistry of these alkyl groups is retained in the oxidized product in nearly all cases, favoring a metal-based, rather than radical-based, oxidative process. Additionally, the high degree of retention of configuration for the diol product confirms that epoxide hydrolysis does not take place to form the diol product, since the latter would give rise to trans-diol.

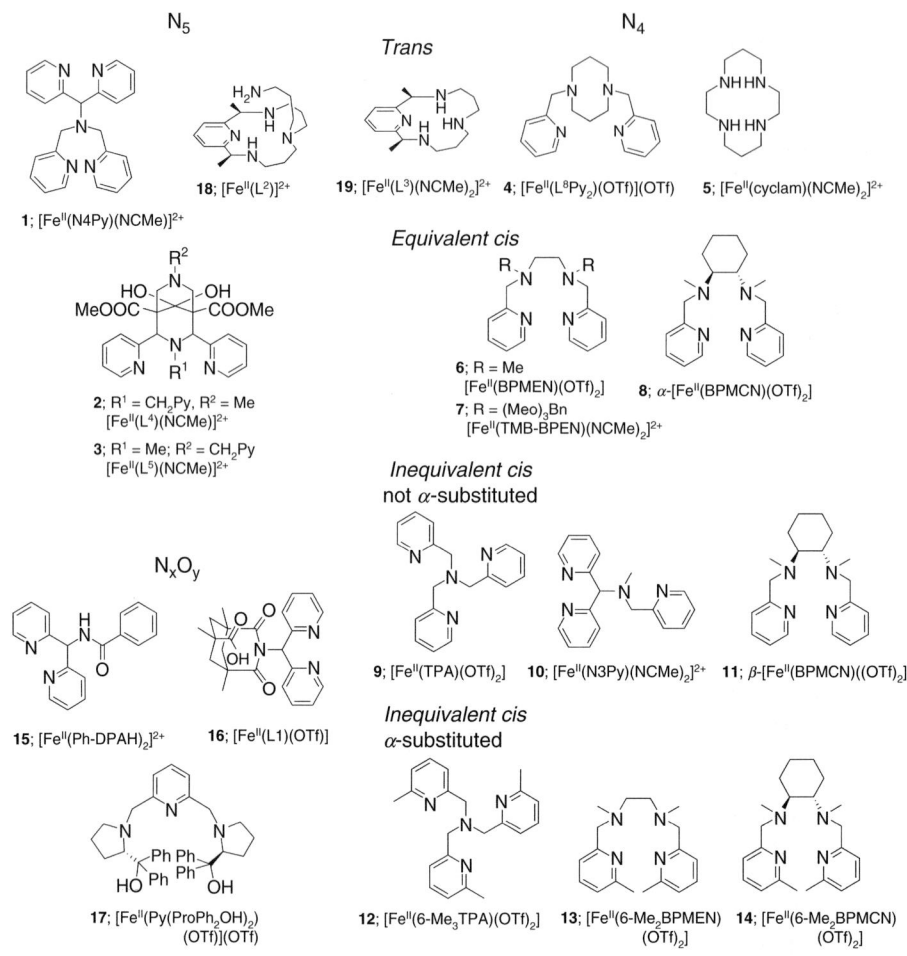

FIGURE 18.2 Ligands for iron-based olefin oxidation catalysts with H_2O_2 oxidant.

Figure 18.3 compares representative structures of nonheme iron olefin oxidation catalysts and their relative oxidative preferences. In many cases, both epoxide and diol are formed as oxidized products. Complexes **1–3**, each supported by a pentadentate ligand, all favor epoxidation but with low stereoselectivity and catalyst efficiency and are thus not good catalysts [28]. Better catalysts are **4** and **5**, which have tetradentate ligands that adopt a *trans* topology resembling heme systems [29,30]. Not surprisingly, these catalysts are highly epoxidation selective.

Also highly selective for epoxidation are complexes with tetradentate ligands that have equivalent *cis*-labile sites (with both being *trans* to tertiary amine groups), **6–8** (Fig. 18.3). However, addition of water can increase the amount of diol product formed [27,31,32]. Catalysts with inequivalent *cis*-labile sites, **9–14** (Fig. 18.3) [27,31,33], favor *cis*-dihydroxylation. Those with no α-methylated pyridyl rings such as **9–11** produce an appreciable amount of epoxide, while those

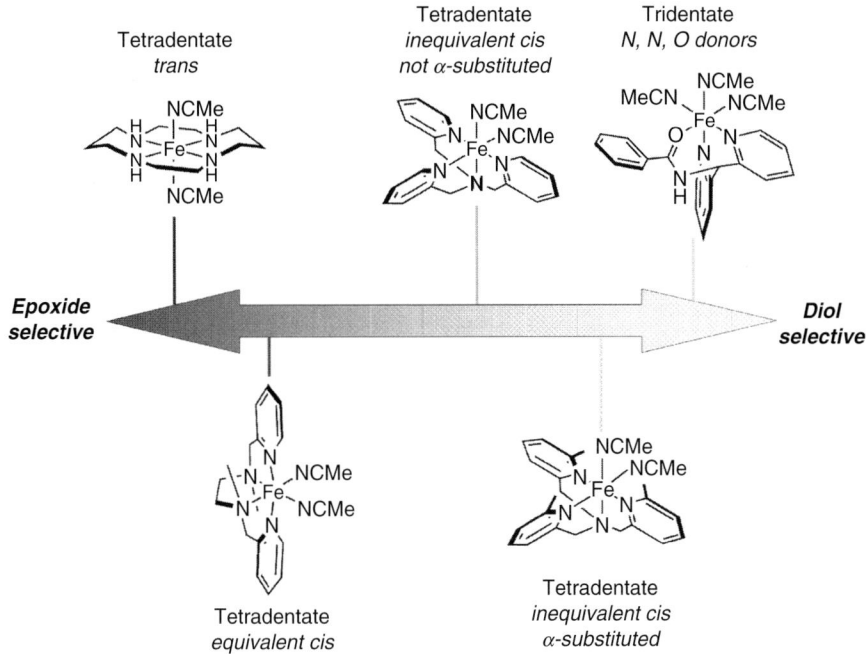

FIGURE 18.3 The epoxidation/*cis*-dihydroxylation selectivity spectrum with representative nonheme iron catalysts that differ in ligand topology.

with α-methylated pyridyl rings, **12–14** [27,34,35], are much more diol selective. From this spectrum of reactivity, it is clear that ligand topology is an important factor that determines the fate of the olefin substrate.

There are a handful of complexes in which oxygen atom donor(s) have been introduced into the polydentate ligand framework. Complexes **15** and **16** have ligands consisting of two pyridines and either a carbonyl oxygen [36] or a carboxylate [37], designed to approximate the 2-His-1-carboxylate ligand set found in the Rieske dioxygenase mononuclear active site. Complex **17**, on the contrary, is a pentadentate N_3O_2 ligand with alcohol functionalities. Complexes **16** and **17** are selective for dihydroxylation but are rather poor catalysts [37,38]. In contrast, **15** exhibits outstanding selectivity for dihydroxylation as well as high oxidative efficiency (Fig. 18.3).

An important feature of the effective catalysts listed in Table 18.1 is the presence of two labile sites. These sites are typically occupied by either weakly coordinating triflate anions or the MeCN solvent. Halide ions are particularly inimical to catalysis. This has been nicely demonstrated in two systematic studies. Ménage and coworkers, using a variation of the BPMEN ligand where the *N*-methyl groups were replaced with 3,4,5-trimethoxybenzyl groups (**7**) [32], synthesized a series of three iron(II) complexes in which the two remaining coordination sites were occupied by two chloride ions, one chloride and one CH_3CN, and two CH_3CN molecules and found that the efficacy of the catalyst depended inversely upon the number of ligated chlorides. The complex with two

TABLE 18.1 Oxidation of cyclooctene catalyzed by nonheme iron complexes in CH_3CN with limiting H_2O_2 oxidant.

	Epoxide: diol	%H_2O_2 converted[a]	Ref
N_5 ligands			
N4Py (**1**)	>50:1	6	[28]
L^4 (**2**)	6:1	10	[28]
L^5 (**3**)	1:1	20	[28]
N_4 ligands			
trans			
$L^8 Py_2$ (**4**)	8.8:1	39	[30]
cyclam (**5**)	34:1	40	[30]
equivalent cis			
BPMEN (**6**)	8.3:1	84	[27]
TMB-BPEN (**7**)	>50:1	31	[32]
α-BPMCN (**8**)	9.8:1	65	[31]
inequivalent cis, not α-substituted			
TPA (**9**)	1:1.2	74	[27]
	1:2	90[b]	[20]
N3Py (**10**)	1:1.6	74	[33]
β-BPMCN (**11**)	1:1.9	77	[31]
inequivalent cis, α-substituted			
6-Me$_3$ TPA (**12**)	1:7	56	[27]
6-Me$_2$BPMEN (**13**)	1:4.3	79	[35]
6-Me$_2$BPMCN[c] (**14**)	1:6.2	47	[35]
N_xO_y[d] ligands			
Ph-DPAH (**15**)	1:14	75	[36]
L1[c] (**16**)	1:6	8	[37]
Py(ProPh$_2$OH)$_2$ (**17**)	1:1.8	17	[38]

[a] The percent of H_2O_2 converted into epoxide and diol products.
[b] Yield in the presence of 2% v/v acetone, which prevented overoxidation of diol.
[c] Substrate = 1-octene.
[d] Refers to polydentate ligands with at least one oxygen atom donor group.

chlorides bound is not at all an olefin oxidation catalyst and reacts with H_2O_2 to produce HO· that hydroxylates the phenyl ring of the ligand. The complex with one chloride is a case where both olefin epoxidation (6–7% conversion efficiency) and ligand oxidation are observed. The complex with no chloride ligands does not effect ligand oxidation at all and does indeed catalyze olefin epoxidation with 30% conversion efficiency.

In a second study, Rybak-Akimova and coworkers investigated the catalytic properties of the dinuclear complex, $[\{Fe^{III}(BPMEN)\}_2(\mu\text{-O})(\mu\text{-OH})]^{3+}$ [39]. This complex is a catalyst comparable to its mononuclear iron(II) version, **6**. The addition of a single equivalent of fluoride resulted in the opening of the

$Fe_2(\mu\text{-}O)(\mu\text{-}OH)$ core to form $[\{Fe^{III}(BPMEN)(F)\}\text{-}(\mu\text{-}O)\text{-}\{Fe^{III}(BPMEN)(OH)\}]^{2+}$, a species with severely diminished catalytic activity. Addition of another equivalent of F^- replaced the remaining –OH group and afforded a complex completely inactive for catalysis. These studies emphasize the need for labile ligands on the iron center to allow efficient access of H_2O_2. Such access promotes inner-sphere oxidation of the iron center to generate the high-valent iron-oxo oxidant, instead of HO• that would be formed by outer-sphere oxidation of the iron center.

3. TOWARD SYNTHETICALLY USEFUL APPLICATIONS

A number of the catalytic systems we have thus far discussed all afford high conversion of oxidant into epoxide and diol products; however, their utility as catalysts is limited because of the requirement for a large excess of substrate. There has been some effort focused on developing nonheme iron complexes to be used as practical catalysts for synthesis, emphasizing conversion of substrate to product(s). Jacobsen and coworkers explored the catalytic activity of **6** in the presence of added acetic acid and found that olefins could be converted into epoxides in high yield with 3 mol% catalyst and 30 mol% HOAc (Table 18.2) [40]. The added acetic acid is clearly important, as reactions under similar reaction conditions but without HOAc afforded a lower yield and selectivity for epoxide [41].

A similar shift in selectivity was observed for the diol-selective catalyst **9** in the presence of acetic acid [42]. In the absence of acetic acid, the epoxide/diol ratio was 1:5.5 in favor of diol; however, the addition of a large excess of acetic acid (20 equiv relative to substrate) shifted the ratio to 1.9:1 in favor of epoxide. In this case, the conversion of substrate into the desired products totaled only 52%, which is synthetically not very useful.

Rybak-Akimova and coworkers also found that adding acid, triflic acid in this case, improved the epoxidation yield of iron complexes with L^2 and L^3, macrocyclic ligands with one pyridine incorporated into the macrocycle [43]. L^2 differs from L^3 in having a pendent amino group that is tethered to the macrocycle (Fig. 18.2). The iron complex with pentadentate L^2, **18**, was relatively ineffective as an olefin oxidation catalyst with H_2O_2. However, addition of 1 equiv of triflic acid gave an epoxide yield of 39%, while 5 equiv further increased the epoxide yield to 86% (Table 18.2). On the contrary, the iron complex of tetradentate L^2, **19**, was effective as a catalyst without triflic acid, affording a 32% epoxide yield, but the addition of 5 equiv of triflic acid improved this yield to 67%. It was shown that the addition of acid to **18** resulted in protonation of the pendent amine, facilitating its dissociation from the iron center. However, the better epoxidation results obtained for **18** in the presence of 5 equiv triflic acid relative to **19** suggested to the authors that the added acid must play a more active role in promoting epoxidation.

The best epoxidation results thus far were reported by Beller and coworkers using an *in situ* generated epoxidation catalyst capable of up to 100% conversion of a range of aryl olefins to epoxides (Table 18.2) [44]. They used a combination of

TABLE 18.2 A Sampling of nonheme Iron oxidation catalysts with H_2O_2 as oxidant with high conversion of substrate to oxidized products

Ligand	Catalyst (mol%)	Additive (additive/substrate)	Substrate	H_2O_2/substrate	Epoxide[a]	Diol[a]	References
BPMEN	3		1-Octene	1.5	73	3	[41]
BPMEN	3	AcOH (0.30)	1-Decene	1.5	85	–	[40]
TPA	3		1-Octene	1.5	6	33	[42]
TPA	3	AcOH (20)	1-Octene	1.5	34	18	[42]
L^3	5		Cyclooctene	1.5	32	–	[43]
L^3	5	HOTf (0.25)	Cyclooctene	1.5	67	–	[43]
L^2	5		Cyclooctene	1.5	9	–	[43]
L^2	5	HOTf (0.05)	Cyclooctene	1.5	39	–	[43]
L^2	5	HOTf (0.25)	Cyclooctene	1.5	86	–	[43]
H_2pydic	5	Pyrrolidine (0.1)	Styrene	2	94	–	[44]

[a] Percentage of substrate converted to indicated product.

FeCl$_3$ H$_2$O, 2 equiv pyridine-2,6-carboxylic acid (H$_2$pydic), and 2 equiv pyrrolidine in *tert*-amyl alcohol solvent, presumably forming a species like [Fe^{III}(pydic)(pyrrolidine)$_{1\ or\ 2}$]$^+$ as the catalyst. The absence of added acid in this system is useful for the formation of acid-sensitive epoxides.

The discovery of iron complexes that can catalyze olefin *cis*-dihydroxylation led Que and coworkers to explore the possibility of developing asymmetric dihydroxylation catalysts. Toward this end, the optically active variants of complexes **11** [(1R,2R)-BPMCN] and **14** [(1S,2S)- and (1R-2R)-6-Me$_2$BPMCN] were synthesized [35]. In the oxidation of *trans*-2-heptene under conditions of limiting oxidant, 1R,2R-**11** was found to catalyze the formation of only a minimal amount of diol with a slight enantiomeric excess (ee) of 29%. However, 1R-2R-**14** and 1S,2S-**14** favored the formation of diol (epoxide/diol = 1:3.5) with ees of ~80%. These first examples of iron-catalyzed asymmetric *cis*-dihydroxylation demonstrate the possibility of developing iron-based asymmetric catalysts that may be used as alternatives to currently used osmium-based chemistry [45].

4. MECHANISTIC LANDSCAPE

The mechanistic landscape described for these nonheme iron catalysts has been developed on the basis of the nature of intermediates that can be trapped at low temperatures in the reaction of iron complexes with H$_2$O$_2$ and complementary H$_2^{18}$O and H$_2^{18}$O$_2$ labeling experiments (Table 18.3). The complexity of this mechanistic landscape is illustrated by the scheme in Fig. 18.4 that summarizes its current picture.

4.1. Fe^{III}–OOH versus Fe^V–O oxidants

Early experiments demonstrated that iron complexes **1** and **9** reacted with H$_2$O$_2$ at low temperature to afford transient species identified as low-spin ($S = \frac{1}{2}$) Fe^{III}-OOH complexes [46–48]. Resonance Raman experiments showed that the coordination of the hydroperoxide to the low-spin iron(III) center resulted in the weakening of the O–O bond and the strengthening of the Fe–O bond [49]; because of this combination of effects, it was suggested that the low-spin iron(III) center served to prime the peroxide O–O bond for cleavage during catalysis [20,27]. Thus, an Fe^{III}-OOH species was implicated in these oxidations. However, experiments monitoring the behavior of the trapped Fe^{III}-OOH intermediates at low temperature in the presence of potential substrates revealed that such species were in fact sluggish oxidants [50], so a further activation step would be required to elicit the observed activity in olefin oxidation.

Subsequent ^{18}O labeling experiments carried out on olefin oxidations by **9** (Table 18.3) provided the first experimental data that established the fate of the peroxide oxygen atoms in the oxidized products [27]. Of particular significance were the *cis*-diol labeling results demonstrating that one diol oxygen derived from H$_2$O$_2$ and the other from H$_2$O. The incorporation of water into the *cis*-diol product required a mechanism in which O–O bond cleavage occurred at least prior to the

TABLE 18.3 Isotope Labeling Results for Cyclooctene Oxidation by H_2O_2 in the Presence of H_2O Catalyzed by Iron Complexes[a]

		Cyclooctene oxide		cis-Cyclooctane-1,2-diol				
		$H_2^{18}O$	$H_2^{18}O_2$	$H_2^{18}O$		$H_2^{18}O_2$		
Complex	Ligand	$^{18}O\%$	$^{18}O\%$	$^{16}O^{16}O\%$	$^{16}O^{18}O\%$	$^{16}O^{18}O\%$	$^{18}O^{18}O\%$	References
4	L8Py2	<1	60	—	—	—	—	[30]
5	Cyclam	<1	88	—	—	—	—	[30]
6	BPMEN	30	73	23	73	58	1	[55]
9	TPA	9	90	13	86	97	3	[27]
11	β-BPMCN	15	66	97	3	4	93	[30]
12	6-Me3TPA	3	54	99	1	4	96	[27]
15[b]	Ph-DPAH	—	—	67	33	36	64	[36]

[a] Reaction conditions for $H_2^{18}O_2$ experiments: 10 equiv of $H_2^{18}O_2$ (diluted from 2% aqueous solution) were added by syringe pump to a solution of catalyst and substrate in CH_3CN. Reaction conditions for $H_2^{18}O$ experiments: same as above except for using $H_2^{16}O_2$ (diluted from 35% aqueous solution) and the addition of 1,000 equiv of $H_2^{18}O$ to this solution.
[b] Substrate = 1-octene.

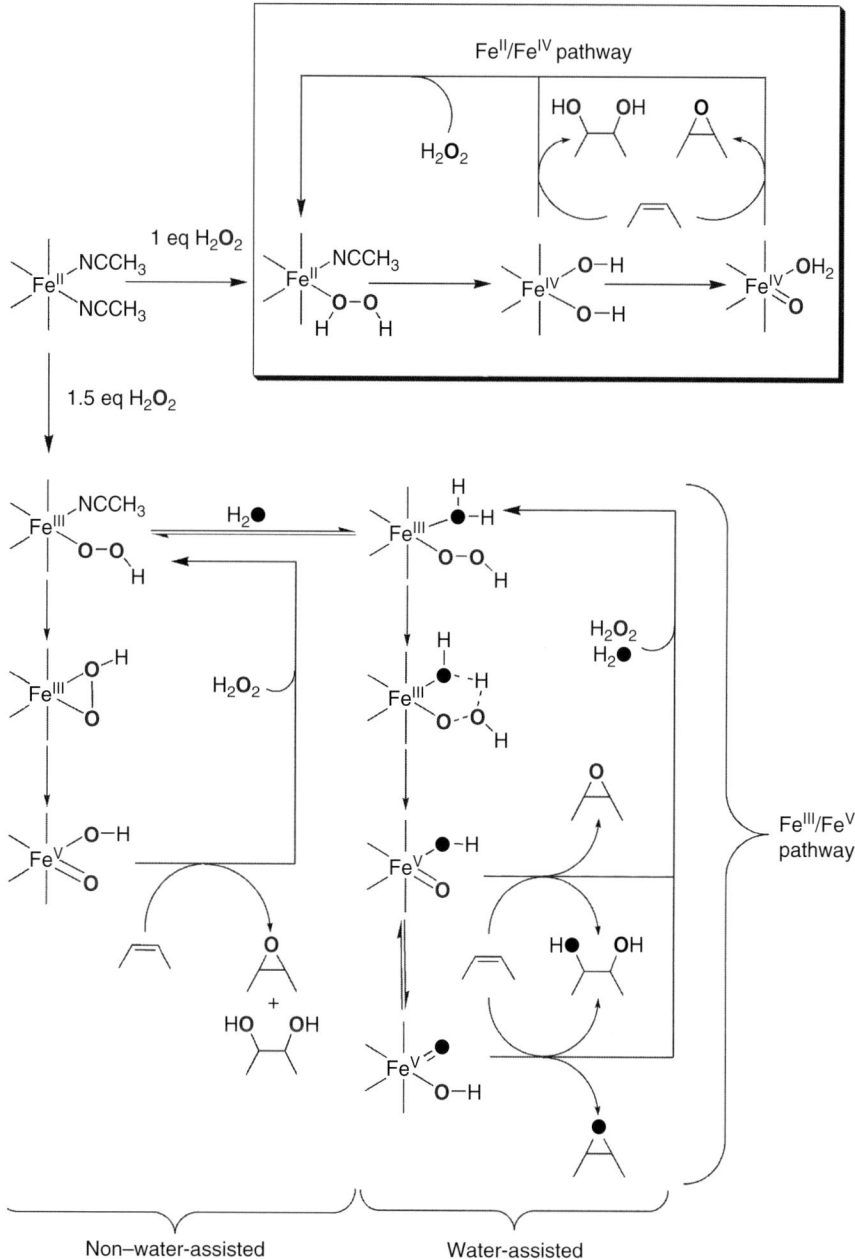

FIGURE 18.4 Mechanistic landscape for olefin oxidation by H_2O_2 catalyzed by nonheme iron complexes. Sources of oxygen atoms include H_2O_2 (unfilled circle) and H_2O (filled circle).

formation of the second C–O bond, if not earlier. The extent of label incorporation from water was also found to depend on the added water concentration. This dependence showed saturation behavior that suggested a pre-equilibrium step involving the added water. Accordingly, a water-assisted mechanism (*wa*) was proposed consisting of the initial formation of a [(TPA)Fe^{III}-OOH(OH$_2$)] species that derives from the [(TPA)Fe^{III}-OOH(NCMe)] intermediate observed at −40 °C [46]. Heterolytic cleavage of the O–O bond in [(TPA)Fe^{III}-OOH(OH$_2$)] is promoted by the coordinated water to generate the HO–Fe^V=O oxidant that adds across the C=C bond to form the *cis*-diol. Unlike for the *cis*-diol, the oxygen atom for the corresponding epoxide product from the same reaction derived mainly from H$_2$O$_2$, with only 10% label incorporated from water. Nevertheless, the H$_2^{18}$O concentration dependence of epoxide labeling exhibited the same saturation behavior as for diol, suggesting that the same oxidant was responsible for epoxidation. An HO–Fe^V=O oxidant can be used to rationalize the epoxide labeling by assuming that the oxo oxygen, which is derived from the cleavage of the –OOH moiety, is the atom transferred to the olefin most of the time. In less than 10% of the time the hydroxo oxygen is transferred. This transfer can take place via an oxo-hydroxo tautomerization that transfers the hydroxo proton intramolecularly to the oxo atom to form an isomeric HO–Fe^V=O oxidant. In the case of **9**, the two labile sites are not equivalent since they are *trans* to different ligands, an amine, and a pyridine [51]. So the two putative HO–Fe^V=O isomers do not necessarily have the same energy and one isomer, presumably the initially formed one, is more favored than the other. This consideration would explain the low level of epoxide labeling from H$_2^{18}$O in the **9**-catalyzed reaction.

Density functional theory (DFT) calculations show that this mechanistic hypothesis is energetically reasonable. The generation of the HO–Fe^V=O oxidant from the [(TPA)Fe^{III}-OOH(OH$_2$)] intermediate was found to have a thermodynamic cost of only 5 kcal/mol and a kinetic barrier of ∼20 kcal/mol (Fig. 18.5) [52]. Furthermore, the HO–Fe^V=O oxidant could carry out either epoxidation or *cis*-dihydroxylation of the olefin, depending on which oxygen atom of the oxidant initiated attack of the substrate [53]. Thus, epoxidation occurs by oxo attack on the olefin, forming the first C–O bond and an intermediate carbon-based radical, which then rebounds to form the second C–O bond. On the other hand, *cis*-dihydroxylation is initiated by hydroxo attack to form the first C–O bond and an intermediate carbon-based radical, followed by rebound with the oxo group to form the second C–O bond.

Subsequent experiments for **6**-catalyzed olefin oxidation also supported the existence of a *wa* pathway. As for **9**, a low-spin Fe^{III}-OOH intermediate could be trapped at low temperature, −50 °C in this case [54]. In the presence of 1,000 equiv H$_2^{18}$O, the *cis*-diol formed exhibited a labeling pattern having one diol oxygen atom from H$_2$O$_2$ and the other from water, while the epoxide showed 30% incorporation of water (Table 18.3) [55]. However, the BPMEN ligand in **6** adopts a *cis*-α topology, such that the two labile sites are both *trans* to tertiary amines and thus are equivalent [56]. So, unlike for **9**, the two possible HO–Fe^V=O isomers that can form are in fact energetically equivalent and would be expected to undergo

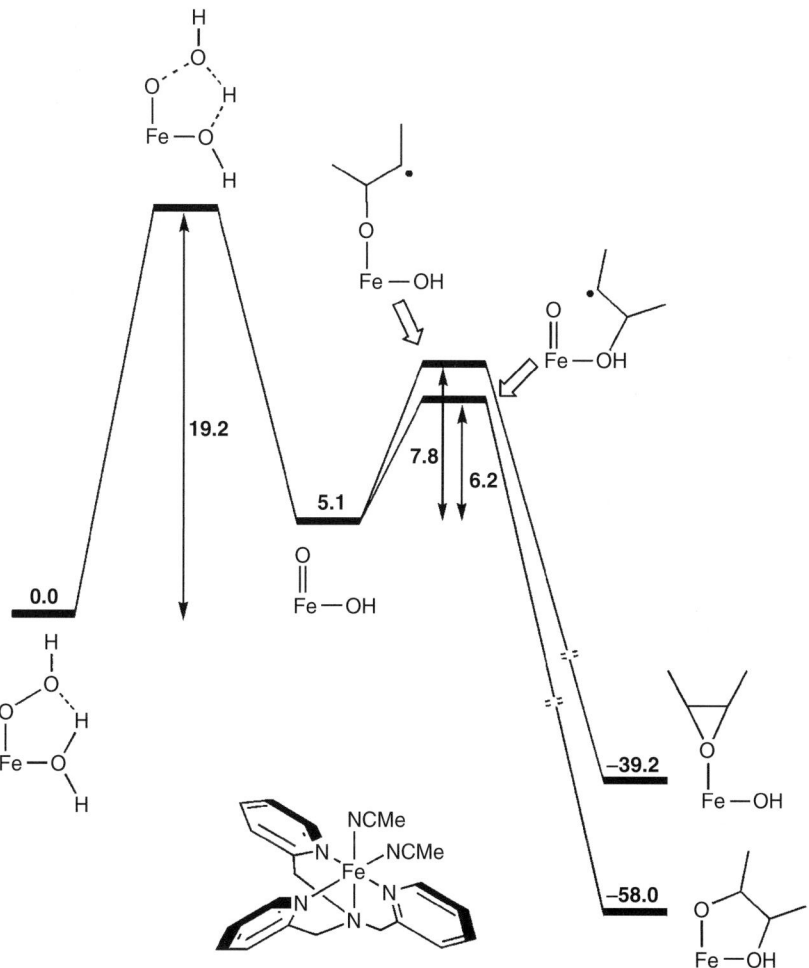

FIGURE 18.5 Energy profile calculated for water-assisted epoxidation and *cis*-dihydroxylation by **9**.

facile oxo-hydroxo tautomerization and give rise to 50% labeled epoxide. This is clearly not the case.

It should be noted that, unlike for **9** [20], the presence of water affects the catalytic chemistry of **6** significantly [55]. The addition of the 1,000 equiv H_2O typically used for labeling experiments resulted in a 35% decrease in yield and dramatically changed the product composition, from an epoxide/diol ratio of 8:1 in the absence of added water to 1.2:1 in its presence, so diol formation was quite enhanced. Adding water clearly favors the *wa* pathway, but the oxidant formed in this reaction at much lower concentrations of water must be different, strongly favoring epoxidation in a non–water-assisted (*nwa*) pathway (Fig. 18.4). DFT calculations have proposed that [(BPMEN)Fe^{III}-OOH(MeCN)] itself can be the epoxidizing agent, with an activation barrier of 21 kcal/mol [55].

Complexes with weaker-field ligands are more diol-selective catalysts and exhibit a different labeling pattern. As exemplified by results for **12**, both oxygen atoms observed in the *cis*-diol product derived from H_2O_2. Furthermore, the use of 50:50 $H_2^{16}O_2 : H_2^{18}O_2$ did not yield any singly labeled diol product, demonstrating that diol is formed from a single molecule of H_2O_2 [27]. Thus, water does not appear to play a role in this mechanism, and an *nwa* pathway is proposed. In this case, it cannot be determined whether O–O bond cleavage occurs prior to olefin attack. Accordingly, either iron-peroxo species or high-valent iron-oxo species can be considered as the oxidant.

The mechanistic picture became more complex as labeling data for more catalysts were obtained. For catalyst **11** [30], ^{18}O label from $H_2^{18}O$ was incorporated into epoxide 15% of the time, a result similar to that found for **9** (Table 18.3). The similar extent of labeling was not surprising, since the *cis*-labile sites of **11** are, like those of **9**, *trans* to an amine and a pyridine and thus inequivalent. However, unlike for the *cis*-diol product from **9**, both oxygen atoms from H_2O_2 were incorporated into the *cis*-diol product of **11**, suggesting that the *cis*-dihydroxylation pathway used an *nwa* mechanism like **12**. These results would appear to require two distinct oxidants: one generated from a *wa* mechanism (HO–Fe^V=O) for epoxidation and another responsible for *cis*-dihydroxylation that is generated via an *nwa* pathway. At the present time, how this situation comes about is not understood.

Labeling experiments for the diol-specific catalyst **15** showed the predominant incorporation of both peroxide oxygen atoms into the diol, but with incorporation from water in a significant fraction of product (Table 18.3) [36]. This pattern was different from that observed for **6** and **9** where one of the diol oxygens derived almost completely from water. Owing to the facial tridentate nature of its polydentate ligand, this complex has three *cis*-labile sites that would allow simultaneous coordination of a water molecule and both oxygen atoms of the hydroperoxo group. Thus, it was proposed that *nwa* O–O bond cleavage resulted in the formation of an Fe^V(O)(OH)(OH$_2$) oxidant that attacked the olefin substrate. Some water incorporation would be observed if intramolecular proton transfer would occur among the oxo, hydroxo, and aqua ligands to effect label scrambling prior to attack of substrate. The incorporation of water label into the *cis*-diol product in this reaction demonstrates for the first time that peroxide O–O bond cleavage must occur prior to the formation of the second C–O bond of the diol product even in the *nwa* pathway that is favored for these weak-field ligand complexes.

4.2. Fe^{IV}=O versus Fe^V=O oxidants

A key observation supporting the Fe(III)–Fe(V) mechanistic scheme presented above was the low-temperature trapping of iron(III)-peroxo species for **6, 9, 11**, and **12** [46,49,54,57,58]. Since the olefin oxidations catalyzed by these complexes are highly stereoselective, it is unlikely that Fe^{III}-OOH species proposed to form in the course of catalysis would undergo O–O homolysis to produce HO•. Instead, it has been argued that the Fe^{III}-OOH species must undergo O–O heterolytic cleavage to form the HO–Fe^V=O oxidant, but direct evidence for the latter has not been obtained. In contrast, Fe^{IV}O species have been generated from **1–3, 6, 9,** and **11** by reaction of the iron(II) starting material with peracids or PhIO [59]. Their relative stability has allowed them to be fully characterized spectroscopically and, in some cases, even crystallized [60,61]. Thus, their existence requires a consideration of whether such species may be involved in the catalytic oxidations discussed in this chapter. The fact that Fe^{IV}=O species are stable enough to be trapped may lead some to argue that such intermediates cannot be the catalytic active species in efficient epoxidation systems. However, arguments to the contrary have been presented recently [62].

The reaction of **2** with H$_2$O$_2$ in aqueous solution to form the corresponding Fe^{IV}=O complex has been reported [63]. The isosbestic behavior observed in the conversion as monitored by UV–vis spectroscopy favored the direct conversion of Fe^{II} to Fe^{IV}=O without the involvement of an Fe(III) intermediate. This observation opens the door for the inclusion of an Fe(II)/Fe(IV) cycle in the mechanistic landscape for olefin oxidation (see box in Fig. 18.4). Indeed, such a cycle was determined to be energetically feasible in a recent computational study of the reaction of H$_2$O$_2$ with a tetradentate ligand analog of **2**. The initial Fe^{II}(HOOH) adduct evolved to form Fe^{IV}(OH)$_2$ after O–O bond cleavage and then Fe^{IV}=O(OH$_2$) subsequent to proton transfer. Upon reaction with an olefin, the former would afford the diol product, while the latter would afford the epoxide product [62].

To date, the observed reactivity of Fe^{IV}=O species toward olefins does not lend unequivocal support for this idea. Although Fe^{IV}=O species derived from **2** or **3** were found to react with cyclooctene generating epoxide in about 50% yield, oxidation of olefins with H$_2$O$_2$ catalyzed by **2** or **3** was not very efficient under argon (20% conversion of oxidant to product or less) and showed evidence for a significant contribution from radical autooxidation when carried out in air [28]. Furthermore, the Fe^{IV}=O complex supported by a 15-membered macrocyclic ligand was inactive in olefin epoxidation, but became quite active only in the presence of excess oxidant (PhIO). A PhIO·Fe^{IV}=O adduct was proposed as the oxidant instead [64].

Fe^{IV}=O species have also been implicated in one recent study [65] to explain the dramatic effect of acetic acid in enhancing the epoxide yield and selectivity of olefin oxidations mediated by **6** and **9** (see Section 3) [40]. NMR evidence was obtained by Talsi and coworkers for the formation of Fe^{IV}=O species from the reaction of **6** or **9** with H$_2$O$_2$ in the presence of acetic acid at −50 °C. The Fe^{IV}=O species may be formed as a consequence of the iron-catalyzed *in situ* formation of peracetic acid as proposed by Fujita *et al.* [42], which has been shown to react with **9** efficiently to form [(TPA)Fe^{IV}O]$^{2+}$ [66]. It remains to be established whether such species indeed participate in epoxidation catalysis at higher temperature.

As indicated by the scheme in Fig. 18.4, the mechanistic landscape for bio-inspired olefin oxidation catalysis by nonheme iron complexes is increasing in complexity. The evidence available to date supports the involvement of the Fe(III)/Fe(V) cycle for **6** and **9** and related complexes having tetradentate ligands with *cis*-labile sites and no α substituents on the pyridines. Much less is known about the mechanisms for other complexes, owing in part to the lack of information about the nature of possible iron-peroxo adducts that must form in order to activate the O–O bond for olefin oxidation. An Fe(II)/Fe(IV) cycle has been proposed for consideration based on DFT calculations [62] and may in fact operate in the mechanisms of these other complexes. Such a cycle is particularly attractive as it would mimic the oxidation state changes that the active site iron undergoes in proposed mechanisms for a number of nonheme iron oxygenases [2,7]. The information currently available in this area emphasizes the important role the ligand plays in modulating the reactivity of the iron center, and much more needs to be learned. In the period following submission of this chapter, several papers have appeared that are quite relevant to the subject of this chapter [67–72].

ACKNOWLEDGMENT

This work was supported by the U.S. Department of Energy (DE-FG02–03ER15455).

REFERENCES

[1] I. Bertini, H. B. Gray, E. I. Stiefel, J. S. Valentine (Eds.), *Biological Inorganic Chemistry*, University Science Books, Sausalito, 2007.
[2] M. Costas, M. P. Mehn, M. P. Jensen, L. Que, Jr., Dioxygen activation at mononuclear nonheme iron active sites: Enzymes, models, and intermediates, *Chem. Rev.* 104 (2004) 939.
[3] P. R. Ortiz de Montellano (Ed.), *Cytochrome P-450: Structure, Mechanism, and Biochemistry*, 3rd ed., Kluwer Academic/Plenum Publishers, New York, 2005.
[4] W. Nam, Comprehensive Coordination Chemistry II, in: J. A. McCleverty, T. J. Meyer (Eds.), Vol. 8, L. Que, Jr., W. B. Tolman (Vol. Eds.), *Cytochrome P-450* p. 281 Elsevier, San Diego, 2004.
[5] H. Fujii, Electronic structure and reactivity of high-valent oxo iron porphyrins, *Coord. Chem. Rev.* 226 (2002) 51.
[6] J. H. Dawson, Probing structure-function relations in heme-containing oxygenases and peroxidases, *Science* 240 (1988) 433.
[7] M. M. Abu-Omar, A. Loaiza, N. Hontzeas, Reaction mechanisms of mononuclear non-heme iron oxygenases, *Chem. Rev.* 105 (2005) 2227.
[8] L. P. Wackett, Mechanism and applications of Rieske non-heme iron dioxygenases, *Enzyme Microb. Technol.* 31 (2002) 577.
[9] D. J. Ferraro, L. Gakhar, S. Ramaswamy, Rieske business: Structure-function of Rieske non-heme oxygenases, *Biochem. Biophys. Res. Commun.* 338 (2005) 175.
[10] B. Kauppi, K. Lee, E. Carredano, R. E. Parales, D. T. Gibson, H. Eklund, S. Ramaswamy, Structure of an aromatic ring-hydroxylating dioxygenase—Naphthalene 1,2-dioxygenase, *Structure* 6 (1998) 571.
[11] K. D. Koehntop, J. P. Emerson, L. Que, Jr., The 2-His-1-carboxylate facial triad: A versatile platform for dioxygen activation by mononuclear non-heme iron(II) enzymes, *J. Biol. Inorg. Chem.* 10 (2005) 87.
[12] S. M. Resnick, D. T. Gibson, Diverse reactions catalyzed by naphthalene dioxygenase from *Pseudomonas* sp. strain NCIB 9816, *J. Ind. Microbiol.* 17 (1996) 438.
[13] D. R. Boyd, G. N. Sheldrake, The dioxygenase catalysed formation of vicinal *cis*-diols, *Nat. Prod. Rep.* 15 (1998) 309.
[14] M. D. Wolfe, J. V. Parales, D. T. Gibson, J. D. Lipscomb, Single turnover chemistry and regulation of O_2 activation by the oxygenase component of naphthalene 1,2-dioxygenase, *J. Biol. Chem.* 276 (2001) 1945.

[15] M. D. Wolfe, D. J. Altier, A. Stubna, C. V. Popescu, E. Münck, J. D. Lipscomb, Benzoate 1, 2-dioxygenase from *Pseudomonas putida*: Single turnover kinetics and regulation of a two-component Rieske dioxygenase, *Biochemistry* 41 (2002) 9611.
[16] A. Karlsson, J. V. Parales, R. E. Parales, D. T. Gibson, H. Eklund, S. Ramaswamy, Crystal structure of naphthalene dioxygenase: Side-on binding of dioxygen to iron, *Science* 299 (2003) 1039.
[17] M. D. Wolfe, J. D. Lipscomb, Hydrogen peroxide-coupled *cis*-diol formation catalyzed by naphthalene 1,2-dioxygenase, *J. Biol. Chem.* 278 (2003) 829.
[18] A. Bassan, M. R. A. Blomberg, P. E. M. Siegbahn, A theoretical study of the *cis*-dihydroxylation mechanism in naphthalene 1,2-dioxygenase, *J. Biol. Inorg. Chem.* 9 (2004) 439.
[19] K. Chen, M. Costas, L. Que, Jr., Spin state tuning of non-heme iron-catalyzed hydrocarbon oxidations: Participation of Fe^{III}-OOH and Fe^{V}=O intermediates, *Dalton Trans.* (2002) 672.
[20] R. Mas-Ballesté, M. Fujita, C. Hemmila, L. Que, Jr., Bio-inspired iron-catalyzed olefin oxidation. Additive effects on the *cis*-diol/epoxide ratio, *J. Mol. Catal. A: Chem.* 251 (2006) 49.
[21] P. D. Oldenburg, L. Que, Jr., Bio-inspired nonheme iron catalysts for olefin oxidation, *Catal. Today* 117 (2006) 15.
[22] X. Wan, G. Doridot, M. M. Joullié, Progress towards the total synthesis of trichodermamides A and B: Construction of the oxazine ring moiety, *Org. Lett.* 9 (2007) 977.
[23] A. Padwa, H. Zhang, Synthesis of some members of the hydroxylated phenanthridone subclass of the *Amaryllidaceae* alkaloid family, *J. Org. Chem.* 72 (2007) 2570.
[24] K. Uchida, S. Yokoshima, T. Kan, T. Fukuyama, Total synthesis of ($\pm$)-morphine, *Org. Lett.* 8 (2006) 5311.
[25] J.-K. Jung, B. R. Johnson, T. Duong, M. Decaire, J. Uy, T. Gharbaoui, P. D. Boatman, C. R. Sage, R. Chen, J. G. Richman, D. T. Connolly, G. Semple, Analogues of Acifran: Agonists of the high and low affinity niacin receptors, GPR109a and GPR109b, *J. Med. Chem.* 50 (2007) 1445.
[26] B. S. Lane, K. Burgess, Metal-catalyzed epoxidations of alkenes with hydrogen peroxide, *Chem. Rev.* 103 (2003) 2457.
[27] K. Chen, M. Costas, J. Kim, A. K. Tipton, L. Que, Jr., Olefin *cis*-dihydroxylation versus epoxidation by nonheme iron catalysts: Two faces of an Fe^{III}-OOH coin, *J. Am. Chem. Soc.* 124 (2002) 3026.
[28] M. R. Bukowski, P. Comba, A. Lienke, C. Limberg, C. Lopez de Laorden, R. Mas-Ballesté, M. Merz, L. Que, Jr., Catalytic epoxidation and 1,2-dihydroxylation of olefins with bispidine-iron(II)/H_2O_2 systems, *Angew. Chem. Int. Ed.* 45 (2006) 3446.
[29] W. Nam, R. Y. N. Ho, J. S. Valentine, Iron-cyclam complexes as catalysts for the epoxidation of olefins by 30% aqueous hydrogen peroxide in acetonitrile and methanol, *J. Am. Chem. Soc.* 113 (1991) 7052.
[30] R. Mas-Ballesté, M. Costas, T. van den Berg, L. Que, Jr., Ligand topology effects on olefin oxidations by bio-inspired [$Fe^{II}(N_2Py_2)$] catalysts, *Chem. Eur. J.* 12 (2006) 7489.
[31] M. Costas, L. Que, Jr., Ligand topology tuning of iron-catalyzed hydrocarbon oxidations, *Angew. Chem. Int. Ed.* 41 (2002) 2179.
[32] Y. Mekmouche, S. Ménage, J. Pécaut, C. Lebrun, L. Reilly, V. Schuenemann, A. Trautwein, M. Fontecave, Mechanistic tuning of hydrocarbon oxidations with H_2O_2, catalyzed by hexacoordinate ferrous complexes, *Eur. J. Inorg. Chem.* (2004) 3163.
[33] M. Klopstra, G. Roelfes, R. Hage, R. M. Kellogg, B. L. Feringa, Non-heme iron complexes for stereoselective oxidation: Tuning of the selectivity in dihydroxylation using different solvents, *Eur. J. Inorg. Chem.* (2004) 846.
[34] K. Chen, L. Que, Jr., *cis*-Dihydroxylation of olefins by a nonheme iron catalyst. A functional model for Rieske dioxygenases, *Angew. Chem. Int. Ed.* 38 (1999) 2227.
[35] M. Costas, A. K. Tipton, K. Chen, D.-H. Jo, L. Que, Jr., Modeling Rieske dioxygenases. The first example of iron-catalyzed asymmetric *cis*-dihydroxylation of olefins, *J. Am. Chem. Soc.* 123 (2001) 6722.
[36] P. D. Oldenburg, A. A. Shteinman, L. Que, Jr., Iron-catalyzed olefin *cis*-dihydroxylation using a bio-inspired N,N,O-ligand, *J. Am. Chem. Soc.* 127 (2005) 15672.
[37] P. D. Oldenburg, C.-Y. Ke, A. A. Tipton, A. A. Shteinman, L. Que, Jr., A structural and functional model for dioxygenases with a 2-His-1-carboxylate triad, *Angew. Chem. Int. Ed.* 45 (2006) 7975.

[38] S. Gosiewska, M. Lutz, A. L. Spek, R. J. M. K. Gebbink, Mononuclear diastereopure non-heme Fe(II) complexes of pentadentate ligands with pyrrolidinyl moieties: Structural studies, and alkene and sulfide oxidation, *Inorg. Chim. Acta* 360 (2007) 405.
[39] S. Taktak, S. V. Kryatov, T. E. Haas, E. V. Rybak-Akimova, Diiron(III) oxo-bridged complexes with bpmen and additional monodentate or bidentate ligands: Synthesis and reactivity in olefin epoxidation with H_2O_2, *J. Mol. Catal. A: Chem.* 259 (2006) 24.
[40] M. C. White, A. G. Doyle, E. N. Jacobsen, A synthetically useful, self-assembling MMO mimic system for catalytic alkene epoxidation with aqueous H_2O_2, *J. Am. Chem. Soc.* 123 (2001) 7194.
[41] J. Y. Ryu, J. Kim, M. Costas, K. Chen, W. Nam, L. Que, Jr., High conversion of olefins to *cis*-diols by non-heme iron catalysts and H_2O_2, *Chem. Commun.* (2002) 1288.
[42] M. Fujita, L. Que, Jr., In situ formation of peracetic acid in iron-catalyzed epoxidations by hydrogen peroxide in the presence of acetic acid, *Adv. Synth. Catal.* 346 (2004) 190.
[43] S. Taktak, W. Ye, A. M. Herrera, E. V. Rybak-Akimova, Synthesis and catalytic properties in olefin epoxidation of novel iron(II) complexes with pyridine-containing macrocycles bearing an aminopropyl pendant arm, *Inorg. Chem.* 46 (2007) 2929.
[44] G. Anilkumar, B. Bitterlich, F. G. Gelalcha, M. K. Tse, M. Beller, An efficient biomimetic Fe-catalyzed epoxidation of olefins using hydrogen peroxide, *Chem. Commun.* (2007) 289.
[45] H. C. Kolb, M. S. VanNieuwenhze, K. B. Sharpless, Catalytic asymmetric dihydroxylation, *Chem. Rev.* 94 (1994) 2483.
[46] C. Kim, K. Chen, J. Kim, L. Que, Jr., Stereospecific alkane hydroxylation with H_2O_2 catalyzed by an iron(II)-tris(2-pyridylmethyl)amine complex, *J. Am. Chem. Soc.* 119 (1997) 5964.
[47] G. Roelfes, M. Lubben, K. Chen, R. Y. N. Ho, A. Meetsma, S. Genseberger, R. M. Hermant, R. Hage, S. K. Mandal, V. G. Young, Jr., Y. Zang, H. Kooijman, A. Spek, L. Que, Jr., B. L. Feringa, Iron chemistry of a pentadentate ligand that generates a metastable Fe^{III}-OOH intermediate, *Inorg. Chem.* 38 (1999) 1929.
[48] G. Roelfes, V. Vrajmasu, K. Chen, R. Y. N. Ho, J.-U. Rohde, C. Zondervan, R. M. la Crois, E. P. Schudde, M. Lutz, A. L. Spek, R. Hage, B. L. Feringa, E. Münck, L. Que, Jr., End-on and side-on peroxo derivatives of non-heme iron complexes with pentadentate ligands: Models for putative intermediates in biological iron/dioxygen chemistry, *Inorg. Chem.* 42 (2003) 2639.
[49] R. Y. N. Ho, G. Roelfes, B. L. Feringa, L. Que, Jr., Raman evidence for a weakened O-O bond in mononuclear low-spin iron(III)-hydroperoxides, *J. Am. Chem. Soc.* 121 (1999) 264.
[50] M. J. Park, J. Lee, Y. Suh, J. Kim, W. Nam, Reactivities of mononuclear non-heme iron intermediates including evidence that iron(III)-hydroperoxo species is a sluggish oxidant, *J. Am. Chem. Soc.* 128 (2006) 264.
[51] A. Diebold, K. S. Hagen, Iron(II) polyamine chemistry: Variation of spin state and coordination number in solid state and solution with iron(II) tris(2-pyridylmethyl)amine complexes, *Inorg. Chem.* 37 (1998) 215.
[52] A. Bassan, M. R. A. Blomberg, P. E. M. Siegbahn, L. Que, Jr., A density functional study of O-O bond cleavage for a biomimetic non-heme iron complex demonstrating an Fe^V-intermediate, *J. Am. Chem. Soc.* 124 (2002) 11056.
[53] A. Bassan, M. R. A. Blomberg, P. E. M. Siegbahn, L. Que, Jr., Two faces of a biomimetic non-heme HO-Fe^V=O oxidant: Olefin epoxidation versus *cis*-dihydroxylation, *Angew Chem. Int. Ed.* 44 (2005) 2939.
[54] E. A. Duban, K. P. Bryliakov, E. P. Talsi, Characterization of low-spin ferric hydroperoxo complexes with N,N´-dimethyl-N,N´-bis(2-pyridylmethyl)-1,2-diaminoethane, *Mendeleev Commun.* (2005) 12.
[55] D. Quiñonero, K. Morokuma, D. G. Musaev, R. Mas-Ballesté, L. Que, Jr., Metal-peroxo versus metal-oxo oxidants in non-heme iron-catalyzed olefin oxidations: Computational and experimental studies on the effect of water, *J. Am. Chem. Soc.* 127 (2005) 6548.
[56] K. Chen, L. Que, Jr., Evidence for the participation of a high-valent iron-oxo species in stereospecific alkane hydroxylation by a non-heme iron catalyst, *Chem. Commun.* (1999) 1375.
[57] M. P. Jensen, M. Costas, R. Y. N. Ho, J. Kaizer, A. Mairata i Payeras, E. Münck, L. Que, Jr., J.-U. Rohde, A. Stubna, High-valent nonheme iron. Two distinct iron(IV) species derived from a common iron(II) precursor, *J. Am. Chem. Soc.* 127 (2005) 10512.

[58] N. Lehnert, R. Y. N. Ho, L. Que, Jr., E. I. Solomon, Electronic structure of high-spin Fe(III)-alkylperoxo complexes and its relation to low-spin analogs: Reaction coordinate of O-O bond homolysis, *J. Am. Chem. Soc.* 123 (2001) 12802.
[59] X. Shan, L. Que, Jr., High-valent nonheme iron-oxo species in biomimetic oxidations, *J. Inorg. Biochem.* 100 (2006) 421.
[60] J.-U. Rohde, J.-H. In, M.-H. Lim, W. W. Brennessel, M. R. Bukowski, A. Stubna, E. Münck, W. Nam, L. Que, Jr., Crystallographic and spectroscopic characterization of a nonheme Fe(IV)=O complex, *Science* 229 (2003) 1037.
[61] E. J. Klinker, J. Kaizer, W. W. Brennessel, N. L. Woodrum, C. J. Cramer, L. Que Jr., Structures of nonheme oxoiron(IV) complexes from X-ray crystallography, NMR spectroscopy, and DFT calculations, *Angew Chem. Int. Ed.* 44 (2005) 3690.
[62] P. Comba, G. Rajaraman, H. Rohwer, A density functional theory study of the reaction of the biomimetic iron(II) complex of a tetradentate bispidine ligand with H_2O_2, *Inorg. Chem.* 46 (2007) 3826.
[63] J. Bautz, M. R. Bukowski, M. Kerscher, A. Stubna, P. Comba, A. Lienke, E. Münck, L. Que, Jr., Formation of an aqueous oxoiron(IV) complex at pH 2–6 from a nonheme iron(II) complex and H_2O_2, *Angew. Chem. Int. Ed.* 45 (2006) 5681.
[64] Y. Suh, M. S. Seo, K. M. Kim, Y. S. Kim, H. G. Jang, T. Tosha, T. Kitagawa, J. Kim, W. Nam, Nonheme iron(II) complexes of macrocyclic ligands in the generation of oxoiron(IV) complexes and the catalytic epoxidation of olefins, *J. Inorg. Biochem.* 100 (2006) 627.
[65] E. A. Duban, K. P. Bryliakov, E. P. Talsi, The active intermediates of non-heme-iron-based systems for catalytic alkene epoxidation with H_2O_2/CH_3COOH, *Eur. J. Inorg. Chem.* (2007) 852.
[66] M. H. Lim, J.-U. Rohde, A. Stubna, M. R. Bukowski, M. Costas, R. Y. N. Ho, E. Münck, W. Nam, L. Que, Jr., An Fe^{IV}=O complex of a tetradentate tripodal nonheme ligand, *Proc. Natl. Acad. Sci. USA* 100 (2003) 3665.
[67] F. G. Gelalcha, B. Bitterlich, A. Anilkumar, M. K. Tse, M. Beller, Iron-catalyzed asymmetric epoxidation of aromatic alkenes using hydrogen peroxide, *Angew. Chem. Int. Ed.* 46 (2007) 7293.
[68] J. Tang, P. Gamez, J. Reedijk, Efficient [bis(imino)pyridine-iron]-catalyzed oxidation of alkanes, *Dalton Trans.* (2007) 4644.
[69] M. S. Chen, M. C. White, A Predictably selective aliphatic C—H oxidation reaction for complex molecule synthesis, *Science* 318 (2007) 789.
[70] R. Mas-Ballesté, L. Que, Jr., Iron catalyzed olefin epoxidation in the presence of acetic acid. Insights into the nature of metal-based oxidant, *J. Am. Chem. Soc.* 129 (2007) 15964.
[71] P. C. A. Bruijnincx, I. L. C. Buurmans, S. Gosiewska, M. A. H. Moelands, M. Lutz, A. L. Spek, G.v. Koten, R. J. M. K. Gebbink, Iron(II) complexes with Bio-inspired N,N,O ligands as oxidation catalysts: Olefin epoxidation and *cis*-Dihydroxylation, *Chem. Eur. J.* 14 (2008) 1228.
[72] K. Suzuki, P. D. Oldenburg, L. Que, Jr., Iron-catalyzed asymmetric olefin *cis*-dihydroxylation with 97% enantiomeric excess, *Angew. Chem. Int. Ed.* 47 (2008) 1887.

CHAPTER **19**

Quantum Chemical Analysis of the Reaction Pathway for Styrene Epoxidation Catalyzed by Mn-Porphyrins

María C. Curet-Arana, Randall Q. Snurr, and **Linda J. Broadbelt**

Contents

1.	Introduction	472
2.	Methodology	473
3.	Results	475
4.	Conclusions	483
	Acknowledgments	483
	References	484

Abstract Embedded quantum/classical calculations were used to study the epoxidation reaction of styrene catalyzed by Mn-porphyrins. Optimized geometries were obtained for the Mn-porphyrin, reaction intermediates, and transition state structures along the proposed reaction path. A polarizable continuum model (PCM) was used to study solvent effects, with dichloromethane as the solvent. While it has been shown previously that the concerted intermediate between the oxidized porphyrin and the alkene is the lowest energy configuration, a transition state to directly form the concerted intermediate without the prior formation of a radical could not be found. A stepwise mechanism, in which a radical intermediate is formed before the concerted intermediate, is proposed.

Key Words: Density functional theory, DFT calculations, Porphyrin, Alkene, Styrene, Epoxidation, Reaction pathway. © 2008 Elsevier B.V.

Department of Chemical and Biological Engineering and Institute for Catalysis in Energy Processes, Northwestern University, Evanston, Illinois 60208

Mechanisms in Homogeneous and Heterogeneous Epoxidation Catalysis © 2008 Elsevier B.V.
DOI: 10.1016/B978-0-444-53188-9.00019-5 All rights reserved.

1. INTRODUCTION

In pursuit of biomimetic catalysts, metalloporphyrins have been extensively studied in attempts to mimic the active site of cytochrome P450, which is an enzyme that catalyzes oxidation reactions in organisms. In recent decades, catalysis of alkene epoxidation with metalloporphyrins has received considerable attention. It has been found that iron [1–3], manganese [4,5], chromium [6], and cobalt porphyrins can be used as model compounds for the active site of cytochrome P450, and oxidants such as iodosylbenzene, sodium hypochlorite [7,8], hydrogen peroxide [9], and peracetic acid [10] have been shown to work for these systems at ambient temperature and pressure. While researchers have learned a great deal about these catalysts, several practical issues limit their applicability, especially deactivation.

The stability of metalloporphyrins during epoxidation reactions depends on the metal used and the environment surrounding it. Manganese porphyrins have enhanced stability compared to iron catalysts, but Mn-porphyrins still deactivate very rapidly. In order to increase the stability of metalloporphyrins, efforts have been made to isolate the catalytic center by adding bulky groups to the sides [11–13]. Immobilization of the catalyst onto a solid support, such as silica [14], and elaborate structures, like the well-known picnic-basket porphyrins created by Collman and coworkers, have also improved the stability of porphyrins [15]. Recently, Nguyen, Hupp and coworkers have used supramolecular chemistry to construct self-assembling systems that contain a regular structure for porphyrin encapsulation [16,17]. These "molecular squares" bind the porphyrin and have been shown to increase the stability of the catalyst.

Some insight into the deactivation behavior of Mn-porphyrins during alkene epoxidation has been derived from an analysis of proposed mechanisms for the reaction. The general catalytic cycle involved in oxidizing alkenes to epoxides is shown in Fig. 19.1. In the first step, the oxygen atom is transferred from the oxidant to the metal-based catalyst, and in the second step, the oxygen atom is transferred to the alkene in order to form the epoxide. While the general features of the catalytic cycle are known, the exact identity of the intermediates is not known. For example, the state of the active –M–[O]– is not fully resolved. A metal-oxo species (Mn=O) has been widely suggested as the active form of the catalyst for epoxidation reactions [18,19], but other species have also been proposed in the literature [20]. Structural and spectral characterization has been obtained for some Mn-oxo complexes (Mn=O), and the ^{1}H NMR spectrum obtained by Groves and coworkers [21–23] and the X-ray absorption spectroscopy obtained by Ayougou et al. [24] provide the strongest evidence for the nature of this species.

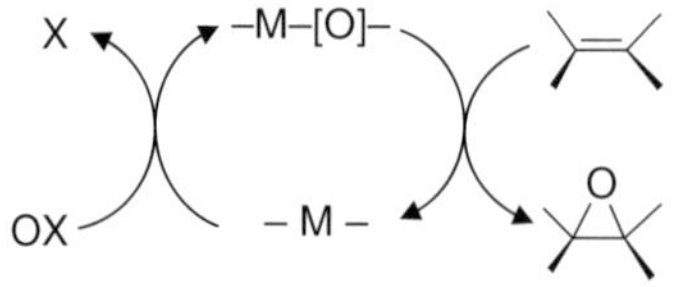

FIGURE 19.1 Schematic diagram of the catalytic cycle involving a metal catalyst that transfers oxygen from an oxidant to an alkene to form an epoxide.

There is also debate about the nature of the intermediates involved in the second step in the catalytic cycle illustrated in Fig. 19.1, that is, the transfer of oxygen to the alkene to form the epoxide. No intermediates have been detected experimentally, but five different possibilities have been proposed in the literature for the alkene complexed to the oxidized porphyrin [11,25–29]. The five proposed intermediates are radical, cation, concerted, metallaoxetane, and pi-radical-cation species. The literature is rather complicated due to the lack of direct experimental observation, and it is not clear that conclusions from, say, iron and chromium porphyrins also apply to manganese porphyrins [28]. Arasasingham *et al.* claim unequivocal evidence for a radical intermediate being involved in the oxidation of alkenes by manganese porphyrins [28]. They also discuss a charge-transfer complex that is similar to the concerted intermediate. Recently, density functional theory (DFT) and quantum mechanics/molecular mechanics (QM/MM) calculations were applied to styrene epoxidation by Mn-porphyrins:

$$\text{(19.1)}$$

These calculations predicted that the radical and concerted intermediates have the lowest energies, again suggesting that their participation in the catalytic cycle is likely for styrene epoxidation by Mn-porphyrins [30]. This study also demonstrated that the QM/MM method is in very good agreement with full DFT calculations and that electronic configurations, geometries, and relative energies between spin states are similar for DFT and QM/MM for all stable intermediates.

The present study builds on the previous DFT and QM/MM results, where stable intermediates were identified [30], to explicitly map the reaction coordinates of individual steps in the catalytic cycle. Transition state structures are identified and characterized, and energy barriers are calculated. The effect of the solvent, dichloromethane, on the energy barriers is also analyzed. Based on the transition state structures identified, a plausible mechanism for styrene epoxidation by Mn-porphyrins is proposed.

2. METHODOLOGY

Integrated QM/MM calculations were carried out using the ONIOM method [31–34] as implemented in Gaussian 98 [35]. The system was divided into two layers, an inner layer, which was treated at both a high level and a low level of theory, and the full system, which was treated at only the low level. The high-level part was analyzed using DFT with the unrestricted PW91 functional [36,37] and the effective core potential LANL2DZ as the basis set. The low-level part was analyzed with the universal force field (UFF) [38]. The specific Mn-porphyrin analyzed is shown in Fig. 19.2 (left). It is a mimic of Mn-tetraphenyl porphyrin, which is commonly studied for epoxidation reactions [39,40]. In Fig. 19.2, the atoms that were treated at the high level are shown with their atomic symbols. For the high-level calculation, hydrogen atoms were used as link atoms when bonds between the high and low-level parts were broken. For ONIOM calculations, the total extrapolated energy is defined as:

$$E_{ONIOM} = E_{(system, low\ level)} + E_{(inner\ layer, high\ level)} - E_{(inner\ layer, low\ level)} \qquad (19.2)$$

The energies of iodosylbenzene and phenyl iodide were calculated using full DFT. For interaction of styrene, styrene oxide, and phenylacetaldehyde with the Mn-porphyrin, the phenyl ring was treated at the low level, and all other atoms were assigned to the inner layer. An example of this partitioning is shown in Fig. 19.3 for the radical intermediate.

The geometries and energies for the stable intermediates were obtained by performing full geometry optimization with no symmetry constraints. Optimization of stable species was verified with frequency calculations. Vibrational frequencies of the molecule were calculated within the rigid-rotor harmonic oscillator approximation. Results were also tested by taking the inner layer of the results optimized using ONIOM and performing DFT single point and stable calculations with PW91/LANL2DZ [41,42]. Transition state geometries were obtained with the quasi-synchronous transit method (QST3). Energies were corrected for basis set superposition error (BSSE) with the counterpoise method [43] when appropriate. Both the triplet and quintet states for all intermediates and transition states were analyzed. Standard statistical mechanics was used to calculate the translational, vibrational, rotational, and electronic contributions to the partition function in order to estimate free energies. Free energies were calculated

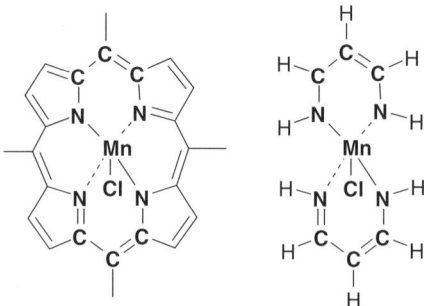

FIGURE 19.2 Inner and outer layers used for the ONIOM calculations. Left: atoms in the inner layer are shown with their atomic symbols and atoms in the outer layer are shown in a stick representation. Right: the inner layer is shown with the hydrogen link atoms.

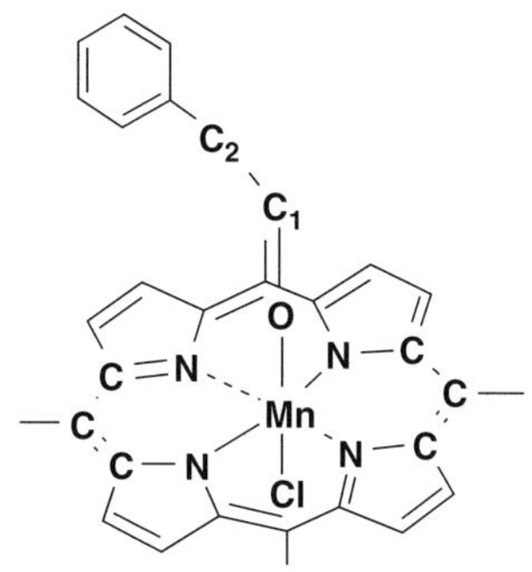

FIGURE 19.3 Inner and outer layers used for ONIOM calculation of the radical intermediate formed by the complexation of styrene with the oxidized porphyrin. Atoms in the inner layer are shown with their atomic symbols.

at 298 K. A polarizable continuum model (PCM) was used to study solvent effects on the energies of stable intermediates and transition states. For these calculations, dichloromethane was used as the solvent, which has a dielectric constant of 8.93.

3. RESULTS

In our previous work, we analyzed different reaction intermediates that have been proposed in the literature for the complexation of alkenes with the oxidized Mn-porphyrin [30]. We found that the concerted intermediate was the lowest energy configuration for the binding of either styrene or propene to the oxidized Mn-porphyrin and that the radical intermediate was 26 kJ/mol higher in energy than the concerted intermediate for the styrene system. Based on these findings, a plausible reaction mechanism involving only the most stable intermediate would involve the complexation of styrene with the oxidized Mn-porphyrin to form a concerted intermediate followed by the desorption of styrene oxide or phenylacetaldehyde. This series of steps is consistent with postulates put forth in the literature [26,27], albeit with the nature of the oxidized Mn-porphyrin and the intermediate involving the alkene still under debate and not directly evaluated via either experiment or theory. We sought transition state structures for these three steps in the mechanism. However, despite the energetic preference for the concerted intermediate, we could not find a transition state to form this intermediate without prior formation of a radical intermediate despite numerous attempts with different initial structures. Therefore, the alternative mechanism we propose based on transition state structures that were located is shown in Fig. 19.4.

The mechanism in Fig. 19.4 is stepwise, in which the radical intermediate is formed before the concerted one. Step 1 involves formation of the metal-oxo intermediate via oxygen transfer to the manganese center from iodosylbenzene. In Step 2, the alkene reacts with the oxidized porphyrin to form the radical intermediate. From this intermediate, the concerted intermediate is formed (Step 3), and finally styrene oxide desorbs in Step 4. The radical intermediate can also

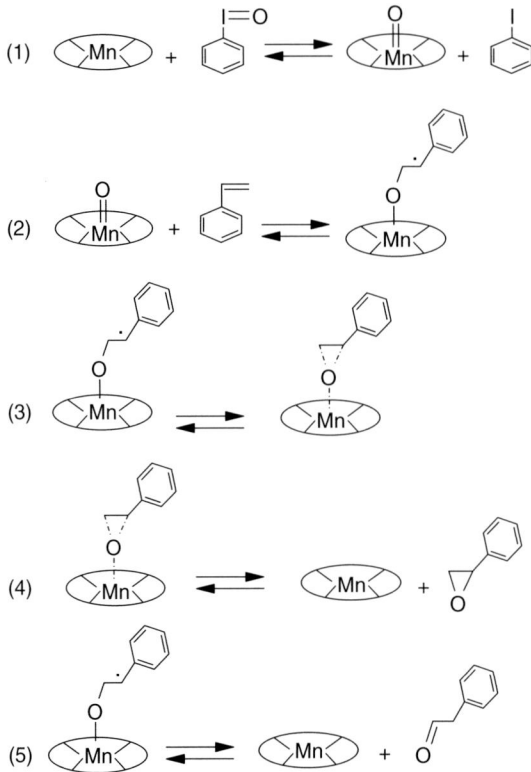

FIGURE 19.4 Mechanism of styrene epoxidation with iodosylbenzene catalyzed by Mn-porphyrins proposed based on transition state structures identified.

react to form the by-product phenylacetaldehyde via Step 5. All steps were analyzed in detail to quantify the electronic energy and free energy barriers of reaction, as described in the following paragraphs.

The electronic energy as a function of the reaction coordinate for the proposed reaction mechanism is shown in Figs. 19.5 and 19.6. For steps that involve a molecule that is not chemically bonded to the porphyrin, we found low energy states in which the two species are weakly associated. For example, if styrene and the oxidized porphyrin are initially very far apart, they form a weakly associated state before reacting. Similarly, the product molecule weakly interacts with the catalyst before becoming completely unassociated. These weakly associated states are stable states with no imaginary frequencies and are ~40 kJ/mol lower in energy than the isolated species. (Basis set superposition error corrections were ~2 kJ/mol.) To probe the nature of these weak interactions, we calculated the Coulombic energy between the two molecules using a very simple model in which a point charge is placed on each atom with the appropriate Mulliken charge. For example, this estimated Coulombic energy between the oxidized Mn-porphyrin and styrene in the weakly associated state is 7 kJ/mol, thereby accounting for 18% of the total interaction energy compared to the isolated species. An average charge

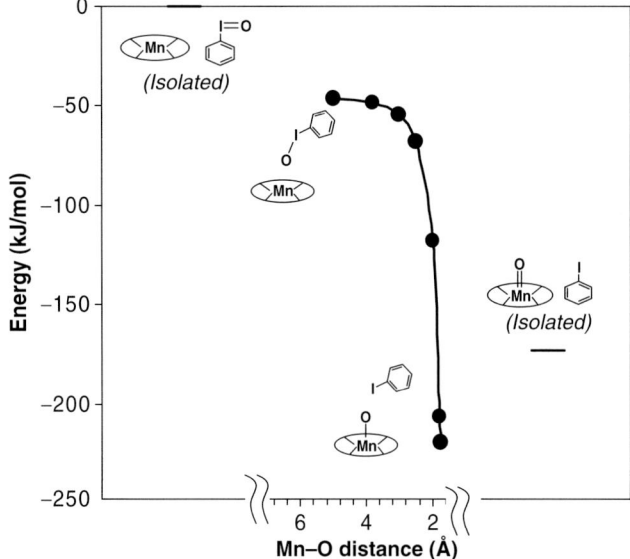

FIGURE 19.5 Electronic energy as a function of the distance between the Mn atom and the oxygen atom as the reaction coordinate for the oxidation of Mn-porphyrin with iodosylbenzene for the quintet spin state.

transfer of $0.03e^-$ between the molecules was observed, indicating some chemical interaction. Single point energy calculations with full DFT (PW91/LANL2DZ) also revealed weakly associated states that were lower in energy than the isolated species.

The first reaction in the mechanism is the oxidation of the Mn-porphyrin with iodosylbenzene. The electronic energy as a function of the reaction coordinate for the quintet spin state is shown in Fig. 19.5, starting and ending with the weakly interacting complexes. The reaction coordinate was taken to be the distance between the manganese and oxygen atoms. This reaction is highly exothermic, with a $\Delta E_{reaction}$ of ~-170 kJ/mol, and proceeds without a barrier. The results for this step on the triplet spin surface were similar.

The energy diagram for the rest of the steps comprising the reaction mechanism is shown in Fig. 19.6, where the energy barriers for both triplet and quintet surfaces are shown. The reference energy for this diagram is the oxidized porphyrin and styrene in the triplet state when both species are infinitely separated. As styrene approaches the oxidized Mn-porphyrin, a stable, weakly associated state is found with the double bond of styrene located ~5 Å from the oxygen atom bonded to Mn. For this state, there is slight charge transfer between the two species, and the triplet state is 24 kJ/mol lower in energy than the quintet state. This step is exothermic by 41.5 and 45.4 kJ/mol for the quintet and triplet spin states, respectively.

The first barrier shown in Fig. 19.6 (TS-2 for Step 2 in Fig. 19.4) involves the transition state to form the radical intermediate. This barrier is 51 kJ/mol for

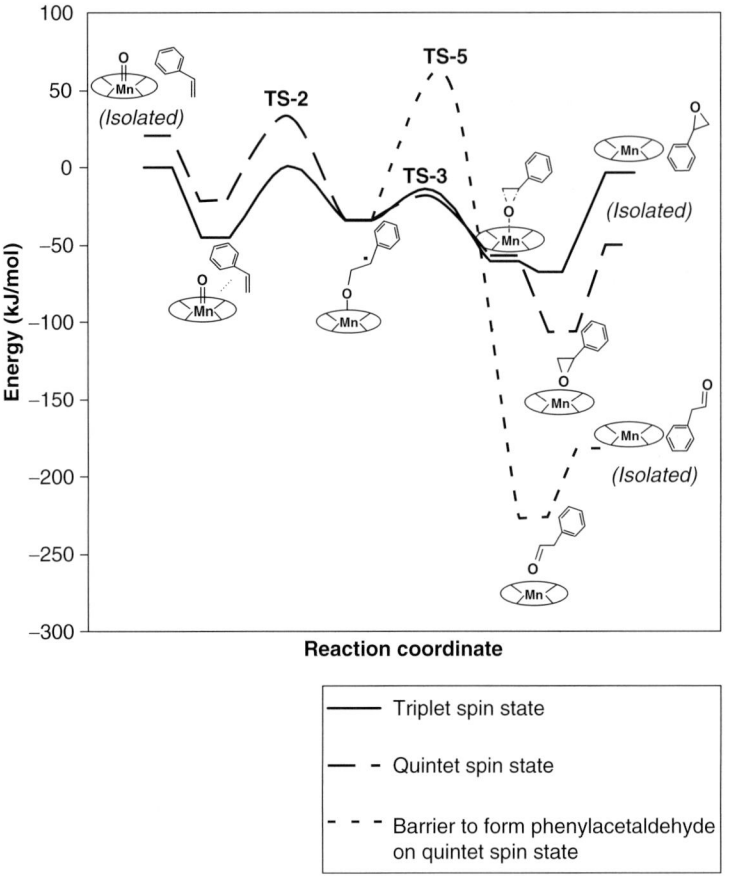

FIGURE 19.6 Electronic energy as a function of the reaction coordinate for the epoxidation of styrene using a Mn-porphyrin catalyst. Numbering of the transition states (TS) corresponds to the steps in Fig. 19.4.

the triplet state and 58 kJ/mol for the quintet state. In the transition state, the distance between Mn and O lengthens from 1.65 to 1.72 Å as shown in Fig. 19.7, and the spin density on C_2 is −0.4 for the triplet state and 0.4 for the quintet state. Once the radical intermediate is formed, there is a single bond between C_1 and C_2, with a length of 1.50 Å, and the spin density on C_2 is −1 for the triplet state and +1 for the quintet state. The second barrier shown in Fig. 19.6 is the transition state to form the concerted intermediate from the radical intermediate (TS-3 for Step 3 in Fig. 19.4). The barrier is ~15 kJ/mol for both spin states. In this transition state, the oxygen atom is interacting with the Mn atom and the two carbon atoms of styrene. As shown in Fig. 19.7, the bond distance between Mn and O lengthens from 1.87 Å in the radical intermediate to 1.94 Å, and the C–O bond distance of the proximate carbon atom changes from 1.45 to 1.49 Å. The second carbon atom, where the

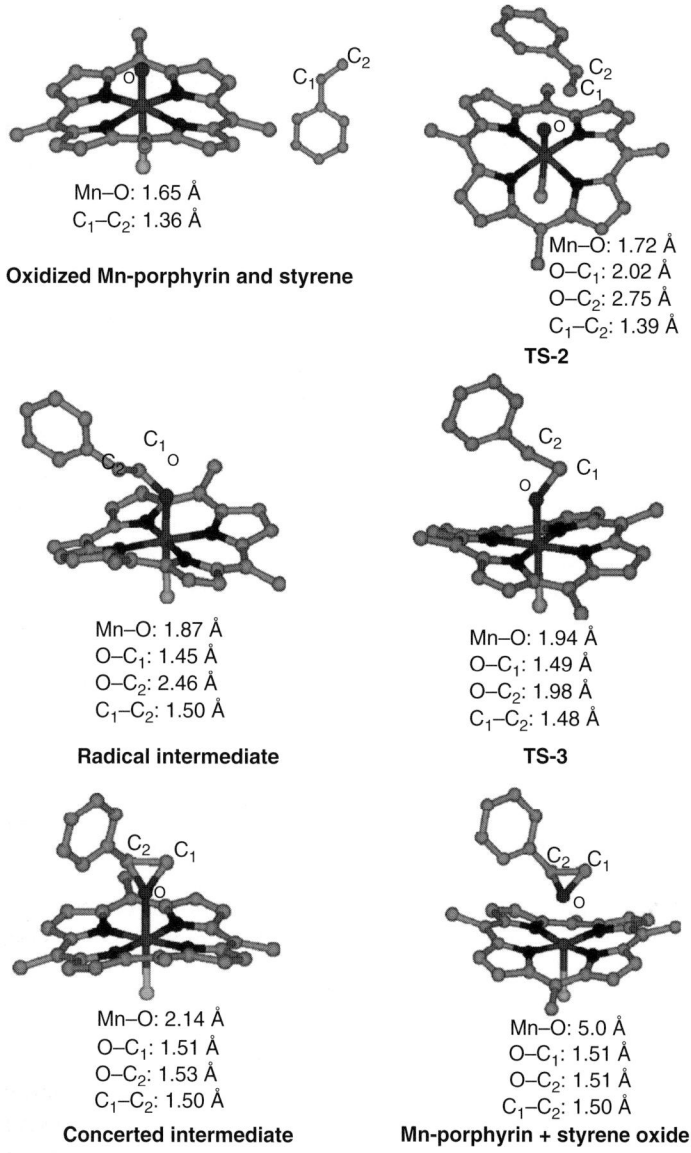

FIGURE 19.7 Geometries for stable intermediates and transition states in the triplet spin state. Hydrogen atoms are not shown for clarity.

radical center was located in the radical intermediate, moves closer to a distance of 1.98 Å away from the oxygen atom. In the concerted intermediate, the Mn–O bond length is 2.14 Å, and the O–C_1 and O–C_2 bond lengths are 1.51 and 1.53 Å, respectively. Spin densities on the Mn atom are 2.00 for the triplet spin state and 3.64 for the quintet state.

It is interesting that the differences in the relative energies between the two different spin states for the radical intermediate, TS-3, and the concerted intermediate are very small. The spin densities on C_2 for TS-2, the radical intermediate, and TS-3 have very similar magnitudes for both multiplicities, but they are opposite in sign. Moreover, the geometries of the species for both spin states are very similar.

Following the formation of the concerted intermediate, the next step (Step 4 in Fig. 19.4) involves breaking the bond between manganese and oxygen to form a weakly interacting state between the epoxide product and the Mn-porphyrin. This step proceeds without an energy barrier and is exothermic for both spin states: 50 kJ/mol for the quintet state and 7 kJ/mol for the triplet state. Finally, styrene oxide and the bare Mn-porphyrin are completely separated. This step is endothermic, and the final state lies 64 and 56 kJ/mol higher in energy than the weakly interacting state for the triplet and quintet spin states, respectively.

The last remaining step in Fig. 19.4, in which the radical intermediate reacts to form phenylacetaldehyde (Step 5), is shown in Fig. 19.6 to pass through TS-5. The reaction coordinate involves the migration of one of the hydrogen atoms on the carbon atom bonded to the oxygen atom (C_1) to the other carbon (C_2). This transition state, which is shown for the quintet state in Fig. 19.6, lies 100 kJ/mol higher in energy than the radical intermediate. The barrier for the triplet state, which is not shown, was 168 kJ/mol. Even though phenylacetaldehyde is more thermodynamically stable than the epoxide, these catalysts are more selective toward styrene oxide, with selectivities higher than 80% for styrene oxide in experiment [44]. The significantly higher predicted barrier for the formation of phenylacetaldehyde is consistent with these observations.

Free energies for the structures identified on the electronic energy surface were calculated using vibrational frequencies and statistical thermodynamics. Fig. 19.8 illustrates the free energy as a function of the reaction coordinate for the lowest activation energy path to form styrene oxide and phenylacetaldehyde. As shown in this plot, the free energies of the weakly associated states and the isolated counterparts are nearly equivalent, thus smoothening the more pronounced ruggedness observed on the electronic energy surface. The lowest barrier to form the radical intermediate, TS-2, is in the triplet state at 58 kJ/mol. Once the radical intermediate is formed, the subsequent steps have the minimum energy barrier on the quintet spin state. Since the triplet and quintet spin states of the radical intermediate are degenerate, and even their geometries are nearly identical on both spin states, the crossing from the triplet surface to the quintet surface should be facile. The lowest barriers to form TS-3 and TS-5 are thus for the quintet state and are 19 and 78 kJ/mol, respectively. The desorption of styrene oxide is endothermic with a barrier of 9 kJ/mol, while the desorption of phenylacetaldehyde is 12 kJ/mol downhill. The difference in free energy between styrene oxide and phenylacetaldehyde is 132 kJ/mol. This difference in free energy results in an equilibrium constant for the isomerization of styrene oxide to phenylacetaldehyde of 1.3×10^{23}. This high equilibrium constant is consistent with the value of 2.0×10^{23} obtained when the free energies were calculated with a group additivity method, which has an accuracy of 2 kJ/mol [45,46].

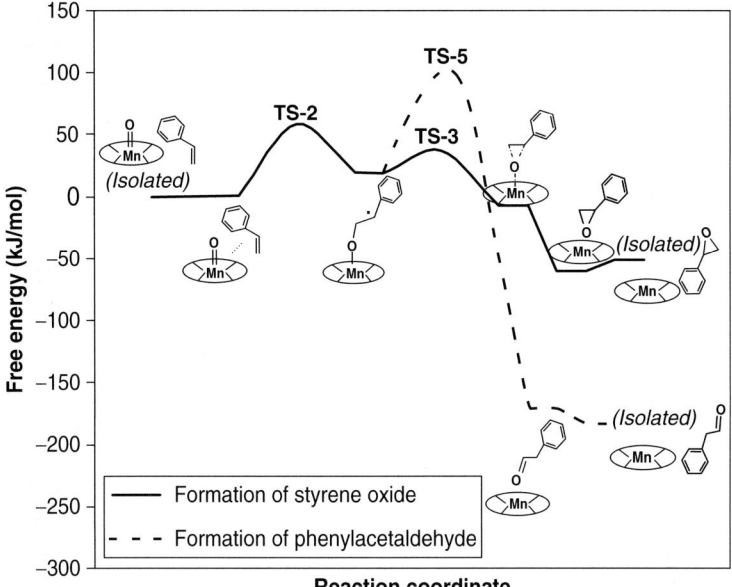

FIGURE 19.8 Free energy as a function of the reaction coordinate for the lowest activation energy path for epoxidation of styrene and formation of phenylacetaldehyde.

The PCM was used to study solvent effects on the energies of stable intermediates and transition states. For these calculations, dichloromethane was used as the solvent. Single point calculations were performed with and without PCM using PW91/LANL2DZ on the optimized structures obtained in vacuum from ONIOM. Figure 19.9a and b illustrates the energies as a function of the reaction coordinate on the quintet spin state for the reaction pathways to form both styrene oxide and phenylacetaldehyde. As seen from these figures, the energies are significantly lower in the presence of the solvent for all intermediates when compared to vacuum results as expected. The main difference between the vacuum and solvent results in the energy diagrams shown in Fig. 19.9a and b is the difference in energy of the oxidized Mn-porphyrin and styrene when they are isolated and when they form the interacting complex. While in vacuum calculations the energy for this weakly interacting state is significantly lower than for the isolated species, in solvent, this difference in energy is not observed. In fact, when the solvent is included in this weakly interacting state, only a small charge transfer is observed between the oxidized porphyrin and styrene ($0.01e^-$), while in vacuum the charge transfer is much higher ($0.06e^-$). For the rest of the states, the energy difference between the results obtained in vacuum and with PCM remained nearly constant along the reaction coordinate, including the transition states. The activation energies in vacuum and in the solvent are quite similar, as shown in Table 19.1. The energy difference between vacuum and solvent calculations in Fig. 19.9 averages 62 kJ/mol. It is interesting, however, that styrene oxide as well as

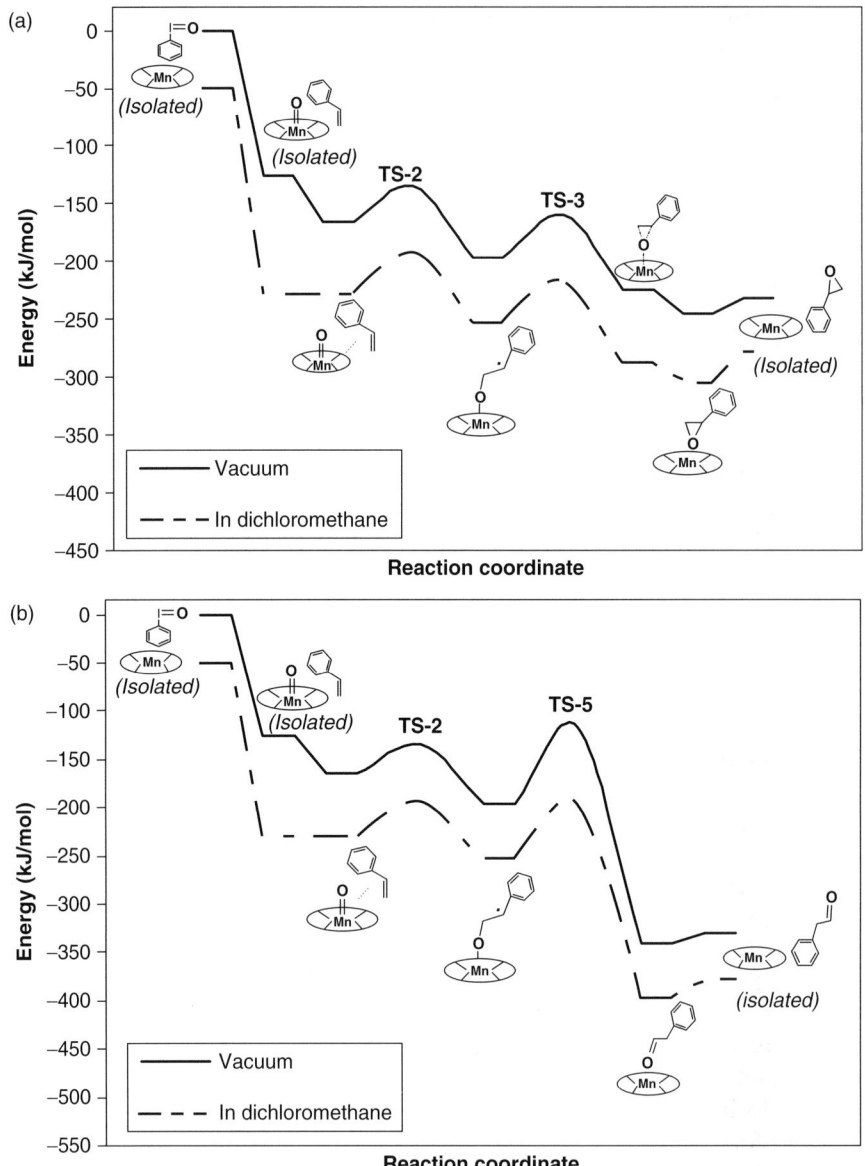

FIGURE 19.9 Electronic energy as a function of the reaction coordinate for (a) epoxidation of styrene and (b) formation of phenylacetaldehyde. Results are shown for the quintet spin state in vacuum and in dichloromethane.

phenylacetaldehyde interact with the Mn-porphryin to form a low energy state even in the solvent. In the presence of the solvent, the charge transfer between the products and the Mn-porphyrin remained the same as in vacuum with an approximate value of $0.03e^-$.

TABLE 19.1 Comparison of the activation energies in vacuum and in dichloromethane solvent

Step in the reaction mechanism			Activation energy in vacuum (kJ/mol)	Activation energy in solvent (kJ/mol)
Mn (with styrene O)	→ TS-2 →	Mn (with radical O)	30.0	37.0
Mn (with radical O)	→ TS-3 →	Mn (with epoxide O)	37.9	35.9
Mn (with radical O)	→ TS-5 →	Mn (with aldehyde O)	80.8	74.3

4. CONCLUSIONS

The epoxidation of styrene and the formation of phenylacetaldehyde using Mn-porphyrin catalysts were analyzed with QM/MM ONIOM calculations. For the epoxidation reaction, the calculations suggest a stepwise mechanism, in which a radical intermediate is formed before the concerted intermediate. The calculations reveal that the oxidation of the porphyrin with iodosylbenzene proceeds without a reaction barrier. The lowest energy barrier for the formation of the radical intermediate is 51 kJ/mol in the triplet state. The lowest energy barrier for the formation of the concerted intermediate from the radical intermediate is 15 kJ/mol in the quintet state. The barrier to form the side product, phenylacetaldehyde, is predicted to be much higher in energy, consistent with the selectivity of this catalyst to form styrene oxide over phenylacetaldehyde. The PCM was used to take the solvent dichloromethane into account. In the presence of solvent, the energy barriers did not change significantly for any of the steps in the mechanism compared to gas-phase calculations.

ACKNOWLEDGMENTS

This research was supported by the Chemical Sciences, Geosciences, and Biosciences Division, Office of Basic Energy Sciences, Office of Science, U.S. Department of Energy, Grant No. DE-FG02–03ERIS457 and by the Illinois Minority Graduate Incentive Program. The authors thank Profs. SonBinh Nguyen and Joseph Hupp for helpful discussions.

REFERENCES

[1] J. T. Groves, T. E. Nemo, R. S. Myers, Hydroxylation and epoxidation catalyzed by iron-porphine complexes–oxygen-transfer from iodosylbenzene, *J. Am. Chem. Soc.* 101 (1979) 1032.
[2] J. T. Groves, T. E. Nemo, Aliphatic hydroxylation catalyzed by iron porphyrin complexes, *J. Am. Chem. Soc.* 105 (1983) 6243.
[3] D. Mansuy, J. F. Bartoli, M. Momenteau, Alkane hydroxylation catalyzed by metalloporphyrins–evidence for different active oxygen species with alkylhydroperoxides and Iodosobenzene as oxidants, *Tetrahedron Lett.* 23 (1982) 2781.
[4] J. T. Groves, W. J. Kruper, R. C. Haushalter, Hydrocarbon oxidations with oxometalloporphinates–isolation and reactions of a (porphinato)manganese(V) complex, *J. Am. Chem. Soc.* 102 (1980) 6375.
[5] C. L. Hill, B. C. Schardt, Alkane activation and functionalization under mild conditions by a homogeneous manganese(III) porphyrin-iodosylbenzene oxidizing system, *J. Am. Chem. Soc.* 102 (1980) 6374.
[6] J. T. Groves, W. J. Kruper, Preparation and characterization of an Oxoporphinatochromium (V) complex, *J. Am. Chem. Soc.* 101 (1979) 7613.
[7] H. Amatsu, T. K. Miyamoto, Y. Sasaki, Olefin epoxidation catalyzed by sterically bulky metalloporphyrins (Metal=Fe,Mn) by use of sodium-hypochlorite as an oxygen source–a kinetic-study, *Bull. Chem. Soc. Jpn.* 61 (1988) 3193.
[8] B. Meunier, E. Guilmet, M. E. Decarvalho, R. Poilblanc, Sodium-hypochlorite – a convenient oxygen source for olefin epoxidation catalyzed by (Porphyrinato)manganese complexes, *J. Am. Chem. Soc.* 106 (1984) 6668.
[9] T. G. Traylor, S. Tsuchiya, Y. S. Byun, C. Kim, High-yield epoxidations with hydrogen-peroxide and tert-butyl hydroperoxide catalyzed by Iron(III) porphyrins – heterolytic cleavage of hydroperoxides, *J. Am. Chem. Soc.* 115 (1993) 2775.
[10] S. Banfi, M. Cavazzini, G. Pozzi, S. V. Barkanova, O. L. Kaliya, Kinetic studies on the interactions of manganese-porphyrins with peracetic acid. Part 1. Epoxidation of alkenes and hydroxylation of aromatic rings, *J. Chem. Soc. Perkin Trans.* 2(4) (2000) 871.
[11] J. Razenberg, A. W. Vandermade, J. W. H. Smeets, R. J. M. Nolte, Cyclohexene epoxidation by the mono-oxygenase model (tetraphenylporphyrinato) manganese(III) acetate sodium-hypochlorite, *J. Mol. Catal.* 31 (1985) 271.
[12] A. N. de Sousa, M. de Carvalho, Y. M. Idemori, Manganese porphyrin catalyzed cyclohexene epoxidation by iodosylbenzene: The remarkable effect of the meso-phenyl ortho-OH substituent, *J. Mol. Catal. A. Chem.* 169 (2001) 1.
[13] M. C. Feiters, A. E. Rowan, R. J. M. Nolte, From simple to supramolecular cytochrome P450 mimics, *Chem. Soc. Rev.* 29 (2000) 375.
[14] F. G. Doro, J. R. L. Smith, A. G. Ferreira, M. D. Assis, Oxidation of alkanes and alkenes by iodosylbenzene and hydrogen peroxide catalysed by halogenated manganese porphyrins in homogeneous solution and covalently bound to silica, *J. Mol. Catal. A. Chem.* 164 (2000) 97.
[15] J. P. Collman, X. M. Zhang, V. J. Lee, E. S. Uffelman, J. I. Brauman, Regioselective and enantio-selective epoxidation catalyzed by metalloporphyrins, *Science* 261 (1993) 1404.
[16] M. L. Merlau, M. D. P. Mejia, S. T. Nguyen, J. T. Hupp, Artificial enzymes formed through directed assembly of molecular square encapsulated epoxidation catalysts, *Angew. Chem., Int. Ed.* 40 (2001) 4239.
[17] M. L. Merlau, W. J. Grande, S. T. Nguyen, J. T. Hupp, Enhanced activity of manganese(III) porphyrin epoxidation catalysts through supramolecular complexation, *J. Mol. Catal. A. Chem.* 156 (2000) 79.
[18] B. Meunier, Metalloporphyrins as versatile catalysts for oxidation reactions and oxidative DNA Cleavage, *Chem. Rev.* 92 (1992) 1411.
[19] M. J. Gunter, P. Turner, Metalloporphyrins as models for the cytochromes-P-450, *Coord. Chem. Rev.* 108 (1991) 115.
[20] S. H. L. Wang, B. S. Mandimutsira, R. Todd, B. Ramdhanie, J. P. Fox, D. P. Goldberg, Catalytic sulfoxidation and epoxidation with a Mn(III) triazacorrole: Evidence for a "third oxidant" in high-valent porphyrinoid oxidations, *J. Am. Chem. Soc.* 126 (2004) 18.

[21] J. T. Groves, J. B. Lee, S. S. Marla, Detection and characterization of an oxomanganese(V) porphyrin complex by rapid-mixing stopped-flow spectrophotometry, *J. Am. Chem. Soc.* 119 (1997) 6269.
[22] N. Jin, J. T. Groves, Unusual kinetic stability of a ground-state singlet oxomanganese(V) porphyrin. Evidence for a spin state crossing effect, *J. Am. Chem. Soc.* 121 (1999) 2923.
[23] N. Jin, J. L. Bourassa, S. C. Tizio, J. T. Groves, Rapid, reversible oxygen atom transfer between an oxomanganese(V) porphyrin and bromide: A haloperoxidase mimic with enzymatic rates, *Angew. Chem. Int. Ed.* 39 (2000) 3849.
[24] K. Ayougou, E. Bill, J. M. Charnock, C. D. Garner, D. Mandon, A. X. Trautwein, R. Weiss, H. Winkler, Characterization of an oxo(porphyrinato)manganese(IV) complex by X-ray-absorption spectroscopy, *Angew. Chem. Int. Ed.* 34 (1995) 343.
[25] J. P. Collman, J. I. Brauman, B. Meunier, T. Hayashi, T. Kodadek, S. A. Raybuck, Epoxidation of olefins by cytochrome-P-450 model compounds–kinetics and stereochemistry of oxygen atom transfer and origin of shape selectivity, *J. Am. Chem. Soc.* 107 (1985) 2000.
[26] J. P. Collman, J. I. Brauman, P. D. Hampton, H. Tanaka, D. S. Bohle, R. T. Hembre, Mechanistic studies of olefin epoxidation by a manganese porphyrin and hypochlorite–an alternative explanation of saturation kinetics, *J. Am. Chem. Soc.* 112 (1990) 7980.
[27] R. J. M. Nolte, A. S. J. Razenberg, R. Schuurman, On the rate-determining step in the epoxidation of olefins by monooxygenase models, *J. Am. Chem. Soc.* 108 (1986) 2751.
[28] R. D. Arasasingham, G. X. He, T. C. Bruice, Mechanism of manganese porphyrin-catalyzed oxidation of alkenes–role of manganese(IV) oxo species, *J. Am. Chem. Soc.* 115 (1993) 7985.
[29] G. X. He, R. D. Arasasingham, G. H. Zhang, T. C. Bruice, The rate-limiting step in the one-electron oxidation of an alkene by oxo meso-tetrakis(2,6-dibromophenyl) porphinato -chromium(V) is the formation of a charge-transfer complex, *J. Am. Chem. Soc.* 113 (1991) 9828.
[30] M. C. Curet-Arana, G. A. Emberger, L. J. Broadbelt, R. Q. Snurr, Quantum chemical determination of stable intermediates for alkene epoxidation with Mn-porphyrin catalysts, *J. Mol. Catal. A. Chem.*, in press.
[31] T. Matsubara, S. Sieber, K. Morokuma, A test of the new "integrated MO + MM" (IMOMM) method for the conformational energy of ethane and n-butane, *Int. J. Quantum Chem.* 60 (1996) 1101.
[32] F. Maseras, K. Morokuma, IMOMM–a new integrated *ab-initio* plus molecular mechanics geometry optimization scheme of equilibrium structures and transition-states, *J. Comput. Chem.* 16 (1995) 1170.
[33] M. Svensson, S. Humbel, R. D. J. Froese, T. Matsubara, S. Sieber, K. Morokuma, ONIOM: A multilayered integrated MO + MM method for geometry optimizations and single point energy predictions. A test for Diels-Alder reactions and Pt(P(t-Bu)(3))(2) + H-2 oxidative addition, *J. Phys. Chem.* 100 (1996) 19357.
[34] S. Humbel, S. Sieber, K. Morokuma, The IMOMO method: Integration of different levels of molecular orbital approximations for geometry optimization of large systems: Test for n-butane conformation and S(N)2 reaction: RCl + Cl, *J. Chem. Phys.* 105 (1996) 1959.
[35] M. J. Frisch, G. W. Trucks, H. B. Schlegel, G. E. Scuseria, M. A. Robb, J. R. Cheeseman, V. G. Zakrzewski, J. A. Montgomery, Jr., R. E. Stratmann, J. C. Burant, S. Dapprich, J. M. Millam, et al. "Gaussian 98," revision A.7, Pittsburgh, PA: Gaussian Inc., 1998.
[36] J. P. Perdew, K. Burke, Y. Wang, Generalized gradient approximation for the exchange-correlation hole of a many-electron system, *Phys.Rev.B.* 54 (1996) 16533.
[37] J. P. Perdew, Y. Wang, Accurate and simple analytic representation of the electron-gas correlation-energy, *Phys. Rev. B.* 45 (1992) 13244.
[38] A. K. Rappe, C. J. Casewit, K. S. Colwell, W. A. Goddard, W. M. Skiff, UFF, a full periodic-table force-field for molecular mechanics and molecular-dynamics simulations, *J. Am. Chem. Soc.* 114 (1992) 10024.
[39] M. Nasr-Esfahani, M. Moghadam, S. Tangestaninejad, V. Mirkhani, Biomimetic oxidation of Hantzsch 1,4-dihydropyridines with tetra-n-butylammonium periodate catalyzed tetraphenyl-porphyrinatomanganese(III) chloride Mn(TPP)Cl, *Bioorg. Med. Chem. Lett.* 15 (2005) 3276.
[40] J. Bernadou, B. Meunier, Biomimetic chemical catalysts in the oxidative activation of drugs, *Adv. Synth. Catal.* 346 (2004) 171.

[41] R. Seeger, J. A. Pople, Self-consistent molecular-orbital methods. 18. Constraints and stability in Hartree-Fock Theory, *J. Chem. Phys.* 66 (1977) 3045.
[42] R. Bauernschmitt, R. Ahlrichs, Stability analysis for solutions of the closed shell Kohn-Sham Equation, *J. Chem. Phys.* 104 (1996) 9047.
[43] S. F. Boys, F. Bernardi, Calculation of small molecular interactions by differences of separate total energies - some procedures with reduced errors, *Mol. Phys.* 19 (1970) 553.
[44] M. C. Curet-Arana, Mechanistic analysis of styrene epoxidation catalyzed by manganese porphyrins and molecular squares, Ph.D. thesis, in Chemical Engineering. Northwestern University, Evanston, Illinois, 2006.
[45] S. W. Benson, *Thermochemical kinetics; methods for the estimation of thermochemical data and rate parameters*. John Wiley, New York, 1976.
[46] S. Stein, J. Rukkers, R. Brown, Structures and Properties Database and Estimation Program. (1991).

INDEX

A

Acetonitrile, 43, 128, 185, 188, 203, 208, 223, 226
Acidic epoxidation systems, 421–425
   allylic strain effects, 423
   conversion and selectivity vs. QCl/W ratio, 422
   of geraniol using $Na_{12}[WZn_3(ZnW_9O_{34})_2]$ solution, 421–422
   optimum α_W angle, 423–424
   selectivity in reactions, using [n-$C_{16}H_{31}$ (HOCH$_2$CH$_2$)NMe$_2$]H$_2$PO$_4$, 424
   toluene addition and QCl/W ratio, 422
   α, β-unsaturated enone formation, 423
   vanadium substituted $Q_{12}[WZn(VO)_2(ZnW_9O_{34})_2]$, use of, 422–423
Activation energy, 234, 236–237, 239–240
Active oxidants, 452
Active oxygen species, 156–157, 159
Active site, catalytic epoxidation
   in catalyst deactivation, 362
   nature of, 358–359
   in sharpless mechanism, 360
Active W-based catalysts, 416
Acyclic allylic alcohols, 423
2-Adamantanone, 168
Adsorption, 235–242, 244, 248, 252–259
Ag/α-Al$_2$O$_3$ catalyst
   with Cl-promoted catalyst, 251–254
   with Cs-promoted catalyst, 249–251
   ethylene desorption, 257–259
   ethylene epoxidation on
      adsorption and desorption of oxygen, 237
      temperature programmed reaction, 244–247
   kinetics of adsorption and desorption of ethylene
      on α_2-O state, 257
      oxidised catalyst, α_1-O state, 255–256
      unoxidised and oxidised catalyst, 254–255
   oxygen dosing on, 247–249
Ag(111) and Ag(110), adsorption of ethylene on, 240
Ag-based bimetallic catalysts. See also $Ag_{14}M_1$ catalysts
   activation energies vs. surface reaction heats, 275–276
   EO and acetaldehyde formation, activation energy for, 272–274
   linear relationships, 275–276
   loading of metal M, 272
   microkinetic model in, 276–277
   OME stability and EO selectivity in, 277–279
   trends for selectivity, $\Delta\Delta E$ variable for, 272–274
Ag catalyst, 5, 251, 259
Ag/γ-Al$_2$O$_3$ catalyst, 287
$Ag_{14}M_1$ catalysts
   ethylene oxide (EO) selectivity
      calculations of, 279
      predictions of, 278
   linear correlations for activation energies, 276
   oxygen heat of adsorption QO and stability of OME, 277
   reaction coordinates, for acetaldehyde and EO formation on, 274
   relative activation energies, 275
Ag–O bond, 235
Aliquat 336, phase-transfer catalyst, 420–421
Alkene. See also Alkene epoxidation
   cis-alkenes epoxidation, 208–210
   chain length, 182
   cycloalkenes, 199
   enantiocontrol of, 185
   oxygen transfer to, 179, 190
   unfunctionalized, 197–198, 203–204, 208
Alkene epoxidation, 193–194, 430
   allyl benzene derivatives of, 107
   catalytic cycle in, 472–473
   catalytic epoxidation, hydrogen peroxide mediated, 107
   chiral catalyst recycling, 110–112
   diastereoselectivity in, 108–109
   enantioselectivity in, 109–110
   hydrogen peroxide (H_2O_2) as oxidant in, 104–105
   kinetic analysis and NMR studies of, 113–114
   mechanistic hypothesis, 114
   Lewis acidity properties in, 106
   of 1-octene, acidity and catalytic activity of complexes, 106
   (P-P)Pt(C_6F_5)(H_2O) catalyst synthesis, 105
   regioselectivity in, 107–108
   stability of metalloporphyrins, 472
Alkylhydroperoxides, 122
Alkyl metaborate esters, 21
Allyl chloride epoxidation

Allyl chloride epoxidation (*Cont.*)
  catalyst recovery in, 443–444
  dynamics of, 442–443
  effect of catalyst amount, 444
  epichlorohydrin yield in, 440–441
  MAS-NMR spectra, 443
  organic solvents in, 441
  reaction time effects in, 442
  temperature effects in, 441–442
Allyl epoxides, 9
Allylic alcohols, 122, 423
  epoxidation reactions and catalysts, 22–23
Alumina
  in olefin epoxidation, 23
  as support material, 13
Amine-mediated epoxidation system, 24–25
Amino acetal catalyst, 191
Aminomethylphosphonic acid, 124
4-Aminophenol, 397
Ammonium nitrate (NH_4NO_3), 321–322
Amphidinolide H (AmpH), 9
Amsterdam density functional (ADF), 267–275, 277
  cluster calculations, 267–268
Anti-hypertensive agent. *See* Levcromakalim
Armstrong's catalyst, 181
Armstrong's intramolecular epoxidation, 183
Aryl selenium compounds, 23
Asymmetric catalysis, 459
Asymmetric epoxidation. *See also* Alkene epoxidation
  achiral catalyst for, 184
  of alkenes, 193–194
  Armstrong's intramolecular, 183
  of benzopyrans, 211
  of cycloalkenes, 199
  of dienes, 112
  of iminium salt, 182
  in ketiminium, 181
  of 4-methyl-1-pentene and terminal alkenes, 111
  of oxaziridinium, 185
  of 1-phenylcyclohexene, 190
    using TPPP, 202, 205, 207–208
  of terminal alkenes, 110
  of unfunctionalized alkenes, 197–198, 203–204
Au–Ag catalyst, 285, 290
Au–Ag/γ-Al_2O_3 catalyst, 289, 291
Au–Ba/Ti-TUD catalyst, 46–47
Au/CeO_2 catalysts, 287, 289, 294
Au surfaces, computational models of, 306
Au/Ti catalysts, 319, 331–332
Au/Ti–MCM-48 catalysts, 320, 326–328
Au/TiO_2 catalysts

Deposition-precipitation (DP) and impregnation methods, 302
mechanistic study of
  Au clusters, models of, 306
  bidentate propoxy species, 304
  inelastic neutron scattering (INS) spectrum of, 305
  temperature-programmed desorption (TPD) experiments, 303
Au/Ti-SiO_2 catalysts, 45–47, 299
Azepinium
  catalysts, 194–196
  enantioselectivity of, 207
  formation of, 194, 197
  substituted salts, 207

B

Bacillus megaterium, 30
Baeyer–Villiger oxidation, 168, 180
  of ketones, 106
Basis set superposition error (BSSE), 474, 476
Benchmarking sandwich POM-catalyzed epoxidations, 418–420
Benzopyran, 210–211
Bimetallic AgM catalyst, 272
Bimetallic catalysts, computational strategies. *See also* Ethylene epoxidation
  application of, 268–269
  cluster calculations, ADF, 267–268
  design, 266, 268–269
  periodic slab calculations, DACAPO, 267
1,1′-Bi-2-naphthol (BINOL), 21, 67, 196, 402–403
Binding energy (Q), 267, 303, 382
Bio-inspired oxidation catalysis, 451–453. *See also* Olefin oxidation catalysis
Biological epoxidation reactions
  approaches, for use of enzymes, 28–30
  catalysts and conditions, 29
Biomimetic catalysts, 5, 71, 156, 451, 472
Biomimetic epoxidation
  ligands used in, 27
  reactions and catalysts in, 26–27
  Ru complexes, 28
4,4′-Bipyridine, 401
Bohe's improved achiral catalyst, 184
Bond dissociation energies (BDEs), 135–136
Bond dissociation enthalpy (BDE)
  in molecule-induced homolysis, 222
  in radical oxidation, 218
  in solvent effect, 226
BPMEN ligand, 454–456, 462
2-(2-Bromoethyl)benzaldehyde, 185–187, 190
Brønsted base, 130
Brunauer-Emmette-Teller (BET), 286–288

t-BuOOH, 62, 122–123, 136, 138, 142–148
Butadiene epoxidation, 6, 41
t-Butanol, 43, 123, 316
tert-Butyl alcohol, 15, 123, 340
tert-Butyl hydroperoxide (TBHP), 16, 123, 374, 376–380, 389

C

Calcium–BINOL complex, 403
Carbamazepine (CBZ), 7, 12
Carbamazepine 10,11-epoxide (CBZ-EP), 7
Carbonyl ruthenium(II) porphyrin, 398
Catalysis, by gold, 316–321
Catalyst A [p-$C_5H_5NC_{16}H_{33}]_3[PW_4O_{16}]$, 432
Catalyst durability, 121, 123, 125, 128
Catalyst systems, 32–34
　rates for, 34
　reactivity patterns, 32–33
Catalytic additives, 458
Catalytic asymmetric anhydrous epoxidation
　asymmetric epoxidations by TPPP
　　reaction solvent, effect of, 207–211
　　temperature studies, 201–206
　TPPP selection of, 200–201
Catalytic asymmetric epoxidation.
　See Asymmetric epoxidation
Catalytic dehydration. *See* Styrene monomer propylene oxide process
Catalytic epoxidation
　by H_2O_2 and Lewis acid, 121
　system, 113
Catalytic intermediates, 459, 462, 465
Cetyltrimethyl ammonium chloride, 432
Chalcone oxide, 403
Charge decomposition analysis (CDA), 48
Chemical vapor deposition (CVD)
　Ti/MCM-41catalyst, 374, 380–385
　Ti/MCM-48 catalyst, 374, 380–383
　Ti/SiO_2 catalyst, 374, 376, 378–380
Chemisorption, 41, 258–260
Chiral TADDOL-derived hydroperoxide, 422
Chloroauric acid, 285
m-Chlorobenzoic acid, 223–224
Chlorohydrin
　method, for pesticide, 439–440
　process for PO, 298
Chloromethylated polystyrene, 396
Chloromethylated poly(styrene-co-DVB), 398
Chloromethylated styrene-DVB copolymer, 395
Chloromethyloxirane, 7
m-Chloroperbenzoic acid (m-CPBA), 25, 107
$[(C_6H_{13})_4N]_3[PO_4[W_4O(O_2)_2]_{24}$, for olefin epoxidation, 431
Chromenes, 393

Cl promotion. *See also* Ag/α-Al_2O_3 catalyst
　in ethylene desorption, 258–259
　oxidation adsorption energy, 252–254
　oxidation desorption energy, 251–253
　selectivity improvement in oxidation, 251
Cobalt(II) acetate, 402
Cobalt(II)-salen, 402
Coldspray ionization mass spectrometry (CSI-MS), 157
Competitive epoxidation, 169
Computational flow dynamics (CFD), 366–367
Concerted oxygen insertion, 221
Copolymerization, acryloyl-substituted manganese-salen complex, 393
Copper (Cu), 266, 272, 274
Corrolazines, 25
Corroles, 25
Cross-bridged cyclam, 134, 138
Cs promotion. *See also* Ag/α-Al_2O_3 catalyst
　in ethylene desorption, 257–258
　O_2 desorption spectrum, 249–250
　temperature programmed reaction spectrum, 250–251
Cu–Ag alloys, 40
Cu–Ag as bimetallic catalyst, 266
Cumene hydroperoxide (CHP), 374, 403
　PO selectivity, 380
　for propylene epoxidation, 376–378
CVD-prepared
　Ti-containing catalysts, 374
　Ti/SiO_2 catalysts, 378
6-Cyanochromene, 394
Cyclisation, 243–244, 246, 250, 254
Cyclobutanone, 168
Cyclohexene epoxidation, 440–441
Cyclohexene oxide, 6–7, 439, 444
Cyclooctene, 395–397
　conversion *vs.* time for epoxidation of, 421
　manganese containing $[WZnMn^{II}_2(ZnW_9O_{34})_2]^{12-}$ for epoxidation, 168
　oxidation, catalyzed by nonheme iron complexes, 456
　reaction profiles of epoxidation, 160–162
Cyclopentanone, 168
Cytochrome P-450, 121, 156, 452, 472

D

Density functional theory (DFT), 162, 274, 303, 319, 321, 328, 462–463, 466, 473
　calculations
　　cluster using ADF, 267–268
　　for oxametallacycle (OME) stability, 276–277
　　periodic slab using DACAPO, 267
　　in propylene epoxidation, 40–41

Density functional theory (DFT) (*Cont.*)
  in Ti catalyst epoxidation, 44
  in Ti-tartrate epoxidation, 55
Deposition-precipitation (DP), 298, 302, 317
Desorption, 235–243, 245, 249–259
3,5-Diamino-1,2,4-triazole, 397
Di-and tetra-nuclear small peroxotungstates, 157
Diastereoselectivity, 108–109
  epoxidation of chiral terminal alkene, 109
Dichloroethane (DCE), 13, 106, 164, 207, 423, 441
Dichloroethylene, 124
cis-and *trans*-1,2-Dichloroethylene, 242
2,6-Dichloropyridine-N-oxide (Cl$_2$pyNO), 398
Dielectric constant, 188
Diels–Alder reactions, 168
Dihydrogen peroxide, 417–418
Dihydroisoquinolinium
  alcohol containing, 189
  asymmetric epoxidation, 195
    using TPPP, 201–203, 207–208
  chair conformation, 190, 192
  ether containing, 190
  formation of, 185–186
  1-phenylcyclohexene, 187
Dihydroisoquinolinium salts, 186
1,2-Dihydronaphthalene, 394, 397
cis-Dihydroxylation, 452–455, 459, 462–464
  catalysts with inequivalent *cis*-labile sites, 454
  of C=C bonds, 452
  development asymmetric dihydroxylation catalysts, 459
  energy profile calculated for water-assisted epoxidation and, 463
  hydroxo attack to form the first C–O bond, 462
  of olefins and osmium-catalyzed, 453
  pathway used *nwa* mechanism, 464
  selectivity spectrum with representative nonheme iron catalysts, 455
β-Diketonate complexes, 389
(μ-η^1:η^1-peroxo)d^0-Dimetal complexes, 159
Dimethyldioxirane (DMD), 48
2,6-Dimethyl-phenyl isocyanide, 106
Dinuclear peroxotungstate, 159
trans-Diol, 453
Dioxiranes
  generating PINO radical, 222
  oxidation by peracids and, 227
Direct propylene epoxidation. *See* Propylene epoxidation
Disilicoicosatungstates, [{γ-SiW$_{10}$O$_{32}$(H$_2$O)$_2$}$_2$(μ-O)$_2$]$^{4-}$, 167
Divinyl benzene (DVB), 389
1-Dodecene, cyclohexene, 396
DVB-methyl methacrylate copolymer, 389

E

Early transition metals, in olefin epoxidation, 120–121
Electron energy loss spectrometry (EELS), 240
Electron paramagnetic resonance (EPR), 47, 223
Eletrophilic oxygen, 235
Enantiomeric excess, 180, 203, 207–208, 210
Enantiopure β-aminoalcohols, 7
Enantioselectivity, 109–110
  case of geraniol, 122
  for dienes, 110, 112
  with surfactants related to triton family, 110–111
  limitation of, 112
Enthalpy, 218
Epichlorohydrin, 7, 441–442. *See also* Allyl chloride, epoxidation
Epoxidation
  activity
    dichloroethane (DCE) as solvent in, 106
    methoxy residues in, 107
  on F$_{20}$TPPFe(III) with H$_2$O$_2$, 66
  manganese as catalyst
    hydrogen abstracting ability, 135–136
    oxidation, 134–135
  manganese as catalyst using hydrogen peroxide, in olefin
    deviation of oxygen content, 139–140
    isotope labeling experiment, 138–139
    manganese oxo species, 137
    mechanism, 142
    oxygen rebound mechanism, 138
    oxygen transfer process, 136–137
    reactions on various olefins, 143–144
    cis-stilbene, 140–141
  manganese catalyst using t-BuOOH, in olefin
    mass spectra study, 147–148
    radical process, 144–146
    reactions on various olefins, 143–144
    cis-stilbene, 146
    ultraviolet-visible spectrophotometry study, 142
  of olefin
    H$_2$O$_2$ oxidation, 121
    iodosylbenzene adducts, 133–134
    Lewis acid catalysation, 121–122
    transition metal complexes, 120–121
    oxidants for, 35–37
  of {PO$_4$[WO(O$_2$)$_2$]$_2$[WO(O$_2$)$_2$(H$_2$O)]}$^{3-}$, 431
  reactions and catalysts, 13
  reactions involved, 121–122
  regio- and enantioselective reactions, 122
Epoxides, 121, 191, 195–196, 201
  in acetonitrile and chloroform, 210
  conversion, 180, 198, 203–204

enantiomer of, 188, 206–207
and endproducts, 11–12
market value of, 5, 9
with pharmacological and medicinal
   properties, 8
ring opening of, 209, 211
threo-Epoxides, 423
Epoxybutene
   applications of, 6
   in butadiene epoxidation, 16
   deuterium-labeled 1,3-butadiene to, 41
Epoxyeicosatrienoic acid, 9
cis-4,5-Epoxy-1-hexene, 107
1,2-Epoxy-3-methyl-3-butanol, 9
1,2-Epoxyoctane synthesis, 378–379
Epoxy resins, 6–9, 156, 397, 407, 430
Erythro-epoxides, 423
Escherichia coli, 29, 34
2-Ethylanthrahydroquinone (EAHQ), 125
   in hydrogen peroxide production, 430
   in propylene epoxidation, 435–437
Ethylanthrahydroquinone/H2 process, 125
2-Ethylanthraquinone (EAQ), 430
Ethylbenzene (EB), 356
   air-oxidation of, 357–358
   in catalytic dehydration, 367
   in trace component chemistry, 363–365
Ethylbenzene hydroperoxide (EBHP), 357, 362,
      364–365. *See also* Chemical vapor
      deposition (CVD)
   conversion, 375
   epoxide yield, epoxide selectivity and turnover
      frequencies, 377
   kinetics of propylene epoxidation with,
      384–385
   pseudo-first-order kinetic model over
      Ti/MCM-41, 384
   $TiCl_4$ deposition
      temperature on, 381
      time on, 376–378
cis- and *trans*-1,2-d_2-Ethylene, 241
Ethylene dichloroethane (EDC), 284
Ethylene epoxidation, 317–318. *See also*
      Ag/α-Al_2O_3 catalyst
   adsorption on oxidised silver, 240–242
   on Ag and bimetallic catalysts
      selectivity in, 268
      transition states in, 269
   Au particle sizes in, 295
   catalyst characterization of
      Au crystallite sizes, 287–288
      BET surface areas in, 287–288
      impregnation method preparation, 288–289
      single-step sol–gel technique, 288–289
      TEM/EDS technique, 288

catalyst preparation
   CeO_2 support synthezise, 286
   impregnation method, 285–286
   incipient wetness method, 285
catalyst selectivity of, 291–292
CO_2 selectivity, 292, 294
desorption of
   Cl promoted silver catalyst, 258–259
   Cs promoted silver catalyst, 257–258
differential flow reactor in, 286–287
ethylene conversion, 289–290
high-surface-area (HSA) γ-alumina, 287–288
kinetics of adsorption and desorption
   on oxidised catalyst, $α_1$-O state, 255–256
   on oxidised catalyst, $α_2$-O state, 257
   on unoxidised and oxidised catalyst, 254–255
microkinetic model for, 276–279
over Ag/γ-Al_2O_3 catalyst, 287
oxygen desorption of, 294
silver (Ag) catalyst in
   Cl promotion, 251–254
   Cs promotion, 249–251
   role of intermediates, 242–244
   subsurface oxygen, 247–249
   temperature programmed reaction,
      244–247
subsurface oxygen, 247–249
support material for, 284
temperature-programmed desorption (TPD),
   291, 293
TiO_2 and CeO_2, 284–285, 294
turnover number in, 290, 293
Ethylene oxidation, 38, 284
Ethylene oxide (EO)
   adsorption on oxidised silver, 240–242
   applications of, 5
   formation by activation energy, 270–272
   linear correlations for, 275–276
   production of
      catalyst used, 12–13
      molecular oxygen in, 38
      space time yield, 13
   role of intermediates, 242–244
   selectivity on bimetallic catalysts, 277–279
   selectivity trends in, 272–274
   silver as catalyst for, 266
2-Ethylhexanoate, 397

F

Fatty acid epoxides, 9
Fatty acid hydroxylase cytochrome P450 BM-3,
   29–30
Fe(bispidine), epoxides and diols with, 70
Fe(II)/Fe(IV) cycle, 465–466
FeOOR(porph) adduct in epoxidation, 65

Fe porphyrin, 25
Fermi temperature (k_BT), 267
Fluorinated acetone
 as catalyst, 21
 structure of, 59
Fluorinated alcohols, 21
2-Fluoropyridine, 128–129
Fourier transform infrared (FTIR), 304–305, 373, 375, 432
Free energy, styrene epoxidation, 480
Fulminating gold, 341
Fumagillin, 8

G

Gas-phase epoxidation
 Au/3D mesoporous Ti-SiO2, catalyst life, 308–309
 Au/TS-1 catalytic performance, 307
 Au with Ag, replacement of, 310
 mechanistic studies, 303–306
Geometry optimization, 474
Geraniol, 422
2,3-Geranyl oxide, 422
Gold, 324, 326–327
 catalyst, 285, 288, 294
 clusters, 306
 nanoparticle catalyst (see Propene epoxidation)
 titania-based catalyst, 340

H

5H-Dibenzepine-5-carboxamide, 7
Heterogeneous catalysis, 234, 359
Heterogeneous epoxidation
 Au/titanosilicate catalysts, 45–47
 epoxidation reactions and catalysts, 17
 silver (Ag) catalyst
  of butadiene, 41–42
  of ethylene, 38–40
  of propylene, 40–41
 titanium (Ti) catalyst, 44
  active solid redox system in, 42
  with hydrogen peroxide, 43–44
  mechanisms of, 44–45
  structure of oxidant formed, 44
  for Ti(η^1-OOH)species, 44–45
Heteropoly anion, [PO44]$^{3-}$, 124–125
Heteropolyphosphotungstate anion, 432–433
1,5-Hexadiene, 109
cis- and trans-1,4-Hexadiene, 107
Hexafluoroacetone (HFA), 218–219
Highest occupied molecular orbital (HOMO), 303, 360
High-resolution electron energy loss spectroscopy (HREELS), 242

High-valent iron, 457, 464
Homogeneous epoxidation
 epoxidation reactions and catalysts, 18–19
 Lewis acid mechanism
  methyltrioxorhenium, 56–57
  molybdenum complexes, 48–52
  polyoxometallates, 56
  titanium complexes, 53–56
  vanadium complexes, 52–53
 main group epoxidation catalysts, 21, 23–25
 redox mechanism
  nonheme Fe complexes, 69–70
  porphyrin complexes, 60–66
  salen complexes, 66–69
 transition metal complexes
  allylic alcohols, epoxidation reactions and, 22–23
  catalyst system, 19
  H_2O_2 epoxidations, 17
  limonene oxide (LO) in, 20
  methyltrioxorhenium, 20
  phase-transfer reagents (PTRs), 17, 19
  polyoxometallates, 19–20
Homolysis, 221–222, 226–227
5-(2-pyridylmethylidene)-Hydantoin, 402
Hydrogen abstraction reaction, 122, 136, 143–144
Hydrogen efficiency, 340–341, 345, 352–353
Hydrogen peroxide (H_2O_2), 156–171, 453–454, 456–462, 464–465
 as adducts with methyltrioxorhenium, 128–129
 in aqueous solutions, 17, 31
 based epoxidation, 431
 in catalytic epoxidation
  advantages of, 430
  of allyl chloride, 441–444
  of cyclohexene, 439–440
  of propylene, 435–439
 formation, 306
 hydroperoxidation process, 298
 methyltrioxorhenium, oxidation of, 128
 as oxidant, 5, 35
 propylene epoxidation with, 15
 propylene oxide synthesis, 46
Hydrogen peroxide/propylene oxide (HPPO), 125
Hydrogen tetrachloroaurate(III) solution (HAuCl$_4$), 342
Hydro-oxidation, 340
Hydroperoxo species, 132, 162
α-Hydroxyketones, 9, 11
N-Hydroxyphthalimide (NHPI), aerobic epoxidation
 butterfly-type transition states in, 221
 methods, 218–219
 molecule-induced homolysis in, 221–222

NHPI and PINO in, 218
of olefins
  acetaldehyde in, 224–225
  epoxides yield, 224
  rate constant, 226
  solvent effect, 226–227
PINO EPR spectrum in, 223
PINO radical generation in, 222
of primary olefins, 219
of propylene, 220–221
reaction system of, 220
Hypervalent Fe and Mn porphyrin, intermediates in epoxidation, 64

I

ICP-AES. *See* Inductively coupled plasma-atomic emission spectrometer
Iminium salt-mediated catalytic asymmetric epoxidation, 184
  alcohol-containing iminium salts, 189–190
  aminoether and aminoacetal precursor, 190–193
  2-(2-bromoethyl)benzaldehyde method, 185–186
  catalysts based
    on binaphthalene structure, 196–199
    on dibenzo[c,e]azepinium salts, 194–196
  reaction parameters, 186–189
  spiro transition state, 194
Impregnation method, 285, 288, 294, 298, 231–302, 380. *See also* Ethylene epoxidation; Propylene epoxidation
Indene, 394
Inductively coupled plasma-atomic emission spectrometer (ICP-AES), 373, 375–376
Inductively coupled plasma (ICP), 432–433
Inelastic neutron scattering (INS), 305
Iodosylbenzene (PhIO), 393, 396
  in Jacobsen–Katsuki epoxidation, 67
  in nonheme Fe complexes, 69–70
  properties of, 36
Iron, 453, 455, 459
Iron-based olefin oxidation catalysts with H_2O_2 oxidant, ligands for, 454
di-Iron containing sandwich-type POM, 169
Ishii–Venturello system, 431
Isoamylene oxide, 6
cis–trans Isomerization, 39, 241
Isopinocampheylamine, 187, 192
Isotopic labeling, 146, 460

J

Jacobsen catalyst, 122, 393
Jacobsen–Katsuki
  epoxidation, of *cis*-stilbene, 67
  Mn(salen) catalyst, 7
  oxidation, 25
Jacobsen Ti-tartrate epoxidation mechanism, 54–55
  DFT calculations, 54–55
  ligand exchange pathway, 54
  structure of, 55

K

Katsuki-type complexes, 403
Keggin and Wells–Dawson-type polyoxometalate, 417
Keggin structure, 440, 444
Keggin-type silicodecatungstate γ-$SiW_{10}O_{36}]^{8-}$, 159
Ketiminium, 181
Kinetic experiments design, 329–330
Kinetic isotope effects (KIE), 133
Knudsen diffusivity, 383
Komatsu's ketiminium salt-mediated epoxidation, 181

L

Lacunary polyoxotungstates, 157, 159, 166
Lacunary POM [γ-$SiW_{10}O_{34}(H_2O)_2]^{4-}$, 20
Lanthanoid-loaded polymers, 402–403
Late transition metal, 122, 130
Layered double hydroxide (LDH), 394
Leaching, 34, 42, 362, 393, 397, 401–402
Levcromakalim, 7, 12, 210–212
Lewis acid
  catalysatation range, 130–131
  epoxidation mechanism (*see* Mid-to late transition metal catalysts)
  properties, 105–106
Lewis acid mechanism, 47. *See also* Homogeneous epoxidation
  in Fe porphyrin system epoxidation, 65–66
  methyltrioxorhenium, 56–57
  molybdenum complexes
    alkylperoxo mechanism of, 50
    disfavored metalladioxolane mechanism, 51–52
    peroxo complex and Sheldon complex, 49
    predominant orbital interactions, 48
  polyoxometallates, 56
  titanium complexes
    kinetic rate expression, 54
    ligand exchange pathway, 54–55
    structure of, 55
  in transition metal ion epoxidation, 72
  vanadium complexes, 52–53
Light olefins, pressure intensified epoxidation process. *See also* Olefin epoxidation

Light olefins, pressure intensified epoxidation process. *See also* Olefin epoxidation (*Cont.*)
  catalyst stability, in five successive run, 129
  CEBC and industrial processes, 126
  equilibrium constants, for media, 129–130
  liquid/gas biphasic catalytic propylene oxide process, 126
  and productivity of PO process, 127
Limonene, 8–9
  epoxidation of, 20
Liquid-phase epoxidation, 300
Low energy electron diffraction (LEED), 239–240
Lowest unoccupied molecular orbital (LUMO), 360

M

Magic angle spinning nuclear magnetic resonance (MAS-NMR), 432
N-(3,4-dihydroxybenzyl)maleimide, 399
Manganese(II) acetate, 389
Manganese(III)-salen complexes, 389
Manganese porphyrins
  alkene epoxidation
    catalytic cycle in, 472–473
    stability of metalloporphyrins, 472
  geometries and energies, 474
  styrene epoxidation mechanism by concerted and radical intermediates, 475
    concerted intermediate formation in, 478–479
    electronic and coulombic energy function in, 476–477
    oxidation with iodosylbenzene, 477
    phenylacetaldehyde formation, 480
Manganese tetra(4-aminophenyl) porphyrin, 395
Manganese tetra(4-pyridyl) porphyrin, 395
Merrifield resin, 394–397
Mesoporous
  Au/Ti-TUD catalyst, 330
  titanium silicates, 311, 322–324
Metal-free NHPI, 219
Metalladioxolane, 51–53
Metallaoxetane, 473
Metalloporphyrins, 430, 472
Methane monooxygenase (MMO), 28
Methanol, 15, 36, 43, 66, 127–128, 130, 300, 347
4,4'-Methylenedianiline, 397
Methylfluorosulphonate, 178
4-Methylhexene, 108
4-Methylpentene, 108
4-Methylpent-2-ene, 395
Methyl phenyl carbinol (MPC), 357, 368
4-Methylstyrene, 394
α-Methylstyrene, 394, 396, 401

cis-2-Methylstyrene, 393–394
cis-β-Methylstyrene, 211
Methyl *tert*-butyl ether (MTBE), 15
Methyltri-n-octylammonium, 418
Methyltrioxorhenium (MTO), 124, 401, 430
  catalytically active intermediates, 128
  disadvantages of, 20
  epoxidation of, 56–57
  with hydrogen peroxide, as adducts
    equilibrium constant, 129
    types of adducts, 128
  ligands, role, 130
  as organometallic catalyst, 124
  solubility of, 127
  stability of, 128
Michaelis–Menten equation, 62, 65
Microkinetic model. *See also* Ag-based bimetallic catalysts
  reversible reactions in, 277
  for selectivity trends for bimetallic catalysts, 277–279
  UBI–QEP method, use of, 277
Mid-to late transition metal catalysts, 131
  iron compounds, 131–133
  manganese in Lewis acid adducts, 133–134
  Mn(IV) catalyst, 134–135
    Mn(Me$_2$EBC)Cl$_2$ catalyst system and olefin epoxidation, 135–142
    with Mn(Me$_2$EBC)Cl$_2$ using t-BuOOH, 142–148
Mimoun complex, 49
Mizuno's catalyst, 125
Mn(IV) catalyst, 134–135, 137–138, 141–142, 144–146
Mn(Me$_2$EBC)Cl$_2$ catalyst system, 135
Mn(salen) system, Lewis acid and redox pathways in epoxidation, 69
Mn-tetraphenyl porphyrin. *See* Manganese porphyrins
Mn(V) oxenoid complex, 133
Mo complexes, in epoxidation, 49
Molecular-induced homolysis
  inhibition of, 226
  in peracids and dioxiranes, 221–222
Molybdenum-catalyzed epoxidation. *See also* Homogeneous epoxidation
  alkylperoxo mechanism, 50
  metalladioxolane mechanism, 51
Molybdenum ethoxide, 397
Molybdenyl acetylacetonate, 397
Monocationic catalysts
  Baeyer–Villiger ketone oxidation in, 106
  synthesis of, 105
Monodentate ligands, 130

N

Nanofiltration
  in POM catalyst recycling
    ceramic membranes in, 425–426
    induction period, 426
Nickel-containing sandwich-type POM, 169
Nickel(II) acetylacetonate, 402
Nickel-substituted quasi-Wells–Dawson-type polyfluorooxometalate, 169
Nonheme, 452–459, 461, 465–466
  iron catalysts, ligand topology, 455
Non-water-assisted *(nwa)* pathway, olefin oxidation, 462–463
Norbornene oxide, 7
Norbornylene, 136
N-(p-acetoxyphenyl)maleimides, 399
N-(p-hydroxyphenyl)maleimide, 399
N-(piperonyl)maleimides, 399
Nuclear magnetic resonance (NMR), 157, 164
Nucleophilic oxygen, 235
Nudged Elastic Band (NEB)
  activation energies for EO and acetaldehyde formation, 270
  calculations over multiple iterations, 269
  convergence criterion of image in ASE/DACAPO version, 272
  for transition state and activation energy barrier, 267
  vs. hybrid approach for rational catalyst design, 268

O

Olefin epoxidation. *See also* Epoxidation
  of allyl chloride, 441–444
  applications of, 431
  of cyclohexene, 439–440
  H_2O_2 oxidation, 121
  iodosylbenzene adducts, 133–134
  Lewis acid catalystation, 121–122
    cytochrome P450, role of, 131
    mechanism in protic and aprotic solvents, 131–132
  manganese catalyst using hydrogen peroxide
    deviation of oxygen content, 139–140
    isotope labeling experiment, 138–139
    manganese oxo species, 137
    mechanism, 142
    oxygen rebound mechanism, 138
    oxygen transfer process, 136–137
    cis-stilbene, 140–141
  manganese catalyst using t-BuOOH
    mass spectra study, 147–148
    radical process, 144–146
    reactions on various olefins, 143–144
    cis-stilbene, 146
    ultraviolet-visible spectrophotometry study, 142
  oxidant advantages of, 430
  by porphyrins, 63
  of propylene, 435–439
  regio- and enantioselective reactions, 122
  of terminal olefins, 440
  transition metal complexes in, 120–121
Olefin oxidation, bio-inspired iron catalyzed. *See also* Olefin epoxidation
  catalytic additives, 458
  $Fe^{III}(BPMEN)_2(\mu-O)(\mu-OH)]^{3+}$ in, 456–457
  Fe^{III}-OOH *vs.* Fe^V-O oxidants
    density functional theory (DFT), 462
    isotope labeling results for, 460
    non-water-assisted *(nwa)* pathway, 463–464
    ^{18}O labeling experiments, 459–460
    water-assisted mechanism *(wa)* energy profile, 462–463
  Fe^{IV}=O *vs.* Fe^V=O oxidants
    acetic acid role, 465
    Fe(II)/Fe(IV) cycle, 465–466
    presence of excess oxidant (PhIO), 465
  high-valent iron-oxo oxidant in, 457, 464
  H_2O_2 oxidant in, 453–454
  intermediates, mechanistic landscape for, 461
  iron-based ligands for
    nitrogen-rich polydentate, 453–454
    tetradentate ligand, 453–454
  nonheme iron catalysts
    acetic acid role, 457
    epoxidation/*cis*-dihydroxylation selectivity spectrum, 455
    epoxide and diol product, 454–455
    H_2O_2 oxidant limiting catalysts, 453, 456
    as practical catalysts, 458
    triflic acid role, 457
  peroxide activation, 452
α-Olefins, 224
Organic hydroperoxides, 374
Organocatalysis, 177
Organophosphoryl polyoxotungstate derivative $[\gamma$-$SiW_{10}O_{36}(PhPO)_2]^{4-}$, 167
Oxametallacycle (OME), 243–244, 247, 266
Oxaziridine
  Armstrong's intramolecular epoxidation, 182–183
  mediated epoxidation, 193–194
  quaternization of, 178
Oxaziridinium salts, 178
  catalytic cycle for, 179, 184–185
  derived from (1S,2R)-(+)-norephidrine, 179
  diastereoisomeric form, for asymmetric induction, 193

Oxaziridinium salts (*Cont.*)
  dihydroisoquinoline derivative, 178
  mediated epoxidation, 179
  si or re face of iminium species, in forming, 185
  (1S,2R)-(+)-norephedrine derivative, 178–179
  structure of, 178
  transferring oxygen to nucleophilic substrates, 180
Oxenoid, 360
Oxometallacycle, 51
Oxone, 184
  tetra-N-butylammonium, 200
  mediated epoxidation, 180–181, 183
  as stoichiometric oxidant, 199
Oxone^{TM}, 23–24
Oxygen-rebound mechanism, 59, 65, 121–122, 132–133, 137–139, 138
Oxygen starvation, 365–366
Oxygen transfer, 360–361

P

Pacific Northwest National Laboratory (PNNL), 267
Packed-bed reactor operation, 378
Pd-based catalysts, 298–299
Pd/TS-1 catalyst, 299, 310
1,4-Pentadiene, 109
Pentafluorophenyl, 104, 106
Peracids
  in alkene epoxidation, 224
  in molecule-induced homolysis, 222
Periodic slab calculations, DACAPO, 267
Peroxide activation, 452
Peroxide adduct, 147–148
μ-η2:η^{2}-Peroxo complexes, 48–50, 156, 162
Peroxolysis, 361–362
Peroxotungstates, 157
  catalytic oxidation, 159, 166
  molecular structure of tungstates, 158
Peroxycarboxylic acid, 59
Peroxyl radicals, 218, 224
Phase-transfer catalyst, 420, 422
Phase-transfer reagents (PTRs), 124
  in anion encapsulation, 17, 19
  in olefin epoxidation, 20
1-Phenylcyclohexene, 190, 205, 207–208, 393
beta-Phenylethanol (BPEA), 364–365
4-Phenylpyridine-*N*-oxide (PPNO), 393–394
Phosphotungstic acid, 157
Phthalimido-*N*-oxyl radical (PINO), 218, 226.
  See also *N*-Hydroxyphthalimide (NHPI)
α-Pinene, 10–11, 401–402
Plasmid pSPZ10, 29
Platinum (Pt(II)) complexes, 104.
  See also Alkene epoxidation

Polarizable continuum model (PCM), 481–482
P. oleovorans GPol, 29
Polyether polyols, 9
Poly(ethylene glycol) monomethylether (MPEG), 399–400
Poly(ethylene glycol)-supported manganese porphyrins, 396
Polymer-supported manganese species, 390–392
Polymer supported metal catalysts
  BINOL-complexes of lanthanoids and calcium
    epoxidation of chalcone by TBHP, 403
    polymer-anchored
      chiral 1,1'-bi-2-naphthol, 402
    preparation of, 402–403
  cobalt catalysts
    aerobic exoxidation of alkenes with an aldehyde, 402
    polyaniline, poly-*o*-toluidine and poly-*o*-anisidine, 402
  manganese complexes, 389
  manganese-porphyrin complexes
    encapsulation of, 396
    extraction of, 395
    sulfonated manganese octabromoporphyrin derivative, 395–396
    supported catalysts for, 395
  manganese-salen complexes
    axial bonding of, 394
    catalyst recycling, 393–394, 404
    for catalytic epoxidation, 389–392
    Jacobsen catalysts in, 393
    with phosphonium groups, 394
    plasma mass spectrometry, 393
  molybdenum catalysts
    synthesizing of cross-linked poly(4-vinyl pyridine-co-styrene), 397
    thermosetting epoxy resins for, 397–398
  nickel catalysts
    heterogenous catalyst, 402
    polybenzimidazole, 402
  platinum catalysts
    dinuclear amidate bridged platinum(III) complex, 402
    low catalytic activity, 402
  rhenium catalysts
    immobilization of methyltrioxorhenium (MTO), 401–402
  ruthenium and iron catalysts
    catalyzed epoxidation, 398, 405
    co-polymerization of vinyl-substituted carbonyl ruthenium(II) porphyrin, 398
  titanium catalysts
    esterification of tartaric acid, 399–400
    inorganic-organic hybrid catalyst system, 400

N-(p-hydroxyphenyl)maleimide and Ti(OiPr)$_4$, 399
  preparation of, 399, 405–406
tungsten catalysts
  epoxidation reaction with aqueous H$_2$O$_2$, 400
  polymer-anchored tungsten carbonyl catalysts, 401
Polyoxometalates (POMs), 156, 417, 422–423, 426, 430–431
  active oxygen species, 157–159
  as biological-type catalyst, 19–20
  catalytic oxidation by
    lacunary polyoxotungstates, 166–168
    peroxotungstates, 166
    transition-metal-substituted polyoxometalates, 168–169
  H$_2$O$_2$-based oxidation, classifications, 157
  hydrogen peroxide, activation of
    lacunary polyoxotungstates, 157, 159–163
    peroxotungstates, 158–159
    transition-metal-substituted, 164–166
  Keggin-type, 157, 159
  in Lewis acid mechanism, 56
  oxidation catalyst, advantages of, 156–157, 166
Poly(4-styrylmethyl-acylchloride), 395
Polytungstozincate anions, 416
Poly(4-vinylpyridine-co-styrene), 397
Poly(4-vinylpyridine-N-oxide), 401
Prileshajev epoxidation, 9
Process improvement. *See* Styrene monomer propylene oxide process
Promoters, 308. *See also* Propylene epoxidation
  alkali earth metals, 324–325
  postsynthesis addition of titanium, 325
  silylation, 326–327
  trimethylamine cofeeding, 327
Promotion. *See also* Ag/α-Al$_2$O$_3$ catalyst
  Cl promotion
    in ethylene desorption, 258–259
    oxidation adsorption energy, 252–254
    oxidation desorption energy, 251–253
  Cs promotion
    in ethylene desorption, 257–258
    O$_2$ desorption spectrum, 249–250
    temperature programmed reaction spectrum, 250–251
Propene epoxidation
  catalyst activity testing, 342–343
  catalyst characterization of, 342–344
  catalyst deactivation
    first order kinetics, 351–352
    hydrogen oxidation in, 350, 352
    particle size distribution, 348–349
    Ti-Si carriers, 352
    vs. propene oxide, 350

catalyst preparation, 341–342
catalytic performance, 345
  Ti-SBA-15 and Ti-SiO$_2$-supported catalysts, 347–348
  titania-supported catalyst, 343
  at two selected temperature, 346
chlorohydrin process *vs.* hydroperoxide processes, 340
scanning electron microscopy (SEM), 342–343
transmission electron microscopy (TEM), 342–343, 349
x-ray photoelectron spectroscopy (XPS), 342, 348
Propene oxide, 340
Propene oxide-*tert*-butyl alcohol (PO-TBA), 340
cis-1-Propenylphosphonic acid (CPPA), 8
Propylene chlorohydrin, 123
Propylene epoxidation
  with aqueous hydrogen peroxide, 438–439
  Au on TI-containing mesoporous supports, 322–324
    Ti-SBA-15 materials, 323–324
    titanosilicates and mixed oxides, 324
    Ti-TUD materials and Ti-MCM-41 materials, 323
  with EBHP
    Ti/MCM-41 and Ti/MCM-48 catalysts, 380–384
    Ti/SiO$_2$ catalyst, 376–378
  gas-phase epoxidation
    Au/3D mesoporous Ti-SiO$_2$, catalyst life of, 308–309
    Au/TiO$_2$ catalysts, mechanistic study of, 303–307
    Au/TS-1, catalytic performance of, 307–308
    Au with Ag, replacement of, 310
  kinetic reactions in, 327–330
    power rate law expression r$_{po}$, 329–330
    sequential *vs.* simultaneous, 330–331
  kinetics of, 384–385
  liquid-phase of, 300–301
  with O$_2$ and H$_2$, catalytic performance data, 299
  over Au/TS-1 catalysts, 318–322
    activity, selectivity and stability of catalyst, 318
    Au (gold) deposition, control of, 320–321
    defects in TS-1, 321–322
    nature of active sites in Au, 319–320
  promotors and postsynthesis treatment in, 322–327
    alkali earth materials, 324–325
    silylation, 326–327
    titanium grafting, 325
    trimethylamine cofeeding, 327

Propylene epoxidation (*Cont.*)
  in situ hydrogen peroxide generation
    catalyst recycling in, 436–437
    EAHQ effects of, 436
    FTIR and ^{31}P MAS-NMR in, 437–438
    net reaction of, 435
    temperature effects of, 435–436
Propylene glycol, 6, 124–125, 130, 316, 374, 436, 439
Propylene oxametallacycle (OMPP), 40
Propylene oxide (PO), 6, 219, 227, 298, 304, 356, 374, 378
  Halcon process, 48
  industrial preparation, 317
    catalysts performance, 126
    catalysts used, 124–125
    chemical process, 126
    chlorohydrin and peroxidation process, 123
    plans and HPPO technology, 125–126
    pressure intensification, 127
    pyridine, role of, 130
  production of, 14–15
  in propylene epoxidation
    with aqueous hydrogen peroxide, 438–439
    with *in situ* hydrogen peroxide, 435–438
  selectivity, 376
  synthesis, 374
    hydrogen-oxygen route in, 45–46
    kinetics, 46
    reaction pathways, 46–47
  Ti/MCM-41 and Ti/MCM-48 catalysts, 380–381
  Ti/SiO$_2$ catalyst, 376
  yield, 376
Proton parking place, 361
Pseudonomas sp. strain VLB120, 29
Pyridine-2,6-dicarboxylic acid (H$_2$pydic), 25
Pyridine-N-oxide (PyNO), 126

Q

[(n-C$_8$H$_{17}$)$_3$NMe]Cl (QCl), 422
QCl/W ratio, 422, 424
Quantum mechanical calculations, 360–361
Quantum mechanics/molecular mechanics (QM/MM), 320–321, 473
Quaternary ammonium cations, 433–434
Quaternary ammonium phosphotungstates, 431

R

Radical intermediate. *See* Styrene epoxidation
Radical oxidations, 121
Reaction-controlled phase-transfer catalysis.
  See also Olefins epoxidation
  for allyl chloride epoxidation, 440–444
  applications of, 430–431
  for cyclohexene epoxidation, 439–440
  heteropolyphosphotungstate on
    characterization of, 432–433
    preparation of, 431–432
  for propylene epoxidation
    with aqueous hydrogen peroxide, 438–439
    with *in situ* hydrogen peroxide, 435–438
  quaternary ammonium cation influence, 433–434
  systems for, 431–432
Reaction-controlled phase-transfer catalyst, 431
Reaction kinetic studies
  Au/TiO$_2$ and Au/TiO$_2$/SiO$_2$ system, 327–328
  oxidation of hydrogen (H$_2$), 328–330
  stable Ba-doped, mesoporous Au/Ti-TUD catalyst, use of, 328–330
Reaction mechanism. *See also* Ag/α-Al$_2$O$_3$ catalyst
  on ethylene epoxidation over silver
    heat of adsorption of intermediates, 244
    intermediate formation, 242–243
    microkinetic model, 243
    temperature programmed reaction, 244–247
Reactive distillation, 368–369
Redox mechanism, 59. *See also* Homogeneous epoxidation
  nonheme Fe complexes in, 69–70
  porphyrin complexes in
    derivatives and reactivity of, 60–61
    hydroperoxides reactions in, 61–62
    intermediates in, 64
    kinetics of, 62–63, 66
    in olefins epoxidation, 63–64
    reaction pathways, 65
  salen complexes in
    Jacobsen–Katsuki epoxidation, 66–67
    stereoselective epoxidation in, 68–69
    vinylcyclopropane probe in, 68
Reflection-absorption infrared spectroscopy (RAIRS), 236, 240
Regioselectivity, 107–108
  in epoxidation of dienes, 108
Rhodococcus rhodochrous, 30
Rieske dioxygenases. *See also* Olefin oxidation, bio-inspired iron catalyzed
  active sites in, 451–452
  nonheme iron catalysts, 452–453
  structure of, 452
Ring-opening metathesis polymerization (ROMP), 389
Ruthenium(II) porphyrins, 398

S

Sandwich polyoxometalates (POMs). *See also* Acidic epoxidation systems
  in catalyst recycling
    ceramic membranes in, 425
    induction period of, 426
    nanofiltration in, 425–426
  characteristics of, 418
  in cyclooctene epoxidation
    cocatalysts for, 420–421
    conversion *vs.* time, 421
  diastereoselectivity of, 423
  in geraniol epoxidation, 421–423
  preparation of, 417–418
  structure of, 417
Scanning electron microscopy (SEM), 342
Schiff bases, 66
Selective oxidation, 234
Selectivity, 242–246, 248–251, 259, 266
Selenium dioxide, 431
Semibatch reactor, 298, 301, 310
Sharpless catalyst, 359, 399
Sharpless epoxidation
  of allyl epoxide, 9
  mechanism, 360–361
  of Ti tartrate catalyst
    ligand exchange pathway, 54–55
    structure and mechanism of, 55–56
Sheldon complex, 49
Shell technology. *See* Styrene monomer propylene oxide (SMPO) process
Silicodecatungstate $[\gamma\text{-SiW}_{10}O_{34}(H_2O)_2]^{4-}$, 160
Silver (Ag), 266–279. *See also* Ethylene epoxidation
  adsorption and desorption of oxygen
    Ag–O bond, 235–236
    fractional coverage, 239
    kinetics of adsorption, 237
    oxygen desorption spectrum, 237–238
    oxygen dosing, 239–240
  ethylene and ethylene oxide adsorption, 240–242
Silver catalyst, 285, 290–291, 294–295
  in heterogeneous epoxidation
    of butadiene, 41–42
    of ethylene, 39–40
    of propylene, 40–41
Silylation, 326–327
Single crystal X-ray structural analysis, 157, 159–160
SiW_{10} catalyst, 125
Slater-type atomic orbitals, 267
SM/PO molar ratio, 364
Solvent effect, 226–227
  for aerobic epoxidation of olefins catalyzed by NHPI, 226

Space time yield, 12
  for Au/titanosilicate catalysts, 45–46
  for limonene oxide (LO), 20
  in production
    of epoxybutene, 16
    of propylene, 15
    of styrene oxide, 29
Space time yield (STY), 298–301
Spiro structure
  of dimethyldioxirane (DMD), 58–59
  of epioxidation intermediates, 57–58
  of Mo complexes, 52–53
Spiro transition state, 193–194
cis-Stilbene, 131
  catalytic t-BuOOH epoxidation of, 146
  Lewis acid epoxidations, 131
  nonradical pathway in, 141
  oxide and trans-stilbene oxide product, 136–137, 139
trans-Stilbene, 16, 394, 401–402
Streptomyces fradiae, 8
Strong metal-support interaction (SMSI), 284–285
Styrene, 393–394
Styrene epoxidation. *See also* Manganese porphyrins
  DFT and QM/MM calculations, 473
    ONIOM method, 473–475
  with iodosylbenzene catalyzed by Mn-porphyrins, 476–477
    activation energies in vacuum and in dichloromethane solvent, 483
    concerted internediate formation in, 478–480
    electronic energy as function, 476–477, 482
    free energies calculation, 480–481
    geometries for stable intermediates and transition states, 479
    Mn-porphyrin oxidation, 477
    phenylacetaldehyde formation in, 480
    polarizable continuum model (PCM) of, 481–482
Styrene monomer propylene oxide (SMPO) process
  catalyst deactivation mechanism, 362
  catalyst dehydration in
    gas-phase processes, 367–368
    liquid-phase processes, 368–369
  catalytic epoxidation
    active site in, 358–359
    molecular orbital analysis in, 360–361
    Sharpless mechanism in, 360
    simplified mechanism of, 361
    titanium peroxide in, 359–360
  computational flow dynamics (CFD) in, 366–367
  flow of, 357

Styrene monomer propylene oxide (SMPO)
  process (*Cont.*)
  reaction process of, 356–358
  specific capital in, 363
  trace component chemistry
    BPEA in, 364–365
    effects of, 363
    oxygen starvation in, 365–366
    phenols formation in, 365
Styrene monomer (SM), 356
Styrene oxide
  applications of, 7
  production of, 29
Subsurface O atoms, 247–249
  for selective oxidation of ethylene, 247
Sulphoxidation, 167, 170
Sumitomo hydroperoxide process, 340
Supercritical CO_2, 300–301
Supported catalysts. *See also* Polymer supported metal catalysts
  BINOL-complexes of lanthanoids and calcium, 402–403
  cobalt, nickel, and platinum catalysts, 402
  manganese catalysts, 389
    manganese-porphyrin complexes, 395–396
    manganese-salen complexes, 389, 393–394
  molybdenum catalysts, 396–398
  rhenium catalysts, 401–402
  ruthenium and iron catalysts, 398–399
  titanium catalysts, 399–400
  tungsten catalysts, 400–401

T

Temperature-programmed desorption (TPD), 236, 238, 245, 248, 291, 293, 303
Terminal alkenes
  assymetric epoxidation of, 110–112
  catalytic system for, 106–107
  electrophilic oxidation in, 104
Terpenes, 8–9
Terpenic compounds, 10
Terpenic substrates, epoxidation of, 8
γ-Terpinene, 401
Tetraalkylammonium heteropolyperoxotungstates, 431
Tetraamido macrocyclic ligand (TAML), 28
Tetracationic and tetraanionic porphyrins, 395
Tetrahydroisoquinoline, 178
Tetrakis(hydroxymethyl)phosphonium chloride (THPC), 302
Tetra-*n*-butylammonium hydroxide (TBAOH), 162
Tetra-*n*-butylammonium (TBA), 159, 166
Tetraphenylphosphonium monoperoxysulfate or bisulfate (TPPP), 24

Tetraphenylphosphonium monoperoxysulphate (TPPP)
  alkenes epoxidation, 209–210
  asymmetric epoxidations using, 201
  and benzopyrans epoxidation, 211
  catalytic asymmetric epoxidation
    of 1-Phenylcyclohexene, 202, 205, 208
    of various alkenes, 209
    of various *cis*-alkenes, 210
  epoxidation of benzopyrans using catalyst 36 and, 211
  equivalents of, 213
  formation of, 200–201
  and NMR data, 212
  and 1-phenylcyclohexene epoxidation, 202, 205, 208
Thermal conductivity detector (TCD), 342, 344
Thermoanaerobium brockii, 36
Thermogravimetric analysis (TGA), 318
Thermosetting epoxy resins, 397
$TiCl_4$, 372. *See also* Ethylbenzene hydroperoxide (EBHP)
  chemical vapor deposition of, 375
  deposition on MCM-41 and MCM-48, 380, 383
  deposition temperature on ethylbenzene hydroperoxide (EBHP) conversion, 381
  deposition time
    on ethylbenzene hydroperoxide (EBHP), 376
    on tert-butyl hydroperoxide (TBHP) conversion, 379
  EBHP conversion and PO yield, 376
  reaction with SiOH, temperature sensitivity, 384
  selectivity *vs.* TiCl4 deposition time, 378
Ti/MCM-41 catalyst. *See also* Chemical vapor deposition (CVD)
  deposition temperature, oxidants, and selectivity, 380
  effective diffusion coefficient and pore size, 383
  kinetic model, 384–385
  for kinetic study., 384
  preparation of, 380–381
  propylene oxide yield for, 381
  tetrahedrally coordinated titanium concentration, 381
  vs. Ti/MCM-48, 381, 383
  XPS and infrared spectra, 382
Ti/MCM-48 catalyst, 383. *See also* Chemical vapor deposition (CVD)
  effective diffusion coefficient, 383
  pore size distribution of, 383
  preparation of, 380–381
  propylene oxide (PO) yield for, 381
  vs. Ti/MCM-41, 383
  XPS and infrared spectra, 382

Ti(OiPr)4, 399
Ti peroxides, 44
Ti-SBA-15
    in catalyst preparation, 341
    mesoporous silica, 341
    propene oxide yields, 346
    SEM and TEM analysis on, 343
Ti/SiO$_2$ catalyst. *See also* Chemical vapor deposition (CVD)
    packed-bed reactor, for stability, 378
    preparation and propylene oxide yield, 376
    titanium, as function of deposition time, 378
    TOF of, 377, 379–380
    types of titanium species in, 379
Titanium, 164, 316, 323–325
Titanium-on-silica catalyst, 362
Titanium peroxide, Ti(OOR), 359, 361
Titanium silicalite-1 (TS-1), 7, 430
    as catalyst in propylene epoxidation
        activity, selectivity and stability of catalyst, 318
        ammonium nitrate treatment, 321–322
        defects in, 321–322
        synthesis of, 317
Ti-tartrate catalyst and intermediate, 55
Trace component chemistry, 363–364. *See also* Styrene monomer propylene oxide (SMPO) process
Transition-metal-substituted POMs, 157, 164, 168
Transmission electron microscopy (TEM), 307–308, 342
1,4,7-Triazacyclononane moieties, 389
Tributylphosphate (TBP), 435
Triferric sandwich-type POM, 169
Trimethylamine, 309
Trimethylsilylation, 308–309
Tungstate catalyzed epoxidation
    catalyst activity in, 420–421
    phase-transfer catalyst in, 420
Tungsten, 158–160, 162, 166–167
Tungsten-based polyoxometallates, 125
Tungsten-phosphor (W/P) ratio, 432
Turnover frequencies (TOFs), 5, 377, 379–380
    for allylic alcohols, 22–23
    for biological catalysts, 29
    for biomimetic catalysts, 26–27
    for commodity catalyst, 13, 15–17
    graphical form of, 34–35
    for homogeneous catalysts, 16, 18–19
    for terpenic compounds, 10

U

Ultrafine titania particles, 362
Ultraviolet photoelectron spectrometry (UPS), 240
Universal force field (UFF), 474

V

L-Valine, 389, 399
Vanadium, 164, 166
Vanadium-based sandwich POM, 423
Vanderbilt ultrasoft pseudopotential, 267
Venturello anion, 401
Venturello compound, 125
Venturello's catalyst, 124
Vernolic acid, 9
Vinylcyclopropane probe, 68
Vosko–Wilk–Nusair (VWN) function, 267

W

Water-assisted mechanism (*wa*), olefin oxidation, 462–463
W-based catalyst systems, 416, 418
Wells–Dawson-type sandwich POM, 169
[WZn$_3$(ZnW$_9$O$_{34}$)$_2$]$^{12-}$, sandwich polyoxometalate anion, 417

X

X-ray fluorescence (XRF), 342–343
X-ray photoelectron spectroscopy (XPS), 300, 342, 373, 375
    Ti/MCM-41and Ti/MCM-48 catalyst, 381–382

Y

Yang's *in situ* iminium salt epoxidation system, 182
Ytterbium–BINOL complex, 403

Z

Zeolite catalysts
    in gas-phase processes, 367–368
    in liquid-phase processes, 368–369